JOURNAL OF CHROMATOGRAPHY LIBRARY — volume 51B

chromatography,
5th edition

fundamentals and applications of chromatography and related differential migration methods

part B: applications

JOURNAL OF CHROMATOGRAPHY LIBRARY — volume 51B

chromatography, 5th edition

fundamentals and applications of chromatography and related differential migration methods

part B: applications

edited by

E. Heftmann

P.O. Box 928, Orinda, CA 94563, U.S.A.

ELSEVIER
Amsterdam — Oxford — New York — Tokyo   **1992**

0463-4822

CHEMISTRY

ELSEVIER SCIENCE PUBLISHERS B.V.
Sara Burgerhartstraat 25
P.O. Box 211, 1000 AE Amsterdam, The Netherlands

Distributors for the United States and Canada:

ELSEVIER SCIENCE PUBLISHING COMPANY INC.
655 Avenue of the Americas
New York, NY 10010, U.S.A.

Library of Congress Cataloging-in-Publication Data

Chromatography : fundamentals and applications of chromatography and
  related differential migration methods / edited by E. Heftmann. --
5th ed.
       p.   cm. -- (Journal of chromatography library ; v. 51A-B)
   Includes bibliographical references and indexes.
   Contents: pt. A. Fundamentals and techniques -- pt. B.
Applications.
   ISBN 0-444-88404-1 (set : alk. paper). -- ISBN 0-444-88236-7 (pt.
A : alk. paper). -- ISBN 0-444-88237-5 (pt. B : alk. paper)
   1. Chromatographic analysis.   I. Heftmann, Erich.   II. Series.
QD79.C4C485 1992
543'.089--dc20   91-35963
                                                         CIP

ISBN 0-444-88237-5 (Part B)
ISBN 0-444-88404-1 (set)

This book is printed on acid-free paper.

Printed in The Netherlands

TO BIBI

CONTENTS

List of Authors

Dr. Hugo A. H. Billiet, Laboratorium voor Analytische Scheikunde, Technische Hogeschool Delft, De Vries van Heystplantsoen 2, NL-2628 RZ Delft, The Netherlands (Part B)

Dr. Phyllis R. Brown, Department of Chemistry, University of Rhode Island, Kingston, RI 02881, USA (Part B)

Dr. Shirley C. Churms, Department of Chemistry, University of Cape Town, Private Bag, Rondebosch 7700, South Africa (Part B)

Dr. Karen M. Gooding, SynChrom, Inc., P.O. Box 310, Lafayette, IN 47902-0310, USA (Part B)

Dr. Paul R. Haddad, Department of Analytical Chemistry, University of New South Wales, P.O. Box 1, Kensington, N.S.W. 2033, Australia (Part B)

Dr. Lars Hagel, Pharmacia LKB Biotechnology, S-751 82 Uppsala, Sweden (Part A)

Dr. Jeffrey B. Harborne, Department of Botany, University of Reading, Reading RG6 2AS, Great Britain (Part B)

Dr. F. Xavier de las Heras, Petroleum Geochemistry Group, Energy Center Building, School of Geology & Geophysics, 100 East Boyd Street, University of Oklahoma, Norman, OK 73019, USA (Part B)

Dr. Robert S. Hodges, Department of Biochemistry, University of Alberta, Edmonton, Alta. T6G 2H7, Canada (Part B)

Dr. Yoichiro Ito, Laboratory of Technical Development, National Heart, Lung, and Blood Institute, National Institutes of Health, 9000 Rockville Pike, Bethesda, MD 20892, USA (Part A)

Dr. Karl Jacob, Institut für Klinische Chemie am Klinikum Grosshadern der Ludwig-Maximilians-Universität München, Postfach 701260, D-8000 München 70, Germany (Part B)

Dr. Josef Janča, Institute of Analytical Chemistry, Czechoslovak Academy of Sciences, Leninova 82, CS-611 42 Brno, Czechoslovakia (Part A)

Dr. Nan-In Jang, Zenith Laboratories, 140 Legrand Ave., Northvale, NJ 07647, USA (Part B)

Dr. Jan-Christer Janson, Pharmacia LKB Biotechnology AB, S-751 82 Uppsala, Sweden (Part A)

Dr. F. W. Karasek, Department of Chemistry, University of Waterloo, Waterloo, Ont. N2L 3G1, Canada (Part B)

Dr. Arnis Kuksis, Banting & Best Department of Medical Research, University of Toronto, Toronto, Ont. M5G 1L6, Canada (Part B)

Dr. Karel Macek, Institute of Physiology, Czechoslovak Academy of Sciences, Budejovicka 1083, CS-142 20 Praha 4, Czechoslovakia (Part B)

Dr. Jan Macek, Research Institute of Rheumatic Diseases, Na slupi 4, CS-128 50 Praha 2, Czechoslovakia (Part B)

Dr. Colin T. Mant, Department of Biochemistry, University of Alberta, Edmonton, Alta. T6G 2H7, Canada (Part B)

Dr. Thomas H. Mourey, Eastman Kodak Research Laboratories, Rochester, NY 14650, USA (Part B)

Dr. K. P. Naikwadi, Department of Chemistry, University of Waterloo, Waterloo, Ont. N2L 3G1, Canada (Part B)

Dr. Szabolcs Nyiredy, Gyógynövény Kutató Intézet, József A. u. 68, H-2011 Budakalász, Hungary (Part A)

Dr. Emilios Patsalides, School of Chemistry, University of Sydney, Sydney, N.S.W. 2006, Australia (Part B)

Dr. Terry M. Phillips, Immunochemistry Laboratory, Ross Hall, George Washington University Medical Center, 2300 Eye Street N.W., Washington, DC 20037, USA (Part A)

Dr. R. P. Philp, Petroleum Geochemistry Group, Energy Center Building, School of Geology and Geophysics, 100 East Boyd Street, University of Oklahoma, Norman, OK 73019, USA (Part B)

Dr. Colin F. Poole, Department of Chemistry, Wayne State University, Detroit, MI 48202, USA (Part A)

Dr. Salwa K. Poole, Department of Chemistry, Wayne State University, Detroit, MI 48202, USA (Part A)

Dr. Hans Poppe, Laboratorium voor Analytische Scheikunde, Universiteit van Amsterdam, Nieuwe Achtergracht 166, NL-1018 WV Amsterdam, The Netherlands (Part A)

Dr. Fred E. Regnier, Department of Biochemistry and Chemistry, Purdue University, West Lafayette, IN 47907, USA (Part B)

Dr. Tadeus Reichstein, Institut für Organische Chemie, Universität Basel, St. Johanns-Ring 19, CH-4056 Basel, Switzerland (Part A)

Dr. Pier Giorgio Righetti, Dip. di Scienze e Tecnologie Biomediche, Sez. Chimica Organica e Biochimica, Università di Milano, Via Celoria 2, I-20133 Milano, Italy (Part A)

Dr. Peter J. Schoenmakers, Philips Natuurkundig Laboratorium, Postbus 8000, NL-5600 JA Eidhoven, The Netherlands (Part A)

Dr. T. C. Schunk, Eastman Kodak Research Laboratories, Rochester, NY 14650, USA (Part B)

Dr. Joseph Sherma, Department of Chemistry, Lafayette College, Easton, PA 18042, USA (Part A)

Dr. Lloyd R. Snyder, LC Resources Inc., 26 Silverwood Ct., Orinda, CA 94563, USA (Part A)

Dr. Louis G. M. Uunk, Philips Natuurkundig Laboratorium, Postbus 8000, NL-5600 JA Eidhoven, The Netherlands (Part A)

Dr. Harold F. Walton, Cooperative Institute for Research in Environmental Sciences, University of Colorado, Campus Box 449, Boulder, CO 80309, USA (Part A)

Dr. Nian En Zou, Department of Biochemistry, University of Alberta, Edmonton, Alta. T6G 2H7, Canada (Part B)

List of Abbreviations

A

A	ampere
Å	ångström $= 10^{-8}$ cm
AA	acetic acid
AAS	atomic absorption spectrometry
AASP	Advanced Automated Sample Processor
Ab	antibody
ABA	abscisic acid
AC	alternating current
Ac	acetyl
acac	acetylacetonate
ACM	acryloylmorpholine
ADAM	1-aminoadamantane
Ade	adenine
Ado	adenosine
ADP	adenosine 5′-diphosphate
AFID	alkali flame-ionization detector
AFS	atomic fluorescence spectrometry
Ag	antigen
ag	attogram $= 10^{-18}$ g
Ala	alanine
AMD	automated multiple development
amol	attomol $= 10^{-18}$ mol
AMP	adenosine 5′-monophosphate
AMPS	2-acrylamido-2-methylpropanesulfonic acid
AOAC	Association of Official Analytical Chemists
APIM	N-(4-anilinophenyl)isomaleimide
APIP	N-(4-anilinophenyl)isophthalimide
aq.	aqueous
Arg	arginine
Asp	aspartic acid
ASTM	American Society for Testing and Materials
Asn	asparagine
atm	atmosphere $= 1$ bar $= 760$ torr $= 14.7$ psi $= 10^5$ Pa
ATP	adenosine 5′-triphosphate
AUFS	absorbance units full scale
AZT	3′-azido-2′,3′-dideoxythymidine

B

BAC	*N,N′*-bisacrylylcystamine
BAMPITC	4-(*t*-butyloxycarbonylaminomethyl)phenyl isothiocyanate

BAP	bisacrylylpiperazine
bar	atmosphere = ca. 14.7 psi
BAW	*n*-butanol/acetic acid/water (4:1:5)
BHT	2,6-di-*t*-butyl-*p*-cresol (butylated hydroxytoluene)
Bis	*N,N'*-methylene bisacrylamide
BN chamber	Brenner/Niederwieser chamber
bp	base pair
br	boiling range
Bu	butyl
BuHCl	*n*-butanol/2 *N* HCl (1:1)
BuSA	butanesulfonic acid

C

C	centigrade, celsius
C_4	leukotriene C_4
CA	carrier ampholyte
ca.	circa
CAD	collision-activated dissociation
cAMP	adenosine 3',5'-cyclic monophosphate
CCC	countercurrent chromatography
CCD	chemical composition distribution; countercurrent distribution
CCK	cholecystokinin
cCMP	cytidine 3',5'-cyclic monophosphate
4CDD	tetrachlorodibenzo-*p*-dioxin
5CDD	pentachlorodibenzo-*p*-dioxin
6CDD	hexachlorodibenzo-*p*-dioxin
7CDD	heptachlorodibenzo-*p*-dioxin
8CDD	octachlorodibenzo-*p*-dioxin
CDP	cytidine 5'-diphosphate
CE	capillary electrophoresis; cholesteryl ester
CEC	cation-exchange chromatography
CER	ceramide
CFFF	concentration field-flow fractionation
CGE	capillary gel electrophoresis
cGMP	guanosine 3',5'-cyclic monophosphate
CHFE	contour-clamped homogeneous field electrophoresis
CI	chemical ionization
CIE	crossed immunoelectrophoresis
cIMP	inosine 3',5'-cyclic monophosphate
CL	cardiolipin
CLC	column liquid chromatography
CLSA	closed-loop stripping analysis
CM	carboxymethyl
cm	centimeter
CMC	critical micelle concentration; 1-cyclohexyl-3-(2-morpholinoethyl)car-bodiimide metho-*p*-toluenesulfonate
CMP	cytidine 5'-monophosphate
conc.	concentrated
cP	centipoise
CPC	centrifugal planar (partition) chromatography; coil planet centrifuge
CPCl	cetylpyridinium chloride
CsA	cephalosporin A

CTAB	cetyltrimethylammonium bromide
Ctd	cytidine
CTP	cytidine 5'-triphosphate
cUMP	Uridine 3',5'-cyclic monophosphate
CV	coefficient of variation
Cys	cysteine
Cyt	cytosine
CZE	capillary zone electrophoresis

D

D	dalton
2-D	2-dimensional
D_4	leukotriene D_4
DABITC	4'-*N,N*-dimethylamino-4-azobenzene isothiocyanate
DABS	4'-*N,N*-dimethylamino-4-azobenzenesulfonyl
DABTH	dimethylaminoazobenzene thiohydantoin
dAdo	2'-deoxyadenosine
dADP	2'-deoxyadenosine 5'-diphosphate
dAMP	2'-deoxyadenosine 5'-monophosphate
DAN	2,3-diaminonaphthalene
DANABITC	4'-*N,N*-dimethylnaphthyl-4-azobenzene isothiocyanate
DAP	2,3-diaminopropionic acid
DAPMP	2,6-diacetylpyridine bis(*N*-methylenepyridiniohydrazone)
DAPMT	2,6-diacetylpyridine bis(*N*-methylene-*N,N,N*-trimethylammonio-hydrazone)
DATD	*N,N'*-diallyltartardiamide
dATP	2'-deoxyadenosine 5'-triphosphate
DC	direct current
DCA	dichloroacetic acid
DCCC	droplet countercurrent chromatography
DCE	1,2-dichloroethane
dCMP	2'-deoxycytidine 5'-monophosphate
DCP	direct-current plasma
DCPAES	directly coupled plasma atomic emission spectroscopy
DCTA	dodecyltrimethylammonium
dCtd	2'-deoxycytidine
dCTP	2'-deoxycytidine 5'-triphosphate
DDC	2',3'-dideoxycytidine
DEAE	diethylaminoethyl
DEDTC	diethyldithiocarbamate
DEDTP	diethyldithiophosphate
DEHPA	di(2-ethylhexyl)phosphoric acid
DG	diacylglycerol
DGDG	digalactosyl diacyl glycerol
DGMG	digalactosyl monoacyl glycerol
dGMP	2'-deoxyguanosine 5'-monophosphate
dGTP	2'-deoxyguanosine 5'-triphosphate
dGuo	2'-deoxyguanosine
DHEBA	*N,N'*-(1,2-dihydroxyethylene)bisacrylamide
disc	discontinuous
DIITC	diphenylindenonyl isothiocyanate
dIno	2'-deoxyinosine

DITH	diphenylindenonyl thiohydantoin
DMA	dimethylaniline
DMAC	*N,N*-dimethylacetamide
dm6Ado	2′-deoxy-N^6-methyladenosine
dm6AMP	2′-deoxy-6-methyladenosine 5′-monophosphate
dm5CMP	2′-deoxy-5-methylcytidine 5′-monophosphate
DMAPI	*p-N,N*-dimethylaminophenyl isothiocyanate
DMF	dimethylformamide
DMOX	dimethyloxazoline
DMP	dimethylphenol
DMSO	dimethylsulfoxide
DNBS	2,4-dinitrobenzenesulfonic acid
DNP	dinitrophenyl
DNS	dimethylaminonaphthalene-5-sulfonyl (dansyl)
Do	dopamine
DP	degree of polymerization
DPA	diphenylamine
DRI	differential refractive index
DRIFT	diffuse reflectance Fourier transform (infrared spectroscopy)
dThd	2′-deoxythymidine
dTMP	2′-deoxythymidine 5′-monophosphate
dTTP	2′-deoxythymidine 5′-triphosphate
dUMP	2′-deoxyuridine 5′-monophosphate
DVB	divinylbenzene
E	
E_4	leukotriene E_4
EA	elemental analysis; ethylamine
ECD	electron-capture detector
ECL	equivalent chainlength
EDA	ethylene diacrylate
EDC	electrically driven chromatography; 1-ethyl-3-(3-dimethylamino-propyl)carbodiimide
EDTA	ethylenediaminetetraacetic acid
EFFF	electrical field-flow fractionation
EFFFF	elutriation focusing field-flow fractionation
EG	ethylene glycol
EGP	ethanolamine glycerophosphatides
EHPA	2-ethylhexylphosphoric acid
EI	electron impact
ELISA	enzyme-linked immunosorbent assay
em	emission
en	ethylenediamine
EP	electrostatic precipitator
EPA	Environmental Protection Agency
Eqn.	Equation
equiv	equivalent
ESA	ethanesulfonic acid
ESD	electronic scanning densitometry
ESR	electron spin resonance
Et	ethyl
ETU	ethylenethiourea

| eV | electron volt |
| ex | excitation |

F

FA	fatty acids
FAAS	flame atomic absorption spectrometry
FAB	fast atom bombardment
FC	free cholesterol
FD	field desorption
FDA	Food & Drug Administration
FFA	free fatty acids
FFF	field-flow fractionation
FFFF	flow field-flow fractionation
FFPC	forced-flow planar chromatography
fg	femtogram = 10^{-15} g
FID	flame-ionization detector
FMOC	9-fluorenylmethyl chloroformate
fmol	femtomol = 10^{-15} mol
fod	2,2-dimethyl-6,6,7,7,8,8,8-heptafluoro-3,5-octanedione
FPD	flame-photometric detector
FPLC	Fast Protein Liquid Chromatography
FSOT	fused-silica open-tubular (column)
ft	foot = 30.48 cm
FT	Fourier transform

G

g	gram; gravity
GC	gas chromatography
GDP	guanosine 5'-diphosphate
GLC	gas/liquid chromatography
Gln	glycine; glutamine
Glu	glutamic acid
Gly	glycine
GM	gangliosides
GMP	guanosine 5'-monophosphate
GPC	gel-permeation chromatography
GSC	gas/solid chromatography
GSL	glycosphingolipid
GTP	guanosine 5'-triphosphate
Gua	guanine
Guo	guanosine

H

h	hour
HAC	hydroxyapatite chromatography
HDPE	high-density polyethylene
HECD	Hall electroconductivity detector
HEDC	bis(2-hydroxyethyl)dithiocarbamate
hep	heptane
HETE	hydroxyeicosatetraenoic acid
HETP	height equivalent to a theoretical plate

hex	hexane
HFAA	hexafluoroacetylacetone
HFBA	hexafluorobutyric acid
HFIP	hexafluoro(2-propanol)
hfa	hexafluoropentane-2,4-dione
HIBA	2-hydroxyisobutyric acid
HIC	hydrophobic-interaction chromatography
HID	helium-ionization detector
HILIC	hydrophilic-interaction chromatography
His	histidine
HMDS	hexamethyldisilazane
HPIAC	high-performance immunoaffinity chromatography
HPLC	high-performance liquid chromatography
HPPC	high-pressure planar chromatography
HPPLC	high-performance precipitation liquid chromatography
HPTLC	high-performance thin-layer chromatography
HR	high-resolution
HSA	human serum albumin; hexanesulfonic acid
HSES	hydrostatic equilibrium system
HSTLC	high-speed thin-layer chromatography
HTMAB	hexadecyltrimethylammonium bromide
Hyp	hypoxanthine

I

IA	indole-3-acetamide
IAA	indole-3-acetic acid
IAN	indole-3-acetonitrile
IBA	indole-3-butyric acid
IC	ion chromatography
ICPAES	inductively coupled plasma atomic emission spectroscopy
ICPMS	inductively coupled plasma mass spectrometry
ID	inside diameter
IDP	inosine 5'-diphosphate
IEC	ion-exchange chromatography
IEF	isoelectric focusing
IEP	immunoelectrophoresis
Ile	isoleucine
IMAC	immobilized-metal affinity chromatography
IMP	inosine 5'-monophosphate; ion-moderated partitioning
in.	inch = 2.54 cm
Ino	inosine
IP	ion pairing
IPG	immobilized pH gradients
IR	infrared
ISRP	internal-surface reversed-phase
ITP	inosine 5'-triphosphate; isotachophoresis
IUPAC	International Union for Pure and Applied Chemistry

K

K	degree kelvin
kb	kilobase

kD	kilodalton = 10^3 D
KDO	3-deoxy-D-*manno*-2-octulosonic acid
kg	kilogram
KP	potassium phosphate

L

l	liter
LALLS	low-angle laser light scattering
LC	liquid chromatography
LCP	liquid-crystal polysiloxane
LDPE	low-density polyethylene
LEC	ligand-exchange chromatography
Leu	leucine
LLC	liquid/liquid chromatography
LLDPE	linear low-density polyethylene
LPC	lysophosphatidylcholine
LPE	lysophosphatidylethanolamine
LSC	liquid/solid chromatography
LSD	lysergic acid diethylamide
LSS	linear solvent strength
LT	leukotriene
LTB$_4$	leukotriene B$_4$
Lys	lysine

M

M	molar
m	meter
mA	milliampere = 10^{-3} A
μA	microampere = 10^{-6} A
m6Ade	N^6-methyladenine
m6Ado	N^6-methyladenosine
MAK	methylated albumin on kieselguhr
MBE	moving-boundary electrophoresis
MCC	microcapillary chromatography
M-chamber	microchamber
MCPC	microchamber centrifugal planar chromatography
m5Ctd	5-methylcytidine
m5Cyt	5-methylcytosine
MDA	3,4-methylenedioxyamphetamine
MDMA	3,4-methylenedioxymethamphetamine
MDNP	methyldinitrophenyl
MDPF	2-methoxy-2,4-diphenyl-3[2H]-furanone
ME	methyl ester
Me	methyl
MEC	micellar electrochromatography
MECA	molecular emission cavity analysis
MEK	methyl ethyl ketone
MES	morpholinoethane sulfonate
Met	methionine
meq	milliequivalent = 10^{-3} equivalent
μeq	microequivalent = 10^{-6} equivalent

MFFF	magnetic field-flow fractionation
MGDG	monogalactosyl diacyl glycerol
MGMG	monogalactosyl monoacyl glycerol
mg	milligram = 10^{-3} g
μg	microgram = 10^{-6} g
m2Gua	N^2-methylguanine
m2Guo	N^2-methylguanosine
m2,2Guo	$N^2{}_2$-dimethylguanosine
MH	Mark-Houwink
MID	multiple-ion detection
MIKES	mass-analyzed ion-kinetic-energy spectra
MIM	metastable-ion monitoring
min	minute
m1Ino	1-methylinosine
MIPAES	microwave-induced plasma atomic emission spectroscopy
MIQ	minimum identifiable quantity
ml	milliliter = 10^{-3} l
μl	microliter = 10^{-6} l
mM	millimolar = 10^{-3} M
mm	millimeter = 10^{-3} m
μm	micrometer = 10^{-6} m
MMA	monomethylarsonate
mmol	millimol = 10^{-3} mol
μmol	micromol = 10^{-6} mol
MO	methyloxime
MOG	Mills-Olney-Gaither (method)
mol.	molecular
mp	melting point
MPa	megapascal (10^6 Pa)
MPI	methylphenanthrene index
MPLC	medium-pressure liquid chromatography
MS	mass spectrometry
MS/MS	tandem mass spectrometry
msec	millisecond = 10^{-3} sec
MSWI	municipal solid-waste incinerator
MTBE	methyl t-butyl ether
mV	millivolt = 10^{-3} V
MW	molecular weight
MWD	molecular weight distribution
m1Xan	1-methylxanthine
m7Xao	7-methylxanthosine

N

N	normal
NAD	nicotinamide adenine dinucleotide
NADH	nicotinamide adenine dinucleotide (reduced form)
NADP	nicotinamide adenine dinucleotide 3'-phosphate
NADPH	nicotinamide adenine dinucleotide 3'-phosphate (reduced form)
NALPE	N-acyl lysophosphatidylethanolamine
NAPE	N-acyl phosphatidylethanolamine
NBDCl	7-chloro-4-nitrobenz-2-oxa-1,3-diazole

NBDF	7-fluoro-4-nitrobenz-2-oxa-1,3-diazole
N chamber	normal chamber
NCI	negative chemical ionization
NCPC	normal-chamber centrifugal planar chromatography
ng	nanogram = 10^{-9} g
NI	negative ion
nl	nanoliter = 10^{-9} l
nm	nm = 10^{-9} m
nmol	nanomol = 10^{-9} mol
NMP	N-methylpyrrolidone
NMR	nuclear magnetic resonance
NOEL	no-observed-effect level
NP	sodium phosphate
NPAH	nitrogen-containing polynuclear aromatic hydrocarbon
NPC	normal-phase chromatography
NPD	nitrogen/phosphorus detector
NSAID	nonsteroidal anti-inflammatory drug
NSO	nitrogen, sulfur, and oxygen compounds
NT	nortestosterone

O

OC	organochlorine
OD	outside diameter
ODS	octadecylsilane; octadecylsilyl
OFAGE	orthogonal-field alternation gel electrophoresis
ON	organonitrogen
OP	organophosphorus
OPA	o-phthaldialdehyde
OPLC	overpressured-layer chromatography
OPMLC	overpressured-multilayer chromatography
ORM	overlapping resolution maps
OSA	octanesulfonic acid
OTLC	open-tubular liquid chromatography

P

P	phosphonyl
PA	phosphatidic acid
Pa	pascal = 10^{-5} bar
PAA	polyacrylamide
PACE	programable, autonomously controlled gel electrophoresis
PAD	pulsed amperometric detector
PAF	platelet activating factor
PAG	polyacrylamide gel
PAGE	polyacrylamide gel electrophoresis
PAH	polycyclic aromatic hydrocarbons
PAM	Pesticide Analytical Manual
PAR	4-(2-pyridylazo)resorcinol
PAS	photoacoustic spectrometry
PBM	probability-based matching
PC	planar chromatography; phosphatidylcholine
PCA	perchloric acid

PCB	polychlorinated biphenyls
PCDD	polychlorinated dibenzo-*p*-dioxins
PCDF	polychlorinated dibenzofurans
PCP	pentachlorophenol
PCRD	postcolumn reaction detection
PDAD	photodiode-array detector
PDMS	poly(dimethyl siloxane)
PE	phosphatidylethanolamine
PEG	polyethylene glycol
PEI	poly(ethylene imine)
PEO	poly(ethylene oxide)
PET	poly(ethylene terephthalate)
PFB	pentafluorobenzoate
PFGE	pulsed-field gel electrophoresis
PFGGE	pulsed-field gradient gel electrophoresis
PFPA	pentafluoropropionic acid
PG	phosphatidylglycerol; prostaglandin
pg	picogram $= 10^{-12}$ g
PGB_2	prostaglandin B_2
PGF	prostaglandin F
PGI_2	prostacyclin
Ph	phenyl
Phe	phenylalanine
PI	phosphatidylinositol
p*I*	isoelectric point
PICS	paired-ion chromatographic system
PID	photoionization detector
PITC	phenylisothiocyanate
pl	picoliter $= 10^{-12}$ l
PLOT	porous-layer open-tubular (column)
PMBP	1-phenyl-3-methyl-4-benzoylpyrazolone
PMD	programed multiple development
pmol	picomol $= 10^{-12}$ mol
POP	1,3-dipalmitoyl-2-oleoylglycerol
PPP	tripalmitoylglycerol
ppb	parts per billion $= 10^{-9}$ parts
ppm	parts per million $= 10^{-6}$ parts
PPO	poly(propylene oxide)
ppt	parts per trillion $= 10^{-12}$ parts
Pr	propyl
PRC	protein reaction cocktail
Pro	proline
PRS	protein reaction system
PrSA	propanesulfonic acid
PS	polystyrene; phosphatidylserine
PSA	pentanesulfonic acid
PSD	poresize distribution
psi	pounds per square inch $= 51.77$ torr
PTC	phenylthiocarbamyl
PTFE	polytetrafluoroethylene (Teflon)
PTH	phenylthiohydantoin

PTV	programed-temperature vaporization
PUFA	polyunsaturated fatty acids
PVC	polyvinyl chloride
Py	pyrolysis

Q

Q	quaternary
QAE	quaternary aminoethyl (*N,N,N*-triethylaminoethyl)acrylamide
QSRR	quantitative structure/retention relationship

R

Ref.	Reference No.
RFGE	rotating-field gel electrophoresis
RI	refractive index
R.I.	Retention Index
RIA	radioimmunoassay
RPC	reversed-phase chromatography
R.P.C.	rotation planar chromatography
rpm	rotations per minute
RRT	relative retention time
RSD	relative standard deviation
RT	retention time

S

S	sulfonyl
SA	sulfonic acid
SAN	styrene/acrylonitrile
satd.	saturated
SAX	strong-anion-exchange (chromatography)
SB	styrene/butadiene
SBCD	short-bed continuous development
SCB	short-chain branching
SCE	standard calomel electrode
S chamber	sandwich chamber
SCOT	support-coated open-tubular (column)
SCX	strong-cation-exchange (chromatography)
SD	standard deviation
SDS	sodium dodecyl sulfate
SE	sulfoethyl
SEC	size-exclusion chromatography; steric-exclusion chromatography
sec	second
SEMA	styrene/ethyl methacrylate copolymer
Ser	serine
SERS	surface-enhanced Raman spectrometry
SF	supercritical fluid
SFC	supercritical-fluid chromatography
SFE	supercritical-fluid extraction
SFFF	sedimentation field-flow fractionation
SFFFFF	sedimentation/flotation focusing field-flow fractionation
SHP	shielded hydrophobic phase
SIM	selected (single)-ion monitoring

SIMS	secondary-ion mass spectrometry
SMA	styrene/methyl acrylate
SMMA	styrene/methyl methacrylate copolymer
SNS	sodium naphthoquinone 4-sulfonate
SP	sulfopropyl
SPAH	sulfur-containing polynuclear aromatic hydrocarbon
SPE	solid-phase extraction
SPH	sphingomyelin
sq.	square
SRM	selected-reaction monitoring

T

t	tertiary
TACT	*N,N',N''*-triallylcitrictriamide
TAFE	transverse alternating field electrophoresis
TBA	tetrabutylammonium
TBAOH	tetrabutylammonium hydroxide
TBDMS	*t*-butyldimethylsilyl
TBDMSO	*t*-butyldimethylsiloxy
TBDPS	*t*-butyldiphenylsilyl
TCA	trichloroacetic acid
TCB	1,2,4-trichlorobenzene
TCD	thermal conductivity detector
TCDD	tetrachlorodibenzodioxins
TCDF	tetrachlorodibenzofuran
TDP	thymidine 5'-diphosphate
TEAA	triethylammonium acetate
TEAF	triethylammonium formate
TEAP	triethylammonium phosphate
TEMED	*N,N,N',N'*-tetramethylethylenediamine
temp.	temperature
TFA	trifluoroacetic acid; trifluoroacetyl
tfa	1,1,1-trifluoropentane-2,4-dione
TFFF	thermal field-flow fractionation
TG	triacyl glycerol
THC	tetrahydrocannabinol
Thd	thymidine
THF	tetrahydrofuran
ThFFF	thermal field-flow fractionation
Thr	threonine
Thy	thymine
TIC	total-ion chromatogram
TID	thermionic ionization detector
TLC	thin-layer chromatography
TMP	thymidine 5'-monophosphate
TMS	trimethylsilyl
TMSO	trimethylsiloxy
TNBS	2,4,6-trinitrobenzenesulfonic acid
TREF	temperature rising elution fractionation
Tris	tris(hydroxymethyl)aminomethane
Trp	tryptophan

TX	thromboxane
TXB	thromboxane B
Tyr	tyrosine

U

U chamber	ultramicro-chamber
UCPC	ultramicro-chamber centrifugal planar chromatography
UDP	uridine 5'-diphosphate
UKDE	United Kingdom Department of the Environment
UM	ultramicro-chamber
UMP	uridine 5'-monophosphate
Ura	uracil
Urd	uridine
UTP	uridine 5'-triphosphate
UV	ultraviolet

V

V	volt
Val	valine
Vis	visible range
vol.	volume
v/v	volume-by-volume

W

WAX	weak-anion-exchange (chromatography)
WCOT	wall-coated open-tubular (column)
WCX	weak-cation-exchange (chromatography)
wt.	weight
w/w	weight-by-weight

X

Xan	xanthine
Xao	xanthosine
XMP	xanthosine 5'-monophosphate

Z

ZE	zone electrophoresis

List of Italic Symbols

A	area
A_s	peak asymmetry factor
$B°$	specific permeability coefficient
$C\%$	grams of crosslinker/$\%T$
C_i	capacity
c	concentration
c_m	concentration in the mobile phase
c_s	concentration in the stationary phase
D	diffusion coefficient
D_m	solute diffusion coefficient in the mobile phase
D_p	solute diffusion rate in the pore
D_s	solute diffusion coefficient in the stationary film
D_T	thermal diffusion coefficient
d	mean layer thickness; Stokes diameter
d_c	column internal diameter
d_f	stationary-phase film thickness
d_o	lumen of open-tubular column
d_p	particle diameter
$d_{p,n}$	number-average particle diameter
$d_{p,w}$	weight-average particle diameter
d_t	internal column diameter
E	separation impedance
E_f	electric field
E_s	solvent/solute interaction energy
F	flowrate
F_f	focusing force
f	solute fraction
g	branching parameter
H	plateheight
h	reduced plateheight
hR_F	$100 \times R_F$
I_R	retention index
K	equilibrium distribution constant; partition coefficient
K_a	acid dissociation constant
K_D	distribution coefficient
$K_R°$	infinite dilution gas/liquid partition coefficient
k	Boltzmann constant
k'	capacity factor; retention factor
$\bar{k}$	average k' in gradient elution
k_a	association constant
k_d	dissociation constant

L	column length
M	mass
M_n	number-average molecular mass
M_p	peakmaximum molecular mass
M_r	molecular mass
M_v	viscosity-average molecular mass
M_w	weight-average molecular mass
M_z	centrifugation-average molecular mass
m_T	electrophoretic mobility
N	number of theoretical plates (platenumber); column efficiency
N_{eff}	effective platenumber
N_{req}	platenumber required for a separation
n	peak capacity; also number
n_{R_S}	number of peaks resolved with resolution R_s
n_s	stroke frequency
ng	centrifugal force
P	pressure
P_c	critical pressure
P_E	effective pore radius
P_i	inlet pressure
P_o	outlet pressure
P_r	reduced pressure
P^+, P^-	ion pair
Q	equilibrium quotient
Q_{inj}	amount injected
Q_{max}	loadability
R	retention; relative zone velocity
$\mathscr{R}$	Rydberg constant
$<R>$	radius of gyration
R_F	migration rate relative to the solvent front
R_h	effective hydrodynamic radius
R_i	resistance to flow
R_M	logarithm of the capacity factor
R_s	peak resolution
RRT	relative retention time
RT	retention time
r	correlation coefficient; radius
r_s	diameter of solute molecule
S_i	sensitivity; response factor
T	absolute temperature
$T\%$	(grams of acrylamide + grams of Bis)/100
T_c	critical temperature
T_g	glass-transition temperature
T_r	reduced temperature
t	time
t_o	column deadtime
t_a	retention time of the first band
t_G	gradient time
t_R	retention time
t'_R	corrected retention time
t_r	delay time required for analysis

t_z	retention time of the last band
$\underline{u}$	flow velocity
$\overline{u}$	average linear velocity
u_e	actual velocity; interstitial velocity
u_{eo}	electro-osmotic velocity
u_{nom}	nominal velocity
u_{opt}	optimum velocity
V^o	channel volume
V_o	void volume
V_C	volumetric flow
V_c	column bed volume
V_d	displacement volume
V_e	elution volume
V_h	hydrodynamic volume
V_i	interstitial volume
V_{inj}	injection volume
V_M	volume of the mixing chamber
V_m	volume of mobile phase; column deadvolume
V_p	porevolume
V_R	retention volume
V_r	chamber volume
V_s	volume of stationary phase
V_t	total liquid volume
$\underline{v}$	reduced velocity
$\overline{v}$	average reduced velocity
v_z	peakwidth variance
W	bandwidth at the baseline
W_h	bandwidth at half peakheight
w	distance between channel walls
Z_f	distance traveled by the solvent front

List of Greek Symbols

α	separation factor; ratio of partition coefficients
β	packing-dependent variable
γ	activity coefficient
γ_m	obstruction factor
γ_t	pore tortuosity factor
ΔG°	standard Gibbs free energy
ΔH°	enthalpy
ΔP	pressure drop
ΔS°	entropy
$\Delta\Phi$	change in mobile-phase composition
δ	Hildebrand solubility parameter
δ_f	reduced film thickness
ε°	solvent strength parameter
ε_e	porosity of packing
ε_i	molar absorptivity
ε_m	column porosity
η	viscosity
$[\eta]$	intrinsic velocity
η_t	carrier gas velocity
λ	obstruction factor; flow inequality parameter
μ	electrophoretic mobility
μ_{eo}	electro-osmotic mobility
ρ_p	density
σ	standard deviation
σ_L	standard deviation in length units
σ_t	standard deviation in time units
σ_v	standard deviation in volume units
τ	pore tortuosity factor
Y	pressure resistance factor
Φ	solvent strength
Φ_A	volume fraction of A
ϕ	shape factor
χ	compressibility
ψ	phase ratio; pseudouridine
ψ_B	association factor
ω	column packing parameter; angular velocity

Chapter 12

Inorganic species

PAUL R. HADDAD and EMILIOS PATSALIDES

CONTENTS

12.1 GAS CHROMATOGRAPHY

This section examines recent developments and applications in inorganic GC, mainly for the period 1982-1989, and covers aspects dealing with elements, binary compounds, coordination compounds, anions, and organometallics. Relevant aspects of the thermo-chromatography and supercritical-fluid chromatography of these groups are also included. Developments in inorganic GC generally parallel developments in column tech-nology, detection systems, reaction chromatography systems, and new methods of derivatization. Much of the recent body of information has been examined in monographs [1-3] and comprehensive general reviews [4-9]. More specialized reviews dealing with gases [10], anions [11-14], and coordination compounds [15-20] have also appeared.

12.1.1 Elemental analysis

Elemental analysis for carbon, hydrogen, nitrogen, etc., employing reaction GC is of interest to inorganic as well as organic chemists, and there is continuing development in methodology [21-29], instrumentation [30-34], and applications [9]. Recent developments include coupled gas chromatography/elemental analysis (GC/EA), which has been re-viewed by Rezl [27], and a new method reported by Sullivan and Grob [29] for the simultaneous determination of up to nine elements in a single sample. In the latter method, hydrogenolysis of the sample at ca. 650°C, aided by powdered magnesium followed by a nickel catalyst, results in quantitative conversion of C, O, N, S, P, Si, Cl, Br, and I to the products CH_4, H_2O, NH_3, H_2S, PH_3, SiH_4, Cl_2, Br_2, and I_2 respectively. These are then separated and determined by GC on graphitized carbon.

Reaction GC provides a rapid means for precise determination of C, H, N, O, S in a variety of metals, specialty inorganic chemicals, semiconductor materials, geological sam-ples, and catalysts. Procedures described by Drugov [9] for the determination of C are based either on combustion to CO_2 or plasmochemical reaction in H_2 at 2000 K to CH_4 and C_2H_2. The latter procedure was said to be capable of detecting levels of carbon close to $10^{-5}\%$ by weight. Detection with the ECD may give similar detection sensitivities. Other applications reported by Drugov [9] include oxygen in silicon and sulfur in special-purity inorganic reagents.

12.1.1.1 Elemental analysis and coupled techniques

GC coupled to highly selective detection systems, based on atomic spectroscopy or mass spectrometry, provides very sensitive methods of quantitation, even for difficult elements, such as H, C, O, Cl, Br, I, S, P, B, and Si. Such methods do not provide molecular weights or structural information though and must therefore be considered to be complementary to conventional GC/MS, GC/FTIR, and other coupled spectroscopic techniques. Coupled element-selective techniques have been reviewed by Ebdon et al. [35] for the period up to 1985. Recent developments include gas chromatography/inductively coupled plasma mass spectrometry (GC/ICPMS) [36], but this technique has yet to be applied to inorganic or organometallic compounds. Another recently introduced technique is gas chromatography/Fourier transform inductively coupled plasma atomic emission spectroscopy (GC/FTICPAES) [37]. This technique currently operates in the red to near-infrared between 15 700 cm^{-1} and 7900 cm^{-1} and can detect and record atomic spectral emissions in this region. The elements C, H, N, O, F, Cl, Br, and S can be detected simultaneously in gas chromatographic effluents. A spectrum for the above wavelength range takes less than 1 sec to record. Other developments in this area include improvements in instrumentation, interfacing to capillary columns, and the development of software for control, data processing, and manipulation [35]. It is not difficult to predict that more complex parallel systems, such as GC/MS with ICPMS, perhaps utilizing only a single mass spectrometer, will be developed in the near future. A summary of the main techniques currently described is given in Table 12.1.

Of the various combinations in Table 12.1, GC/AAS and GC/MECA are probably the simplest to assemble, and relatively compact working systems could readily be assembled from available GC, AAS, or flame photometer instrumentation. However, elements such as B, C, H, P, S, Si, and the halides, as well as many refractory metals, cannot be detected with high sensitivity by AAS, and the linear calibration range is usually limited to 1-2 orders of magnitude. GC/MECA is a versatile flame molecular emission technique, which, under suitable conditions, can detect S, P, As, Sb, B, Si, Ge, Sn, C, and N, though detection limits at the picogram level have yet to be achieved.

At present, GC/MIPAES is the most common combination technique for the selective quantification of inorganic and organometallic compounds. MIPAES is highly compatible with GC, as it can operate at ambient pressure and the He or Ar carrier gas can also serve as plasma gas. Other claimed features include picogram level sensitivity for S, P, halogens, and other nonmetals without the need for a purged or vacuum spectrometer, high sensitivity for many metallic species, low operating costs, and long-term stability and reproducibility [43]. Importantly, this technique has been successfully interfaced with capillary columns without loss of sensitivity or resolution [44,45]. The high specificity of such spectrometric detectors, and their ability to operate in combination with capillary columns, means that very complex mixtures of organic, organometallic, and inorganic compounds

References on p. B61

TABLE 12.1

SPECTROSCOPIC AND OTHER ELEMENT-SELECTIVE DETECTION SYSTEMS FOR GC

Technique	Combination technique	Elements detected	Approx. detection limit (ng)	Linear range (order of magnitude)	Features	Refs.
Microwave-induced plasma atomic emission spectroscopy	GC/MIPAES	Most metallic and nonmetallic elements	0.001-10	3-4	Suitable for use with capillary columns. Commercial systems available. Intolerant of large sample peaks, e.g. solvent.	35
Inductively-coupled plasma atomic emission spectroscopy	GC/ICPAES	Most metallic and nonmetallic elements	0.01-100	3-4	Some loss of sensitivity since ICPs are not currently designed to handle GC effluents efficiently. Use of capillary columns recently described.	35
Directly coupled plasma atomic emission spectroscopy	GC/DCPAES	Most metallic and nonmetallic elements	0.01-100	4	Same as GC/ICPAES.	35
Atomic absorption spectroscopy (i) flame (ii) furnace	GC/AAS	Many metallic elements	1-100 0.01-10	1-2 1-2	Typically limited to single-element monitoring. Furnace system recently coupled to a capillary column.	35 38

Atomic fluorescence spectroscopy (i) flame (ii) furnace	GC/AFS	Many metallic and metalloid elements	0.1-10	3-4 3-4	Multi-element capability. Dispersive or nondispersive mode.	35 39, 40
Inductively coupled plasma mass spectrometry	GC/ICPMS	Most metallic and nonmetallic elements		3	Not yet applied to inorganics or organometallics. Isotope ratio measurements.	36
Molecular emission cavity analysis	GC/MECA	S, P, As, Sb, B, Si, Ge, Sn, C, N. Possibly other elements.	2-100	1-2	Related to flame-photometric detector.	41, 42

References on p. B61

can be monitored for a number of elements simultaneously. The great majority of applications of combination techniques deal with either organic compounds containing P, S, Si, B, or halogens or with organometallic compounds of Hg, As, Se, Pb, and Sn. For the latter group, speciation and analyses of a broad range of sample types, including biological and environmental materials, air, water, fuels, and standard reference materials, have been carried out. Volatile compounds can be determined directly following extraction, cleanup, and preconcentration. Complex, volatile mixtures, e.g., pyrolyzates or synthetic fuels originating from coal or shale oil, containing compounds of S, P, Se, As, and Si, are particularly suitable for analysis by these techniques. Involatile species can be converted to volatile alkyl, silyl, or hydride derivatives prior to analysis. Alternatively, such samples can be analyzed by cognate combination techniques utilizing HPLC or SFC [46]. Purely inorganic compounds or metal chelates have been studied to a much lesser degree. Indeed, where total element determinations are required, there may be no advantage in GC separation, as determinations can be effected spectroscopically.

12.1.1.2 Elemental forms

The GC determination of an element in its *elemental* form(s) represents a particular case of element speciation. Thus, the determination of residual chlorine in water requires the determination of Cl_2, as distinct from aqueous chloride. In principle, Cl_2 can be determined directly by GC, but in practice, derivatization is more appropriate. Ellis and Brown [47], e.g., have determined residual Cl_2, by derivatization to 4-chloro-2,6-dimethyl-phenol and determination of the latter, following silanization, by GC/FID. Chlorine concentrations in the range 0.01-8.6 mg/l were determined with relative standard deviations below 1%.

Other elements determined by GC (apart from the element gases) include sulfur, phosphorus, and iodine. Sulfur can be determined directly by GC [48], although a number of peaks attributable to the molecular species S_2-S_{16} were observed in the chromatogram. Alternatively, derivatization to triphenylphosphine sulfide [49] or conversion to H_2S by reaction GC hydrogenation [50] can be utilized and applied to the determination of sulfur in antimony [51], sediments [50], and Kraft pulping liquors [49]. In the last application, elemental and polysulfide sulfur were converted to the triphenylphosphine sulfide derivative, and a toluene extract was analyzed on a packed OV-17 column with temperature programing between 120 and 280°C and flame-ionization detection. Polysulfide sulfur concentrations in the range 0.3-17.1 g/l agreed closely with results obtained by the standard amalgam method [49].

Elemental phosphorus (as P_4) has been determined in samples from an aquatic environment [52]. In the reported procedure, phosphorus was extracted with benzene, and after cleanup, the extract was analyzed on a packed OV-1 column at 210-230°C using an AFID. In water, P in the range 0.08-1.30 μg/l was determined.

The separation of elemental iodine from other forms of this element in emissions from nuclear facilities has been attempted by GC [53]. However, tracer studies indicated that as much as 20% of elemental iodine introduced into the column was strongly retained. A derivatization similar to that described above for chlorine may, therefore, be required.

Other elements, with the exception of Hg, may be too reactive or involatile for conventional GC. However, a variety of metallic elements (Na, K, Cs, Ba, Eu, Yb, Tm, Tl, Pb, Bi, Po, Am, Cf, Fm, Md, Lr, etc.) [54-58] can be chromatographed by thermochromatography (Section 12.1.2.3) at 600-1400 K in titanium, graphite, or quartz columns. Thermochromatography includes any GC technique that operates above the maximum temperature of conventional GC, ca. 400°C. Typically, this will involve the use of tube furnaces with packed or unpacked quartz or other tubing as the chromatographic column. Although this technique does not at present appear to be suitable for analytical element separations, it does allow the adsorption and volatilization behavior of metallic elements to be studied and has been useful for the characterization of heavy actinide elements.

12.1.2 Binary compounds of metallic and nonmetallic elements

The main groups to be considered here comprise monomeric hydrides, halides, and oxides sufficiently volatile to be determined by conventional GC or by thermochromatography. Gaseous species, e.g., CH_4, NH_3, H_2S, CO, CO_2, and SO_2 are not discussed.

12.1.2.1 Water

For the determination of water, GC is rapidly evolving as a very selective and sensitive method. It can replace, or complement, established methods, based on oven drying, Karl Fischer titrations, infrared absorption, and thermogravimetry [59,60].

For determining water in liquids and solids above ca. 0.1%, established procedures based on chromatographic separation on porous polymers and thermal conductivity detection are suitable, provided water is well separated from other components in the chromatogram. Sorbents described for this purpose are based on polystyrenes, including Porapaks Q, N, T, Super Q, Chromosorbs 101, 102, 104, Polysorb 1, GPX-103 [61-65]; methacrylate polymers and copolymers [66,67], and carbon molecular sieves, such as Carbosieve S and Carbopack C [61]. On many porous polymers tailing of the water peak occurs. This can be reduced by the addition of polar compounds, such as methanol, to the carrier gas [68]. Porous polymers coated with polar stationary phases have also been enlisted to decrease tailing and retention and to improve selectivity [69]. Recent applications of such columns are described for water determination in vitamin C [70], liquid ketones [66,71], crude mineral oil [64], pharmaceuticals [65], liquid petroleum [72], and high-purity hydrides [73]. Because of the limited sensitivity of this procedure, water concentrations below 0.1% require a preconcentration step [63], which increases analysis time.

Low concentration (<0.1%) of water in solids, liquids, and gases can be determined indirectly by reaction GC. In these methods water is converted to hydrogen, methane, acetylene, or other organic compounds [9]. Their advantage is that they permit the use of more sensitive detectors, such as the FID. The method most widely favored requires the conversion of water to acetylene [74] according to Eqn. 1.

$$CaC_2 + H_2O \rightarrow CaO + C_2H_2 \tag{1}$$

For solids and liquids this method is normally combined with headspace gas analysis [75-77]. Acids interfere, as do high concentrations of low-molecular-weight alcohols, since these react with calcium carbide to generate additional acetylene. Such headspace procedures have been used to determine the moisture content of coal [78], of chemical and pharmaceutical products [76], and of liquid fuels [79] down to about 40 ppm. Methods for water in natural gas [80], ammonia [81], and nitrogen [82] at ppm levels have also been reported.

The recent application of open-tubular columns to water determination is of great significance, since such columns can be utilized for rapid determination of ppm levels of water with high specificity. Thus, a procedure for determining water below 2 ppm in gases or liquids by means of a fused-silica Carbowax 20M column without preconcentration and with the helium ionization detector has been reported by Andrawes et al. [61,62]. Significantly, the determination limit was controlled by moisture contamination rather than limitations in sensitivity of the detection system. Megabore open-tubular columns (DBWAX, DB-210, or DB-17) together with a TCD have also been found suitable for determining moisture levels in organic solvents below 10 ppm [83]. Use of the ECD with selective electron-capture sensitization may also be useful for this application [84]. The utilization of such systems will almost certainly largely replace those based on porous polymers and indirect procedures.

12.1.2.2 Other hydrides

Other hydrides studied by GC include those of boron and element groups IVA, VA, VIA, and VIIA. Compounds of group VIIA, i.e. hydrogen halides, are highly reactive and corrosive and are best converted to suitable derivatives prior to GC (Section 12.1.4). The hydrides of C, Si, Ge, and B produce extensive homologous series: alkanes, silanes, germanes, boranes, and carboranes, carbosilanes, etc. Their GC characterization is exceedingly useful. The chromatographic properties of these and other hydrides are well documented in earlier monographs [3, 85] and are not re-examined here. It is sufficient to state that such compounds can be chromatographed on conventional nonpolar or porous polymer columns at 50-200°C without difficulty.

Recent interest in metalloid hydrides stems from development of the hydride generation technique and its utilization in trace analysis and speciation studies. In this technique, hydrides of inorganic or organic compounds of As, Sb, Bi, S, Se, Te, Ge, and Sn are generated in aqueous solution by treatment with sodium or potassium borohydride, and the volatile hydrides so generated are analyzed spectroscopically or by GC [86]. Preconcentration by means of cryogenic trapping or sorbents can also be used. Detection can be effected with the gold gas-porous electrode [88], PID [89,90], ECD [91], or element-selective detectors, based on AAS, AES, or AFS. Fry et al. [90] developed a procedure based on the PID for simultaneously determining As, Sn, Sb, and Se in water with determination limits as low as 0.001 ppb for a 28-ml sample. However, H_2Se was difficult to elute at subnanogram levels, and careful choice of columns and column conditioning was required. Similar procedures have been described for trace determination of As in food-grade phosphoric acid and for Ge, Sn, Pb, Sb, Bi, Se, and Te in environmental samples [93].

Phosphorus can also be converted to its hydride, but more drastic reaction conditions are required [94]. The sample containing phosphate or other form of phosphorus is mixed with aq. $NaBH_4$ in a quartz boat and dried at 40°C. The residue is then flash-heated at 460°C in a tube furnace, and the generated phosphine is analyzed by GC/FPD. This method has been applied to the determination of phosphorus in pond and seawater [95] and is suitable for levels below 0.1 ppm. Phosphorus in the form of phosphine is used as a fumigant for stored cereals. A GC method for determining PH_3 residues below 0.001 ppb in such samples with the AFID has been reported [96] and seems suitable for routine use.

12.1.2.3 Halides

Halides studied by GC include those of element groups IIIA, IVA, VA, VIA, the interhalogens ClF_2, ClF_3, BrF_3, BrF_5, etc. and the halides of transition metals, such as Ti, V, Mo, and W. The GC of these compounds and the requirements for inert chromatographic systems have been examined in great detail by Guiochon and Pommier [85]. Only recent aspects need to be examined here.

Many reactive halides are utilized industrially on a large scale and, despite the problems, GC is usually the most suitable method of analysis [85]. Usually, the determination of impurities, decomposition or byproducts is of interest. Methods for impurity determinations have been developed for PCl_3 and $POCl_3$ [97], $SiCl_4$ [90], $TiCl_4$ [89], SF_6 [100,101], $SnCl_4$ [102], and HCl [103].

Not all halides are highly reactive. Compounds, such as CF_4, CCl_4, C_2Cl_6, C_3Cl_8, C_4Cl_6, C_6Cl_6, and SF_6 (but not other sulfur halides) are quite stable and be can chromatographed on conventional columns without special precautions. Sulfur hexafluoride (SF_6), in fact, is so stable that its industrial use is resulting in a steady concentration increase in the atmosphere [104]. A highly sensitive GC method for determining SF_6 in air is based on a Molecular Sieve 5A column and the ECD at 300°C. A linear calibration curve for the range 0-0.3 pg SF_6 was obtained [104], and it was possible to determine < 0.1 ppt SF_6 in air. GC methods for toxic and other breakdown products of SF_6 generated in electrical or welding environments (including CF_4, C_2F_6, C_3F_8, C_4F_{10}, C_7F_{14}, SO_2, SOF_2, SO_2F_2, SF_4, CF_3OSF_5, and $S_2F_{10}O$), have also been reported [101,105].

Halides of low volatility, including those of Zn, Cd, Np, Tc, Nb, Zr, Re, Os, Hg, Ir, U [106,107], and those of the lanthanides [108,109], have been analyzed at temperatures up to 1200°C by thermochromatography. This is carried out in graphite or silica columns with or without suitable alkali metal halides as stationary phase. Elution is facilitated by

References on p. B61

TABLE 12.2

DERIVATIZING REAGENTS FOR THE GAS CHROMATOGRAPHY OF METAL IONS

Ligand	Structure	Selectivity	Fluorinated derivatives studied[*]	Ions detected at pg level[**]	Refs.
β-Diketone	I	broad spectrum	yes	Be(II), Al(III), Cr(III) Cr(VI), Rh(III)	4, 16
Monothio-β-diketone	II	class b ions	yes	Ni(II), Pd(II)	16, 116
Dithio-β-diketone	III	class b ions	yes		117, 118
β-Ketoenamine	IV	Cu(II), Ni(II), Pd(II)	yes		119
β-Ketoenamine (tetradentate)	V	bivalent ions of coordination no. 4	yes	Ni(II), Cu(II), Pd(II), Co(II), V(IV)O	4, 120, 121
β-Ketoenamine (hexadentate)	VI	trivalent ions of coordination no. 6	no		122
β-Thionoenamine	VII	class b ions	yes	Ni(II)	123
β-Thionoenamine (tetradentate)	VIII	class b bivalent ions of coordination no. 4	yes		124, 125

N,N-Dialkyldithiocarbamate	IX	mainly class b ions	yes	Ir(III), Ni(II), Rh(III) Co(III), Pd(II), Cr(III) Cr(VI), Pt(II)	126-131
O,O'-Dialkyldithiophosphate	X	class b ions	no	Zn(II), Ni(II), Pd(II)	132, 133
Dialkyldithiophosphinate	XI	class b ions	no		134, 135
Porphyrin		bivalent ions of coordination no. 4	no		136

* For literature to 1988.

** Refers to ECD, except for dialkyldithiophosphate complexes where the FPD was used.

References on p. B61

addition of suitable reactive components (e.g., Cl_2 or CCl_4) to the carrier gas. By incorporating vapor-phase complexing agents (BCl_3, $AlCl_3$, $FeCl_3$) it has been possible to elute such halides at considerably lower column temperatures (200-600°C). Thermochromatography of halides has been found useful for very rapid preseparation and preconcentration of actinides and radioisotopes of other elements prior to spectrometric analysis [110]. An application of this method is the determination of plutonium in soil [111].

12.1.2.4 Other binary compounds

Apart from common gases, such as CO, CO_2, N_2O, NO_2, and SO_2, other binary compounds sufficiently volatile for GC comprise mainly an odd collection of oxides (N_2O_3, N_2O_5, N_4S_4, P_2O_3, P_2O_5, SeO_2, SeO_3, TeO_2, As_2O_3, and MoO_3), sulfides (e.g., CS_2, SiS_2, AsS_2, P_2S_3, P_2S_5), and selenides (C_2Se_4, Se_4S_4, Se_2S_6). Of these, only CS_2 is of common interest and has been determined at ppt levels in air by GC/MS [112]. Metallic oxides of Cm, Am, Pu, Np, U, Ir, Pt, Re, Tc, and Os have also been chromatographed at temperatures up to 1200°C on metallic surfaces [113-115]. The sublimation behavior of the oxides was found to be strongly affected by the nature of the metal surface. Heats of adsorption and some separations were reported.

12.1.3 Coordination compounds

The GC of coordination complexes encompasses trace determinations of metals, physicochemical measurements, the study of on-column reactions, the separation of isomers, and the utilization of complexes as components of the stationary phase for enhancing selectivity.

12.1.3.1 Trace determination of metals

The determination of metal complexes by GC, in the broadest sense, is difficult and often frustrating. At present, GC procedures for the trace determination of entire metal groups, such as the groups IA, IIA, lanthanides, and actinide elements are, simply, non-existent. On the other hand, elements such as Be, Al, Cr, Co, Rh, Cu, Ni, Pd, and V can be determined individually or in small groups at concentrations which compare favorably with other sensitive methods of metal determination. A list of ligand types examined to date and the applicability of these in trace determinations is given in Table 12.2.

The β-diketones and their analogs comprise the largest and most extensively studied GC reagents for metal ions. The parent β-diketones are broad-spectrum reagents, forming complexes with nearly all metallic ions [137,138]. Unfortunately, methods for trace determinations based on those reagents are limited essentially to the ions in Table 12.2. For such methods the fluorinated ligand 1,1,1-trifluoropentane-2,4-dione (Hfta) is mainly used, together with solvent extraction and electron-capture detection. Other ligand analogs of the β-diketones, such as those represented by Structures II-IV and VII, are more selective and have only been applied to the trace determination of Ni(II) [16,117,119,122].

Tetradentate ligands of the type represented in Structure V, due to the greater stability of the complexes, have been more successful. Trace determinations with fluorinated reagents of this type have been developed for Cu(II), Ni(II), Co(II), Pd(II), and V(IV)O [4]. However, complexes of the corresponding β-thionoenamines, represented by Structure VIII, even fluorinated types, have been disappointing [123,125]. Hexadentate Schiff bases, indicated in Structure VI, are selective for trivalent transition ions, such as Fe(III), Co(III), V(III), and Cr(III) and form derivatives with excellent chromatographic properties. Unfortunately, the latter do not appear to to be readily formed in derivatization reactions in aqueous solution [123].

Another important class of derivatizing reagents comprises the sulfur donor ligands N,N-dialkyldithiocarbamates, O,O'-dialkyldithiophosphates, and dialkyldithiophosphinates, as represented in Structures IX, X, and XI, respectively. Complexes of related ligands, such as xanthates, thioxanthates, and dithioalkylates may also exhibit favorable GC properties. These reagents differ from the β-diketone group in the size of the chelate rings (4- rather than 6-membered). The stability of their complexes is probably due to strong dπ-dπ metal → sulfur back-bonding. Fluorinated complexes of the dialkyldithiocarbamates (IX) have been studied by GC and exhibit greatly enhanced volatility and ECD response relative to the corresponding nonfluorinated complexes [128]. However, fluorinated complexes of the dialkyldithiophosphates (X) have yet to be successfully synthesized [141]. At least 17 derivatives of the fluorinated dialkyldithiocarbamate (IX, where R = CF_3CH_2-) can be chromatographed satisfactorily at the nanogram level. Detection, at subnanogram levels, of the complexes of ligands IX to XI can be effected with the ECD, FPD (sulfur mode), or AFID [132,133]. Applications of these reagents in trace determinations are not extensive at present [131,142,143]. Many labile derivatives cannot be satisfactorily chromatographed at picogram levels, even on fused-silica capillary columns. Also, on-column interelement effects, reminiscent of the β-diketonates, are operative [131], and this may limit the wider application of these reagents.

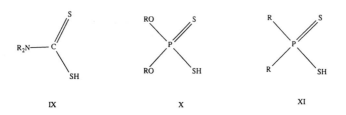

An example of macrocyclic reagents that have been utilized in GC are the porphyrins [136]. Derivatives of these reagents (metalloporphyrins) can be satisfactorily chromatographed at elevated temperatures (300-400°C) on fused-silica columns. However, studies of such complexes seem to be aimed chiefly at identifying natural metalloporphyrins in crude petroleum oils. Eglinton et al. [144], for example, have utilized high-temperature GC/MS to identify vanadyl porphyrins in shale oil. These complexes were eluted from a glass capillary column, coated with the OH-terminal polysiloxane phases PS086 or OV-225-OH at 400°C.

The broader application of coordination complexes for the trace determination of metallic species by GC is largely limited by the adverse on-column behavior exhibited by many metal complexes even on fused-silica capillary columns. Nonideal behavior includes elevated baselines, peak broadening and tailing, reversible and irreversible adsorption, and displacement effects [16]. These are, in turn, attributed to a variety of factors, including homolytic and heterolytic dissociation, adsorption, chemisorption, oxidation, hydrolysis, dehydrogenation, and catalytic decomposition [16]. The observed behavior depends not only on the metal ion and ligand, but, obviously, also on stationary-phase polarity, support, and plugging material (for packed columns), construction materials, temperature, and history of the column. Standardized procedures for detecting and quantifying nonideal behavior have been developed [26] and may be useful for testing new derivatives or columns. The chief problem in this area is the lability of coordination complexes. This has become evident from the behavior of a variety of different chelate systems and from similar problems experienced in HPLC [145,146] and SFC [147]. One way to overcome this problem is to incorporate ligand vapor in the carrier gas [148-150]. However, this method is feasible only when the ligand is volatile. The reaction of ligand with metallic surfaces in the syringe, injection system, or detector, and the problems of detection in the presence of a vast excess of ligand represent further problems. For example, it is not possible to use the ECD effectively when ligand is present in the carrier gas. Although this problem can be overcome by utilizing element-selective monitoring

[151], the inability to use the ECD is a distinct disadvantage. Ultimately, the best way to overcome the problem of lability may be to utilize macrocyclic, or macrobicyclic ligands to enhance chelate stability. Ligands of this type are known to give highly stable derivatives, even with ions that normally give highly labile complexes [152-154]. However, the very high selectivity of such ligands means that a particular reagent may be suitable for only a few metal ions.

12.1.3.2 Coordination complexes as components of the stationary phase

By virtue of their Lewis acidity, coordinatively unsaturated complexes can alter stationary-phase selectivity in GC. Such complexes can therefore be very useful for selective GC separations. As illustrated in Table 12.3, the complexes can be incorporated by adsorption or chemical bonding to the support, they can be dispersed or dissolved in the stationary phase or used as polymeric sorbents. In all cases, the special interactions exploited are Lewis acid/base interactions, with the metal ion acting as the Lewis acid. In a broader sense, such specific interactions are the basis of complexation GC, reviewed by Szczepaniak et al. [155,169].

For coordination complexes or other metallic additives, two types of specific interactions are usually exploited: hard acid/base and soft acid/base interactions. An example of selectivity based on hard acid/base interactions is illustrated in Fig. 12.1 for a packed column containing Eu(fod)$_3$, dissolved in squalane as stationary phase. It can be seen

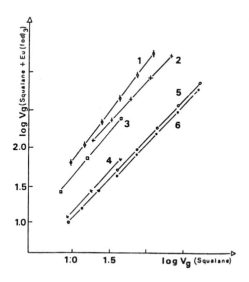

Fig. 12.1. Log/log plot of specific retention (V_g) on a squalane column impregnated with Eu(fod)$_3$ versus a squalane-only column, both at 120°C. 1 = C$_3$-C$_8$ n-alcohols; 2 = methyl esters of C$_4$-C$_8$ n-carboxylic acids; 3 = diethyl, di-n-propyl, di-n-butyl ketones; 4 = diethyl, di-n-propyl, di-n-butyl ethers; 5 = C$_2$-C$_{12}$ n-alkanes; 6 = C$_8$-C$_{12}$ n-1-alkenes; fod represents the ligand anion of the β-diketone 2,2-dimethyl-6,6,7,7,8,8,8-heptafluoro-3,5-octanedione (Hfod). (Reproduced from Ref. 157 with permission.)

TABLE 12.3

COORDINATION COMPOUNDS AS SORBENTS AND STATIONARY-PHASE ADDITIVES

Complex	Column	Selectivity and application	Refs.
Dithiophosphinates Zn(II), Cd(II), Co(II), Ni(II), Pd(II)	polymeric sorbents	various nucleophiles	155
Phosphinates Cu(II), Mg(II), Co(II)	polymeric sorbents	nucleophiles and olefins	155
Schiff base complex Co(II)	immobilized and activated by addition to polystyrene/vinyl pyridine copolymer sorbent	selectively retains oxygen; applied to analysis of gas mixtures at ambient temperature	156
β-Diketonates Eu(III)	ca. 1% by wt. in squalane	nucleophiles; alcohols > methyl esters > ketones > ethers	157
Bisdiphenylphosphine halides Cu(II), Co(II)	metal halides, anchored to silica-bonded phosphinosilanes	selective for olefins and other compounds with π bonds	158-161
β-Diketonates La(III), Cu(II), Ni(II), Zn(II)	polymeric sorbents	nucleophiles; utilized in precolumns	162
β-Diketonates Eu(III)	polymeric sorbents	nucleophiles; utilized in precolumns	163

Fluorinated β-diketonates Eu(III), Pr(III)	dissolved in SE-30	nucleophiles; nitrogen bases > alcohols > ketones > ethers	164
Fluorinated β-diketonates La(III)	dissolved in SE-30	nucleophiles; aldehydes, ketones, esters, ethers, alcohols, etc., separated from hydrocarbon and halogenated compounds	165
3-Heptafluorobutanoyl-1R-camphorate Mn(II), Co(II), Ni(II)	coated onto open-tubular column in admixture with squalane	nucleophiles; separation of enantiomers of ethers, ketones, alcohols, esters, etc.	166, 167
Polycomplexonates of vinyl pyridine, styrene, and glycidyl methacrylate copolymers Cu(II), Ag(I), Hg(II)	used as sorbents	alkenes, aromatics; halogen, nitrogen and sulfur compounds	168

References on p. B61

that the presence of a complex has little effect on the retention of alkanes, alkenes, and ethers, but causes pronounced retention of stronger nucleophiles, such as alcohols, carboxylic acids, and ketones [157]. An application of polymeric lanthanide β-diketonate sorbents is in precolumns for selectively abstracting and preconcentrating nucleophiles in complex volatile mixtures [163,164]. Hydrocarbons and halogenated hydrocarbons are not retained on such sorbents and can be adsorbed and preconcentrated on a second analytical column by thermal focusing prior to analysis. Nucleophilic molecular species retained on the first column can then be desorbed and analyzed as a separate fraction. Another application is the separation of enantiomeric mixtures, based on coordination complexes containing chiral centers, such as the β-diketonate represented in Structure XIV.

XIV

These stationary-phase additives have been useful for separating mixtures of racemic Lewis bases, such as ethers, ketones, and alcohols, where the center of asymmetry in the molecules in close to a Lewis base site. An example of the separation routinely possible is shown in Fig. 12.2. Nonetheless, the active chiral complexes are relatively volatile, and column temperatures above 100°C are not suitable. An example of column selectivity, based on soft acid/base interaction is shown in Fig. 12.3. Here, Cu(II), present as bisdi-phenylphosphinedichlorocopper(II) interacts selectively with compounds containing π bonds (alkenes, aromatics, etc.). Although similar effects can be obtained with inorganic salts, such as $CuCl_2$, CuCl, $AgNO_3$ [169], the coordination complex is stereoselective [159] due to the coordinating ligands, which moderate the metal/solute interactions. Thus, the order of elution of unsaturated C_5 hydrocarbons is Z-2-pentene > 1-pentene > E-2-pentene.

12.1.3.3 Physicochemical and related aspects of the GC of coordination compounds

Methods for determining the heats of solution of volatile complexes, based on the procedure described by Littlewood [170], can be utilized to determine the nature of intermolecular forces experienced by coordination compounds in the stationary phase and to study changes brought about by fluoroalkyl substitution in the complex. Based on the similarity of the heats of solution of hexafluoropentane-2,4-dione (hfa) complexes in SE-30 and Apiezon L, for example, Wolf et al. [171] concluded that interactions of the

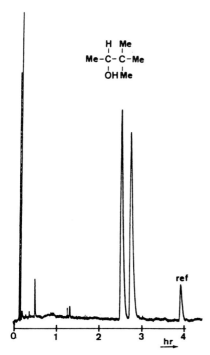

Fig. 12.2. Separation of the enantiomers of 3,3-dimethyl-2-butanol on a squalane column containing the 1R-enantiomer of the nickel complex indicated in Structure XIV (ca. 0.2 M solution). Nickel capillary column (100 m x 0.5 mm ID) at 70°C, 3.0 ml/min, N_2 (split ratio 1:50). The assignment of each peak to a particular enantiomer was not indicated. (Reproduced from Ref. 166 with permission.)

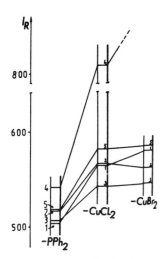

Fig. 12.3. Retention indices (I_R) of C_5 hydrocarbons: 1 = 1-pentene; 2 = Z-2-pentene; 3 = E-2-pentene; 4 = 1-pentyne; 5 = 2-methyl-1,3-butadiene. -PPh₂ = diphenylphosphineorganosilane column only; -CuCl₂ = -PPh₂ column containing coordinated CuCl₂; -CuBr₂ = -PPh₂ column containing coordinated CuBr₂. (Reproduced from Ref. 159 with permission.)

References on p. B61

complexes were similar in both phases. However, a distinction between the behavior of octahedral and square planar complexes was evident. Whereas nonspecific dispersion forces were observed for octahedral complexes, specific, presumably Lewis acid/base interactions, were indicated for the square planar complexes. Thus, square planar complexes are more sensitive to stationary-phase composition, and this has been exploited in GC separations [172]. Separation of octahedral complexes would thus seem to be dependent on small differences in dispersion forces and may impede their GC separation. Complexes derived from ligands of Structure V exhibit strong stationary-phase selectivity effects, and this can also be exploited in GC separations [178].

The separation of reaction products of volatile coordination compounds is possible when the former are thermally stable and volatile [174,175]. However, the isomeric species must be stable with respect to isomerization. For example, among the β-diketonates, where geometrical isomerism and optical isomerism are both possible, separation of isomeric species has been reported only for the substitution-inert complexes of Cr(III), Rh(III), and Mo(III) [176,177]. However, even these complexes exhibit on-column isomerization at the temperatures (100-200°C) required for elution. Complexes of quadridentate Schiff bases indicated in Structure V exhibit ligand isomerism which does not result in on-column isomerization, even at elevated temperatures, and isomers of these complexes can readily be separated [178].

Isomerization, decomposition, and other reactions of volatile complexes can be studied by on-column methods [121,179] or by the more versatile pulsed GC reactor method of Dyagileva et al. [180]. Marriott and Lai [179], using the on-column method, have shown that at 398-458 K isomerization of the geometrical isomers of Cr(tfa)$_3$ proceeds first-order with respect to chelate concentration. The conversion of the cis- to the trans-isomer was found to be more facile than that of trans- to cis-isomer. Using a similar technique, Patsalides and Robards [121] have shown that β-thionoenamine complexes undergo pseudo-first-order decomposition on fused-silica open-tubular columns at 190-220°C. The halflives for the Zn(II), Cu(II), and Ni(II) complexes were calculated to be 0.72, 3.1, and 29.6 min, respectively, at 190°C. GC studies of the thermal decomposition of lanthanide acetylacetonates at 150°C were reported by Khalmurzaev et al. [181]. Decomposition was found to proceed essentially as shown in Eqn. 2. A side-reaction is shown in Eqn. 3.

$$Ln(acac)_3 \cdot H_2O \rightarrow LnOH(acac)_2 + Hacac \tag{2}$$

$$Ln(acac)_3 \cdot H_2O \rightarrow Ln(CH_3CO_2)(acac)_2 + CH_3COCH_3 \tag{3}$$

Pyrolysis GC studies of coordination complexes are, surprisingly, relatively rare. Uden et al. [182] utilized pyrolysis GC/MS and pyrolysis GC/IR to identify the organic pyrolysis products of the Cu(II) and Ni(II) complexes of the Schiff base N,N'-ethylenebis(salicylaldimine). Similar studies by Bratspies et al. [183,184] of dihalodiethyldithiocarbamate complexes of Sn(IV) indicated that the major reaction products (CS$_2$, Et$_2$NH, EtNCS, SnS, SnS$_2$, Et$_2$NCSNEt$_2$) are formed by a combination of ionic and free-radical mechanisms.

12.1.4 Anions

The determination of inorganic anions by GC can be considered to be complementary to that by ion chromatography and conventional HPLC. This makes chromatography powerful and versatile for trace anion determination. The basic requirement for anion determination by GC is the facile conversion of an anion to a neutral, volatile derivative. Typically, the latter is selected to have favorable detection properties as well. Whereas in the GC of metal ions, the volatile species is a coordination compound, in anion derivatization a volatile derivative is usually produced by the formation of a relatively stable covalent bond. Indeed, such bonds can be formed by a variety of methods, and there is no doubt that many more types of derivatives will be examined in the future. In the simplest case, anions of weak acids are converted by acidification to the corresponding conjugate acids or acid anhydrides, and the gaseous products (e.g., CO_2, H_2S, HCN, and SO_2) are determined. A more general procedure involves nucleophilic displacement, where the analyte anion (A^-) displaces a ready leaving group, L^-, from the neutral reagent, RL (Eqn. 4), to give the derivative AR.

$$A^- + R\text{--}L \rightarrow A\text{--}R + L^- \tag{4}$$

Other derivatization reactions include electrophilic substitution, condensations, and 1,2- or 1,4-additions. A summary of derivatizing reagents and recent application in GC analyses is given in Table 12.4.

Indirect methods for determining inorganic anions as volatile β-diketonates have been described by Kito and Komatsu [214,215]. In the procedures described for fluoride and sulfide, reduction in the GC response for suitable Be(II) and Cu(II) β-diketonates, respectively, were ascribed to the following reactions, where L = β-diketonate ligand:

$$BeL_2 + 4F^- + 2H^+ \rightarrow BeF_4{}^{2-} + 2HL$$

$$CuL_2 + S^{2-} + 2H^+ \rightarrow CuS + 2HL$$

Similar reactions can be applied for determining other strong-complexing anions, such as CN^-, CNS^-, or I^- by using suitable substrate β-diketonates or other reagents.

Other indirect methods are based on homogeneous catalysis. For example, CN^- has been determined at concentrations below 0.1 μg/ml in water by its catalytic effect on the methanolysis of benzoin to benzaldehyde [216,217]. However, such methods appear less desirable than conventional, direct methods.

For derivatization, either broad-spectrum reagents or nearly specific reagents can be used. Broad-spectrum reagents include silylating reagents [203] and alkylating reagents [187-195]. Whereas the former generally give hydrolytically unstable derivatives, alkylating reagents have been used successfully with aqueous systems involving phase-transfer catalysis to facilitate extraction and derivatization. With pentafluorobenzylmethanesulfonate it was also possible to alkylate Br^-, I^-, and CNS^- directly in aqueous solutions,

TABLE 12.4

DERIVATIZING REAGENTS FOR THE DETERMINATION OF ANIONS BY GAS CHROMATOGRAPHY

Reagent	Anions	Reaction type (selectivity*)	Derivative	Refs.
1,2-Aryldiamine	SeO_3^{2-} NO_2^-	condensation (HS)	piaselenol benzotriazole	14 185
N-Alkyl-1,2-aryldiamine	NO_2^-	condensation (HS)	N-alkylbenzotriazole	185, 186
N-Butyl-p-toluenesulfonate	Br^-, I^-, CNS^-, NO_3^-	nucleophilic displacement (LS)	n-butyl derivative	187
Pentafluorobenzyl-p-toluenesulfonate	Br^-, I^-, CN^-, SCN^-		pentafluorobenzyl derivative	188, 189
Pentafluorobenzylmethane-sulfonate	Br^-, I^-, CN^-, SCN^-		pentafluorobenzyl derivative	190
Pentafluorobenzyl bromide	CN^-, SCN^-, NO_2^-	nucleophilic displacement (LS)	pentafluorobenzyl derivative	191, 192
Methyl sulfate	CN^-, SCN^-, NO_2^-, I^-	nucleophilic displacement (LS)	methyl derivative	193, 194
Ethyl sulfate	CN^-, SCN^-, NO_2^-, I^-		ethyl derivative	195
Trimethylchlorosilane	F^-	nucleophilic displacement (LS)	trimethylfluorosilane	196, 197
	SiO_4^{4-}, $Si_2O_7^{6-}$, $Si_3O_{10}^{8-}$, etc.	nucleophilic displacement (LS)	trimethylsilylsilicates	198
Benzene	NO_3^-	aromatic electrophilic substitution (HS)	nitrobenzene	199

Reagent	Species	Reaction	Product	Ref.
1-Hydroxypropane-2,3-dithiol	AsO_2^-, AsO_3^-	condensation (S)	dithiolato complex	200
1,2-Ethanedithiol	MoO_4^{2-}	condensation (S)	dithiolato complex	201
Phenylmercury(II) nitrate	Cl^-, Br^-, I^-	nucleophilic substitution (S)	aryl halide	202
N-Methyl-N-(t-butyldimethyl-silyl)trifluoroacetamide	BO_3^-, SO_4^{2-}, PO_3^-, AsO_2^-, VO_3^-, PO_4^{3-}, AsO_3^-	acid catalyzed nucleophilic displacement (LS)	t-butyldimethylsilyl derivative	203
Primary aryl amine	I^-, CN^-	decomposition of phenyl diazonium salt (S)	aryl iodide or cyanide	204
Sodium borohydride	SeO_3^{2-}, AsO_2^-, AsO_3^-, PO_4^{3-}	reduction (S)	hydride	205, 206
Styrene	Br^-	1,2-addition of Br_2 (S)	styrene dibromide	207
Ethylene oxide	Cl^-, Br^-	1,2-addition of HX (S)	haloalcohol	208
Acetone, 3-pentanone	Br^-, I^-	base-catalyzed substitution of X_2 (S)	α-haloketone	209, 210
H^+	CN^-, CO_3^{2-}, SO_3^{2-}, S^{2-}	acid/base reaction (S)	acid or acid anhydride	211, 212
Ce(IV)	N_3^-	redox reaction (HS)	N_2	213

* HS = high selectivity, S = selective, LS = low selectivity.

References on p. B61

though such derivatization seems to be restricted to species which are stronger nucleophiles than water. Broad-spectrum reagents, despite their obvious appeal, have not been widely applied in actual analytical procedures.

Anions for which highly selective procedures have been developed and utilized include SeO_3^{2-}, NO_3^-, NO_2^-, F^-, CN^-, Cl^-, Br^- and I^-. The determination of selenium by GC is very selective and can be used to determine SeO_3^{2-}, SeO_4^{2-}, or total selenium by control of reaction conditions. Many derivatives (piaselenols) have been prepared and are easily synthesized from the corresponding phenylenediamines in acid medium. The most suitable reagents at present appear to be 5-nitro-1,2-diaminobenzene, 5-trifluoromethyl-1,2-diaminobenzene [14], and 3-bromo-5-trifluoromethyl-1,2-diaminobenzene [219]. For the analysis, conventional nonpolar columns and electron-capture detection are used. Much of the background chemistry and earlier work have recently been reviewed [14].

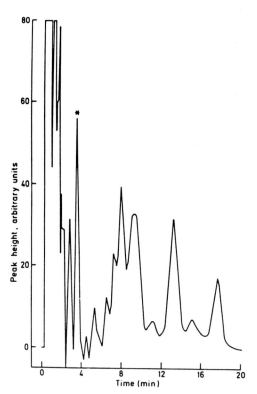

Fig. 12.4. Chromatogram showing the peak (marked with an asterisk) for the 5-nitropia-selenol derivative of selenium, obtained for a sample of milk (following digestion, derivatization, and extraction). Column, 1 m x 4 mm ID, packed with 15% SE-30 on 60- to 80-mesh Chromosorb W (200°C); electron-capture detection. (Reproduced from Ref. 14 with permission.)

The GC procedure for Se, in contrast to related photometric procedures, is remarkably free of interference; even 10 000 ppm of 25 common metal ions will not interfere in the determination of 1 ppm or less of Se [14]. Nonetheless, a multitude of peaks in addition to that of the analyte is usually observed in chromatograms with the FID or ECD. These peaks are, at times, so numerous as to complicate quantification (see Fig. 12.4). They are due to compounds which arise from (i) incomplete or partial oxidation of the sample; (ii) impurities in the reagent; and (iii) the presence of nitrous acid or other oxidizing agents, which react with the reagent to produce a variety of products [185]. These problems have been overcome, to some extent, by the addition of urea following acid digestion (presumably to destroy excess nitrous acid) and acid washing of the piaselenol extract [218]. Generally, this appears to be sufficient for most analyses, but a greater degree of specificity could be attained by using spectrometric detection, e.g. with MIPAES.

The GC procedure for selenium has been applied to a very wide variety of samples, including inorganic reagents, biological specimens, water, and metals [14]. Concentrations close to 5 ppb Se have been determined in milk and saliva [218], and the accuracy of the method has been established by analysis of standard reference materials [218,219].

Another highly selective derivatization reaction involves the reaction of fluoride with trimethylsilanol, obtained from the hydrolysis of trimethylchlorosilane in water. The selectivity of this reaction appears to be the extraordinary stability of the Si-F bond, being of the order of 590 kJ/mol. Detection has so far been confined to the FID, although the use of the ECD with suitable derivatives may greatly improve detection sensitivity. The GC procedure for F^- is said to be more reproducible but slower than the potentiometric analysis with an ion-selective electrode [220]. With this derivatization method, fluoride has been determined in biological samples [196,197], ambient air [221], urine [220], and blood [222,223].

Methods for other halide ions are based on a variety of derivatizing reactions (see Table 12.4). Bromine has been determined in vegetables [207,224], cereals [210], and plasma [225]; iodide has been determined in eggs [226] and blood [227], and chloride has been determined in adipose tissue following conversion to the phenylchloromercury(II) derivative [202]. The method for determining NO_3^- by GC with selective electron-capture detection is almost exclusively based on derivatization to nitrobenzene, as originally described by Tesch et al. [199]. This method is nearly specific for NO_3^- and has been applied to the trace determination of NO_3^- in biological materials [228,229], urine [228], soil [229], air [230], and sulfuric acid [231]. Nitrite can also be determined by this procedure following the oxidation of NO_2^- to NO_3^-. A highly selective derivatization for NO_2^- is based on conversion to N-alkylbenzotriazole in weakly acidic medium [185,186]. Only one analytical application of this reaction has been reported to date [186].

Cyanide has been determined as cyanogen bromide in baijui (a Chinese alcoholic beverage) [232] or as HCN following acidification and headspace analysis of blood [233,234] and bean jam [235].

Other inorganic anions for which GC procedures have been described include CO_3^{2-} [236], SO_3^{2-} [237], S^{2-} [236,238], and N_3^- [213]. An improved derivatization procedure for silicate anions in minerals (but not aqueous silicate) has also appeared [198].

12.1.5 Organometallics

Organometallics studied by GC to date include various alkyl, aryl, vinyl, and silyl compounds of Be, IIIA, IVA, VA, IIB element groups, Si compounds (silanes, chloroalkyl and chloroaryl silanes, silatranes), carbonyl, arylcarbonyl, and trifluorophoshinocarbonyl complexes of transition metals, metallocenes and their substituted derivatives [3-5,8,85]. With few exceptions, organometallics exhibit limited stability with respect to one or more of thermolysis, hydrolysis, oxidative or photooxidative decomposition, and catalytic decomposition. These limitations dominate the GC properties of such compounds. For the less stable compounds, great care in all aspects of their GC, including sampling, transfer, injection, column selection, and deactivation is required. The alkyl compounds Me_2Be, Me_3Al, Me_3Ga, and Me_2Zn, e.g., are so air- and water-sensitive that extraordinary means of purifying the carrier gas are required [239-241]. Solvents should be deaerated, nonreactive toward the solute, and free of reactive impurities, such as peroxides. Special procedures for sampling from vaccum/inert gas lines, inert gas boxes, Schlenk flasks, or other sealed containers may be required for introducing samples into the GC column. Some methods of accomplishing this are described in Refs. 239-243.

The choice of GC columns for organometallics can be crucial, and it is generally safer to avoid metal-lined injection systems and metallic columns. The suitability of a column for particular organometallics can be determined by procedures described by Kirsch et al. [244] and by Patsalides et al. [245]. These procedures, in effect, determine how the response for an organometallic compound varies relative to an inert reference compound, e.g., an n-alkane, with respect to (i) dilution of the solution and (ii) length of time on the column. Variations observed are usually due to adsorption and/or decomposition. A feature of the GC of many organometallics is the irreversible deterioration of column characteristics following their introduction into the column. This can arise from catalytic decomposition of the stationary phase, bonding of organometallics to the stationary phase, or deposition of metallic decomposition products in the column. Such changes can be monitored by standard procedures, described by Grob et al. [246] as applied to aryltricarbonylchromium(0) complexes [245]. A method for regenerating columns involves treatment with a continuous stream of trifluoroacetylacetone or other suitable scavenging ligands to remove metal or oxide deposits. This treatment obviously excludes chromatographic systems with exposed metal parts.

The GC separation of organometallics is usually not problematic, but problems can arise in the separation of closely similar isomeric species. Hwang et al. [247] found it necessary to utilize serially coupled capillary columns with different stationary phases to separate all complexes of the series $M(PF_3)_n(CO)_{n-6}$ (n = 0 to 6, M = Cr, Mo, W), including isomeric forms. Furthermore, a different combination of columns was required for complexes of each of the above elements.

The nature of the detector is not critical, unless either highly complex mixtures of organics/organometallics are to be analyzed or trace determination is required. Thus, for the detection of organometallic mixtures or reaction products at levels of ca. 1% or higher in solution, the TCD or FID are usually adequate. A FID selective for silicon compounds and other organometallics has been described [248,249] and may be suitable in some applications. More selective and sensitive detection is based on the ECD or element-selective detection, viz. AAS, AFS, or AES.

Where organometallics are involatile or highly reactive, derivatization by alkylation, silylation, hydridization, coordination, or other reactions can be utilized. Other, less selective procedures, based on hydrolysis or pyrolysis have been used, mainly for assays of organometallic or metalloorganic reagents, such as Grignard or alkyllithium reagents [3].

12.1.5.1 Physicochemical studies

Physicochemical studies of organometallics include reactions involving pyrolysis, alkyl group exchange, oxidation, isomerization, or catalysis of organic reactions. Both on-column and off-column studies are possible. Systems involving homogeneous catalysis, reviewed by Pscheidl et al. [250] are of particular importance because of their practical applications.

Heats of solution, Kováts indices and Incremental indices have been used to determine solute/stationary-phase interactions [245-254]. These data have been applied to clarify the electronic structure of organometallic solutes, such as aryltricarbonylchromium(0) complexes [245], haloalkylsilanes [254], and silatranes and related compounds [251-253]. In the silatranes, for example, anomolously high Kováts indices have been interpreted in terms of appreciable transannular Si→N bonding in the silatrane structure [252], consistent with the conclusions reached on the basis of dipole moment, spectroscopic, and X-ray structural data.

12.1.5.2 Environmental organometallics

Though organometallics as a class may seem no more than laboratory curiosities, an increasing number, mainly certain compounds of Pb, Hg, and Sn, can now be determined at very low concentrations in polluted environments. Research into biomethylation has also led to the realization that some organometallics, e.g., of Pb, Hg, As, Sn, Se, Tl, and Sb, can originate from biochemical conversions of inorganic substrates [255,256]. The speciation and determination of organometallics or moieties at ppm-ppt levels is usually carried out by GC with the ECD or other sensitive detector after selective extraction, derivatization, and preconcentration. Derivatization may be important in such procedures, since a variety of species of a particular element can be determined simultaneously. Usually, the species of interest represent various stages of alkylation or dealkylation of the original organometallic compounds due to weathering or biological transformation. The simultaneous determination of such species is of value in environmental studies. Contemporary methods of derivatization involve alkylation, hydridization (see Table 12.5),

TABLE 12.5

SUMMARY OF METHODS FOR THE TRACE DETERMINATION OF ENVIRONMENTAL ORGANOMETALLICS BY GAS CHROMATOGRAPHY

Species	Preseparation/extraction	Derivatization	GC column***	Detection	Refs.
PbMe4, PbMe3Et, PbMe2Et2*, PbMeEt3, PbEt4	solvent extraction or gas strip/trap/purge	none	nonpolar silicone column, e.g., OV-1	ECD, AAS (283.3 nm)	257
PbMe3+, PbMe2 2+ etc.*	solvent extraction as chloride or diethyl-dithiocarbamate complex	none or alkylation	as above	ECD, AAS (283.3 nm)	258, 259
Me2Sb(OH)2, Me2SbO(OH)**	gas strip/trap/purge	hydridization	OV-3	AAS (217.6 nm)	260
Me2Hg**	solvent extraction	none or to MeHgCl or Hg	nonpolar, Chromosorb 101	AAS, AFS, AES (253.7 nm)	261
MeHg+**	gas strip/trap/purge	to MeHgCl or Hg	deactivated polar column, e.g., DEGS	ECD, AAS, AFS, AES (253.7 nm)	262-264
Me3As**, Me2AsO(OH), MeAsO(OH)2, Me3AsO	solvent extraction or gas stripping	hydridization	DC-550 on Chromosorb W Chromosorb 101 coated with 5% PEG-20M	AAS (193.7 nm)	265-267

Me₂Se[*,***], Me₂Se₂, Me₂SeO₂	gas stripping or methanol extraction	QF-1 or XE-60 column	AAS, AFS, AES (196.0 nm)	268-270
R₄Sn, R₃SnX, R₂SnX₂, RSnX₃ R₂SnOSnR₂	extraction from aq. HCl/NaCl. Cleanup on silica gel or Florisil column optional	nonpolar packed or capillary column	hydridization or alkylation	FPD (600 nm) ECD AAS, AFS AES (224.6 nm)	271-277

R = Me, Et, n-Pr,
n-Bu, n-Oct, cyclohexyl,
phenyl
X = Cl, OAc, OH, etc.

[*] Anthropogenic in origin.

[**] From biomethylation.

[***] Use of columns without metal surfaces is assumed.

References on p. B61

arylation [258], and silylation [262]. Such reactions can also be used for the quantitation of inorganic ions, such as Hg^{2+} [262], Sn^{2+}, and Pb^{2+}. An example of the derivatization of organotin species that is also applicable to other organometalloids is shown in Scheme 12.1. A summary of GC methods for organometallics of current interest is given in Table 12.5. Surprisingly, many environmental organometallics of Pb, Hg, and Sn can be determined at ng-pg levels without problems of decomposition, irreversible adsorption, or alkyl group exchange, a fact which reflects the relatively high stability of these species. Adsorption on packed columns has been encountered mainly with polar halide species, such as $MeHgCl$, R_3SnCl, R_2SnCl_2, and $RSnCl_3$ when these are chromatographed without derivatization

R_4Sn		R_4Sn		R_4Sn
R_3SnR'	alkylation	R_3SnX	hydridization	R_3SnH
R_2SnR_2'	$\longleftarrow$	R_2SnX_2	$\longrightarrow$	R_2SnH_2
$RSnR_3'$	$R'MgBr$	$RSnX_3$	$NaBH_4$	$RSnH_3$
SnR_4'		SnX_4		SnH_4

Scheme 12.1

However, even in these cases effective column deactivation is possible. Column deactivation for MeHgCl determination can be effected by addition of $HgCl_2$ to the stationary phase [279], while addition of traces of HCl to the carrier gas is effective for determining organotin chlorides [280]. Other problems reported are oxidation of tetraalkyllead species during adsorptive preconcentration from air [281] and alkyl group exchange during alkylation of organolead species [282].

12.1.5.3 Organolead species

Although organolead species can be formed by biomethylation [283,284], the major environmental source of such species is of anthropogenic origin. The species, originating mainly from leaded gasoline, are Me_4Pb, Et_4Pb, Et_3MePb, Me_3EtPb, and Et_2Me_2Pb. These can be determined directly by GC after extraction and preconcentration. Methods have been described for the separation and trace determination of these species in air [281,285,286], water, sediments, and biological material [257] by GC with ECD, furnace AAS, or MIPAES as detectors. The ECD is characteristically less selective than other detectors, but it is possible to determine co-extractives, such as dichloroethane, at the same time. For solids and liquids, a single extraction with hexane or benzene is sufficient, and the extract can be analyzed without preconcentration. The analysis of biological material is usually the most difficult due to the poor recoveries (typically <75%) of tetraalkyllead species [257]. Since tetraalkylleads are volatile, samples should be extracted immediately after collection [257]. The determination of alkyllead compounds in air in-

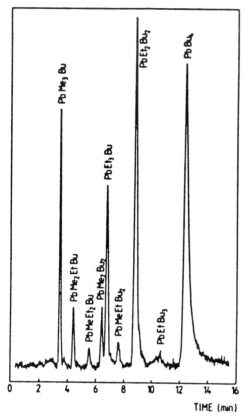

Fig. 12.5. Chromatogram of butylated lead species, following the extraction of inorganic and ionic organometallic lead species from rainwater (AAS detection). (Reproduced from Ref. 289 with permission.)

volves trapping at $-80°C$ [285] or adsorption on porous polymer [281,286]. Oxidative decomposition of alkyllead species by ozone in adsorption analyses can be avoided by passing sampled air over ferrous sulfate. As little as 0.2 ng/m^3 of organic lead can be determined, based on 1-h sampling times. Thus, GC is well suited for the determination of organic lead species in urban air.

Recent progress in this area involves the development of methods for the simultaneous trace determination of ionic alkylleads R_3Pb^+ and R_2Pb^{2+}, including mixed alkyl species [282]. Such species can be determined directly as the chlorides [288]. However, a more suitable procedure, involving selective extraction of the species as their diethyldithiocarbamate complexes and conversion to tetraalkylleads, is favored. A typical chromatogram for this type of derivatization is shown in Fig. 12.5. Procedures based on this approach have been developed to determine ionic alkylleads in water down to 1-5 ng/l, sediment, biological samples (7-10 ng/g), and aerosols (2 pg/m^3) [259,289,290].

12.1.5.4 Organomercury species

Of the various forms of mercury frequently determined (total, elemental, and organomercury) only organomercury, mainly methylmercury, is commonly determined by GC, although all three forms can be so determined [259]. The methodology for extracting methylmercury is well established and generally involves the Westoo double-extraction procedure [291] or a related procedure [262]. The extracted methylmercury(II) chloride can then be quantified by GC, using the ECD, furnace-AAS [292], or other element-selective detector [261,294-297]. As previously indicated, column deactivation and the avoidance of metal columns are necessary in trace determinations. Recent applications of the determination of methylmercury are described for biological material [292], including fish tissue [279,297], blood [298], urine [263], atmosphere [261], sediments [299], and natural waters [300].

12.1.5.5 Organotin species

Organotin compounds of types R_4Sn, R_3SnX, R_2SnX_2, and R_3SnX are found in the environment largely as a result of their use in agriculture, wood preservation, and marine antifouling paints [301,302]. They are analogous to the organolead species, but are considerably more stable, both chromatographically and environmentally. In particular, species $RSnX_3$ are stable under ambient conditions, whereas the corresponding $RPbX_3$ species are not. Although alkyltin halides can be determined directly by GC [280] following extraction, current practice involves their conversion to either alkyl derivatives [271,275,303] or hydride derivatives [272-274,277,304,305], as shown in Scheme 12.1. A significant feature of these derivatization reactions is that they are facile and occur without displacement of the original alkyl groups. GC conditions for such derivatives do not appear to be critical, although metal columns and metal-lined injection systems should be avoided. Columns of methyl- or methylphenylpolysiloxanes are commonly used with detection by ECD [272], FPD [306-309], or nonflame-AAS [271,286,306-308]. The ECD has adequate sensitivity to both alkyl and hydride species and is convenient [272], but poor specificity may require cleanup of extracts to remove interfering electron-capturing organics. Element-selective detectors, on the other hand, are both sensitive and selective, but are not commonly available for routine use.

The general method for recovery of organotins involves either purge-and-trap, in the case of alkyltin hydrides, or extraction with hexane or benzene [271,305], which can be utilized for both the tetraalkyl and hydride derivatives. A method for the simultaneous extraction of both organic and inorganic tin from aqueous solution prior to derivatization involves extraction with a benzene solution of tropolone [271]. Sensitivity in such procedures can usually be improved by preconcentrating extracts to a small volume in, e.g., a Kuderna-Danish apparatus.

Recent applications involve methods for the determination of organotins in vegetables [275], fish tissue to ca. 10 ppb [304,310], water to ca. 5 ng/l [274], environmental samples [303], agricultural products [311], and PVC products [312].

12.2 LIQUID CHROMATOGRAPHY

12.2.1 Introduction

Until 1980, the development of liquid chromatographic (LC) methods for the separation and determination of inorganic species was much less rapid than for the application of these techniques to organic species. The advent of ion chromatography (IC) and its subsequent widespread application has provided the stimulus for an acceleration of interest in LC methods for inorganic species. In the time period covered by this review (mainly 1982-1989), more than 1500 papers have appeared on this topic. Because this volume of literature precludes comprehensive coverage, this review will focus on some of the more significant developments and applications, further representative examples of the main approaches being provided in tabular form. Further detail may be found in other review articles on specific aspects of LC of inorganic species, and these will be mentioned as each topic is discussed.

LC methods will be considered to include only modern, high-performance techniques, such as IC (which, for simplicity, will be interpreted to represent *only* ion-exchange methods), HPLC (both normal- and reversed-phase), ion-pair chromatography and ion-exclusion chromatography. Detailed descriptions of the operating principles of these techniques may be found in Part A, particularly Chapter 5. In the interests of brevity, TLC and electrophoretic techniques will not be discussed. Further, it will be convenient to subdivide the subject by considering separately each broad class of solute species, namely inorganic anions, inorganic cations, organometallic species, and coordination compounds.

Section 12.1 demonstrates the widespread utility of GC for inorganic analysis. However, the use of GC requires that the solute be volatile and thermally stable. Most inorganic species do not meet these criteria, unless they are bound to an organic moiety. Ionic solutes, such as free inorganic anions and cations, must therefore be derivatized to produce volatile compounds, and this process may perturb the distribution of solute species in a sample. In contrast, LC methods permit the direct determination of these ionic solutes, without the requirement for derivatization. Moreover, many of the volatile organometallic species analyzed by GC are also suitable for determination by LC.

Detection of eluted solutes is often a problem in GC, where detection must usually be based on a measurable property of the derivatizing agent, rather than a property of the inorganic solute itself. The existence of sensitive general-purpose and selective detectors for LC, especially conductivity detectors, means that detection of inorganic solutes is often relatively straightforward. However, it is also fair to say that the detectors used in LC lack the sensitivity exhibited by the ECD, widely employed in GC determinations of inorganic solutes.

12.2.2 Inorganic anions

12.2.2.1 Ion chromatography of anions

Ion chromatography on low-capacity, high-efficiency ion-exchange materials has provided the mainstay for inorganic anion determinations. IC can be performed either in the suppressed or nonsuppressed modes (Chapter 5). Suppressed IC involves passage of the eluent and solute anions through a suitable ion-exchange column and thence through a suppressor device in order to reduce the background conductance of the eluent. In most cases, this reduction is achieved in the suppressor by dynamic exchange of eluent cations for hydrogen ions, with subsequent protonation of weak acid eluent anions. This process does not affect those solute anions which are the conjugates of strong acids. On the other hand, for nonsuppressed IC a very dilute eluent is used with a column of low ion-exchange capacity (typically 10-100 μeq/g), so that the eluted solutes can be detected readily without the requirement for a suppressor.

IC has the distinction of being one of the most commonly reviewed topics in the general field of liquid chromatography. In the past six years, eight books [313-320] and more than 70 reviews have appeared. Many of these reviews are concerned chiefly with the instrumentation and fundamental applications of IC, while others treat the subject in more depth [321-328].

12.2.2.1.1 Stationary phases

The special ion-exchange stationary phases used in IC have undergone remarkable development over the past six years. The agglomerated resins used for suppressed IC, which are formed by electrostatic binding of aminated latex particles onto a sulfonated core particle, have shown a trend towards smaller particle sizes for both the core particle and the latex particles. Modern materials are now typically produced from highly cross-linked 5- to 10-μm microporous polystyrene/divinylbenzene (PS/DVB) as the core particles, with functionalized latex particles in the size range 60-300 nm serving as the actual ion-exchange sites [329]. The result of this is that chromatographic efficiency has been improved dramatically over that exhibited by earlier materials, so that the separation of seven common anions in about 6 min is now a routine exercise. It has been shown recently that very efficient agglomerated anion exchangers can be produced either by simple hydrophobic binding of aminated latex particles onto a neutral, unfunctionalized core particle [330], or by mechanical binding of the latex with a suitable glue [331,332]. Regardless of the manner in which they have been produced, agglomerated anion-exchange resins exhibit excellent chromatographic efficiency.

Stationary phases for nonsuppressed IC have also undergone rapid development. The commercially available silica-based anion exchangers used to introduce the technique have now been largely superseded by resin-based materials in which the base polymer used to carry the quaternary ammonium functionatily is either PS/DVB or polymethacrylate. The latter polymer, which is hydrophilic in nature, has the advantage that fouling of the column by organic sample components is reduced, whereas the PS/DVB materials

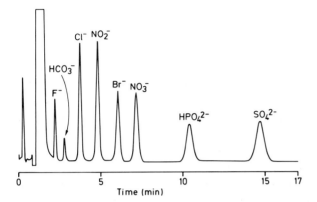

Fig. 12.6. Separation of anions on a polymethacrylate-based surface-aminated anion exchanger. A Waters IC-Pak HR column was used with a gluconate/borate eluent and conductivity detection. (Chromatogram courtesy of Waters.)

offer the advantages of pressure stability and tolerance towards organic solvents. Fig. 12.6 shows a separation of anions on a polymethacrylate anion exchanger.

Alumina is an amphoteric oxide which can act as a cation exchanger in alkaline solution or an anion exchanger in acidic solution. Alumina has been used as a chromatographic material for organic compounds, and more recently its use for the separation of inorganic anions has been revived [333,334]. The retention order for these solutes on 5-μm alumina with an acetate buffer as eluent is as follows:

$$F^- > SO_4^{2-} > Cr_2O_7^{2-} > HCOO^- > ClO_2^- > BrO_3^- > Cl^- > NO_2^- > NO_3^- > Br^-$$
$$> ClO_3^- > SCN^- > I^- > ClO_4^-$$

This order is roughly opposite to that observed on conventional exchangers with quaternary ammonium functionalities. The halide elution sequence is consistent with the aluminium-halide formation constants, and this suggests that anion interaction occurs at the Al atom. The value of the overall formation constant (log β_6) for AlF_6^{3-} is 19.84 [335], which explains why fluoride cannot be eluted from an alumina column, even with a strong eluent. Phosphate (not listed in the above selectivity order) shows similar behavior.

12.2.2.1.2 Eluents

A very restricted range of eluents was used in early IC separations, a sodium carbonate/bicarbonate buffer being used typically for suppressed IC and potassium hydrogen phthalate for nonsuppressed IC. The range of eluents used in both modes of IC has shown a dramatic increase over the period of this review. For example, sodium tetraborate [336], sodium hydroxide [337], sodium p-cyanophenate [338], tyrosine [339], and zwitterionic species [340] have all been employed successfully with suppressed IC. Similarly, aliphatic or aromatic carboxylate anions (such as citrate [341], tartrate [342], succinate [343], trimesate [344], pyromellitate [345], dihydroxybenzoate [346], and salicylate [347]), aliphatic and aromatic sulfonates (such as methanesulfonate [348,349] and p-toluenesulfonate [350]), sodium hydroxide [351], complexes of borate with polyhydroxy compounds (such as gluconate [352] or mannitol [353,354]) and inorganic species (such as nitric acid [355] and sodium perchlorate [356]) have all proved suitable for anion separations by nonsuppressed IC.

12.2.2.1.3 Detection

Conductivity detection remains the most widely employed detection method for IC. Its utility for suppressed IC has been enhanced considerably by the introduction of highly efficient suppressors which use ion-exchange membranes to facilitate the replacement of eluent cations by hydrogen ions from the suppressor. A cation-exchange membrane is used to separate the eluent from a solution of a suitable acid. This membrane may take the form of a hollow fiber [331,357-360] or flat sheets [336] (Chapter 5). In the latter case, the resultant suppressor is called a micromembrane suppressor, and such a device can completely suppress an eluent consisting of 100 mM NaOH, at a flowrate of 1 ml/min [338]. This high suppression capacity has, for the first time, permitted the use of gradient elution in IC with conductivity detection [336]. Thus, a linear gradient from 25 to 100 mM NaOH permits the separation of 36 inorganic and organic anions in 30 min. Successful gradients have also been achieved in nonsuppressed IC with isoconductive eluents [361,362]. These eluents are formed with the same eluent anion, but with different cations, in each eluent (Fig. 12.7).

Spectrophotometric detection at wavelengths in the range 200-220 nm is suitable for a wide range of anions [326], particularly nitrate [363,364], nitrite [365,366], bromide [367], bromate [368], iodate [369], thiocyanate [367], and thiosulfate [370]. Moreover, use of a UV-absorbing eluent, such as phthalate, permits the indirect detection of anions which themselves do not absorb at the detection wavelength. This technique, called indirect photometric chromatography [371], is widely applied. Aromatic carboxylate [372] or aromatic sulfonate [350] anions are suitable eluent competing ions. Atomic spectroscopy has also been coupled to IC for the detection of inorganic anions. Examples include the detection of selenite and selenate by atomic absorption spectroscopy [373] and the detection of arsenate and arsenite by DCPAES [374]. The chief problem encountered in these methods is matching the eluent flowrate with the nebulizer uptake rate on the spectrometer. A new detection technique, called replacement IC, has been reported [375] in which the cations in the eluted sample band are replaced with lithium ions on a cation-exchange column in the lithium form, mounted after the analytical column. The lithium ions are then detected by flame-emission spectroscopy.

Electrochemical detection is utilized in IC for situations demanding extreme sensitivity or special selectivity. Amperometry, polarography, coulometry, and potentiometry have all been employed for this purpose. Amperometric detection is most often applied to the detection of cyanide [376,377], sulfide [377,378], bromide [379], nitrite [380], and thiosulfate [381], by way of oxidative reactions at platinum or silver working electrodes. Indirect amperometric detection of electroinactive anions has been shown to be possible through utilization of the pH change accompanying elution of anions from a suppressed IC system [382]. Polarography [383] and coulometry [384] are less frequently used than amperometry. Potentiometry is utilized for the selective detection of anions, usually with an ion-selective electrode as the indicator electrode. This approach is applicable to the detection of halides and pseudo-halides by means of solid-state [385-388] or coated-wire [356,389] electrodes. More general detection is possible with the use of a metallic copper indicator electrode [390-394]. This electrode will respond directly to any solute which can participate in reactions with copper metal, cuprous ions, or cupric ions, through oxidation, reduction, or complexation mechanisms. Moreover, indirect detection can be accomplished by incorporating a copper-complexing ligand (such as phthalate) as the competing anion in the eluent. Changes in the phthalate concentration accompanying solute elution are sensed by the copper indicator electrode, so that all solutes are detected.

Post-column reactions have been used to a limited extent for the detection of inorganic anions. Most applications relate to the detection of phosphates [395], polyphosphates [396], and phosphonates [396], generally through the use of molybdate reagents, leading to the formation of the heteropoly blue complex. A more generally applicable postcolumn reagent is iron(III) perchlorate, which forms colored complexes with more than 20 inorganic anions [397]. This reagent has been employed not only for the detection of common anions, but also for sulfur oxo-anions [398] and polyphosphates [399,400].

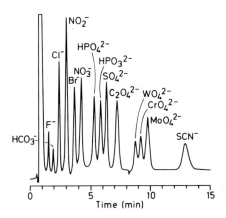

Fig. 12.7. Anion-exchange separation with an isoconductive gradient. A Waters IC Pak Anion HR column was used. The column was equilibrated with Eluent A (8.25 mM boric acid/1.11 mM gluconic acid/3.08 mM cesium hydroxide/0.48 mM glycerin/12% acetonitrile) and a step gradient to Eluent B (12.65 mM boric acid/1.70 mM gluconic acid/4.72 mM lithium hydroxide/0.75 mM glycerin/12% acetonitrile) was initiated at the moment of injection. Note that with the pump used, there was a considerable lag before the gradient reached the column. (Reproduced from Ref. 362 with permission.)

12.2.2.1.4 Sample handling

Early IC methods for the determination of inorganic anions showed excellent results when applied to relatively simple sample matrices, such as waters, but were less reliable with more complex sample types. The subject of sample handling in IC has therefore received considerable attention in recent years. The emphasis has been on the analysis of strongly alkaline samples (such as those found in the Bayer and Kraft processes), the determination of anions at trace levels, and the extension of IC methods to include the analysis of solids, gases, aerosols, and organic compounds.

Strongly alkaline samples usually produce chromatograms with uneven baselines because of the destabilizing effect of the injected hydroxide ions on the eluent/stationary phase equilibria. Simple neutralization of the sample with acid is unsuitable because of the resultant contamination of the sample by the acid anion. In such cases, pH adjustment can be performed by batchwise addition of a cation-exchange resin in the hydrogen form [401]. Alternatively, ion-exchange replacement of sodium ions in the sample with hydrogen ions can be accomplished with the aid of a cation-exchange membrane, which separates the sample from either an acid solution [402], or some cation-exchange resin in the hydrogen form [403,404]. This procedure is identical in nature to the suppression of eluent conductance with ion-exchange membranes, as used in suppressed IC. A commercially available device (Milli-Trap, from Millipore) is recommended for sample treatment in IC. The sample is passed by syringe through a hollow-fiber cation-exchange membrane, immersed in an acid solution.

Routine detection limits for IC with conductivity measurement usually fall in the 500 ppb – 1 ppm range for most anions when an injection volume of 100 μl is used. These detection limits may be decreased by increasing the injection volume (up to 50 ml [364,405,406]), but the limitation to this procedure is the size of the resultant solvent peak, which may eventually obscure the solute peaks. It is interesting to note that there is little significant increase in peakwidth for eluted anions when the injection volume is increased in this way. The reason for this is that solute anions in the sample become bound at the head of the column during sample injection and do not begin to traverse the column until contacted by eluent ions following the injected sample. Trace enrichment of anions is also commonly achieved in an anion-exchange precolumn ("concentrator column"), through which the sample is pumped. Solute ions become trapped in that column and, at the conclusion of sample loading, are transferred to the IC column for separation and detection [407-409]. This process is attractive, because it is simple and convenient to apply, amenable to automation, offers high enrichment factors, and is less prone to sample contamination effects than alternative methods. Trace enrichment of inorganic anions on concentrator columns has been investigated systematically in a series of papers [349,350,405,410,411], in which such factors as the chromatographic hardware requirements, the type of eluent, nature of the packing material in the concentrator column, and effects of sample loading parameters are discussed. Trace enrichment of samples for IC analysis can also be achieved by dialytic techniques for the transfer of solute anions through an anion-exchange membrane into a small volume of a suitable receiver solution of high ionic strength [412-414].

Solid samples, such as soils, rocks, and plastics require specialized sample treatment prior to analysis by IC. Acid dissolution is inappropriate, because high levels of acid anion are introduced into the sample. In most cases, combustion techniques offer the most suitable means for sample destruction and conversion of sample components into a form suitable for measurement by IC. This approach is especially convenient for the determination of the levels of halogens, phosphorus, and sulfur in such samples. During the combustion process, these species are oxidized to produce volatile compounds that are then collected in an appropriate trapping solution to form inorganic anions, which are then separated and determined by IC. The type of combustion procedure used (Schoeniger flask, Parr bomb, or furnace methods) depends on the nature of the sample, while the type of trapping solution (alkaline, oxidizing, or reducing) depends on the particular species being analyzed. Thus, Cl, Br, P, and S in organic reagents may be determined by Schoeniger flask combustion, with collection in an alkaline solution of hydrogen peroxide [415], while F, Cl, and S may be determined in geological samples after pyrohydrolysis in an induction furnace, with collection in carbonate buffer solution [416]. Alkali fusion methods are also suitable for the dissolution of solid samples in IC, sodium carbonate [417,418] and sodium hydroxide [419] being typical fluxing materials. This approach is typically applied to the analysis of geological materials and glasses.

IC is rapidly becoming the method of choice for the determination of airborne pollutants (gases, aerosols, and particulates) in environmental analysis and occupational hygiene. The sample can be collected in a combination of a filter medium (to trap particulates), a system of denuders (to trap gases), and an impregnated filter (to trap aerosols) [420]. Alternatively, impingers [421] or adsorption columns [422,423] can be employed for sample collection. In each of these approaches, the sample components are ultimately converted to inorganic anions, which are then determined by IC. This method has been applied to the determination of sulfur dioxide [421], nitrogen dioxide [424], hydrogen sulfide [425], and acid vapors [426,427].

12.2.2.1.5 Applications

IC has found extensive applications over a wide range of areas. Examination of the literature suggests that environmental analysis, particularly of waters, is the most commonly used application of IC, followed by industrial applications, food and plant analysis, and clinical and pharmaceutical analysis. Table 12.6 lists some of these applications, the examples being selected to provide an insight into the types of samples analyzed and the separation and detection methods used. It should be stressed that Table 12.6 provides a partial listing only of the IC applications published over the period covered by this review.

12.2.2.2 Ion-pair chromatography of anions

Ion-pair or paired-ion chromatography, also known as ion-interaction chromatography, offers a useful alternative to ion exchange for the separation of inorganic anions. This approach is especially attractive to those users wishing to extend the capabilities of a conventional HPLC instrument to include anion analysis. When applied to this task,

References on p. B61

TABLE 12.6

TYPICAL DETERMINATIONS OF INORGANIC ANIONS BY LIQUID CHROMATOGRAPHY

Solutes	Sample	Stationary phase	Eluent[*]	Detection mode[**]	Detection limit	Refs.
$C_2O_4^{2-}$	urine	Waters IC-Pak A	0.7 mM phthalate	P (Cu electrode)	0.5 ppm	428
Cl^-, Br^-, NO_3^-	rainwater	Wescan 269-001	2 mM methanesul-fonic acid	S (214 nm)	2-30 ng	348
Cl^-, CN^-, CO_3^{2-}	sea water	Waters IC-Pak A	5 mM KOH	C, A	10 ppb	429
Cl^-, NO_3^-, PO_4^{3-}, SO_4^{2-}	soils	C_{18}	0.5 mM TBA phthalate	IS (257 nm)	50 ppb	430
Cl^-, NO_2^-, NO_3^-, SO_4^{2-}	organic com-pounds	Hamilton PRP-X 100	2 mM phthalate in 10% acetone	C	0.1 ppm	431
Cl^-, NO_2^-, NO_3^-, SO_3^{2-}, SO_4^{2-}	ambient air	Dionex AS-4 latex agglomerate	2.0 mM Na_2CO_3/0.75 mM $NaHCO_3$	C	< 1 ppb	432
Cl^-, PO_4^{3-}, SO_4^{2-}, $C_2O_4^{2-}$	Bayer liquors	Dionex AS-3 latex agglomerate	1.8 mM Na_2CO_3/2.25 mM $NaHCO_3$	C	0.2 ppm	433

Anion	Matrix	Column	Mobile phase	Detection	Detection limit	Ref.
CN^-, CO_3^{2-}, S^{2-}, SO_3^{2-}, SO_4^{2-}, SCN^-, $S_2O_3^{2-}$	polysulfide solution	TSK-gel IC Anion PW	1.3 mM gluconate/1.3 mM borate	C, S (220 nm)	0.8-68 μM	434
CN^-, SCN^-	plasma	TSK Gel LS-22	0.1 M acetate/0.2 M NaClO$_4$	PCR (König reaction)	<1 pmol	435
F^-	wastewaters	Brownlee Polypore H	2.0 mM H$_2$SO$_4$	C	0.2 ppm	436
NO_2^-, NO_3^-	meats	Econosil C$_{18}$	2 mM DCTA phosphate	A, S (220 nm)	10 ppb	380
PO_4^{3-}, $P_2O_7^{4-}$, $P_3O_{10}^{5-}$, SO_4^{2-}	detergents	Waters IC-Pak A	HNO$_3$ gradient	PCR with Fe(ClO$_4$)$_3$	50 ppb	399
SO_3^{2-}	foods	Wescan anion-exclusion	5 mM H$_2$SO$_4$	C, A (0.4 V)	10 ppb	437
WO_4^{2-}, MoO_4^{2-}	plants, soils	Cosmosil C$_{18}$	1.5 mM Tiron/30 mM TBA Br/acetate buffer	S (315 nm)	50 ppb	438

* TBA = tetrabutylammonium, DCTA = dodecyltrimethylammonium, Tiron = 1,2-dihydroxybenzene-3,5- disulfonic acid.

** P = potentiometry, S = spectrophotometry, C = conductivity, A = amperometry, IS = indirect spectrophotometry, PCR = postcolumn reaction.

References on p. B61

ion-pair chromatography has been utilized in two distinct modes. The first of these involves the addition to the eluent of a relatively lipophilic cationic species (such as tetrabutylammonium), called the ion-pairing reagent, which is maintained as an eluent component throughout the chromatographic analysis. This method, known as "dynamic coating ion-pair chromatography", parallels the conventional use of this method for the separation of organic anions. Retention of solute anions is considered to arise from the formation of an electrical double layer of the ion-pairing reagent and its countercation at the stationary-phase surface [439,440]. The second operational mode of ion-pair chromatography involves preliminary treatment of a reversed-phase column with a dilute solution of a very lipophilic ion-pairing reagent (such as tridodecylmethylammonium), generally dissolved in a mixed water/organic solvent, the reagent being omitted from the eluent in the subsequent separation of anions. This approach is known as "permanent-coating ion-pair chromatography", since the reversed-phase column is considered to retain a stable layer of the lipophilic ion-pairing reagent cation, thereby converting it effectively into an anion-exchange material. Separations can then be performed in the same manner as used in ion-exchange applications, but permanent-coating ion-pair chromatography has the advantage that the effective ion-exchange capacity of the column can be regulated easily by controlling the amount of adsorbed ion-pairing reagent.

12.2.2.2.1 Stationary phases

Ion-pair chromatography of anions has been successfully performed on a wide range of stationary phases, including neutral PS/DVB polymers [441,442] and bonded silica materials with C_{18} [443], C_8 [444], phenyl [445], and cyano [446] groups as the chemically bound functionality. Each of these stationary phases gives satisfactory retention of anionic solutes, provided the eluent composition is such that an appropriate amount of the ion-pairing reagent is adsorbed. The choice between stationary phases is usually based on such considerations as chromatographic efficiency [447], pH stability [448], and particle size [449], rather than on differences in chromatographic selectivity. However, it has been noted [450] that the elution order for anionic solutes can vary when the nature of the stationary phase used to support the reagent is altered.

Further factors to be considered in the selection of stationary phases for ion-pair chromatography are specific interactions existing between the stationary phase and either the reagent or the solutes, and the role of residual silanol groups on silica-based stationary phases. Some solutes (e.g., iodide) show particularly strong adsorption on PS/DVB stationary phases, and this has been attributed to the occurrence of π-π interactions with the aromatic moiety of the polymer [447]. The majority of ion-pair separations of inorganic anions are performed on conventional C_{18} silica-based reversed-phase materials or on neutral PS/DVB polymers (such as Hamilton PRP-1, Rohm & Haas XAD-2, or Dionex MPIC columns).

12.2.2.2.2 Eluents

The most important component of the eluent in ion-pair chromatography is the ion-pairing reagent itself. In dynamic coating, the lipophilicity of the reagent governs the degree to which it is adsorbed on the stationary phase, and this, in turn, controls the density of the electrical double layer and, hence, the retention of solute anions. Similarly, the lipophilicity of the reagent in the permanent-coating method influences the ultimate ion-exchange capacity of the coated column. Moreover, the amount of organic modifier present in the eluent can also be used to vary the amount of adsorbed reagent in both the dynamic and permanent-coating methods. It may be noted here that the eluents used in permanent-coating ion-pair chromatography (and indeed the detection methods also) are identical to those discussed earlier for use with fixed-site, quaternary ammonium ion exchangers and therefore need not be considered further.

The counterion fills a very important role in dynamic-coating ion-pair chromatography of anionic solutes. This counterion usually acts as an ion-exchange competing anion and is responsible for the elution (and in many cases also the detection) of the solute anions. Typical counterions are: hydroxide [450], fluoride [441], chloride [451], perchlorate [452], bromide [453], phthalate [454], citrate [455], and salicylate [456]. The nature of the counterion determines the type of separation achieved. The following strengths of counterions in reducing the retention of anionic solutes have been reported for a PRP-1 column with a quaternary ammonium salt as the ion-pairing reagent [453,457]:

$$ClO_4^- > I^- > NO_3^- > Br^- > NO_2^- > Cl^- > \text{citrate} > \text{formate} > PO_4^{3-} > SO_4^{2-} > F^- > OH^-$$

It is not essential for the counterion to act as the ion-exchange competing anion. An alternative approach is to use a separate eluent component, such as phosphate [458], citrate [455], oxalate [459], or phthalate [440] for this purpose. This method is sometimes used to assist in the elution of strongly retained ions.

12.2.2.2.3 Detection

The counterion of the ion-pairing reagent also influences the detection modes applicable to a particular separation. This occurs in exactly the same manner as applies in ion-exchange chromatography with fixed-site exchangers. Thus, counterions, such as citrate, phthalate, and hydroxide are suitable for conductivity detection; hydroxide, fluoride, and chloride are suitable for direct spectrophotometric detection: and phthalate is suitable for indirect spectrophotometric detection.

The majority of dynamic-coating ion-pair chromatographic separations of anions utilize direct spectrophotometric detection and thus they are applied to anions, such as nitrate [363,460], nitrite [363,441], and iodide [461,462]. Indirect spectrophotometric detection is also possible if the counterion is strongly absorbing (or if such a species is added separately to the eluent as a competing anion). For example, an eluent containing tetra-butylammoniumphthalate has been used for the separation and indirect detection of common anions in ion-pair chromatography with a C_{18} column [430,463]. An alternative approach to indirect detection is to use an ion-pairing reagent which is itself strongly absorbing [464,465]. This method can be described as UV visualization and enables UV-transparent solutes to be detected by peaks induced by adsorption or desorption of the reagent from the column, as a result of the passage of the sample band along the column. UV-absorbing ion-pairing reagents used for this type of detection include benzyl-tributylammonium [466], naphthylmethyltributylammonium [467], and inorganic complexes, such as iron(II) phenanthroline [468,469].

12.2.2.2.4 Applications

The applications of ion-pair chromatography to the quantitation of inorganic anions are very similar to those described earlier for IC. Some applications representative of typical stationary phases, eluents, and sample types are included in Table 12.6.

12.2.2.3 Ion-exclusion chromatography of anions

Ion-exclusion chromatography involves the separation of weak acid anions on a column packed with sulfonated cation-exchange resin and with an acid solution as eluent. The acid concentration in the eluent determines the effective charge on the solute. Several factors are known to contribute to the retention of solutes [470,471]. Neutral species can undergo liquid/liquid partitioning between the eluent and occluded water, trapped within the resin bead, while anionic species are excluded from the resin due to electrostatic repulsion from the sulfonate functional groups on the resin surface. Hydrophobic adsorption on the nonfunctionalized areas of the stationary-phase resin may also contribute to the overall retention mechanism. Ion-exclusion chromatography has been used chiefly for the separation of organic acids, and it is only in recent years that it has also been applied to inorganic anions.

12.2.2.3.1 Stationary phases and eluents

The most commonly used packing material for ion-exclusion chromatography of inorganic anions is a microporous or gel-type, fully sulfonated cation exchanger in the hydrogen form. A typical column, packed with resin of this type, is the Bio-Rad HPX-87H. More rigid, crosslinked, macroporous cation exchangers have been introduced recently [472,473]. They have greater pressure stability than the microporous material, but also show stronger hydrophobic interactions with solutes.

Typical eluents for ion-exclusion chromatography are dilute solutions of mineral acids, especially sulfuric acid. If conductivity detection is to be used, a less conductive acid, such as succinic acid [474] or carbonic acid [475], may be substituted.

12.2.2.3.2 Detection

Direct spectrophotometric detection is the normal operating mode for ion-exclusion chromatography, but this has very limited applicability for the detection of inorganic anions, since very few of the UV-absorbing anions fall into the category of weak acids suitable for separation by ion-exclusion chromatography. As stated above, direct conductivity detection is suitable only if the background conductance of the eluent is kept low through the use of a weakly conducting acid. Alternatively, the background conductance of a strong acid eluent can be lowered with a membrane suppressor, with tetrabutyl-ammonium hydroxide as the scavenger solution. For example, passage of a hydrochloric acid eluent through such a suppressor results in exchange of hydrogen ions in the eluent for tetrabutylammonium ions and leads to the formation of weakly conducting tetrabutyl-ammonium chloride [476].

Since solutes are chromatographed as partially ionized species in ion-exclusion chromatography, their detection by conductivity measurements is often rather insensitive. One means of overcoming this difficulty is with the use of an "enhancement" column to increase the conductance of the sample band. An example of this approach is the conductivity detection of carbonate after separation by ion-exclusion chromatography with water as eluent [477]. The eluted sample zone is passed through a cation-exchange enhancement column in the potassium form, and then through an anion-exchange enhancement column in the hydroxide form. The outcome of these steps is that the weakly conducting carbonic acid eluted from the ion-exclusion column is converted to the highly conducting potassium hydroxide. An alternative approach is to increase the conductance of the sample by adding a suitable reagent to the eluent; e.g., the addition of mannitol or fructose to the eluent aids in the conductivity detection of boric acid and germanic acid, due to the formation of conducting complexes [478,479].

Amperometry has also been used extensively with ion-exclusion chromatography, especially for the detection of sulfite and sulfide [437].

12.2.2.3.3 Applications

Ion-exclusion chromatography is particularly suited to the determination of inorganic weak acid anions, such as sulfite, sulfide, and borate, in very complex samples. The reason for this is that strong acid anions are generally unretained on an ion-exclusion column and are eluted as a band with the void volume, so that little interference with retained solutes results. Thus, ion-exclusion chromatography has been applied to the determination of bicarbonate in biological fluids [480], sulfite in foods and beverages [437], and boric acid in nuclear reactor cooling water [481]. The last of these applications combines the determination of boric acid by ion-exclusion chromatography with a simultaneous determination of strong-acid inorganic anions by IC. This is achieved by collecting the inorganic anions in the void volume of the ion-exclusion system and then passing them directly to the IC system. The chromatograms obtained with this coupled system are shown in Fig. 12.8. Further representative applications of the determination of inorganic anions by ion-exclusion chromatography are included in Table 12.6.

12.2.3 Inorganic cations

12.2.3.1 Ion chromatography of cations

Both suppressed and nonsuppressed IC are applied extensively to the separation of inorganic cations, especially alkali-metal and alkaline-earth cations. Many of the developments discussed earlier for IC of anions also apply to cation determinations, but it is fair to

References on p. B61

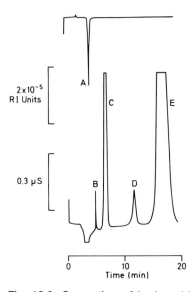

Fig. 12.8. Separation of boric acid and inorganic anions by coupled ion-exclusion and ion-exchange chromatography. A Waters Fast Fruit Juice column was used for ion-exclusion chromatography, with 1.25 mN sulfuric acid as eluent and refractive index detection. The detector output is shown as the upper tracing. A Waters IC Pak A column was used for ion-exchange chromatography, with 3 mM sodium octanesulfonate as eluent and conductivity detection. The detector output is shown as the lower tracing. Peak identities: A = boric acid, B = chloride, C = nitrate, D = iodide, E = sulfate. (Reproduced from Ref. 481 with permission.)

say that the rate of progress of IC has been somewhat slower for cations than for anions. The underlying reason for this is that IC stands alone as the premier method for anion analysis, but represents just one of a number of useful techniques for cation analysis. Nevertheless, the ability of IC to provide rapid and sensitive determination of multiple solutes in a single chromatogram will ensure its continued application in cation analysis.

12.2.3.1.1 Stationary phases

Agglomerated cation-exchange materials have evolved in parallel with anion exchangers of the same type. However, the first commercial resins of this kind became available only relatively recently and are manufactured from aminated PS/DVB core particles, 10-13 μm in diameter, and sulfonated latex particles in the size range 100-250 nm. It is interesting to note that incomplete coverage of the core particle by the latex results in a column that exhibits both anion- and cation-exchange properties. Cation exchange occurs at the sulfonic acid functional groups on the latex particles, while anion exchange occurs at the quaternary ammonium groups on the core particles. This behavior is illustrated in Fig. 12.9.

Nonsuppressed IC (and often also suppressed IC) of cations is generally performed on low-capacity, surface-sulfonated PS/DVB materials. Confinement of the sulfonic acid

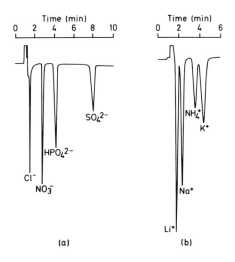

Fig. 12.9. Anion- and cation-exchange characteristics of an anion agglomerate column (Dionex HPIC-CAS1). Eluents: (a) 0.50 mM phthalic acid/0.61 mM sodium tetraborate, (b) 0.50 mM CuCl$_2$. Indirect spectrophotometric detection at (a) 254 nm and (b) 215 nm was used in each case. (Reprinted from Ref. 482 with permission.)

groups to a thin layer (extending radially about 200 Å into the particle [483]) at the surface of the resin bead gives optimal mass-transfer characteristics. An interesting polymer-coated silica cation-exchange material has been synthesized, in which poly(butadiene-maleic acid) functions as the coating polymer [484]. This is a weak-acid cation-exchange material and requires that the eluent pH be > 3.0 to establish cation-exchange sites on the column.

12.2.3.1.2 Eluents

Monovalent cations are eluted easily with dilute solutions of strong acids, usually nitric acid. The hydrogen ion serves as an effective competing cation, such that alkali-metal ions and ammonia can be separated in less than 10 min. Aromatic bases, such as dimethylpyridine and benzylamine [485], and inorganic species, such as copper(II) [486] and cerium(III) [487,488], have recently been shown to be effective as eluents for mono-valent cations.

The strong affinity of divalent cations for sulfonated cation exchangers means that these solutes are difficult to elute by straightforward ion-exchange chromatography, un-less a divalent competing cation (such as ethylenediammonium) is used in the eluent. An alternative approach is to moderate the solute charge by complexation with a suitable ligand. Tartrate, citrate, oxalate, or 2-hydroxyisobutyric acid (HIBA) [489,490] can be employed for this purpose. Solute retention times are governed by the eluent pH and the concentration of the eluent ligand, which in turn determine the conditional formation constant of the solute/ligand complex [491]. A gradient formed by varying the concentra-

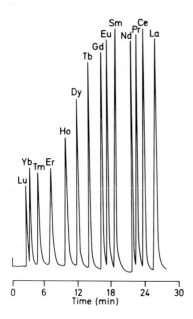

Fig. 12.10. Gradient elution in cation-exchange chromatography with postcolumn reaction detection. A Nucleosil 10SA column was used with an eluent formed from a linear gradient from 10 to 40 mM 2-methyllactic acid (pH 4.6) over a 30-min period. (Reproduced from Ref. 492 with permission.)

tion of ligand in the eluent can be used to improve the separations achieved in the ion-exchange mode [492], as shown in Fig. 12.10.

Combination of a complexing species and a competing cation in the eluent permits simultaneous separation of monovalent and divalent cations [484]. This has proved to be a long-standing challenge for IC. Some ligands, such as pyridine-2,6-dicarboxylic acid [407] and EDTA [493], can form anionic complexes with the solute cations, enabling them to be separated on an anion exchanger.

12.2.3.1.3 Detection

Conductivity detection is particularly suitable for the detection of monovalent cations after separation by IC with nitric acid as eluent. In the suppressed mode, the nitric acid eluent can be converted effectively to water by exchange of nitrate ions in the eluent for hydroxide ions from the suppressor. On the other hand, nitric acid is also well-suited for the nonsuppressed mode, since the very high limiting equivalent ionic conductance of the hydrogen ion offers sensitive, indirect detection of the less-conducting solute cations [494]. Conductivity detection may also be utilized for alkaline-earth cations.

In contrast, conductivity detection is rarely employed for the detection of transition-metal cations, because of the relative insensitivity of this method. Postcolumn reaction with color-forming reagents is more suitable. The most popular reagents used for this purpose are 2,7-bis(2-arsonophenylazo)-1,8-dihydroxynaphthalene-3,6-disulfonic acid (Arsenazo III) and PAR. PAR is the preferred reagent for transition-metal ions, whereas Arsenazo III is preferred for lanthanide ions. Detection limits for postcolumn reaction with PAR are typically in the range between 0.5 and 5 ng for transition-metal ions [408]. Detection limits can be enhanced if an equimolar solution of PAR and Zn(II)-EDTA complex is used as the postcolumn reagent. Eluted metal ions react with the Zn(II)-EDTA to release zinc ions, which then complex with PAR. Detection is therefore based only on the Zn(II)-PAR complex [495-497].

12.2.3.1.4 Applications

As indicated above, the prime application of IC in inorganic cation analysis is the determination of monovalent cations, for which it provides a rapid and sensitive determination. Ion exchange does not offer exceptionally high separation efficiencies for transition metals, unless a pellicular type of ion exchanger (e.g., agglomerated resins or polymer-coated silica) is employed. Table 12.7 shows some representative applications of IC for the determination of inorganic cations.

12.2.3.2 Ion-pair chromatography of cations

Ion-pair chromatography finds extensive use for the separation of transition metals and lanthanides, which it separates with unrivaled efficiency. The dynamic-coating method is the preferred approach, and the ion-pairing reagent is usually an aliphatic sulfonate, such as hexanesulfonate [459] or octanesulfonate [509,510]. The choice of the counterion is of much less importance than in the case of ion-pair chromatography of anions, since ion-exchange plays only a minor role in the elution process for cations. Solute elution is dominated by complexation effects due to eluent ligands, such as tartrate. Therefore, the ligand concentration and the eluent pH are of particular importance in the control of retention. Recent studies have shown that there is a linear relationship between the logarithm of the solute capacity factor and the logarithm of the fraction of total metal present as the free cation [511]. It may also be noted that anionic complexes of metal ions with suitable ligands (especially cyanide) are readily separated by ion-pair chromatography with chromatographic efficiencies well in excess of those attainable by anion exchange [512-515]. In this application, direct spectrophotometric detection at 214 nm is suitable.

Again, postcolumn reaction with PAR or Arsenazo III is the most sensitive detection method, and the combination of dynamic-coating ion-pair chromatography with this mode of detection yields some excellent analytical results for metal ions. Gradient elution by changing the concentration of the eluent ligand is again applicable. These techniques yield optimal results when applied to the determination of lanthanides, so that ion-pair chromatography represents on outstanding analytical method for these species. Fig. 12.11 shows the separation of 14 lanthanides in less than 9 min. Ion-pair chromatography has been used for the determination of lanthanides in such complex samples as uranium

References on p. B61

TABLE 12.7

TYPICAL DETERMINATIONS OF INORGANIC CATIONS BY LIQUID CHROMATOGRAPHY

Solutes	Sample	Stationary phase	Eluent*	Detection mode**	Detection limit	Refs.
Na^+	Kraft black liquor	Waters IC Pak C	2 mM HNO_3	IC	10 ppb	498
Na^+, K^+, NH_4^+	shellfish	Dionex AS-3	10 mM HCl	C	5 ppm	499
Na^+, NH_4^+, K^+, Mg^{2+}, Ca^{2+}	vegetables	Zipax SCX	2.5 mM $CuSO_4$	IS (220 nm)	1 ppm	500
Na^+, K^+, Rb^+, Cs^+, Mg^{2+}, Ca^{2+}	water	Interaction ION 210	0.1 mM $Ce_2(SO_4)_3$	IS (254 nm)	4 ppb	487
Na^+, K^+, Rb^+, Cs^+, NH_4^+	–	aters IC Pak C	2 mM dimethylpyridine	IS (276 nm)	10 ppb	501
Ca^{2+}, Mg^{2+}	sea water	Dionex CS-3	48 mM HCl/8 mM DAP	C	1-10 ppm	502
Fe^{3+}, Zn^{2+}, Pb^{2+}, Ni^{2+}, Co^{2+}	wastewaters	Aminex A5	0.35 M to 0.5 M tartrate (pH 3.4) gradient	PCR, S (540 nm)	10 ppb	503
Fe^{3+}, Cu^{2+}, Ni^{2+}, Zn^{2+}	effluents	TSK silica cation exchanger	1 g/l citric acid/7.5 g/l tartaric acid/120 µl/l EDA	S (546 nm)	50 ppb	504
Fe^{3+}, Pb^{2+}, Zn^{2+}, Ni^{2+}, Co^{2+}, Cd^{2+}, Fe^{2+}, Mn^{2+}	soil	µBondapak C_{18}	2 mM octanesulfonate/50 mM tartrate (pH 3.4)	PCR, S (520 nm)	10 ppb	505

Lanthanides	rocks	Supelcosil LC-18	10 mM octanesulfonate/0.1-0.4 M HIBA gradient	PCR, S (495 nm)	50 ppb	506
Cu^{2+}, Zn^{2+}, Fe^{2+}, Mn^{2+}, Mg^{2+}, Ca^{2+}	wine	Nucleosil 10-SA	1 mM oxalic acid/2.5 mM EDA (pH 3.5)	PCR, S (490 nm)	20 ppb	507
Cyano complexes of Ni^{2+}, Co^{2+}, Fe^{3+}, Au(I), Fe^{2+}, Au(III)	plating solutions	Dionex AS-4	2 mM TBAOH/0.2 mM Na$_2$CO$_3$	C	250 ppb	508

* DAP = diaminopropionic acid, TBAOH = tetrabutylammonium hydroxide, EDA = ethylenediamine, HIBA = α-hydroxyisobutyric acid.

** S = spectrophotometry, C = conductivity, IC = indirect conductivity, IS = indirect spectrophotometry, PCR = postcolumn reaction.

References on p. B61

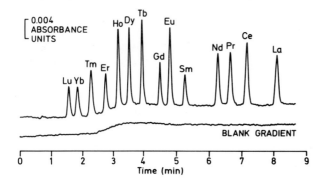

Fig. 12.11. Ion-pair chromatography of lanthanides by gradient elution. A Supelco LC18 column was used. The eluent was a linear gradient at pH 4.6 from 0.05 M HIBA to 0.40 M HIBA over 10 min at 2.0 ml/min, with a constant concentration of 10 mM 1-octanesulfonate in the eluent. Detection by postcolumn reaction with Arsenazo III at 635 nm. (Reprinted from Ref. 517 with permission.)

dioxide fuels [516,517] and rocks [518]. Further applications of this method to inorganic cations are included in Table 12.7.

12.2.3.3 Separation of cations on stationary phases with immobilized ligands

Metal ions can be separated on stationary phases on which a suitable ligand is immobilized by chemical bonding to a support material (usually silica). Many immobilized ligands have been used for this purpose, including 8-hydroxyquinoline [519], β-diketones [519,520], dithizone [519], and 2-pyridinecarboxylaldehyde phenylhydrazone [521]. An alternative approach is to use a crown ether, usually 18-crown-6 (18-crown-6(1,4,7,10,13,16-hexaoxacyclooctadecane)) or its derivatives, as the immobilized ligand. This method has been studied extensively [523-528] and shows particular promise for the separation of alkali-metal ions. It is noteworthy that an anion is bound along with the cation in order to preserve electroneutrality, so that crown ether stationary phases can also be employed for the chromatographic determination of inorganic anions [523,524].

12.2.4 Coordination compounds and organometallics

12.2.4.1 Coordination compounds

Many inorganic cations can be determined by LC as their coordination compounds. The presence of a suitable ligand in the complex permits separation to be achieved by conventional reversed-phase or normal-phase HPLC methods, and often also assists in the detection of the eluted metal complex. Several recent reviews on LC of coordination compounds are available [529-534]. The ligand and chelate should have a number of properties, including the following [530-532]:

(i) The ligand should form neutral complexes with a large number of metals, prepared by relatively simple methods.

(ii) The complexes formed should be coordination-saturated, since this gives the greatest probability of separation of complexes formed from different metals. Moreover, the ligand should not be too large, so that specific properties of the central metal atom are maintained.

(iii) The donor atoms in the ligand should have low total electronegativity to minimize adsorption effects on silica-based reversed phases. Preferred donor atoms are N, O, and S.

(iv) Ligand substituents should not have large induction or steric effects, and electronegative atoms should preferably exist in close proximity to the chelate ring to increase separation selectivity.

(v) The complexes should have high stability, good detectability, and high solubility in nonpolar organic solvents.

Many ligands have found application in LC of metal chelates. These ligands include: dithiocarbamates [535], 8-hydroxyquinoline [536,537], β-diketones [538], 4-(2-pyridyl-azo)naphthol [539], 4-(2-pyridylazo)resorcinol [540], dialkyldithiophosphates [541,542], xanthates [543], 2,3-diaminonaphthalene [544], pyrazolones [545], and hydrazones [521,546]. No single ligand is suitable for all metal ions, and typically only a few metals are determined in a single chromatographic separation. In most cases, water-insoluble chelates are formed, and these must be extracted into a suitable organic solvent prior to chromatographic separation. This sometimes involves extraction with solvents that cannot be injected directly into a reversed-phase HPLC system, so that evaporation and redissolution become necessary. Alternatively, complexes can be formed in situ by injecting metal ions into a mobile phase containing the ligand and an appropriate buffer [547] or through the use of solid-phase reaction on a suitable precolumn [542]. The stability of the metal complexes is also of great importance, because these complexes are generally injected at very low concentrations and are therefore prone to dissociation as they traverse the chromatographic system. This is particularly true of complexes that may undergo ligand-exchange reactions at the surfaces of metallic chromatographic components, such as the injector, interconnecting tubing, and the inlet and outlet frits in the column. Kinetic stability is of more importance than thermodynamic stability, since kinetically inert complexes are more likely to pass intact through the chromatographic system.

Once again, the large volume of literature on HPLC of metal chelates precludes a comprehensive discussion of this topic. Table 12.8 provides a selected listing of some applications of HPLC of metal chelates. To illustrate the utility of this technique, we will focus attention only on the use of dithiocarbamate ligands, since these illustrate most of the properties of other ligands.

12.2.4.1.1 Dithiocarbamate complexes

Alkyldithiocarbamate ligands form complexes with a very wide range of metal ions and therefore offer the potential for the separation of perhaps a larger number of metal ions than any other ligand. Diethyldithiocarbamate (DEDTC) complexes were studied exten-

TABLE 12.8

TYPICAL DETERMINATIONS OF METAL CHELATES BY LIQUID CHROMATOGRAPHY

Solutes	Ligand[*]	Stationary phase	Mobile phase [**]	Detection mode [***]	Detection limit [§]	Refs.
Al(III), In(III)	PMBP	Shim-Pack CLC-ODS	MeCN-MeOH	S (290 nm)	21-121 ng	545
As(III), Sb(III), Bi(III)	DEDTP	Hypersil ODS	10 mM DEDTP in MeCN	S (280)	2 ng	542
Cd(II), Pb(II), Ni(II), Co(III), Hg(II), Cr(III), Se(IV), Cu(II), Te(IV)	DEDTC	Waters C_{18} Rad-Pak	MeOH/MeCN/water (40:35:25)	S (254 nm)	0.5 μg	535
Cd(II), Co(II), Pb(II), Ni(II), Cu(II)	HEDC	μBondapak C_{18}	0.1mM HEDC in 40% aq. MeOH/0.1 mM Zn^{2+}	S (300 nm)	7-53 ppb	548
Co(II), Cu(II), Hg(II), Ni(II)	HEDC	Supelco C_{18}	25 mM triethylammonium acetate	S (255 nm)	5 ppb	549
Cr(VI), Mo(VI), Co(II), Cu(II), Mn(II), Mn(III)	Oxine	C_8 reversed phase	1mM oxine in borate buffer (pH 9)	S (254 nm)	n.s.	537
Cu(II)	HFAA	Silica	100% CH_2Cl_2	DCPAES	n.s.	550
Cu(II), Fe(III)	Oxine	C_{18} reversed phase	10mM oxine, MeCN, and acetate buffer (pH 6)	S (400 nm)	100 ppb	536

Pb(II), Hg(I), Hg(II)	DEDTC	Hypersil ODS	0.05% DEDTC in water/MeOH/CHCl$_3$ (70:20:10)	S (350 nm)	0.5 ppm	547
Se(IV)	DAN	μBondapak C$_{18}$	60% aq. MeOH	S (254 nm), Fluor	10 ppb	544
Ti(IV), Fe(III), U(VI), V(V)	DAPMP, DAPMT	Polymer Labs PLPR-S	1%-40% aq. MeCN gradient	S (340 nm)	n.s.	546

*PMBP = 1-phenyl-3-methyl-4-benzoyl-5-pyrazolone, DEDTP = diethyldithiophosphate, DEDTC = diethyldithiocarbamate, HEDC = bis(2-hydroxyethyl)dithiocarbamate, Oxine = 8-hydroxyquinoline, HFAA = hexafluoroacetylacetone, DAN = 2,3-diaminonaphthalene, DAPMP = 2,6-diacetylpyridine bis(N-methylenepyridiniohydrazone), DAPMT = 2,6-diacetylpyridine bis(N-methylene-N,N,N-trimethylammoniohydrazone).

** MeOH = methanol, MeCN = acetonitrile.

*** S = spectrophotometry, Fluor = fluorimetry, DCPAES = direct current plasma atomic emission spectrometry.

§ n.s. = not stated.

References on p. B61

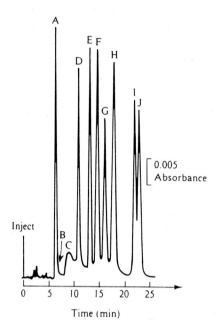

Fig. 12.12. Separation of a mixture of diethyldithiocarbamate complexes. Column, Waters C$_{18}$ Rad-Pak; mobile phase, methanol/acetonitrile/water (40:35:25); flowrate, 2.0 ml/min; detection, 254 nm. Peak identities: A = disulfiram, B = Cd(II), C = Pb(II), D = Ni(II), E = Co(III), F = Cr(III), G = Se(IV), H = Cu(II), I = Hg(II), J = Te(IV). (Reproduced from Ref. 535 with permission.)

sively prior to 1982 and this activity has continued over the time period covered by this review. Fig. 12.12 shows a typical separation of metal-DEDTC complexes achieved on a reversed-phase column with a ternary eluent of water, methanol, and acetonitrile. Two important characteristics are evident from this chromatogram. First, there is a peak due to the ligand oxidation product, bis(diethylthiocarbamyl)disulfide (usually referred to as disulfiram), which results either from excess ligand in the extracting solution, or from ligand produced by dissociation of labile complexes. Second, the peak shape for Pb(II) is very poor, due to the low kinetic stability of this complex and the resultant likelihood of dissociation or ligand-exchange reactions. This behavior occurs with other unstable complexes, such as those of Cd(II), Fe(III), and Zn(II), and is especially evident when columns with stainless-steel frits are used. The porous nature and high surface areas of these frits provide ideal sites for ligand-exchange reactions to occur between the injected metal complexes and metal ions produced from the stainless-steel components, especially nickel. These reactions can be minimized, either by using columns without porous metallic frits, such as radial compression columns [535], by deactivating the frits with an organosilane [551], or by addition of EDTA to the mobile phase [535].

Precolumn formation of dithiocarbamate complexes can be used for the determination of metal ions (as in the above examples) or for the determination of amines. In the latter case, the amines are allowed to react with carbon disulfide and a suitable metal ion (often Ni(II)) to form a dithiocarbamate complex. Because these complexes incorporate two ligand molecules, the possibility exists to separate isomers of the parent amine. For example, if the parent amine exists in two isomeric forms, L_1 and L_2, the resultant dithiocarbamate complexes produced with nickel are $Ni(L_1)_2$, $Ni(L_2)_2$, and NiL_1L_2.

An alternative method for the formation of the dithiocarbamate complexes is to inject the metal ions into a mobile phase containing the ligand. On-column complex formation provides a potentially quick and easy method for multi-element identification and determination. When DEDTC is added to the mobile phase, Cd(II), Pb(II), Co(II), Hg(II), and Cu(II) can be separated [547,552]. The chief problem encountered with on-column complexation is the high background detector signal produced by the presence of the ligand in the mobile phase. This requires that a selective wavelength be used in the case of spectrophotometric detection, or alternatively, amperometric detection must be used [553]. A further problem is the poor solubility of many dithiocarbamate complexes in typical mobile phases for reversed-phase HPLC, but this may be overcome either by the addition of a small amount of chloroform to the mobile phase [547] or through the use of a ligand that forms water-soluble complexes [548]. An example of the latter approach is the use of bis(2-hydroxyethyl)dithiocarbamate, in which the hydroxy groups on the ligand cause metal complexes to be water-soluble at low concentrations [549].

12.2.4.2 Organometallic compounds

One of the factors that limit the applicability of HPLC analysis of metal chelates is the need for forming the chelate. This limitation does not exist for many of the organometallic species. They are subject to chromatographic analysis, since they are often found in a wide range of samples. The more important organometallic species that can be analyzed by HPLC are alkyllead, alkylmercury, alkylarsenic, and alkyltin compounds. Of these, the organoarsenic species are the most widely studied. Monomethylarsonate, dimethylarsinate, and phenylarsonate (as well as the inorganic ions arsenate and arsenite) are formed by the action of many common yeasts, fungi, and bacteria on arsenic present in soils. The high toxicities of these compounds necessitate their accurate determination, especially in water samples. Separation is usually accomplished by anion-exchange chromatography [554,555], but ion-pair chromatography in both the dynamic-coating [556] and permanent-coating [557] modes has also been employed. Organotin compounds find industrial use as stabilizers for PVC, as catalysts, and in marine anti-fouling paints, and can therefore be present in natural waters. Once again, ion-exchange [558] and ion-pair [560] chromatography are the preferred separation modes.

Table 12.9 lists some chromatographic details and applications of HPLC of organometallic compounds.

References on p. B61

TABLE 12.9

TYPICAL DETERMINATIONS OF ORGANOMETALLIC SPECIES BY LIQUID CHROMATOGRAPHY

Solutes[*]	Sample	Stationary phase	Eluent[**]	Detection mode[§]	Detection limit	Refs.
Tributyltin	natural waters	Partisil 10 SCX	70% aq. MeOH/0.1 M ammonium acetate	FAAS	200 ng	558
Tributyltin	estuarine waters	Partisil 10 SCX	80% aq. MeOH/0.15 M ammonium acetate	Fluor	16 ng	559
CH_3Sn^{3+}, $(CH_3)_2Sn^{2+}$, $(CH_3)_3Sn^+$	seawater, clam juice, tuna	Hamilton PRP-1	3 mM hexanesulfonic acid/3 mM KF/0.02 N H_2SO_4/2.5% acetic acid	Hy-DCPAES	40 ppb	560
MMA, DMA, arsenobetaine	-	Hamilton PRP-X100, Vydac 201TP	3 mM $NH_4H_2PO_4$ (pH 6)	ICPAES	n.s.[§§]	561
MMA, DMA, phenylarsonate	-	Hamilton PRP-1	2mM HTMAB (pH 9.6)/1% CH_3COOH/10% DMF[***]	ICPAES	130 ppb	557
MMA	soil-pore waters	Zipax SAX and Benson BAX10	0.1 mM H_2SO_4 then 0.1 M $(NH_4)_2CO_3$[***]	Hy-AAS	10 ppb	555
MMA, phenylarsonate	shale retort waters	Dionex AS-1	20% aq. MeOH to 15% aq. MeOH with 0.02 M $(NH_4)_2CO_3$[***]	GFAAS	1 pp	554

MMA, DMA	air	Dionex AS-1	2.4 mM NaHCO$_3$ + 1.9 mM Na$_2$CO$_3$ + 1 mM Na$_2$B$_4$O$_7$/5 mM Na$_2$B$_4$O$_7$ [***]	Hy-AAS	10 ppb	562
Methylmercury, ethylmercury	tuna	Waters Pico Tag C$_{18}$	60 mM ammonium acetate/0.005% mercapto-ethanol	ICPAES/MS	1 ppb	563

[*] MMA = monomethylarsonate, DMA = dimethylarsinate.

[**] HTMAB = hexadecyltrimethylammonium bromide, DMF = dimethylformamide.

[***] A step gradient between the indicated eluents was used.

[§] FAAS = flame atomic absorption spectrometry, Fluor = fluorimetry, Hy = hydride generation, DCPAES = direct-current plasma atomic emission spectrometry, ICPAES = inductively coupled plasma atomic emission spectrometry, GFAAS = graphite furnace atomic absorption spectrometry, MS = mass spectrometry.

[§§] n.s. = not stated.

References on p. B61

12.2.4.3 Detection methods

In some cases, metal chelates and organometallic compounds may be detected spectrophotometrically. However, when this detection mode is applied to metal chelates, it is often complicated by the presence of excess ligand in the sample. This may require the use of a wavelength at which detection sensitivity is diminished. For this reason, atomic spectroscopic detection is very attractive, and considerable effort has been expended in successful coupling of HPLC instruments with atomic absorption spectrometers (AAS), or direct-current plasma (DCP) and inductively coupled plasma (ICP) atomic emission spectrometers. This subject has been reviewed recently [564].

The problems involved in these methods include matching the chromatographic flow-rate and the uptake rate of the aspiration device on the spectrometer, and the presence of organic solvents in the eluents. The organic solvent affects nebulization efficiency and also acts as a secondary fuel if supplied to an AAS flame. It is therefore appropriate to use pulse nebulization [558], or to incorporate an intermediate step between the HPLC and the nebulizer, such as a fraction collector [565] or a hydride generator [555]. Electrothermal AAS offers high sensitivity, but the sample must be dried and ashed, and thus only indirect coupling of HPLC to AAS is possible. On the other hand, DCP and ICP nebulizers can usually be coupled directly to the output of a HPLC instrument [566], even when microbore columns are used [556].

12.2.5 Determination of water

The determination of water is a very important and frequently encountered analytical problem. LC may be applied to this determination if the water is first converted to a suitable derivative. For example, phenyl isocyanate reacts with water to give *N,N'*-diphenylurea, which can then be determined by LC [567]. A more efficient approach is to use ion-exclusion chromatography to separate water from other components, and this method has been utilized with a hydrogen-form cation-exchange column and an eluent containing very dilute sulfuric acid in methanol [568]. In this case, conductivity detection was employed, and the water peak gave a reduced conductivity relative to that of the background eluent. The main drawback with this method was the variability in the detector response as the concentration of water in the sample was altered. The rate of exchange of protons between a methanol molecule and a water molecule is less than that between two molecules of water or two molecules of methanol. This factor may contribute to the process whereby water is retained on the ion-exclusion column.

Fritz and coworkers [569,570] have reported an alternative ion-exclusion method, in which the water is isolated on a cation-exchange column in either the lithium or hydrogen form, with cinnamaldehyde in methanol as the eluent. Detection is accomplished by spectrophotometric monitoring of the equilibrium existing between cinnamaldehyde and cinnamaldehyde dimethylacetal:

$$2CH_3OH + \text{cinnamaldehyde} \leftrightarrow H_2O + \text{cinnamaldehyde dimethylacetal}$$

This reaction does not occur to any appreciable extent until an acid catalyst is present. This may be the hydrogen-form cation-exchange resin in the column, or an acid can be added to the eluent. After catalysis, the equilibrium lies well to the right. When water is injected, there is a small shift in the equilibrium towards the formation of cinnamaldehyde. This change can be detected spectrophotometrically at 300 nm. Excellent sensitivity is achieved by this method, and the analysis can be completed in less than 2 min.

REFERENCES

1 G. Schwedt, *Chromatographic Methods in Inorganic Analysis*, Huethig Verlag, Heidelberg, 1981.
2 D.N. Sokolov, *Gas Chromatography of Volatile Metal Complexes*, Nauka, Moscow, 1981.
3 T.R. Crompton, *Gas Chromatography of Organometallic Compounds*, Plenum Press, New York, 1982.
4 P.C. Uden, *J. Chromatogr.*, 313 (1984) 3.
5 P.C. Uden, *Chem. Anal.*, 78, Chap. 5 (1985) 229.
6 D.N. Sokolov, *Usp. Khim.*, 57 (1988) 57.
7 K. Robards, S. Clarke and E. Patsalides, *Analyst (London)*, 113 (1988) 213.
8 Yu.S. Drugov, *Zh. Anal. Khim.*, 35, No. 3 (1980) 559.
9 Yu.S. Drugov, *Zh. Anal. Khim.*, 40, No. 4 (1985) 585.
10 K. Robards, V. Kelly and E. Patsalides, in press.
11 F. Xu, *Huanjing Huaxue*, 5, No. 1 (1986) 66; *C.A.*, 105: 107364k.
12 K. Funazo and M. Tanaka, *Bunseki*, No. 2 (1983) 135; *C.A.*, 99: 15520n.
13 M. Tanaka and T. Shono, *Kagaku (Kyoto)*, 36, No. 2 (1981) 149; *C.A.*, 95: 72429f.
14 S. Dilli and I. Sutikno, *J. Chromatogr.*, 300 (1984) 265.
15 R. Neeb, *Pure Appl. Chem.*, 54, No. 4 (1982) 847.
16 K. Robards, E. Patsalides and S. Dilli, *J. Chromatogr.*, 411 (1987) 1.
17 M. Miyazaki, *Kagaku no Ryoiki Zokan*, 132 (1981) 143; *C.A.*, 95: 161289c.
18 T. Kuwamoto and N. Matsubara, *Bunseki*, (1982) 310; *C.A.*, 97: 103358e.
19 J. Maslowska and S. Starzynski, *Wiad. Chem.*, 36 (1982) 669; *C.A.*, 99: 98238a.
20 V.Y. Golubtsova, I.A. Muraveva and L.I. Martynenko, in V.I. Spitysyn and L.I. Martynenko (Editors), *Prikl. Khim. β-Diketonatov Met.*, Nauka, Moscow, 1985, pp. 173-187; *C.A.*, 103: 204894z.
21 E. Pella, L. Bedoni, B. Colombo and G. Giazzi, *Anal. Chem.*, 56 (1984) 2504.
22 P. Mazzeo and A. Mazzeo-Farina, *Microchem. J.*, 28 (1983) 137.
23 Z. Vecera, J. Uhdeová and V. Rezl, *Microchem. J.*, 30 (1984) 369.
24 V. Rezl and J. Uhdeová, *Zh. Anal. Chem.*, 39, No. 12 (1983) 2210.
25 Y. Seo, *Bunseki Kagaku*, 33 (1984) 252.
26 V. Rezl, *Mikrochim. Acta*, 2 (1982) 107.
27 V. Rezl, *J. Chromatogr.*, 251 (1982) 35.
28 T. Hara, T. Fujinaga, F. Okui and A. Arai, *Bull. Chem. Soc. Jpn.*, 56 (1983) 3615 and previous papers of this series.
29 J.F. Sullivan and R.L. Grob, *J. Chromatogr.*, 268 (1983) 219.
30 B. Waskowski, R., Gondko and J. Kaczmerek, *Chem. Anal. (Warsaw)*, 31, No. 2 (1986) 305.
31 V. Rezl and J. Uhdeová, *Zh. Anal. Khim.*, 39, No. 12 (1984) 2210.
32 J. Franc, *Czech. Pat. 204,139*, 1983; *C.A.*, 101: 239642b.
33 B. Waskowski, *Pol. Pat. 124,992*, 1984; *C.A.*, 101: 203677c.
34 V. Rezl, *Czech. Pat. 197,679*, 1983; *C.A.*, 99: 224460g.
35 L.E. Ebdon, S. Hill and R.W. Ward, *Analyst (London)*, 111 (1986) 1113.
36 N.S. Chong and R.S. Houk, *Appl. Spectrosc.*, 41, No. 1 (1987) 66
37 D.E. Pivonka, W.G. Fateley and R.C. Fry, *Appl. Spectrosc.*, 40, No. 3 (1986) 291.
38 D.S. Forsyth, *Anal. Chem.*, 59 (1987) 1742.

39 V.I. Rigin, *Zh. Anal. Khim.*, 41, No. 1 (1985) 46.
40 V.I. Rigin, *Zh. Anal. Khim.*, 35, No. 1 (1980) 64.
41 R. Belcher, S.L. Bogdanski, M. Burguera, E. Henden and A. Townshend, *Anal. Chim. Acta*, 100 (1978) 515.
42 E. Henden, *Anal. Chim. Acta*, 173 (1985) 89.
43 C.S. Cerbus and S.J. Gluck, *Spectrochim. Acta*, 38B (1983) 387.
44 D.B. Quimby, P.C. Uden and R.M. Barnes, *Anal. Chem.*, 50 (1978) 2112.
45 K.B. Olsen, D.S. Sklarew and J.C. Evans, *Spectrochim. Acta*, 40B (1985) 357.
46 L.E. Ebdon, S. Hill and R.W. Ward, *Analyst (London)*, 112 (1987) 1.
47 J. Ellis and P.L. Brown, *Anal. Chim. Acta*, 124 (1981) 431.
48 K.Y. Chen, M. Moussavi and A. Sycip, *Environ. Sci. Technol.*, 7, No. 10 (1973) 949.
49 L.G. Borchardt and D.B. Easty, *J. Chromatogr.*, 299 (1984) 471.
50 C. Heim, I. Devai and J. Harangi, *J. Chromatogr.*, 295 (1984) 259.
51 L. He, Y. Huang, L. Cao and G. Gao, *Fenxi Huaxue*, 14, No. 2 (1986) 139; *C.A.*, 105: 34755b.
52 Y. Zhuang, X. Chen, Y. Yuan, Y. Zhang, Q. Ye and Z. Su, *Huanjing Huaxue*, 4, No. 1 (1985) 67; *C.A.*, 103: 97991e.
53 S.J. Fernandez, L.P. Murphy and R.A. Rankin, *Anal. Chem.*, 56 (1984) 1285.
54 B. Grapengiesser and G. Rudstam, *Radiochim. Acta*, 20 (1973) 85.
55 G. Rudstam and B. Grapengiesser, *Radiochim. Acta*, 20 (1973) 97.
56 S. Hübener and I. Zvára, *Radiochim. Acta*, 27 (1980) 157.
57 D.T. Jost, H.W. Gäggeler, Ch. Vogel, M. Schädel, E. Jäger, B. Eichler, K.E. Gregorich and D.C. Hoffman, *Inorg. Chim. Acta*, 146 (1988) 255.
58 B. Eichler, G.V. Buklanov and S.N. Timokhin, *Kernenergie*, 30 (1987) 469; *C.A.*, 108: 157797m.
59 C. Harris, *Talanta*, 19 (1972) 1523.
60 E. Scholz, *Karl Fischer Titration*, Springer-Verlag, Berlin, Heidelberg, 1984.
61 F.F. Andrawes, *J. Chromatogr.*, 290 (1984) 65.
62 F.F. Andrawes, *Anal. Chem.*, 55 (1983) 1869.
63 K. Kato, K. Sato, H. Tomita and T. Maeda, *Bunseki Kagaku*, 36 (1987) 267.
64 W.K. Al-Thamir, *J. Chromatogr.*, 403 (1987) 288.
65 I.M. Mogilevich and L.E. Mal'tseva, *Khim.-Farm. Zh.*, 22, No. 1 (1988) 118; *C.A.*, 108: 137977s.
66 R. Komers and Z. Sir, *Collect. Czech. Chem. Commun.*, 42 (1977) 3576.
67 Z. Yu and Y. Long, *Gaoding Xuexiao Huaxue Xuebao*, 8, No. 11 (1987) 990; *C.A.*, 108: 1790336n.
68 C.P. A'Campo, S.M. Lemkovitz, P. Verbrugge and P.J. van den Berg, *J. Chromatogr.*, 203 (1981) 271.
69 A.M. Awwad and T.M. Sarkissian, *J. Chromatogr. Sci.*, 16 (1978) 166 and references therein.
70 H. Ge, M. Zhu, G. Gu and J. Fan, *Huadong Huagong Xueyuan Xuebao*, 11, No. 3 (1985) 419; *C.A.*, 105: 85296k.
71 M. Gaza, M. Blebea, L. Dan, H. Kolck and I. Ianculov, *Rev. Chim. (Bucharest)*, 37, No. 1 (1986) 65.
72 K.M. Amirov, E.K. Chepegina and N.V. Zakharova, *Zh. Anal. Khim.*, 41, No. 3 (1986) 420.
73 G.E. Snopartin, A.E. Ezheleva, L.S. Malygina and Z.A. Klemina, *Zh. Anal. Khim.*, 39, No. 3 (1984) 433.
74 A. Goldup and M.T. Westaway, *Anal. Chem.*, 38 (1966) 1657.
75 J.M. Loeper, *Diss. Abstr. Int. B*, 48, No. 8 (1988) 2294.
76 J.M. Loeper and R.L. Grob, *J. Chromatogr.*, 463 (1989) 365.
77 J.M. Loeper and R.L. Grob, *J. Chromatogr.*, 457 (1988) 247.
78 D.R. Jenke and M.R. Hannifan, *Anal. Chem.*, 54 (1982) 843.
79 A. Konopczinski and A. Siedlecki, *Chem. Anal.*, 25 (1980) 777.
80 G.D. Sarafyan, A.M. Ovsepyan, E.S. Ter-Stepanyan and S.T. Ter-Gevorkyan, *Prom-st. Stroit. Arkhit. Ar.*, 8 (1988) 67; *C.A.*, 110: 117893z.
81 G. Kiss, H. Vanco,, H. Krug and H. Sturtz, *Chem. Techn.*, 33 (1981) 147.

82 S. Latif, J.K. Haken and M.S. Wainwright, *J. Chromatogr.*, 258 (1983) 233.
83 R. Dguchi, K. Yamaguchi and T. Shibamoto, *J. Chromatogr. Sci.*, 26 (1988) 588.
84 M.A. Wizner, S. Singhawangcha, R.M. Barkley and R.E. Sievers, *J. Chromatogr.*, 239 (1982) 145.
85 G. Guiochon and C. Pommier, *Gas Chromatography in Inorganics and Organometallics*, Ann Arbor Science Publishers, Ann Arbor, MI, 1973.
86 J. Agterdenbos and D. Bax, *Fresenius Z. Anal. Chem.*, 323 (1986) 783.
87 Z. Weidenhoffer, Z. Plzak and J. Dolansky, *J. Chromatogr.*, 350 (1985) 324.
88 P.R. Gifford and S. Bruckenstein, *Anal. Chem.*, 52 (1980) 1028.
89 A.E. Ezheleva, L.S. Malygina and V.A. Krylov, *Vysokochist Veshchestva*, 3 (1987) 214; *C.A.*, 108: 30915t.
90 S.H. Vien and R.C. Fry, *Anal. Chem.*, 60 (1988) 465.
91 G.C. Huston, R.R. Romanosky, Jr. and J.K. Wachter, *J. Chromatogr. Sci.*, 24 (1986) 458.
92 T. Ding, Z. Wu, Z. Dong, S. Li and J. Wei, *Sepu*, 5, No. 3 (1987) 197; *C.A.*, 107: 152970g.
93 V.I. Rigin, *Zh. Anal. Khim.*, 41, No. 1 (1986) 46.
94 S. Hashimoto, K. Fujiwara and K. Fuwa, *Anal. Chem.*, 57 (1985) 1305.
95 S. Hashimoto, K. Fujiwara and K. Fuwa, *Limnol. Oceanogr.*, 32, No. 3 (1987) 729.
96 K.A. Scudamore and G. Goodship, *Pestic. Sci.*, 37 (1986) 385.
97 V.A. Krylov, G.V. Lazarev, S.G. Sokolova and S.G. Krasotskii, *Vysokochist Veshchestva*, 2 (1987) 159; *C.A.*, 107: 189782y.
98 Y. Li, Y. Yang, J. Zhang and L. Wang, *Fenxi Huaxue*, 12, No. 8 (1984) 763; *C.A.*, 102: 39058g.
99 L.M. Gurevich, R.V. Poponova, I.L. Agafonov and G.G. Volkova, *Zh. Anal. Khim.*, 41, No. 12 (1986) 2249.
100 W. Cheng, G. Ma and X. Zhu, *Fenxi Huaxue*, 15, No. 7 (1987) 588; *C.A.*, 107: 189882f.
101 Z.T. Wang, F.P. Yus and S.K. Hsia, *Zh. Anal. Khim.*, 41, No. 4 (1985) 511.
102 V.A. Krylov, V.A., Gonina, S.K. Barushkov, S.G. Krasotskii, A.N. Gur'yanov, V.F. Khopin and N.N. Vechkanov, *Zh. Anal. Khim.*, 39, No. 6 (1983) 1112.
103 L.N. Morozova, V.I. Maiorov and A.D. Molodyk, *Vysokochist Veshchestva*, 6 (1987) 144; *C.A.*, 108: 215613m.
104 M. Hirota and H. Muramatsu, *Bull. Chem. Soc. Jpn.*, 59 (1986) 329.
105 V.G. Berezkin and Yu.S. Drugov, *Zh. Anal. Khim.*, 39, No. 7 (1984) 1249.
106 J. Rudolph, K. Bachmann, A. Steffan and S. Tsalas, *Mikrochim. Acta*, No. 1 (1978) 471.
107 S. Tsalas and K. Bachmann, *Anal. Chim. Acta*, 98 (1978) 17.
108 F. Dienstbach and K. Bachman, *Radiochim. Acta*, 34 (1983) 215.
109 C.C. Nguyen, A.E. Novgorodov, M. Kaskevich, A. Kolaczkowski and V.A. Khalkin, *Radiochimiya*, 26, No. 1 (1984) 60.
110 B. Grapengiesser and G. Rudstam, *Radiochim. Acta*, 20 (1973) 83
111 F. Dienstbach and K. Bachmann, *Anal. Chem.*, 52 (1980) 620.
112 G.A. Cutter and T.J. Oatts, *Anal. Chem.*, 59 (1987) 717.
113 A. Steffen and K. Bachmann, *Talanta*, 25 (1978) 677.
114 V.P. Domanov, B. Eichler and I. Zuana, *Radiokhimiya*, 26, No. 1 (1984) 60; *C.A.*, 101: 103103u.
115 B.L. Zhuikov, T. Rectz and I. Zuara, *Radiokhimiya*, 28, No. 2 (1986) 246; *C.A.*, 105: 31575g.
116 R.S. Barratt, R. Belcher, W.I. Stephen and P.C. Uden, *Anal. Chim. Acta*, 59 (1972) 59.
117 P.C. Uden, K.A. Nonnemaker and W.E. Geiger, *Inorg. Nucl. Chem. Lett.*, 14 (1978) 161.
118 E. Patsalides and K. Robards, unpublished data.
119 S. Dilli and A.M. Maitra, *J. Chromatogr.*, 358 (1986) 337.
120 S. Dilli and A.M. Maitra, *J. Chromatogr.*, 254 (1983) 133.
121 E. Patsalides and K. Robards, *J. Chromatogr.*, 357 (1986) 49.

122 S. Dilli and E. Patsalides, *J. Chromatogr.*, 270 (1983) 354.
123 L.N. Bazhenova, V.E. Kirichenko, K.I. Pashkevich and L.P. Alesenko, *Zh. Anal. Khim.*, 42, No. 12 (1986) 2205.
124 E. Patsalides, B.J. Stevenson and S. Dilli, *J. Chromatogr.*, 173 (1979) 321.
125 S. Dilli and C. Chepeta, unpublished data.
126 G. Hartmetz, G. Scollary, H. Meierer, A. Meyer and R. Neeb, *Fresenius Z. Anal. Chem.*, 313 (1982) 309.
127 N. Häring, M. Zell and K. Ballschmiter, *Fresenius Z. Anal. Chem.*, 305 (1981) 285.
128 R. Neeb, *Pure Appl. Chem.*, 54, No. 4 (1982) 847 and references therein.
129 M.J. Riekkola, *Mikrochim. Acta*, No. 1 (1982) 327.
130 K. Kritsotakis and H.J. Tobschall, *Fresenius Z. Anal. Chem.*, 320 (1985) 152.
131 H. Schaller and R. Neeb, *Fresenius Z. Anal. Chem.*, 327 (1987) 170.
132 P.J. Marriott and T.J. Cardwell, *J. Chromatogr.*, 234 (1982) 157.
133 W. Jennings, *J. High. Resolut. Chromatogr. Chromatogr. Commun.*, 3 (1980) 452.
134 A. Kleinmann and R. Neeb, *Naturwissenschaften*, 60 (1973) 201.
135 B. Flegler, V. Gemmer-Colos, A. Dencks and R. Neeb, *Talanta*, 26 (1979) 761.
136 P.J. Marriott, P. Gill and G. Eglinton, *J. Chromatogr.*, 236 (1982) 403.
137 J.P. Fackler, *Progr. Inorg. Chem.*, 7 (1966) 361.
138 K.C. Joshi and V.N. Pathak, *Coord. Chem. Rev.*, 22 (1977) 37.
139 S. Dilli and E. Patsalides, *Anal. Chim. Acta*, 128 (1981) 109.
140 S. Dilli, unpublished data.
141 E. Patsalides, unpublished data.
142 R.A. Zabairova, G.N. Bortnikov, F.A. Gorina and K.M. Samarin, *Zavod. Lab.*, 47, No. 3 (1981) 22.
143 A. Radecki, J. Halkiewicz, J. Grzybowski and H. Lamparczyk, *J. Chromatogr.*, 151 (1978) 259.
144 W. Blum, W.J. Richter and G. Eglinton, *J. High Resolut. Chromatogr. Chromatogr. Commun.*, 11 (1988) 148.
145 T.H. Risby and C.A. Tollinche, *J. Chromatogr. Sci.*, 16 (1978) 448.
146 R.C. Gurira and P.W. Carr, *J. Chromatogr. Sci.*, 20 (1982) 461.
147 M. Ashraf-Khorassani, J.W. Hellgeth and L.T. Taylor, *Anal. Chem.*, 59 (1987) 2077.
148 T. Fujinaga, T. Kuwamoto and K. Sugiura, *Bull. Inst. Chem. Res. Kyoto Univ.*, 58 (1980) 201.
149 V.I. Spitsyn, I.A., Muraveva, L.I. Martynenko, D.N. Sokolov and V.I. Golubatsova, *Zh. Neorg. Khim.*, 27 (1982) 853.
150 K.W. Siu, M.E. Fraser and S.S. Berman, *J. Chromatogr.*, 256 (1983) 455.
151 H. Tao, A. Miyazaki and K. Bansho, *Anal. Sci.*, 4 (1988) 299; *C.A.*, 109: 43229t.
152 A.H. Alberts and D.J. Cram, *J. Am. Chem. Soc.*, 99 (1977) 3880.
153 Y. Ito, T. Sugaya, M. Nakatsuka and T. Sqegusa, *J. Am. Chem. Soc.*, 99 (1977) 8366.
154 I. Tabushi, Y. Kobuke and T. Nishiya, *Nature (London)*, 280 (1979) 665.
155 W. Szczepaniak, J. Nawrocki and W. Wasiak, *Chromatographia*, 12, No. 8 (1979) 559 and references therein.
156 J.N. Gills, R.E. Sievers and G.E. Pollack, *Anal. Chem.*, 57 (1985) 1572.
157 E.T. Kowalska and W.J. Kowalski, *Chromatographia*, 19 (1984) 301.
158 W. Wasiak and W. Szczepaniak, *Chromatographia*, 18 (1984) 205.
159 W. Wasiak, *Chromatographia*, 22 (1986) 147.
160 W. Wasiak, *Chromatographia*, 23 (1987) 261.
161 W. Wasiak, *Chromatographia*, 23 (1987) 427.
162 T.J. Wenzel, L.W. Yarmaloff, L.Y. St. Cyr, L.J. O'Meara, M. Donatelli and R.W. Bauer, *J. Chromatogr.*, 396 (1987) 51.
163 J.E. Picker and R.E. Sievers, *J. Chromatogr.*, 217 (1981) 289.
164 J.E. Picker and R.E. Sievers, *J. Chromatogr.*, 203 (1981) 29.
165 W.J. Kowalski, *J. Chromatogr.*, 349 (1985) 457.
166 V. Schurig and W. Bürkle, *J. Am. Chem. Soc.*, 104 (1982) 7573.
167 V. Schurig, *J. Chromatogr.*, 441 (1987) 135.

168 K.I. Sakodynskii, L.I. Panina, Z.A. Reznikova, V.B. Kargman and N.B. Galitskaya, *J. Chromatogr.*, 364 (1986) 455.
169 W. Szczepaniak, J. Nawrocki and W. Wasiak, *Chromatographic*, 12 (1979) 484.
170 A.B. Littlewood, *Gas Chromatography - Principles, Techniques and Applications*, Academic Press, New York, 2nd Edn., 1970, pp. 112, 113.
171 W.R. Wolf, R.E. Sievers and G.H. Brown, *Inorg. Chem.*, 11 (1972) 1995.
172 E. Patsalides and K. Robards, *J. Chromatogr.*, 350 (1985) 353.
173 E. Patsalides, unpublished data.
174 T.J. Cardwell, T.H. Larman and Z.Z. Feng, *J. Chromatogr.*, 358 (1986) 187.
175 T.J. Cardwell and T.H. Lorman, *Inorg. Chim. Acta*, 120 (1986) 25.
176 F.D. Hileman, R.E. Sievers, G.G. Hess and W.D. Ross, *Anal. Chem.*, 45 (1973) 1126.
177 G.R. Brubaker and J.A. Romberger, *Inorg. Chem.*, 19 (1980) 1087.
178 A.M. Maitra and E. Patsalides, *Chromatographia*, in press.
179 P.J. Marriott and Y.H. Lai, *Inorg. Chem.*, 25 (1986) 3680.
180 L.M. Dyagileva, E.I. Tsyganova and Yu.A. Aleksandrov, *Zh. Fiz. Khim.*, 58 (1984) 1030.
181 N.K. Khalmurzaev, I.A. Murav'eva, L.I. Martynenko and V.I. Spitsyn, *Zh. Neorg. Khim.*, 21 (1976) 1635.
182 P.C. Uden, D.E. Henderson, F.P. Di Sanzo, R.J. Lloyd and T. Tetu, *J. Chromatogr.*, 196 (1980) 403.
183 G.K. Bratspies, J.F. Smith, J.O. Hill and R.J. Magee, *Thermochim. Acta*, 19 (1977) 335.
184 G.K. Bratspies, J.F. Smith, J.O. Hill and R.J. Magee, *Thermochim. Acta*, 19 (1977) 361.
185 S. Dilli and E. Patsalides, *J. Chromatogr.*, 280 (1983) 59.
186 T. Kishimoto, Y. Shimoishi and K. Tôei, *Bunseki Kagaku*, 34 (1985) 197.
187 K. Funazo, H.L. Wu, K. Morita, M. Tanaka and T. Shono, *J. Chromatogr.*, 319 (1985) 143.
188 K. Funazo, M. Tanaka, K. Morita, M. Kamino and T. Shono, *J. Chromatogr.*, 354 (1986) 259.
189 T. Tanaka, Y. Yasaka, M. Kamino, T. Shono, K. Funazo and H.L. Wu, *J. Chromatogr.*, 438 (1988) 253.
190 M. Tanaka, H. Takigawa, Y. Yasaka, T. Shono, K. Funazo and H.S. Wu, *J. Chromatogr.*, 404 (1987) 175.
191 H.S. Wu, S.H. Chen, S.J. Lin, W.R. Hwang, K. Funazo, M. Tanaka and T. Shono, *J. Chromatogr.*, 269 (1983) 183.
192 S.H. Chen, H.L. Wu, M. Tanaka, T. Shono and K. Funazo, *J. Chromatogr.*, 396 (1987) 129.
193 K. Funazo, M. Tanaka and T. Shono, *J. Chromatogr.*, 211 (1981) 361.
194 N. Sugiyama, K.I. Saito, A. Nomato, T. Hanabusa, H. Sato and M. Kawai, *Bunseki Kagaku*, 34 (1985) 335.
195 M. Tanaka, K. Funazo, T. Hirashima and T. Shono, *J. Chromatogr.*, 234 (1982) 373.
196 R. Ikenishi, T. Kitagawa, M. Nishiuchi, K. Takehara, H. Yamada, I. Nishino, T. Umeda, K. Iwatani, Y. Nakagawa, M. Sawai and T. Yamashita, *Chem. Pharm. Bull.*, 36, No. 2 (1988) 662.
197 R. Ikenishi and T. Kitagawa, *Chem. Pharm. Bull.*, 36, No. 2 (1988) 810.
198 G.V. Kalmychkov, *Zh. Anal. Khim.*, 37, No. 7 (1982) 1247.
199 J.W. Tesch, W.R. Rehg and R.E. Sievers, *J. Chromatogr.*, 126 (1976) 743.
200 K.W.M. Siu, S.Y. Roberts and S.S. Berman, *Chromatographia*, 19 (1984) 398.
201 E. Patsalides, unpublished data.
202 B.G. Osborne and K.H. Willis, *Analyst (London)*, 110 (1985) 1037.
203 T.P. Mawhinney, *Anal. Lett.*, 16, No. 2 (1983) 159.
204 W. Zhai and X. Qia, *Huaxue Shijie*, 27, No. 3 (1986) 113.
205 S. Hashimoto, K. Fujiwara and K. Fuwa, *Anal. Chem.*, 57 (1985) 1305.
206 S.H. Vien and R.C. Fry, *Anal. Chem.*, 60 (1988) 465.
207 N. Oyamada, S. Veno and K. Kubota, *Ishizaki Shokuhin Eiseigaku Zasshi*, 26, No. 1 (1985) 13; *C.A.*, 103: 177066x.

208 L.J. Anthony and B.E. Prescott, *J. Chromatogr.*, 264 (1983) 405.
209 L. Maros, M. Kaldy and S. Igaz, *Anal. Chem.*, 61 (1989) 733.
210 T. Mitsuhashi, K. Adachi and Y. Kaneda, *Shokuhin Eiseigaku Zasshi*, 28, No. 2 (1987) 130; *C.A.*, 107: 174514p.
211 G. Castello, A.M. Beccaria and G. Poggi, *Analyst (London)*, 112 (1987) 1093.
212 J.F. Lawrence and R.K. Chadha, *J. Chromatogr.*, 298 (1987) 355.
213 E. Kubaszewski, Z. Kurzawa and M. Lozynski, *Anal. Chim. Acta*, 196 (1987) 267.
214 A. Kito and K. Komatsu, *Bunseki Kagaku*, 31 (1982) 438.
215 A. Kito and K. Komatsu, *Bunseki Kagaku*, 32 (1982) E273.
216 Y. Chen, Q. Li, M. Zhou and H. Zhang, *Sepu*, 2, No. 6 (1985) 335; *C.A.*, 104: 218251v.
217 Y. Guo, W. Tan and Y. Li, *Huanjing Huaxue*, 3, No. 3 (1984) 57; *C.A.*, 101: 115955m.
218 S. Dilli and I. Sutikno, *J. Chromatogr.*, 298 (1984) 21.
219 A.F. Al-Attar and G. Nickless, *J. Chromatogr.*, 440 (1988) 333.
220 A. Sahui-Gnassi, H.C. Pham, G. Dumontier, S. Raspaud, P.L. Nguyen and M. Harron, *Falsif. Expert. Chim. Toxicol.*, 80 (1987) 203; *C.A.*, 107: 230602z.
221 H. Nishikawa and T. Hayakawa, *Taiki Osen Gakkaishi*, 23, No. 1 (1988) 7; *C.A.*, 109: 10938v.
222 T. Sato, T. Kimura and T. Ando, *Koku Eisei Gakkai Zasshi*, 37, No. 4 (1987) 506.
223 N. Rizov, I. Benchev and A. Kolarska, *Khig. Zdraveopaz*, 30, No.. 1 (1987) 67; *C.A.*, 107: 110533m.
224 H. Beernaert and A. Vandezande, *Mededel. Fac. Landbouwwet. Rijksuniv. Gent*, 51, No. 2A (1986) 191; *C.A.*, 106: 48741b.
225 Y. Yamano, I. Ito, N. Nagao and S. Ishizu, *Sangyo Igaku*, 29, No. 3 (1987) 192; *C.A.*, 107: 182625r.
226 S.J. Lin and H.L. Wu, *Kao-hsiung I Hsueh K'o Hsueh Tsa Chih*, 3, No. 4 (1987) 283; *C.A.*, 107: 38190c.
227 D.J. Doedens, *J. Anal. Toxicol.*, 9, No. 3 (1985) 109; *C.A.*, 103: 31891r.
228 U. Duesedau and H.J. Felgentraeger, *Z. Gesamte Hyg. Ihre Grenzgeb.*, 31, No. 4 (1985) 232; *C.A.*, 103: 84361z.
229 M.T. Dmitrieu and G.P. Zarubin, *Gig. Sanit.*, 2 (1985) 51; *C.A.*, 102: 130533m.
230 F. Lindquist, *J. Air Pollut. Control Assoc.*, 35, No. 1 (1985) 19; *C.A.*, 102: 136862q.
231 J. Wang, *Huaxue Shii*, 9 (1987) 38; *C.A.*, 108: 87159d.
232 H. Ni and C. Huang, *Zhonghua Yufangyixui Zazhi*, 19, No. 2 (1985) 100; *C.A.*, 103: 213290b.
233 J. Zamecnik and J. Tam, *Anal. Toxicol.*, 11, No. 1 (1987) 47; *C.A.*, 106: 97488d.
234 T. Shinohara and Y. Seto, *Kagaku Keisatsu Kenkyusho Hokoku Hokagaku Hen*, 40, No. 3 (1987) 154; *C.A.*, 108: 70242c.
235 T. Chikamoto and T. Maitani, *Eisei Kogai Kenkyusho Nenpo*, 29 (1984) 147; *C.A.*, 103: 36248w.
236 G. Castello, A.M. Beccaria and G. Poggi, *Analyst (London)*, 112 (1987) 1093.
237 J.F. Lawrence and R.K. Chadha, *J. Chromatogr.*, 298 (1987) 355.
238 G.A. Cutter and T.J. Oatts, *Anal. Chem.*, 59 (1987) 717.
239 N.T. Ivanova, N.N. Khromykh and V.N. Bochkarev, *Zh. Anal. Khim.*, 40, No. 2 (1985) 280.
240 P.J. Baugh, A. Casson, M.W. Jones and A.C. Jones, *J. Chromatogr.*, 411 (1987) 445.
241 Yu.M. Salganskii and A.N. Moiseev, *Vysokochist. Veshchestva*, 6 (1987) 5; *C.A.*, 108: 231158x.
242 L.G. Novotorova, Yu. Novotorov and I.L. Agafonov, *Zh. Anal. Khim.*, 40, No. 4 (1985) 669.
243 N.T. Karabanov, V.Ya. Abakumov, A.Yu. Devyat'yarov, Z.P. Vetrova and N.M. Olefirenko, *Zavod. Lab.*, 50, No. 1 (1984) 16; *C.A.*, 100: 202785d.
244 S.I. Kirsh, N.T. Karabanov and A.V. Mamaeva, *Zh. Anal. Khim.*, 40, No. 12 (1985) 2231.
245 E. Patsalides, S.J. Pratten and K. Robards, *J. Organomet. Chem.*, 364 (1989) 169.
246 K. Grob, G. Grob and K. Grob, Jr., *J. Chromatogr.*, 219 (1981) 13.

247 W.H. Hwang, R.J. Clark and W.T. Cooper, *J. High Resolut. Chromatogr. Chromatogr. Commun.*, 10 (1987) 504.
248 M.A. Osman and H.H. Hill, *J. Chromatogr.*, 213 (1981) 397.
249 D.R. Hansen and H.H. Hill, Jr., *J. Chromatogr.*, 303 (1984) 331.
250 H. Pscheidl, E. Mueller and D. Haberland, *Wiss Z. Ernst-Moritz-Arndt-Univ. Greifsw. Math-Naturwiss. Reihe*, 35, No. 1 (1986) 67; *C.A.*, 106: 195719d.
251 T.I. Rybinka, E.A. Kirichenko, V.A. Kochetov and V.M. Kopylov, *Zh. Anal. Khim.*, 61, No. 12 (1986) 2254.
252 V.D. Shatz, V.A. Belikov, G.I. Zelchan, I.I. Solomennikova and E. Lukevics, *J. Chromatogr.*, 174 (1979) 83.
253 V.D. Shatz, V.A. Belikov, G.I. Zelchan, I.I. Solomennikova, N.P. Yerchak, O.A. Pudova and E. Lukevics, *J. Chromatogr.*, 200 (1980) 105.
254 A.I. Ermakov, E.A. Zharikova and E.A. Kirichenko, *Zh. Fiz. Khim.*, 62 (1988) 1260.
255 F.E. Brinkman and J.M. Bellama (Editors), *Organometals and Organometalloids - Occurrence and Fate in the Environment*, ACS Symposium Series 82, American Chemical Society, Washington, DC, 1978.
256 J.S. Thayer and F.E. Brinkman, *Adv. Organomet. Chem.*, 20 (1982) 313.
257 Y.K. Chau, P.T.S. Wong, G.A. Bengert and O. Kramer, *Anal. Chem.*, 51 (1979) 186.
258 D.S. Forsyth and W.D. Marshall, *Anal. Chem.*, 55 (1983) 2132.
259 Y.K. Chau, P.T.S. Wong, G.A. Bengert and J.L. Dunn, *Anal. Chem.*, 56 (1984) 271.
260 M.O. Andreae, J.F. Asmode, P. Foster and L. Van't dack, *Anal. Chem.*, 53 (1981) 1766.
261 D.S. Ballantine and W.H. Zoller, *Anal. Chem.*, 56 (1984) 1288.
262 J.A. Rodriquez-Vasquez, *Talanta*, 25 (1978) 299.
263 M.A. Seckin, S. Aygun and O.Y. Ataman, *Int. J. Environ. Anal. Chem.*, 26 (1986) 1.
264 N. Bloom and W.F. Fitzgerald, *Anal. Chim. Acta*, 208 (1988) 151.
265 Y. Talmi and D.T. Bostick, *Anal. Chem.*, 47 (1975) 2145.
266 M.O. Andreae, *Anal. Chem.*, 49 (1977) 820.
267 Y. Odanaka, N. Tsuchiya, O. Matano and S. Goto, *Anal. Chem.*, 53 (1983) 929.
268 D.C. Reamer and W.H. Zoller, *Science*, 208 (1980) 500.
269 L. Barkes and R.W. Fleming, *Bull. Environ. Contam. Toxicol.*, 12 (1974) 308.
270 Y.K. Chau, P.T.S. Wong and P.D. Goulden, *Anal. Chem.*, 47 (1975) 2279.
271 Y.K. Chau, P.T.S. Wong and G.A. Bengert, *Anal. Chem.*, 54 (1982) 246.
272 Y. Hattori, A. Kobayashi, S. Takemoto, K. Takami, Y. Kuge, A. Sugimae and M. Nakamoto, *J. Chromatogr.*, 315 (1984) 341.
273 A. Woollins and W.R. Cullen, *Analyst (London)*, 109 (1984) 1527.
274 C.L. Matthias, J.M. Bellama, G.J. Olson and F.E. Brinkman, *Environ. Sci. Technol.*, 20 (1986) 609.
275 H.H. van den Broek, G.B.M. Hermes and C.E. Goewie, *Analyst (London)*, 113 (1988) 1237.
276 J.J. Sullivan, J.D. Torkelson, M.M. Wekell, T.A. Hollingworth, W.L. Saxton, G. Miller, K.W. Panaro and A.D. Uhler, *Anal. Chem.*, 60 (1988) 626.
277 S. Clark, J. Ashby and P.J. Craig, *Analyst (London)*, 112 (1987) 1781.
278 J.E. O'Reilly, *J. Chromatogr.*, 238 (1982) 433.
279 G.H. Alvarez, S.C. Hight and S.G. Capar, *J. Assoc. Off. Anal. Chem.*, 67 (1984) 715.
280 W.A. Aue, B.J. Flinn, C.G. Flinn, V. Paramasigamani and K.A. Russell, *Can. J. Chem.*, 67 (1989) 402.
281 T. Nielson, H. Egsgaard, E. Larsen and G. Schroll, *Anal. Chim. Acta*, 124 (1981) 1.
282 R.J.A. Van Cleuvenbergen, D. Chakraborti and F.C. Adams, *Anal. Chim. Acta*, 182 (1986) 239.
283 V. Schmidt and F. Huber, *Nature (London)*, 259 (1976) 157.
284 P.T.S. Wong, Y.K. Chau and P.L. Luxon, *Nature (London)*, 253 (1975) 263.
285 W.R.A. De Jonghe, D. Chakraborti and F.C. Adams, *Anal. Chem.*, 52 (1980) 1974.
286 C.N. Hewitt and R.M. Harrison, *Anal. Chim. Acta*, 167 (1985) 277.
287 G.W. Rice, J.J. Richard, A.P. D'Silva and V.A. Fassel, *Anal. Chem.*, 53 (1981) 1519.
288 S.A. Estes, P.C. Uden and R.M. Barnes, *Anal. Chem.*, 53 (1981) 1336.

289 R.J.A. Van Cleuvenbergen, D. Chakraborti and F.C. Adams, *Environ. Sci. Technol.*, 20 (1986) 589.
290 D. Chakraborti, R.J.A. Van Cleuvenbergen and F.C. Adams, *Int. J. Environ. Anal. Chem.*, 30 (1987) 233.
291 G. Westoo, *Acta Chem. Scand.*, 20 (1966) 2131.
292 M. Filippelli, *Anal. Chem.*, 59 (1987) 116.
293 Official Methods of Analysis of the Association of Official Analytical Chemists, 14th Ed., A.O.A.C., Washington, DC 1984.
294 R.B. Costanzo and E.F. Barry, *Anal. Chem.*, 60 (1988) 826.
295 G.W. Rice, J.J. Richard, A.P. Silva and V.A. Fassel, *J. Assoc. Off. Anal. Chem.*, 65, No. 1 (1982) 14.
296 K.W. Panaro, D. Erickson and I.S. Krull, *Analyst (London)*, 112 (1987) 1097.
297 K.I. Fujiwara, Y. Tamara and T. Katsura, *Bunseki Kagaku*, 33 (1984) T87.
298 A.G.F. Brooks, E. Bailey and R.T. Snowden, *J. Chromatogr.*, 374 (1986) 289.
299 X. Zhang, J. Qi, G. Zhong and F. Niu, *Huanjing Huaxue*, 5, No. 1 (1986) 1; *C.A.*, 105: 29637k.
300 X. Zhang and X.E. Ren, *Huan Ching K'O Hsueh*, 1, No. 6 (1980) 34; *C.A.*, 94: 108987g.
301 P. Smith and L. Smith, *Chem. Brit.*, 11 (1975) 208.
302 J.J. Zuckermann (Editor), *Organotin Compounds: New Chemistry and Applications*, ACS Symposium Series 157, American Chemical Society, Washington, DC, 1976.
303 M.D. Mueller, *Anal. Chem.*, 59 (1987) 617.
304 J.J. Sullivan, J.D. Torkelson, M.M. Wekell, T.A. Hollingwort, W.L. Saxton, G.A. Miller, K.W. Panaro and A.D. Uhler, *Anal. Chem.*, 60 (1988) 626.
305 R.J. Maguire, *Environ. Sci. Technol.*, 18 (1984) 291.
306 S. Kapila and C.R. Vogt, *J. Chromatogr. Sci.*, 18 (1980) 144.
307 W.A. Aue and C.G. Flinn, *J. Chromatogr.*, 153 (1978) 305.
308 D.R. Hansen, T.J. Gilfoil and H.H. Hill, Jr., *Anal. Chem.*, 53 (1981) 857.
309 C.G. Flinn and W.A. Aue, *Can. J. Spectrosc.*, 25 (1980) 141.
310 J.F. Short, *Bull. Environ. Contam. Toxicol.*, 39 (1987) 412.
311 A.F. Shushvnova, V.V. Kutsovskaya, M.V. Skovorodina and G.I. Makin, *Zh. Anal. Khim.*, 44 (1989) 745.
312 K. Takami, T. Okumura, H. Yamasaki and M. Nakamoto, *Bunseki Kagaku*, 37 (1988) 117.
313 F.C. Smith and R.C. Chang, *The Practice of Ion Chromatography*, Wiley, New York, 1983.
314 J.S. Fritz, D.T. Gjerde and C. Pohlandt, *Ion Chromatography*, Huethig Verlag, Heidelberg, 1982.
315 J.S. Fritz and D.T. Gjerde, *Ion Chromatography*, 2nd Ed., Huethig Verlag, Heidelberg, 1987.
316 J.G. Tarter (Editor), *Ion Chromatography*, Dekker, New York, 1987.
317 J. Weiss, *Handbook of Ion Chromatography*, Dionex Corporation, Sunnyvale, CA, 1986.
318 R.E. Smith, *Ion Chromatography Applications*, CRC Press, Boca Raton, FL, 1988.
319 O.A. Shpigun and Yu.A. Zolotov, *Ion Chromatography in Water Analysis*, Ellis Horwood, Chichester, 1988.
320 P.R. Haddad and P.E. Jackson, *Ion Chromatography: Principles and Applications*, Elsevier, Amsterdam, 1990.
321 R.M. Cassidy and B.D. Karcher, in I.S. Krull (Editor), *Reaction Detection in Liquid Chromatography*, Dekker, New York, 1986, p. 129.
322 S. Rokushika and H. Hatano, in M.V. Novotny and D. Ishii (Editors), *Microcolumn Separations*, Journal of Chromatography Library, Vol. 30, Elsevier, Amsterdam, 1985, p. 277.
323 R.A. Wetzel, C.A. Pohl, J.M. Riviello and J.C. MacDonald, *Chem. Anal. (N.Y.)*, 78 (1985) 355.
324 J.S. Fritz, *Anal. Chem.*, 59 (1987) 335A.
325 J.S. Fritz, *J. Chromatogr.*, 439 (1988) 3.

326 P.R. Haddad and A.L. Heckenberg, *J. Chromatogr.*, 300 (1984) 357.
327 P. Jandik, P.R. Haddad and P.E. Sturrock, *CRC Crit. Rev. Anal. Chem.*, 20 (1988) 1.
328 H. Small, *Anal. Chem.*, 55 (1983) 235A.
329 R.W. Slingsby and C.A. Pohl, *J. Chromatogr.*, 458 (1988) 241.
330 L.M. Warth, J.S. Fritz and J.O. Naples, *J. Chromatogr.*, 462 (1989) 165.
331 Y. Hanoaka, T. Murayama, S. Muramoto, T. Matsuura and A. Nanba, *J. Chromatogr.*, 239 (1982) 537.
332 S. Haldna, R. Palvadre, J. Pentshuk and T. Kleemeier, *J. Chromatogr.*, 350 (1985) 296.
333 G.L. Schmitt and D.J. Pietrzyk, *Anal. Chem.*, 57 (1985) 2247.
334 T. Takeuchi, E. Suzuki and D. Ishii, *Chromatographia*, 25 (1988) 480.
335 D.G. Peters, J.M. Hayes and G.M. Hieftje, *Chemical Separations and Measurements*, Saunders, Philadelphia, PA, 1974, p. A13.
336 J. Stillian, *LC*, 3 (1985) 802.
337 S.A. Bouyoucos, *J. Chromatogr.*, 242 (1982) 170.
338 R.D. Rocklin, C.A. Pohl and J.A. Schibler, *J. Chromatogr.*, 411 (1987) 107.
339 Y.A. Zolotov, O.A. Shpigun, Y.E. Pazukhina and I.N. Voloshik, *Int. J. Environ. Anal. Chem.*, 31 (1987) 99.
340 J.P. Ivey, *J. Chromatogr.*, 287 (1984) 128.
341 J.S. Fritz, D.L. DuVal and R.E. Barron, *Anal. Chem.*, 56 (1984) 1177.
342 T. Okada and T. Kuwamoto, *J. Chromatogr.*, 284 (1984) 149.
343 V. Kordorouba, A. Balikungeri, M. Pelletier and W. Haerdi, *Analusis*, 12 (1984) 364.
344 B.E. Andrew, *J. High Resolut. Chromatogr. Chromatogr. Commun.*, 8 (1985) 189.
345 A. Diop, A. Jardy, M. Caude and R. Rosset, *Analusis*, 14 (1986) 67.
346 R. Golombek and G. Schwedt, *J. Chromatogr.*, 367 (1986) 69.
347 G. Horvai, J. Fekete, Z. Niegreisz, K. Toth and E. Pungor, *J. Chromatogr.*, 385 (1987) 25.
348 J.P. Ivey, *J. Chromatogr.*, 267 (1983) 218.
349 P.R. Haddad and A.L. Heckenberg, *J. Chromatogr.*, 318 (1985) 279.
350 P.E. Jackson and P.R. Haddad, *J. Chromatogr.*, 355 (1986) 87.
351 T. Okada and T. Kuwamoto, *Anal. Chem.*, 55 (1983) 1001.
352 G. Schmuckler, A.L. Jagoe, J.E. Girard and P.E. Buell, *J. Chromatogr.*, 356 (1986) 413.
353 T. Okada, *J. Chromatogr.*, 403 (1987) 27.
354 J.E. Girard, N. Rebbani, P.E. Buell and A.H.E. Al- Khalidi, *J. Chromatogr.*, 448 (1988) 355.
355 A.W. Fitchett and A. Woodruff, *LC*, 1 (1983) 48.
356 J.E. Lockridge, N.E. Fortier, G. Schmuckler and J.S. Fritz, *Anal. Chim. Acta*, 192 (1987) 41.
357 T.S. Stevens, J.C. Davis and H. Small, *Anal. Chem.*, 53 (1981) 1488.
358 S. Rokushika, Z.Y. Qiu and H. Hatano, *J. Chromatogr.*, 260 (1983) 81.
359 T.S. Stevens, *Res. Dev.*, September (1983) 96.
360 T.S. Stevens, G.L. Jewett and R.A. Bredewe, *Anal. Chem.*, 54 (1982) 1206.
361 W.R. Jones and P. Jandik, *Res. Dev.*, September (1988) 92.
362 W.R. Jones, P. Jandik and A.L. Heckenberg, *Anal. Chem.*, 60 (1988) 1977.
363 P.E. Jackson, P.R. Haddad and S. Dilli, *J. Chromatogr.*, 295 (1984) 471.
364 T. Okada, *Bunseki Kagaku*, 36 (1987) 702.
365 L. Eek and N. Ferrer, *J. Chromatogr.*, 322 (1985) 491.
366 J. Osterloh and D. Goldfield, *J. Liquid Chromatogr.*, 7 (1984) 753.
367 Y. Michigami, T. Takahashi, F. He, Y. Yamamoto and K. Ueda, *Analyst (London)*, 113 (1988) 389.
368 P.R. Haddad and P.E. Jackson, *Food Tech. Aust.*, 37 (1985) 305.
369 R.J. Williams, *Anal. Chem.*, 55 (1983) 851.
370 I. Vins and L. Kabrt, *Collect. Czech. Chem. Commun.*, 52 (1987) 1167.
371 H. Small and T.E. Miller, Jr., *Anal. Chem.*, 54 (1982) 462.
372 A. Jardy, M. Caude, A. Diop, C. Curvale and R. Rosset, *J. Chromatogr.*, 439 (1988) 137.

373 D. Chakraborti, D.C.J. Hillman, K.J. Irgolic and R.A. Zingaro, *J. Chromatogr.*, 249 (1982) 81.
374 I.T. Urasa and F. Ferede, *Anal. Chem.*, 59 (1987) 1563.
375 D.J. Freed, *Anal. Chem.*, 47 (1975) 186.
376 W.F. Koch, *J. Res. Nat. Bur. Std.*, 88 (1983) 157.
377 R.D. Rocklin and E.L. Johnson, *Anal. Chem.*, 55 (1983) 4.
378 L.R. Goodwin, D. Francom, A. Urso and F.P. Dieken, *Anal. Chem.*, 60 (1988) 216.
379 G.S. Pyen and D.E. Erdmann, *Anal. Chim. Acta*, 149 (1983) 355.
380 M. Lookabaugh and I.S. Krull, *J. Chromatogr.*, 452 (1988) 295.
381 T. Kawanishi, T. Togawa, A. Ishigami, S. Tanabe and T. Imanari, *Bunseki Kagaku*, 33 (1984) E295.
382 J.G. Tarter, *J. Liquid Chromatogr.*, 7 (1984) 1559.
383 A.M. Bond, I.D. Heritage, G.G. Wallace and M.J. McCormick, *Anal. Chem.*, 54 (1982) 582.
384 Z. Hu and Y. Tang, *Analyst (London)*, 113 (1982) 179.
385 M.P. Keuken, J. Slanina, P.A.C. Jongejan and F.P. Bakker, *J. Chromatogr.*, 439 (1988) 13.
386 J. Slanina, F.P. Bakker, P.A.C. Jongejan, L. van Lamoen and J.J. Mols, *Anal. Chim. Acta*, 130 (1981) 1.
387 E.C.V. Butler and R.M. Gershey, *Anal. Chim. Acta*, 164 (1984) 153.
388 H. Mueller and R. Scholz, in E. Pungor and I. Buzas (Editors), *Ion-selective Electrodes*, Anal. Chem. Symp. Ser., Vol. 22, Elsevier, Amsterdam, 1985, p. 553.
389 H. Hershcovitz, C. Yarnitzky and G. Schmuckler, *J. Chromatogr.*, 244 (1982) 217.
390 P.W. Alexander, P.R. Haddad and M. Trojanowicz, *Anal. Chem.*, 56 (1984) 2417.
391 P.W. Alexander, M. Trojanowicz and P.R. Haddad, *Anal. Lett.*, 17 (1984) 309.
392 P.R. Haddad, P.W. Alexander and M. Trojanowicz, *J. Chromatogr.*, 315 (1984) 261.
393 P.R. Haddad, P.W. Alexander and M. Trojanowicz, *J. Chromatogr.*, 321 (1985) 363.
394 P.R. Haddad, P.W. Alexander and M. Trojanowicz, *J. Chromatogr.*, 324 (1985) 319.
395 Y. Hirai, N. Yoza and S. Ohashi, *J. Chromatogr.*, 206 (1981) 501.
396 E. Vaeth, P. Sladek and K. Kenar, *Fresenius Z. Anal. Chem.*, 329 (1987) 584.
397 T. Imanari, S. Tanabe, T. Toida and T. Kawanishi, *J. Chromatogr.*, 250 (1982) 55.
398 J.N. Story, *J. Chromatogr. Sci.*, 21 (1983) 272.
399 Waters IC Lab. Report No. 309.
400 Dionex Application Note 44R.
401 R.A. Hill, *J. High Resolut. Chromatogr. Chromatogr. Commun.*, 6 (1983) 275.
403 J.A. Cox and N. Tanaka, *Anal. Chem.*, 57 (1985) 383.
404 J.A. Cox and N. Tanaka, *Anal. Chem.*, 57 (1985) 385.
405 A.L. Heckenberg and P.R. Haddad, *J. Chromatogr.*, 299 (1984) 301.
406 T. Okada and T. Kuwamoto, *J. Chromatogr.*, 350 (1985) 317.
407 Dionex Application Note 26.
408 R.M. Cassidy and S. Elchuk, *J. Chromatogr. Sci.*, 18 (1980) 217.
409 R.M. Cassidy and S. Elchuk, *J. Chromatogr. Sci.*, 19 (1981) 503.
410 A.L. Heckenberg and P.R. Haddad, *J. Chromatogr.*, 330 (1985) 95.
411 P.E. Jackson and P.R. Haddad, *J. Chromatogr.*, 389 (1987) 65.
412 J.E. DiNunzio and M. Jubara, *Anal. Chem.*, 55 (1983) 1013.
413 J.A. Cox and N. Tanaka, *Anal. Chem.*, 57 (1985) 2370.
414 J.A. Cox and G.R. Litwinski, *Anal. Chem.*, 55 (1983) 1640.
415 J.F. Colaruotolo and R.S. Eddy, *Anal. Chem.*, 49 (1977) 884.
416 K.L. Evans, J.G. Tarter and C.B. Moore, *Anal. Chem.*, 53 (1981) 925.
417 S.A. Wilson and C.A. Gent, *Anal. Lett.*, 15 (1982) 851.
418 S.A. Wilson and C.A. Gent, *Anal. Chim. Acta*, 148 (1983) 299.
419 C. McCrory-Joy, *Anal. Chim. Acta*, 181 (1986) 277.
420 D.W. Mason and H.C. Miller, in E. Sawicki and J.D. Mulik (Editors), *Ion Chromatographic Analysis of Environmental Pollutants*, Vol. II, Ann Arbor Science Publishers, Ann Arbor, MI, 1979, p. 193.

421 J.D. Mulik, G. Todd, E. Estes, R. Puckett, E. Sawicki and D. Williams, in E. Sawicki and J.D. Mulik (Editors), *Ion Chromatographic Analysis of Environmental Pollutants,* Vol. I, Ann Arbor Science Publishers, Ann Arbor, MI, 1978, p. 23.

422 M.E. Cassinelli and D.G. Taylor, in G. Choudhary (Editor), *Chemical Hazards in the Work Place: Measurement and Control,* American Chemical Society, Washington, DC, 1981, Chap. 10, p. 137.

423 D.V. Vinjamoori and C.-S. Ling, *Anal. Chem.,* 53 (1980) 1689.

424 J.H. Margeson, J.E. Knoll, R.M. Midgett, G.B. Oldaker, III and W.E. Reynolds, *Anal. Chem.,* 57 (1985) 1586.

425 D.D. Siemer, *Anal. Chem.,* 59 (1987) 2439.

426 B. Dellinger, G. Grotecloss, C.R. Fortune, J.L. Cheney and J.B. Homolya, *Environ. Sci. Technol.,* 14 (1980) 1244.

427 T.W. Dolzine, G.G. Esposito and D.S. Rinehart, *Anal. Chem.,* 54 (1982) 470.

428 M.Y. Croft and P.R. Haddad, in D. Sampson (Editor), *Australian Association of Clinical Biochemists Monograph Series,* Australian Association of Clinical Biochemists, Sydney, 1986, p. 138.

429 P. Jandik, D. Cox and D. Wong, *Int. La.,* June (1986) 66.

430 E.G. Bradfield and D.T. Cooke, *Analyst (London),* 110 (1985) 1409.

431 J. Schaefer, J. Burmicz and D. Palladino, *Am. Lab.,* February (1989) 70.

432 Y. Nishikawa and K. Taguchi, *J. Chromatogr.,* 396 (1987) 251.

433 K. The and R. Roussel, *Light Met. (Warrendale, Pa),* (1984) 115.

434 S. Ikeda, H. Satake and H. Segawa, *Nippon Kagaku Kaishi,* 9 (1985) 1704.

435 T. Toida, T. Togawa, S. Tanabe and T. Imanari, *J. Chromatogr.,* 308 (1984) 133.

436 R.E. Hannah, *J. Chromatogr. Sci.,* 24 (1986) 336.

437 H.-J. Kim and Y.-K. Kim, *J. Food Sci.,* 51 (1986) 1360.

438 H. Yamada and T. Hattori, *J. Chromatogr.,* 411 (1987) 401.

440 Q. Xianren and W. Baeyens, *J. Chromatogr.,* 456 (1988) 267.

441 R.L. Smith, Z. Iskandarani and D.J. Pietrzyk, *J. Liquid Chromatogr.,* 7 (1984) 1935.

442 Dionex Application Note 41.

443 B.B. Wheals, *J. Chromatogr.,* 262 (1983) 61.

444 G. Schwedt, *Chromatographia,* 12 (1979) 613.

445 J. Crommen, G. Schill, D. Westerlund and L. Hackzell, *Chromatographia,* 24 (1987) 252.

446 R.N. Reeve, *J. Chromatogr.,* 177 (1979) 393.

447 R.M. Cassidy and S. Elchuk, *J. Chromatogr. Sci.,* 21 (1983) 454.

448 D.L. DuVal, J.S. Fritz and D.T. Gjerde, *Anal. Chem.,* 54 (1982) 830.

449 M.C. Gennaro, *J. Chromatogr.,* 449 (1988) 103.

450 Y. Hirai, N. Yoza and S. Ohashi, *Anal. Chim. Acta,* 115 (1980) 269.

451 F.G.P. Mullins and G.F. Kirkbright, *Analyst (London),* 109 (1984) 1217.

452 M. Dreux, M. Lafosse and M. Pequignot, *Chromatographia,* 15 (1982) 653.

453 Z. Iskandarani and D.J. Pietrzyk, *Anal. Chem.,* 54 (1982) 1065.

454 B.E. Andrew, *LC.GC,* 4 (1986) 1026.

455 W. Mingjia, V. Pacakova, K. Stulik, G.A. Sacchetto, *J. Chromatogr.,* 439 (1988) 363.

456 B.A. Bidlingmeyer, C.T. Santasania and F.V. Warren, Jr., *Anal. Chem.,* 59 (1987) 1843.

457 Z. Iskandarani and D.J. Pietrzyk, *Anal. Chem.,* 54 (1982) 2427.

458 M. Kalbasi and M.A. Tabatabai, *Commun. Soil Sci. Plant. Anal.,* 16 (1985) 787.

459 A.D. Kirk and A.K. Hewavitharana, *Anal. Chem.,* 60 (1988) 797.

460 Z. Iskandarani and D.J. Pietrzyk, *Anal. Chem.,* 54 (1982) 2601.

461 I. Molnar, H. Knauer and D. Wilk, *J. Chromatogr.,* 201 (1980) 225.

462 M. Lookabaugh, I.S. Krull and W.R. LaCourse, *J. Chromatogr.,* 387 (1987) 301.

463 M. Cooke, *J. High Resolut. Chromatogr. Chromatogr. Commun.,* 7 (1984) 515.

464 J.J. Stranahan and S.N. Deming, *Anal. Chem.,* 54 (1982) 1540.

465 W.E. Barber and P.W. Carr, *J. Chromatogr.,* 301 (1984) 25.

466 W.E. Hammers, C.N.M. Aussems and M. Janssen, *J. Chromatogr.,* 360 (1986) 1.

467 W.E. Barber and P.W. Carr, *J. Chromatogr.,* 260 (1983) 89.

468 P.G. Rigas and D.J. Pietrzyk, *Anal. Chem.,* 58 (1986) 2226.

469 P.G. Rigas and D.J. Pietrzyk, *Anal. Chem.,* 60 (1988) 454.
470 T. Jupille, M. Gray, B. Black and M. Gould, *Am. Lab.,* 13 (1981) 80.
471 B.K. Glod and W. Kemula, *J. Chromatogr.,* 366 (1986) 39.
472 D.P. Lee and A.D. Lord, *LC-GC,* 5 (1987) 261.
473 P. Walser, *J. Chromatogr.,* 439 (1988) 71.
474 K. Tanaka and J.S. Fritz, *J. Chromatogr.,* 361 (1986) 151.
475 H. Itoh and Y. Shinbori, *Chem. Lett.,* (1982) 2001.
476 Dionex Technical Note 17.
477 K. Tanaka and J.S. Fritz, *Anal. Chem.,* 59 (1987) 708.
478 J.P. Wilshire and W.A. Brown, *Anal. Chem.,* 54 (1982) 1647.
479 T. Okada and T. Kuwamoto, *Fresenius Z. Anal. Chem.,* 325 (1986) 683.
480 J.R. Kreling and J. DeZwaan, *Anal. Chem.,* 58 (1986) 3028.
481 W.R. Jones, A.L. Heckenberg and P. Jandik, *J. Chromatogr.,* 366 (1986) 225.
482 Dionex Application Note 52.
483 T.S. Stevens and H. Small, *J. Liquid Chromatogr.,* 1 (1978) 123.
484 P. Kolla, J. Kohler and G. Schomburg, *Chromatographia,* 23 (1987) 465.
485 P.R. Haddad and R.C. Foley, *Anal. Chem.,* 61 (1989) 1435.
486 M. Miyazaki, K. Hayakawa and S.-D. Choi, *J. Chromatogr.,* 323 (1985) 443.
487 J.H. Sherman and N.D. Danielson, *Anal. Chem.,* 59 (1987) 490.
488 J.H. Sherman and N.D. Danielson, *Anal. Chem.,* 59 (1987) 1483.
489 J.-M. Hwang, F.-C. Chang and Y.-C. Yeh, *J. Chinese Chem. Soc.,* 30 (1983) 167.
490 H. Saitoh and K. Oikawa, *J. Chromatogr.,* 329 (1985) 247.
491 G.J. Sevenich and J.S. Fritz, *Anal. Chem.,* 55 (1983) 12.
492 W. Wang, Y. Chen and M. Wu, *Analyst (London),* 109 (1984) 281.
493 K. Hayakawa, T. Sawada, K. Shimbo and M. Miyazaki, *Anal. Chem.,* 59 (1987) 2241.
494 R.C.L. Foley and P.R. Haddad, *J. Chromatogr.,* 366 (1986) 13.
495 D. Yan and G., Schwedt, *Fresenius Z. Anal. Chem.,* 320 (1985) 325.
496 D. Yan and G. Schwedt, *Anal. Chim. Acta,* 178 (1985) 347.
497 G. Schwedt, *GIT Fachz. Lab.,* 7 (1985) 697.
498 D. Cox, P. Jandik and W. Jones, *Pulp Pap. Canada,* 88 (1987) T318.
499 H. Saitoh, K. Oikawa, T. Takano and K. Kamimura, *J. Chromatogr.,* 281 (1983) 397.
500 K. Hayakawa, R. Ebina, M. Matsumoto and M. Miyazaki, *Bunseki Kagaku,* 33 (1984) 390.
501 R. Reiffenstuhl and G. Bonn, *Fresenius Z. Anal. Chem.,* 332 (1988) 130.
502 Dionex Application Note 3.
503 R.M. Cassidy, S. Elchuk and J.O. McHugh, *Anal. Chem.,* 54 (1982) 727.
504 Waters IC Lab. Report No. 250.
505 Waters IC Lab. Report No. 272.
506 R.M. Cassidy, F.C. Miller, C.H. Knight, J.C. Roddick and R.W. Sullivan, *Anal. Chem.,* 58 (1986) 1389.
507 D. Yan, E. Stumpp and G. Schwedt, *Fresenius Z. Anal. Chem.,* 322 (1985) 474.
508 M. Nonomura, *Met. Fin.,* 85 (1987) 15.
509 R.L. Smith and D.J. Pietrzyk, *Anal. Chem.,* 56 (1984) 1572.
510 D.J. Barkley, M. Blanchette, R.M. Cassidy and S. Elchuk, *Anal. Chem.,* 58 (1986) 2222.
511 P.R. Haddad and R.C. Foley, *J. Chromatogr.,* 500 (1990) 301.
512 D.F. Hilton and P.R. Haddad, *J. Chromatogr.,* 361 (1986) 141.
513 P.R. Haddad and N.E. Rochester, *Anal. Chem.,* 60 (1988) 536.
514 P.R. Haddad and N.E. Rochester, *J. Chromatogr.,* 439 (1988) 23.
515 B. Grigorova, S.A. Wright and M. Josephson, *J. Chromatogr.,* 410 (1987) 419.
516 R.M. Cassidy, S. Elchuk, N.L. Elliot, L.W. Green, C.H. Knight and B.M. Recoskie, *Anal. Chem.,* 58 (1986) 1181.
517 C.H. Knight, R.M. Cassidy, B.M. Recoskie and L.W. Green, *Anal. Chem.,* 56 (1984) 474.
518 R.M. Cassidy, *Chem. Geol.,* 67 (1988) 185.
519 K.H. Faltynski and J.R. Jezorek, *Chromatographia,* 22 (1986) 5.
520 K.T. DenBleyker, J.K. Arbogast and T.R. Sweet, *Chromatographia,* 8 (1983) 449.

521 N. Simonzadeh and A.A. Schilt, *Talanta,* 35 (1988) 187.
522 E. Blasius and K.P. Janzen, *Israel J. Chem.,* 26 (1985) 25.
523 M. Igawa, K. Saito, M. Tanaka and T. Yamabe, *Bunseki Kagaku,* 32 (1983) E137.
524 M. Igawa, K. Saito, J. Tsukamoto and M. Tanaka, *Anal. Chem.,* 53 (1981) 1942.
525 T. Iwachido, H. Naito, F. Samukawa, K. Ishimaru and K. Toei, *Bull. Chem. Soc. Jpn.,* 59 (1986) 1475.
526 M. Lauth and P. Gramain, *J. Chromatogr.,* 395 (1987) 153.
527 M. Lauth and P. Gramain, *J. Liquid Chromatogr.,* 8 (1985) 2403.
528 M. Nakajima, K. Kimura and T. Shono, *Bull. Chem. Soc. Jpn.,* 56 (1983) 3052.
529 G. Nickless, *J. Chromatogr.,* 313 (1985) 129.
530 A.R. Timerbaev, O.M. Petrukhin and Yu.A. Zolotov, *Fresenius Z. Anal. Chem.,* 327 (1987) 87.
531 B. Steinbrech, *J. Liquid Chromatogr.,* 10 (1987) 1.
532 J.W. O'Laughlin, *J. Liquid Chromatogr.,* 7 (1984) 127.
533 J.C. MacDonald, in J.C. MacDonald (Editor), *Inorganic Chromatographic Analysis,* Wiley-Interscience, New York, 1985, p. 285.
534 I.S. Krull, in M. Bernhard, F.E. Brinkman and P.J. Sadler (Editors), *The Importance of Chemical "Speciation" in Environmental Processes,* Springer-Verlag, Berlin, Heidelberg, 1986, p. 579.
535 S.R. Hutchins, P.R. Haddad and S. Dilli, *J. Chromatogr.,* 252 (1982) 185.
536 J.P. Mooney, M. Meaney, M.R. Smyth, R.G. Leonard and G.G. Wallace, *Analyst (London),* 112 (1987) 1555.
537 B.W. Hoffman and G. Schwedt, *J. High Resolut. Chromatogr. Chromatogr. Commun.,* 5 (1982) 439.
538 R.C. Gierira and P.W. Carr, *J. Chromatogr. Sci.,* 20 (1982) 461.
539 G. Schwedt and R. Rudde, *Chromatographia,* 15 (1982) 527.
540 D.A. Roston, *Anal. Chem.,* 56 (1984) 241.
541 T.J. Cardwell and D. Caridi, *J. Chromatogr.,* 193 (1980) 53.
542 H. Irth, E. Brouwer, G.J. de Jong, U.A.Th. Brinkman and R.W. Frei, *J. Chromatogr.,* 439 (1988) 63.
543 H. Eggers and H.A. Russel, *Chromatographia,* 17 (1983) 486.
544 G.L. Wheeler and P.F. Lott, *Microchem. J.,* 19 (1974) 390.
545 A. Tong, Y. Akama and S. Tanaka, *J. Chromatogr.,* 478 (1989) 408.
546 M.V. Main and J.S. Fritz, *Anal. Chem.,* 61 (1989) 1272.
547 R.M. Smith, A.M. Butt and A. Thakur, *Analyst (London),* 110 (1985) 35.
548 H. Ge and G.G. Wallace, *Anal. Chem.,* 60 (1988) 830.
549 J.N. King and J.S. Fritz, *Anal. Chem.,* 59 (1987) 703.
550 D.J. Mazzo, W.G. Elliott, P.C. Uden and R.M. Barnes, *Appl. Spectrosc.,* 38 (1984) 585.
551 Y-T. Shih and P.W. Carr, *Talanta,* 28 (1981) 411.
552 R.M. Smith and L.E. Yankey, *Analyst (London),* 107 (1982) 744.
553 A.M. Bond, R.W. Knight, J.B. Reust, D.J. Tucker and G.G. Wallace, *Anal. Chim. Acta,* 182 (1986) 47.
554 R.H. Fish, F.E. Brinkman and K.L. Juwett, *Environ. Sci. Technol.,* 16 (1982) 174.
555 S.J. Haswell, P. O'Neill and K.C.C. Bancroft, *Talanta,* 32 (1985) 69.
556 K.L. Lawrence, G.W. Rice and V.A. Fassel, *Anal. Chem.,* 56 (1984) 289.
557 K.J. Irgolic, R.A. Stockton, D. Chakraborti and W. Beyer, *Spectrochim. Acta,* 38B (1983) 437.
558 L. Ebdon, S.J. Hill and P. Jones, *Analyst (London),* 110 (1985) 515.
559 L. Ebdon and J.I.G. Alonso, *Analyst (London),* 112 (1987) 1551.
560 I.S. Krull and K.W. Panaro, *Appl. Spectrosc.,* 39 (1985) 960.
561 G.K.-C. Low, G.E. Batley and S.J. Buchanan, *J. Chromatogr.,* 386 (1986) 423.
562 G.R. Ricci, L.S. Shepard, G. Colovos and N.E. Hester, *Anal. Chem.,* 53 (1981) 610.
563 D.S. Bushee, *Analyst (London),* 113 (1988) 1167.
564 L. Ebdon, S. Hill and R.W. Ward, *Analyst (London),* 112 (1987) 1.
565 K. Yoshimura and T. Taratani, *J. Chromatogr.,* 237 (1982) 89.
566 W.R. Biggs, J.T. Guno and R.J. Brown, *Anal. Chem.,* 56 (1984) 2653.

567 B. Bjorkqvist and H. Toivonen, *J. Chromatogr.,* 178 (1979) 271.
568 T.S. Stevens, K.M. Chritz and H. Small, *Anal. Chem.,* 59 (1987) 1716.
569 N.E. Fortier and J.S. Fritz, *J. Chromatogr.,* 462 (1989) 323.
570 J. Chen and J.S. Fritz, *J. Chromatogr.,* 482 (1989) 279.

Chapter 13

Amino acids and peptides

COLIN T. MANT, NIAN E. ZHOU and ROBERT S. HODGES

CONTENTS

13.1 AMINO ACIDS

13.1.1 Introduction

A knowledge of the amino acid composition of a polypeptide or protein can be of critical importance for its identification, an understanding of its properties, an assessment of its purity, and in establishing its concentration. Other areas where there is a requirement for accurate analysis of the presence and levels of amino acids include peptide synthesis, food and beverage analysis, and the analysis of physiological fluids, e.g., the profiling of amino acids in body fluids [1]. Much work has also been directed towards the separation of enantiomers of amino acids [2-8]. Unlike peptides, which have the advantage of strong peptide bond absorbance at wavelengths <220 nm, most free amino acids have low UV extinction coefficients. Thus, current detection methods for amino acids depend upon the reaction of the primary amino group of the amino acid to yield a colored or fluorescent derivative, this derivatization taking place either before or after the separation of the amino acids has been effected. The most commonly determined amino acids and their 3- and 1-letter designations are listed in Table 13.1.

The first automated system for the analysis of amino acid mixtures was described in 1958 by Spackman, Stein, and Moore [9]. This method was based on the ion-exchange separation of amino acids, followed by their on-line quantitation by reaction with ninhydrin and subsequent detection in the visible range of wavelengths. Since that time, there has been a dramatic increase in the need for faster and more sensitive separations. This challenge has been met by the development of high-performance stationary phases (most notably, hydrophobic reversed-phase sorbents for the precolumn derivatization approach), coupled with the introduction of UV-absorbing or fluorescing derivatization reagents.

TABLE 13.1

THREE- AND ONE-LETTER DESIGNATIONS OF THE 20 AMINO ACIDS FOUND IN PROTEINS

Alanine	Ala	A
Arginine	Arg	R
Asparagine	Asn	N
Aspartic acid	Asp	D
Cysteine	Cys	C
Glutamic acid	Glu	E
Glutamine	Gln	Q
Glycine	Gly	G
Histidine	His	H
Isoleucine	Ile	I
Leucine	Leu	L
Lysine	Lys	K
Methionine	Met	M
Phenylalanine	Phe	F
Proline	Pro	P
Serine	Ser	S
Threonine	Thr	T
Tryptophan	Trp	W
Tyrosine	Tyr	Y
Valine	Val	V

As pointed out by Ogden and Földi [10], it would be inaccurate to claim that there is a perfect analytical method for each amino acid and its metabolites, of which more than 300 are known. These authors also point out that the variety and flexibility of amino acid analysis techniques ensure that each separation problem can be solved by choosing the optimal method.

In contrast to peptides, the potential approaches to the separation of amino acids by HPLC are somewhat limited. The smaller variety of types of amino acids compared to peptide structures makes the only HPLC modes available to the researcher for amino acid separations ion-exchange chromatography (IEC), generally cation-exchange chromatography (CEC), and reversed-phase chromatography (RPC). The choice of HPLC mode is further limited by the choice of derivatization reagent and/or decision to use either pre- or postcolumn derivatization. In fact, some of the most interesting research carried out in the separation of amino acids by HPLC lies in the development and optimization of novel derivatizing reagents.

Much of the current research on amino acid separations by HPLC is carried out by companies which market automated amino acid analyzers. Literature put out by these companies, together with some excellent recent reviews on the subject of the current status of amino acid analysis [10-14], serve to cover the general aspects of HPLC of amino acids very thoroughly. For this reason, instead of merely rehashing these admirable

References on p. B140

sources of information (which include model separations of standard mixtures of amino acids), the current review will focus on examining the advantages of various derivatizing reagents as well as the relative merits of pre- and postcolumn derivatization.

13.1.2 Derivatization reagents

13.1.2.1 Ninhydrin

The automated system for amino acid analyses first described in 1958 [9], based on the cation-exchange chromatographic separation of amino acids and followed by reaction with ninhydrin, remains to this day the most reliable methodology for amino acid analysis of peptides and proteins [10,14-22]. Commercial analyzers based on this methodology are available from such companies as Beckman Instruments and Pharmacia. At high temperatures ($\geq 100°C$), all primary amino acids react with two molecules of ninhydrin to form a chromophore (Ruhemann's purple) with maximal absorption at 570 nm (Fig. 13.1A). Proline and hydroxyproline, which are secondary (imino) acids, react with ninhydrin to form a product with an absorption spectrum distinctly different from that formed from primary amino acids (Fig. 13.1B). For this reason, amino acid separation and quantitation are monitored at two wavelengths, normally 440 nm and 570 nm. In addition, cystine forms both a blue and a yellow chromophore with ninhydrin. In fact, the ratio of the absorbances at 440 nm and 570 nm is a useful diagnostic feature in the identification of amino acids [10].

All primary amino acids form the same complex following reaction with ninhydrin, making this reagent unsuitable for precolumn derivatization. Most cation-exchange HPLC columns utilized for ninhydrin-based postcolumn amino acid analysis still employ sulfonated polystyrene/divinylbenzene resins (5- to 10-μm particle size) [13]. Silica-based ion-exchange packings, popular for peptide separations, have poor resistance to the high pH values, ionic strength, and elevated temperature required to elute free amino acids.

Modern automated amino acid analyzers based on postcolumn ninhydrin derivatization are capable of providing highly reliable and reproducible analyses. Ninhydrin methodology is limited to a sensitivity of ca. 100 pmole for the amino acid present in lowest abundance in a peptide or protein, a somewhat lower sensitivity than can be achieved with other derivatization reagents. However, at this lower sensitivity range, the postcolumn ninhydrin methodology has the clear advantage of being the most versatile in terms of its tolerance to the presence of impurities and in its proven accuracy and reproducibility, based on many years of methodology development [14,21].

13.1.2.2 o-Phthaldialdehyde

o-Phthaldialdehyde (OPA) was originally introduced as an alternative to ninhydrin in postcolumn derivatization and provided a significant increase in sensitivity [23,24]. OPA reacts in alkaline medium with primary amines and, while OPA itself has no inherent fluorescence, its reaction products exhibit very high fluorescent yields (Fig. 13.1C). At high

A   Ninhydrin/primary amino acid derivative
(Ruhemann's purple)
λ_{max} = 570 nm

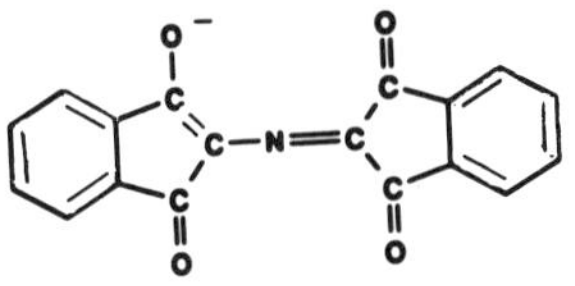

B   Ninhydrin/proline derivative

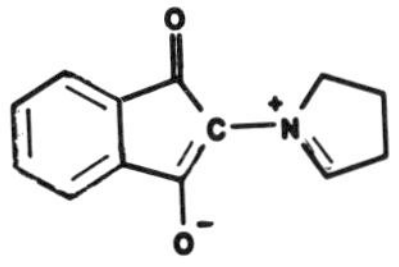

C   OPA/amino acid derivative

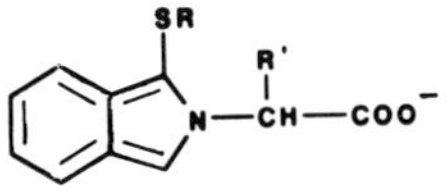

D   PITC/amino acid derivative (PTC derivative)
λ_{max} = 244 nm

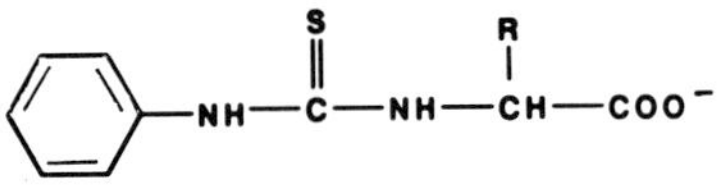

E   DABS-Cl/amino acid derivative
λ_{max} = 420 nm

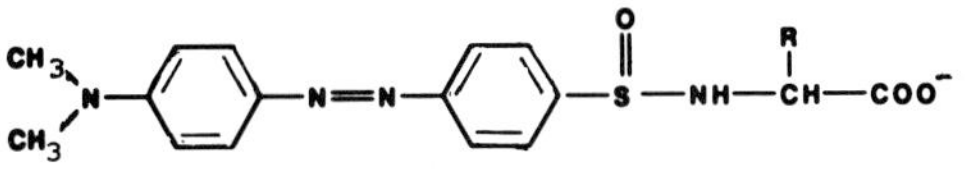

F   FMOC-Cl/amino acid derivative

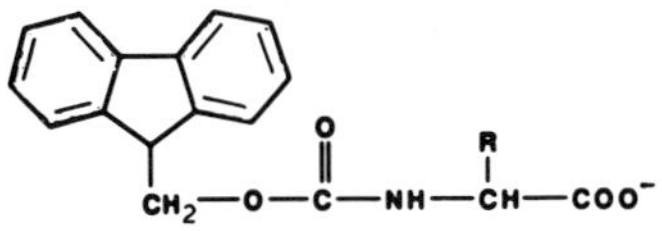

Fig. 13.1. Products of the reaction between amino acids and major derivatization reagents employed in automated amino acid analysis systems. The R groups denote the amino acid sidechains. The R' group shown in the OPA/amino acid product (C) is derived from the thiol reagent required for the reaction to proceed; OPA reacts with primary amines only.

References on p. B140

pH and in the presence of a thiol, such as 2-mercaptoethanol (although other thiols, such as ethanedithiol or 3-mercaptopropionic acid, can be used), the reaction proceeds almost instantaneously at room temperature to form an isoindole product, in which both the amino acid and thiol are incorporated. Mono derivatives are apparently obtained for all amino acids, except lysine, having two primary amino groups, and cysteine, having both amino and sulfhydryl groups, which form bis products. Although the reaction products of cysteine and cystine exhibit only a weak fluorescence, this situation can be improved by alkylation of cysteine with iodoacetic acid prior to derivatization with OPA [25] or by oxidation to cysteine sulfonic acid [26]. Direct comparisons between postcolumn derivatization with ninhydrin and OPA clearly demonstrated the gain in detection sensitivity with fluorescence over colorimetric detection [27,28].

Although OPA was originally applied only to postcolumn derivatization, still a common application [27-40], it is now also widely used as a reagent for precolumn derivatization [41-59]. While separation of amino acids following precolumn derivatization is carried out by RPC (generally, silica-based stationary phases), both RPC [31,34-36] and IEC (specifically, strong-cation-exchange chromatography) [27-30,32,33,37-40] may be employed when OPA is the postcolumn derivatizing reagent. The cation-exchange columns are typically not silica, but generally polystyrene-based, with sulfonate functional groups, while the reversed-phase columns are usually silica-based C_{18} packings, although a polystyrene-based packing has also been employed [34].

One disadvantage of OPA lies in the relative instability of OPA derivatives, perhaps leading to quantitation problems, although this is not a major difficulty in the postcolumn derivatization technique. However, in precolumn derivatization, and following lowering of the pH to 7.2 from a typical reaction pH of 9.5, the products must be immediately applied to the chromatographic system. Lowering the pH both quenches the reaction and serves to stabilize slightly the reaction products. The major disadvantage of OPA is its failure to react with secondary amines (imino acids; proline and hydroxyproline), although this can be circumvented by their on-line oxidation with chloramine-T or hypochlorite [60]. A strategy combining the use of OPA and FMOC (Section 13.1.2.6) has also recently been introduced to overcome this deficiency [58].

The OPA method, specifically the precolumn derivatization method, is frequently selected for amino acid analysis because it is more sensitive and takes less time than other HPLC methods. This technique has also gained significant popularity in the clinical field [10,37,41,42,44,47,48,50,52-55,57,59].

13.1.2.3 Fluorescamine

Fluorescamine (or fluram) was also introduced as a possible alternative to ninhydrin in postcolumn derivatization [61]. At high pH, fluorescamine reacts rapidly with primary amines to form fluorescent products [62], with a concomitant 10- to 100-fold increase in detection sensitivity over that of ninhydrin [63], although fluorescamine itself does not fluoresce. Amino acid derivatives, following fluorescamine labeling, achieve maximum fluorescence at pH 9 or above, showing only very little fluorescence at pH 7.4.

Fluorescamine was briefly popular as an alternative to ninhydrin for postcolumn derivatization (it is still often used for peptide and protein labeling), despite the fact that, like OPA, it fails to react with secondary amines. As in the case of OPA, this problem can be circumvented by the use of oxidizing agents, such as chloramine-T [64], hypochlorite [63], or N-chlorosuccinimide [65].

The use of fluorescamine for precolumn derivatization of amino acid analysis is limited, since the silica-based columns generally used in reversed-phase chromatography are not stable at the high pH values required for satisfactory fluorescence of the amino acid derivatives.

A description of fluorescamine-based HPLC analyzers for amino acids and sugars can be found in Refs. 66 and 67, and Ref. 68 supplies a more general discussion regarding the use of fluorescamine for HPLC in protein chemistry.

13.1.2.4 Phenylisothiocyanate

Phenylisothiocyanate (PITC), also known as Edman's reagent, has long been used for the sequencing of polypeptides and proteins, and was introduced for the analysis of amino acids in the early Eighties [69-72]. It is currently the most extensively used reagent for precolumn derivatization, followed by RPC, in amino acid analysis, being employed for amino acid mixtures from a wide variety of sources [51,73-88].

Instead of being analyzed as phenylthiohydantoin (PTH) derivatives, as occurs during peptide or protein sequencing, amino acids are analyzed as phenylthiocarbamyl (PTC) derivatives (Fig. 13.1D). Thus, at alkaline pH, both primary and secondary amino acids produce PTC derivatives, which can be monitored by UV absorbance (240-255 nm) with a detection limit in the 5- to 50-pmol range. All amino acids form mono-PTC derivatives, except lysine and cysteine, which yield the bis-PTC forms. The products are relatively stable, although some conversion to the PTH derivatives can occur if the pH is not adequately controlled.

Although the PITC methodology can be considered superior to that of other derivatization techniques in a number of respects [14], it has a major disadvantage in that yields of some PTC amino acid derivatives are markedly affected by the presence of some salts, divalent cations, metals, and buffer ions [89]. Thus, while ammonium acetate, sodium chloride, and sodium borate have been found to have no effect on PTC amino acid yields, other salts, such as ammonium acetate, sodium phosphate, and sodium bicarbonate may be highly deleterious. In addition, trace metal ion contamination has also been found to reduce markedly the yields of PTC amino acids. Improved yields have been noted by Dupont et al. [89] and Mora et al. [90] when ethylenediamine tetraacetic acid (EDTA) was added to the derivatization buffer.

There are a number of recent reviews covering the status of PITC in amino acid analysis [91-93], a derivatization technique which has had considerable commercial success (e.g., Waters Pico Tag and Applied Biosystems).

References on p. B140

13.1.2.5 Dansyl chloride and dabsyl chloride

Dansyl chloride (1-dimethylaminonaphthalene-5-sulfonyl chloride; DNS-Cl) was initially introduced for *N*-terminal amino acid analysis of peptides and proteins [94], a purpose for which it is still employed. It reacts with both primary and secondary amines to produce fluorescent derivatives. The application of dansyl chloride suffers from the presence of numerous side products and the formation of multiple derivatives of certain amino acids. Added to this is the requirement of a long reaction time. Thus, although dansyl chloride is frequently employed by researchers for derivatization [95-100] prior to RPC, this method has never been widely accepted for automated amino acid analysis.

Dabsyl chloride (dimethylaminoazobenzenesulfonyl chloride; DABS-Cl), Fig. 13.1E, a reagent closely related to dansyl chloride, was introduced and developed by Chang and colleagues [101-105] and others [106-109] for the detection of amino acids in the visible wavelength range and at the pmol level. It reacts with both primary and secondary amino acids to produce derivatives that have a high absorbance in the visible range, its major advantage. The reaction proceeds at ca. 70°C and pH 8-8.5, producing highly stable derivatives in 10-12 min. Both mono- and bis-derivatives of lysine, tyrosine, and histidine are formed. Despite its advantages over dansyl chloride as a derivatization reagent, the dabsyl chloride technique has a limited tolerance to the presence of salts [105] and it is not yet clear how detection at high levels of sensitivity (10- to 20-pmol range) would be affected by the presence of metals, salts, buffer ions, and other contaminants.

Although not as widely used in automated amino acid analysis as ninhydrin, OPA, and PITC, a commercial adaptation of the dabsyl chloride method is available from Beckman Instruments.

13.1.2.6 9-Fluorenylmethyl chloroformate

9-Fluorenylmethyl chloroformate (FMOC-Cl) reacts with amino groups to yield strongly fluorescent and stable carbamate derivatives (Fig. 13.1F). This highly reactive reagent, originally and still used as an amino-protecting group in peptide synthesis [110], was first applied as a precolumn derivatizing reagent for amino acid analyses by Einarsson et al. [111-113] and others [114,115]. FMOC-Cl reacts with all amino acids at alkaline pH and room temperature to produce highly stable amino acid derivatives, which have a fluorescent yield comparable to OPA derivatives. Both mono- and bis-derivatives can be formed with histidine, tyrosine, and lysine. The relative proportions of these derivatives vary as a function of pH and of the ratio of reagent to amino acid.

During the reaction with FMOC-Cl, hydrolysis products of the reagent are formed that have fluorescent properties very similar to those of the derivatized amino acids. These products interfere in the reversed-phase chromatographic separation, unless they are removed by organic solvent (e.g., pentane) extraction prior to sample application to the RPC column. An automated version of this derivatization procedure has been developed [116], and an evaluation of a commercial instrument (Varian Amino Tag) has recently been reported [117].

Betnér and Földi [118] circumvented the problem of interference by the elution of large peaks of FMOC-Cl and its hydrolysis product, FMOC-OH, which are eluted in the same chromatographic region as several of the FMOC amino acids, by derivatizing the bulk of excess FMOC-Cl with a hydrophobic amine (1-aminoadamantane; ADAM). The resulting derivative is eluted late in the chromatogram after all the FMOC amino acids. A commercial adaptation (Pharmacia) of the FMOC/ADAM system is described in Ref. 22.

As mentioned above, a combined OPA/FMOC approach has been described [58], which uses a commercially available analyzer (Hewlett-Packard Amino Quant). Reaction of primary amino acids is carried out initially with OPA, followed by reaction of proline (or hydroxyproline) with FMOC. Fluorescence monitoring for the OPA primary amino acids and FMOC-proline derivatives is then carried out at appropriate wavelengths.

13.1.2.7 7-Chloro- and 7-fluoro-4-nitrobenz-2-oxa-1,3-diazole

Derivatization reagents based on the structure of nitrobenzofurazan, namely 7-chloro- and 7-fluoro-4-nitrobenz-2-oxa-1,3-diazole (NBD-Cl and NBD-F, respectively) have proved useful for specific applications in analysis of all amino acids.

NBD-Cl was introduced by Ghosh and Whitehouse [119] as a fluorigenic reagent for amines, including amino acids, and further developed by Roth [120] as a useful tool for amino acid analysis, particularly because of its enhanced reactivity with secondary amines. Since that time, this reagent has seen occasional employment for detection of amino acids, with particular emphasis on the detection of proline and hydroxyproline [121-124]. NBD-Cl has been applied to both precolumn [121,122,124] and postcolumn [120,123] derivatization, the detection limit in the latter being ca. 1 nmol for proline and hydroxyproline and the detection limit in the former being somewhat lower. Although NBD-Cl is stable and enables sensitive detection of amino acids, it reacts somewhat slowly with amines and has not seen widespread use as a general analytical reagent for amino acids.

NBD-F has proven more reactive than the chloro analog, enabling faster derivatization [125], coupled with a greater fluorescence yield for proline and hydroxyproline [125,126]. In a manner similar to NBD-Cl, NBD-F has been applied in both precolumn [125-130] and postcolumn [128,131] derivatization. The detection limits for precolumn derivatization are reportedly in the range of 5-15 fmol for most amino acids (ca. 1000-fold greater than postcolumn derivatization). Again like NBD-Cl, NBD-F has not seen widespread use as an analytical method for amino acids, although specific researchers do appear to employ the latter reagent on a regular basis.

13.1.2.8 Other isothiocyanate reagents

4'-N,N-Dimethylamino-4-azobenzeneisothiocyanate (DABITC) has been used successfully in the manual sequencing and N-terminal analysis of peptides and proteins [132-137]. The Edman-type degradation of peptides and proteins effected by this reagent produces colored thiohydantoin derivatives (DABTH), which can then be separated by RPC. The degradation method generally involves a double coupling, first with DABITC and then with PITC. The 4-*N,N*-dimethylnaphthylazobenzene homolog of DABITC (DANABITC) has also been applied successfully, albeit to a lesser extent than DABITC [138,139].

Diphenylindenonylisothiocyanate (DIITC) has been reported to be useful when mass spectrometry by chemical ionization is coupled with previous identification by RPC. The detection of thiohydantoin derivatives (DITH) is carried out by UV monitoring (low-pmol range). Electrochemical detection is a useful alternative, considering the nature of the aromatic moiety [140].

Other precolumn derivatization isothiocyanate reagents have included 4-(*t*-butyloxycar-bonylaminomethyl)phenyl isothiocyanate (BAMPITC; fluorescence detection) [141], trimethylsilyl isothiocyanate for *C*-terminal sequence analysis (UV detection) [142] and *p-N,N*-dimethylaminophenyl isothiocyanate (DMAPI) as an electrochemical label for amino acid analysis [143].

13.1.2.9 Miscellaneous reagents

Other reagents are continually being developed and employed for amino acid detection. Some have quite specific and limited applications, while others are treated as possible contenders for widespread use in automated systems. Such reagents include 2,4,6-trinitrobenzenesulfonic acid (TNBS) and its 2,4-dinitrobenzene analog (DNBS) [144], *N,N*-diethyl-2,4-dinitro-5-fluoroaniline [145], fluoro-2,4-dinitrobenzene [146], 1,1-diphe-nylboronic acid [147], and 1-fluoro-2,4-dinitrophenyl-5-L-alanine amide (Marfey's reagent) [148,149]. Fluorescence detection is possible when employing such reagents as pentane-2,4-dione/formaldehyde [150], 9-anthryldiazomethane [151], laryl chloride [152], 2,3-naphthalenedicarboxyaldehyde (plus cyanide) [153], succinimide α-naphthylcarba-mate [154], 3-benzoyl-2-quinolinecarboxaldehyde [155], 9-hydroxymethylanthracene and fluorescein isothiocyanate [156], and 1-naphthyl isocyanate [157]. A chemiluminescent reagent, 4-isocyanatophthalhydrazide, has also been reported [158]. With the exception of Ref. 150, where the reported reagent was used in postcolumn derivatization following cation-exchange HPLC of amino acids, all of the above reagents were employed for derivatization prior to RPC.

13.1.3 Precolumn versus postcolumn derivatization

13.1.3.1 Advantages and disadvantages of pre- and postcolumn derivatization

The opinions of researchers as to the relative merits of derivatization before or after HPLC of amino acids will, of course, be determined by the requirements of the specific application. Factors such as required detection sensitivity, amount of available sample, sample type (e.g., size of peptide/protein and presence of unusual amino acids), sample source, analysis speed and reproducibility, and even economic considerations (e.g., costs of instrumentation and reagents) will influence the choice between pre- and postcol-umn derivatization of amino acids.

13.1.3.1.1 Postcolumn derivatization

Derivatization following cation-exchange chromatography of free amino acids has distinct advantages, provided that the instrumentation yields reproducible flowrates and reaction temperatures [10]. Moreover, there is no need for derivatization to proceed to completion, and problems of derivative stability are not significant. Other advantages are that automation is easy and that time-consuming and loss-causing sample preparation are unnecessary [11]. Detectors in series can also be used. Most importantly, perhaps, a vast amount of experience has been gained from the traditional cation-exchange/ninhydrin method, which has proven to be reliable and accurate and capable of excellent separation of complex mixtures of amino acids and their metabolites.

A disadvantage of postcolumn derivatization with a conventional HPLC system is that more complex instrumentation is needed than for the precolumn method. Postcolumn derivatization requires the addition of one or more pumps for the derivatizing reagents and solvents [13]. It also requires the addition of a postcolumn reactor, which may lead to additional bandbroadening and, hence, a decrease in detection sensitivity. In addition, the column eluate is continuously diluted by the derivatizing reagent, further decreasing the detection sensitivity. Thus, the maximum sensitivity attainable by postcolumn derivatization will be lower than that obtained with precolumn derivatization, an evident conclusion when detection sensitivities in post- and precolumn derivatization with OPA are compared in Table 13.2.

13.1.3.1.2 Precolumn derivatization

Precolumn derivatization, coupled with RPC instead of IEC, was originally introduced in response to the ever-increasing demand for higher sensitivity and greater speed of analyses. Indeed, advantages of precolumn derivatization methodology include simplicity, speed, and high sensitivity. Thus, if maximum detection sensitivity is required, a fluorescent reagent, combined with precolumn derivatization, is the method of choice.

Disadvantages of the precolumn method include the need to ensure complete reaction of the derivatization reagent and the possibility of interference with the separation by excess reagent, the reaction medium, the formation of artefacts, or the production of several derivatives from one component [11]. In addition, derivative stability may be an important factor during precolumn derivatization, the delay between derivatization and injection becoming critical to the results obtained (e.g., in precolumn derivatization with OPA) [10].

13.1.3.2 Detection characteristics of major derivatization reagents

Table 13.2 lists the detection characteristics of derivatization reagents employed in commercial automated systems. An important point to note here is that the suggested detection limits for each type of derivatization methodology, summarized from various reviews and manufacturers' literature, are more than likely better than those achievable on a regular, everyday basis. Thus, at the present time, the practical detection limits of precolumn derivatization methods are of the order of 10-20 pmol, i.e., a 5- to 10-fold

TABLE 13.2

DETECTION CHARACTERISTICS OF DERIVATIZATION REAGENTS EMPLOYED IN COMMERCIAL AUTOMATED SYSTEMS

	Precolumn derivatization (RPC)				Postcolumn derivatization (IEC)		
	OPA	PITC	DABS-Cl	FMOC	OPA	OPA/chloramine-T	Ninhydrin
Detection of secondary amino acids	No	Yes	Yes	Yes	No	Yes	Yes
Detection method*	UV or F	UV	Vis	UV or F	F	F	Vis
Detection limits (pmol)	1000 (UV) <1.0 (F)	5-50	2-10	<1.0 (F)	3-5	10-100	50-100

* UV, F, and Vis denote ultraviolet, fluorescence, and visible detection, respectively.

increase in sensitivity over the classical postcolumn/ninhydrin method, even though, in theory, the sensitivity with the fluorescent reagents is much higher (low fmol range). The detection limit tends to be dictated by the variable background levels of contamination present in reagents and hardware. As Smillie and Nattriss [14] put it, "in spite of heroic efforts", it has not yet been possible to reduce this background to levels below a few picomoles. In addition, these authors point out that, while in some cases these precolumn derivatization methods compare reasonably well with the postcolumn ninhydrin proce- dures at a level > 100 pmol, the precision and accuracy achievable in the low-pmol range are significantly less.

Smillie and Nattriss [14] suggest that, in assessing the advantages and disadvantages of pre- and postcolumn derivatization methods, it is convenient to consider these at a sensitivity range of > 100 pmol and in the range of 10-100 pmol. Thus, in situations where the amount of sample is not a limiting factor, e.g., in the quality assessment of the purity of synthetic peptides, the postcolumn/ninhydrin methodology remains the most reliable and satisfactory method. At the higher sensitivity level, i.e., when one of the precolumn deriva- tization methods is applied at sensitivity levels < 100 pmol, increased sensitivity tends to come at the expense of progressively decreasing reliability of the data. Thus, while anal- yses at the 20-pmol level may be adequate for composition analysis of a 20- to 30-residue peptide, the value and reliability of compositional data of such an analysis on a large protein are more questionable [14].

13.2 PEPTIDES

13.2.1 Introduction

13.2.1.1 Characterization of peptides

The distinction between a peptide, polypeptide, and protein, in terms of the number of peptide residues they contain, is somewhat arbitrary. However, peptides are usually defined as containing 50 amino acid residues or less. Although molecules of more than 50 residues usually have a stable three-dimensional structure in solution, and are referred to as proteins, conformation can be an important factor in peptides as well as proteins. Secondary structure, e.g., α-helix, is generally absent, even under benign aqueous condi- tions for small peptides (up to ca. 10 residues). However, the potential for a defined secondary or tertiary structure increases with increasing peptide chainlength and, for peptides of more than 20-30 residues, folding of the peptide chain to internalize nonpolar residues is likely to become an increasingly important conformational feature. In addition, the presence of disulfide bridge(s) would be expected to affect conformation, and thus, the retention behavior of peptides in HPLC may differ from that in the fully reduced state. Thus, conformation should always be a consideration in choosing conditions for chroma- tography.

References on p. B140

13.2.1.2 Chromatographic modes used for peptide separations

Since amino acids are the fundamental units of peptides, the chromatographic be-havior of a particular peptide will be determined by the number and properties (polarity, charge potential) of the residue sidechains it contains. There is a fundamental difference in the type and scope of problem facing the researcher when comparing the separation of amino acids with that of peptides. This difference is most clearly obvious when one considers the potential complexity of peptide mixtures in terms of peptide length, amino acid composition and sequence, as well as the relative number and amounts of peptides present. It has already been stated in Section 13.2.1.1 that peptides are usually defined as containing 50 amino acid residues or less. Thus, if one just considers the 20 amino acids found in proteins, the potential number of different 10-residue peptides, for example, which could be formed from these amino acids is vast (20^{10}).

The aforementioned complexity of peptide mixtures affords researchers the option of using several modes of HPLC for peptide separations. The main modes of chromatog-raphy employed in peptide separations, to be covered in depth in this review, take advantage of differences in peptide size (size-exclusion chromatography, SEC), net charge (ion-exchange chromatography, IEC) or hydrophobicity (reversed-phase chroma-tography, RPC). Within these modes, mobile-phase conditions may be manipulated to maximize the separation potential of a particular column. Other, lesser-used peptide sepa-ration techniques include affinity chromatography [159] and, more recently, hydrophobic-interaction chromatography (HIC) [160,161] and hydrophilic-interaction chromatography (HILIC) [162]. Several useful articles and reviews on HPLC of peptides can be found in Refs. 12 and 163-169. For an extensive source of information on the early development of HPLC of peptides, the reader is directed to Ref. 170. A more current and comprehensive overview on this topic is supplied by Ref. 171.

Capillary electrophoresis (CE), a relatively new separation method, is capable of isolat-ing and identifying picomole quantities of peptides [172-174]. As an electrophoretic technique, it is a separation method complementary to HPLC. At present, it does not have the same capability as RPC to resolve peptides of similar charge but differing hydropho-bicities. RPC has been extensively investigated over a long period of time, and this has resulted in the development of a wide range of mobile phases and stationary phases to meet the vast majority of separation requirements. In contrast, buffer systems in CE for the separation, based on solute hydrophobicity, of peptides and proteins are still under development. It remains to be seen whether CE can achieve the same resolving capa-bilities on a routine basis for peptides and proteins of varying size, charge, and hydro-phobicity.

The eventual success of a particular peptide separation is inextricably bound up with the correct choice of column(s) and chromatographic conditions. The proper selection of the column will simplify optimization of the chromatographic conditions. The major limita-tion to mobile-phase conditions for peptide separations lies in the nature and stability of the column packing material. Silica-based packings are still the most widely used. The rigidity of microparticulate silica enables the use of high linear flow velocities of mobile phases. In addition, favorable mass transfer characteristics allow rapid analyses to be performed. However, most silica columns are limited to a pH range of 2.0-8.0. The silica matrix rapidly dissolves in the presence of basic eluents, and the hydrophobic bonded phase is progressively cleaved at low pH. Column packings based on organic polymers are used increasingly in all modes of chromatography (particularly SEC and IEC) of peptides. Examples of peptide separations on these pH-stable packing materials are included in this review.

13.2.1.3 Peptide detection

Ultraviolet (UV) detectors are, by far, the most commonly used detectors for peptide analysis (and in almost all other applications for that matter). Peptide bonds absorb light strongly in the far ultraviolet (<220 nm), providing a convenient means of detection (usually at 210-220 nm). In addition, the aromatic side chains of tyrosine, phenylalanine, and tryptophan absorb light in the 250- to 290-nm ultraviolet range.

Fluorescence [175-178] and electrochemical [178-182] techniques have also been employed for specific peptide analyses. Fluorimeters, in particular, have proved to be very sensitive and selective detectors for HPLC. The average research laboratory is much more likely to possess a UV detection system for peptide HPLC than either fluorescence or electrochemical detection systems. Indeed, Lockhart et al. [178], comparing the three types of detector, determined that, for routine peptide applications, the UV detector is superior to electrochemical and fluorescence detectors with respect to convenience of operation. It can be assumed that all of the peptide applications reported in this review utilized UV detection systems, either alone or as a complement to other systems.

13.2.1.4 Separation goals and multistep separation

HPLC has proved very versatile in the isolation of peptides varying widely in their sources, quantity, and complexity. Peptides from various sources differ in size, net charge, and polarity, and the approach to their separation must be tailored to the separation goals. Thus, purification of a single peptide from a complex peptide mixture (e.g., the separation of a synthetic peptide from impurities due to solid-phase peptide synthesis) will require an approach different from that necessary for separating all components of a complex mixture (e.g., peptide fragments resulting from tryptic cleavage of a protein). The former may require the application of only a single HPLC technique, i.e., taking advantage of only one property (size, charge, or polarity) of the peptide of interest. In contrast, the latter goal will require a combination of separation techniques for efficient resolution of all desired peptides. A brief review of approaches to multistep HPLC of peptides can be found in Ref. 183.

References on p. B140

13.2.1.5 Use of peptide standards

Common to all applications of HPLC is the need to choose the proper column(s) and the most appropriate mobile phase. The logical approach to this is the use of standards to test the suitability of HPLC columns and conditions. The value of standards in monitoring solute retention behavior in HPLC as well as HPLC column and instrument performance cannot be overestimated. Peptide standards structurally similar to the sample of interest are best. They allow the researcher to:

    (a) identify nonspecific interactions of peptides with the column packing (nonideal behavior);

    (b) monitor column performance (efficiency, selectivity, resolution);

    (c) monitor interassay reproducibility of peptide separations;

    (d) monitor column aging;

    (e) monitor the effects of mobile-phase composition and pH;

    (f) monitor the effect of varying parameters (e.g., flowrate in SEC, IEC, RPC; gradient slope in IEC, RPC; or temperature);

    (g) monitor the effect of changing column dimensions;

    (h) monitor the effect of particle and pore sizes;

    (i) compare packing materials (same materials from different manufacturers; different batches of the same packing material from the same manufacturer);

    (j) monitor sensitivity of detector response;

    (k) monitor effect of instrument variations.

13.2.2 Size-exclusion chromatography

The practical value of SEC columns is at present somewhat limited. Complete resolution of peptide mixtures cannot be obtained due to the lack of commercial packings with the correct fractionation range (i.e., ca. 100-6000 D). The fractionation range needed tends to be at the low end of the fractionation ability of most current columns, which were designed mainly for protein separations. The rapid loss of peptide resolution with column age at this extreme end of their fractionation range and their high cost tend to make their purchase risky.

Although the production of a size-exclusion column designed specifically for peptides has yet to be achieved, likely requiring pore diameters less than the lowest currently available, this by no means implies that SEC has no useful role in peptide separations. Among other uses, for example, size-exclusion columns are of value for peptide/protein separations in the early stages of a peptide purification. Selected applications of SEC to peptides are presented in Table 13.3. Common to all of the applications are the particle size of the packings (10 μm in the vast majority of cases) and the range of pore sizes (60-240 Å). Silica-based packings are prevalent in most applications, although excellent results have also been obtained with various non-silica-based columns [184-189]. Column dimensions for analytical size-exclusion columns are generally in the range of 25-60 cm x 4.5-8.0 mm ID. Columns at the upper end of this ID range are often referred to as "semipreparative", which is a misnomer, considering the small capacity of these columns for preparative applications.

13.2.2.1 Mobile-phase selection

Currently available packings for aqueous SEC are generally surface-modified silicas and hydrophilic crosslinked organic polymers (typically polyether-, polyester-, or agarose-type materials, carrying hydroxyl functions). Separation of peptides by a mechanism based solely on peptide size, i.e., ideal SEC, occurs only when there is no interaction between the solutes and the column matrix. Although high-performance size-exclusion columns are designed to minimize nonspecific interactions, most modern SEC columns, whether silica-based or polymer-based, are weakly anionic (negatively charged) and slightly hydrophobic. This results in deviations from ideal size-exclusion behavior [190], due to, respectively, ionic and hydrophobic solute/packing interactions. Peptides of a wide range of net charge and polarity may be expected to be particularly sensitive to both types of nonspecific interactions. Thus, if ideal (and predictable) peptide retention behavior is required, e.g., for molecular weight determination, such nonspecific interactions must be recognized and suppressed.

Electrostatic effects are minimized above an eluent ionic strength of about 0.05 M, and aqueous phosphate or Tris buffers, often containing 0.1-0.4 M salts, are commonly employed as the mobile phase for SEC of peptides. Other, nonvolatile systems have included the use of sodium acetate buffers (with sodium sulfate [191]) and triethylammonium formate buffers [192]. Hydrophobic interactions are promoted under conditions of high salt concentrations, and ionic strengths above 0.6 M should generally be avoided [190,193]. Agents, such as guanidine hydrochloride GuaHCl), urea, and sodium dodecyl-sulfate (SDS) have also been employed in mobile-phase solvents, both to suppress nonspecific interactions and to denature solutes for estimation of molecular weights [184,190,194-197]. Optimum flowrates for analytical size-exclusion columns are generally in the range of 0.2-1.0 ml/min for the best compromise between separation time and resolution. The sample volume must be kept as small as possible.

The major disadvantage of nonvolatile eluents is that they must be removed prior to further purification or analysis. In addition, the presence of denaturing compounds, such as GuaHCl and SDS, precludes the use of low UV wavelengths (210 nm) for monitoring the column effluent. A volatile eluent enables lyophilization of peptide fractions prior to analysis or their direct application to ion-exchange or reversed-phase columns. Swergold and Rubin [185] suggested that the ideal system for SEC of peptides should satisfy at least five criteria: (1) the linear log MW vs. elution time or volume relationship should extend to small peptides; (2) few peptides should have anomalous retention times; (3) complex peptide mixtures should be resolved; (4) the solvent system should be transparent to short UV radiation; and (5) the solvent system should be volatile and easily removed. The term "ideal" is possibly a slight overstatement, since the optimum mobile-phase conditions for a particular application may not require that all the above criteria be met. For example, Criteria 1 and 2 may not be relevant when molecular weight determi-

TABLE 13.3

SELECTED APPLICATIONS OF SIZE-EXCLUSION CHROMATOGRAPHY TO PEPTIDES

Stationary phase	S or NS*	Particle (μm)/pore diameter (Å)	Column dimensions (mm x mm ID)	Sample	MW range of peptides (daltons)	Mobile phase	V or NV**	Remarks	Ref.
TSK-150 TSK-250 (in series)	S S	10/130 10/240	NG*** NG	Atrial natriuretic peptides in neonatal rat heart	ca. 3000	0.1% aq. TFA/32% MeCN	V	Separation of 28-residue peptide from 126-residue protein	198
TSK-250	S	10/240	600 x 7.5	CNBr fragments of rabbit skeletal troponin I	1400-4000	0.1% aq. TFA	V	First step in multi-dimensional HPLC	199
TSK G2000SW	S	10/130	600 x 7.5	Tryptic digest of aldehyde dehydrogenase	NG	0.1 M aq. ammonium bicarbonate	V	Fragment separation	200
TSK G2000SW TSK G3000SW	S S	10/130 10/240	600 x 7.5 600 x 7.5	Synthetic hydrophobic peptide polymers	1600-3400	0.1% aq. TFA/50% MeCN	V	Non-ideal SEC:hydrophobic interactions	201
TSK G2000SW	S	10/130	600 x 7.5	Synthetic peptide polymers	900-4000	50 mM KH$_2$PO$_4$ (pH 7.0) + 1.1 M KCl ± 8 M urea	NV	Separation/MW determination	197

Packing			Dimensions	Sample	MW range	Mobile phase		Remarks	Ref.
TSK G2000SW	S	10/130	NG	Atrial natriuretic peptides in human coronary sinus plasma	ca. 3000	0.1% aq. TFA/20% MeCN	V	Purity checks following extractions	202
TSK G2000SW	S	10/130	300 x 7.5 (2 in series)	CNBr digest of immunoglobulin light chain	ca. 3000-12000	1.0% aq. TFA/35% MeCN	V	Separation/MW determination	203
TSK G2000SW	S	10/130	300 x 7.5	Synthetic heptapeptide immunogen	ca. 800	50 mM aq. TFA	V	Monitoring of synthetic peptide coupling to carrier protein	204
TSK-Gel 2000SW	S	10/130	600 x 7.5	Biologically active peptides	1000-6500	50 mM sodium phosphate (pH 7.2) + 0.3% (w/v) SDS	NV	First step in multi-dimensional HPLC of artificial peptide mixture	194
TSK-Gel 2000SW	S	10/130	600 x 7.5	Peptides in uremic serum	<500-1000	50 mM sodium phosphate (pH 7.2) + 0.3% (w/v) SDS	NV	Two-step (SEC→RPC) separation of peptides	195
TSK-Gel G2000SW	S	10/130	600 x 7.5	Mixtures of biologically active peptides	200-10 000	150 mM phosphate (pH 7.4) + 1 M NaCl + 20% methyl cellosolve + 1% SDS	NV	MW determination; tested several mobile phases	196
TSK SW2000	S	10/130	600 x 7.5	Mixtures of standard peptides	200-6700	50 mM phosphate (pH 5.0) + 35% methanol + 0.1% TFA	NV	MW determination; tested various mobile phases for ideal SEC	205

(Continued on p. B94)

TABLE 13.3 (continued)

Stationary phase	S or NS*	Particle (µm)/pore diameter (Å)	Column dimensions (mm x mm ID)	Sample	MW range of peptides (daltons)	Mobile phase	V or NV**	Remarks	Ref.
Spherogel TSK G2000SW	S	10/130	300 x 7.5	Synthetic size-exclusion peptide standards	800-4000	Various mobile phases, based on 0.1% aq. TFA or phosphate buffers (pH 6.5), containing KCl and/or MeCN; 50 mM KH$_2$PO$_4$ (pH 6.5) + 0.5 M KCl + 8 M urea	V and NV	Demonstration of nonideal vs. ideal SEC; MW determination of myoglobin fragments	184
SynChropak GPC60	S	10/60	300 x 7.8						
Superose 12	NS	10/NG	300 x 10						
TSK G3000SW	S	10/240	600 x 7.5 (2 in series)	Peptide digests of human IgG	NG	50 mM sodium acetate (pH 5.0) + 0.1 M sodium sulfate	NV	Fragment separation; monitoring course of digestion	191
TSK G3000SW	S	10/240	600 x 7.5	Synthetic peptide polymers	900-4000	0.1% aq. TFA; 0.1% aq. TFA, containing 50% MeCN or 50% trifluoroethanol	V	Monitoring solvent effects on quaternary structure of a model protein	206
TSK Spherogel 3000 SW	S	10/240	300 x 4.6	Angiotensin II-like material from rat brain	ca. 5000-7000	12 mM Tris-HCl (pH 8.5) + 0.2 M NaCl	NV	Purification/MW determination	207

Column	[*]	[**]/pore	Dimensions	Sample	MW range	Mobile phase	[**] V/NV	Application	Ref.
PAC I-125	S	10/125	NG	Mixtures of peptide standards	300-5700	TEAP and TEAF buffers (pH <3.0) with various MeCN concentrations	V and NV	Nonideal vs. ideal SEC; effect of temperature on resolution	192
TSK-G3000 PW	NS	13-15/200	300 x 7.5 (2 in series)	Mixtures of standard peptides	300-6500	0.1% aq. TFA + range of MeCN concentrations	V	Nonideal vs. ideal SEC	185
TSK gel G3000PW$_{XL}$	NS	6/NG	300 x 7.8 (2 in series)	Mixture of standard peptides	300-6500	0.1% aq. TFA + 45% MeCN	V	Ideal SEC in volatile solvents	186
Asahipak GS-320	NS	NG	500 x 7.6	Mixtures of standard peptides; tryptic and chymotryptic fragments of RNase F$_1$	<500-5500	50 mM sodium phosphate at various pH values or with 5-40% MeCN; 0.1% aq. TFA /5% MeCN; 0.1 M acetic acid	V and NV	Nonideal vs. ideal SEC; use of non-ideal SEC; influence of pH and temperature on separations	187
Asahipak GS-320H	NS	NG	250 x 7.6						
Asahipak GS-320	NS	NG	500 x 7.6	Tryptic and chymotryptic fragments of RNase F$_1$	<3000	0.1 M acetic acid	V	Fractionation of peptide fragments; nonideal SEC	188
Ultrastyragel 1000 Å	NS	<10/1000	300 x 7.8	Self-associating hydrophobic peptide (gramicidin A)	ca. 1700	Tetrahydrofuran	V	Monomer/dimer equilibrium studies	189

[*] S and NS denote, respectively, silica- and non-silica-based stationary phases.

[**] V and NV denote, respectively, volatile and nonvolatile mobile phases.

[***] NG denotes not given.

References on p. B140

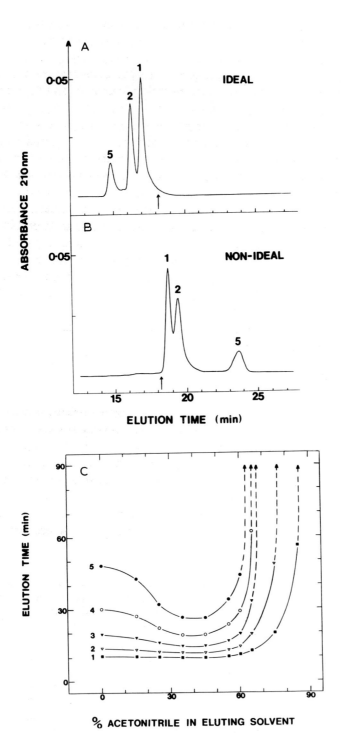

nations or absolute predictability of peptide retention behavior are unnecessary, i.e., when the goal is simply to resolve a peptide mixture. In fact, as discussed in Section 13.2.2.3, in many cases, a mixed-mode effect may be advantageous. However, solvent volatility is a clear asset, particularly when SEC is one step in a multistep purification procedure. Many researchers have demonstrated the utility of 0.1% aq. TFA with or without additions (acetonitrile, trifluoroethanol) for peptide separations [184-187,198,199,201-204,206]. Apart from their low-UV transparency, the presence of these organic modifiers decreases nonspecific (hydrophobic) adsorption and increases the overall solubility of very nonpolar peptides. Resins (both silica-based and polymer-based) manufactured by the Toyo Soda Company (TSK columns) have been, and still are, very widely used size-exclusion packings (reflected in Table 13.3), and mobile phases of TFA/organic modifier have been frequently, and successfully, employed with these packings [184-186,198,199,201-204,206]. A volatile aq. triethylammonium formate (TEAF) buffer containing 30% acetonitrile has also proved useful for peptide separations [192], although the presence of formate necessitates peptide detection at high UV wavelengths (272 nm in this case), i.e., detection depends on aromatic residues being present. Other volatile mobile phases have included 0.1 M ammonium bicarbonate [200] and 0.1 M acetic acid [187,188]. An interesting application by Bañó et al. [189] is the use of tetrahydrofuran as the mobile phase for conformational studies of hydrophobic peptides on a polymer-based SEC column (an uncommon example of nonaqueous SEC of peptides).

13.2.2.2 Peptide standards

In order to maximize the (at present) somewhat limited potential of SEC as a separation technique for peptides, the researcher must have some means of comparing the resolving power of different size-exclusion columns. In addition, since nonspecific solute/packing interactions may be more pronounced with one size-exclusion column than with another, a logical approach to the selection of the correct mobile-phase conditions for a particular application is required. The mobile phase may be either selected to ensure ideal size-exclusion behavior of the peptides or manipulated to take advantage of any potential peptide-resolving capabilities resulting from mixed-mode effects (i.e., nonideal SEC).

Fig. 13.2. Use of synthetic peptide standards to monitor nonspecific interactions in SEC. Column: SynChropak GPC-60 (300 x 7.8 mm ID, 10-μm particle size, 60-Å pore size; SynChrom Inc.). Panel A: ideal SEC of peptide standards; mobile phase, 50 mM KH$_2$PO$_4$/100 mM KCl (pH 6.5); flowrate, 0.5 ml/min. Panel B: nonideal SEC of peptide standards; mobile phase, 5 mM KH$_2$PO$_4$/50 mM KCl (pH 6.5); flowrate, 0.5 ml/min. Panel C: effect of acetonitrile concentration on elution time of peptide standards; mobile phase, 0.1% aq. TFA (pH 2.0), containing 0-85% acetonitrile; flowrate, 1 ml/min; temperature, 26°C; absorbance at 210 nm. The peptide standards have the sequence Ac-(G-L-G-A-K-G-A-G-V-G)$_n$-amide, where n denotes 1, 2, 3, 4, and 5 (10, 20, 30, 40, and 50 residues and +1, +2, +3, +4, and +5 net charge, respectively). The standards were obtained from Synthetic Peptides Inc. The arrows denote the elution time for the total permeation volume of the column. The 1- and 3-letter codes for the amino acids are shown in Table 13.1.

Peptide standards designed specifically for assessing size-exclusion columns should be capable of monitoring both ideal and nonideal behavior. To this end, Mant et al. [184] developed a polymer series of five synthetic peptide standards, Ac-(Gly-Leu-Gly-Ala-Lys-Gly-Ala-Gly-Val-Gly)$_n$-amide, where n = 1-5. The increasing size of the peptide standards (10, 20, 30, 40, and 50 residues; 800-4000 D) enables the accurate molecular-weight calibration of a column in ideal SEC; the increasingly basic character of the standards (1-5 positively charged residues) makes them sensitive to the anionic character of a size-exclusion column; and the increasing hydrophobicity of the standards enables a determination of column hydrophobicity. In addition, the high glycine content of the standards minimizes or eliminates any tendency towards secondary structure. The basic character of the peptide standards also ensures good solubility in aqueous solvents. Fig. 13.2 illustrates the efficacy of the peptide standards in highlighting the properties of a SynChropak GPC-60 silica-based column. The top panels show elution profiles of Peptide Standards 1, 2, and 5 (10, 20, and 50 residues, respectively) with aq. mobile-phase buffers of 50 mM KH$_2$PO$_4$/100 mM KCl (pH 6.5) (Panel A) or 5 mM KH$_2$PO$_4$/50 mM KCl (pH 6.5) (Panel B). In Panel A, the three peptides are eluted in the order of decreasing size, as would be expected under ideal SEC conditions; in addition, the three peptides exhibited a linear log MW vs. elution time relationship. In contrast, in Panel B, the column clearly exhibits nonspecific interactions between the peptides and column matrix at this lower phosphate and KCl concentration, the smallest peptide (Peptide Standard 1; 10 residues) being eluted first and the largest peptide (Peptide Standard 5; 50 residues) being eluted last. In addition, all three peptides are being retained longer than the total permeation volume of the column (denoted by an arrow). By definition, under ideal SEC conditions, no molecule will be retained beyond the total permeation volume of the column. The column is, in fact, behaving like a cation-exchange column [184,190,208-211], the peptides being eluted in the order of increasing positive charge (Peptide Standards 1, 2, and 5 possess a +1, +2, and +5 net charge, respectively) instead of decreasing size. With an increase in the ionic strength of the mobile phase, these electrostatic effects are suppressed to produce the peptide chromatogram based on an ideal size-exclusion mechanism (Panel A). Fig. 13.2C shows the effect of increasing acetonitrile concentration on the retention times of the five peptide standards on the GPC-60 column when 0.1% aq. TFA is used as the mobile-phase eluent. The peptides clearly show nonideal retention behavior in the absence of acetonitrile (the smallest peptide being eluted first). As the concentration of the organic modifier increases, the retention times of all five peptides decreases to a minimum (at ca. 35-45% acetonitrile), presumably as hydrophobic interactions are overcome, and then proceed to increase again, possibly due to the promotion of ionic interactions with the column material. It should be noted that the results of Mant et al. [184] suggested that the addition of salt to the mobile phase disrupted hydrophobic as well as electrostatic peptide/packing interactions.

The commercially available synthetic peptide standards described here have proved extremely beneficial in enabling rapid development of the optimal conditions for SEC of peptides [167,183,184,212] and for monitoring the resolving power of various size-exclusion columns [184]. In addition, they have played an important role in the development of a computer program designed to simulate peptide elution profiles in HPLC [213].

13.2.2.3 Nonideal size-exclusion chromatography

The majority of reports in the literature concerning SEC of peptides (and proteins) tend to describe chromatographic conditions designed to ensure a pure size-exclusion process. It is often overlooked that the nonideal properties of size-exclusion columns can be advantageous in the separation of peptides [184,211]. Excellent examples of this were reported by Yasukawa et al. [187], who separated various hydrophilic and hydrophobic dipeptides on a non-silica-based column by isocratic elution with 50 mM sodium phosphate buffer (pH 7.0). The peptides, retained by adsorption, were eluted in the order of increasing hydrophobicity. Even sequence isomers, such as Trp-Glu and Glu-Trp were successfully separated. In another application by these researchers, fragments of RNAase derived from tryptic and chymotryptic digestion were separated from the proteases and undigested RNAase F by isocratic elution with 0.1 M acetic acid. The proteins were eluted first by a size-exclusion mechanism, and the peptides were then separated by adsorption chromatography. In another example, Mant et al. [184] demonstrated that the separation of five peptide standards of 10, 20, 30, 40, and 50 residues (Section 13.2.2.2) was markedly better on a SynChropak GPC-60 column (Fig. 13.2) when advantage was taken of the nonspecific adsorptive properties of the column than when the column was utilized under conditions of ideal SEC.

13.2.2.4 Ideal size-exclusion chromatography

The ability to predict the position and/or elution order of peptides in SEC of a peptide mixture would greatly simplify preliminary separation of peptides in a chemical or proteolytic protein digest. Under conditions of ideal SEC, large peptide fragments, resulting from incomplete protein digestion, can be quickly identified and removed. Thus, it is important to achieve a linear relationship between log MW and retention time or volume over a wide molecular weight range. This linear relationship has been obtained by using volatile solvents and TSK G2000SW (peptide molecular weight range of 3000-12000 D) [203], TSK G3000SW (900-8000 D) [206], or TSK G3000PW (300-134 000 D [185] and 300-7000 D [186,214]) columns. Similar results for peptides in the 1000- to 44 000-D range were obtained by Rivier [192] on a Waters PAC-I125 column by using a nonvolatile aq. triethylammonium phosphate (TEAP) buffer (0.25 M, pH 2.5), containing 15-30% acetonitrile. It should be noted that among these examples of ideal SEC of peptides and proteins, all but one involved the separation of synthetic peptides [206] or the separation of artificial peptide mixtures [185,186,192,214]. The one exception was the ideal SEC behavior on a TSK G2000SW column of cyanogen bromide fragments of an immunoglobulin light chain, with 1% aq. TFA/35% acetonitrile as the mobile phase, reported by Iadarola et al. [203]. Similar results were reported by Mant and Hodges [199] in SEC of

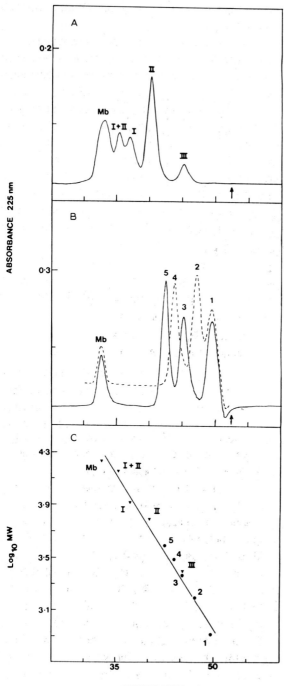

cyanogen bromide fragments of rabbit skeletal troponin I (1400-4000 D) on a TSK-250 column with 0.1% aq. TFA as the mobile phase.

These results and the size-exclusion chromatogram of the peptide standards, shown in Fig. 13.2A, were obtained under nondenaturing conditions. Many proteins and large peptides may deviate from ideal size-exclusion behavior due to differences in conformation. Thus, the tendency of peptides or protein fragments to maintain or reform a particular conformation in nondenaturing media, as opposed to their random coil configuration in denaturing media, will complicate retention time prediction. Thus, Mant et al. [184] could obtain a linear log MW vs. peptide retention time relationship for myoglobin and its cyanogen bromide fragments (2500-17000 D), together with the five peptide polymer standards described previously, only under highly denaturing conditions [50 mM KH_2PO_4 (pH 6.5)/0.5 M KCl/8 M urea] (Fig. 13.3). If the conformational character of a peptide/protein mixture in a particular mobile phase is uncertain and ideal size-exclusion behavior is desired, SEC should always be carried out under highly denaturing conditions.

Although more common in protein applications, molecular weight determinations in the presence or absence of denaturants in SEC can, in fact, provide a means of demonstrating tertiary or quaternary structures in peptides. Thus, Lau et al. [197] demonstrated the use of SEC in the formation of two-stranded α-helical coiled coils from 29- and 36-residue peptides.

13.2.3 Ion-exchange chromatography

Although generally overshadowed in the past by RPC as the preferred HPLC mode for peptide separations, IEC has more recently begun to come into its own as a recognized asset for many applications. Commercially available high-performance ion-exchange packings, capable of retaining both highly charged and weakly basic or acidic peptides, are being introduced. In addition, apart from the many silica-based ion-exchange columns available (silica still being the most common support for ion-exchange sorbents), polymer-based packings exhibiting a broad pH tolerance and satisfactory resistance to compression have also provided excellent resolution of peptide mixtures. This is in contrast to RPC, where, for the most part, the practical advantage of polymer-based over silica-based packings has yet to be demonstrated convincingly. Column dimensions for most analytical ion-exchange columns are in the range of 150-250 x 4.0-4.6 mm ID. Aside from the greater resolving power of RPC, one reason why less attention has been paid to IEC for peptide separations is probably the need to desalt samples prior to further analysis.

Fig. 13.3. Ideal SEC of protein fragments and of a mixture of synthetic peptide standards. Column, Altex Spherogel TSK G2000SW (300 x 7.5 mm ID, Beckman Instruments); mobile phase, 50 mM KH_2PO_4/0.5 M KCl/8 M urea (pH 6.5); flowrate, 0.2 ml/min; temperature, 26°C. Panel A: Elution profile of horse-heart myoglobin (Mb) and its cyanogen bromide cleavage fragments (I, II, I + II, III). Panel B: Elution profile of horse-heart Mb and synthetic peptide standards 1-5. Panel C: Plot of log MW vs. retention time of Mb, cyanogen bromide fragments of Mb and the five synthetic peptide standards. Peptide Standards 1, 2, 3, 4, and 5 contain 10, 20, 30, 40, and 50 residues, respectively (Fig. 13.2). The arrows denote the elution time for the total permeation volume of the column (A and B). (Reprinted from Ref. 184, with permission.)

References on p. B140

TABLE 13.4

SELECTED APPLICATIONS OF ION-EXCHANGE CHROMATOGRAPHY TO PEPTIDES

Stationary phase	Mode[*] S or NS[**]	Particle (μm)/pore diameter (Å)	Column dimensions (mm x mm ID)	Sample	MW range of peptides (daltons)	Mobile phase	V or NV[***]	Remarks	Ref.
Cation-exchange chromatography									
Syn-Chropak S300	SCX	S	6.5/300	250 x 4.1	Mixtures of standard peptides, including CNBr fragments of rabbit skeletal troponin I	ca. 900-4000	5 mM KH$_2$PO$_4$ to 5 mM KH$_2$PO$_4$/KCl (pH 3.0 and 6.5); 5-10 mM salt/min	NV	Column characterization; compared SCX and SAX; examined non-ideal IEC (pH effect)	215
Syn-Chropak S300	SCX	S	6.5/300	250 x 4.1	CNBr fragments of rabbit skeletal troponin I	1400-4000	See Ref. 215; 5 mM salt/min	NV	Second step in multistep HPLC	199
Syn-Chropak S300	SCX	S	6.5/300	250 x 4.1	Synthetic undecapeptide CEX standards	1100	5 mM KH$_2$PO$_4$ to 5 mM KH$_2$PO$_4$/NaCl (pH 3.0 and 6.5); 5-20 mM salt/min	NV	Development of computer program to predict peptide elution profiles in HPLC	213

Packing	Mode			Column (mm)	Sample	MW range	Conditions	Det.	Comments	Ref.
Syn-Chropak S300	SCX	S	6.5/300	250 x 4.1	Synthetic undecapeptide CEX standards and synthetic peptide polymers	ca. 400-6000	5 mM KH$_2$PO$_4$ to 5 mM KH$_2$PO$_4$/0.5 M NaCl (pH 3.0 and pH 6.5), both buffers containing 0-40% MeCN; 20 mM salt/min	NV	Nonideal vs. ideal IEC (pH and hydrophobicity effects); effect of peptide chain length and charge density on retention behavior	216
PolySE aspartamide	SCX	S	5/300	200 x 4.6						
Mono S HR 5/5	SCX	NS	10/NG§	50 x 5						
SE aspartamide	SCX	S	5/300	200 x 4.6	Mixtures of standard peptides	ca. 500-2000	5 mM phosphate to 5 mM phosphate/0.5 M NaCl (pH 3.0), both buffers containing 25% MeCN; 8.3 mM salt/min	NV	Column characterization; identification of peptide contaminants in solid-phase synthesis	217
PolySE aspartamide	SCX	S	5/300	200 x 4.6	Mixtures of standard peptides	NG	5 mM potassium phosphate to 5 mM potassium phosphate/KCl (pH 3.0), both buffers containing 0-50% MeCN; 5-10 mM salt/min, isocratic elution of dipeptide	NV	Column characterization; mixed-mode effects	218
SE aspartamide	SCX	S	5/300	200 x 4.6	Staphylococcus aureus V8 protease protein digest fragments	ca. 600-6000	5 mM phosphate to 5 mM phosphate/0.5 M NaCl (pH 3.0), both buffers containing 25% MeCN; 6.7-8.3 mM salt/min	NV	Peptide mapping of standard proteins; comparison with RPC	219

(Continued on p. B104)

TABLE 13.4 (continued)

Stationary phase	Mode[*]	S or NS[**]	Particle (μm)/pore diameter (Å)	Column dimensions (mm x mm ID)	Sample	MW range of peptides (daltons)	Mobile phase	V or NV[***]	Remarks	Ref.
Partisil SCX	SCX	S	NG/NG	250 x 4.6	Mixtures of standard peptides	500-6500	5 mM pyridine/0.04 M HOAC; 3.0 M pyridine/0.5 M HOAC; step gradients	V	Final step in multi-step HPLC of artificial mixture of biologically active peptides	194
Mono S HR 5/5	SCX	NS	10/NG	50 x 5	Plasma β-endorphin	ca. 1700	0.05 M NH$_4$OAc to 0.50 M NH$_4$OAc (pH 5.5), both buffers containing 20% MeCN; 5.9% B/min	V	Required MeCN for good separation of opioid peptides	220
Mono S HR 5/5	SCX	NS	10/NG	50 x 5	Guinea pig pancreatic polypeptide	ca. 4000	0.1% aq. TFA to 0.1% aq. TFA/0.5 M NaCl (pH 2.0), both solvents containing 20% MeCN; 10 mM salt/min	NV	Second step in multistep purification of peptide	221
Bio-10 and Bio-20	SCX SCX	NS NS	NG/NG NG/NG	100 x 3 100 x 3	Tryptic peptides and other small peptides	NG	pH and salt step gradients and isocratic separations using various citrate phosphate, borate, formate buffers; pH 1.7-10.0; 23-60°C	NV	Column characterization; IEC of very acids and basic peptides	222

Column	Type	S/NS	Capacity	Dimensions (mm)	Sample	MW	Eluent	V/NV	Comments	Ref.
Protein Pak SP-5PW	SCX	NS	10/1000	75 x 7.5	Proteolytic fragments of bovine neurophysins	NG	0.02 M NH$_4$OAc (pH 4.3) to 0.5 M NH$_4$OAc (pH 7.5) in 30 min	V	Disulfide bond assignment by HPLC, including WAX and RPC	223
Hitachi-Gel 3013C	NG	NS	6/NG	250 x 4.0	Tryptic digests of Bence-Jones proteins	NG	Water to 0.4 M ammonia containing 50% MeCN and 25% 2-propanol (adjusted to pH 6.2 with methanesulfonic acid); 0.5 ml/min; 70°C	V	Peptide mapping on a macroreticular cation-exchange resin; following lyophilization, small amount of residual salt removed by RPC	224
Hitachi-Gel 3013C	NG	NS	6/NG	250 x 4.1	Tryptic digest of lambda light chain of human IgD	NG	Water to 0.4 M ammonia/5 mM acetic acid containing 50% MeCN and 25% 2-propanol (adjusted to pH 6.2 with methanesulfonic acid); 70°C	V	First step in successive IEC; RPC purification of tryptic peptides	225
Syn-Chropak CM300	WCX	S	6.5/300	250 x 4.1	Various standard peptides	ca. 700-2300	50 mM KH$_2$PO$_4$ to 50 mM KH$_2$PO$_4$/1 M KCl (pH 4.5); 8 mM salt/min. Isocratic separations at pH 4.5 and 6.5	NV	Column characterization; isocratic separations of similarly charged peptides	226
Bio-Sil TSK CM-35W	WCX	S	NG/NG	NG	Bovine pituitary peptides	ca. 200-5000	50 mM NH$_4$OAc to 1 M NH$_4$OAc (pH 5.5), step gradient	V	Batch extraction of peptides; WCX followed by WAX column in series	227

(Continued on p. B106)

TABLE 13.4 (continued)

Stationary phase	Mode[*]	S or NS[**]	Particle (μm)/pore diameter (Å)	Column dimensions (mm × mm ID)	Sample	MW range of peptides (daltons)	Mobile phase	V or NV[***]	Remarks	Ref.
NG-carboxylate functionalities	WCX	S	5/NG	250 × 4.6	Angiotensin peptides	ca. 400-1000	10 mM ammonium formate (pH 4.2) containing 10% MeCN to 50 mM ammonium formate (pH 4.2), containing 20% MeCN	V	Satisfactory separation not accomplished without MeCN gradient	228
Ultropac TSK 535 CM	WCX	NG	NG/NG	150 × 7.5	Tryptic digest of aldehyde dehydrogenase	NG	20 mM NH$_4$OAc to 20 mM NH$_4$OAc/0.1 M NaCl (pH 5.0), both buffers containing 8 M urea; ca. 2 mM salt/min	NV	Fragment separations; compares soft gel with HPLC columns	200

Anion-exchange chromatography

Syn-Chropak Q300	SAX	S	6.5/300	250 × 4.1	CNBr fragments of rabbit skeletal troponin I	ca. 2000-4000	5 mM KH$_2$PO$_4$ to 5 mM KH$_2$PO$_4$/1 M KCl (pH 6.5); 5 mM salt/min	NV	Compared SAX with SCX of peptides	215
TSK gel QAE-2SW	SAX	S	NG/NG	250 × 4.6	Tryptic digest of phosphorylated peptide analog of ribosomal protein S6	NG	50 mM NH$_4$HCO$_3$ to 300 mM NH$_4$HCO$_3$ (pH 7.5)	V	Further purification of RPC peptide fractions	229

Mono Q	SAX	NS	NG/NG	50 x 5	Insulin and insulin derivatives	ca. 5000	10 mM Tris-HCl to 10 mM Tris-HCl/0.3 M NaCl (pH 8.6), both buffers containing 7 M urea	NV	Demonstrated rapid fractionation of crystalline insulin	230
Syn-Chropak AX100	WAX	S	5/100	250 x 4	Ribonuclease S-peptide methyl esters	ca. 2000	MeCN to 10 mM NH$_4$OAc (pH 6.0)	V	Triethylammonium acetate (pH 4.4) gave better resolution than NH$_4$OAc; latter preferred for preparative work because it is readily lyophilized	231
TSK IEX-540 DEAE	WAX	S	NG/NG	NG	Tryptic digest of human ceruloplasmin	NG	20 mM Tris/acetic acid (pH 8.0) to 0.4 M NH$_4$OH, methanesulfonic acid, 20 mM Tris/acetic acid (pH 8.0); step gradient	NV	Automated tandem HPLC of peptides (WAX followed by RPC); ca. 260 peaks resolved	232
Ultropac TSK 545 DEAE	WAX	S	NG/NG	150 x 7.5	See Ref. 200	See Ref. 200	10 mM NH$_4$HCO$_3$ to 0.2 M NH$_4$HCO$_3$ (pH 7.5)	V	See Ref. 200	200
Bio-Sil TSK DEAE-3SW	WAX	S	NG/NG	NG	See Ref. 227	See Ref. 227	See Ref. 227	V	See Ref. 227	227

(Continued on p. B108)

TABLE 13.4 (continued)

Stationary phase	Mode*	S or NS**	Particle (µm)/pore diameter (Å)	Column dimensions (mm x mm ID)	Sample	MW range of peptides (daltons)	Mobile phase	V or NV***	Remarks	Ref.
MicroPak AX-10	WAX	S	10/NG	300 x 4	Various peptide mixtures (from dipeptides up)	NG	10 mM triethylammonium acetate (TEAA) (pH 3-5) containing MeCN (isocratic); MeCN to 10 mM TEAA gradients; 40 mM HCOOH (pH 2.6) (isocratic). Various temperatures (30-60°C)	V	Review of many applications on this column, including separations of isomeric dipeptides, acidic peptides, standard peptides and tryptic digests	233
Protein Pak DEAE-5PW	WAX	NS	10/1000	75 x 7.5	See Ref. 223	See Ref. 223	0.02 M NH$_4$OAc (pH 7.5) to 0.5 M NH$_4$OAc (pH 4.3) in 30 min	V	Disulfide bond assignment by HPLC, including SCX, RPC	223

*SCX denotes strong-cation-exchange HPLC; WCX, weak-cation-exchange HPLC; SAX, strong anion-exchange HPLC; WAX, weak-anion-exchange HPLC.

**S and NS denote, respectively, silica- and non-silica-based stationary phase.

***V and NV denote, respectively, volatile and nonvolatile mobile phase.

§ NG denotes not given.

However, RPC and IEC are often complementary, i.e., their combined use can provide optimal separation of a peptide mixture or assess the purity of a peptide preparation. IEC, followed by RPC, allows both rapid desalting of the IEC fractions and peptide separation and is the preferred order for multicolumn separations [167,183,199].

13.2.3.1 Anion-exchange versus cation-exchange chromatography

The ion-exchange properties of a particular packing are determined by the specific ionic groups incorporated into the bonded phase. Weak-cation-exchange functionalities are usually carboxymethyl (CM), and strong-cation-exchange groups can be sulfonyl (S), sulfoethyl (SE), sulfopropyl (SP), or phosphonyl (P) ionic groups. Weak-anion-exchange packings contain diethylaminoethanol (DEAE), ethylamine (EA), or polyethyleneimine (PEI) functional groups, while strong-anion-exchange sorbents are quaternary amines (Q). A common misapprehension is that the term "weak" and "strong" refer to the strength of binding of solutes to an ion-exchange sorbent. In fact, these terms describe the variation in ion-exchange capacity of a packing with mobile-phase pH. Thus, the tertiary amines of the DEAE or PEI weak-anion-exchange sorbents assume a charge according to the mobile-phase pH (the lower the pH, the more highly positively charged the sorbent and the higher the ion-exchange capacity); the overall negative charge on a cation-exchange sorbent containing the weak CM functionality declines with a decrease in pH below a value of ca. 4.0, lowering the ion-exchange capacity of the sorbent. In contrast, due to a relatively permanent positive (anion exchangers; Q) or negative (cation exchangers; S, SP, P) charge, the ion-exchange capacity of strong-ion-exchange sorbents remains high and constant over a broad pH range.

Selected applications of IEC to peptides are presented in Table 13.4. Features common to most of the applications include the particle size of the packings (5-10 μm) and the 300-Å pore size, the latter being a good compromise for IEC of both peptides and proteins [215]. It can be seen that peptide separations have been achieved on all four types of ion exchangers, i.e., by weak- and strong-cation-exchange chromatography (WCX and SCX, respectively) and weak- and strong-anion-exchange chromatography (WAX and SAX, respectively). As a general rule, there is no real advantage in favoring weak-ion-exchange packings over strong-ion-exchange packings for peptide separations. In the case of AEX, if predictable retention behavior of peptides is desired, the constant net positive charge over a broad pH range on a strong-anion-exchange column is an obvious necessity. The utility of strong-cation-exchange sorbents lies in their ability to retain, unlike weak-cation-exchange sorbents, their negatively charged character in the acidic to neutral pH range. In addition, the use of strong cation exchangers at two pH values (for example, pH 3.0 and pH 7.0) can have dramatic effects on the peptide separation, since the net charge is affected by protonation of the sidechain functional groups of aspartic acid and glutamic acid residues at pH 3.0 (thus emphasizing any basic, positively charged character of the peptide) or ionization of these groups at pH 7.0, which decreases the net positive charges on the peptides. The solubility of peptides will also be affected by the pH of the medium. It is quite possible (in fact, likely) that all or most of the

References on p. B140

applications reported for weak-anion-exchange and -cation-exchange packings (Table 13.4) could have been successfully carried out on the corresponding strong-ion-exchange columns. The bottom line here is that if a choice has to be made between purchasing a weak- or a strong-ion-exchange column for peptide applications, opt for the latter. Taking this a step further, if a choice has to be made between buying a strong-cation-exchange column and a strong-anion-exchange column for general peptide applications, purchase the former. Among other things, the use of acidic eluents for SCX (as mentioned above, a major advantage of this type of packing) helps to extend the lifetime of silica-based packings, still the most widely used material for peptide separations. Extensive use of silica-based columns with aqueous solvents (pH 6-8) will accelerate silica dissolution. In addition, SCX is probably the most useful ion-exchange mode for multistep peptide separations.

13.2.3.2 Mobile-phase selection

Elution of solutes from ion-exchange columns is effected by salt counterions displacing similarly charged solute ions from the charged sites on the stationary phase. The retention time of a peptide in anion- or cation-exchange chromatography will depend on a number of factors, including buffer pH and the nature and ionic strength of the anion or cation employed for displacement of acidic or basic peptides, respectively. Gooding et al. [234] reported certain general guidelines concerning the efficacy of salt counterions in displacing proteins from ion-exchange columns, and these guidelines are likely to be also applicable to peptides: thus, divalent ions tend to be stronger displacers than monovalent species and to produce lower retention times; in addition, smaller ions tend to have a higher elution strength than larger ions of the same group. Arranged in order of decreasing solute retention, the reported approximate sequence for cations of commonly used salts is: $K^+ < Na^+ < NH_4^+ < Ca^{2+} < Mg^{2+}$; and for anions: $CH_3COO^- < Cl^- < HPO_4^{2-} \leq SO_4^{2-}$.

For most ion-exchange separations carried out with nonvolatile mobile phases, sodium or potassium ion is used as the cationic counterion and chloride ion as the anionic counterion, and this is reflected in Table 13.4 [199,200,213,215-219,221,226,230]. It has been suggested that continual use of high concentrations of halide ions may have a deleterious effect on stainless-steel surfaces [235], and therefore acetate is occasionally the anionic counterion of choice [235,236]. However, with careful washing of the pump and lines with water following ion-exchange experiments, the constant use of halide ions should cause no problems. In addition, stainless-steel components and lines can be passified periodically with dilute nitric acid. Other anions employed for specific peptide separations have included bicarbonate, borate, and formate [222,237]. Careful control of mobile-phase pH, which may affect the overall net charge on the peptide as well as the ion-exchange capacity of weak-ion exchangers, is effected through the use of Tris [230,232], citrate [222], or – most commonly – phosphate [199,213,215-219,222,226] buffers.

As mentioned previously for SEC of peptides, many researchers choose to take advantage of volatile mobile-phase systems for IEC applications. These systems have generally taken the form of linear or step ionic-strength gradients of ammonium acetate [220,223,227,231] or ammonium bicarbonate [200,229]. Examples of other, lesser used, volatile systems are also included in Table 13.4 [194,224,225,228,233]. It should be noted that the use of pyridine/acetic acid [194,238] or formate [222,228] buffers in the mobile phase precludes the use of spectrophotometry below 250 nm.

A feature of many ion-exchange applications is the addition of an organic modifier, frequently acetonitrile [216-221,224,225,228,231,233], to the mobile phase. The utility of such nonpolar mobile-phase additives for IEC of peptides is discussed in Section 13.2.3.4.

13.2.3.3 Elution mode

Peptides may be removed from an ion-exchange sorbent by either (linear or step) gradient or isocratic elution. Linear gradient elution is generally the elution mode of choice when attempting to resolve mixtures of peptides with a wide range of net charges. A linear NaCl or KCl gradient in 5-50 mM KH_2PO_4 buffer at pH 6.5 (anion- or cation-exchange chromatography) or at pH 3.0 (SCX) constitutes suitable elution conditions for most separations and is certainly a good place to start. Care should be taken in the choice of ionic strength of the starting buffer. If it is too high, weakly acidic or basic peptides, which may otherwise be retained by anion- or cation-exchange sorbents, respectively, may be eluted with unretained compounds. The necessity of further mobile-phase modifications, e.g., the addition of an organic modifier to suppress potential nonspecific hydrophobic interactions, may be gauged following a pilot experiment with peptide standards specifically designed to monitor ion-exchange column performance (Section 13.2.3.5). Flowrates of 0.5-2.0 ml/min are favored for analytical ion-exchange separations. The choice of gradient rate (increasing concentration of counterion per unit time) will be dictated by the complexity and charge distribution of the peptide mixture to be resolved, but an increase of 5-20 mM salt/min is suitable for most analytical purposes. As a general rule, the optimum separation of two peptides by gradient elution on cation- or anion-exchange columns will be obtained when there is a net charge difference of at least 1 unit between the two peptides. However, the relative polypeptide chainlengths of two peptides of the same net charge can also have a profound effect on their resolution [213,216] (Section 13.2.5).

Linear ionic-strength gradients of buffers (e.g., 0.05-0.5 M phosphate buffer) are occasionally employed for peptide separations [226,230,235,239], although these types of gradients are more common in volatile mobile-phase systems [194,200,220,223,227-229,231].

A somewhat different form of gradient elution has been described by Dizdaroglu [233] in some interesting separations of acidic, neutral, and mildly basic peptides on a weak-anion-exchange column (MicroPak AX-10). Its stationary phase carries the difunctional group $-CH_2-CH_2-CH_2-\overset{+}{N}H_2-CH_2-CH_2-\overset{+}{N}H_3$. The mixed ionic/hydrophobic properties of this functional ligand enables peptides to be separated by taking advantage of both their charge and hydrophobic characteristics, i.e., by a multimodal mechanism. Dizdaroglu employed mixtures of triethylammonium acetate (TEAA) and acetonitrile as eluents, acetonitrile promoting ionic interaction of peptides with the anion exchanger. Peptides were eluted with increasing concentration of TEAA in the mobile phase. Ref. 233 is a review of many applications of this weak-anion-exchange column, including separations of isomeric dipeptides, acidic peptides, standard peptides, and tryptic digests (Table 13.4).

Although less common, a pH gradient instead of the usual salt gradient may be used for the separation of peptides [222,233,230]. When the pH is linearly increased or decreased, the overall charge on a peptide will change as ionizable groups are protonated or deprotonated. This leads to differential elution of peptides, depending on their overall acidic or basic residue composition. Obviously, the effect of pH changes on the differential ionization of stationary-phase functional groups must be considered when employing weak-ion-exchange columns [223]. In addition, instances have been reported [215,216] where the ion-exchange capacity of a strong-cation-exchange column decreased with decreasing pH due to progressive ionization of unwanted functional groups on the support. The ensuing increasing positive charge on the support served to reduce the overall net negative charge (due to sulfonate functional groups) of the stationary phase.

Although not employed as extensively as linear gradient elution, step gradient elution has been the method of choice for specific peptide applications [194,222,227,232].

Optimal separation conditions for isocratic elution of peptides are generally more difficult to develop than those for gradient elution. An interesting example of isocratic peptide elution from a weak-cation-exchange column was reported by Cachia et al. [226], who were able to use this elution mode to separate peptides of similar net charge, which therefore are poorly resolved by gradient elution. These researchers employed isocratic elution at pH 6.5 (50 mM KH_2PO_4 buffer, containing 0.3 or 0.4 M KCl) to resolve almost completely two basic peptides of similar size and identical net charge (+6) on a SynChropak CM 300 column. The TEAA/acetonitrile systems of Dizdaroglu [233] also produced several good isocratic separations of dipeptides and other larger, closely related peptides on the MicroPak AX-10 weak-anion exchanger. Other examples of isocratic peptide elution are given in Refs. 218 and 222.

13.2.3.4 Nonspecific hydrophobic interactions in ion-exchange chromatography

Although, as the name implies, the separation mechanism of IEC is electrostatic in nature, ion-exchange packings may also often exhibit significant hydrophobic characteristics, giving rise to mixed-mode contributions to solute separations [240,241]. As pointed out by Rounds et al. [242], a small amount of hydrophobic character in an ion exchanger is not necessarily detrimental to solute separations and may even enhance resolution by mixed-mode effects. However, when only the predominant, i.e. ionic, stationary-phase/solute interaction is required (ideal ion-exchange behavior), the mobile phase must be manipulated so as to minimize nonspecific interactions, e.g., by the addition of nonpolar organic solvents, such as 2-propanol or acetonitrile, to the mobile-phase buffers to suppress hydrophobic interactions between the solute and the ion-exchange packing.

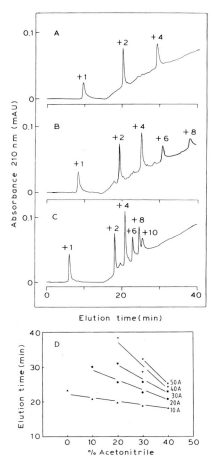

Fig. 13.4. Strong-cation-exchange chromatography of synthetic peptide polymers. Column, SynChropak S-300 (250 x 4.1 mm ID, 6.5-μm particle size, 300-Å pore size, SynChrom Inc.). (A) to (C) effect of nonspecific, hydrophobic interactions of peptide elution profiles; conditions, linear AB gradient (20 mM salt/min following 10-min isocratic elution with Buffer A). Buffer A = 5 mM KH$_2$PO$_4$ (pH 6.5), Buffer B = Buffer A + 0.5 M NaCl; both buffers contain (A) 10, (B) 20, or (C) 40% acetonitrile (v/v); flowrate, 1 ml/min; temperature, 26°C. (D) Plot of peptide polymer elution time vs. percentage of acetonitrile in mobile-phase buffers. The sequence of the peptides was Ac-(L-G-L-K-A)$_n$-amide, where n = 1, 2, 4, 6, 8, 10 (+1, +2, +4, +6, +8, +10 net charge, respectively; peptides 5A, 10A, 20A, 30A, 40A, 50A, respectively). Peptide 5A was not eluted by the salt gradient, being eluted during the initial isocratic elution with Buffer A, and is therefore not included in the plot in (D). The 1- and 3-letter codes for the amino acids are shown in Table 13.1.

The hydrophobic character of high-performance ion-exchange packings has long been recognized. However, only recently has a systematic approach to examining the effect and magnitude of the hydrophobicity of these packings in IEC of peptides been reported. Burke et al. [216] synthesized series of basic peptide polymers (5-50 residues) of varying hydrophobicity and subjected them to SCX on both silica-based (SynChropak S-300,

PolySulfoethyl A) and non-silica-based (Mono S HR 5/5) columns. The sequence of one of these polymer series was Ac-(Leu-Gly-Leu-Lys-Ala)$_n$-amide, where $n = 1, 2, 4, 6, 8, 10$ (5 to 50 residues, $+1$ to $+10$ net change). Both mobile-phase buffers [Buffer A $= 5$ mM KH$_2$PO$_4$ (pH 6.5), Buffer B $=$ Buffer A $+ 0.5$ M NaCl] contained 0, 10, 20, 30, or 40% acetonitrile (v/v). Fig. 13.4 shows elution profiles of the peptide polymers on the S-300 column in the presence of 10, 20, or 40% acetonitrile (Fig. 13.4, Panels A-C, respectively) in the mobile-phase buffers. Fig. 13.4D shows a graphic representation of the effect of acetonitrile concentration on peptide retention time. In the absence of acetonitrile, only the 5- and 10-residue peptides ($+1$ and $+2$ net charge, respectively) were eluted, with reasonable peak shape, by a sodium chloride gradient up to a concentration of 0.5 M. The 20-residue peptide ($+4$ net charge) appeared as a late-eluted, very broad, badly skewed peak. The more hydrophobic 30-, 40-, and 50-residue peptides ($+6$, $+8$, and $+10$, respectively) were not eluted by 0.5 M sodium chloride. Mant and Hodges [215] had previously shown that a mixture of peptides of average hydrophobicity and a range of net charges from $+2$ to $+8$ at pH 6.5 was easily removed from the S-300 column in the absence of an organic solvent by a salt gradient up to a concentration of only 0.4 M. Thus, the observed results for the peptide polymer series suggested that, in addition to ionic-solute/packing interactions, nonspecific hydrophobic interactions were also affecting the retention behavior of the peptides. With the addition of 10% acetonitrile to the mobile phase (Fig. 13.4A), designed to help overcome any hydrophobic (as opposed to ionic) interactions, the 5-, 10-, and 20-residue peptides ($+1$, $+2$, and $+4$ net charge, respectively) were now eluted with good peak shapes within a concentration range of 0-0.3 M NaCl. As the concentration of acetonitrile was increased further to 20% (Fig. 13.4B), 30%, and 40% (Fig. 13.4C) acetonitrile, the more hydrophobic 30-, 40-, and 50-residue peptides ($+6$, $+8$, and $+10$ net charge, respectively) were also eluted from the column. In addition, the retention times of all peptides decreased with increasing levels of acetonitrile (Fig. 13.4A-D). Similar results were obtained with the PolySulfoethyl A and Mono S columns.

Refs. 217, 218, 220, and 228 (Table 13.4) show how the presence of acetonitrile in the mobile phase enhanced the efficiency of peptide separations by IEC; these observations were said to suggest the possibility of mixed-mode (hydrophobic and ionic interactions) characteristics. Although the addition of a nonpolar organic solvent, generally acetonitrile, is featured in a considerable number of papers on IEC of peptides, there is a consistent failure on the part of many of these articles to justify the presence of these solvents. Presumably, these solvents were employed for reasons similar to those stated above.

The work of Burke et al. [216] and others [217,218,220,228,243] suggests that the addition on a regular basis of a low level of acetonitrile (e.g., 10-20% v/v) to mobile-phase buffers in IEC of peptides is worthwhile to suppress any hydrophobic interactions with the ion-exchange packing and thereby to ensure efficient elution of all peptides in a mixture. However, it is also worthwhile to keep in mind a cautionary note by Nowlan and Gooding [244] that the addition of organic solvents, such as methanol, acetonitrile, or 2-propanol to the mobile phase should be undertaken judiciously, because they may adversely affect the biological activity of proteins or cause salt precipitation.

13.2.3.5 Peptide standards

The value of peptide standards in IEC lies in allowing a quick comparison of the performance characteristics of different ion-exchange columns or assessment of the performance characteristics of a specific cation-exchange column:

(a) Standards will confirm that the column can, indeed, retain charged species (the lower the net charge on the peptide that can be retained, the more versatile the column). In addition, a mixture of standards with a range of net charges will reveal whether the column retains charged species systematically, i.e. peptide retention time generally increases with increasing net charge.

(b) Standards can confirm that a particular mobile phase will elute charged species from an ion-exchange column.

(c) Standards can be used to assess the effect of pH variations on the resolving capability and load capacity of an ion-exchange column. This is particularly important in SCX, where the manipulation of mobile phases throughout the acidic to neutral pH range is frequently employed for peptide separations.

(d) Standards can be used to determine the presence and extent of any nonspecific, i.e. hydrophobic, characteristics of a column packing.

A series of four synthetic undecapeptide standards has recently been developed to meet the above requirements for assessing cation-exchange columns [212,216]. The sequences of the four standards are

(C1) Ac-Gly-Gly-Gly-Leu-Gly-Gly-Ala-Gly-Gly-Leu-Lys-amide,

(C2) Ac-Lys-Tyr-Gly-Leu-Gly-Gly-Ala-Gly-Gly-Leu-Lys-amide,

(C3) Ac-Gly-Gly-Ala-Leu-Lys-Ala-Leu-Lys-Gly-Leu-Lys-amide,

(C4) Ac-Lys-Tyr-Ala-Leu-Lys-Ala-Leu-Lys-Gly-Leu-Lys-amide.

Peptides C1-C4 contain, respectively, 1-4 basic residues (lysine residues) and no acidic residues. Thus, over the pH range used for the majority of cation-exchange separations [pH 3.0 to 7.0 (SCX) or pH 4.5 to 7.0 (WCX)], the net charges of $+1$ to $+4$ for peptides C1-C4, respectively, do not change. The hydrophobicity of the standards increases from C1 to C4, with a concomitant increase in peptide sensitivity to potential hydrophobic-solute/packing interactions. Two examples of the use of these standards in detecting nonideal column behavior are described below.

Fig. 13.5 shows the gradient elution chromatograms at pH 6.5 (top) and pH 3.0 (bottom) of Peptide Standards C1-C4 on a silica-based strong-cation-exchange column. There was an initial 5-min isocratic elution with the starting buffer prior to the start of the gradient. At pH 6.5 (Fig. 13.5, top), Standards C2, C3, and C4 ($+2$, $+3$, and $+4$ net charge, respectively) were removed by gradient elution, while Peptide C1 ($+1$ net charge) was eluted during the initial isocratic elution. In contrast, at pH 3.0 (Fig. 13.5, bottom), only Peptides C3 and C4 were eluted by the gradient, while both Peptides C1 and C2 were eluted during the initial isocratic elution. In addition, the retention times of Peptides C3 and C4 were reduced considerably. Ideally, there should have been no change in elution time of these peptides with a change in buffer pH. The observed effects apparently

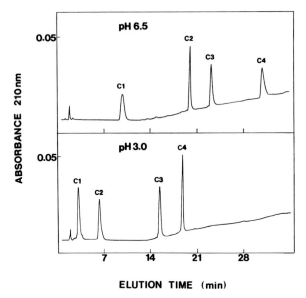

Fig. 13.5. Strong-cation-exchange chromatography of synthetic peptide standards. Conditions: linear AB gradient (20 mM salt/min), following 5-min isocratic elution with Buffer A. Buffer A = 5 mM KH$_2$PO$_4$ [pH 6.5 (top) or pH 3.0 (bottom)], Buffer B = Buffer A + 1 M NaCl; flowrate, 1 ml/min; temperature, 26°C. The peptide standards have the sequences (C1) Ac-G-G-G-L-G-G-A-G-G-L-K-amide, (C2) Ac-K-Y-G-L-G-G-A-G-G-L-K-amide, (C3) Ac-G-G-A-L-K-A-L-K-G-L-K-amide, and (C4) Ac-K-Y-A-L-K-A-L-K-G-L-K-amide. Peptide standards C1, C2, C3, and C4 contain net charges of +1, +2, +3, and +4, respectively. The standards were obtained from Synthetic Peptides Inc. The 1- and 3-letter codes for the amino acids are shown in Table 13.1.

resulted from a reduction in the capacity of the column to retain charged species as the eluent became more acidic.

Fig. 13.6 shows elution patterns of the peptide standards obtained on a second silica-based strong-cation-exchange column at pH 6.5. The mobile-phase buffers were identical to those shown in Fig. 13.5, except for the addition of 5% (top) or 10% (bottom) acetonitrile (v/v) to both mobile-phase buffers. The inclusion of acetonitrile in the mobile phase was found to be necessary for the elution of the four peptides within a reasonable time and with acceptable peak shape. In the absence of acetonitrile, neither Peptide C3 nor C4 (+3 and +4 net charge, respectively) were eluted from the column, while C2 (+2 net charge) was eluted as a broad, skewed peak. The improvement in the chromatogram of the four standards on addition of 5% acetonitrile (Fig. 13.6, top) and the further improvement by a level of 10% acetonitrile (Fig. 13.6, bottom) indicated that, in its absence, the column packing did exhibit hydrophobic, in addition to ionic, characteristics.

From results such as those described above (Figs. 13.5 and 13.6), the experimenter can decide on the column best suited to his needs and/or the most appropriate mobile-phase conditions for the separation of a particular peptide mixture. Results such as these

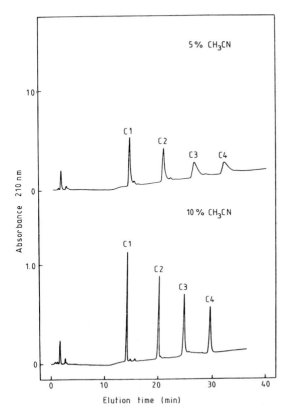

Fig. 13.6. Strong-cation-exchange chromatography of synthetic peptide standards. Conditions: linear AB gradient (20 mM salt/min, following 5-min isocratic elution with Buffer A). Buffer A = 5 mM KH$_2$PO$_4$ (pH 6.5), Buffer B = Buffer A + 1 M NaCl. Both buffers contain 5% (top) or 10% (bottom) acetonitrile (v/v); flowrate, 1 ml/min; temperature, 26°C. The sequences of Peptide Standards C1, C2, C3, and C4 (+1, +2, +3, and +4, net charge, respectively) are shown in the legend to Fig. 13.5.

also highlight the need for a similar set of standards to assess the suitability of anion-exchange columns for peptide applications.

13.2.3.6 Effect of polypeptide chainlength and charge density on retention behavior

In addition to overall net charge, Burke et al. [216] demonstrated that other factors that affect the retention behavior of peptides in IEC include polypeptide chainlength and charge density. This is illustrated in Figs. 13.7A and 13.7B, which show chromatograms of, respectively, a mixture of five synthetic peptide size-exclusion standards (10, 20, 30, 40, and 50 residues; +1, +2, +3, +4, and +5 net charge, respectively) and a mixture of four synthetic peptide cation-exchange standards (11 residues; +1, +2, +3, and +4 net charge) on a strong-cation-exchange column. Mobile-phase conditions were designed to

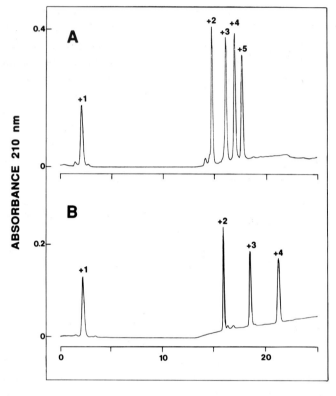

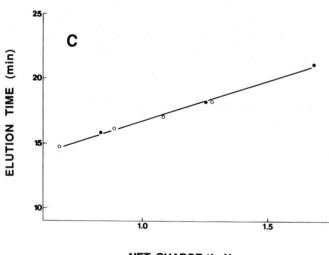

ensure the suppression of any nonideal behavior [216]. From Fig. 13.7, it is clear that similarly charged species are not necessarily eluted at similar times. For instance, the 50-residue peptide (+5 net charge; Fig. 13.7A) was not retained as long as 11-residue peptides with net charges of +3 or +4 (Fig. 13.7B). Similarly, the 40-residue peptide (+4 net charge; Fig. 13.7A) was eluted prior to an 11-residue peptide with a net charge of +3 (Fig. 13.7B). The two series of peptide standards differed significantly in their range of both peptide chainlength (10-50 residues in Fig. 13.7A; 11 residues in Fig. 13.7B) and charge density (+1 net charge per 10 residues for the peptides in Fig. 13.7A; +1 to +4 net charge per 11 residues for the peptides in Fig. 13.7B). Burke et al. [216] rationalized results such as these by demonstrating a linear relationship between peptide elution time and net charge/ln N (where N is the number of residues the peptide contains), such as that shown in Fig. 13.7C.

Linearization of peptide retention behavior in IEC is useful in correlating the overall net charge on a peptide at a given pH with its amino acid composition. In addition, this linearization has proved vital during development of an expert system [213], designed to simulate peptide elution patterns in HPLC (Section 13.2.5.2).

13.2.4 Reversed-phase chromatography

The excellent resolving power of RPC has made it the predominant HPLC technique for peptide separations. Its ability to separate peptides of closely related structures makes it extremely useful for high-speed, analytical, and preparative applications. RPC is not only usually superior to other modes of HPLC with respect to both speed and efficiency, but it also offers the widest scope for manipulation of mobile-phase characteristics to improve peptide separations.

Favored RPC sorbents for the vast majority of peptide separations continue to be silica-based packings containing n-alkyl (e.g., octyl, octadecyl) functionalities. Of several hundred published papers on RPC of peptides collected for this review, ca. 80% reported using C_{18} (octadecyl) columns, ca. 10% C_8 (octyl) columns, and ca. 5% C_4 (butyl) columns. Other packings included silica-based materials containing C_3 (propyl) or phenyl functional groups. Dimensions of analytical columns generally range from 100 to 250 x 4 to 4.6 mm ID. Although non-silica (usually polystyrene/divinylbenzene)-based supports are available, their value remains largely untested (with some exceptions [245-248]) in peptide and protein applications. Similar to IEC columns, analytical RPC columns used in

Fig. 13.7. Strong-cation-exchange chromatography of synthetic peptides. Column, Mono S HR 5/5 (50 x 5 mm ID, 10-μm particle size; Pharmacia). Conditions, linear AB gradient (20 mM salt/min, following 10-min isocratic elution with Buffer A). Buffer A = 5 mM KH$_2$PO$_4$ (pH 6.5), Buffer B = Buffer A plus 0.5 M NaCl. Both buffers contain 40% acetonitrile (v/v); flowrate, 1 ml/min; temperature, 26°C. Panel A: Mixture of five synthetic size-exclusion standards (10-50 residues; +1 to +5 net charge, respectively). The sequences of the peptides are shown in the legend to Fig. 13.2. Panel B: Mixture of four synthetic undecapeptide cation-exchange standards (+1 to +4 net charge). Sequences of the peptides are shown in the legend to Fig. 13.5. Panel C: plot of observed peptide elution time vs. peptide net charge, divided by the logarithm of the number of residues (net charge/ln N).

References on p. B140

peptide separations commonly have a particle size of 5-10 μm and a pore size of 300 Å. Matrices with 80- to 100-Å pore size do not provide optimum resolution and recovery of peptides larger than 30 residues [206]. In general, for peptides and proteins (30-150 residues), the 300-Å pore size matrices gives better resolution and recovery [190,249,250]. Although not yet in extensive use, the nonporous packings (both silica- and non-silica-based) described by Horváth [248,251-253] show promise for specific applications in RPC of peptides and proteins.

13.2.4.1 Mobile-phase selection

Reversed-phase silica-based columns may contain surface silanols which act as weak acids and are ionized above pH 3.5. The weak acids may interact with the basic residues of peptides chromatographed on reversed-phase columns and have an adverse effect on resolution, characteristically producing long retention times and peak broadening. Although excellent resolution of peptide mixtures may be obtained with acidic or neutral eluents, the majority of researches have been carried out by RPC at pH values < 3.0. Apart from the suppression of silanol ionization under these conditions, silica-based sorbents are more stable at low pH. Whatever the pH employed, suppression of nonideal (ionic) interactions with the hydrophobic stationary phase is always an important concern.

13.2.4.1.1 Organic modifiers

Acetonitrile is the favored organic solvent for the great majority of peptide separations in RPC. This is highlighted by the fact that, in 90-95% of all published methods for RPC of peptides, acetonitrile was used as the organic modifier. This is due to several advantageous characteristics of this solvent, including low viscosity, relatively high volatility, low UV transparency, and a high degree of selectivity. Other organic solvents have also proved useful for specific applications. Wilson et al. [254] noted that the hydrophobic strength of organic solvents in the mobile phase decreases in the order 1-propanol $\cong$ 2-propanol > acetonitrile $\gg$ methanol. Thus, 2-propanol (or 1-propanol), a less polar solvent than acetonitrile, has proved useful for the separation of hydrophobic peptides [201,225,254-263], whereas methanol, a more polar solvent than acetonitrile, may be required for separation of hydrophilic peptides [207,254,256,259-261,263-274]. Ethanol has also found occasional use as an organic modifier in RPC of peptides [275,276]. In addition, mixtures of these organic solvents (usually two) have been used, for both isocratic [277,278] and gradient [203,256,264,279] elution. Better peptide separations can occasionally be obtained with a mixture of organic solvents rather than with a single organic modifier. The high viscosity of 2-propanol, in particular, may produce undesirably high column backpressures, unless flowrates are reduced relative to those typically used in analytical RPC with acetonitrile (1-2 ml/min). Alternatively, the temperature can be raised, although this can cause deterioration of column packing or acceleration of column aging [280].

13.2.4.1.2 Ion-pairing reagents

Most of the flexibility and resolving power of RPC is derived from the use of ion-pairing reagents. Peptides are charged molecules at most pH values, and the presence of different counterions may have a profound influence on their chromatographic behavior. Differences in the polarities of peptides in a mixture can be maximized through careful choice of ion-pairing reagents. Favored models for the mechanism of ion-pair separations involve either formation of ion pairs with the solutes, followed by their retention on a reversed-phase sorbent [281,282], or a dynamic ion-exchange event in which the ion-pairing reagent is first retained by the reversed-phase column and then solute molecules exchange ions with the counterions that are associated with the sorbed ion-pair reagent [283-286]. Both models lead to similar predictions concerning separation as a function of experimental conditions. Whatever the mechanism, the resolving power of ion-pairing reagents is solely due to their interaction with the ionized groups of peptides [287]: anionic counterions will interact with the protonated basic residues (arginine, lysine, histidine) of a peptide and with protonated free N-terminal amino groups; cationic counterions will show an affinity for ionized acidic residues (glutamic and aspartic acid) and for ionized free C-terminal carboxyl groups.

The effect of a particular ion-pairing reagent on the retention time of a peptide depends upon the hydrophobicity of the ion-pairing reagent. The homologous series of alkylsulfonates (e.g., butanesulfonate, hexanesulfonate) [246,252,262,268,275,276,288-290] and volatile perfluorinated carboxylic acids [e.g., trifluoroacetic acid (TFA) and heptafluorobutyric acid (HFBA)] [291-293] offer a range of anionic ion-pairing reagents with graded hydrophobicity. Quaternary ammonium ions are useful as cationic ion-pairing reagents [246,274,294-296], particularly tetrabutylammonium phosphate, which is a strongly hydrophobic cationic counterion. Although tertiary alkylamines have been employed as cationic ion-pairing reagents [295], the main value of, e.g., triethylamine lies in its use as a major buffer component (Section 13.2.4.1.4) rather than merely an additive.

Both anionic and cationic ion-pairing reagents are generally used in the 2- to 20-mM concentration range. It is important to be consistent in the level of counterion used in the mobile-phase solvents. There is a concentration dependence effect on peptide retention [287,292,297,298], and peptide elution patterns can vary quite markedly over a narrow range of reagent concentrations. This effect will vary with increasing numbers of positively charged or negatively charged residues in the case of anionic and cationic counterions, respectively. Varying the concentration of a particular ion-pairing reagent, instead of changing the reagent, can occasionally be of use in improving the resolution of peptide mixtures [287].

13.2.4.1.3 Aqueous trifluoroacetic acid/acetonitrile mobile phases

Most experiments on ion-pair RPC for peptide separations take advantage of the excellent resolving power and selectivity of the anionic ion-pairing properties of TFA, particularly in the use of TFA/water to TFA/acetonitrile gradients. The low pH (ca. 2.0) of this solution (0.05%-0.1% TFA, v/v) ensures protonation of hydrophilic carboxylic acid groups, which increases the hydrophobicity of the peptides, thereby increasing the

TABLE 13.5

SELECTED APPLICATIONS OF REVERSED-PHASE CHROMATOGRAPHY TO PEPTIDES

	Stationary phase [*]	Particle (μm)/pore diameter (Å)	Column dimensions (mm × mm ID)	Sample	MW range of peptides (daltons)	Mobile phase [**],[***]	V or NV[§]	Remarks	Ref.
(i)	SynChropak RP-P C$_{18}$	10/300	50, 100 and 250 × 4.6	Cyanogen bromide fragments of human hemoglobin	NG[§§]	0.1% TFA, PFPA or HFBA/1-PrOH; 0.1% TFA/2-PrOH or EtOH; 1%/min	V	Varied hydrophobicity of anionic IP reagent to resolve peptides	257
(ii)	SynChropak RP-P C$_{18}$ Aquapore RP-300 C$_8$	6.5/300 7/300	250 × 4.1 220 × 4.6	Synthetic peptide mixtures	ca. 500-1500	0.1% H$_3$PO$_4$, TFA or HFBA/MeCN; 0.02%-0.8% TFA/MeCN; 1%/min	V and NV	Varied hydrophobicity or concentration of anionic IP reagent to resolve peptides; predicted effects of varying counterion	287
(iii)	μBondapak C$_{18}$	NG/NG	NG	Major peptide components of rat neurointermediary lobe	ca. 900-4500	0.1% TFA or 0.13% HFBA/MeCN; 0.33%/min	V	Derived amino acid residue retention coefficients in systems containing anionic IP reagents of different hydrophobicities	321

	Column	Particle size/porosity	Dimensions (mm)	Sample	M_r range	Mobile phase	Mode	Remarks	Ref.
(iv)	μBondapak C$_{18}$	NG/NG	NG	Rat pituitary peptides	NG	0.1 M H$_3$PO$_4$ (pH 1.0), 0.01 M TEAF (pH 3.0), 0.01 M TEAA (pH 4.5), 0.01 M TEAA (pH 5.5), 0.01 M TEAP (pH 7.0), 0.1% TFA or 0.13% HFBA/MeCN; 0.3%-0.7%/min	V and NV	Improved resolution of peptide mixtures by manipulating pH and nature of IP reagent	294
(v)	Spherisorb C$_6$, C$_8$ and C$_{18}$	5/NG	150 × 4.6	Substance P and fragments	ca. 850-1250	0.1 M NaH$_2$PO$_4$ (pH 3.0) + 0.01 M-0.03 M ESA, PrSA, BuSA or PSA/EtOH or MeCN; isocratic	NV	Examined effect of increasing hydrophobicity of anionic ion-pairing reagents (alkane sulfonates) on peptide retention behavior	276
(vi)	Micropellicular C$_8$ and C$_{18}$	2/non-porous	30 × 4.6	Tryptic, chymotryptic and cyanogen bromide fragments of proteins	NG	0.1% TFA/MeCN; 50 mM NaH$_2$PO$_4$ (pH 2.8) + 5 mM HSA or 1 mM HSA or 1 mM OSA; 0.1% TFA + 5 mM HSA; rapid gradient-rates and flowrates; high temperatures	V and NV	Development of RPC techniques for rapid peptide mapping using anionic IP reagents	252
(vii)	Zorbax ODS C$_{18}$	NG/NG	250 × 4.6	Protected synthetic peptides	ca. 400-3000	0.005 M TFABr, 0.005 M TBAHSO$_4$, 0.01 M TBABr or 0.01 M TBAHSO$_4$/MeOH or MeCN; isocratic	NV	Examined effect of type and concentration of hydrophobic cationic IP reagents (tetrabutylammonium salts) on peptide retention behavior	296

(Continued on p. B124)

TABLE 13.5 (continued)

Stationary phase*	Particle (µm)/pore diameter (Å)	Column dimensions (mm x mm ID)	Sample	MW range of peptides (daltons)	Mobile phase**, ***	V or NV§	Remarks	Ref.
(viii) µBondapak C_{18} Partisil ODS-3 C_{18} MicroPak MCH-5 Lichrosorb RP-8 C_8 Vydac 218TP	10/NG 10/NG 5/NG 7/NG 10/300	NG NG 150 x NG NG NG	Peptide hormones	ca. 500-2000	0.1% TFA/MeOH and (or) MeCN; binary and ternary solvent systems; linear gradients	V	Comparison of MeOH and MeCN as organic modifiers; demonstrated how ternary (as opposed to binary) solvent systems may improve peptide resolution	264
(ix) Altex Ultrapore RPSC C_3	5/300	75 x 4.6	Hydrophobic synthetic peptide polymers	1600-3400	0.1% TFA/MeCN or 2-PrOH; 1.0%-2.0%/min	V	2-PrOH preferred organic modifier for hydrophobic peptides	201
(x) µBondapak C_{18}	NG/NG	30 x 3.9	Chymotryptic digests of mutant and wild-type cytochrome c	ca. 200-1600	0.1% TFA or 100 mM K_2HPO_4 (pH 7.2)/MeCN; ca. 0.2%-0.6%/min	V and NV	Peptide mapping by RPC at different pH values	362
(xi) Brownlee RP-300 C_8	7/300	220 x 4.6	Endo- and exo-protease fragments of proteins	NG	0.01 M NH$_4$OAc (pH 6.04) or 0.1% TFA/MeCN; 0.67%-0.8%/min	V	Peptide mapping by RPC at different pH values	335

No.	Column	Particle size (μm) / Pore size	Dimensions (mm)	Sample	MW range	Mobile phase / conditions	Detection	Remarks	Ref.
(xii)	Lichrosorb RP-18 C_{18}	7/NG	250 × 4.0	Synthetic peptides and isopeptides	ca. 200–700	0.1% H_3PO_4 (± 0.1 M $NaClO_4$; pH 2.2), 0.05% TFA (pH 2.3), 0.01 M NH_4OAc (pH 5.7) or 0.005 M NaH_2PO_4 (pH 7.4)/MeCN; 1.5%/min	V and NV	Separation of closely related peptides by RPC at different pH values	315
(xiii)	Hypersil ODS C_{18} / Merck C_{18}	5/NG, 10/NG	NG, NG	Tryptic peptides of ribosomal proteins from E. coli	ca. 200–2500	0.5% TFA (pH 2.0), NH_4F (pH 4.4) or NH_4F (pH 7.8)/MeCN or MeOH; 0.4%–2.0%/min	V	Separation of peptides by RPC at different pH values; some peptides insoluble at low pH	340
(xiv)	μBondapak C_{18}	NG/NG	30 × 3.9	Tryptic digest of bovine β-casein	ca. 200–6000	10 mM K phosphate (pH 6.5)/MeCN; 1.0%–2.0%/min	NV	pH values above 6.0 required due to precipitation of some peptides at lower pH	367
(xv)	LiChrosorb RP-18 C_{18}	5/NG	150 × 4.0	Angiotensins and other peptides	ca. 300–1250	0.05 M phosphate (pH 8.5)/MeCN or MeOH; isocratic	NV	Benzoin-derivatized peptides fluoresced most intensely in weakly alkaline medium	365
(xvi)	LiChrosorb RP-18 C_{18}	NG/NG	250 × 4.0	Immunostimulant peptide and diastereomeric impurity	ca. 400	0.015 M Na_2HPO_4 (pH 9.0); isocratic	NV	Separation of two diastereomers due to deprotonation of α-NH_2 group at pH 9.0	370

(Continued on p. B126)

References on p. B140

TABLE 13.5 (continued)

Stationary phase*	Particle (µm)/pore diameter (Å)	Column dimensions (mm x mm ID)	Sample	MW range of peptides (daltons)	Mobile phase**, ***	V or NV§	Remarks	Ref.
(xvii) SynChropak RP-4 C$_4$	6.5/300	250 x 4.1	Synthetic octapeptide analogs and synthetic peptide standards	ca. 850-1200	0.1% TFA/MeCN; 10 mM (NH$_4$)$_2$HPO$_4$ (pH 7.0) + 0.1 M NaClO$_4$/MeCN; 1%/min	V and NV	Derivation of amino acid residue retention coefficients in acidic and neutral media	261
SynChropak RP-8 C$_8$	6.5/300	250 x 4.1						
SynChropak RP-P C$_{18}$	6.5/300	250 x 4.1						
SynChropak RP-P C$_{18}$	6.5/300	50 x 4.1						
Ultrapore RPSC C$_3$	5/300	75 x 4.6						
Partisil 5 C$_8$	5/60	250 x 4.6						
(xviii) Hamilton PRP-1 (non-silica)	10/NG	150 x 4.1	Tryptic digests of proteins plus various other peptides	ca. 200-4000	0.1% TFA or 5 mM NH$_4$HCO$_3$ (pH 8.0, 9.6 or 11.0)/MeCN; 2%/min	V	Characterized non-silica-based column for peptide separations over wide pH range; derived amino acid residue retention coefficients at pH 2.0 and 8.0	247

	Packing	Particle/pore size	Column (mm)	Sample	MW	Mobile phase	V/NV	Comments	Ref.
(xix)	Polystyrene microspheres (non-silica)	3/non-porous	30 × 4.6	Tryptic digest of human growth hormone	NG	0.2% TFA (pH 2.0) or 3 mM Na$_3$PO$_4$ + 0.5 mM decyltrimethylammonium bromide (pH 11.0); rapid linear gradients and flowrates; 80°C	V and NV	Rapid peptide mapping in acidic and highly alkaline medium	248
(xx)	C$_{18}$ C$_8$ C$_8$	5/300 7/300 6.5/300	250 × 4.6 220 × 4.6 75 × 4.6	Synthetic peptide standards	ca. 1200	0.05% TFA (pH 2.0), 10 mM TEAP (pH 4.5) or 10 mM (NH$_4$)$_2$HPO$_4$ (pH 7.0) + NaClO$_4$/MeCN; mixed MeCN (1%/min) and salt gradients	V and NV	Development of procedure for monitoring free silanols on silica-based packings	347
(xxi)	Supelcosil LC-8-DB C$_8$ LiChrosorb RP-8 C$_8$	NG/100 NG/NG	250 × 4.6 250 × 4.6	Opioid peptides	ca. 500-1450	0.1% TFA, 50 mM TEAP (pH 3.0) or 50 mM NH$_4$OAc (pH 5.0)/MeCN; 0.25%-0.5%/min	V and NV	Demonstrated best separation in TEAP system; silanol interactions observed in TFA system were blocked in TEAP system	334
(xxii)	μBondapak C$_{18}$ Nucleosil C$_{18}$	10/NG 7/NG	300 × 4.0 300 × 4.0	Neuropeptides	NG	TFA (pH 2.1) or 10 mM NH$_4$TFA (pH 2.0)/MeCN; 10 mM NH$_4$OAc (pH 4.5 or 5.4)/MeOH; 10 mM TEAP (pH 3.0)/2-PrOH; 0.3%-1.4%/min	V and NV	Mobile phases of different selectivities enabled high resolution of a variety of peptides	259

(Continued on p. B128)

TABLE 13.5 (continued)

Stationary phase [*]	Particle (μm)/pore diameter (Å)	Column dimensions (mm x mm ID)	Sample	MW range of peptides (daltons)	Mobile phase [**], [***]	V or NV[§]	Remarks	Ref.
(xxiii) Several stationary phases, mainly C_4, C_8, C_{18}	3 to 10/NG	50 to 250 x 3.0 to 5.0	Insulin and insulin derivatives	ca. 5800	0.1% TFA, 0.25 M TEAP (pH 3.0) or 0.25 TEAF (pH 6.0)/MeCN or 2-PrOH; 0.04%–0.4%/min; isocratic elution	V and NV	Investigated effects of different mobile phases and packings on separation of closely related peptides; silanol interactions observed in TFA system were blocked in TEAP system	345
(xxiv) ODS-120T C_{18}	NG/NG	300 x 7.5	Peptic haemoglobin lysate	ca. 150-6500	10 mM NH_4OAc (pH 6.1)/MeCN; 0.5%/min	V	Peptide mapping and separation of peptides from SEC fractions	336
(xxv) μBondapak C_{18}	NG/NG	NG	Various oligopeptides, mainly peptide hormones	NG	0.04 M TEAF (pH 3.15)/MeCN; isocratic elution	V	Demonstrated how dilute and volatile formate buffer system allows detection at low UV wavelengths (200 nm)	352

(xxvi)	Vydac Tp 214 C$_4$	NG/NG	250 × NG	Synthetic membrane-spanning peptides	NG	0.1% TFA + 0.08% morpholine + 0.1% n-octyl β-D-glucopyranoside/MeCN + 2-PrOH; 2%/min	NV	Presence of nonionic detergent enabled efficient purification of hydrophobic peptides	279
(xxvii)	Zorbax C$_{18}$	6/100	250 × 4.0	Various peptides and polypeptides	≤ 14 000	50 mM KH$_2$PO$_4$ or NaH$_2$PO$_4$ (pH 2.3) + varying concentrations of Triton X-100 or Brij 35/MeCN; 0.6%/min; isocratic elution	NV	Examined effect of presence and concentration of nonionic detergents in RPC of peptides	364

* Unless noted otherwise, all stationary phases are silica-based.

** All mobile phases are aqueous-based; unless noted otherwise, linear gradient (%/min denotes % organic modifier/min) elution conditions were employed.

*** TFA, PFPA, and HFBA denote trifluoroacetic, pentafluoropropionic, and heptafluorobutyric acid, respectively; TEAA, TEAP, and TEAF denote triethylammonium acetate, phosphate, and formate, respectively; NH$_4$OAc, NH$_4$F, and NH$_4$TFA denote ammonium acetate, formate, and trifluoroacetate, respectively; ESA, PrSA, BuSA, PSA, HSA, and OSA denote ethane-, propane-, butane-, pentane-, hexane-, and octanesulfonic acid, respectively; TBA denotes tetrabutylammonium; MeCN, MeOH, EtOH, 1-PrOH, and 2-PrOH denote acetonitrile, methanol, ethanol, 1-propanol, and 2-propanol, respectively; IP = ion pairing.

§ V and NV denote volatile and non volatile, respectively.

§§ NG denotes not given.

References on p. B140

interaction of peptides with the reversed-phase sorbent. In addition, the suppression of surface silanols in silica-based packings at low pH (<4.0) decreases ionic interactions with the hydrophobic stationary phase. TFA is an excellent solvent for most peptides, is completely volatile and, due to its low UV transparency, enables detection at wavelengths below 220 nm. The volatility of the aq. TFA/acetonitrile mobile phase eliminates the need for subsequent sample desalting, making RPC ideal as the final step in a multistep purification procedure [183,199]. Finally, the best resolution of a peptide mixture is usually obtained between 15% and 40% of the organic solvent in the gradient [250], and this is generally achieved with acetonitrile. The powerful peptide resolving capabilities of the aq. TFA/acetonitrile system are particularly highlighted in peptide mapping studies, where chemical or enzymatic cleavage of proteins frequently produces a large number of peptide fragments [299-306]. In addition, this solvent system can be used to resolve peptides of closely related structures [260,307-311], including diastereomers [307,308,311].

The recommended initial approach to most analytical peptide separations is the use of elution gradients from aq. TFA (0.05%-0.1%) to TFA (0.05%-0.1%)/acetonitrile at room temperature at a flowrate of 0.5-2.0 ml/min. Peptide resolution can than be optimized by varying the steepness of the acetonitrile gradient (generally 0.5-2.0% acetonitrile/min). Numerous mobile-phase systems, other than aq. TFA/acetonitrile, have been reported for a host of applications. However, in many instances, the reasoning behind the use of a particular mobile phase is somewhat obscure. The advantages of the aq. TFA/acetonitrile system are so apparent that there ought to be ample justification for not using it for peptides, at least as a starting point. Alternative RPC conditions are reviewed in Section 13.2.4.1.4, and specific applications of many of these mobile-phase systems are presented in Section 13.2.4.1.5 and Table 13.5.

13.2.4.1.4 Alternative mobile phases

Simple acidic solutions have often been employed in peptide separations. One of the first solvents used in RPC of peptides was, in fact, dilute phosphoric acid [255,287, 294,297,312-318]. Despite being nonvolatile, H_3PO_4, owing to its UV transparency, enables sensitive detection of the peptide bond. In addition, the low pH of 0.1% aq. H_3PO_4 suppresses ionization of silanol groups and, hence, enhances peptide elution profiles. HFBA (0.05-0.1%) has frequently been used as an alternative to TFA as a more strongly hydrophobic anionic ion-pairing reagent [221,257,272,287,291-294,297,319-324]. High concentrations (0.1-2.0 M) of acetic or formic acid, which are effective peptide solvents, have also been employed for RPC [250,325,326]. The addition of pyridine to these acids, though occasionally useful for some peptide separations, precludes UV detection below 250 nm [250,254,255]. In addition, the deleterious effect of high concentrations of acids on silica-based reversed-phase packings prohibits excessive use of these systems [287]. Lesser-used acidic mobile phases include dilute HCl [314,327-329], dilute $HClO_4$ [318], dilute acetic acid [330,331] and a TFA/acetic acid mixture [332].

A volatile ammonium acetate system has proved useful for the separation of peptide mixtures [207,259,263,265,270-273,277,311,315,333-336] over a wide range of pH values (generally pH 3.0-7.0); a volatile ammonium formate system has also seen some use

[337,338]. In addition, 0.1 M ammonium bicarbonate [247,256,339-341] and 0.02 M ammonium trifluoroacetate [259] have been employed in volatile mobile phases in the pH range 7-11.

Amines are particularly effective in deactivating interactions between silanol groups and sample molecules, and the development of buffers based on triethylamine, particularly triethylamine phosphate (TEAP) [259,262,294,311,333,334,339,342-351] and TEAF [275,294,345,351-356], has enabled the effective resolution of peptide mixtures in the pH 2-7 range. The ionic strength of these triethylamine-based buffers, when employed for peptide separations, has ranged between 10 mM and 0.25 M, depending on the application. TEAF has the advantage over TEAP of being volatile, although it does not have the UV transparency of the latter buffer, particularly so at high ionic strengths. However, the UV transparency of the TEAF system may be improved, if dilute solutions (10-40 mM) are used [294,352]. The addition of morpholine (0.05-0.15%), a heterocyclic secondary amine, has also been an occasional feature of peptide separations [279,357,358].

Nonvolatile phosphate buffers (generally sodium or potassium phosphate) have also frequently been employed for reversed-phase separations of peptides, enabling precise pH control over a wide pH range (typically, pH 2.0-8.0) [252,261,267-270,275,276,315, 347,349,359-371].

If it is desirable to carry out RPC at pH values higher than 3.5, the addition of almost any salt to the mobile phase will block interactions of peptides with silanol groups present on the reversed-phase sorbent [295]. However, it should be noted that above a pH of ca. 4.0 (and occasionally below this value), the presence of buffer salts may not be sufficient to eliminate silanol interactions entirely. Hence, it is often necessary to increase the ionic strength of the aqueous buffers by the further addition of 0.01-0.2 M salts. Thus, the addition of salts, such as chlorides [289,332,369] and sulfates [349,368,372], or of a chaotropic reagent, such as sodium perchlorate [255,261,315,347,359,360,373-375], is a feature of some peptide separations, thereby improving the elution pattern.

13.2.4.1.5 Specific applications of the alternative mobile phases

While by no means exhaustive, the list of selected applications in Table 13.5 covers a wide range of mobile-phase systems other than the prevalent aq. TFA/acetonitrile system. There are two major advantages of deviating from that system: firstly, change of the ion-pairing reagent (either from an anionic to a cationic reagent, or a variation in counter-ion hydrophobicity) can drastically affect the separation of a peptide mixture; and, secondly, a change in pH may have a profound effect on the charge characteristics of ionizable peptide sidechains and endgroups, with subsequent (often dramatic) change in peptide elution behavior. In addition, there is an obvious link between these two advantages, since, if a change from an anionic to a cationic counterion is desired, an increase in pH (to above pH 4.5) is required to ensure complete ionization of potentially negatively charged functional groups (C-terminal α-COOH; sidechain carboxyls of Asp and Glu). In contrast, if a change from a cationic to an anionic counterion is desired, the pH must be decreased (< pH 4.0) to ensure protonation of these potentially negatively charged

References on p. B140

groups and, thus, to take full advantage of the basic character of the peptide (i.e., the presence of His, Lys, or Arg residues or of the N-terminal α-NH$_2$ group).

Several of the applications selected for Table 13.5 describe the effectiveness of varying counterion hydrophobicity for resolving peptide mixtures. These include the use of anionic ion-pairing reagents (either volatile perfluorinated carboxylic acids [257,287,321] or non-volatile alkylsulfonates [252,276]) or cationic ion-pairing reagents [296]. Guo et al. [287], in fact, developed a method for predicting the change in peptide retention times with variations in hydrophobicity of anionic counterions. In addition, changes in counterion hydrophobicity have been employed for peptide mapping studies [252].

Changes in the selectivity of RPC at different pH values (often concomitant with a change in ion-pairing reagent) have been frequently advantageous in peptide applications, including peptide mapping studies [248,335,362], separation of closely related peptides [315,345], separation of diastereomers [370], and as a general aid in the isolation and purification of peptides [259,294,334,340]. The popularity of ammonium acetate-based buffers [259,315,334-336] lies in the wide pH range over which these buffers may be employed, which complements their volatility as well as the high solubility of many peptides in these buffers. This is also true of triethylammonium formate (TEAF)-based buffers [294,345,352], particularly when the ionic strength of these buffers is low enough to allow low-wavelength UV detection of peptides [294,352]. Recognition of the different column selectivities resulting from pH variations has also prompted the derivation of amino acid sidechain hydrophobicity coefficients in acidic and neutral (or close to neutral) medium [247,261] (Section 13.2.4.3). Peptide solubility is, of course, important in RPC. Highly acidic peptides are frequently somewhat insoluble in acidic media, where they tend to precipitate. Thus, under these circumstances, pH values closer to 7 are the norm [340,367].

In several applications presented in Table 13.5, solvents other than acetonitrile were used in order to take advantage of variations in the nonpolar nature of the organic modifier to effect a particular peptide separation [201,257,259,264,276,279,340,345,365]. The application with perhaps the clearest requirement for a more hydrophobic organic modifier (2-propanol) than acetonitrile was that of Taneja et al. [201] in RPC of extremely hydrophobic, leucine-rich, synthetic peptide polymers. Another approach to the separation of hydrophobic peptides by RPC has been the addition of nonionic detergents, such as n-octyl-β-D-glucopyranoside [279], Triton X-100, or Brij 35 [364] to the mobile phase when solubilized, hydrophobic peptides, such as membrane-spanning peptides [279], bind irreversibly to reversed-phase sorbents. These detergents appear to act by binding to the column packing, thereby reducing the number of potential sites of interaction on the stationary phase and, hence, the overall hydrophobicity of the packing.

13.2.4.2 Isocratic elution

Gradient, rather than isocratic, elution is by far the preferred elution mode for RPC of peptides and proteins, due to the mainly adsorption/desorption mechanism by which these solutes interact with the hydrophobic stationary phase [257,376]. The limited partitioning of peptides with the reversed-phase packing, a consequence of multisite binding, makes the use of isocratic conditions impractical for separating mixtures of peptides exhibiting a wide range of hydrophobicities [376]. However, isocratic elution has proved effective for the separation of peptides closely related in hydrophobicity. Szókan et al. [270], for instance, employed isocratic elution as a means of monitoring synthetic stages during solid-phase peptide synthesis. This was a form of purity control of peptides to detect synthetic impurities, formed as a consequence of side-reactions, such as oxidation, desulfation, racemization, transpeptidation, alkylation, and decomposition. Isocratic elution may also be used to separate synthetic impurities (e.g., deletion or terminated peptides) from the product of interest [275]. Other interesting applications of isocratic elution of peptides closely related in hydrophobicity have included: separations of small peptides [289], angiotensin peptides [374], Substance P and fragments [276], and synthetic peptide analogs [278]. The employment of isocratic conditions has also been a common feature of separations of iodinated derivatives of peptide hormones, such as somatostatin [343] and insulin [344,345,349,351,368]. Hruby and co-workers [307,308,311] and others [370] have highlighted the potentially powerful resolving capabilities of isocratic elution by its use for the separation of peptide diastereomers. Another example of isocratic separation of peptide isomers involved the resolution of α- and β-aspartyl peptides [277]. Thus, although optimal conditions for isocratic RPC separations of peptides are generally more difficult to develop than those for gradient elution, there is no denying the effectiveness of the former elution mode for specific applications.

13.2.4.3 Peptide standards

During an extensive review of the literature concerning the use of RPC for peptide separations, one common trend of which one becomes aware is that the stationary phases chosen by many researchers for their particular applications are unsuitable. This generally manifests itself in two ways: firstly, various (usually nonvolatile) mobile phases at low pH other than aq. TFA/acetonitrile systems are used, apparently due to undesirable silanol activity with the latter mobile phase; and, secondly, even in the absence of any nonideal behavior, a particular reversed-phase packing may simply not have the resolving power required for the intended application. In addition, some combination of the above is often apparent. If care is taken to select the appropriate stationary phase, in terms of both resolving power and minimal nonideal (ionic) characteristics (obviously the latter will have a profound effect on the former), the subsequent choice and optimization of elution conditions will be greatly simplified. Peptide standards designed specifically to aid the researcher in choosing the appropriate RPC column are described below.

Table 13.6 shows the sequences of six synthetic decapeptide standards (I1, I2, S2, S3, S4, and S5), designed to monitor RPC column performance [377]. The hydrophobicity of the peptide analogs increases only slightly between S2 and S4. Between S2 and S3 there is a change from an α-H to a β-CH$_3$ group; between S3 and S4 there is a change from a β-CH$_3$ group to two methyl groups, attached to the β-CH group; and between S4 and S5 there is a change from an α-H to an isopropyl group, attached to the α-carbon. The

References on p. B140

TABLE 13.6

SYNTHETIC PEPTIDE STANDARDS FOR MONITORING THE RESOLVING POWER OF RPC COLUMNS

Peptide	Sequence[*]
I1	Ac-Arg-Gly-Gly-Gly-Gly-**Ile**-Gly-**Ile**-Gly-Lys-amide
I2	Ac-Arg-Gly-Gly-Gly-Gly-**Ile**-Gly-Leu-Gly-Lys-amide
S2[**]	Ac-Arg-Gly-Gly-Gly-Gly-Leu-Gly-Leu-Gly-Lys-amide
S3	Ac-Arg-Gly-**Ala**-Gly-Gly-Leu-Gly-Leu-Gly-Lys-amide
S4	Ac-Arg-Gly-**Val**-Gly-Gly-Leu-Gly-Leu-Gly-Lys-amide
S5	Ac-Arg-Gly-**Val**-**Val**-Gly-Leu-Gly-Leu-Gly-Lys-amide

[*] Ac = N^{α}-acetyl; amide = C^{α}-amide.

[**] Residues in bold denote substitutions in the sequence of Peptide Standard S2.

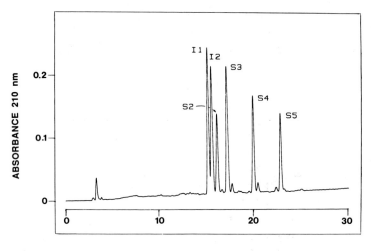

ELUTION TIME (min)

Fig. 13.8. Reversed-phase chromatography of synthetic peptide standards. Column, Syn-Chropak RP-P C_{18} (250 x 4.6 mm ID, 6.5-μm particle size, 300-Å pore size; SynChrom Inc.). Conditions, linear AB gradient (1% B/min), where Eluent A is 0.1% aq. TFA and Eluent B is 0.1% TFA in acetonitrile; flowrate, 1 ml/min; temperature, 26°C. The sequences of the six peptide standards are shown in Table 13.6. The standards were obtained from Synthetic Peptides Inc.

hydrophobicity variations between I1, I2, and S2 are even more subtle. There is a change of only an isoleucine to a leucine residue between I1 and I2, and between I2 and S2. Guo et al. [261] demonstrated that Leu is slightly more hydrophobic than Ile, although these residues contain the same number of carbon atoms. Since Ile is β-branched, the β-carbon is close to the peptide backbone and not as available for interaction with the hydrophobic stationary phase compared to the conformation of the Leu sidechain. Thus, this peptide mixture enables a very precise determination of the resolving power of a reversed-phase column (demonstrated in Fig. 13.8 for an analytical SynChropak RP-P C_{18} column).

A series of synthetic undecapeptide standards has also been developed for monitoring the extent of silanol activity on silica-based reversed-phase packings [347]. The sequences of these peptides, which also serve as cation-exchange standards, are shown in the legend to Fig. 13.5 [note that the peptides are designated as C1, C2, C3, and C4 (IEC) or 1, 2, 3, and 4 (RPC), depending on the mode of application]. Peptides 1-4 contain,

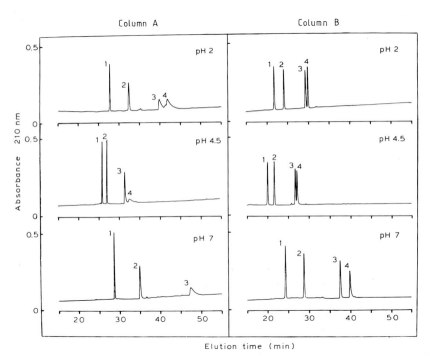

Elution time (min)

Fig. 13.9. Monitoring of free silanols on commercial reversed-phase columns with peptide standards. Column A: C_{18} column; 250 x 4.6 mm ID; 5-μm particle size; 300-Å pore size. Column B: C_8 column; 220 x 4.6 mm ID; 7 μm; 300 Å. Conditions: pH 2.0, linear AB gradient (1% B/min), where Eluent A is 0.05% aq. TFA and Eluent B is 0.05% TFA in acetonitrile; pH 4.5, linear AB gradient (2% B/min, equivalent to 1% acetonitrile/min and 1 mM TEAP/min), where Eluent A is aq. 10 mM TEAP (pH 4.5) and Eluent B is 50% acetonitrile, containing 60 mM TEAP. Eluent B was prepared by mixing equal volumes of acetonitrile and an aq. 120 mM solution of TEAP (pH 4.5); pH 7.0, linear AB gradient (1.67% B/min, equivalent to 1% acetonitrile/min and 1.67 mM NaClO$_4$/min), where Eluent A is aq. 10mM (NH$_4$)$_2$HPO$_4$ (pH 7.0) and Eluent B is 60% aq. acetonitrile, containing 100 mM NaClO$_4$. Flowrate, 1 ml/min; temperature, 26°C. The numbers 1-4 denote Peptide Standards 1-4, respectively. The sequences of the standards are shown in the legend to Fig. 13.5 (Peptides C1-C4 in Fig. 13.5 denote Peptides 1-4 in Fig. 13.9). (Reprinted from Ref. 347, with permission.)

respectively, 1-4 basic residues (Lys), no acidic residues being present. Thus, over the pH range used for the majority of reversed-phase separations (pH 2.0 to 7.0), the net charges of +1 to +4 for Peptides 1-4, respectively, do not change. The basic, positively charged character of the peptides ensures their sensitivity to any ionic, as opposed to hydrophobic, stationary-phase characteristics; this sensitivity increases with increasing basic character (Peptide 1 to Peptide 4) of the standards. The hydrophobicity of the standards also increases from Peptide 1 to Peptide 4. Fig. 13.9 demonstrates the application of these standards to monitoring silanol activity on two different silica-based columns at pH values of 2.0, 4.5, and 7.0. The contrast in performance between the two columns is striking. Column A exhibited significant ionic interactions with the basic peptide standards (apparent mainly from the retention behavior of Peptides 3 and 4) over the entire pH range (pH 2.0-7.0) available to silica-based reversed-phase columns, even under conditions designed to suppress or block such interactions, i.e., low pH or the presence of amines or salts in the mobile phase. Peptides with charges of +3 and +4 are not uncommon and, thus, this column is clearly unsuitable for most peptide applications. In contrast, the elution profiles obtained on Column B were excellent at all three pH values. Some ionic interaction was apparent at pH 7.0, even on this column, stressing once again the advantage of emploing mobile phases at pH values <4.0, if at all possible.

13.2.4.4 Prediction of retention times

Several research groups have determined sets of amino acid residue hydrophobicity coefficients for predicting peptide retention times in RPC, on the assumption that the chromatographic behavior of a peptide is mainly or solely dependent on amino acid composition, and this assumption holds well enough for small peptides (up to ca. 15 residues) [247,254,261,280,321,359-361,378,379]. Retention values have generally been obtained by computer-calculated regression analyses of the retention times of a wide range of mainly unrelated peptides of varied composition [247,254,321,359-361]. However, the most precise coefficients currently available are likely those of Guo et al. [261], who rejected the regression-analysis approach in favor of measuring the contributions of individual amino acids, at pH 2.0 and pH 7.0, to the retention time of a synthetic model peptide at a given chainlength. The octapeptide sequence, Ac-Gly-X-X-(Leu)$_3$-(Lys)$_2$-amide, was substituted at position X by all 20 naturally occurring amino acids. The predicted retention time (τ) of a peptide in RPC was then equal to the sum of the retention coefficients (ΣR_c) for the amino acid residues in the peptide, plus the time correction for an internal standard.

Several researchers have noted that peptides longer than 15-20 residues tended to be eluted more rapidly than predicted from a hydrophobic consideration alone [176,254,255,361,378-382]. The effect on peptide retention time of increasing peptide length is clearly illustrated in Fig. 13.10A, which shows the elution profile of a series of five synthetic peptide size-exclusion standards (10-50 residues) on a C$_{18}$ column. There is, in fact, an exponential relationship between peptide chainlength and peptide retention time in

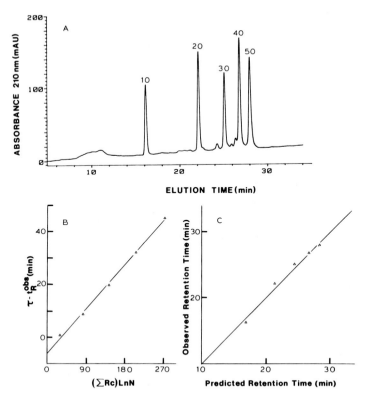

Fig. 13.10. RPC of a mixture of synthetic peptide polymers. Column: SynChropak RP-P C$_{18}$ (250 x 4.6 mm ID, 6.5-μm particle size, 300-Å pore size; SynChrom Inc.). Conditions: linear AB gradient (1% B/min), where Eluent A is 0.1% aq. TFA and Eluent B is 0.1% TFA in acetonitrile; flowrate, 1 ml/min; temperature, 26°C. Panel A: elution profile of five peptide polymers (10-50 residues). Panel B: plot of predicted minus observed retention time $(\tau - t_R^{obs})$ vs. the sum of the retention coefficients (ΣR_C) of Guo et al. [261] times the logarithm of the number of residues (ln N). Panel C: correlation of predicted and observed retention times of the peptide polymers (derived from the plot shown in Panel B, as described by Mant et al. [382]). The sequences of the peptide polymers are shown in the legend to Fig. 13.2. (Reprinted from Ref. 213, with permission.)

RPC [167,206,213]. Thus, in Fig. 13.10A, the effect on peptide retention of increasing length of the peptide polymers decreases progressively with each 10-residue addition.

The intimate relationship between peptide hydrophobicity and chainlength and their combined effect on peptide retention behavior in RPC was detailed by Mant et al. [382]. These researchers demonstrated a linear relationship between predicted (τ) minus observed (t_R^{obs}) retention time vs. the product of peptide hydrophobicity (expressed as ΣR_C, the sum of the coefficients of Guo et al. [261]) and the logarithm of the number of residues, ln N (Fig. 13.10B). Using the slope and intercept from such a plot, the retention behavior of peptides up to 50 residues in length can be predicted. Fig. 13.10C shows the good correlation between predicted and observed retention times for the peptides shown in Fig. 13.10A, once peptide chainlength has been taken into account. This predictive process subsequently formed the basis for an expert system [213], capable of simulating reversed-phase peptide elution profiles.

13.2.4.5 Reversed-phase chromatography as a physicochemical model of biological systems

One of the most difficult and important challenges facing biochemists today is understanding protein folding, specifically, how the amino acid sequence of a protein determines its three-dimensional structure and the pathway of folding. The goal of predicting polypeptide and protein conformation from primary structure information is being pursued by many researchers, using a variety of methodologies. A very promising HPLC-based approach to the problem of predicting higher orders of protein conformation from amino acid sequence information only is the study of the retention behavior of peptides or protein fragments instead of the complete protein molecule. RPC, in particular, is becoming the favored approach in much of this work. The hydrophobic interactions between peptides and the nonpolar stationary phase upon which this technique depends may well reflect similar intraprotein interactions between the nonpolar residues that stabilize the folded or three-dimensional structure of the native protein molecule.

An excellent example of the value of RPC as a model of biological systems was recently reported by Zhou et al. [383], who employed synthetic peptides to study the retention behavior of amphipathic α-helices. A nonpolar environment, such as a hydrophobic stationary phase, may induce helical structures in potentially helical molecules [167,384]. Further, if a molecule becomes helical on binding and contains a preferred binding domain, as in the case of an amphipathic helix, then obviously some residues may not be contributing to the same extent to the overall hydrophobicity of the peptide. Through the use of two sets of model synthetic peptide analogs, one set representing amphipathic α-helical structures ([Ac-Lys-Cys-Ala-Glu-Leu-Glu-Gly-(Lys-Leu-Glu-Ala-Leu-Glu-Gly)$_n$-amide] where n = 1-4) and the other set representing nonamphipathic α-helical structures ([Ac-Lys-Cys-Ala-Glu-Gly-Glu-Leu-(Lys-Leu-Glu-Ala-Gly-Glu-Leu)$_n$-amide], where n = 1-4), Zhou et al. [383] demonstrated that it is possible not only to predict the retention behavior of amphipathic α-helices during RPC, but also to deduce the presence of an amphipathic α-helical structure in peptides based upon their retention data. This predictive method was a refinement of previous work by Guo et al. [261] and Mant et al. [382] in predicting reversed-phase retention behavior of peptides. The importance of understanding the interactions of the hydrophobic domains of amphipathic α-helical structures should not be underestimated. These structures may play an important role in protein folding, for instance. Amphipathic α-helical structures are also important in the binding of peptide hormones to their biological receptors [385]. Thus, the interaction of amphipathic α-helices with a hydrophobic surface during RPC is likely to be a good model for their binding to receptors or packing with other α-helices in the folded protein. In addition, if it is predicted that fragments of proteins bind to reversed-phase sorbents as amphipathic helices, they may be in this conformation in the native protein.

13.2.5 Expert systems

13.2.5.1 Introduction

Peptide mixtures derived from different sources differ widely in complexity and quantity, as well as in characteristics (length, net charge, hydrophobicity) of the individual peptides. In addition, within the three main modes of HPLC employed for peptide separations [SEC, IEC (specifically, cation-exchange chromatography) and RPC], mobile-phase conditions may be manipulated in many ways to maximize the separation potential of a particular HPLC column. Although a desired peptide separation may be obtained by trial and error, this may take many attempts, with consequent loss of time and sample. Obviously, any methodology that can aid in selecting a procedure with minimum expenditure of both time and sample would be welcome. Two types of computer-assisted HPLC method development systems have evolved to fulfill these requirements: (i) a system which predicts elution profiles from scratch, i.e. solely on the basis of peptide sequence information; (ii) optimization programs useable following one or a few initial chromatographic experiments with the peptide sample of interest.

13.2.5.2 Simulation of chromatograms from sequence information alone

A computer program, called ProDigest-LC, has been developed to assist scientists in devising methods of SEC, IEC, and RPC for the analytical separation and purification of biologically active peptides and peptide fragments from enzymatic and chemical digests of proteins [213,386]. This program predicts peptide elution profiles, at varying flowrates (SEC, IEC, RPC) and gradient rates (IEC, RPC), requiring no information about the peptides except their amino acid composition.

The program was developed [213] from principles of predictive behavior of peptides in HPLC, described in Sections 13.2.2.4 (SEC), 13.2.3.6 (IEC), and 13.2.4.4 (RPC). Specifically, development of the program was based upon the work of Mant et al. (SEC) [184], Burke et al. (IEC) [216], Guo et al. (RPC) [261], and Mant et al. (RPC) [382]. The use of peptide standards plays a key role in the operation of ProDigest-LC, both as a means of identifying and suppressing any nonideal behavior (a prerequisite for accurate retention prediction) and for calibration of HPLC columns and instrumentation.

The experiments simulated by the program eliminate the time-consuming trial-and-error approach to peptide purification. In addition, ProDigest-LC is also a teaching aid for chromatographers, designed to help students or researchers to select the appropriate conditions for chromatography (HPLC mode, column, and mobile phase) and giving them the option of examining the effect of varying flowrate, gradient rate, sample size, and sample volume on the separation. Thus, ProDigest-LC simulates peptide elution profiles without the need for actual experiments with the sample of interest. Having gone through the program until the desired separation has been simulated, the researcher may then carry out the experiment. On the basis of the observed chromatogram, the researcher may then embark on further optimization of the chromatographic conditions, perhaps through the use of a complementary optimization program. Future developments of ProDigest-LC include refinements in the program to take into account the presence of preferred binding domains in peptides [383] as well as the extension of the IEC predictive facility to anion-exchange chromatography.

It is always more accurate to simulate the effect of changes in parameters on observed elution patterns then to base these simulations on previously predicted profiles, and a commercially available optimization computer program, DryLab G, has been designed specifically for this purpose [387-392]. The program requires a minimum of two experiments, following which the effects on the peptide elution pattern of manipulating parameters, such as gradient time, gradient shape (multisegmented gradients), and flow-rate may be simulated until the desired resolution is achieved. For a recent review of optimization programs, in addition to other aspects of computer-assisted method development in chromatography, the reader is directed to Ref. 393.

ACKNOWLEDGEMENTS

This work was supported by research grants from the Medical Research Council of Canada and by equipment grants, a postdoctoral fellowship (N.E.Z.), and a research allowance from the Alberta Heritage Foundation for Medical Research.

REFERENCES

1 Z. Deyl, J. Hyánek and M. Horáková, *J. Chromatogr.*, 379 (1986) 177.
2 S. Einarsson, B. Josefsson, P. Möller and D. Sanchez, *Anal. Chem.*, 59 (1987) 1191.
3 D.S. Dunlop and A. Neidle, *Anal. Biochem.*,165 (1987) 38.
4 P. Masia, I. Nicoletti, M. Sinibaldi, D. Attanasio and A. Messina, *Anal. Chim. Acta*, 204 (1988) 145.
5 M. Maurs, F. Trigalo and R. Azerad, *J. Chromatogr.*, 440 (1988) 209.
6 E. Armani, L. Barazzoni, A. Dossena and R. Marchelli, *J. Chromatogr.*, 441 (1988) 287.
7 P. E. Hare, in M. Zief and L.J. Crane (Editors), *Chromatographic Chiral Separations*, Chromatographic Science Series, Vol. 40, Dekker, New York, 1988, p. 165.
8 A. Duchateau, M. Crombach, M. Aussems and J. Bongers, *J. Chromatogr.*, 461 (1989) 419.
9 D.H. Spackman, W.H. Stein and S. Moore, *Anal. Chem.*, 30 (1958) 1190.
10 G. Ogden and P. Földi, *LC.GC*, 5 (1987) 28.
11 H. Engelhardt, in H. Engelhardt (Editor), *Practice of High Performance Liquid Chromatography: Applications, Equipment and Quantitative Analysis*, Springer-Verlag, Heidelberg, 1986, p. 409.
12 T.E. Hugli (Editor), *Techniques in Protein Chemistry*, Academic Press, New York, 1989.
13 C. Lazure, J.A. Rochemont, N.G. Seidah and M. Chretien, in F.E. Regnier and K.M. Gooding (Editors), *HPLC of Biological Macromolecules: Methods and Applications*, Dekker, New York, 1990, p. 263.

14 L.B. Smillie and M. Nattriss, in C.T. Mant and R.S. Hodges (Editors), *HPLC of Peptides and Proteins: Separation, Analysis and Conformation,* CRC Press, Boca Raton, FL, 1991.
15 W. Blom and J. Huijmans, *Sci. Tools,* 32 (1985) 10.
16 R.P. Andrews and N.A. Baldar, *Sci. Tools,* 32 (1985) 44.
17 J. Macek, L. Miterová and M. Adam, *J. Chromatogr.,* 364 (1986) 253.
18 S. Al-Najafi, C.A. Wellington, A.P. Wade, T.J. Sly and D. Betteridge, *Talanta,* 34 (1987) 749.
19 D.I. Seracu, *Anal. Lett.,* 20 (1987) 1417.
20 T. Hori and S. Kihara, *Fresenius Z. Anal. Chem.,* 330 (1988) 627.
21 V. Arrizon-Lopez, R. Biehler, J. Cummings and J. Harbaugh, in C.T. Mant and R.S. Hodges (Editors), *HPLC of Peptides and Proteins: Separation, Analysis and Conformation,* CRC Press, Boca Raton, FL, 1991.
22 H. Krapf, in C.T. Mant and R.S. Hodges (Editors), *HPLC of Peptides and Proteins: Separation, Analysis and Conformation,* CRC Press, Boca Raton, FL, 1991.
23 M. Roth, *Anal. Chem.,* 43 (1971) 880.
24 M. Roth and A. Hampai, *J. Chromatogr.,* 83 (1973) 353.
25 J.D.H. Cooper and D.C. Turnell, *J. Chromatogr.,* 227 (1982) 158.
26 K.S. Lee and D.G. Drescher, *J. Biol. Chem.,* 82 (1977) 250.
27 R.L. Cunico and T. Schlabach, *J. Chromatogr.,* 266 (1983) 461.
28 M.W. Dong and J.R. Gant, *J. Chromatogr.,* 327 (1985) 17.
29 G.J. Hughes and K.J. Wilson, *J. Chromatogr.,* 242 (1982) 337.
30 T.N. Ferraro and T.A. Hare, *Anal. Biochem.,* 143 (1984) 82.
31 N. Seiler and B, Knodgen, *J. Chromatogr.,* 341 (1985) 11.
32 I.R. Elrifi, D.B. Layzell, B.J. King, G.E. Weagle and D.H. Turpin, *J. Liq. Chromatogr.,* 9 (1986) 2199.
33 Y. Hashimoto, S. Yamagata and T. Hayakawa, *Anal. Biochem.,* 160 (1987) 362.
34 P.G. Rigas, S.J. Arvanitis and D.J. Pietrzyk, *J. Liq. Chromatogr.,* 10 (1987) 2891.
35 J. Haginaka and J. Wakai, *J. Chromatogr.,* 396 (1987) 297.
36 J. Haginaka and J. Wakai, *Anal. Biochem.,* 171 (1988) 398.
37 M. Griffin, J. Leah, N. Mould and G. Compton, *J. Chromatogr.,* 431 (1988) 285.
38 Y. Hashimoto, *J. Chromatogr.,* 462 (1989) 341.
39 A. Fiorino, G. Frigo and E. Cucchetti, *J. Chromatogr.,* 476 (1989) 83.
40 V. Barkholt and A.L. Jensen, *Anal. Biochem.,* 177 (1989) 318.
41 W. Rajendra, *J. Liq. Chromatogr.,* 10 (1987) 941.
42 P. Schrynemackers-Pitance and S. Schoos-Barbette, *Clin. Chim. Acta,* 166 (1987) 91.
43 J. Chow, J.B. Orenberg and K.D. Nugent, *J. Chromatogr.,* 386 (1987) 243.
44 G.A. Qureshi and P. Södersten, *J. Chromatogr.,* 400 (1987) 247.
45 D.E. Willis, *J. Chromatogr.,* 408 (1987) 217.
46 M. Eslami, J.D. Stuart and K.A. Cohen, *J. Chromatogr.,* 411 (1987) 121.
47 M. Farrant, F. Zia-Gharib and R.A. Webster, *J. Chromatogr.,* 417 (1987) 385.
48 J. Schmidt and C.J. McClain, *J. Chromatogr.,* 419 (1987) 1.
49 D.J. Aberhart, *Anal. Biochem.,* 169 (1988) 350.
50 H.M.H. van Eijk, M.A.H. van der Heijden, C.L.H. van Berlo and P.B. Soeters, *Clin. Chem.,* 34 (1988) 2510.
51 G. McClung and W.T. Frankenberger, Jr., *J. Liq. Chromatogr.,* 11 (1988) 613.
52 L.L. Brown, P.E. Williams, T.A. Becker, R.J. Ensley, M.E. May and N.N. Abumrad, *J. Chromatogr.,* 426 (1988) 370.
53 T.A. Durkin, G.M. Anderson and D.J. Cohen, *J. Chromatogr.,* 428 (1988) 9.
54 R. Schuster, *J. Chromatogr.,* 431 (1988) 271.
55 M.R. Euerby, *J. Chromatogr.,* 454 (1988) 398.
56 P. Brunet, B. Sarrobert and N. Paris-Pireyre, *J. Chromatogr.,* 455 (1988) 173.
57 G.A. Qureshi and M.S. Baig, *J. Chromatogr.,* 459 (1988) 237.
58 D.T. Blankenship, M.A. Krivanek, B.L. Ackermann and A.D. Cardin, *Anal. Biochem.,* 178 (1989) 227.
59 G.A. Qureshi and A.R. Qureshi, *J. Chromatogr.,* 491 (1989) 281.

B142

60 P. Boehlen and M. Mellet, *Anal. Biochem.,* 94 (1979) 313.
61 S. Udenfriend, S. Stein, P. Böhler, W. Dairman, W. Leingruber and M. Weigele, *Science,* 178 (1972) 871.
62 J.V. Castell, M. Cervera and R. Marco, *Anal. Biochem.,* 99 (1979) 379.
63 P.A. St.-John, *Aminco Lab. News,* 31 (1975) 1.
64 D.G. Drescher and K.S. Lee, *Anal. Biochem.,* 84 (1978) 559.
65 M. Weigele, S. de Bernardo and W. Leingruber, *Biochem. Biophys. Res. Commun.,* 50 (1973) 352.
66 M.M. Tikhomirov, A.Y. Khorlin, W. Voelter and H. Bauer, *J. Chromatogr.,* 167 (1978) 197.
67 S. Stein and L. Brink, *Methods Enzymol.,* 79 (1981) 20.
68 S. Stein and S. Udenfriend, *Anal. Biochem.,* 136 (1984) 7.
69 D.R. Koop, E.T. Morgan, G.E. Tarr and M.J. Coon, *J. Biol. Chem.,* 257 (1982) 8472.
70 G.E. Tarr, D.D. Black, V.S. Fujita and M.J. Coon, *Proc. Natl. Acad. Sci. US,* 80 (1983) 6552.
71 B.A. Bidlingmeyer, S.A. Cohen and T.L. Tarvin, *J. Chromatogr.,* 336 (1984) 93.
72 R.L. Heinrikson and S.C. Meredith, *Anal. Biochem.,* 136 (1984) 65.
73 K.L. Rogers, R.A. Philibert, A.J. Allen, J. Molitor, E.J. Wilson and G.R. Dutton, *J. Neur. Methods,* 22 (1987) 173.
74 R.G. Elkin and A.M. Wasynczuk, *Cereal Chem.,* 64 (1987) 226.
75 W.E. Schmidt, J.M. Conlon, V. Mutt, M. Carlquist, B. Gallwitz and W. Creutzfeldt, *Eur. J. Biochem.,* 162 (1987) 467.
76 J.L. Glajch and J.J. Kirkland, *J. Chromatogr. Sci.,* 25 (1987) 4.
77 M.M.T. O'Hare, O. Tortora, U. Gether, H.V. Nielsen and T.W. Schwartz, *J. Chromatogr.,* 389 (1987) 379.
78 J.L. Tedesco and R. Schafer, *J. Chromatogr.,* 403 (1987) 299.
79 P.R. Young and F. Grynspan, *J. Chromatogr.,* 421 (1987) 130.
80 R. Murphy, J.B. Furness and M. Costa, *J. Chromatogr.,* 408 (1987) 388.
81 H.P.J. Bennett and S. Solomon, *J. Chromatogr.,* 359 (1986) 221.
82 D.J. Strydom, *Anal. Biochem.,* 174 (1988) 679.
83 A.S. Inglis, N.A. Bartone and J.R. Finlayson, *J. Biochem. Biophys. Methods,* 15 (1988) 249.
84 B.C. Pramanik, C.R. Moomaw, C.T. Evans, S.A. Cohen and C.A. Slaughter, *Anal. Biochem.,* 176 (1989) 269.
85 R. Gupta and N. Jentoft, *J. Chromatogr.,* 474 (1989) 411.
86 K. Hayakawa and J. Oizumi, *J. Chromatogr.,* 487 (1989) 161.
87 D. Atherton, in T.E. Hugli (Editor), *Techniques in Protein Chemistry,* Academic Press, New York, 1990, p. 273.
88 K.A. West and J.W. Crabb, in T.E. Hugli (Editor), *Techniques in Protein Chemistry,* Academic Press, New York, 1990, p. 295.
89 D.R. Dupont, P.A. Keim, A. Chui, R. Bello, M. Bozzini and K.J. Wilson, in T.E. Hugli (Editor), *Techniques in Protein Chemistry,* Academic Press, New York, 1990, p. 284.
90 R. Mora, K.D. Berndt, H. Tsai and S.C. Meredith, *Anal. Biochem.,* 172 (1988) 368.
91 S.A. Cohen and D.J. Strydom, *Anal. Biochem.,* 174 (1988) 1.
92 G.E. Tarr, in J.E. Shively (Editor), *Methods of Protein Microcharacterization,* Humana Press, Clifton, NJ, 1986, p. 155.
93 B.A. Bidlingmeyer, T.L. Tarvin and S.A. Cohen, in K. Walsh (Editor), *Methods in Protein Sequence Analysis,* Humana Press, Clifton, NJ, 1986, p. 229.
94 W. Tapuhi, D.E. Schmidt, W. Lindner and B.L. Karger, *Anal. Biochem.,* 115 (1981) 123.
95 L. Zecca and P. Ferrario, *J. Chromatogr.,* 337 (1985) 391.
96 A.Ph. Lobazov, V.A. Mostovnikov, S.V. Nechaev, B.G. Belenkii, J.J. Kever, E.M. Korolyova and V.G. Maltsev, *J. Chromatogr.,* 365 (1986) 321.
97 F.J. Márquez, A.R. Quesada, F. Sánchez-Jiménez and J. Núñez de Castro, *J. Chromatogr.,* 380 (1986) 275.
98 A. Hernndobler, *J. High Resolut. Chromatogr. Chromatogr. Commun.,* 9 (1986) 602.
99 J.C. Rutledge and J. Rudy, *Am. J. Clin. Pathol.,* 87 (1987) 614.

100 A. Negro, S. Garbisa, L. Gotte and M. Spina, *Anal. Biochem.*, 160 (1987) 39.
101 J.-K. Lin and J.-Y. Chang, *Anal. Chem.*, 47 (1975) 1634.
102 J.-Y. Chang, R. Knecht and G. Braun, *Biochem. J.*, 199 (1981) 547.
103 J.-Y. Chang, R. Knecht and D.G. Braun, *Biochem. J.*, 203 (1982) 803.
104 J.-Y. Chang, R. Knecht and D.G. Braun, *Methods Enzymol.*, 91 (1983) 41.
105 J.-Y. Chang, R. Knecht, P. Jenoe and S. Vekemans, in T.H. Hugli (Editor), *Techniques in Protein Chemistry*, Academic Press, New York, 1989, p. 305.
106 J. Vendrell and F. Avilés, *J. Chromatogr.*, 358 (1986) 401.
107 G.J. Hughes, S. Frutiger and C. Fonck, *J. Chromatogr.*, 389 (1987) 327.
108 S. Odani, N. Kenmochi and K. Ogata, *J. Biochem.*, 103 (1988) 872.
109 H.J. Schneider, *Chromatographia*, 28 (1989) 45.
110 L.A. Carpino and G.Y. Han, *J. Org. Chem.*, 37 (1972) 3404.
111 S. Einarsson, B. Josefsson and S. Lagerkvist, *J. Chromatogr.*, 282 (1983) 609.
112 S. Einarsson, *J. Chromatogr.*, 348 (1985) 213.
113 S. Einarsson, S. Folestad, B. Josefsson and S. Lagerkvist, *Anal. Chem.*, 58 (1986) 1638.
114 I. Betnér and P. Földi, *Chromatographia*, 22 (1986) 381.
115 T. Näsholm, G. Sandberg and A. Ericsson, *J. Chromatogr.*, 396 (1987) 225.
116 R. Cunico, A.G. Mayer, C.T. Wehr and T.L. Sheehan, *BioChromatography*, 1 (1986) 6.
117 A.J. Smith, J.M. Presley and W. McIntyre, in T.E. Hugli (Editor), *Techniques in Protein Chemistry*, Academic Press, New York, 1989, p. 255.
118 I. Betnér and P. Földi, *LC.GC*, 6 (1988) 832.
119 P.B. Ghosh and M.W. Whitehouse, *Biochem. J.*, 108 (1968) 155.
120 M. Roth, *Clin. Chim. Acta*, 83 (1978) 273.
121 M. Ahnoff, I. Grundevik, A. Arfwidsson, J. Fonselius and B.A. Persson, *Anal. Chem.*, 53 (1981) 485.
122 H. Umagat, P. Kucera and L.F. Wen, *J. Chromatogr.*, 239 (1982) 463.
123 H. Yoshida, T. Sumida, T. Masujima and H. Imai, *J. High Resolut. Chromatogr. Chromatogr. Commun.*, 5 (1982) 509.
124 W.J. Linblad and R.F. Diegelman, *Anal. Biochem.*, 138 (1984) 390.
125 K. Imai and Y. Watanabe, *Anal. Chim. Acta*, 130 (1981) 377.
126 Y. Watanabe and K. Imai, *Anal. Biochem.*, 116 (1981) 471.
127 Y. Watanabe and K. Imai, *J. Chromatogr.*, 239 (1982) 723.
128 K. Imai, Y. Watanabe and T. Toyo'oka, *Chromatographia*, 16 (1982) 214.
129 Y. Watanabe and K. Imai, *J. Chromatogr.*, 309 (1984) 279.
130 H. Kotaniguchi and M. Kawakatsu, *J. Chromatogr.*, 420 (1987) 141.
131 Y. Watanabe and K. Imai, *Anal. Chem.*, 55 (1983) 1786.
132 J.-Y. Chang, *Methods Enzymol.*, 91 (1983) 455.
133 B. Oray, H.S. Lu and R.W. Gracy, *J. Chromatogr.*, 270 (1983) 253.
134 A. Lehmann and B. Wittman-Liebold, *FEBS Lett.*, 176 (1984) 360.
135 J.-Y. Chang, *Biochem. J.*, 199 (1981) 557.
136 C.-Y. Yang and S.J. Wakil, *Anal. Biochem.*, 137 (1984) 54.
137 J.-Y. Chang, *Anal. Biochem.*, 170 (1988) 542.
138 J.-Y. Chang, D. Brauer and B. Wittman-Liebold, *FEBS Lett.*, 93 (1978) 205.
139 J.-Y. Chang and E.H. Creaser, *J. Chromatogr.*, 132 (1977) 303.
140 I.V. Nasimov, N.B. Levina, V.V. Shemyakin, B.V. Rosynov, I.A. Bogdanova and V.G. Merimson, in C. Birr (Editor), *Methods in Peptide and Protein Sequence Analysis*, Elsevier, Amsterdam, 1980, p. 475.
141 J.J. L'Italien and S.B.H. Kent, *J. Chromatogr.*, 283 (1984) 149.
142 D.H. Hawke, H.-W. Lahm, J.E. Shively and C.W. Todd, *Anal. Biochem.*, 166 (1987) 298.
143 T.J. Mahachi, R.M. Carlson and D.P. Poe, *J. Chromatogr.*, 298 (1984) 279.
144 S.V. Vitt, M.M. Vorob'ev, E.A. Paskonova and M.B. Saporovskaya, *J. High Resolut. Chromatogr. Chromatogr. Commun.*, 6 (1983) 158.
145 I. Fermo, F.M. Rubino, E. Bolzacchini, C. Arcelloni, R. Paroni and P.A. Bonini, *J. Chromatogr.*, 433 (1988) 53.

146 R.C. Morton and G.E. Gerber, *Anal. Biochem.,* 170 (1988) 220.
147 C.J. Strang, E. Henson, Y. Okamoto, M.A. Paz and P.M. Gallop, *Anal. Biochem.,* 178 (1989) 276.
148 G. Szókáu, G. Mezö and F. Hudecz, *J. Chromatogr.,* 444 (1988) 115.
149 S. Kochhar and P. Christen, *Anal. Biochem.,* 178 (1989) 17.
150 K. Kakehi, T. Konishi, I. Sugimoto and S. Honda, *J. Chromatogr.,* 318 (1985) 367.
151 T. Yoshida, A. Uetake, H. Murayama, N. Nimura and T. Kinoshita, *J. Chromatogr.,* 348 (1985) 425.
152 R.M. Metrione, *J. Chromatogr.,* 363 (1986) 337.
153 B.K. Matuszewski, R.S. Givens, K. Srinivasachar, R.G. Carlson and T. Higuchi, *Anal. Chem.,* 59 (1987) 1102.
154 K. Iwaki, N. Nimura, Y. Hiraga, T. Kinoshita, K. Takeda and H. Ogura, *J. Chromatogr.,* 407 (1987) 273.
155 S.C. Beale, J.C. Savage, D. Wiesler, S.M. Wietstock and M. Novotny, *Anal. Chem.,* 60 (1988) 1765.
156 C.M.B. van den Beld, H. Lingeman, G.J. van Ringen, U.R. Tjaden and J. van der Greef, *Anal. Chim. Acta,* 205 (1988) 15.
157 A. Neidle, M. Banay-Schwartz, S. Sacks and D.S. Dunlop, *Anal. Biochem.,* 180 (1989) 291.
158 S.R. Spurlin and M.M. Cooper, *Anal. Lett.,* 19 (1986) 2277.
159 G. Fassina, M. Zamai, M. Brigham-Burke and I.M. Chaiken, *Biochemistry,* 28 (1989) 8811.
160 A.J. Alpert, *BioChromatography,* 2 (1987) 131.
161 A.J. Alpert, *J. Chromatogr.,* 444 (1988) 269.
162 A.J. Alpert, *J. Chromatogr.,* 499 (1990) 177.
163 R.H. Burdon and P.H. van Knippenberg (Editors), *Laboratory Techniques in Biochemistry and Molecular Biology,* Vol. 17, Elsevier, Amsterdam, 1987.
164 P.T. Matsudaira (Editor), *A Practical Guide to Protein and Peptide Purification for Microsequencing,* Academic Press, San Diego, CA, 1989.
165 C. Fini, A. Floridi, V.N. Finelli and B. Wittman-Liebold (Editors), *Laboratory Methodology in Biochemistry,* CRC Press, Boca Raton, FL, 1990.
166 J.J. Villafranca (Editor), *Current Research in Protein Chemistry,* Academic Press, San Diego, CA, 1990.
167 C.T. Mant and R.S. Hodges, in K.M. Gooding and F.E. Regnier (Editors), *HPLC of Biological Macromolecules: Methods and Applications,* Dekker, New York, 1990, p. 301.
168 M.T.W. Hearn (Editor), *HPLC of Proteins, Peptides and Polynucleotides,* VCH, Weinheim, in press.
169 S. Blackburn (Editor), *Handbook of Chromatography: Peptides,* CRC Press, Boca Raton, FL, 1986.
170 W.S. Hancock (Editor), *Handbook of HPLC for the Separation of Amino Acids, Peptides and Proteins,* Vols. I and II, CRC Press, Boca Raton, FL, 1984.
171 C.T. Mant and R.S. Hodges (Editors), *HPLC of Peptides and Proteins: Separation, Analysis and Conformation,* CRC Press, Boca Raton, FL, 1991.
172 A.S. Cohen and B.L. Karger, *J. Chromatogr.,* 397 (1987) 409.
173 P.D. Grossman, K.J. Wilson, G. Petrie and H.H. Lauer, *Anal. Biochem.,* 173 (1988) 265.
174 J.C. Colburn, P.D. Grossman, S.E. Moring and H.H. Lauer, in C.T. Mant and R.S. Hodges (Editors), *HPLC of Peptides and Proteins: Separation, Analysis and Conformation,* CRC Press, Boca Raton, FL, 1991.
175 T.D. Schlabach and C.T. Wehr, *Anal. Biochem.,* 127 (1982) 222.
176 C.T. Wehr, L. Correia and S.R. Abbott, *J. Chromatogr. Sci.,* 20 (1982) 114.
177 T.D. Schlabach, *J. Chromatogr.,* 266 (1983) 427.
178 K.L. Lockhart, R.A. Kenley and M.O. Lee, *J. Liq. Chromatogr.,* 10 (1987) 2999.
179 M.W. White, *J. Chromatogr.,* 262 (1983) 420.
180 L.H. Fleming and N.C. Reynolds, *J. Liq. Chromatogr.,* 7 (1984) 793.
181 A. Sauter and W.J. Frick, *J. Chromatogr.,* 297 (1984) 215.

182 A.F. Spatola and D.E. Benovitz, *J. Chromatogr.,* 327 (1985) 165.
183 C.T. Mant and R.S. Hodges, *J. Liq. Chromatogr.,* 12 (1989) 139.
184 C.T. Mant, J.M.R. Parker and R.S. Hodges, *J. Chromatogr.,* 397 (1987) 99.
185 G.D. Swergold and C.S. Rubin, *Anal. Biochem.,* 131 (1983) 295.
186 H. Sasaki, T. Matsuda, O. Ishikawa, T. Takamatsu, K. Tanaka, Y. Kato and T. Hashimoto, *Science Report of Toyo Soda, Mfg. Co., Ltd.,* 29 (1985) 37.
187 K. Yasukawa, M. Kasai, Y. Yanagihara and K. Noguchi, *J. Chromatogr.,* 332 (1985) 287.
188 H. Yoshida and S. Naijo, *Anal. Biochem.,* 159 (1986) 273.
189 M.C. Bañó, L. Braco and C. Abad, *J. Chromatogr.,* 458 (1988) 105.
190 F.E. Regnier, *Methods Enzymol.,* 91 (1983) 137.
191 T. Tomono, T. Suzuki and E. Tokunaga, *Anal. Biochem.,* 123 (1982) 394.
192 J.E. Rivier, *J. Chromatogr.,* 202 (1980) 211.
193 Y. Kato, K. Komiya, H. Sasaki and T. Hashimoto, *J. Chromatogr.,* 193 (1980) 29.
194 H. Mabuchi and H. Nakahashi, *J. Chromatogr.,* 213 (1981) 275.
195 H. Mabuchi and H. Nakahashi, *J. Chromatogr.,* 228 (1982) 292.
196 Y. Shioya, H. Yoshida and T. Nakajimi, *J. Chromatogr.,* 240 (1982) 341.
197 S.Y.M. Lau, A.K. Taneja and R.S. Hodges, *J. Biol. Chem.,* 259 (1984) 13253.
198 P.P. Shields and C.C. Glembotski, *Biochem. Biophys. Res. Commun.,* 146 (1987) 547.
199 C.T. Mant and R.S. Hodges, *J. Chromatogr.,* 326 (1985) 349.
200 H. von Bahr-Londström, U. Moberg, J. Sjödahl and H. Jörnvall, *Bioscience Reports,* 2 (1982) 803.
201 A.K. Taneja, S.Y.M. Lau and R.S. Hodges, *J. Chromatogr.,* 317 (1984) 1.
202 T. Yandle, I. Crozier, G. Nicholls, E. Espiner, A. Carne and S. Brennan, *Biochem. Biophys. Res. Commun.,* 146 (1987) 832.
203 P. Iadarola, M.C. Zapponi, M. Stoppini, M.L. Meloni, L. Minchiotti, M. Galliano and G. Ferri, *J. Chromatogr.,* 443 (1988) 317.
204 U. Bläsi, R.P. Linke and W. Lubitz, *J. Immunol. Methods,* 108 (1988) 209.
205 M.A. Vijayalakshmi, L. Lemieux and J. Amiot, *J. Liq. Chromatogr.,* 9 (1986) 3559.
206 S.Y.M. Lau, A.K. Taneja and R.S. Hodges, *J. Chromatogr.,* 317 (1984) 129.
207 M. Pohl, A. Carayon, F. Cesselin and M. Hamon, *J. Neurochem.,* 51 (1988) 1407.
208 E. Pfannkoch, K.C. Lu and F.E. Regnier, *J. Chromatogr. Sci.,* 18 (1980) 430.
209 H. Engelhardt and D. Mathes, *Chromatographia,* 14 (1981) 325.
210 H. Engelhardt, G. Ahr and M.T.W. Hearn, *J. Liq. Chromatogr.,* 4 (1981) 1361.
211 W. Kopaciewicz and F.E. Regnier, *Anal. Biochem.,* 126 (1982) 8.
212 C.T. Mant and R.S. Hodges, in C.T. Mant and R.S. Hodges (Editors), *HPLC of Peptides and Proteins: Separation, Analysis and Conformation,* CRC Press, Boca Raton, FL, 1991.
213 R.S. Hodges, J.M.R. Parker, C.T. Mant and R.S. Sharma, *J. Chromatogr.,* 458 (1988) 147.
214 G.D. Swergold, O.M. Rosen and C.S. Rubin, *J. Biol. Chem.,* 257 (1982) 4207.
215 C.T. Mant and R.S. Hodges, *J. Chromatogr.,* 327 (1985) 147.
216 T.W.L. Burke, C.T. Mant, J.A. Black and R.S. Hodges, *J. Chromatogr.,* 476 (1989) 377.
217 D.L. Crimmins, J. Gorka, R.S. Thoma and B.D. Schwartz, *J. Chromatogr.,* 443 (1988) 63.
218 A.J. Alpert and P.C. Andrews, *J. Chromatogr.,* 443 (1988) 85
219 D.L. Crimmins, R.S. Thoma, D.W. McCourt and B.D. Schwartz, *Anal. Biochem.,* 176 (1989) 255.
220 U.-H. Stenman, T. Laatikainen, K. Salminen, M.-L. Huhtala and J. Leppäluoto, *J. Chromatogr.,* 297 (1984) 399.
221 J. Eng, C.-G. Huang, Y.-C.E. Pan, J.D. Hulmes and R.S. Yalow, *Peptides,* 8 (1987) 165.
222 P. Mychack and J.R. Benson, *LC.GC,* 4 (1986) 463.
223 S. Burman, E. Breslow, B.T. Chait and T. Chaudhary, *J. Chromatogr.,* 443 (1988) 285.

224 T. Isobe, T. Takayasu, N. Takai and T. Okuyama, *Anal. Biochem.*, 122 (1982) 417.
225 N. Takahashi, Y. Takahashi and F.W. Putnam, *J. Chromatogr.*, 266 (1983) 511.
226 P.J. Cachia, J. Van Eyk, P.C.S. Chong, A. Taneja and R.S. Hodges, *J. Chromatogr.*, 266 (1983) 651.
227 S. James and H.P.J. Bennett, *J. Chromatogr.*, 326 (1985) 329.
228 P.A. Doris, *J. Chromatogr.*, 336 (1984) 392.
229 Y. Sakanoue, E. Hashimoto, K. Mizuta, H. Kondo and H. Yamamura, *Eur. J. Biochem.*, 168 (1987) 669.
230 B.S. Welinder and S. Linde, in W.S. Hancock (Editor), *Handbook of HPLC for the Separation of Amino Acids. Peptides and Proteins*, Vol. II, CRC Press, Boca Raton, FL, 1984, p. 357.
231 M.A. Jimenez, M. Rico, J.L. Nieto and A.M. Gutierrez, *J. Chromatogr.*, 360 (1986) 288.
232 N. Takahashi, N. Ishioka, Y. Takahashi and F.W. Putnam, *J. Chromatogr.*, 326 (1985) 407.
233 M. Dizdaroglu, *J. Chromatogr.*, 334 (1985) 49.
234 K.M. Gooding, M.N. Schmuck and M.P. Nowlan, *ACS Meeting, Miami, FL, September, 1989.*
235 J.R. McDermott and A.M. Kidd, *J. Chromatogr.*, 296 (1984) 231.
236 K.M. Gooding and M.N. Schmuck, *J. Chromatogr.*, 327 (1985) 139.
237 R.L. Patience and L.H. Rees, *J. Chromatogr.*, 352 (1986) 241.
238 H. Nika and T. Hultin, *Methods Enzymol.*, 91 (1983) 359.
239 J. Gariépy, B.D. Sykes and R.S. Hodges, *Biochemistry*, 22 (1983) 1765.
240 W. Kopaciewicz, M.A. Rounds, J. Fansnaugh and F.E. Regnier, *J. Chromatogr.*, 266 (1983) 3.
241 L.A. Kennedy, W. Kopaciewicz and F.E. Regnier, *J. Chromatogr.*, 359 (1986) 73.
242 M.A. Rounds,, W.D. Rounds and F.E. Regnier, *J. Chromatogr.*, 397 (1987) 25.
243 P.C. Andrews, *Peptide Research*, 1 (1988) 93.
244 M.P. Nowlan and K.M. Gooding, in C.T. Mant and R.S. Hodges (Editors), *HPLC of Peptides and Proteins: Separation, Analysis and Conformation*, CRC Press, Boca Raton, FL, 1991.
245 Z. Iskandarini and D.J. Pietrzyk, *Anal. Chem.*, 53 (1981) 489.
246 Z. Iskandarini, R.L. Smith and D.J. Pietrzyk, *J. Liq. Chromatogr.*, 7 (1984) 111.
247 T. Sasagawa, L.E. Ericsson, D.C. Teller, K. Titani and K.A. Walsh, *J. Chromatogr.*, 307 (1984) 29.
248 V.-F. Maa and Cs. Horváth, *J. Chromatogr.*, 445 (1988) 71.
249 K.J. Wilson, E. van Wieringen, S. Klauser, M.W. Berchtold and G.J. Hughes, *J. Chromatogr.*, 237 (1982) 407.
250 M. Hermodson and W.C. Mahoney, *Methods Enzymol.*, 91 (1983) 352.
251 K. Kalghatgi and Cs. Horváth, *J. Chromatogr.*, 398 (1987) 335.
252 K. Kalghatgi and Cs. Horváth, *J. Chromatogr.*, 443 (1988) 343.
253 K. Kalghatgi and Cs. Horváth, in C.T. Mant and R.S. Hodges (Editors), *HPLC of Peptides and Proteins: Separation, Analysis and Conformation*, CRC Press, Boca Raton, FL, 1991.
254 K.J. Wilson, A. Honegger, R.P. Stötzel and G.J. Hughes, *Biochem. J.*, 199 (1981) 31.
255 K.J. Wilson, A. Honegger and G.J. Hughes, *Biochem. J.*, 199 (1981) 43.
256 D.R. Knighton, D.R.K. Harding, J.R. Napier and W.S. Hancock, *J. Chromatogr.*, 249 (1982) 193.
257 W.C. Mahoney, *Biochim. Biophys. Acta*, 704 (1982) 284.
258 P.A. Schueler, R.P. Eide, W.S. Herman and W.C. Mahoney, *J. Neurochem.*, 47 (1986) 133.
259 K. Hermann, R.E. Lang, Th. Unger, C. Bayer and D. Ganten, *J. Chromatogr.*, 312 (1984) 273.
260 C.T. Mant and R.S. Hodges, *LC.GC*, 4 (1986) 250.
261 D. Guo, C.T. Mant, A.K. Taneja, J.M.R. Parker and R.S. Hodges, *J. Chromatogr.*, 359 (1986) 499.

262 E.M. Dotimas, K.R. Hamid, R.C. Hider and U. Ragnarsson, *Biochim. Biophys. Acta,* 911 (1987) 285.
263 R. Corder, R.C. Gaillard and P. Böhlen, *Regul. Pept.,* 21 (1988) 253.
264 L. Tan, *J. Chromatogr.,* 266 (1983) 67.
265 A.F. Spatola and D.E. Benovitz, *J. Chromatogr., 327 (1985) 165.*
266 R.E. Koeppe, II, J.H. Haw and J.A. Paczkowski, *FEBS Lett.,* 183 (1985) 313.
267 W.O. Richter and P. Schwandt, *J. Neurochem.,* 44 (1985) 1697.
268 P.D. Marley, K.I. Mitchelhill and B.G. Livett, *Brain Res.,* 363 (1986) 10.
269 D.C. Liebisch, E. Weber, B. Kosicka, C. Gramsch, A. Herz and B.R. Seizinger, *Proc. Natl. Acad. Sci. US,* 83 (1986) 1936.
270 G. Szókan, A. Török and B. Penke, *J. Chromatogr., 387 (1987) 267.*
271 H.P.J.M. Noteborn, I. Ebels, A.C. Reinharz, P. Pévet, B. Benson and C.A. Salemink, *J. Pineal Res.,* 5 (1988) 573.
272 M. Fenger and A.H. Johnsen, *J. Endocrinol.,* 118 (1988) 329.
273 L. Lozzi, M. Rustici, A. Santucci, L. Bracci, S. Petreni, P. Soldani and P. Neri, *J. Liq. Chromatogr.,* 11 (1988) 1651.
274 G.R. Rhodes, M.J. Rubenfield and C.T. Garvie, *J. Chromatogr.,* 488 (1989) 456.
275 B. Fransson, L. Grehn and U. Ragnarsson, *J. Chromatogr.,* 328 (1985) 396.
276 B. Fransson, *J. Chromatogr.,* 361 (1986) 161.
277 L. Gozzini and P.C. Montecucchi, *J. Chromatogr.,* 362 (1986) 138.
278 K.D. Lork, K.K. Unger, H. Brückner and M.T.W. Hearn, *J. Chromatogr.,* 476 (1989) 135.
279 J.M. Tomich, L. Wulf Carson, K.J. Kanes, N.J. Vogelaar, M.R. Emerling and J.H. Richards, *Anal. Biochem.,* 174 (1988) 197.
280 D. Guo, C.T. Mant, A.K. Taneja and R.S. Hodges, *J. Chromatogr.,* 359 (1986) 519.
281 Cs. Horváth, W. Melander and I. Molnár, *J. Chromatogr.,* 125 (1976) 129.
282 Cs. Horváth, W. Melander, I. Molnár and P. Molnár, *Anal. Chem.,* 49 (1977) 2295.
283 N.E. Hoffmann and J.C. Liao, *Anal. Chem.,* 49 (1977) 2231.
284 J.C. Kraak, K.M. Jonker and J.F.K. Huber, *J. Chromatogr.,* 142 (1977) 671.
285 P.T. Kissinger, *Anal. Chem.,* 49 (1977) 883.
286 J.L.M. van de Venne, J.L.H.M. Hendrikx and R.S. Deelder, *J. Chromatogr.,* 167 (1978) 1.
287 D. Guo, C.T. Mant and R.S. Hodges, *J. Chromatogr.,* 386 (1987) 205.
288 E. Spindel, D. Pettibone, L. Fisher, J. Fernstrom and R. Wurtman, *J. Chromatogr.,* 222 (1981) 381.
289 J. Ishida, M. Kai and Y. Ohkura, *J. Chromatogr.,* 356 (1986) 171.
290 S. Kimura, Y. Sugita, I. Kanazawa, A. Saito and K. Goto, *Neuropeptides,* 9 (1987) 75.
291 H.P.J. Bennett, C.A. Browne and S. Solomon, *J. Liq. Chromatogr.,* 3 (1980) 1353.
292 W.M.M. Schaaper, D. Voskamp and C. Olieman, *J. Chromatogr.,* 195 (1980) 181.
293 D.R.K. Harding, C.A. Bishop, M.F. Tartellin and W.S. Hancock, *Int. J. Peptide Protein Res.,* 18 (1981) 314.
294 H.P.J. Bennett, *J. Chromatogr.,* 266 (1983) 501.
295 W.S. Hancock and D.R.K. Harding, in W.S. Hancock (Editor), *Handbook of HPLC for the Separation of Amino Acids, Peptides and Proteins,* Vol. II, CRC Press, Boca Raton, FL, 1984, p.3.
296 E.V. Titova, M.A. Chlenov and L.I. Kudrjashov, *J. Chromatogr.,* 364 (1986) 209.
297 A.N. Starratt and M.E. Stevens, *J. Chromatogr.,* 194 (1980) 421.
298 M.T.W. Hearn and B. Grego, *J. Chromatogr.,* 203 (1981) 349.
299 P.A. Hartman, J.D. Stodola, G.C. Harbour and J.G. Hoogerheide, *J. Chromatogr.,* 360 (1986) 385.
300 H. Kawasaki, S. Imajoh, S. Kawashima, H. Hayashi and K. Suzuki, *J. Biochem.,* 99 (1986) 1525.
301 M. Tamura, Y. Yubisui, M. Takeshita, S. Kawabata, T. Miyata and S. Iwanaga, *J. Biochem.,* 101 (1987) 1147.
302 R.A. Bank, B.C. Crusins, T. Zwiers, S.G.M. Meuwissen, F. Arwert and J.C. Pronk, *FEBS Lett.,* 238 (1988) 105.
303 E.E. Huston, J.-C. Grammer and R.G. Yount, *Biochemistry,* 27 (1988) 8945.

304 R.F. Ebert and C.H. Schmelzer, *J. Chromatogr.,* 443 (1988) 309.
305 M.R. Mauk and A.G. Mauk, *J. Chromatogr.,* 439 (1988) 408.
306 J.I. Rushbrook, A.G. Wadewitz, M. Elzinga, T.-T. Yao and R.G. Somes, Jr., *Biochemistry,* 27 (1988) 8953.
307 M. Lebl, W.L. Cody and V.J. Hruby, *J. Liq. Chromatogr.,* 7 (1984) 1195.
308 W.L. Cody, B.C. Wilkes and V.J. Hruby, *J. Chromatogr.,* 314 (1984) 313.
309 M.A. Heinemann, J.N. Whitaker and J.M. Seyer, *J. Chromatogr.,* 329 (1985) 295.
310 E.M. Brown, H.J. Dower and R. Greenberg, *J. Chromatogr.,* 443 (1988) 247.
311 V.J. Hruby, A. Wawasaki and G. Toth, in C.T. Mant and R.S. Hodges (Editors), *HPLC of Peptides and Proteins: Separation, Analysis and Conformation,* CRC Press, Boca Raton, FL, 1991.
312 W.S. Hancock, C.A. Bishop, R.L. Prestidge, D.R.K. Harding and M.T.W. Hearn, *J. Chromatogr.,* 153 (1978) 391.
313 B. Grego, F. Lambrou and M.T.W. Hearn, *J. Chromatogr.,* 266 (1983) 89.
314 T. Imito and H. Yamada, *Mol. Cell. Biochem.,* 51 (1983) 111.
315 H. Gaertner and A. Puigserver, *J. Chromatogr.,* 350 (1985) 279.
316 M.T.W. Hearn, A.N. Hodder and M.-I. Aguilar, *J. Chromatogr.,* 327 (1985) 47.
317 G. Flouret, T. Majewski, D.R. Peterson, A.J. Kenny and F.A. Carone, *Am. J. Physiol.,* 252 (1987) E320.
318 H.-P. Fiedler, T. Hörner and A. Wörn, *Chromatographia,* 24 (1987) 433.
319 M. van der Rest, H.P.J. Bennett, S. Solomon and F.H. Glorieux, *Biochem. J.,* 191 (1980) 253.
320 H.P.J. Bennett, C.A. Browne and S. Solomon, *Biochemistry,* 20 (1981) 501.
321 C.A. Browne, H.P.J. Bennett and S. Solomon, *Anal. Biochem.,* 124 (1982) 201.
322 A.W. Burgess, J. Knesel, L.G. Sparrow, N.A. Nicola and E.C. Nice, *Proc. Natl. Acad. Sci. US,* 79 (1982) 5753.
323 K. Yoshizaki, V. de Bock, I. Takai, N.-S. Wang and S. Solomon, *Regul. Pept.,* 14 (1986) 11.
324 M.C. Cappell, K.B. Brosnihan, W.R. Welches and C.M. Ferrario, *Peptides,* 8 (1987) 939.
325 V. Di Marzo, A. Etienne, G. Marino, H.R. Morris and A. Palmisano, *Neuropeptides,* 6 (1985) 53.
326 A. Palmisano, G. Marino, V. Di Marzo, H.R. Morris, T.A. Howlett and S. Tomlin, *Neuropeptides,* 7 (1986) 281.
327 H. Yamada, T. Imoto and S. Noshita, *Biochemistry,* 21 (1982) 2187.
328 D.G. Smyth, *Anal. Biochem.,* 136 (1984) 127.
329 M. Hayakawa, I. Kudo, M. Tomita, S. Nojima and K. Inoue, *J. Biochem.,* 104 (1988) 767.
330 B. Giros, C. Llorens-Cortes, C. Gros and J.-C. Schwartz, *Peptides,* 7 (1986) 669.
331 C.V. Yang, W.A. Gonnerman, L. Taylor, R.B. Nimberg and P.R. Polgar, *Endocrinology,* 120 (1987) 63.
332 A.I. Smith and J.R. McDermott, *J. Chromatogr.,* 306 (1984) 99.
333 J.E. Rivier, R. McClintock, R. Galyean and H. Anderson, *J. Chromatogr.,* 288 (1984) 303.
334 T. Kanmera and R.P. Sequeria, in R.S. Rapaka and R.L. Hawks (Editors), *Opioid Peptides: Molecular Pharmacology, Biosynthesis and Analysis,* Research Monograph series 70, National Institute on Drug Abuse, Rockville, MD, 1986, p. 319.
335 M. Castagnola, L. Cassiano, R. DeCristofaro, R. Landolfi, D.V. Rossetti and G.B.M. Bettolo, *J. Chromatogr.,* 440 (1988) 231.
336 J.M. Piot, D. Guillochon and D. Thomas, *Chromatographia,* 25 (1988) 307.
337 M.T.W. Hearn, M. Guthridge and J. Bertolini, *J. Chromatogr.,* 397 (1987) 371.
338 P.G. Stanton, B. Grego and M.T.W. Hearn, *J. Chromatogr.,* 296 (1984) 189.
339 M.T.W. Hearn, B. Grego and C.A. Bishop, *J. Liq. Chromatogr.,* 4 (1981) 1725.
340 R.M. Kamp, Z.-J. Yao and B. Wittmann-Liebold, *Hoppe-Seyler's Z. Physiol. Chem.,* 364 (1983) 141.
341 G.E.D. Jackson and N. M. Young, *Anal. Biochem.,* 162 (1987) 251.
342 J.E. Rivier, *J. Liq. Chromatogr.,* 1 (1978) 343.

343 H. Antoniotti, P. Fagot-Revurat, J.P. Esteve, D. Fourmy, L. Pradayrol and A. Ribet, *J. Chromatogr.*, 296 (1984) 181.
344 B.S. Welinder, S. Linde, B. Hansen and O. Sonne, *J. Chromatogr.*, 298 (1984) 41.
345 B.S. Welinder, S. Linde and B. Hansen, *J. Chromatogr.*, 348 (1985) 347.
346 G.H. Fridland and D.M. Desiderio, *J. Chromatogr.*, 379 (1986) 251.
347 C.T. Mant and R.S. Hodges, *Chromatographia*, 24 (1987) 805.
348 A. Vandermeers, P. Gourlet, M.-C. Vandermeers-Piret, A. Cauvin, P. De Neef, J. Rathe, M. Svoboda, P. Robberecht and J. Christophe, *Eur. J. Biochem.*, 164 (1987) 321.
349 B.S. Welinder, H.H. Sorensen and B. Hansen, *J. Chromatogr.*, 398 (1987) 309.
350 F.-S. Shen and I. Lindberg, *Neuropeptides*, 13 (1989) 17.
351 S. Linde and B.S. Welinder, in C.T. Mant and R.S. Hodges (Editors), *HPLC of Peptides and Proteins: Separation, Analysis and Conformation*, CRC Press, Boca Raton, FL, 1991.
352 D.M. Desiderio and M.D. Cunningham, *J. Liq. Chromatogr.*, 4 (1981) 721.
353 D. Liu and D.M. Desiderio, *J. Chromatogr.*, 422 (1987) 61.
354 A. Cupo, J.P. Niel, J.P. Miolan, Y. Jule and Th. Jarry, *Neuropeptides*, 12 (1988) 257.
355 H. Ong, A. De Léan and C. Gagnon, *Clin. Chem.*, 34 (1988) 2775.
356 T.-T. Nguyen, A. De Léan and H. Ong, *Anal. Biochem.*, 179 (1989) 24.
357 M.A. Stadalius, M.A. Quarry and L.R. Snyder, *J. Chromatogr.*, 327 (1985) 93.
358 J.L. Glajch, M.A. Quarry, J.F. Vasta and L.R. Snyder, *Anal. Chem.*, 58 (1986) 280.
359 J.L. Meek, *Proc. Natl. Acad. Sci. US*, 77 (1980) 1632.
360 J.L. Meek and Z.L. Rossetti, *J. Chromatogr.*, 211 (1981) 15.
361 S.-J. Su, B. Grego, B. Niven and M.T.W. Hearn, *J. Liq. Chromatogr.*, 4 (1981) 1745.
362 W.H. Vensel, V.S. Fujita, G.E. Tarr, E. Margoliash and H. Kayser, *J. Chromatogr.*, 266 (1983) 491.
363 M.T.W. Hearn and B. Grego, *J. Chromatogr.*, 296 (1984) 61.
364 M.T.W. Hearn and B. Grego, *J. Chromatogr.*, 296 (1984) 309.
365 M. Kai, T. Miyazaki, Y. Sakamoto and Y. Ohkura, *J. Chromatogr.*, 322 (1985) 473.
366 J.J. L'Italien, *J. Chromatogr.*, 359 (1986) 213.
367 C. Carles and B. Ribadeau-Dumas, *J. Dairy Res.*, 53 (1986) 595.
368 B.S. Welinder, H.H. Sorensen and B. Hansen, *J. Chromatogr.*, 408 (1987) 191.
369 R.E. Carraway and S.P. Mitra, *Regul. Pept.*, 18 (1987) 139.
370 S. Görög, B. Herényi, O. Nyéki, I. Schön and L. Kisfaludy, *J. Chromatogr.*, 452 (1988) 317.
371 W.C. Duckworth, F.G. Hamel, J. Liepnicks, B.H. Frank, C. Yagil and R. Rabkin, *Endocrinology*, 123 (1988) 2701.
372 R.H. Snider, C.F. Moore, E.S. Nylen and K.L. Becker, *Horm. Metab. Res.*, 20 (1988) 254.
373 K.J. Wilson and G.J. Hughes, *Chimia*, 35 (1981) 327.
374 L.B. Klickstein and B.U. Wintroub, *Anal. Biochem.*, 120 (1982) 146.
375 R.L. Emanuel, G.H. Williams and R.W. Giese, *J. Chromatogr.*, 312 (1984) 285.
376 C.T. Mant, T.W.L. Burke and R.S. Hodges, *Chromatographia*, 24 (1987) 565.
377 T.W.L. Burke, C.T. Mant and R.S. Hodges, in C.T. Mant and R.S. Hodges (Editors), *HPLC of Peptides and Proteins: Separation, Analysis and Conformation*, CRC Press, Boca Raton, FL, 1991.
378 T. Sasagawa, T. Okuyama and D.C. Teller, *J. Chromatogr.*, 240 (1982) 329.
379 C.T. Mant and R.S. Hodges, in M.T.W. Hearn (Editor), *HPLC of Proteins, Peptides and Polynucleotides*, VCH, Weinheim, in press.
380 M.J. O'Hare and E.C. Nice, *J. Chromatogr.*, 171 (1979) 209.
381 E.C. Nice, M.W. Capp, N. Cooke and M.J. O'Hare, *J. Chromatogr.*, 218 (1981) 569.
382 C.T. Mant, T.W.L. Burke, J.A. Black and R.S. Hodges, *J. Chromatogr.*, 458 (1988) 193.
383 N.E. Zhou, C.T. Mant and R.S. Hodges, *Peptide Res.*, 3 (1990) 1.
384 C.T. Mant and R.S. Hodges, in C.T. Mant and R.S. Hodges (Editors), *HPLC of Peptides and Proteins: Separation, Analysis and Conformation*, CRC Press, Boca Raton, FL, 1991.

385 E.T. Kaiser and F.J. Kézdy, *Science,* 223 (1984) 249.
386 C.T. Mant, T.W.L. Burke, N.E. Zhou, J.M.R. Parker and R.S. Hodges, *J. Chromatogr.,* 485 (1989) 365.
387 J.W. Dolan and L.R. Snyder, *LC.GC,* 5 (1987) 971.
388 B.F.D. Ghrist, B.S. Cooperman and L.R. Snyder, *J. Chromatogr.,* 459 (1988) 1.
389 B.F.D. Ghrist and L.R. Snyder, *J. Chromatogr.,* 459 (1988) 25.
390 B.F.D. Ghrist and L.R. Snyder, *J. Chromatogr.,* 459 (1988) 43.
391 L.R. Snyder, J.W. Dolan and D.C. Lommen, *J. Chromatogr.,* 485 (1989) 65.
392 J.W. Dolan, D.C. Lommen and L.R. Snyder, *J. Chromatogr.,* 485 (1989) 9.
393 R.W. Giese, J.K. Haken, K. Macek and L.R. Snyder (Editors), Computer assisted method development in chromatography, *J. Chromatogr.,* 485 (1989).

Chapter 14

Proteins

FRED E. REGNIER and KAREN M. GOODING

CONTENTS

14.1 INTRODUCTION

Chromatographic fractionation of proteins began three decades ago with the introduction of carbohydrate gel-type ion-exchange and size-exclusion media [1,2]. During the ensuing twenty years the number of separation modes available for proteins grew through the addition of gel-type affinity [3], hydroxyapatite [4], and hydrophobic-interaction chromatography (HIC) [5] columns. A typical column of this era was packed with a soft carbohydrate gel of >70-μm particle diameter and operated at a linear velocity of <60 cm/h. One or more of these columns was used in the purification of most proteins before 1980.

As separation technology advanced, it became apparent that chromatographic materials of >50-μm particle diameter had serious mass-transfer limitations. The slow rate of diffusion of macromolecules into such large-particle-diameter materials limited both their resolution and separation speed. This led to the production of rigid, 10-μm-particle-diameter sorbents for proteins and the introduction of HPLC [6]. To achieve the high degree of mechanical strength required for operation at mobile-phase velocities of 300-

500 cm/h, HPLC supports are very rigid materials, such as silica or highly crosslinked organic resins. A secondary property of these rigid materials is that they do not shrink or swell. This is an enormous advantage in reversed-phase chromatography (RPC) where it is necessary to use organic solvent gradients.

The molecular size and structural lability of proteins are significant considerations in column and mobile-phase selection. Because proteins may range from ten to several hundred kilodaltons in size, it is possible that large proteins may be sterically excluded from porous sorbents unless the pores are very large. Pores larger than 300 Å generally allow the penetration of most proteins [7]. However, it should be appreciated that solute penetration of the pore matrix is not the only consideration. When the pore diameter of the column packing approaches the size of the protein, diffusion into the pore matrix will be restricted and produce extensive bandspreading. It is preferable that the pore diameter be larger than five times the molecular dimensions of the solute to preclude diffusion restrictions in the pore matrix. An alternative solution to restricted pore diffusion is to eliminate the pores. Separations with nonporous sorbents are routinely achieved in 1-5 min [8]. Unfortunately, these materials are of much lower surface area and of little utility in semi-preparative and preparative separations.

Since the biological activity of proteins is closely related to their three-dimensional structure and this structure is labile, special consideration must be given to the selection of columns and operating conditions. For example, adsorption of proteins on sorbents that are not specifically designed for protein separations may lead to denaturation and poor recovery of biological activity. Packing materials for the separation of proteins generally have an external hydrophilic layer to which the stationary phase is attached, even in the case of HIC materials. Whenever proteins are so strongly adsorbed on packings that nonphysiological conditions must be used to achieve desorption, there is some risk of denaturation. This is particularly true in the case of RPC where very acidic mobile phases and organic solvents are required for elution.

This chapter will examine protein separations from three points of view: contributions to the separation process that are the result of protein structure, contributions that are related to the separation mode, and unique problems associated with the separation of certain classes of proteins.

14.2 PROTEIN STRUCTURE CONTRIBUTIONS

Primary, secondary, tertiary, and quaternary structure play a major role in the chromatographic behavior of proteins. In this discussion these levels of structure will be reduced to two components: (a) amino acid composition and (b) the distribution of these amino acids in the three-dimensional structure.

14.2.1 Chromatographic behavior based on amino acid composition

The twenty-one amino acids generally found in proteins may be divided into four categories: (a) basic, (b) acidic, (c) neutral hydrophobic, and (d) neutral hydrophilic. On the basis of this classification it is then possible to differentiate between proteins, depending on the number and ratio of various types of amino acids. The chances that any two proteins will have the same net charge, hydrophobicity, size, and affinity for chelated metals are extremely low.

14.2.1.1 Ion-exchange chromatography

Ion-exchange or electrostatic-interaction chromatography (IEC) is one of the most effective forms of chromatography for the separation of proteins. Through the immobilization of charged groups on a hydrophilic support matrix, it is possible to create sorbents that separate proteins predominantly on the basis of charge. Proteins are adsorbed on ion-exchange columns at low (0.01-0.05 M) salt concentration and are eluted with an ascending gradient of salt ranging up to 1.0 M [9]. Sulfonic acid and quaternary amine groups are the most widely used stationary phases for the preparation of strong-cation- and -anion-exchange sorbents, whereas carboxyl groups and tertiary amines are most frequently used in the preparation of the corresponding weak-cation- and -anion-exchange materials [10].

Optimization of separations on IEC columns is achieved by varying pH, ionic strength, the type of displacing ion, and gradient slope [11,12]. Proteins differ in the composition and ratio of charged amino acids, the microenvironment surrounding charged amino acids, and the degree of intramolecular interactions between charged groups. Consequently, there is generally a pH value at which the charge of any given protein differs from that of other species in the sample. Frequently, the optimum pH and ionic strength are found by trial and error, but pH maps, such as those seen in Fig. 14.1, can indicate optimum operating conditions [13].

The nature of the salt used for IEC has a large effect on selectivity, as can be seen in Fig. 14.2 [14]. The effects appear to be due to ionic interactions with the protein as well as with the functional groups on the sorbent because both anions and cations can change selectivity on any IEC column [11,12].

Secondary interactions also play a role in IEC. In particular, hydrophobic interactions can contribute to the retention mechanism when salts with a high molal salting-out coefficient (e.g., sodium sulfate or ammonium sulfate) are added to the mobile phase [15]. This is most important when there are hydrophobic moieties at the surface of the protein and the sorbent has some significant hydrophobic character.

Hydroxyapatite chromatography (HAC) might also be considered to be a form of IEC. The mineral matrix of hydroxyapatite is unique in that it is amphoteric and probably separates proteins by some combination of simultaneous cationic and anionic interactions [16].

References on p. B168

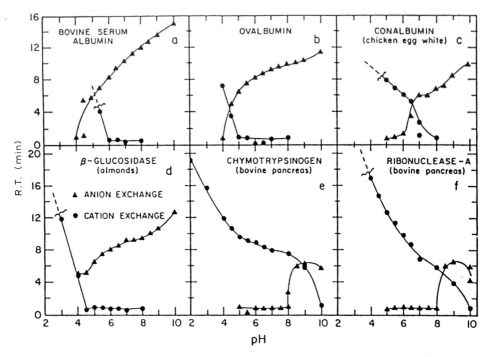

Fig. 14.1. Protein ion-exchange retention maps. Columns, Mono S or Mono Q, 50 x 5 mm ID; buffers were 0.01 *M* and were appropriate for the pH conditions; 20-min linear gradient from 0 to 0.5 *M* NaCl; flowrate, 1 ml/min. (Reprinted from Ref. 13, with permission.)

14.2.1.2 Reversed-phase chromatography

RPC of polypeptides on alkylsilane-derivatized silica was introduced in the Seventies [17]. Because RPC materials contain residual silanols that interact with basic amino acids, they can produce extensive peak trailing. This problem has been overcome by eluting columns with acidic mobile phases to repress the ionization of silanols [18]. In addition to repressing silanol ionization, acidic additives form ion pairs with amino groups in polypeptides. Essentially, these ion-pairing agents form dynamic derivatives with polypeptides and either decrease or increase chromatographic retention, depending on whether the ion-pairing agent is hydrophilic or hydrophobic. The incremental change in retention is determined by the number of amino groups in the molecules. Through the use of a variety of ion-pairing agents, it is now possible to purify almost any polypeptide to homogeneity on a single RPC column [19]. RPC is usually the method of choice for peptide analysis due to its superb resolution. Fig. 14.3 illustrates selectivity and peak shapes in the separation of angiotensin variants and other peptides.

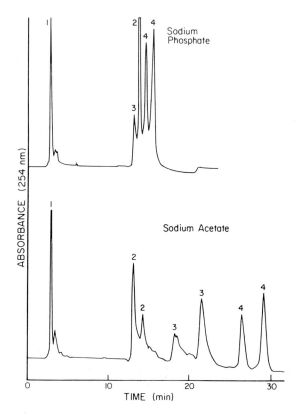

Fig. 14.2. Comparative selectivities for a four-protein mixture with two different salts in IEC. Column, SynChropak AX 300, 250 x 4.6 mm ID; buffer, 0.02 M Tris (pH 7); 30-min linear gradient from 0 to 1 M salt, as indicated; flowrate, 1 ml/min. Sample: 1 = myoglobin, 2 = conalbumin, 3 = ovalbumin, 4 = β-lactoglobulins B and A.

The hydrophobic interaction between polypeptides and RPC packings is so strong that organic solvents are generally required for elution. Unfortunately, these solvents alter the three-dimensional structure of most proteins and diminish their biological activity. For this reason, RPC is of lesser utility in the fractionation of large peptides and multimeric proteins that are difficult to refold [19].

14.2.1.3 Hydrophobic-interaction chromatography

HIC separates proteins by a hydrophobic-interaction mechanism under milder conditions than RPC [20]. The interaction between HIC columns and proteins is not strong enough to produce adsorption in buffers of less than 0.1 M concentration. Adsorption on these columns occurs only when salts with a high molal salting-out coefficient, such as ammonium sulfate, are added to increase the surface tension of the mobile phase. In the

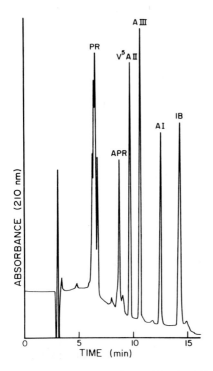

Fig. 14.3. Analysis of peptides by RPC. Column, SynChropak RP-P, 250 x 4.6 mm ID; mobile phase, A, 0.1% TFA in water; B, 0.1% TFA in 2-propanol; 30-min linear gradient from 5 to 100% B; flowrate, 1 ml/min. Sample: PR = protamine sulfate, APR = aprotinin, V⁵AII = Val⁵ angiotensin II, AIII = angiotensin III, AI = angiotensin I, IB = insulin B.

presence of 1 *M* salt, the solvophobic effect is sufficiently strong to make polypeptides interact with the HIC column hydrophobically, thus minimizing their surface area of contact with the polar mobile phase. Elution is achieved with a descending salt gradient, usually to 0.10 or 0.05 *M* salt. The nature of the salt can alter selectivity; however, the effect is less pronounced than in IEC [21,22]. The excellent selectivity which can be achieved by HIC is illustrated in the separation of calcitonin analogs, seen in Fig. 14.4 [23].

Although secondary, pH also plays a role in HIC separations [22]. Carboxyl and amino groups at the surface of a protein may be included in the hydrophobic interface between the solute and ligands on the support. When these groups are ionized, higher salt concentrations are required to keep the solute hydrophobically adsorbed.

14.2.1.4 Immobilized-metal affinity chromatography

IMAC was introduced by Porath et al. [24]. In this type of chromatography, a metal is immobilized at the surface of a sorbent through the aid of a chelating agent. Separations are based on interactions that occur between electron donors (histidine, tryptophan, and

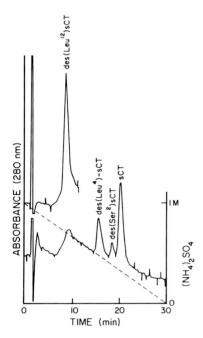

Fig. 14.4. HIC of calcitonin analogs. Column, SynChropak Propyl, 100 x 4.6 mm ID; buffer, 0.02 *M* phosphate (pH 7); 30-min gradient from 1 to 0 *M* ammonium sulfate; flowrate, 1 ml/min [23].

cysteine) on the polypeptide surface and immobilized metal atoms [Cu(II), Ni(II), Ca(II), Zn(II), Fe(II), and Fe(III)], which act as electron acceptors. Selectivity is influenced by the number of accessible electron-donating groups on the protein surface, the type of metal immobilized, and the mobile-phase pH [25]. Strongest binding of a polypeptide is achieved through histidine residues with immobilized Cu(II) at basic pH. The affinity of proteins for immobilized metal ions is highest for Cu(II) and increases in proportion to the number of surface histidine residues [25]. Elution is achieved by either a descending pH gradient or an ascending gradient of a displacing agent, such as imidazole at pH 5-8.

14.2.1.5 Affinity chromatography

Affinity chromatography (Chapter 7) was first used as a distinct separation method in the late Sixties [3]. Although "affinity" is a fundamental component of the retention process in all surface-mediated separations, this form of chromatography is generally based on the affinity of an immobilized biological species for another biological species being transported through a column. In most cases this technique exploits interactions between proteins and stationary ligands for purification. The ligand is chemically tethered to the surface of a support and specifically captures the protein. Two negative features of affinity chromatography are that different columns are required for different separations and that the capturing agents must be chemically immobilized on the support. This inconvenience has been ameliorated in recent years by the commercial introduction of preactivated, prepacked columns on which capturing agents can be immoblized in situ by brief incubation.

Desorption of bound materials from affinity columns is achieved by either specific displacement or partial denaturation of the protein. Specific displacement is most effective with weak, reversibly associating complexes. If the binding is very strong, extremes in pH and/or the use of organic solvents generally dissociate ligand/protein complexes.

14.2.2 Chromatographic behavior based on three-dimensional structure

Very early in the evolution of chromatography of polypeptides it was recognized that chromatographic behavior could not be predicted by amino acid composition alone; some proteins of identical amino acid composition are separable [26]. During the last decade, it became clear that the three-dimensional structure of a protein plays an important role in its interaction with surfaces and, therefore, in its chromatographic properties [29,30].

14.2.2.1 Size-exclusion chromatography

Size-exclusion chromatography (SEC) is based on the hydrodynamic volumes and thus on the tertiary structure of proteins. Although it does not have the resolution potential of interactive modes of chromatography, it is usually one of a series of steps in protein purification. It is a fast method of separating proteins from molecules dissimilar in molecular weight and of changing their environment by desalting [27].

Sorbents for HPSEC of proteins are designed to have a neutral carbohydrate-like layer which will have minimal interaction with proteins and will not render them biologically inactive [27]. The support matrices are composed of controlled-porosity silica or polymers with large porevolume to maximize peakcapacity.

Mobile phases for SEC generally contain 0.05-0.1 M salt to minimize ionic interactions or exclusion caused by free silanols on silica matrices or by carboxyl groups on polymers. All SEC supports have some ionic and hydrophobic characteristics. Hence, mobile phases must be formulated to eliminate any interactions [28]. For certain proteins, additives are necessary to ensure solubility or minimize adsorption. These additives include glycerol, SDS, surfactants, and organic solvents.

14.2.2.2 Role of surface amino acids on retention

Natural globular proteins have several structural features that set them apart from synthetic polymers. The most dramatic difference is that natural proteins may be equivalent in amino acid content and size but different in three-dimensional structure [30]. This is due to variations in the sequence of the amino acids and therefore in the folding of the polypeptide chains. Amino acid distribution within the three-dimensional structure of natural polypeptides is somewhat predictable. Polar amino acids tend to be more numerous at the external surface and in contact with water or polar solvents, whereas hydrophobic amino acids have a higher probability of being located in the interior of the molecule [31]. Additionally, the distribution of amino acids at the surface of proteins is not uniform. For example, there is a region on the surface of lysozyme, opposite the catalytic cleft, that is more hydrophobic than the rest of the surface [32]. Cytochrome c has a similar region on its surface in which negatively charged amino acids are concentrated [33]. The tetrameric protein, lactate dehydrogenase, has an unequal charge distribution, in that its H subunits are much more electronegative than the M subunits [29].

A second structural feature that sets proteins apart from synthetic polymers is that the polypeptide structure is semirigid under near-physiological conditions. Although the three-dimensional structure of a protein may be disrupted by organic solvents or extremes of pH, ionic strength, or heat, these polypeptides generally exist in a specific conformational state [31]. It is assumed that the tertiary structure of proteins is preserved when they are adsorbed on surfaces under physiological conditions, because it has been demonstrated that the biological activity of enzymes and antibodies is generally retained. In contrast, the structure of most synthetic polymers is a random coil, and they tend to conform to the surface on which they are adsorbed, crawling along the surface by a process called "reptation".

Another chromatographically important feature of proteins is their size relative to the dimensions of both the pore and the solvated bonded phases at the surface of the sorbents. As the solute size approaches the pore diameter, diffusion is severely restricted and bandspreading becomes extensive [7].

The structural features of proteins will have the following effects on chromatographic behavior:

(1) The size and structural rigidity of proteins make it impossible for all amino acids in the molecule to interact simultaneously with the stationary phase. By necessity, a small number of amino acids will determine the chromatographic behavior [29]. The contribution of any single amino acid is most likely the average of its contact time with the surface [29].

(2) Amino acids in the interior of a protein will not come in direct contact with the stationary phase and will have little impact on chromatographic behavior. This is particularly important in hydrophobic interactions, where the forces that determine chromatographic retention are of very short range. Electrostatic interactions, in contrast, have a longer range and can be effective even when a charged amino acid is not in direct contact with the surface [29].

(3) An asymmetric distribution of amino acids within the three-dimensional structure of proteins can cause one portion of the molecule to dominate the chromatographic behavior. In the cases of lysozyme, cytochrome c, and lactate dehydrogenase, cited above, certain portions of the surfaces have been found to be totally responsible for their chromatographic retention [29]. The most extreme case of this phenomenon is found in affinity chromatography. The active site of an enzyme or the 8-12 amino acid epitope of a protein antigen will totally determine its retention properties.

These multiple features of the interactions of semirigid macromolecular solutes with

References on p. B168

solid surfaces have led to the concept that there is a specific region(s) of a protein that interacts with a sorbent. This area of contact has been designated the "chromatographic contact region" [29] and may actually be composed of several regions on the protein surface that are not necessarily contiguous. Contact with these multiple regions may or may not be simultaneous. It may thus be concluded that only the amino acids in the chromatographic contact region make a contribution to retention. This means that differences between two molecular species will not be sensed by the chromatographic system if they are outside of the contact region.

The chromatographic contact region in one separation mode may be different from that in another mode. It was mentioned in Section 14.2.1 that different amino acids are involved in the various separation modes and that the structural distribution of amino acids responsible for chromatographic behavior varies greatly among proteins [29]. This explains why one separation mode can discriminate between protein variants when another mode fails. Separation cannot occur if the chromatographic contact regions are not the areas where the two molecules differ.

It is widely accepted that, although protein structure is semirigid, "proteins are not rocks" and there is great concern that protein structure may be altered by either the chromatographic process or some treatment to which a protein was exposed during purification [29]. If such structural changes alter the chromatographic contact region, there is a high probability that the change will be sensed by the chromatographic system. Conversely, structural alterations that do not impact that region will seldom be detected. This is particularly significant in biotechnology, where faulty protein biosynthesis, chemical alterations caused by purification procedures, or incorrect protein folding must be detected and rectified [19]. It is evident from this discussion that the three-dimensional structure of proteins is intimately tied to chromatographic behavior.

For the purpose of discussion we have assumed that the structure of proteins is not altered during the chromatographic process. This may not be true, particularly in the case of RPC. The use of organic solvents under acidic conditions can induce substantial alteration of protein structure [34,35]. When this happens, there is also a high probability that the chromatographic contact region will be altered, with a concomitant change in chromatographic behavior. It is important to consider where these structural alterations are occurring, the rate at which they are occurring during elution, the extent to which the column catalyzes conformational changes, and the degree of refolding that may occur after elution. Although most frequently associated with RPC, alteration of protein structure on a column can occur in any chromatographic mode. If the tertiary structure can be easily recovered, as in the case of small peptides, or if the purpose of chromatography is to acquire analytical data, on-column structural changes may be of no concern. However, in the preparative isolation of high-molecular-weight species with retention of biological activity, on-column structural alterations are definitely undesirable.

14.3 CHROMATOGRAPHIC BEHAVIOR OF SELECTED PROTEIN CLASSES

14.3.1 Hemoglobin

Human hemoglobin is a tetrameric protein composed of two sets of globin chain subunits, which may be modified by glycosylation or acetylation. Four globin chains are commonly found, but there are numerous others, which differ from them by one amino acid. By 1988, the number of known variants of human hemoglobin was 514 [36]. The excellent resolution of many different hemoglobin moieties by ion-exchange chromatography raised considerable interest in the use of HPLC because of its speed and potential for automation [37,38]. High-performance anion-exchange chromatography was used initially [39,40], but the superior resolving power of high-performance cation-exchange chromatography for hemoglobins has subsequently preempted anion exchange and occasioned a total changeover in techniques [41].

Identification and quantitation of hemoglobins that differ by a globin chain or an amino acid substitution (variants) or by the degree of glycosylation or acetylation are very important clinically. The identification of hemoglobin variants is necessary for diagnosis of such conditions as sickle cell anemia and thalassemia. Quantitation of these variants is also useful for monitoring clinical procedures, such as blood transfusions. Fig. 14.5 illustrates the excellent resolution of hemoglobin variants that can be achieved by HPLC. Hb S differs from Hb A_0 by only a single amino acid substitution in two of the four subunits. This minor difference affects the tertiary structure, chromatographic properties, and oxygen-binding characteristics of Hb S (sickle cell hemoglobin).

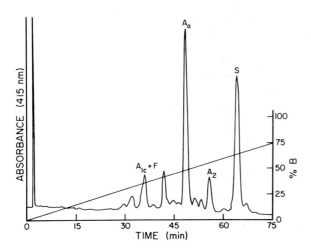

Fig. 14.5. Separation of hemoglobin variants by HPIEC. Column, SynChropak CM300, 250 x 4.6 mm ID; buffer, 0.03 M Bis-Tris/0.0015 M KCN (pH 6.4); 100-min linear gradient from 0 to 0.15 M sodium acetate; flowrate, 1 ml/min.

References on p. B168

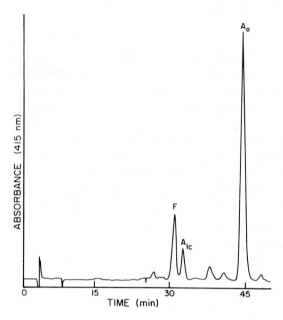

Fig. 14.6. Separation of Hb A_{1c} from other hemoglobins by HPIEC. Column, SynChropak CM300, 250 x 4.6 mm ID; buffer, 0.035 M Bis-Tris/0.0015 M KCN (pH 6.4); 83-min linear gradient from 0.003 M ammonium acetate to 0.017 M ammonium acetate/0.15 M sodium acetate; flowrate, 0.8 ml/min.

The level of glycosylated hemoglobin (Hb A_{1c}) in diabetic patients indicates the efficacy of their insulin and dietary regimens by reflecting the average glucose level of the preceding weeks. Glycosylated hemoglobin is found in normal blood cells at levels of 4-6%, but the percentage is increased in diabetes. The only difference between Hb A_0 and Hb A_{1c} is that the N-terminal valine of two of the chains are covalently bound to glucose. Fig. 14.6 shows the separation of Hb A_{1c} from Hb F, which allows the former to be accurately quantitated. It can be seen that Hb A_{1c} is chromatographically more similar to Hb F, which has two different subunit chains, than it is to Hb A_0, which differs only by two glucose units. This resolution of Hb A_{1c} and Hb F requires ammonium sulfate in the mobile phase and therefore is probably based on a mixed ion-exchange and hydrophobic-interaction mechanism.

Although cation-exchange chromatography has been the most effective technique for hemoglobin analysis, RPC has been used in a complementary role to identify and quantitate globin chains [42]. Certain variants cannot be differentiated when the intact proteins are analyzed by cation-exchange chromatography, but can be identified by RPC of their globin chains. This is an example of "silent" differences, which are not located in the chromatographic contact region when the protein is in its intact globular conformation. These different amino acids become exposed during denaturation and separation of the chains under the RPC conditions.

14.3.2 Cereal proteins

Cereal proteins are among the most difficult groups of proteins to analyze, due to their general insolubility in water or salt solutions, their tendency to aggregate, and their heterogeneity. Nevertheless, it is necessary to separate and identify cereal proteins for agricultural studies [43], including general protein characterizations [44], varietal identification by "fingerprinting" [45], genetic studies [46], prediction of grain quality [47,48], and assessment of effects of processing [49].

Although the insolubility of cereal proteins in water and salt solutions makes them difficult to analyze by traditional size-exclusion and ion-exchange techniques, their solubility in acid and organic solvents has made them well-suited for hydrophobic discrimination by RPC. Alkyl reversed-phase ligands (C_3-C_{18}) with standard mobile phases containing trifluoroacetic acid and acetonitrile have been used successfully to resolve cereal proteins. Profiles, such as that seen in Fig. 14.7 [50], allow one to compare

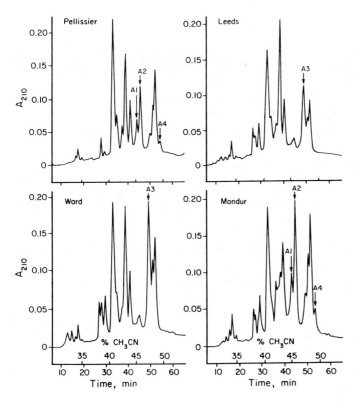

Fig. 14.7. Comparison of gliadin proteins from four wheat varieties by RPC. Column, SynChropak RP-P, 250 x 4.1 mm ID; Eluent A, 0.1% trifluoracetic acid/15% acetonitrile in water; Eluent B, 0.1% trifluoroacetic acid/80% acetonitrile in water; 55-min linear gradient from 20 to 55% B; flowrate, 1 ml/min. (Reprinted from Ref. 50, with permission.)

References on p. B168

protein components of different varieties of a specific grain or to identify varieties. The resolution of both wheat gliadins and corn zeins [51] can be enhanced by the use of elevated temperatures, which possibly cause disruption of aggregates. Because most of these studies are analytical rather than preparative in nature, disruption of tertiary structure caused by RPC is of no great consequence.

The difficulties in handling cereal proteins make auxiliary techniques, such as sample preparation, critical to successful and reproducible analyses. The cereal proteins must be extracted with a solvent which does not extract other, related classes of proteins. For example, a gliadin fraction should not include glutenins [52]. Additionally, the extraction solvent must not cause early elution in HPLC. Solvents that extract nonprotein classes of compounds, such as starches, must be avoided, since they may foul the column and necessitate cleaning it with solvents, such as DMSO [53].

SEC has had limited success in the analysis of cereal proteins. The use of detergents, such as SDS, in the mobile phase allows cereal proteins to be eluted according to their molecular weights. Although SEC cannot produce fingerprints or resolve as many different cereal proteins as RPC, it has been shown useful in predicting the breadmaking quality of wheat [47,54].

14.3.3 Glycoproteins

Glycoproteins are proteins containing covalently bound sugars. In these proteins, several different carbohydrates may be bound to many different amino acid locations in varying amounts, and this results in extensive microheterogeneity. For example, IgD is composed of more than 1000 different species, which differ only in carbohydrate content [55]. Because chromatographic systems do not have the capability of totally resolving so many closely related peaks, it is frequently better to use HPLC methods which do not resolve this carbohydrate-based microheterogeneity.

Ion-exchange methods can discriminate between proteins that differ by carbohydrate linkages. For certain simple glycoproteins, such as glycosylated hemoglobins and ovalbumin, this can result in excellent separations. This is the basis on which Hb A_{1c} is separated from Hb A_0, as shown in Fig. 14.6. However, more complex glycoproteins, such as IgD, hemopexin, and β-2-glycoprotein I (B-2-I), give no distinct separation between any of the protein forms on an anion-exchange column [56]. Broad and tailing peaks are obtained, because the columns slightly resolve the different forms. Fig. 14.8A illustrates the chromatogram of B-2-I [56].

Although HAC seemingly operates like IEC, it does not discriminate between small differences in the carbohydrate sidechains of glycoproteins (Fig. 14.8B) [56]. Generally, separations appear to be dependent on conformational differences rather than the quantity or position of carbohydrate groups [55]. Conformational differences may be due to the presence or absence of a carbohydrate sidechain.

In RPC, only a minimal effect on separations is caused by carbohydrate sidechains [57]. It is an excellent choice for those glycoproteins which are not adversely affected by the acidic and organic solvents used as eluents. Ceruloplasmin and IgD are examples of

β_2 - glycoprotein I

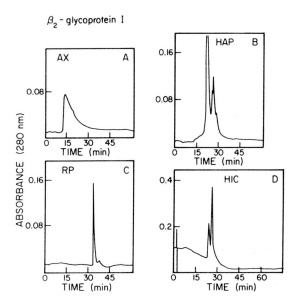

Fig. 14.8. Chromatographic characterization of human β_2-glycoprotein I by four modes of HPLC. Columns: (A) anion-exchange HPLC on a 25 x 0.4-cm-ID SynChropak AX 300 column, (B) 10 x 0.75-cm-ID hydroxyapatite, (C) reversed-phase HPLC on a 25 x 0.4-cm-ID SynChropak RP-P column, (D) hydrophobic-interaction chromatography on a 7.5 x 0.75-cm-ID TSK Phenyl 5PW column. In each case, the eluent was monitored by absorbance at 280 nm for protein. In A, the column was eluted at a flowrate of 1.0 ml/min with a linear gradient for 60 min from 0.02 M Tris/acetic acid buffer (pH 8.0) to 0.02 M Tris/acetic acid buffer (pH 8.0), containing 1.0 M sodium acetate. In B, elution was at a flowrate of 0.8 ml/min for 60 min with a linear gradient from 0.01 M sodium phosphate buffer (pH 6.8)/0.3 mM calcium chloride to 0.5 M sodium phosphate buffer (pH 6.8)/0.01 M calcium chloride. In C, the column was eluted at a flowrate of 1.0 ml/min with a linear gradient from 0.1% trifluoroacetic acid to 60% acetonitrile/0.1% trifluoroacetic acid during 60 min. In D, elution was at a flowrate of 1.0 ml/min with a gradient of decreasing salt concentration from 1.7 M ammonium sulfate/0.1 M sodium phosphate buffer (pH 7.0) to 0.1 M sodium phosphate (pH 7.0). Because of the low grade of ammonium sulfate used, the baseline decreased gradually with decrease in salt concentration. (Reprinted from Ref. 56, with permission.)

glycoproteins that are denatured or precipitated under RPC conditions [55]. Although carbohydrates increase the hydrophilicity of proteins slightly, separations are primarily based on amino acid composition. Fig. 14.8C shows that B-2-I [56] yields only one major peak in RPC.

HIC, likewise, does not resolve glycoproteins that differ only in carbohydrate content. Because it is milder than RPC, it does not disrupt the tertiary structure of proteins. It can be used to analyze many proteins on the basis of the hydrophobicity of surface amino acids [58]. HIC separates B-2-I into two species, one of which contains a lipid (Fig. 14.8D) [56]. This discrimination was not observed in RPC (Fig. 14.8C), because the lipid component was stripped off during chromatography [55].

References on p. B168

14.3.4 Immunoglobulins

Immunoglobulins are antibodies produced by B-lymphocytes when the immune system is challenged by an antigenic substance. Immunoglobulins may be divided into five classes, based on differences in molecular mass and chemical properties. All immunoglobulins are multimeric, being composed of two heavy (H) and two light (L) chains, joined by disulfide bridges. Immunoglobulin G (IgG) is a molecule of approximately 150 kD that is composed of two 50-kD H chains and two 25-kD L chains. Immunoglobulins E (IgE) and D (IgD) have molecular weights of 200 and 280 kD, respectively, and are similar to IgG in being single structural units, consisting of two H and two L chains. Immunoglobulins M (IgM) and A (IgA) are more complex. IgM is a pentamer of 900 kD. IgA exists in both monomeric and multimeric forms. The 165-kD monomeric form found in serum consists of two L chains and two H chains, whereas IgA in mucous secretions exists as a dimer.

The two major problems encountered in immunoglobulin purification are (a) the separation of target antibodies from contaminating proteins and (b) the resolution of a mixture of antibody species. Because of the very large differences between antibodies, reports of single-step purification procedures are probably more the exception than the rule. Six general types of chromatography have been reported in the purification of immunoglobulins: SEC, anion-exchange and cation-exchange chromatography, HAC, HIC, and various affinity chromatography methods. The type of chromatographic system selected depends primarily on the type of antibody purification [59].

SEC can be very useful in the fractionation of immunoglobulins when the species differ in molecular weight by at least a factor of 2. Examples in which SEC has been used effectively are: the separation of IgM, IgA, and IgG [60]; IgA from the Fc and F(ab) fragments [61]; enzyme/antibody conjugates from underivatized antibody and enzyme [62]; IgG aggregates from IgG [63]; albumin aggregates from IgG [63], and IgG from albumin and transferrin [64].

Anion-exchange chromatography has been the most widely used chromatographic method for purifying monoclonal and polyclonal IgG species, which have pI values ranging from 5.2 to 8 [65,59]. Polyclonal IgG species from different mammals can generally be separated by anion-exchange chromatography [65], but the resolution of IgG subclasses within a species is much more difficult. Under optimum conditions, IgG_{2b} and IgG_3 from mouse serum can be resolved, whereas IgG_1 and IgG_{2a} cannot.

Cation-exchange chromatography has been effectively used to separate monoclonal antibodies at pH 4.5-5.0, where they are positively charged [66]. On a mixed-bed column with cation- and anion-exchange sorbents, IgG was isolated from blood serum, using a buffer at its pI [67].

Purification of IgG from serum by HAC generally eliminates more than 95% of the contaminant proteins and has yielded 75% or higher recovery [16,68]. Sixteen monoclonal antibodies from all subclasses of IgG and IgM have been separated on a hydroxyapatite column, eluted with a gradient from 0.005 to 0.3 M phosphate at pH 7.2, as seen in Fig. 14.9 [16]. Broad peaks of polyclonal antibodies obtained with this column were attributed to polymorphism.

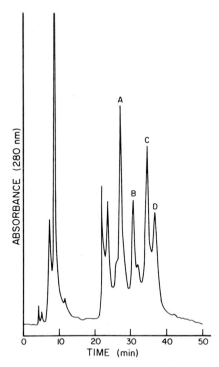

Fig. 14.9. Resolution of six antibodies from mouse ascites fluid by hydroxyapatite HPLC [16]. Column, TAPS, 100 x 7.5 mm ID; 30-min linear gradient from 0.005 to 0.3 M sodium phosphate (pH 6.8); flowrate, 0.5 ml/min. Sample: A = IgG_1, IgG_2; B = IgM; C = IgG_{2b}, IgG_3; D = IgG_{2a}.

IEC has been most effective when used in combination with some other fractionation technique, such as ammonium sulfate precipitation [68], anion- or cation-exchange chromatography [65], HIC [68], or SEC [69,70]. HIC is generally as effective as IEC [68] and, in those cases where the pl of an IgG species is <7.2, it can be superior.

Affinity chromatography of antibodies on a column of immobilized microbial binding proteins, such as Protein A or Protein G, has been achieved by adsorption at neutral pH and elution with a descending pH gradient. When a Protein A column was eluted with a pH gradient ranging from 9 to 2, IgG subclasses could be analyzed with high efficiency [71]. Of the two proteins, Protein G shows the broader binding activity for IgG species.

REFERENCES

1 H.A. Sober and E.A. Peterson, *J. Am. Chem. Soc.*, 76 (1954) 1711.
2 J. Porath and P. Flodin, *Nature (London)*, 183 (1959) 1657.
3 P. Cuatrecasas, M. Wilchek and C.B. Anfinsen, *Proc. Natl. Acad. Sci.*, 61 (1968) 636.
4 A. Tiselius, S. Hjertén and O. Levin, *Arch. Biochem. Biophys.*, 65 (1956) 132.
5 H.P. Jenissen, *Protides Biol. Fluids*, 23 (1975) 675.

6 S.H. Chang, K.M. Gooding and F.E. Regnier, *J. Chromatogr.,* 125 (1976) 103.
7 G. Vanecek and F.E. Regnier, *Anal. Biochem.,* 109 (1980) 345.
8 G.P. Rozing and H. Goetz, *J. Chromatogr.,* 476 (1989) 3.
9 F.E. Regnier, *Anal. Biochem.,* 126 (1982) 1.
10 F.E. Regnier and R.M. Chicz, in K.M. Gooding and F.E. Regnier (Editors), *HPLC of Biological Macromolecules: Methods and Applications,* Dekker, New York, 1990, pp. 77- 93.
11 W. Kopaciewicz and F.E. Regnier, *Anal. Biochem.,* 133 (1983) 251.
12 K.M. Gooding and M.N. Schmuck, *J. Chromatogr.,* 296 (1984) 321.
13 W. Kopaciewicz and F.E. Regnier, *J. Chromatogr.,* 266 (1983) 3.
14 M.P. Nowlan and K.M. Gooding, in R. Hodges and C. Mant (Editors), *HPLC of Peptides and Proteins,* CRC Press, Boca Raton, FL, 1991.
15 M.L. Heinitz, L. Kennedy, W. Kopaciewicz and F.E. Regnier, *J. Chromatogr.,* 443 (1988) 173.
16 Y. Yamakawa and J. Chiba, *J. Liq. Chromatogr.,* 11 (1988) 665.
17 W.S. Hancock, C.A. Bishop, R.L. Prestidge, D.R.K. Harding and M.T.W. Hearn, *J. Chromatogr.,* 153 (1978) 391.
18 W.C. Mahoney and M.A. Hermodsen, *J. Biol. Chem.,* 255 (1980) 11199.
19 J. Frenz, W.S. Hancock, W.J. Henzel and Cs. Horváth, in K.M. Gooding and F.E. Regnier (Editors), *HPLC of Biological Macromolecules: Methods and Applications,* Dekker, New York, 1990, pp. 145-177.
20 R.E. Shansky, S.-L. Wu, A. Figueroa and B.L. Karger, in K.M. Gooding and F.E. Regnier (Editors), *HPLC of Biological Macromolecules: Methods and Applications,* Dekker, New York, 1990, pp. 95-144.
21 W.R. Melander, D. Corradini and Cs. Horváth, *J. Chromatogr.,* 317 (1984) 67.
22 M.N. Schmuck, M.P. Nowlan and K.M. Gooding, *J. Chromatogr.,* 371 (1986) 55.
23 M.L. Heinitz, E. Flanigin, R. C. Orlowski and F.E. Regnier, *J. Chromatogr.,* 443 (1988) 229.
24 J. Porath, J. Carlsson, I. Olsson and G. Belfrage, *Nature (London),* 258 (1975) 598.
25 Z. El Rassi and Cs. Horváth, in K.M. Gooding and F.E. Regnier (Editors), *HPLC of Biological Macromolecules: Methods and Applications,* Dekker, New York, 1990, pp. 179-213.
26 D.L. Brautigan, S. Ferguson-Miller and E. Margoliash, *J. Biol. Chem.,* 253 (1978) 130.
27 K.M. Gooding and F.E. Regnier, in K.M. Gooding and F.E. Regnier (Editors), *HPLC of Biological Maacromolecules: Methods and Applications,* Dekker, New York, 1990, pp. 47- 75.
28 E. Pfannkoch, K.C. Lu, F.E. Regnier and H.G. Barth, *J. Chromatogr. Sci.,* 18 (1980) 430.
29 F.E. Regnier, *Science,* 238 (1987) 319.
30 F.E. Regnier, *LC,* 1 (1983) 350.
31 A.L. Leninger, *Biochemistry,* Worth Publishers, New York, 1971, pp. 109-128.
32 J.L. Fausnaugh and F.E. Regnier, *J. Chromatogr.,* 359 (1986)131.
33 J. Fausnaugh-Pollitt, G. Thevenon, L. Janis and F.E. Regnier, *J. Chromatogr.,* 443 (1988) 221.
34 R.H. Ingraham, S.Y.M. Lau, A.K. Taneja and R.S. Hodges, *J. Chromatogr.,* 327 (1985) 77.
35 S.-L. Wu, A. Figueroa and B.L. Karger, *J. Chromatogr.,* 371 (1986) 3.
36 International Hemoglobin Information Center Variant List, *Hemoglobin,* 12 (1988) 209.
37 J.B. Wilson, in K.M. Gooding and F.E. Regnier (Editors), *HPLC of Biological Macromolecules: Methods and Applications,* Dekker, New York, 1990, pp. 457-472.
38 U.H. Stenman, in K.M. Gooding and F.E. Regnier (Editors), *HPLC of Biological Macromolecules: Methods and Appalications,* Dekker, New York, 1990, pp. 473-486.
39 K.M. Gooding, K.C. Lu and F.E. Regnier, *J. Chromatogr.,* 164 (1979) 506.

40 T.H.J. Huisman, M.B. Gardiner and J.B. Wilson, in S.M. Hanash and G.J. Brewer (Editors), *Advances in Hemoglobin Analysis*, Alan R. Liss, New York, 1981, pp. 69-82.
41 J.B. Wilson, M.E. Headless and T.H.J. Huisman, *J. Lab. Clin. Med.*, 102 (1983) 174.
42 J.B. Shelton, J.R. Shelton and W.A. Schroeder, *J. Liq. Chromatogr.*, 7 (1984) 1969.
43 J.A. Bietz, in K.M. Gooding and F.E. Regnier (Editors), *HPLC of Biological Macromolecules: Methods and Applications*, Dekker, New York, 1990, pp. 429-455.
44 H. Wieser and H.D. Belitz, *J. Cer. Sci.*, 9 (1989) 221.
45 P.K.W. Ng, M.G. Scanlon and W. Bushuk, *Cereal Res. Comm.*, 17 (1989) 5.
46 L.V.S. Sastry, J.W. Paulis, L.A. Cobb, J.S. Wall and J.D. Axtell, *J. Agric. Food Chem.*, 34 (1986) 1061.
47 J.A. Bietz, *Bakers Digest*, 58 (1984) 15.
48 T. Burnouf and J.A. Bietz, *Seed Sci. Technol.*, 15 (1987) 79.
49 M. Menkovska, G.L. Lookhart and Y. Pomeranz, *Cereal Chem.*, 64 (1987) 311.
50 T. Burnouf and J.A. Bietz, *J. Cer. Sci.*, 2 (1984) 3.
51 J.A. Bietz and L.A. Cobb, *Cereal Chem.*, 62 (1985) 332.
52 J.A. Bietz, T. Burnouf, L.A. Cobb and J.S. Wall, *Cereal Chem.*, 61 (1984) 124.
53 T. Burnouf and J.A. Bietz, *J. Chromatogr.*, 299 (1984) 185.
54 T. Burnouf and J.A. Bietz, *J. Chromatogr.*, 327 (1985) 333.
55 F.W. Putnam and N. Takahashi, in K.M. Gooding and F.E. Regnier (Editors), *HPLC of Biological Macromolecules: Methods and Applications*, Dekker, New York, 1990, pp. 571-584.
56 F.W. Putnam and N. Takahashi, *J. Chromatogr.*, 443 (1988) 267.
57 N. Takahashi, Y. Takahashi, T.L. Ortel, J.N. Lozier, N. Ishioka and F.W. Putnam, *J. Chromatogr.*, 317 (1984) 11.
58 J.L. Fausnaugh, L.A. Kennedy and F.E. Regnier, *J. Chromatogr.*, 317 (1984) 141.
59 S.I. Sivakoff, in K.M. Gooding and F.E. Regnier (Editors), *HPLC of Biological Macromolecules: Methods and Applications*, Dekker, New York, 1990, pp. 487-528.
60 G. Sann, G. Schneider, S. Loeke and H.W. Doerr, *J. Immunol. Methods*, 59 (1983) 121.
61 S.B. Mortensen and M. Kilian, *J. Chromatogr.*, 296 (1984) 257.
62 S.I. Sivakoff, *BioChromatography*, 1 (1986) 42.
63 J.K. Lee, F.J. Deluccia, E.L. Kelly, C. Davidson and F.R. Borger, *J. Chromatogr.*, 444 (1988) 141.
64 S.W. Burchiel, J.R. Billman and T.R. Alber, *J. Immunol. Methods*, 69 (1984) 33.
65 D.R. Nau, *BioChromatography*, 4 (1989) 4.
66 M. Carlsson, A. Hedin, M. Inganas, B. Harfast and F. Blomberg, *J. Immunol. Methods*, 79 (1985) 89.
67 R.W. Stringham and F.E. Regnier, *J. Chromatogr.*, 409 (1987) 305.
68 B. Pavlu, V. Johansson, C. Nyhlen and A. Wichman, *J. Chromatogr.*, 359 (1986) 449.
69 P. Clezardin, J.L. McGregor, M. Monach, H. Boukerche and M. Dechavanne, *J. Chromatogr.*, 319 (1985) 67.
70 M.J.F. Schmerr, K.R. Goodwin, H.D. Lehmkuhl and R.C. Cutlip, *J. Chromatogr.*, 326 (1985) 225.
71 S. Ohlson and J. Wieslander, *J. Chromatogr.*, 397 (1987) 207.

Chapter 15

Lipids

A. KUKSIS

CONTENTS

15.1 INTRODUCTION

Since the publication of the last edition of this book [1] the practice of lipid chromatography has continued along both established and innovative lines. In this revised chapter, emphasis has been placed upon the newer developments and techniques. In selecting the material for inclusion preference has been given to advances that have resulted in significant breakthroughs in practical applications, while reference citations have been selected to reflect typical and not necessarily original applications of the techniques. For more detailed coverage of recent advances in the chromatography of lipids reference may be made to various reviews and monographs about a single or limited number classes of compounds, e.g., GLC of fatty acids [2-5], GLC [6] and HPLC [7] of prostaglandins, GLC of neutral lipids [8-10], HPLC of lipids [11,12], HPLC of polar lipids [13-15], or GC/MS [5,6,10,11,16], LC/MS [17-19] and MS/MS [19] of lipids and related compounds. Other specific reviews are noted in the text.

15.2 PREPARATION OF LIPID EXTRACTS

Traditional solubilization and extraction of neutral and weakly acidic lipids by liquid/liquid partition with chloroform and methanol [20,21], chloroform and 2-propanol for plant tissues [22], and more recently, the less toxic 2-propanol and hexane [23] is time-consuming and involves large volumes of solvents, which are expensive and hazardous to health. The conventional extracts are also contaminated to various extents with nonlipid components, which must be removed by adsorption [24] or reversed-phase [25] chromatography. Liquid/solid extraction, which greatly reduces the solvent requirements and provides cleaner samples, has therefore gained in popularity [26], but it is not a replacement for liquid/liquid extraction. Considerable progress has been made in the automation of sample preparation [27].

15.2.1 Liquid/liquid extraction

Mixtures of chloroform and methanol continue to be widely used for lipid extraction, although the limitations of these procedures have been extensively reviewed [1,5]. They must be applied with special care when preparing extracts of cell membranes. Saito et al. [28] have examined the effect of these extraction techniques on the recovery of the individual lipids from erythrocytes. They showed that chloroform/methanol extraction resulted in losses of acidic lipids, which ranged from 8 to 12% for phosphatidylserine (PS) and phosphatidic acid (PA) under the experimental conditions. Freyburger et al. [29] have shown that the yield of lipids extracted by chloroform/methanol (2:1) from human erythrocytes decreases as a function of the relative content of water added to or present in the erythrocyte pellet prior to lipid extraction. As the volume of added water increases, the recovery of PS drops dramatically and approaches zero, while the yield of other phospholipids remains unchanged. In contrast, extraction of the lipids with 2-propanol/chloroform was not affected by water and resulted in complete recovery of all phospholipids at all dilutions with water. This is consistent with the earlier observations of Rose and Oaklander [30], who had shown that a mixture of 2-propanol/chloroform (11:7) leads to highly reproducible lipid yields from erythrocytes, following lysis of the packed erythrocytes with an equal volume of distilled water. Presumably, a solvent system that can deal effectively with the erythrocyte lipids is likely to produce excellent results with the lipids of other membranes. For the recovery of the more acidic phospholipids, e.g. poly-phosphoinositides, an initial extraction with neutral solvents is followed by an extraction with acidic solvents [31]. This approach allows the recovery of any plasmalogens that may be present without exposing them to acids, which degrades them to lysophospholipids [32].

15.2.2 Liquid/solid extraction

Much more effective is liquid/solid extraction, which may be accomplished by passing a dilute solution of body fluids or solubilized lipid mixture through a solid adsorbent bed. Liquid/solid extraction with radially compressed minicolumns requires less solvent and yields greater recoveries and accuracy because of fewer transfers. It results in savings of time because of fewer steps, as well as in greater safety and less exposure of personnel to solvents. Furthermore, it eliminates emulsion formation and minimizes sample peroxidation. Hamilton and Comai [33] demonstrated that small silica Sep-Pak columns could be used to separate neutral and polar lipids with > 96% recovery when combinations of hexane, methyl t-butyl ether (MTBE), and acetic acid (AA) are used. Cholesteryl esters (CE) (hexane/MTBE, 200:3, 12 ml), triacyl glycerols (hexane/MTBE 96:4, 12 ml), free fatty

References on p. B218

acids (hexane/MTBE/AA, 100:2:0.2, 12 ml), free cholesterol and partial acyl glycerols (MTBE/AA 100:0.2, 12 ml) were eluted separately and were separated from polar lipids, which were retained by the column under these conditions. Polar lipids are eluted from the column by a combination of MTBE/methanol and ammonium acetate. Phosphatidyl-ethanolamine (PE) and phosphatidylinositol (PI) are eluted together in one fraction, while phosphatidylcholine (PC), sphingomyelin (SPH), and lysophosphatidylcholine (LPC) are recovered in another fraction. Recoveries of each phospholipid were > 98% when as-sayed by means of radioactive tracers. This method differs from a previous routine [34] by the substitution of MTBE for chloroform. The processing of 10 samples can be accom-plished in less than 1 h. Prior to use, each Sep-Pak column was washed with 4 ml of hexane/MTBE (96:4), followed by 12 ml of hexane. This washing procedure removes substances that interfere with the HPLC procedure used to monitor lipids.

Kaluzny et al. [35] and Figlewicz et al. [36] have reported the successful use of both silica and aminopropyl-bonded-phase Bond Elut cartridges to separate lipid mixtures into individual classes in high yield and purity. The Bond Elut system [35] may be operated with a special apparatus. The lipids are eluted with solvents of progressively decreasing polarity. It should be noted that the solvent composition and volumes are critical and that there is no continuous monitoring of the solvent by UV. Figlewicz et al. [36] used the reversed-phase columns for the separation of complex lipids, such as PC, ceramides, sulfatides, and gangliosides from the respective water-soluble radioactive precursors after in vitro biosynthesis. The lipids extracted from the incubation mixtures were eluted with chloroform/methanol (2:1). The coefficients of correlation (r) between the Bond Elut and conventional methods of recovery of radioactive PC, sulfatides, and gangliosides were 0.99. Recently, Egberts and Buiskool [37] employed the aminopropyl solid-phase column for an efficient isolation of the acidic phospholipid phosphatidylglycerol (PG) from natural pulmonary surfactant.

Powell [38] has employed octadecylsilylsilica (Sep-Pak C_{18} cartridges) to extract fatty acids and prostaglandins from urine and other biological fluids. Solutes are retained on the basis of hydrophobic interactions with the stationary phase. Substances more polar than these can be eluted from the column with aqueous solvents containing small amounts of miscible organic solvents (e.g. methanol, ethanol, or acetonitrile). The desired material is then eluted with a solvent containing a higher concentration of the organic component. Body fluids, such as urine and plasma, can be applied directly to the cartrid-ges after acidification. The cartridge is then washed with water, followed by aq. ethanol. The solutes are eluted with the desired combination of ethanol and water. However, sample acidification may lead to isomerization of leukotrienes. Powell [38] obtained > 90% recovery of prostaglandin standards and ca. 59% recovery of 15-hydroxy-5,8,11,13-eicosatetraenoic acid (15-HETE) from biological fluids. Eskra et al. [39] have developed an extraction procedure with better recovery of HETEs and other lipoxygenase pathway products by minimizing the number of processing steps, transfers, and evapora-tions of the sample. Acetonitrile extraction and fractionation in Sep-Pak C_{18} cartridges met these criteria. Acetonitrile (66%) was an effective protein precipitant, allowing a satisfactory and highly reproducible recovery of leukotrienes, HETEs, and PGB_2 from plasma. Steinhilber et al. [40] have briefly reviewed the solid-phase extraction methods for lipoxins and leukotrienes and have proposed procedures for improved recovery of both. The Radial-Pak cartridge allowed sufficient separation of the lipoxin isomers and a baseline resolution of LTB_4 and its isomers, including $5S,12S$-diHETE within 8 min.

Saito et al. [28] have used DEAE-Sephadex columns (A-25, acetate form, 0.64 x 3.6 cm) to isolate phospholipids from erythrocytes. The lipid extract prepared according to Rose and Oaklander [30] in 1 ml of chloroform/methanol (1:1) was applied to the DEAE-Sephadex column, following equilibration with chloroform/methanol (1:1). The neutral lipids, including choline and ethanolamine phosphatides and sphingomyelin, were eluted with 4 ml of chloroform/methanol (1:1). The column was then washed with 1 ml of chloroform and 8 ml of chloroform/acetic acid (3:1), which removed free fatty acids and pigments. The acidic phospholipids, serine, inositol, and glycerol phosphatides, were eluted with 4 ml of chloroform/methanol/aq. 3 M acetate (pH 7.0) (50:50:8). The acidic phospholipids were recovered along with sulfatides and gangliosides. The gangliosides were removed in the upper phase by partitioning with 1.37 ml of 0.9% NaCl, added to the eluate. The sulfatides were recovered by subsequent TLC. Good recoveries were claimed.

15.3 GENERAL CONDITIONS OF LIPID CHROMATOGRAPHY

The primary methods of lipid analysis continue to be normal-phase TLC and HPLC for class separations, and GLC and reversed-phase HPLC for the resolution of molecular species [1]. Recently, chiral-phase HPLC has been introduced for the separation of enantiomeric hydroxy acids, secondary alcohols, and mono- and diradyl glycerols [41]. Each of the techniques is available in a variety of refinements, which are important for specific applications and optimum resolution of each sample.

15.3.1 Methods of separation

Normal-phase TLC on plain silica gel is commonly employed for the separation of the major neutral and polar lipid classes. For the separation of mono- and diradyl glycerols it is essential to use normal-phase TLC in combination with boric acid-treated silica gel [1]. Likewise, normal-phase TLC in combination with boric acid-treated silica gel improves the separation of phosphatidylinositol from phosphatidylserine (PS) in a polar solvent system [1]. Silica gel, impregnated with silver nitrate, continues to provide effective means for the normal-phase TLC resolution of saturated and unsaturated lipid classes. Excellent resolution has recently been obtained by Breuer et al. [42], using precoated alumina plates, but silver ions bound to ion-exchange HPLC columns provide advantages [43,44]. Normal-phase TLC on extra-finely ground silica gel is now being widely used to obtain maximum separation of lipid classes on miniplates in the form of high-performance TLC (HPTLC) [45]. In other instances, quartz rods have been substituted for the open plates as the adsorbent surface [46]. These rods provide separations very similar to those obtained on plates, but they can be quantitatively assayed by passing them through a specially constructed FID (latroscan). Both HPTLC plates [45] and quartz rods [46] can be used with the adsorbent modifiers developed for coated plates. Neither the rods nor the HPTLC plates are well suited for sample collection, although this has been attempted.

The general conditions for HPLC of lipids have recently been discussed by Christie [11] and Shukla [12]. Normal-phase HPLC is commonly performed on silicic acid-based packing materials, but in special instances other sorbents have also been employed. Thus, chiral-phase columns have been used for the resolution of enantiomeric lipid molecules [41] and ion-exchange columns with immobilized silver ions for saturated and unsaturated neutral lipids [43,44]. The separation of phospholipid classes by normal-phase HPLC has proved to be more difficult than anticipated, but successful applications are being reported [47-49], including preparative applications [50,51]. The HPLC columns are utilized with a variety of solvent systems, as required for the solubilization of the solutes and for their effective desorption or displacement from the sorbent surface. Both isocratic and gradient elution have been utilized. The substitution of MTBE for chloroform is advantageous for both separation and UV detection (206 nm) of the solutes.

Reversed-phase HPLC is utilized mainly for resolving molecular species of neutral lipids [11,12], but notable resolution has also been obtained for the molecular species of intact glycerophospholipids [52]. Reversed-phase columns, singly or in pairs, yield excellent resolution of various neutral lipids on the basis of carbon atom and double-bond number, or equivalent chainlength. A variety of solvent systems are employed to ensure compatibility with the lipid class solubility, polarity, and the nature of the detector employed for qualitative and quantitative monitoring of the eluate.

High-performance centrifugal partition chromatography has been introduced to overcome some of the distinct disadvantages of both normal- and reversed-phase HPLC in the separation, isolation, and purification of lipids of biological importance [53]. Specifically, this process has been utilized in the preparation of highly purified phospholipids from egg yolk and of long-chain polyunsaturated fatty acids from fish oils.

The separation of fatty acid derivatives obtained by supercritical-fluid chromatography is also of great interest. This has been monitored by UV (phenyl esters) [54], on-line FTIR (free fatty acids) [55], and EI-MS (triacyl glycerols) [56].

Most GLC separations are currently performed on capillary columns containing either nonpolar or polar stationary phases [5]. The nonpolar phases are generally employed for carbon-number or molecular-weight resolution, but certain separations based on unsaturation and the shape of the molecule are also possible on these more efficient columns [8,57]. Nonpolar capillary columns possess high temperature stability and can be operated up to 350°C without significant loss of stationary phase. The columns can be used in various lengths with hydrogen as the carrier gas. Since the introduction of aluminum cladding, fused-silica columns can now be operated in excess of the 370°C limit imposed by the polyimide coating. In order to utilize the high-temperature capability of these columns, a nonpolar stationary phase, known as HT5, has been developed and successfully employed for the resolution of hydrocarbons having up to 90 carbons and even more, over a temperature range of 200-480°C [58]. The column was capable of effective resolution of butterfat triacyl glycerols in the 200-370°C range. At the 10-ng level, only a 5% decrease in peak area was observed between C_{38} and C_{56} triacyl glycerol standards.

Capillary GLC on polar stationary phases is employed mainly for the resolution of fatty acid methyl esters. The TMS and TBDMS derivatives of mono- and diradyl glycerols can also be separated on these columns at somewhat higher temperatures [5,10]. However, with the currently available polar liquid phases, the column temperature may not exceed 270°C. Because the polar capillary columns possess considerable vapor pressure at elevated temperatures, they can be used for temperature programing only over a limited temperature range [59]. Even polyunsaturated (up to 12 double bonds) diacyl glycerol species could be eluted isothermally, but this required over 90 min.

Capillary GLC with polarizable phenylmethylsilicone stationary phases is possible at temperatures up to 350°C [60]. These liquid phases acquire considerable polarity above 300°C and can be used for the resolution of natural triacyl glycerols on the basis of carbon and double-bond number. This liquid phase behaves very much like a nonpolar phase below 280°C, where the free fatty acids and monoacyl glycerols are eluted. These columns also give excellent resolution of mixtures of neutral lipids, ranging from diacyl glycerols and ceramides, as TMS or TBDMS ethers, to cholesteryl esters and triacyl glycerols [61]. The cholesteryl esters must be recovered below 310°C, as these compounds tend to decompose at higher temperatures.

Both normal- and reversed-phase HPLC have been combined with on-line CI-MS [16-19] by using a variety of instrumental set-ups, while GLC on both polar and nonpolar capillary columns has been combined with EI- and CI-MS [5,10,17,19] by direct introduction of the capillary into the ion source. The process dubbed MS/MS or tandem mass spectrometry permits sequential analysis of the ionized analyte for its unequivocal identification. Model applications have been reported in the lipid field [62-65].

15.3.2 Quantitation

Chromatographic peak quantitation is simple with detectors that give either mass or molar response. Thus, the FID commonly employed in GLC responds correctly to the carbon mass in the solute [66]. It is very sensitive and possesses a wide dynamic range. It is not affected by temperature programing. It has also been employed with success for the quantitation of the solutes separated on Chromarods [46,67]. In a few instances, flame-ionization detection has also been combined with HPLC [68-70] via a special moving-wire or moving-belt arrangement, which allows the removal of the solvent prior to passing through the FID. These detection systems usually do not require extensive calibration.

The UV-absorbing derivatives prepared from various lipids and employed in HPLC provide a correct molar response, as each molecule, regardless of its molecular weight, carries one UV-absorbing element [71]. The only requirement is that the UV-absorbing derivatives can be quantitatively prepared and are stable under separation conditions. Thus, the diradyl glycerols have been analyzed as the benzoyl [72] and dinitrobenzoyl [73] derivatives, while the gangliosides [74,75] and their degradation products [76] have been determined as the perbenzoyl derivatives. A significantly higher response is seen for the fluorescent naphthyl urethanes [77]. These derivatives do not require extensive calibration of the detection system.

Although lipid molecules do not possess significant UV absorption of their own, they can be detected in the low-UV range (190-220 nm) due to absorption by double bonds and carbonyl functions [22]. The HPLC system must be operated with solvents that do not possess significant optical density in the corresponding UV absorption range. However, absorbance varies from one molecular species to the next and must be calibrated for quantitative work. Nevertheless, useful results have been obtained for the diradyl glycerol acetates [78-80] and intact glycerophospholipids [33,49,52,81]. However, conjugated double bonds and lipid peroxides do absorb strongly in the UV [82,83].

HPLC can also be combined with refractometry for triacyl glycerol peak detection and quantitation [84]. This method requires isocratic operation as well, as it imposes certain limits on the choice of solvents. The method is not as sensitive as UV detection, and quantitation is difficult at best and impossible in some instances. In supercritical-fluid chromatography, FTIR detection has been effective in the identification and quantitation of fatty acids [55]. Promising results for universal lipid detection in HPLC have been obtained with a light- or laser light-scattering detector, which is also referred to as a mass detector [11]. Although the response is nonlinear and less sensitive than UV or flame-ionization detection, it is applicable to substances that cannot be derivatized and do not possess UV absorption of their own. Originally employed for neutral lipid detection [85], it has now been successfully applied to monitoring phospholipids in the HPLC column effluents [86].

Mass spectrometry can also be employed for quantitation of lipids with both GLC and HPLC, especially for the estimation of the relative concentrations of overlapping components, where other methods are not applicable [87-89]. However, the ion yields are variable for both parent and fragment ions, and extensive calibration is necessary for accurate work. The use of stable-isotope-labeled internal standards offers special advantages for peak identification and quantitation [18,19].

Palumbo and Zullo [90] have described quantitative TLC estimation of lipids by staining with iodine under controlled conditions. After exposing the developed plate to iodine, it is sprayed with a suitable solvent to prevent halogen evaporation, and the stained lipids are collected by scraping the spots off the plate. The absorbed iodine is determined by measuring spectrophotometrically (at 410 nm) the rate of decolorization of a solution of Ce(IV) by As(III) in strong acid. Nagata et al. [91] have described a relatively simple procedure for quantitation of lipid concentrations on TLC plates by densitometry of transparent copies. After resolution in an appropriate solvent, the lipids are visualized by staining with iodine, ninhydrin, or molybdate and are then photocopied onto transparent sheets, using a standard office copier. The density of each spot on the photocopy is determined with a simple densitometer and related to sample concentration to give a direct measure of each lipid, applied to the plate over a range up to 6- to 8-fold. It should be noted, that iodine is a powerful catalyst for transesterification, and iodine-stained glycerolipid spots may not be suitable for subsequent analyses of molecular species.

Fowler et al. [92] have reviewed the advantages of fluorometric scanning of thin-layer chromatograms and have proposed the use of Nile Red (ex: 580 nm; em: 649 nm) as a general reagent for the rapid in situ quantitation of lipids. Wood et al. [93] have reassessed the densitometric HPTLC method of Macala et al. [94] in terms of linearity of standards in the nanogram range and have concluded that this method is efficient for examining different lipid classes in samples where the lipid content is low. Poole and Poole [95] have reviewed the principal methods for obtaining quantitative information from separations by PC techniques.

15.3.3 Peak identification

Peak identification has become the most difficult part of chromatographic analysis of lipids because of the recent emphasis on minor components. Calculations of relative retention times based on the behavior of a few standards are not adequate for the prediction of retention times of molecules for which there are no standards. Calculations of equivalent carbon numbers and partition coefficients are only of limited help, for the same reasons. Interlaboratory comparisons are complicated by the variability in the stationary phase and column preparation, and by changes in column performance with aging [96]. Since peak collection is not practical, peak identification must remain tentative, unless extensive prefractionation, fraction collection, and independent characterization has been undertaken. However, a systematic analysis on different chromatographic phases can be helpful, and analytical strategy becomes of utmost importance for the identification of unknown components. In fact, an appropriate strategy [24] must be devised and adopted for each type of analysis, keeping in mind the origin of a clinical sample, e.g., diet, disease, drug treatment, and method of sample preparation and storage. Christie [97] has recently reconfirmed the value of the equivalent chainlength (ECL) in the GLC identification of unknown fatty acids by using stationary phases of selected types, while Beaumelle and Vial [98] have provided a theoretical explanation of the phenomenon on the basis of the shapes of molecules and opportunities for interaction between double bonds and the walls of the capillary column.

On-line FTIR detection has provided peak identification in various biological applications of GLC [99]. It complements GC/MS, but is not as sensitive. It is best used in combination with supercritical-fluid chromatography [55]. Computer-aided graphics also help to visualize the UV signal. Using a photodiode-array detector, Greenspan et al. [100] identified cholesteryl esters, triacylglycerols, ubiquinone, α-tocopherol, dolichol, cholesterol, 7-dehydrocholesterol, and retinol in the 0.1% 2-propanol/heptane isocratic effluents of normal-phase HPLC. Radio-GLC has received a boost from the recent development [101] of the synchronized accumulating radioactivity detector, while the sensitivity of radio-HPLC detection of solutes has been increased by the use of arrays of detectors [102]. These detectors complement stable-isotope detection by GC/MS and LC/MS, respectively [17].

A technique that cuts across many of the uncertainties and allows a direct identification of the components in any chromatographic peak is MS. It can provide the needed confirmation of the identity of known components and information about unknowns that is frequently sufficient for entertaining reasonable guesses about their potential identity, at least for components that are fully resolved. MS can also identify overlapping known components or molecular species. In the case of overlapping unknowns, MS must be preceded by appropriate chromatographic separation. Both EI- and CI-MS have been used to identify the native molecules or special derivatives, prepared to yield the most informative fragment ions [18,19,88,89]. Preparation of appropriate derivatives is essential for the GC/MS determination of double-bond distribution in polyunsaturated fatty acids [5,103-107].

15.4 SEPARATION OF LIPID CLASSES

Both analytical and preparative separations of lipid classes are necessary for the determination of the composition of lipids from natural sources. The methods differ, depending on the nature and complexity of the sample. A preliminary segregation of the lipid extract into nonacidic and acidic components by means of a Sephadex column can be of immense help in the later identification of both classes and molecular species of lipids [28,93,94].

15.4.1 Resolution of nonpolar lipids

The use of adsorption columns [1] for the separation and isolation of different neutral lipid classes has been largely replaced in recent years with TLC, HPTLC, and normal-phase HPLC, which are much faster and give more complete separation. However, for certain applications, the use of the adsorption column has been essential, e.g., resolution of neutral and acidic glycosphingolipids [108,109]. A kind of adsorption column chromatography has been recently reintroduced in the form of adsorbent cartridges [33-38]. The columns are effective for the separation of lipid classes differing greatly in polarity, but not for closely related lipid classes, which tend to overlap.

15.4.1.1 Thin-layer and high-performance thin-layer chromatography

Beesley [110] has reviewed the instrumentation and methodology currently available to increase sample throughput and enhance sensitivity, reproducibility, and reliability. Special emphasis was placed upon methods of sample application, which are critical for good TLC performance, especially in HPTLC. The use of TLC has continued to provide rapid, effective, and inexpensive means for neutral lipid separation. The extent of lipid class resolution has been greatly increased by the new high-performance and reversed-phase layers, the efficiency of which has rivaled that of normal-phase HPLC. Macala et al. [94] have described a HPTLC densitometry method for resolving the major neutral lipids, isolated from brain tissue by DEAE Sephadex column chromatography, but the method has wide application. Excellent separations are obtained of the nonacidic lipids of rat brain by HPTLC by using two developments. The first, chloroform/methanol/acetic acid/formic acid/water (35:15:6:2:1) (to about 4.5 cm) separates choline and ethanolamine phospholipids and cerebrosides. A second development, with hexane/diisopropyl ether/acetic acid

(65:35:2) (to top of plate), separates cholesterol, oleyl alcohol (internal standard), triacyl glycerols, and cholesteryl esters. The compounds are located by charring the plates with 3% cupric acetate in 8% phosphoric acid. Wood et al. [93] have recently reexamined the method and have recommended it for application in the detection of lipids in the nanogram range.

There has been a heavy emphasis on the use of TLC/FID for industrial applications [111]. The method allows a rapid and unambiguous quantitation of lipid classes in crude materials and in finished products. It is not suitable for measuring low lipid levels, unless a variability of 5-8% is acceptable. However, applications in biomedical research have also been numerous [111]. Sebedio et al. [112] have continued to employ with good results the latroscan TLC/FID system for the study of the composition of commercial frying oil samples, previously isolated by column chromatography on hydrated silicic acid, according to standard methods. Likewise, Okumura et al. [67] estimated the free diacyl glycerol content in the rat heart by the latroscan TLC/FID method.

HPTLC has been used [113] to compare the oxidation products of authentic polyunsaturated fatty acids, their esters, and fatty alcohols. Monohydroperoxy and highly polar products were separated from linoleic, linolenic, and arachidonic acids.

Nikolova-Damyanova and Amidzhin [114] have obtained effective separation of triacyl glycerols by means of reversed-phase TLC. Triacyl glycerol groups with ECL numbers from 30 to 52 in different natural mixtures gave essentially baseline resolution. The Silica Gel G plates for this purpose were silanized with dimethyldichlorosilane, and a mobile phase consisting of acetone/acetonitrile/water was employed in various ratios, depending on the composition of the sample.

15.4.1.2 High-performance liquid chromatography

Although the substitution of normal-phase HPLC for normal-phase TLC as a method for neutral lipid class isolation was proposed many years ago [1], the lack of general detection techniques and the greater expense of the HPLC system dampened enthusiasm. However, mobile phases have been developed with minimal UV absorption so that lipid separations can be monitored in the 190- to 210-nm range [33,34,115].

Originally, Hamilton and Comai [34] described a HPLC separation and quantitative recovery of cholesteryl esters and triacyl glycerols with solvent systems containing chloroform. More recently, these authors [49] have replaced chloroform with MTBE, which permits detection of solutes at 206 nm. The elution is isocratic and thus requires only one pump rather than a complex gradient system. Using hexane/MTBE/AA (100:5:0.02) as the mobile phase, serum cholesteryl esters emerged at 1.4-2.0 min and triacyl glycerols between 4.8 and 6.0 min. Radioactive cholesteryl oleate was recovered in 100% yield, and radioactive triacyl glycerol in 98% yield. A second mobile phase, hexane/MTBE/AA (70:30:0.02) effectively separated free fatty acids (FFA) and free cholesterol (FC) with quantitative recovery of each. The sn-1,2-diacyl glycerols overlapped the FC, but could be separated from it with a third mobile phase, made up of hexane/2-propanol/acetic acid (100:2.0:0.02), which resolved cholesteryl esters plus triacyl glycerols, FFA, 1,3-diacyl glycerols, FC, and *sn*-1,2-diacyl glycerols. Carbon chainlength and number of double bonds had little influence on the separation. In both instances, the column was μPorasil (30 mm x 3.9 mm).

15.4.2 Resolution of polar lipids

Intact polar lipids can be separated by either one-, two- or three-dimensional TLC methods, which have been successfully employed by many investigators [1]. More recently, normal-phase HPLC has been effectively substituted for conventional TLC, but not for HPTLC, which, in combination with in situ densitometry, has remained unsurpassed as a method of lipid class separation [115].

15.4.2.1 Thin-layer and high-performance thin-layer chromatography

Individual phospholipid classes continue to be isolated by TLC as the method of choice. A major shortcoming of the conventional TLC systems is that complete separation of the acidic phospholipids is not obtained due to overlap. Fine and Sprecher [116] have described a unidimensional TLC system for the separation of phospholipids on boric acid-impregnated plates. The plates were impregnated by dipping them upside-down into a solution of 1.2% boric acid in 50% ethanol. After air-drying for 15 min, the plates were reactivated at 100°C for 60 min. The samples were either spotted or streaked on a preadsorbent area, then developed in chloroform/methanol/water/ammonium hydroxide (120:75:6:2). The method resolved eight phospholipid standards, including PI, PS, and PG. When present in large excess, SPH and PS overlapped. The problem could be overcome by increasing the concentration of boric acid, although this caused the LPC and PS standards, as well as PG and SPH standards, to remain unresolved. Mitchell et al. [117] have reported the separation of phosphoinositides and other phospholipids from human erythrocytes, platelets, and murine lymphoma cells by two-dimensional TLC. Ca^{2+}-free lipid samples were loaded on Silica Gel HL plates and developed, first in chloroform/methanol/water/conc. ammonia (48:40:7:5), and then in chloroform/methanol/formic acid (55:25:5).

Macala et al. [94] and Wood et al. [93] have given detailed accounts of HPTLC systems for the determination of the neutral and acidic phospholipids in rat brain. These methods are equally well suited for the analysis of phospholipid classes from other sources. Excellent separations of the choline and ethanolamine phospholipids were obtained in a dual development with a neutral lipid system, as explained in Section 15.4.1.1. The acidic serine and inositol phospholipids and sulfatides were resolved by a dual development with an acidic solvent system. In the first development, chloroform/methanol/acetic acid/formic acid/water (35:15:6:2:1) was allowed to run ca. 6 cm to separate the inositol phosphatides, serine phosphatides, and sulfatides. In the second development, hexane/diisopropyl ether/acetic acid (65:35:2) was allowed to ascend to the top of the plate, to separate the internal standard from the free fatty acids. The resolved lipids were located by charring with 3% cupric acetate in 8% phosphoric acid. These and similar resolutions have been described in greater detail elsewhere [115]. Saito et al. [28] have described a densitometric method for the determination of minor acidic phospholipids from erythrocytes, liver, and kidney. Total lipids were separated into neutral and acidic fractions by DEAE-Sephadex column chromatography. A clear-cut separation of the different acidic phospholipids was achieved by HPTLC with chloroform/acetone/acetic acid/formic acid/water (60:60:4:10:3). The plate was sprayed with 0.25% potassium dichromate in 15% sulfuric acid.

Heape et al. [118] have described an improved HPTLC method for the analysis of total natural phospholipids. The lipid extracts are spotted or banded onto HPTLC plates (Silica Gel 60F-254) and chromatographed with the following solvent systems: (i) methyl acetate/1-propanol/chloroform/methanol/0.25% aq. KCl (25:25:28:10:7) for 6 cm; (ii) hexane/diethyl ether/acetic acid (90:10:2) for an additional 2 cm. This method gives excellent resolution of sphingomyelin, the choline, serine, and ethanolamine phosphatides, and monogalactosyl diacyl glycerol in the order of increasing R_F values. This method has been utilized for the resolution of red blood cell phospholipids by Freyburger et al. [29]. For TLC densitometry, the plate was prewashed with chloroform/methanol/water (65:35:8), followed by heating at 125°C for 5 min. After application of the solutes, the plate was developed to 7 cm with chloroform/acetone/acetic acid/formic acid/water (60:60:4:10:3), dried, and redeveloped to the top with n-hexane/diisopropyl ether/acetic acid (65:35:2). The sulfatides that overlapped the serine and inositol phosphatides could be separated from PS with another solvent system, chloroform/acetone/methanol/acetic acid/formic acid (30:30:3:1:5), but the separation of PA and cardiolipin (CL) decreased. Another minor band, observed on these chromatograms between PA and CL, was tentatively identified as cholesteryl sulfate.

Banerjee et al. [119] have obtained improved resolution of phospholipids on Chromarods, treated with oxalic acid, without interfering with Iatroscan TLC/FID detection and measurement. However, PI and PS were not resolved in the three different solvent systems tried.

15.4.2.2 High-performance liquid chromatography

A number of excellent solvent systems have been devised for the separation of phospholipid classes in normal-phase HPLC [34,47-52]. Hamilton and Comai [49] have reported an improved HPLC method for the separation of common phospholipids in less than 10 min, compared to 60-120 min in most published works [34,47,48,52]. Using MTBE/methanol/ammonium acetate (pH 8.6) (5:8:2) as the eluting solvent, the inositol and ethanolamine phosphatides were separated within 1-2 min, followed at 4 min by PC, at 5 min by SPH, and at 6 min by LPC. There was a 15% overlap of SPH and PC. Excellent recoveries were claimed in all instances (greater than 98%). PG and CL emerged in the PI fraction. The procedure has also been used to separate lipids from human serum and rat liver. Homan and Pownall [120] have reported an effective resolution of all major classes of both neutral lipids and phospholipids containing pyrene-labeled fatty acids, on a silica column by gradient elution with a ternary solvent system, beginning

with isooctane/tetrahydrofuran (99:1) and continuing with increasing amounts of 2-pro-
panol/dichloromethane (4:1) and 2-propanol/water (1:1), passed through the column at
1.7 ml/min. The cholesteryl esters (CE), triacyl glycerols (TG), diacyl glycerols (DG),
monoacyl glycerols (MG), FFA, PG, PE, PA, PS, PC, and LPC lipid classes were eluted in
this order.

Independently, promising results have been obtained with a DIOL column [121-123].
Andrews [121] reported good separation of PE, PG, PI, PS, PC, and SPH on a 12.5 cm x
0.46-cm column, packed with 5-μm Lichrosorb DIOL, with a 3 cm x 0.46-cm guard
column, packed with 5-μm SI 60 silica. The oven temperature was 38°C, and detection
occurred at 201 nm. The elution was performed with a linear gradient from 88% Eluent A
(acetonitrile) and 12% Eluent B [acetonitrile/water (7:2)] to 25% Eluent A and 75% Eluent
B in 8 min. The column effluent was monitored at 203 nm. Kuhnz et al. [122] improved the
separation of the phospholipids, particularly PS and PI, by substituting 0.005 M phosphate
buffer for water. Heinze et al. [123] revised the method of Andrews [120] and proposed a
specific and sensitive clinical routine method for critical lung maturity cases. A linear
gradient elution program permits quantitation of phospholipids in amniotic fluid, with lyso
PC as an internal standard. After elution with 88% Eluent A (acetonitrile) and 12% Eluent B
[acetonitrile/water (4:1)] for 5 min, a rise to 23% B was effected from 5 to 8 min after
starting; then from 8 to 12 min, a rise to 70% B, and after 13 min up to 75% B. After
keeping Eluent B constant at 75% until 18 min, it was linearly decreased between 18 and
19 min to 12% B. The oven temperature was 55°C and detection was at 201 nm. In the UV
recording, the water gradient produces a constant rise in the baseline, which must be
corrected. The recoveries ranged from 89% to 107% and were independent of the composition of the mixture. The method showed splitting of SPH and PE peaks and resolved
additional new peaks, which were not identified. The absolute values of PC analyzed by
HPLC were considerably lower than those from the enzymatic method, although the
correlations between the two methods were excellent. If phosphate buffers are used in
analyses, the solvents must be prepared fresh daily.

Nakamura et al. [31] have reported a simple HPLC method for the determination of
phosphatidylinositol 4-phosphate and phosphatidylinositol 4,5-bisphosphate in brain. The
phosphatides are derivatized with 9-anthryldiazomethane to yield the corresponding
mono- and di-(9-anthryl) derivatives, which are resolved with an isocratic solvent system
of dioxane/methanol/water/conc. phosphoric acid (40:60:4:0.1) on a Nucleosil 7C18 column (30 cm x 4.6 mm).

A sensitive HPLC method for the separation of major phospholipid classes from skeletal and cardiac muscle samples has been described by Seewald and Eichinger [124]. The
method is based on the simultaneous use of a pH gradient and a polarity gradient. The
complex elution system included acetonitrile, 0.2% phosphoric acid, and methanol. A
baseline separation was obtained for CL, PI, PS, PE, PC, SPH, lyso PE, and lyso PC in
less than 50 min. The eluted lipids were detected at 205 nm.

A novel plasmalogen, believed to be a precursor of the fecapentaenes, has been
isolated and purified by normal-phase HPLC from the colonic microflora of humans and
pigs by Van Tassell et al. [125]. Initial separations were made on silica gel with chloro-

form/methanol (94:6), while final resolution was obtained by normal-phase HPLC on μBondapakNH$_2$ with chloroform/methanol/conc. ammonium hydroxide (60:40:1). The peak detection was at 340 nm, where the conjugated double bond system shows a strong absorption.

Moschidis and Andrikopoulos [126] have reported an improved separation by HPLC of the phosphono analogs of PC and PE from related phospholipids.

15.4.3 Resolution of glycolipids

The natural distribution of glycoglycerolipids in animal tissues has been recently reviewed by Slomiany et al. [127]. Glycolipids are commonly isolated by means of adsorption chromatography as a separate acetone fraction [1]. If a DEAE-Sephadex column is used to effect the initial lipid class separation, the acidic gangliosides and sulfatides are isolated along with the acidic phospholipids (PS and PI), from which the gangliosides are recovered by solvent partitioning [28]. The sulfatides are resolved by TLC.

15.4.3.1 Thin-layer and high-performance thin-layer chromatography

Adsorption chromatography has been used to obtain either a total neutral glycosphingolipid (GSL) fraction by elution with acetone/methanol (9:1) or by stepwise elution with linearly increasing amounts of methanol in chloroform to separate GSL into fractions with different polarities. The final purification is usually accomplished by preparative TLC [128]. Strasberg et al. [109] have described a method for isolating milligram quantities of the four neutral glycosphingolipids, glucocerebroside, lactosylceramide, triaosylceramide, and globoside from human placental tissue. The method involves the use of a continuous chloroform/methanol gradient from 19:1 to 4:1 on a silicic acid column (5 g silica), which yields 1 to 3 mg of GL-1a to -4a each from 933 g of placental pellet. Aliquots of every third fraction from the column were compared by TLC in chloroform/methanol/water (14:6:1) with appropriate standards. Plates were sprayed with 50% sulfuric acid and charred at 100°C to locate the GSL bands. Young and Borgman [129] have used continuous development to increase the separation of glycosphingolipids by HPTLC. HPTLC plates have the disadvantage that the bands, although clearly resolved, may be very closely spaced, so that individual bands cannot be scraped off the plate. Increased separation was achieved for short- and long-chain neutral glycolipids (Solvents A and B), acetylated neutral glycolipids and gangliosides (Solvents C and D), and human meconium neutral glycolipids (Solvents E and F). The solvents and development times were:

A, chloroform/methanol/water (62:30:6), 30 min;

B, chloroform/methanol/water (75:18:2.5), 2.5 h;

C, chloroform/methanol/0.25% aq. CaCl$_2$ (60:40:9), 37 min;

D, chloroform/methanol/0.25% aq. CaCl$_2$ (62.5:30:6), 2 h;

E, chloroform/methanol/water (55:40:11), 47 min; and

F, chloroform/methanol/water (60:35:8), 45 min.

Antibody staining of TLC plates was performed as follows. After drying, the plates were dipped in hexane containing 0.05% polyisobutylmethacrylate, blocked with 5% bovine serum albumin in phosphate-buffered saline, and then incubated in succession with anti-Lewis a antibody CF4C4 and iodinated staphylococcal protein A.

Ariga et al. [130] have used the neutral and acidic lipid fractions isolated from silicic acid or DEAE-Sephadex column chromatography for HPTLC analysis of individual gly-colipid classes. The analytical methods were similar to those described by Macala et al. [94] and Saito et al. [28]. The isolated neutral and acidic lipids (excluding neutral and acidic glycerolipids) were quantified by HPLC according to the internal standard method of Macala et al. [94]. To detect the asialo-GM_1 (gangliotetrose) structures, the various ganglioside fractions were examined by HPTLC immunostaining [131]. Shimamura et al. [132] purified individual glycolipids from PC12 cells by developing HPTLC plates with chloroform/methanol/0.02% aq. $CaCl_2$ (65:25:4). Ladisch et al. [133] separated ganglio-sides from human neuroblastoma tissue on precoated Silica Gel 60 HPTLC plates with chloroform/methanol/0.25% aq. $CaCl_2$ (60:40:9). Gangliosides were visualized as purple bands with resorcinol/HCl reagent.

Ogawa et al. [134] have reported an improved method for HPTLC separation of the neutral glycosphingolipids GlcCer, GalCer, LacCer, and Ga2Cer. Silica Gel G 60 HPTLC plates were developed with 2-propanol/15 M ammonia/methyl acetate/water (75:5:0:25, 75:5:5:25, or 75:10:5:15, as required).

Alvarez and Touchstone [135] have described the separation of acidic and neutral glycolipids on a thin-layer plate having an amino-modified layer (5 cm) adjacent to a silica gel layer (5 cm), continuously developed with chloroform/methanol/water (65:35:8). The sample was applied to the amino layer, 0.5 cm from the bottom.

15.4.3.2 High-performance liquid chromatography

Ladisch et al. [133] have used normal-phase HPLC to separate semipurified total gangliosides according to their carbohydrate content. The system, originally developed by Gazotti et al. [136], consisted of a 25-cm Hibar RT Lichrosorb Si 1000 NH_2 column (4 mm ID and 7 μm average particle size) and a complex gradient of acetonitrile and Sorensen's buffer. Eluent A [acetonitrile/5 mM Sorensen's phosphate buffer (pH 5.6) (83:17)] and Eluent B [acetonitrile/20 mM Sorensen's phosphate buffer (pH 5.6) (1:1)] were used as follows: 100% A for 7 min, then a linear gradient from 100% A to A/B (66:34) over 53 min, and finally a linear gradient to A/B (36:64) over 20 min. The elution was monitored at 215 nm. Fractions of the gangliosides from the normal-phase HPLC columns (GM_2) were separated into molecular species of different ceramide structures by reversed-phase HPLC [137], using a Hibar RT Lichrosorb Si 100 RP-8 column and a linear gradient of 1:1 to 3:2 acetonitrile in sodium phosphate buffer. Elution was monitored at 195 nm. Hoving et al. [138] have noted that silica HPLC of phospholipids and sphingolipids with mobile phases containing phosphoric acid leads to gradual hydrolysis of plasmalogens during passage through the column. Whalen et al. [139] have used a commercially available RadialPak NH_2 column to separate underivatized gangliosides by ion-exchange HPLC.

The procedure separates neutral glycolipids from gangliosides, and the gangliosides into classes, based on their number of sialic acid residues. For this purpose, the column is modified by protonation to pH 5.4, so that it exhibits both ion-exchange and adsorption properties. The separation was accomplished by gradient elution with 1 N NaCl in methanol solution of geometrically increasing salt content. The eluate peaks were determined at 210 nm. An effective separation of a complex mixture of gangliosides has been obtained by Watanabe and Tomono [140], using three DEAE controlled-pore-size glass columns in sequence. Good resolution was effected for mono-, di-, and trisialogangliosides as well as for the neutral glycolipids.

There have been only a few applications of HPLC to the analysis of plant lipids. A range of complex plant lipids has been separated, using gradients of hexane/2-propanol/water [141,142] and isocratic elution with acetonitrile/methanol/sulfuric acid [143] as the mobile phases with columns of silica gel. Similar conditions have been used in combination with an amine-bonded phase [144]. These procedures have been reviewed by Christie [11]. Christie and Morrison [145] have recently described an improved solvent system for the resolution of the galactosyl diacyl glycerols and glycerophospholipids. It consists of a linear gradient, generated by three reservoirs in a ternary solvent mixer, containing hexane/2-butanone/acetic acid (35:65:0.4), hexane/chloroform/2-propanol/aq. buffer (42:5:45:3) and hexane/chloroform/2-propanol/aq. buffer (32:5:50:8). The aq. buffer was 0.5 mM serine, adjusted to pH 7.5 with ethylamine. A baseline separation was obtained for a glycolipid-rich fraction from wheat flour. The mono- and digalactosyl di- and monoacyl glycerols preceded the N-acyl PE and the N-acyl lyso PE. Quantitative analyses of the cereal lipids indicated that the response of the light-scattering detector to glycolipids resembled that for phospholipids.

15.5 SEPARATION OF ENANTIOMERS

In recent years, effective resolution of enantiomeric lipid molecules has been obtained by several chromatographic methods, which have utilized two principally different approaches. In one instance, conventional chromatographic columns have been used for the resolution of diastereomers, prepared by complexing racemic lipid molecules with enantiomeric complexing agents [146]. In the other approach, racemic lipid molecules in the form of appropriate chemical derivatives are resolved into pure enantiomeric species by chromatography on chiral stationary phases [147,148].

15.5.1 Long-chain alcohols and hydroxy acids

A complete resolution of chiral alcohols and hydroxy fatty acids of long chainlengths as the 3,5-dinitrophenylurethane derivatives has been obtained on (S)-2-(4-chlorophenyl)isovaleric acid and its amide derivatives [147] and on N-(R)-1-(α-naphthyl)ethylaminocarbonyl(S)-valine [149], bonded to silica gel. The enantiomers of a wide variety of chiral secondary alcohols have been resolved as the α-naphthylurethane derivatives by HPLC on a chiral stationary phase, containing a conformationally restricted β-amino acid and

90% aq. methanol as the mobile phase [150]. For preparative isolation of these alcohols, a convenient procedure, involving a mild trichlorosilane cleavage, affords the unracemized alcohol from the urethane in high yield [151]. Itabashi et al. [152] have now reported the resolution of enantiomeric aliphatic diols as the dinitrophenylurethanes on N-(S)-2-(4-chlorophenyl)isovaleroyl-D-phenylglycine, chemically bonded to γ-aminopropylsilanized silica, with n-hexane/1,2-dichloroethane/ethanol (40:12:3) as the eluent. A method capable of resolving subnanogram amounts of 12-HETE enantiomers as the pentafluorobenzoate ester derivatives has been developed [153] by ionically coupling two Baker-Bond dinitrobenzoylphenylglycine columns and employing hexane/2-propanol (1000:8) as eluent. The ^{18}O-labeled pentafluorobenzoate of 12-HETE was used as internal standard for stable-isotope-dilution GLC with negative-chemical-ionization mass spectrometry (GC/NCI-MS).

15.5.2 Monoradyl glycerols

A complete separation of monoalkyl glycerols as the diastereomeric 1-(1-naphthyl)ethylmethane derivatives has been obtained by HPLC on an achiral stationary phase [146]. Takagi and Itabashi [154,155] have demonstrated complete resolution of the monoalkyl glycerol enantiomers on a chiral liquid phase, as described for the aliphatic diols [152]. The sn-1 enantiomer is eluted first and the sn-3 enantiomer last. Since the 3,5-dinitrophenylurethanes absorb in the UV, the peaks are readily detected and quantified [155]. Similar separations are obtained with the enantiomeric sn-1- and sn-3-acyl glycerols [156]. However, since the chiral column provides only a limited resolution based on molecular weight of the solutes, it is necessary to collect the peak fractions and to determine their hydrocarbon chain composition. The fatty acids can be identified by transmethylation and GLC of the fatty acids, while the alkyl glycerols must be identified as the alkyl glycerol derivatives by GLC of appropriate derivatives on a polar capillary column. Presumably, the dinitrophenylurethane groups can be removed from the glycerol molecules by reaction with trichlorosilane at elevated temperature (Section 15.5.3). Alternatively, the HPLC peaks may be identified by mass spectrometry in an LC/MS system. Since mixtures of the sn-(3)- and sn-2-monoradyl glycerols of medium-chainlength derivatives may overlap during chiral-phase HPLC, it is mandatory that the sn-1(3)- and sn-2-isomers as well as monoalkyl and monoalkenyl glycerols be first resolved into their chemical classes and positional isomers by normal-phase TLC or normal-phase HPLC (Section 15.4.1).

15.5.3 Diradyl glycerols

Michelsen et al. [146] have reported complete separation of enantiomeric diradyl glycerols and partial separation of diacyl glycerols as their diastereomeric 1-(1-naphthyl)ethylurethane derivatives by HPLC on an achiral phase. However, the incompletely resolved diacyl glycerols could not be identified with optically pure samples. The natural diradyl glycerols occur as sn-1,2(2,3)- and X-1,3-positional isomers. The sn-1,2(2,3)-diradyl glycerols can exist as enantiomers, and each enantiomer in both reverse isomer forms, provided two different hydrocarbon chains are present. The sn-1,2(2,3)- and X-1,3-isomers are readily separated by GLC and HPLC, but the enantiomeric sn-1,2- and sn-2,3-diacyl glycerols have been resolved only recently [157] as the 3,5-dinitrophenylurethane derivatives by HPLC, using a chiral-phase column, N-(R)-1-(α-naphthyl)ethylaminocarbinol-(S)-valine, chemically bonded to γ-aminopropyl silanized silica, with a gradient of hexane/dichloroethane/ethanol (80:20:1) to (250:20:1) for elution. The sn-1,2-enantiomers emerge clearly ahead of the sn-2,3-enantiomers, when only one type of hydrocarbon chain is present. Similar separations have been obtained [157] with the dialkyl glycerols, containing a single type of substituent. In case of different radyl substituents, the enantiomers may overlap, although this effect can be partly minimized by appropriate selection of eluents. The X-1,3-diradyl glycerols, if also present, are eluted ahead of the sn-1,2-enantiomers, which are eluted ahead of the sn-2,3-enantiomers.

The individual components can be identified by determining the fatty acid composition of the peaks, which can be collected. However, the 3,5-dinitrophenylurethanes can also be degraded to the free diacyl glycerols, which can then be trimethylsilylated and separated into molecular species on polar capillary columns, as described in Ref. 158. In this manner, it is possible to obtain resolution of natural diacyl glycerols on the basis of chirality and molecular association of fatty acids in the glycerol molecule. The reverse isomers are not resolved, unless they differ greatly in the length of the hydrocarbon chains occupying the primary and secondary positions of the glycerol molecule. Alternatively, the peaks emerging from the chiral HPLC column can be identified by CI-MS, which has been shown to yield characteristic parent and fragment ions [159]. Overlapping also occurs, when mixtures of diradyl glycerols are chromatographed at the same time. Therefore, it is necessary to separate the dialkyl, alkylacyl, alkenylacyl and diacyl glycerols into the individual chemical classes prior to the preparation of the 3,5-dinitrophenylurethanes and their resolution on the chiral column.

Some reverse isomers of diacyl glycerols can be identified by mass spectrometry [88] on the basis of the $[CH_2-O-OCR]^+$ fragment, which can be produced only from the sn-2-position of the glycerolipid molecule, provided such a fragment is formed in sufficient yield under the chemical ionization conditions.

15.5.4 Triradyl glycerols

Although triradyl glycerols can also occur as enantiomers, they cannot be resolved by the present methods of enantiomer separation. The best one can do is to degrade the triradyl glycerols to diradyl glycerols by means of the Grignard reaction and then to determine the composition of the fatty acids in the enantiomeric diacyl glycerols [24].

15.6 RESOLUTION OF MOLECULAR SPECIES

Resolution of molecular species is the ultimate goal of many lipid analyses. Only a few lipid classes can be completely resolved into individual species. Usually, combinations of two or more complementary analytical systems must be utilized for this purpose. Detailed

References on p. B218

accounts of the analytical approaches explored until 1982 may be found in the previous edition of this book [1].

15.6.1 Fatty acids and fatty acid dimers

Fatty acids are conveniently separated into molecular species, or small groups thereof, by means of GLC on polar liquid phases, following their conversion to methyl or other alkyl esters [2-5]. In combination with flame-ionization detection, the method also provides excellent quantitation. However, many natural mixtures of fatty acids are so complex that it is not possible to identify all peaks on the basis of relative retention times alone [5]. For the isolation and further characterization of unknown fatty acids it is useful to effect an initial fractionation of the acids into groups of uniform degree of unsaturation and geometric configuration by argentation chromatography [44,160]. Final identification of unknown fatty acids may be obtained by GC/MS or LC/MS, following preparation of specialized derivatives, which yield characteristic ion fragments in the mass spectrometer [5,11].

15.6.1.1 Argentation high-performance liquid chromatography

Argentation chromatography is usually performed in combination with TLC and, in isolated instances, on silver nitrate-coated quartz rods [46]. However, HPLC with silver ion complexation, in which the silver ions are bound to an ion-exchange resin, provides a simple method for the isolation of molecular fractions differing in the degree of unsaturation [38,44]. The utilization of a column (250 x 4.6 mm) of Nucleosil 5SA has been described in detail [11].

Using both AgNO$_3$-TLC and AgNO$_3$-Chromarods, separations of lipids with 0-4 double bonds are usually obtained [1]. With the silver-loaded ion-exchange column, the methyl esters of fatty acids of bovine testis lipids have been resolved [44] into well-defined peaks, containing from 0 to 6 double bonds, by applying a linear gradient of methanol to methanol/acetonitrile (9:1) over 30 min at a flowrate of 0.75 ml/min. When analyzed by GLC, each of the peaks was shown to contain only the expected fatty acids, without overlap. Detection was by light scattering [86]. Attempts to use the column for the separation of picolinyl ester derivatives of fatty acids, which are valuable for GC/MS analyses of double-bond distribution, were not successful, possibly because of an appreciable amount of transesterification due to residual sulfonic acid groups in the stationary phase and methanol in the mobile phase [44].

15.6.1.2 Reversed-phase high-performance liquid chromatography and liquid chromatography with mass spectrometry

Numerous methods have been developed for the rapid separation by HPLC of derivatives of fatty acids which either absorb UV or fluorescence [1]. These are only updated here. Quilliam and Yaraskavitch [161] have investigated the *t*-butyldiphenylsilyl derivatization for the purpose of improving the HPLC analysis of fatty acids with UV and/or mass spectrometric detection. The esters proved stable under the HPLC conditions and could be separated on a RP-8 reversed-phase column, using acetonitrile as the mobile phase. The derivatives proved to be specifically suitable for EI-MS, and their nonpolar nature facilitated LC/MS with a moving-belt interface. Korte et al. [162] have refined the *p*-bromophenacyl method of reversed-phase HPLC for the rapid quantitative determination of fatty acids as esters with C_{10}-C_{22} carbons. The compounds are detected at 254 nm, and the method is applicable to extracts of tissues and cells in monolayer culture.

Juengling and Kammermeier [163] have developed a one-vial procedure for quantifying free fatty acids in human blood serum and heart tissue. The free acids were extracted with Freon 11 and then derivatized to the fluorescent coumarin esters prior to HPLC. The esters were resolved into molecular species by isocratic elution with 83% aq. acetonitrile on RP-8 Lichrosorb Supersphere. Fatty acids with longer chain lengths (e.g., 22:0) were chromatographed with 90% aq. acetonitrile. The method showed high reproducibility (less than 4% RSD) and good recovery (91-105%) in the concentration range 20 pmol to 20 nanomol.

15.6.1.3 Gas/liquid chromatography and gas chromatography with mass spectrometry

GLC on polar liquid phases continues to reign as the most powerful single method for separating complex mixtures of natural fatty acids. Since reliable correlation exists between the chromatographic retention time and structure of an acid [2,4], the method is useful for tentative identification of fatty acids. Christie [97] has recently reassessed the value of GC equivalent chainlengths (ECL) for the identification of unknown fatty acids by taking advantage of the availability of a large number of different isomeric C_{18} fatty acids, including most of the isomeric *cis*-octadecenoates and methylene-interrupted *cis*-,*cis*-octadecadienoates, and some *cis*-,*cis*-octadecadienoates with more than one methylene group between the double bonds, along with polyunsaturated fatty acids of cod liver oil and pig testis lipids. The liquid phases tested were available as wall-coated open tubular (WCOT) fused-silica capillary columns (Carbowax 20M, Silar 5CP and CP-Sil 84, and 5% phenylmethylsilicone). In general, the data were in good agreement with those published earlier [41], including the differences between the actual and calculated ECL values. Ackman [4] has proposed that liquid phases of the Carbowax 20M type should be utilized in the standard reference WCOT column for interlaboratory studies, because they permit the elution of the most unsaturated species of each carbon number before the fully saturated species of the next highest carbon number emerges. However, commercial preparations of Carbowax columns tend to vary in their properties [96]. The availability of high-quality polar capillary columns has greatly facilitated both separation and quantitation of fatty acids by GLC [2-5].

The certainty of GLC identification of fatty acids can be greatly increased by preliminary segregation of the acids on the basis of unsaturation or molecular weight [1]. Independent chromatographic confirmation of the identity of the fatty acids may be obtained by GLC examination of special chemical derivatives. Thus, a micromethod has been developed [164] for a quantitative determination of double-bond positions in unsaturated fatty acid methyl esters differing in position, number of double bonds, and chainlength. The method involves ozonolysis in the presence of 1,3-propanediol, reduction of the hydroperoxides with dimethylsulfide, and conversion of aldehyde and aldester fragments to stable derivatives, followed by GLC on a polar liquid phase.

Dutton et al. [165] have described an analytical approach to the determination of 55 possible nonconjugated positional isomers of octadecenoic acid by ozonolysis, GLC on polar capillary columns (free fatty acids), and computer solution of linear equations. The procedure was tested by ozonolyzing a synthetic seven-component mixture of cis-,cis-octadecenoates, converting it to aldehydes, aldehyde esters, and dialdehydes. The fragments were separated by GLC. Equations for an arbitrarily constructed 12 x 15 matrix of linear equations and a computer solution of the matrix provided the composition of the final octadecadienoate mixture. Practical analyses were made of the hydrazine reduction of γ-linolenic acid and the diene products from the biological desaturation of isomeric monoenes.

A combination of GLC with MS provides a more complete identification of the oxygenation products [166-168], but this approach is suitable only for oligounsaturated, not polyunsaturated fatty acids. Alternatively, the fatty acids may be converted to N-acyl pyrrolidides [167,169] or picolinyl esters [169,170], which easily stabilize the ions containing double bonds. Although the pyrrolidides give distinctive modes of fragmentation, the interpretation of spectra becomes more difficult if the number of double bonds is greater than 4 [169]. A preliminary derivatization of double bonds, using nascent deutero-diimide, followed by conversion to pyrrolidides and mass spectrometric analysis of the resulting derivatives, has allowed the correct assignment for up to six double bonds, but the results were found to be variable with different types of instruments [171].

Christie et al. [172] have compared the usefulness of the pyrrolidide and picolinyl ester derivatives for identification of fatty acids in natural lipid samples rich in unsaturated fatty acids, e.g., pig testis and cod liver oil. They obtained satisfactory resolution on capillary columns of fused silica, coated with stationary phases of varying polarity. The picolinyl esters, when subjected to GC/MS on a column containing crosslinked methylsilicone, gave distinctive mass spectra, which could be interpreted in terms of number and location of double bonds. The GC/MS was performed on fused-silica capillary columns coated with a crosslinked (5% phenylmethyl)silicone, and with helium as carrier gas. The column temperature was programed from 60 to 220°C at 50°C/min, then to 250°C at 1°C/min [173]. In a subsequent study, Christie et al. [174] examined the GC/MS spectra of the complete series of isomeric methylene-interrupted octadecadienoates and of octadec-9-ynoate. However, difficulties of interpretation arose, when the double bonds were close to the carboxyl groups and the isomers were not separable. Isomers with double bonds located centrally were generally separable when they were 2 carbons apart. Christie et al. [174] have also examined the GC/MS spectra of the picolinyl ester derivatives of 5,12-, 6,12-, 7,12-, 8,12-, 6,10-, and 6,11-octadecadienoic acids. In each spectrum, features were found that were diagnostic of the position of the terminal double bond. However, authentic spectra were required for positive identification of unknowns. By a combination of AgNO₃-HPLC and GLC on polar capillary columns of methyl esters and GC/MS of picolinyl esters on nonpolar columns, Christie et al. [175] identified complex fatty acid mixtures, including fatty acids with mono- and multi-methylbranched isomers, mono- and polyunsaturated fatty acids of the n-1, n-3, n-4, n-5, n-6, n-7, n-8, n-9, n-11, and n-13 families, and dienoic acids with several methylene groups between the double bonds, with chainlengths from C_{13} to C_{22}. Christie et al. [176] have examined the usefulness of the picolinyl esters for the location of triple bonds in some isomeric dimethylene-interrupted octadecadiynoic acids. In the electron-impact mass spectra of each isomer there were always present molecular ions and ions characteristic of the pyridine ring. The triple bond most remote from the carboxyl group was identifiable because of the presence of diagnostic ions, but the proximal triple bond was not. Therefore, the picolinyl esters are not as useful for the location of triple bonds. The picolinyl esters were identified by mass spectrometry, as described in detail elsewhere [172-175]. The rest of the fractions were hydrolyzed to free fatty acids with 1 M ethanolic KOH before conversion to the picolinyl ester derivatives by an improved procedure, described by Christie and Stefanov [177]. In this procedure, the mixture of fatty acid anhydrides with trifluoroacetic acid anhydride is allowed to react with 3-(hydroxymethyl)pyridine in the presence of 4-dimethylaminopyridine.

Zhang et al. [178] have proposed the formation of 2-alkenyl-4,4-dimethyloxazolines (DMOX) by condensation of the long-chain polyunsaturated fatty acids with 2-amino-2-methylpropanol. These derivatives exhibit clean and regular fragmentation patterns that allow easy discrimination of positional isomers and assignment of double-bond location in the chain. The derivatives are readily recovered and resolved on nonpolar capillary columns under conditions similar to those employed for fatty acid methyl esters. The usefulness of the method has been demonstrated [179] by analyzing the polyunsaturated fatty acids from fish oils and rat testis lipids. The GLC analysis was performed on a 10 m x 0.28-mm glass capillary column, coated with SE-54 (crosslinked and bonded). The unsaturated fatty acid derivatives were eluted before the saturated ones. Furthermore, monoenoic acid derivatives with double bonds near the carboxyl group emerged before those having a double bond remote from the polar end.

There is a close resemblance between the mass spectra of the DMOX derivatives and those of the corresponding N-acyl pyrrolidides in the region above m/z 100. However, the DMOX furnish much more prominent fragment ions at the high-mass end. Only three types of ions are present in the spectra: the McLafferty rearrangement product (m/z 113, usually as the base peak of 2-unsubstituted fatty acid series), the ion due to cyclization/displacement (m/z 126, usually as the second-largest peak in the spectrum), and a series of homologous ions, containing the heterocyclic ring (m/z 126 + 14x, where x = 1, 2, 3, ...). The double-bond position can be deduced easily by using the empirical "12-mass-difference rule" proposed by Zhang et al. [178]. The presence of a double bond at carbon n of the fatty acid is indicated by a gap of 12 mass units in the homologous series instead of the normal 14 mass units between ions containing n - 1 and n carbon atoms of the original acid moiety. As confirmed on model compounds [178], peaks indicating the location of double bonds in spectra of polyunsaturated fatty acid (PUFA) derivatives at C_4

(m/z 138), C_5 (m/z 152), and C_6 (m/z 166) are accompanied by a strong odd-mass ion at m/z 139, 153, and 167, respectively. The DMOX method is suitable for the GC/MS of naturally occurring long- or very-long-chain (greater than 22 C) PUFA. Yu et al. [180] have extended the application of this method to branched-chain fatty acids. Balazy and Nies [181] have reported a rapid, mild, and quantitative derivatization method for preparation of picolinyl esters of epoxides derived from 18:2, 20:4, and 22:6 fatty acids. These esters yielded characteristic spectra, including the molecular ions and a sequence of peaks with abundant ions for cleavage of the carbon/carbon bonds at the oxirane ring. Crabtree et al. [182] prepared the PFB derivatives of hydroxy fatty acid methyl esters for detection by GLC/ECD and GC/NICI-MS. The 2-hydroxy carboxylic acids from C_{12}-C_{26} were studied on the high-temperature phase Poly-S 179. A series of α,ω-dihydroxy, monohydroxy, dihydroxy, and keto fatty acids were identified as the TMS derivatives by Gulacar et al. [183], who used conventional GC/MS for the examination of the lipid extracts from a 4000-year-old Nubian burial ground.

For identification of the normal-chain saturated fatty acids and for a preliminary characterization of branched-chain saturated and unsaturated components, the methyl esters may be admitted to the mass spectrometer directly from the polar capillary column to obtain the molecular weight [19]. However, many polar-phase columns produce too much column bleed, and cannot be used for effective mass spectrometric analyses.

Aveldano [184] has identified a complete series of very-long-normal-chain polyunsaturated fatty acids in the phosphatidylcholines of vertebrate retina, using established chromatographic methods. These fatty acids have 4, 5, or 6 double bonds and belong to homologous series of even-carbon polyenes, having up to 36 carbon atoms. Aveldano and Sprecher [185] used oxidative ozonolysis to show that the very-long-chain tetraenes belong to the n-6 series, hexaenes to the n-3, and major pentaenes to the n-3 series of fatty acids. Molecular ions were obtained by EI-MS of methyl esters.

A capillary GC/MS method for the quantitation of very-long-chain (C_{22} to C_{26}) fatty acids in microliter samples of plasma has been described Aubourg et al. [186]. The 17:0 and 27:0 acids were used as internal standards. The TMS esters of the fatty acids were resolved on a nonpolar capillary column, the peaks were selectively normalized and integrated, and the fatty acid composition was calculated from calibration standards.

Janssen et al. [187] have demonstrated that EI-MS spectra of poly(trimethylsiloxy) derivatives of polyenoic long-chain carboxylic acids, containing a conjugated diene unit, clearly locate all double bonds. Abundant, diagnostic fragment ions arise from cleavage of the bonds at the positions of the original double bonds. The method was applied to several alkadienoic, trienoic, and tetraenoic acids containing a conjugated diene unit. These methods of double-bond location were similar to those employed earlier for the location of double-bond positions in molecular ionization and hydrogenation products of α- and γ-linolenic acids [188]. Baba et al. [189] have described detailed analyses of mycolic acid molecular species from eight strains of *Mycobacterium smegmatis*. The GC/MS was performed on the TMS ether derivatives of the acids, recovered from TLC plates. The α'-mycolic acids were monoenoic acids, ranging from C_{60} to C_{66} and possessing an α-unit of $C_{24:0}$; α-mycolates were dienoic acids, ranging from C_{75} to C_{79} and possessing an α-unit of $C_{24:0}$.

Le Quere et al. [190] have described a simple, on-line hydrogenation method for GC/MS analysis of unsaturated fatty acid esters, in which a fused-silica tube (60 cm x 0.32 mm) coated with palladium acetylacetonate is used. Structures of cyclic fatty acid esters, isolated from heated linseed oil, were elucidated. Frank et al. [191] have described a mass spectrometric method for the determination of dimers of PUFA, found in hepatic endoplasmic reticulum of rats after inhalation of carbon tetrachloride. For this purpose, the sample was hydrogenated and the fatty acids converted to PFB esters, which were analyzed by direct-probe NICI-MS. The PFB esters of the fatty acid dimers started to vaporize at about 260°C after the normal fatty acid PFB esters had evaporated. The most abundant ions, corresponding to the loss of the PFB moiety from the molecular ion, were quantified in relation to that of a tricontanoic acid internal standard.

Christopoulou and Perkins [192] have prepared synthetic dimers, structurally representative of those produced during thermal-oxidative reactions in fats and oils as standards for the development of chromatographic procedures for their separation. These compounds, including dihydroxy, tetrahydroxy, and diketo dimers of methyl stearate, were analyzed by packed and capillary GLC and by reversed-phase HPLC and TLC. Significant difficulties were experienced in developing satisfactory GLC separations of the high-molecular-weight and complex structures. The best GLC separations of the various polar and nonpolar dimers within each class were obtained with the OV-17 phase. More satisfactory separations were obtained when reversed-phase HPLC was used with acetonitrile/acetone (1:1) as the mobile phase and refractometry as the detection method. The separations proceeded according to the polarity of the various dimers, and complete resolution of all dimers, except those of the thermal dimer of methyl linoleate and the dehydro dimer of methyl oleate, were obtained. Subsequently, these researchers [193] isolated dimers by reversed-phase HPLC from partially hydrogenated soybean oil used for frying. The identification of specific products was confirmed by GC/MS. Evidence was presented for the occurrence of the monohydroxy, dihydroxy, and keto groups in the dimers of linoleic acid as well as the dehydrodimer, the bicyclic, tricyclic, and thermal dimer of methyl linoleate, and the dehydrodimer of methyl oleate. The GC/MS analyses were performed on the TMS ethers by CI-MS. Grandgirard et al. [194] have detected by means of GC/FTIR and GC/MS the mono-*trans* geometric isomers of 20:5n3 and 22:6n3 acids in the livers of rats, fed heated linseed oil. The fatty acids were isolated as the methyl esters by preparative reversed-phase HPLC, followed by AgNO$_3$-TLC.

Jensen and Gross [195] have shown that branched fatty acids can be distinguished from isomeric straight-chain fatty acids by collisionally activating the [M-H]⁻ ions desorbed by using fast-atom bombardment mass spectrometry (FAB-MS). Mixtures of homologs and isomers could be investigated by using the combination of FAB and tandem mass spectrometry (MS/MS). Tomer et al. [196] have demonstrated that a substituent, such as an acyl branch, hydroxyl group, cyclopropene ring, cyclopropane ring, or epoxide ring, interrupts the normal fragmentation pattern of saturated fatty acids or collisionally activated gas phase in a characteristic fashion that permits identification of the substituent

References on p. B218

and location of its position on the hydrocarbon chain. The method is suitable for analysis of mixtures of carboxylic acids.

15.6.2 Prostaglandins and other oxo acids

The analysis of molecular species of prostaglandins and related substances has been discussed in great detail in recent reviews by Birkle et al. [197], Pace-Asciak [6], and Powell [7,198]. Individual molecular species or small groups of them can be readily detected and quantified by radioisotope [197] and enzyme-linked [199] immunoassays. These methods have been combined with HPLC to increase the selectivity of the identification [200,201] or to use the antibody precipitation as a means of class isolation for subsequent analysis by HPLC [202]. GLC and HPLC methods are capable of analyzing many components at a time and, in combination with MS, they can provide specific and selective detection of all or most species present in the sample [6]. However, specific chromatographic systems alone have also given effective separations and identifications of molecular species [7].

15.6.2.1 Normal-phase high-performance liquid chromatography

The usefulness of normal-phase HPLC for the separation of prostaglandins and related substances was discussed at some length in the previous edition of this book [1]. Powell [7] has provided a recent review of the application of normal-phase HPLC to the analysis of the lipoxygenase and cyclooxygenase products of arachidonic acid (20:4) metabolism by this method and has tabulated a series of useful solvent systems. Thus, the products of 20:4 oxidation by lipoxygenase could be effectively resolved on silica (5-μm RoSil), using a series of linear gradients of hexane/2-propanol/acetic acid (99.4:0.6:0.1) to hexane/2-propanol/acetic acid (85:15:0.1). This system separated 5S,12S-dh-20:4; iso-1,6-$trans$-LTB$_4$; iso-2,6-$trans$-12-epi-LTB$_4$, LTB$_4$, PGD$_2$, PGE$_2$, PGF$_{2\alpha}$, and 20-OH LTB$_4$ (in the order of increasing retention time) over a period of 120 min [7]. The thromboxane-containing products of 20:4 oxidation by cyclooxygenase were more satisfactorily resolved by means of a series of linear gradients from hexane/toluene/acetic acid (50:50:0.5) to toluene/ethyl acetate/acetonitrile/methanol/acetic acid (30:40:30:2:0.5). This system separated 15h-15-HETE, 12h-12-HETE, 11h-11-HETE, and 60-F$_{1\alpha}$-6-oxoPGF$_{1\alpha}$ (in the order of increasing retention time) over a period of 120 min. The products were detected by measuring radioactivity, as this solvent system did not permit peak detection by UV absorbance. According to Powell [198], care must be exercised in interpreting the monohydroxy region of chromatograms in cases where the injected solution contained significant amounts of polar solvents in addition to the mobile phase. Thus, addition of small amounts of 2-propanol to the mobile phase in normal-phase HPLC can markedly improve the recovery of monohydroxy metabolites, but 2-propanol also affects the chromatographic behavior of these solutes. Recent studies have shown that normal-phase HPLC is not suitable for the separation of peptido-leukotrienes [7].

Argentation HPLC on silica has proven difficult to adopt directly from argentation TLC, which has been widely used in the past to separate prostaglandins on the basis of the number of double bonds [1]. Powell [7] has found that silver-ion-loaded cation-exchange columns [43] can be used to separate a wide variety of metabolites of both 20:4 and 20:3 acids. The order of elution varies with the exact composition of the liquid phase, because the solutes interact with the stationary phase by olefin/silver ion interactions and by polar/polar interactions. If the mobile phase contains relatively high concentrations of acetonitrile (18%), the order of retention times resembles that in normal-phase HPLC (i.e., 20:4, 12-HETE, PGE_2, PGE_1 and PGE_2, $PGF_{1\alpha}$ and $PGF_{2\alpha}$. When the concentration of acetonitrile is reduced to only 1% or less and the solvent consists almost entirely of methanol, the retention times are consistent with the number of double bonds (e.g., 20:4 and 5-HETE, 20:3, PGE_2 and $PGF_{2\alpha}$, and PGE_1 and $PGF_{1\alpha}$). Powell [7] has applied this method to various problems in prostaglandin biochemistry, including the identification of metabolites, purification of isotopically labeled metabolites, and isolation of compounds for subsequent mass spectrometry. Powell [203] has listed the retention times of several labeled and unlabeled metabolites of 20:4 on silver-ion-loaded cation-exchange columns, eluted with a variety of mobile phases.

15.6.2.2 Reversed-phase high-performance liquid chromatography and liquid chromatography with mass spectrometry

Reversed-phase HPLC can be used for the separation of all classes of prostaglandins and related compounds, including peptido-leukotrienes. Most of the solvent systems used for reversed-phase HPLC of prostaglandins consist of aq. methanol or aq. acetonitrile to which various acids or buffers have been added. These solvent mixtures are transparent in the UV and therefore compatible with monitoring the UV absorbance of column effluents. Do et al. [204] have developed a general method for the separation and determination of the specific activity of eicosanoids without derivatization. A combination of a flow-through radioactivity detector and a variable-wavelength UV detector with a micro-bore, reversed-phase HPLC column was used to detect, separate, and quantify 16-keto $PGF_{1\alpha}$, TXB_2, $PGF_{2\alpha}$, PGE_2, PGD_2, 11,9-epoxymethanoprostaglandin H_2, and arachidonic acid in a reproducible manner at high sensitivity.

Powell [7] has discussed the reversed-phase HPLC of the dihydroxy and mono-hydroxy metabolites of 20:4 and has pointed out the advantages and disadvantages of specific solvent combinations. In addition, Powell [7] has examined in great detail the reversed-phase HPLC conditions for the separation of the peptido-leukotrienes. An effective system for their separation consists of a 0.0008 to 0.2% trifluoroacetic acid (TFA) gradient in 70% acetonitrile and an Ultrasphere ODS column. Thus, the 5-lipoxygenase products of 20:4 oxidation are eluted in the order of increasing retention times, as follows: leukotrienes B_4, C_4, E_4, and D_4 [205]. Aside from TFA, acetic and phosphoric acids give good results for the analysis of peptido-leukotrienes and other 20:4 metabolites. However, phosphoric acid is not volatile, and special efforts are necessary to remove it.

References on p. B218

Powell [205] has succeeded in designing solvent gradients that can be used for reversed-phase HPLC separation of complex mixtures of 20:4 metabolites, e.g., prosta-glandins, thromboxanes, and ω-oxidation products of LTB$_4$, dihydroxy products of lipoxygenase oxidation, monohydroxy products, and peptido-leukotrienes. A simple, linear gradient between water/acetonitrile/phosphoric acid (75:25:0.025) and methanol/aceto-nitrile/TFA (60:40:0.0016) gave an effective separation of a mixture of 20:4 metabolites on an end-caped Ultrasphere ODS column [205]. In the order of increasing retention time they were: 6-keto-F$_{1\alpha}$, 20h-B$_4$, TXB$_2$, F$_{2\alpha}$, E$_2$, D$_2$, PGB$_2$, followed by the four 5,12-di-hydroxy metabolites (iso-1, iso-2, LTB$_4$, 5S,12S-dh) HHT, followed by monohydroxy derivatives (15h, 11h, 12h, 5h), the calcium ionophore A23187, and the leukotrienes (C$_4$, E$_4$, and D$_4$). A similar system, but consisting of three solvents, could be used to separate 20:4 metabolites on a non-endcapped Spherisorb ODS-2 column, with a linear gradient from water/acetonitrile/TFA (75:25:0.001) to methanol/acetonitrile/TFA (60:40:0.001) over a 45-min period, followed immediately by a second gradient for 15 min, in which the concentration of TFA was increased to ca. 0.009%. Except for the peptido-leukotrienes, which are eluted at the end of the gradient, the chromatogram was very similar to that obtained on the capped Ultrasphere ODS-silica column.

HPLC can be used for the quantitation of prostaglandins and related compounds, on the basis of either their UV absorbance or their radioactivity. Sugata et al. [206] have determined the fatty acid free radical species, generated by soybean lipoxygenase from 18:2, 18:3, and 20:4 acids, by means of a combination of reversed-phase HPLC with electron spin resonance (ESR) spectroscopy. The use of UV detection together with HPLC/ESR resulted in a 500-fold increase in sensitivity of detection of the radical species. Powell [7] has reported the use of a 0.2-ml cell, packed with cerium-coated glass beads for monitoring the radioactivity of the prostaglandins in the effluents of the HPLC columns. Borgeat and Picard [207] have proposed the use of 19-hydroxyprostaglandin B$_2$ as a polar internal standard, which they consider essential for avoiding variable recoveries of polar compounds in the on-line extraction/reversed-phase HPLC analysis of lipoxygenase products.

The potential use of LC/MS in the identification and quantitative assay of prostaglan-dins and related substances is an attractive prospect, but the present technology is still mostly at the theoretical stage, the practical attempts failing due to low sensitivity. Ther-mospray LC/MS has recently been successfully applied in the structural analysis of 5-HETE and LTB$_4$ and of the di- and trihydroxy metabolites of 22:6n3, produced by brain homogenates [208]. A similar technique was employed to confirm the formation of 11-, 9-, and 5-HETEs by human inflammatory cells [209].

15.6.2.3 Gas/liquid chromatography and gas chromatography with mass spectrometry

Several groups of investigators have reported systematic studies on the GLC and GC/MS analysis of prostaglandins and related compounds [1]. The use of capillary GLC is mandatory for the analysis of the stable cyclooxygenase metabolites of arachidonic acid (PGF$_{2\alpha}$, PGE$_2$, PGD$_2$, 6-keto-PGF$_{1\alpha}$, thromboxane B$_2$), because these compounds cannot be separated in conventional packed columns [210]. For this purpose, the pentaflu-orobenzoates of the methoxime trimethylsilyl (PFB-MO-TMS) derivatives are best suited. The mass spectra of these derivatives give significant, intense ions in the high-mass range, so that specific and sensitive responses are obtained when the single-ion monitor-ing (SIM) technique is used. Chiabrando et al. [211] have described a one-step extraction procedure for urinary metabolites of thromboxane A_2 for MS. TXB_2, 2,3-dinor-TXB_2, and their deuterium-labeled internal standard (TXB_2-D_8) are simultaneously extracted from urine by an antibody raised against TXB_2 (cross-reacting with 2,3-dinor-TXB_2) and coupled to Sepharose. Samples are then derivatized and directly analyzed by capillary GC/NCI-MS. Chiabrando et al. [212] have described a further advance in terms of rapidity and simplicity of assay of prostanoids, such as 6-keto-$PGF_{1\alpha}$, or TXB_2, based on immu-noaffinity extraction, stable-isotope dilution, and GC/MS detection for profiling urinary metabolites of prostacyclin and thromboxane. The lipid extract was directly derivatized to form PFB ester, MO, and TMS ether derivatives. The method was aplied to evaluate the urinary excretion of 6-keto-$PGF_{1\alpha}$, 2,3-dinor-6-keto-$PGF_{1\alpha}$, TXB_2 and 2,3-dinor-6-keto-$PGF_{1\alpha}$. Thromboxane TXA_3 and prostaglandin PGI_3 in man after ingestion of 20:5 acid have been identified and quantified by GC/EI-MS [213].

Powell and Gravelle [214,215] have extensively utilized GC/MS for the identification of various metabolites of the leukotrienes by polymorphonuclear leukocytes of rat, man, and pig. In earlier studies, Powell [216] had used GC/MS to study the ω-oxidation products of a variety of dihydroxyicosatetraenoic acids and had identified the LTB_4 ω-oxidation prod-uct as 5S,12S-dh-20:4.

Hartzell and Andersen [217] have reported a systematic study of the mass spectral fragmentation of the methyl ester (ME)-MO-TMS derivatives of D and E prostaglandins and selected ω-chain analogs. The EI mass spectra of ME-TMS derivatives of $PGF_{2\alpha}$ and the ME-MO-TMS derivatives of PGE_2, PGD_2, 6-oxo-$PGF_{1\alpha}$, and 2,3-dinor-6-oxo-$PGF_{1\alpha}$ [218] and of thromboxanes [219] have been investigated, and collisionally activated de-composition mass spectra of the most intense parent ions in the high-mass region have been determined. It was hoped that characteristic daughter ions of high intensity would allow reliable quantitation of prostaglandins in biological fluids and a reduction of sample cleanup. In a parallel study [220] NCI mass spectra of selected prostaglandins and thromboxanes as the PFB-MO-TMS derivatives were assessed, and the collisionally acti-vated spectra of the prostanoid PFB/MO/TMS derivatives were recorded.

Gut et al. [221] have reported an improved direct CI mass spectrometric technique, in which a polyimide-coated fused-silica fiber is used as an extended probe tip, to obtain molecular ions and diagnostic fragment ions of underivatized arachidonic acid metabolites involved in leukotriene biosynthesis.

Hughes et al. [222] have described a method for a quantitation of lipid peroxidation products in total hepatic lipid by GC/MS. Three isomers (11-, 12-, and 15-HETE) acid were quantitated, using SIM techniques based on the internal standard methyl 15-hydroxyara-chidate. The advantage of the proposed methodology over existing techniques for measuring lipid peroxidation was primarily one of specificity.

A stable-isotope dilution assay for the simultaneous determination of two metabolites of prostacyclin, 6-oxo-PGF$_{1\alpha}$ and 2,3-dinor-6-oxo-PGF$_{1\alpha}$, in human seminal fluid and human urine has been described by Fischer and Meese [223]. The synthesis of a new deuterated internal standard, 18,18,19,19-^{2}H$_4$-2,3-dinor-6-oxo-PGF$_{1\alpha}$, enabled specific and sensitive quantitation, based on NCI-MS of the MO/PFB/TMS ether derivatives in SIM, by registering the [M-181]$^-$ fragments. The detection limit was 10 pg per sample. A highly sensitive and specific assay for the quantitation of thromboxane B$_2$ in human urine was described by Meese et al. [224]. The method is based on the use of a low-blank tetra-deuterated internal standard (18,18,19,19-^{2}H$_4$-thromboxane B$_2$), which the authors prepared and then purified by HPLC. The analyses were made with the MO/PFB/TMS ethers of the unknowns and internal standards in the GC/NICI-MS system. The detection limit by SIM in the NCI-MS mode was 10 pg and 30 pg per sample for pure standards and biological samples, respectively.

Pace-Asciak [6] has recently reviewed the application of GC/MS techniques to the analysis of prostaglandins and related substances. His approach to the problem of analyzing as many products as possible in the eicosanoid family has been to keep all products together during sample purification and to make use of GLC to separate the products, with MS acting as a specific and selective detector. This technique was suitable for the analysis of the leukotrienes, lipoxins, hepoxilins, prostaglandins, prostacyclins, thromboxanes, as well as the epoxy-prostaglandins and diols [225-228]. Pace-Asciak [6] has described detailed procedures for the workup of samples for GC/MS, which involves ethyl acetate extraction, C$_{18}$-Sep-Pak purification, PFB derivatization, MO derivatization, diethyl ether extraction, C$_{18}$-Sep-Pak separation, TMS derivatization, and hexane extraction, preliminary to GC/MS analysis. The latter was performed on a 60-m fused-silica column, containing DB-1 methyl silicone (0.25-μm film thickness) as the stationary phase, by splitless injection. The carrier gas was hydrogen, and the column temperature was programed from 100 to 280°C at 30°C/min, where it was held until the end of the analysis. Prostaglandins were eluted from the column at 280°C in the order of increasing retention time as follows: α-dinor 6KF$_{1\alpha}$, α-dinor 15KD-6KF$_{1\alpha}$, F$_{2\alpha}$, 15K-F$_{2\alpha}$, 15KD-F$_{2\alpha}$, D$_2$, E$_2$, 15KD-E$_2$, TXB$_2$, 6KF$_{1\alpha}$, 15K-6KF$_{1\alpha}$, 15KD-6KF$_{1\alpha}$, 6KE$_1$, C$_{22}$F$_{2\alpha}$, and 15K-6KE$_1$. The peaks were identified by SIM on the basis of the [M-PFB] fragments of the d_0 and d_4 derivatives by NCI with methane as the reactant gas. Pace-Asciak [6] has recently redefined the conditions for proper use of the deuterated products as internal standards for qualitative identification and quantitation of prostaglandins, including the criteria of purity and homogeneity of the standard. An alternative to the use of deuterated products is the use of ^{18}O-labeled carboxyl oxygens [229]. Neither of these stable isotope derivatives of prostaglandins is available commercially, and this is a serious handicap. Pace-Asciak [6] has proposed the use of C$_{22}$-PGF$_{2\alpha}$, an analog of PGF$_{2\alpha}$ in which the chainlength has been extended by two carbon atoms, as an internal standard for products unavailable in deuterated form. The current sensitivity of the GC/MS system is in the low-picogram and high-femtogram range. Pace-Asciak [6] has made extensive use of the GC/NCI-MS methodology in the assay of biological systems, including the release of prostaglandins from the isolated, perfused rat kidney, homogenates of rat lung, and the determination of prostaglandins and their metabolites in murine and human blood.

Rosenfeld et al. [230] have recently described an integrated procedure for the isolation and derivatization of prostaglandins for GLC analysis with ECD and NCI-MS. The procedure involves the ion-exchange resin XAD-2, which retains PGE_2 and acts as a support for the oximation at the carbonyl group and as a catalyst for the PFB preparation.

15.6.3 Neutral glycerolipids

A complete structural analysis of a lipid mixture requires that it be separated into molecular species and that the composition of specific acyl moieties be determined in all the relevant portions of the molecule. With neutral lipid molecules containing one or two acyl groups this is now possible by chromatographic means. With neutral lipids that contain three hydrocarbon chains, physical separation by chromatographic means is often impossible, and the molecular species present in the unresolved mixture must be identified and quantitated by other methods. Likewise, chromatographic resolution of molecular species of polar lipids containing two or more acyl chains is frequently impossible by chromatographic means, although certain exceptions are known. It is usually necessary to convert the polar lipids into neutral lipids prior to chromatographic resolution and quantitation of the component species. The most successful methods of separation and quantitation of molecular species have proved to be capillary GLC on polar liquid phases and reversed-phase HPLC, along with MS.

15.6.3.1 Capillary gas/liquid chromatography and gas chromatography with mass spectrometry on nonpolar liquid phases

The early GLC and GC/MS work with packed columns containing nonpolar liquid phases was extensively reviewed in the preceding edition of the book [1] as well as elsewhere [5,19] and will not be repeated here. Instead, the discussion will be confined to the use of nonpolar capillary columns in GLC or in GC/MS. The capillary GLC separations are usually accomplished with temperature programing and hydrogen as the carrier gas [5,8]. Mares and Husek [231] have studied in detail the factors affecting the quantitation of intact triacyl glycerols by the use of nonpolar capillary columns. Column performance was found to depend on such factors as the injection technique, column quality, flowrate of the carrier gas, the amount of the solute, and its molecular weight. However, if sufficient care is taken in optimizing the system and in measuring the weight correction factors, excellent reproducibility can be achieved [231].

As with packed columns, the most difficult resolutions are those involving triacyl glycerols, which are separated according to molecular weight, but some resolution is also seen between saturated and unsaturated species of the same carbon number [8]. Components differing by one methylene unit are clearly resolved, even on relatively short capillary columns (6 m). Similar resolution of natural triacyl glycerols has been obtained by others with the triacyl glycerols of seed oils [57], human plasma [232], and other animal sources

[233]. The method has been employed in comparative studies of triacyl glycerol and alkyl diacyl glycerol composition in human milk [234] and of short-chain triacyl glycerols and chloropropanediol diesters in ruminant milk fats [235]. Increasing the length of the capillary column permits significant resolution based on both chainlength and unsaturation [236].

Myher et al. [237] have demonstrated that this resolution is due to a progressively longer retention of triacyl glycerols containing capric, caprylic, caproic, butyric, and acetic acids within a carbon number. Geeraert and Sandra [238] have obtained a partial resolution of saturated and unsaturated triacyl glycerols of coffee oil on a 15-m glass capillary, coated with OV-101, using temperature programing from 310 to 330°C. Up to four fractions could be recognized, appearing in the order: UUU, UUS, USS, and SSS (where S is a saturated and U an unsaturated fatty acid residue). The slightly earlier elution of the unsaturated components is responsible for the peak broadening and overlap with the odd-carbon-number species usually encountered in natural triacyl glycerol samples on nonpolar capillary GLC.

The nonpolar capillary columns are also suitable for the resolution of cholesteryl esters according to carbon number, provided the column temperature is kept sufficiently low (below 310°C) to avoid decomposition of the more highly unsaturated species [61,232,239-242].

Comparable separations on the nonpolar capillary columns have been obtained for diacyl glycerols. They also permit the separation of X-1,2- and X-1,3-isomers [1]. For this purpose, the diacyl glycerols are analyzed as the TMS or TBDMS ethers, or as acetates. The separation of the saturated and unsaturated species within a carbon number can be greatly improved by increasing the column length [8], but no practical application of this increased power of resolution has been reported.

The nonpolar capillary columns are also suitable for the separation of free sterols, monoacyl glycerols, and free fatty acids. Thus, cholesterol, campesterol, and sitosterol are readily resolved as the TMS ethers [1]. The monoacyl glycerols are resolved on the basis of the position of the fatty acid in the molecule, the secondary esters emerging ahead of the primary esters. The free fatty acids are recovered as the TMS or TBDMS esters, with partial separation between the saturated and oligounsaturated species.

The nonpolar capillary columns are eminently suited for GC/MS. The highly stable stationary phase provides a low-background bleed, even at relatively high temperatures. The capillary columns can be threaded directly into the ion source, eliminating the need for complex interfaces and molecular separators, which cause problems with the recovery of high-molecular-weight components. In addition, MS allows the identification of the unresolved components, thus greatly improving the information available from either GLC or MS alone. With EI spectra, the triacyl glycerol peaks are identified on the basis of the retention time and the nature of the diacyl glycerol type of fragment ions, although small molecular ions may also be observed [19]. The softer CI provides substantial molecular or quasi-molecular ions, but yield only small diacyl glycerol fragment ions, which may not provide sufficient information for complete identification of the molecular species present. Both ammonia [244] and methane [245,246] have been successfully employed as reagent gases in the past for positive CI. By this means, triacyl glycerols have been quantitated in a variety of natural lipid extracts, including butterfat [245] and algae [246]. Alkyl diacyl glycerols give mass spectra similar to those of triacyl glycerols [247]. The molecular ion is small and that for [M-18]$^+$ is smaller than that in triacyl glycerols.

The TMS and TBDMS ethers of diacyl glycerols yield ion fragments which are very similar to those obtained upon triacyl glycerol fragmentation [88]. The diacyl glycerol species are readily identified from the [M-ROO]$^+$ and [M-57]$^+$ ions. However, their yield varies somewhat with the nature of the fatty acid substituents. Similarly, both EI- and CI-MS have been effectively utilized for the identification of the steryl esters in the effluents of nonpolar capillary columns [241-243,245].

Ratnayake et al. [248] reported the applicability of the Finnigan MAT Ion-Trap Detector mass spectrometer for the structure determination of novel glyceryl ethers, isolated from cod flesh. The glyceryl ethers derived from diacyl monoalkyl glycerols were resolved as isopropylidene derivatives on capillary columns, coated with OV-101 methylsilicone. Bossant et al. [249] employed GLC on nonpolar columns with ECD and MS/MS to establish the absence of 1-O-alkyl-2-propionyl-sn-glycerol moieties in natural extracts of the platelet-activating factor in human polymorphonuclear neutrophils. The compounds were resolved on nonpolar capillary columns. Nes [250] has described the GC/MS of phytophthorols, which comprise a series of saturated and unsaturated odd-chain fatty alcohols (C_{17}-C_{21}), esterified with myristic acid, each possessing a 1,2-diol group located near the ester terminus. These compounds are produced by *Phytophthora cactorum*, a fungus. The phenolic lipids of plants, which possess a phenolic in place of a carboxyl or glyceryl group, have been investigated by Tyman [251] by TLC/GLC, TLC/MS, and HPLC and were identified as anacardic acid, cardol, and cardanol. These lipids have methylene-interrupted double bonds in the fatty acid chains.

15.6.3.2 Capillary gas/liquid chromatography and gas chromatography with mass spectrometry on polar liquid phases

Polar capillary GLC has been extensively utilized for the separation of naturally occurring monoacyl and diacyl glycerols and the monoacyl and diacyl glycerol moieties of natural triacyl glycerols and glycerophospholipids. Similar resolutions have been recorded for the naturally occurring and derived alkyl acyl and alkenyl acyl glycerols, although they have been analyzed less frequently [1,24,252].

Specifically, molecular species of diacyl glycerols as the TMS or TBDMS ethers are resolved [253,254] on the basis of their molecular weight and degree of unsaturation on the SP-2330 column (10 m x 0.25 mm ID) employed in fatty acid analyses, provided the oven temperature is raised to about 250 or 260°C. Complete resolution was obtained for nearly all molecular species, ranging from 16:0-16:0 to 18:1-22:6. However, the 18:0-18:2 and 16:0-20:4 species partially overlapped, as the retention ratio was only 1.06, despite a ratio of 1.19 calculated from the relative retention times of the component fatty acids. Apparently, the polarity of the column at the low temperature used for fatty acid separation is different from that at the high temperature employed for diacyl glycerol resolution. Other

References on p. B218

overlaps were seen for minor diacyl glycerol species. This GLC column has been utilized for the resolution of the diacyl glycerol moieties of linseed oil triacyl glycerols, derived by Grignard degradation [253], the diacyl glycerol moieties of the phosphatidylcholines in rat liver [253], the diacyl glycerol moieties of the phosphatidylinositols in soybean [255], the diacyl and alkenyl acyl glycerol moieties in the heart and kidney of rats [254], and the diradyl glycerol moieties in the choline and ethanolamine phosphatides of Ehrlich ascites cells and their fat granule membranes [256]. In all instances, the TMS ethers gave better resolution and higher recoveries than the TBDMS ethers, which required somewhat higher elution temperatures than the TMS ethers. Improved baseline stability and peak resolution was obtained for the diradyl glycerols on a fused-silica column (15 m x 0.32 mm ID), coated with cross-bonded RTx 2330. The diradyl glycerol TMS ethers were separated isothermally at 250°C, using split injection (split ratio 7:1) and H_2 carrier gas at 3 psi head pressure. The new column was employed for the resolution of the diradyl glycerol moieties of the choline, ethanolamine, inositol, and serine phosphatides of the red blood cells [257] and plasma [258] of man and of the diradyl glycerol moieties of the choline and ethanolamine phosphatides of rat intestinal mucosa and the lymph chylomicrons of rats [259]. Over 200 molecular species were identified and quantitated in representative samples of human red blood cells, plasma, and rat intestine. Other recent applications have been reported in the analysis of monoacyl glycerols derived from triacyl glycerols by digestion with pancreatic lipase [260], the monoalkyl glycerols derived from red cell phosphatidylcholines, and in the analyses of the monoalkenyl glycerols from red blood cell phosphatidylethanolamines [257,258].

The fused-silica capillary columns containing the bonded RTx 2330 stationary phase have also been utilized for the resolution of the highly complex diacyl glycerol moieties derived from menhaden oil triacyl glycerols by Grignard degradation. Isothermal chromatography at 260°C was utilized to extrapolate the relative retention times for the many unknown species from various secondary and a few primary diacyl glycerol standards [261] to obtain the tentative identities of the major diacyl glycerol moieties of menhaden oil during a stereospecific analysis [262].

The potential usefulness of the polarizable methylphenylsiloxane liquid phase for the resolution of the diacyl glycerol mixtures has been demonstrated by Kuksis et al. [61]. Capillary columns (25 m x 0.25 mm ID), prepared with 50 and 65% phenylsilicone content, gave comparable resolution in the temperature range 320-340°C used for triacyl glycerol resolution. However, resolution and recovery of the more highly unsaturated species remained incomplete. Since these columns are relatively nonpolar below 310°C, the separations must be performed above this temperature, where partial degradation takes place. However, higher flowrates and lower oven temperatures may permit higher recoveries and better resolution during isothermal operation, which thus far has not been tried for this separation.

The usefulness of the polarizable capillary columns for the separation of natural triacyl glycerols of both plant and animal origin has been demonstrated by Geeraert [57] and Geeraert and Sandra [60]. In case of bovine milk fat, which represents one of the most complex natural triacyl glycerol mixtures, complete resolution was obtained for the higher-molecular-weight species, while the lower-molecular-weight species overlapped extensively. The long-chain species were identified on the basis of the retention times of various secondary and a few primary standards [60]. The short-chain triacyl glycerols have been subsequently identified by a combination of the polarizable capillary GLC and MS [237]. Oshima et al. [263] have recently employed the polarizable 65% phenylmethylsilicone stationary phase to separate triacyl glycerol species and to identify them by EI-MS. For this purpose, they monitored the $[RCO]^+$ and the $[M - RCOO]^+$ ions. At the present stage of development, the polarizable capillary columns do not give adequate recoveries of the long-chain polyunsaturated fish oil triacyl glycerols, which apparently decompose at the elevated temperatures and long retention times necessary to effect their elution [60]. Possibly, higher flowrates and shorter columns, permitting elutions at lower temperatures and over shorter periods of time, may result in more satisfactory resolution and recovery of polyunsaturated triacyl glycerols.

The polar capillary columns also allow effective separation of mixtures of cholesteryl and plant steryl esters. Thus, the different fatty acid esters of cholesterol, campesterol, and sitosterol, isolated from a patient with phytosterolemia, have been resolved on the basis of molecular weight and degree of unsaturation of the fatty acid, except for the few instances where the molecular weights coincide [243].

15.6.3.3 Reversed-phase high-performance liquid chromatography and liquid chromatography with mass spectrometry

The general usefulness of reversed-phase HPLC for the resolution of molecular species of neutral lipids was discussed in the previous edition of this book [1] and many specific examples have been given more recently by Christie [11] and Shukla [12]. A major problem affecting all HPLC systems is the lack of a simple quantitative detector, although the light-scattering detector appears promising [11]. Other detectors for HPLC have been discussed elsewhere [12,264].

Payne-Wahl and Kleiman [265] quantitated estolide triacyl glycerols in *Sapium* seeds by reversed-phase HPLC, using an IR detector. They also used the IR detector for quantitation of free acids, mono-, di-, and triacyl glycerols. Phillips et al. [226] have used reversed-phase HPLC in combination with a FID for the quantitation of triacyl glycerols in cocoa butter, soybean oil, and olive oil. Differences in the response of individual species were small and did not require the use of response factors. Foglia et al. [267] used HPLC in combination with an IR detector to analyze the 1-alkyl-3-acyl and 1-alkyl-2-acyl products of detritylation of 1-alkyl-2-acyl-trityl glycerols, cholesterol serving as reference standard.

Stolyhwo et al. [268] demonstrated the potential of using a laser light-scattering detector for the quantitation of triacyl glycerols of complex mixtures with few calibration factors. However, Robinson et al. [269] showed that the light-scattering detector (also called mass detector) gave a nonlinear response for triacyl glycerols with a detection limit of less than 1 μg. The mass detector was found to be highly satisfactory for efficient monitoring of triacyl glycerol resolution on a 3-μm, 15 cm x 4.6-mm-ID Spherisorb ODS column by gradient elution with dichloroethane and acetonitrile [270]. The nonaqueous reversed-phase HPLC analysis of triacyl glycerols has been reviewed by Barron and Santa-Maria [271].

Frede [272] and Singleton and Pattee [273] attest to the importance of temperature programing in the HPLC of triacyl glycerols. Singleton and Pattee [274] have used a micellar mobile phase to decrease the elution time of peanut oil triacyl glycerols in reversed-phase HPLC. The overall pattern of the triacyl glycerol species recognized in the peanut oil was comparable to that obtained by Bezard and Ouedraogo [275], who used acetone/acetonitrile (70:30) as the mobile phase. Sempore and Bezard [84] have reported further qualitative and quantitative analyses of peanut oil triacyl glycerols by reversed-phase HPLC with a RI detector, including the collection of fractions as an aid in peak identification.

Frede [272] has described three new methods of improving triacyl glycerol separation by reversed-phase HPLC with propionitrile as eluent in combination with a RI detector. A high temperature was indispensable for saturated long-chain compounds, which tend to crystallize on the column. As a result of temperature programing, all chromatograms showed higher resolution than those obtained at constant temperature. Specifically, butterfat yielded more peaks than previously observed, while soybean oil revealed the presence of a series of ten long-chain compounds, following 1,3-dipalmitoyl 2-oleoyl glycerol (POP). Frede and Thiele [276] have subsequently adopted this system to improve the HPLC resolution of butterfat triacyl glycerols. The improvements were achieved by decreasing the sample load, recycling the solvent over longer periods of time, and increasing the detector sensitivity by employing an interferential refractometer in connection with a thermostat to stabilize the temperature of the entire HPLC system. Gilkinson [277] has described a gradient reversed-phase system which is compatible with UV detection at 220 nm and is suitable for fats containing a wide range of carbon number. Cis- and trans-isomers were well resolved and better precision was obtained for polyunsaturated fats by reversed-phase HPLC than by capillary GLC. The type of unsaturation present in the eluted triacyl glycerols was obtained from the absorbance ratio at 220/225 nm, monitored by a diode-array detector during isocratic elution. The method was illustrated by analyses of butterfat and soybean oil triacyl glycerols.

Shukla and Spener [278] have used reversed-phase HPLC and UV detection at 220 nm to separate triacyl glycerols of Flacourtiaceae seed oils, which contain cyclopentenyl fatty acids (chaulmoogric oils). A comparison of retention volumes of triacyl glycerols containing cyclopentenyl fatty acids with those of straight-chain fatty acids shows that CCC is equal to OOO, PCC to POO, and PPC to PPO, where C is chaulmoogric; O, oleic; and P, palmitic acid. These authors [278] also showed that nonaqueous reversed-phase systems possess the advantage of resolving triacyl glycerol critical pairs that are not separated by TLC and GLC techniques. A new postcolumn reactor for the analysis of triacyl glycerols with high sensitivity and selectivity has been developed by Kondoh and Takano [279]. In this reactor, triacyl glycerols in the HPLC effluent are hydrolyzed with KOH, and the resulting glycerol is oxidized to formaldehyde with periodic acid. The formaldehyde is made to react with acetylacetone in the presence of ammonium acetate, and the reaction product is detected colorimetrically at 410 nm. The authors reported a linear working range of 0.3 to 50 nmol trilauroylglycerol. The usefulness of the detector was illustrated by analyses of coconut oil and synthetic triacyl glycerols.

Kuksis et al. [280] and Marai et al. [281] have reported similar LC/MS separations and identifications for corn oil, stripped lard, and human plasma triacyl glycerols by means of a gradient of 30-90% propionitrile in acetonitrile at 30°C. The LC/MS combination was achieved through the direct liquid inlet interface, which, however, permits the admission to the mass spectrometer of only 1-2% of the total column effluent, and therefore requires overloading of the HPLC column in order to obtain sufficient mass of minor components. This results in peak skewing and a partial overlap of any minor peaks by major preceding peaks that may tail. This LC/MS system has been applied to the resolution and identification of the molecular species of triacyl glycerols in bovine milk fats and butter oil distillates [280]. Due to the much greater variety of fatty acids represented, none of these peaks contained single components, most of them being constituted of several major and many minor species. Kuksis et al. [282] have subsequently completed a detailed assessment of the composition of the butterfat triacyl glycerol peaks by LC/MS of molecular distillates of butteroil. A comparable complexity was demonstrated by LC/MS for the triacyl glycerols of goat milk fat [234,235].

The reversed-phase LC/MS system has been shown [283] to be suitable for the separation and identification of menhaden oil triacyl glycerols. Due to the presence of a large number of polyunsaturated fatty acids, this oil is much more complex than the vegetable oils and even ruminant milk fats. The longer-retained, more saturated species are well resolved, and the spectra easily interpreted. The peaks in the early portion of the chromatogram are very closely spaced, due to the great variety of the polyunsaturated species having identical ECL. Since the intensities of the "diacyl glycerol" ions are not representative of their true proportions in the parent triacyl glycerols, it is not possible to calculate the peak composition. Therefore, there is great interest in the detection of triacyl glycerol species by chloride-attachment NCI-MS, which yields only the pseudomolecular ion for each triacyl glycerol species [284]. The ions are produced by addition of 1% dichloromethane to the acetonitrile/propionitrile gradient used as eluent. This method shows promise for quantitation of triacyl glycerols overlapping in any of the HPLC peaks, provided that they differ in molecular weight. Quantitation of triacyl glycerol species in positive CI requires extensive calibration of the LC/MS system [89].

Reversed-phase HPLC methods have been extensively utilized for the separation of diradyl glycerol acetates. Nakagawa and Horrocks [79] reported a method for the resolution of individual molecular species of diacyl, alkylacyl and alkenylacyl glycerols derived from glycerophospholipids. The alkylacyl and alkenylacyl glycerols as the acetates were eluted with acetonitrile/2-propanol/MTBE/water (63:28:7:2), while the diacyl glycerol acetates were separated by acetonitrile/2-propanol/MTBE/water (72:18:8:2). The separations were accomplished at 30°C with UV detection at 205 nm. By this method, Nakagawa et al. [285] demonstrated high amounts of alkyl arachidonoylglycerol in rabbit alveolar macrophages. Nakagawa and Waku [286] studied the de novo biosynthesis of dipalmitoylglycerol moieties in the choline phosphatides of these cells, and Ojima-Uchiyama et al. [287] observed the high levels of alkylacyl glycerol moieties in the choline

References on p. B218

phosphatides of human eosinophils. The acetate derivatives were used by Choe et al. [288] for the identification of long-chain diacyl glycerols derived from the ethanolamine phosphatides of frog retina. In this instance, acetonitrile/2-propanol (70:30) was employed as the eluent, with detection at 210 nm. Nakagawa and Waku [289] have presented a detailed account of the reversed-phase HPLC separation of the arachidonoyl species of glycerophospholipids in alveolar macrophages.

Reversed-phase HPLC has been very widely used for the separation of molecular species of diradyl glycerols, following the preparation of UV-absorbing derivatives, such as benzoates, p-nitrobenzoates, and dinitrobenzoates [1,11,72]. However, a complete resolution of molecular species is usually not obtained with a single solvent system. Takamura et al. [73] have accomplished the complete separation of the diacyl glycerol moieties of the common glycerophospholipids as the dinitrobenzoates by HPLC, using a combination of two solvent systems. The peaks that were not resolved by acetonitrile/2-propanol (4:1) were separated by methanol/2-propanol (19:1). This method was subsequently applied [290] to the separation of molecular species of alkenylacyl and alkylacyl subclasses of human platelet phospholipids. The dinitrobenzoates were first separated into the diradyl glycerol subclasses by either TLC or by normal-phase HPLC. The methods of Takamura et al. [73] were employed by Poulos et al. [291] for the resolution of the long-chain diacyl glycerol species, derived from Zellweger brain phosphatidylcholine and by Bell [292] for the resolution of the polyunsaturated species of diacyl glycerols, derived from PC, PE, and PI of ripe roes of cod. A third, complementary solvent system for use with a C_8 reversed-phase column was methanol/water/acetonitrile (93:5:2), which resolved the species rich in n-3 polyunsaturates. Ramesha et al. [293] have reported a sensitive HPLC method for the separation and quantitation of glycerophospholipid subclasses and molecular species as the diradyl glycerol 1-anthroyl derivatives. The anthroyl derivatives were first resolved into the alkenylacyl, alkylacyl and diacyl glycerol subclasses, either by normal-phase HPLC or by TLC. The individual molecular species were separated by reversed-phase HPLC with acetonitrile/2-propanol (70:30) as the mobile phase. Rustow et al. [77] have reported the HPLC analysis of the fluorescent naphthylurethane derivatives of diradyl glycerols. The naphthylurethanes can be detected at significantly lower concentrations, but high temperature (85°C) and long derivatization times (2 h) are required for their preparation. This routine has been subsequently applied [294] to the separation of the molecular species of the alkylacyl and diacyl glycerol subclasses derived from the ethanolamine phosphatides of bovine erythrocytes. The diradyl glycerol subclasses were resolved by normal-phase HPLC. Binaglia et al. [295] have proposed the preparation of the naphthoates as fluorescent derivatives for diradyl glycerol analysis by normal-phase TLC. There would appear to be no reason why these derivatives could not be employed for more effective analyses by reversed-phase HPLC.

Pind et al. [296] have resolved molecular species of diradyl glycerol moieties of natural glycerophospholipids as the TBDMS ethers by reversed-phase HPLC, using a linear gradient of 30-90% propionitrile in acetonitrile over a period of 30 min. The species were detected and identified by CI-MS after the column effluent had been admitted to a quadrupole mass spectrometer via a direct liquid inlet interface. The TBDMS ethers are stable to moisture and yield excellent resolution. The method was later used [17] for the detection and quantitation of the deuterium-labeled diacyl glycerol moieties of rat liver PC and PE. The use of the TBDMS ethers for the reversed-phase HPLC separation of diradyl glycerols has been discussed in greater detail elsewhere [17]. Haroldsen and Murphy [297] have analyzed the molecular species of the rat lung as the dinitrobenzoyl derivatives by ECD and CI-MS, while Ramesha and Pickett [298] have analyzed the diradyl glycerol PFB esters by NCI-MS, which permits detection and quantitation in femtomole quantities. The monoradyl glycerols making up the fecapentaenes 12 and 14 have been characterized by Peters et al. [299], using reversed-phase HPLC, radio-isotopic dilution analysis, and detection at 340 nm. Appropriate reference lipids were synthesized in the laboratory and resolved with a gradient of 0.01 M sodium phosphate in acetonitrile/methanol (4:1).

15.6.4 Ceramides

The TMS and TBDMS ethers of ceramides are readily resolved on nonpolar capillary GLC columns to give a complete resolution of the odd- and even-carbon-number species and a partial separation of the saturated and monounsaturated species, which emerge ahead of the saturated species of the same carbon number [1]. The ceramides may be resolved on the basis of both carbon and double-bond number, using a polar capillary column (RTx 2330), but the temperature required for this purpose is close to the limit of thermal stability of the stationary phase. Because of their higher temperature limit, the polarizable capillary columns (50- 65% phenylmethylsilicone) are particularly well suited for the resolution of the molecular species of natural ceramides [61]. Separations are obtained on the basis of the number of carbons as well as the number of double bonds in the chains of both the fatty acid and the nitrogenous base. On the 65% phenylmethylsilicone column the various species were recovered in quantitatively correct proportions. Marai et al. [300] have used GC/CI-MS to characterize the TMS and TBDMS ethers of ceramides derived from plasma sphingomyelin. The ceramides were analyzed as part of the plasma total lipid profile, determined by nonpolar capillary GLC.

Nilsson and Zopf [301] have described a procedure for the release of ceramides from glycosphingolipids by trifluoroacetylation. Only the sphingosine ceramides, which have a double bond, were cleaved. The TFA derivatives of the ceramides were separated and identified by GC/MS on a nonpolar liquid phase. The recently discovered ceramide glycanase allows a general and simple derivatization of the ceramide portion of glycosphingolipids. Hansson et al. [302] have taken advantage of this capability to develop a novel method for structural characterization of the released ceramides as the permethylation products, by nonpolar capillary GC/MS.

The early HPLC separations of various ceramide derivatives are discussed by Christie [11]. More recently, molecular species of the ceramide moieties of egg yolk and human plasma sphingomyelin have been resolved as the TBDMS ethers by reversed-phase HPLC, with a linear gradient of 30-90% propionitrile in acetonitrile containing 1% dichloromethane and detection by chloride-attachment NCI-MS [303].

References on p. B218

15.6.5 Glycerophospholipids and sphingomyelins

Although molecular species of the glycerophospholipids and sphingomyelins are best chromatographed following removal of the polar head groups [1,5,10,52] (see also Section 15.4.4), an extensive resolution of molecular species is also obtained with intact phospholipids [52,81] or with phospholipids having chemically modified polar head groups [304-306].

15.6.5.1 Argentation chromatography

The separation of unmodified PC, PE, and PI by conventional AgNO$_3$-TLC was discussed in the previous edition of this book [1]. More recent studies have been confined to the AgNO$_3$-TLC resolution of the methylated dinitrobenzoyl derivatives of PE [304,306]. The dimethyl esters of phosphatidic acid have been resolved by argentation HPTLC and by argentation HPLC [305]. Other phospholipids have been converted to the phosphatidic acids by hydrolysis with phospholipase D and methylation with diazomethane.

15.6.5.2 High-performance liquid chromatography

Although the major glycerophospholipids have been separated intact into 30 to 35 molecular species by reversed-phase HPLC [52,81], the method has not enjoyed universal acceptance. The major problem is the necessity to quantitate the peaks by short-wave UV, which does not yield readily comparable signals for all species. Alternatively, the peaks must be collected and quantitated by GLC of the component fatty acids or diacyl glycerols, or by phosphorus analyses. In such a case, it is more convenient to perform a chemical or enzymatic dephosphorylation of the parent compounds and to convert the diradyl glycerols into derivatives that are more readily resolved and quantitated. Nevertheless, there have been numerous successful applications of reversed-phase HPLC in the speciation of intact glycerophospholipids [307-311]. Furthermore, the intact glycerophospholipids do not allow an effective resolution of the alkylacyl, alkenylacyl and diacyl species or subclasses [31]. Satsangi et al. [312] have reported a method for direct conversion of the 1-alkyl-2-acyl glycerophosphocholine to the 1-alkyl-2-acyl 3-heptafluorobutyrate, without isomerization. The reaction with heptafluorobutyric anhydride was 90% complete after 4 h at 35°C, and 100% complete after incubation overnight at room temperature. An analogous reaction was performed with pentafluorobenzoyl chloride to give the PFB derivative. These compounds are well suited for reversed-phase HPLC and for LC/MS with NCI. However, these methods are not suitable for work with plasmalogens, which are readily degraded in the presence of acids. The plasmalogens are best dephosphorylated with phospholipase C. For these reasons, many workers, including Patton and Robins [81], have resorted to degradation of the glycerophospholipids to diradyl glycerols and the preparation of the benzoates or similar derivatives for subsequent HPLC resolution and sensitive UV detection and quantitation.

Alternatively, some of the glycerophospholipids can be converted intact to UV-absorbing derivatives with improved chromatographic properties [1]. Recently, Nakagawa and Waku [306] have described a HPLC procedure for the separation of the molecular species of alkylacyl and diacyl subclasses of the ethanolamine phosphatides, following dinitrophenylation and methylation. The derivatives were first separated into the alkenyl-acyl, alkylacyl, and diacyl subclasses by TLC, and then each subclass was separated into individual molecular species by reversed-phase HPLC. Alkylacyl and diacyl derivatives were resolved to show 15 to 20 peaks in a single chromatogram. The methyl dinitrophenyl derivatives of egg yolk PE were dissolved in methanol for the resolution of the molecular species by HPLC on a reversed-phase column (LiChrosorb RP-18, 25 cm x 4 mm ID). Samples were eluted isocratically with acetonitrile/2-propanol/methanol/water (85:4:3:2) at a flowrate of 1 ml/min.

Terao et al. [82] have developed a method for the detection and determination of phospholipid peroxidation products in biological systems by reversed-phase HPLC. The hydroperoxy and hydroxy derivatives of PC and PE were eluted with a linear gradient of 5 to 20% water in acetonitrile.

15.6.5.3 Liquid chromatography with mass spectrometry and direct-probe mass spectrometry

Sugnaux and Djerassi [313] have described the advantages and problems of on-line LC/MS with direct liquid introduction interface and have compared it with off-line desorption/chemical ionization, including the determination of the structure of phospholipids. However, only flow injection was used to demonstrate the feasibility of the approach. Jungalwala et al. [314] used the moving-belt interface for coupling the HPLC column to the mass spectrometer for the LC/MS analysis of phospholipids. Positive and negative MS of various phospholipids, such as PC, PE, PS, PI, PG, and SPH were obtained in the chemical ionization mode with ammonia or methane as the reagent gas. About 5 μg of individual phospholipids were required for analysis by LC/MS, but with selected-ion detection subnanogram levels could be monitored. Characteristic ions were obtained for the polar head groups and molecular species. Erdahl and Privett [315] have described the combination and use of a HPLC system in conjunction with a computer-controlled CI mass spectrometer. It is based on the continuous conversion of the effluents to volatile products by reduction with hydrogen prior to their introduction into the mass spectrometer. These LC/MS analyses of pure samples of PC showed that the acyl groups are split from the backbone structure to yield $[RCOOH + 1]^+$ and $[RCO]^+$ ions; a strong ion with m/z 72 arises from the choline moiety, which can be used to detect and identify PC. Thomas et al. [316] used ammonia-desorption CI of ether-based phospholipids. Deuterated ammonia was used for structure elucidation. Much more extensive use has been made of FAB-MS in the analysis of the various glycerophospholipid and sphingomyelin classes. Thus, Gross [317] employed this method to characterize the ethanolamine phospholipids and sphingomyelins from sarcolemma, while Varenne et al. [318] used it for characterizing the platelet-activating factor (PAF) and related phospholipids. Others have

References on p. B218

employed secondary-ion MS for the analysis of crude [319] or isolated [320,321] phospholipids. Benfenati and Reginato [319] described negative- and positive-mode secondary-ion MS and field-desorption (FD) spectra of the molecular species of various choline, ethanolamine, inositol, and glycerol phosphatides. In some cases, the glycerophospholipids were found to lose the chain at the *sn*-2-position selectively, and this allows the nature and position of both fatty acid chains to be established. The possibility of identifying the position of the two chains in PC by the analysis of mass-spectrometric fragments has been suggested previously [320], but no practical applications have been made.

Sherman et al. [322] have shown that FAB produces useful spectra of the three phosphoinositides and the metabolically related phospholipids, lyso PI and PA. Analysis of the $[M-H]^+$ ions for the fatty acid ester composition by mass-analyzed ion-kinetic-energy spectra (MIKES) was shown to be inadequate to resolve fatty acyl daughter ions when the parent ion contains isobaric species. In contrast, analysis on a triple-sector instrument with and without collisional activation did provide complete compositional information. A complication in quantitative analyses was the dissimilar ion yield from fatty acyl-bearing fragments originating from compositionally different parent ions. Jensen et al. [323] have extended the previous investigations on the application of negative-ion FAB-MS/MS methods to the study of PI, PE, PG, CL, PA, and PAF and its analogs. What was found especially significant was the formation of $[M - H]^-$ ions and ions that correspond to the carboxylate portions of these molecules.

Haroldsen and Gaskell [65] described a procedure for the determination of PAF by stable-isotope dilution and FAB-MS/MS. Low-energy collisional activation of the $[M + H]^+$ ion yielded a single daughter ion of *m/z* 184, characteristic of phosphocholine derivatives. For precise and accurate quantitation, the internal standard was [^{2}H3]acetylhexadecylglycerolphosphocholine, which yields an analogous daughter ion of *m/z* 185. Secondary-ion MS was employed by Itabe et al. [324] to identify 2-azelaoylphosphatidylcholine as one of the cytotoxic products generated during oxyhemoglobin-induced peroxidation of PC. The oxidation products were isolated by both normal-phase and reversed-phase HPLC. The FAB spectra of the three phospholipids containing a secondary, tertiary, and quaternary nitrogenous base, attached to a 1,2-dipalmitoyl-*sn*-glycero-3-phosphate skeleton each, showed an ion derived from the intact nitrogenous base [325]. The collisional-activation MIKES, produced from these ions, was shown to be unique for each particular base.

15.6.6 Glycolipids

The TLC and GLC methods of resolution of the simple galactosyl acyl glycerols were discussed in the previous edition of this book [1]. A reversed-phase HPLC separation and quantitation by flame-ionization detection of the molecular species of PG, and the mono- and digalactosyl diacyl glycerols (MGDG and DGDG, respectively) from plant chloroplasts has been described by Smith et al. [326]. Molecular species of DGDG, ranging from 18:3-16:4 to 18:1-16:0 were resolved with 96% aq. methanol. Rezanka and Podojil [327] used reversed-phase HPLC to resolve the molecular species of the major polar lipid classes of algae, followed by an identification and quantitation of the species by nonpolar capillary GC/MS of the derived diacyl glycerol TMS ethers. The molecular species of DGMG and DGDG were resolved by using an isocratic system of 20 mM choline chloride in methanol/water/acetonitrile (90.5:7:2.5). The MGDG and DGDG in plant lipid extracts have been resolved by a ternary HPLC system, described by Christie and Morrison [145].

Detailed analyses of various complex glycolipids, including gangliosides, by TLC and HPTLC methods have been discussed and illustrated by Ando and Saito [115]. Kubo and Hoshi [328] have reported a simple and effective procedure for eliminating silica gel from gangliosides after preparative TLC. Gangliosides were extracted from the scraped-off gel with chloroform/methanol/water (10:10:3) and dried. After filtration through a sintered-glass funnel, the supernatant was applied to a short reversed-phase C_{16}-Sep-Pak cartridge column. The gangliosides were eluted from the cartridge with methanol in yields exceeding 90%. Excellent separations of bovine brain gangliosides on HPTLC plates were obtained, following consecutive development with chloroform/methanol/12 mM $MgCl_2$/ 15 M NH$_4$OH (60:35:7.5:3) and chloroform/methanol/12 mM $MgCl_2$ (58:40:9). GM1, GD1a, GD2, GD1b, and GT1b, were detected by a Cu^{2+}/resorcinol spray. Hirabayashi et al. [329] have used borate-treated and borate-free HPTLC for the separation of bovine brain cerebrosides, previously isolated on reversed-phase HPLC columns. Clausen et al. [330] have reported HPTLC of isolated galactosyl A^a and other glycolipids, following acetylation, with dichloroethane/acetone/water (50:50:1) as the eluent. The glycolipids were located with a primulin spray and were recovered by sonication in chloroform/methanol/water (2:1:0.15) and deacetylated. The purification was at each step monitored by TLC immunostaining with monoclonal antibodies. The HPTLC/immunostaining analysis of total glycolipid extracts and purified components was performed, following the development of the plates with chloroform/methanol/water (5:4:1), containing 0.05% $CaCl_2$. Clausen et al. [331] utilized these methods for a successful isolation and characterization of novel glycolipids with blood group A-related structures, e.g., galactosyl-A and sialosyl-galactosyl-A. The glycolipids were further characterized by direct-probe EI- and FAB-MS. Ostrander et al. [332] have used HPTLC for the separation of purified gangliosides from sole liver. The plates were developed with chloroform/methanol/water (60:40:9), containing 0.02% $CaCl_2$, and were stained with resorcinol. The glycolipids resolved on the HPTLC plates were also subjected to immunostaining, following desialylation. The final fractions were characterized by direct-probe EI- and FAB-MS.

The glycoglycerolipids for the TLC and HPTLC analyses are usually isolated by normal-phase HPLC on silicic acid or Iatrobead [15,108], or on DEAE-cellulose [333] columns with chloroform/methanol (1:1) and ammonia or ammonium acetate to provide appropriate polarity. Prior to normal-phase HPLC and TLC, the glycosphingolipids are isolated as a purified glycolipid class by DEAE-Sephadex column chromatography [15,108]. By using a high-carbon (20%) C_{18}-silica bead column, Hirabayashi et al. separated monohexosylceramides isolated from bovine brain [329] and Japanese quail intestine [334] into 42 and 56 peaks, respectively, which could be subsequently recovered for further analysis. Each of the subclasses was then analyzed by HPTLC and GC/MS.

References on p. B218

The combined methods allowed the complete identification of molecular species of glyco-sphingolipids.

Bonafede et al. [335] have described the isolation and characterization of a unique ganglioside, 9-O-acetyl-GD$_3$. The compound was isolated in large quantities by several conventional ion-exchange and silica gel column chromatographic procedures. The iso-lated 9-O-acetyl-GD$_3$ was characterized on the basis of its TLC behavior and immuno-reactivity with a specific monoclonal antibody [336] and by conversion to authentic GD$_3$ by mild base treatment. The HPTLC and immunostaining was performed by developing the plates with chloroform/methanol/aq. 0.02% CaCl$_2$ (50:45:10). The HPTLC behavior of the unique butterfat ganglioside was compared to that of a large selection of other gangliosides, which were effectively resolved under the selected working conditions. Young and Borgman [337] have described a continuous-development TLC method for improved separation of short- and long-chain neutral glycolipids, acetylated neutral gly-colipids, and gangliosides. Gottfries et al. [338] have elaborated a complete strategy for the isolation of individual mono- and disialogangliosides. These authors used normal-phase HPLC or partitioning to obtain a crude ganglioside fraction. The reaction was peracetylated and passed through a second normal-phase HPLC column. After anion-ex-change chromatography, the gangliosides were resolved by normal-phase HPLC with chloroform/methanol/water mixtures as eluents. The method served to prepare ganglio-sides for subsequent structural characterization by a TLC/enzyme-linked immunosorbent assay (ELISA), FAB-MS and GC/MS. The method was illustrated by a separation of GalNAc-II^{3}NeuAc-GgOse$_4$Cer from GalNAc-isoII^{3}NeuAc-GgOse$_4$Cer.

Ariga et al. [130] reported extensive HPTLC separations of the neutral glycolipids from PC 12 pheochromocytoma cells and have characterized a globoside as a major neutral lipid component. The neutral and acidic lipid fractions were isolated by column chroma-tography on DEAE-Sephadex and analyzed by HPTLC and by FAB-MS. The brain-type gangliosides of the ganglio series were also detected by a HPTLC/immunostaining method. Shimamura et al. [132] have used similar methods for the characterization of another major glycolipid in PC 12 cells as III^3 gal$_\alpha$-globotriaosylceramide. It was isolated by HPTLC and characterized by FAB-MS, the saccharide sequence having been estab-lished by a new method involving the use of endoglycoceramidase. Degrandis et al. [339] used a TLC binding assay to show that pig edema disease toxin recognizes globo-tetraosylceramide. The glycolipids were separated by TLC with chloroform/methanol/water (65:25:4). The plates were blocked by incubation in 3% gelatin and, after washing, incubated with the toxin. Ladisch et al. [133] have used HPTLC and a combination of negative- and positive-ion FAB-MS and collisionally activated dissociation tandem MS to identify the underivatized gangliosides. The semipurified total gangliosides were separated and recovered according to their carbohydrate composition by normal-phase HPLC [325]. Marked ceramide heterogeneity was evidenced by the reversed-phase HPLC res-olution of 18 individual ceramide species of neuroblastoma GM$_2$, which were collected for further analysis. Previously, Ladisch [340] had employed similar methodology for human brain gangliosides.

The methods of structure elucidation of glycosphingolipids and gangliosides, involving high-performance tandem MS have been reviewed by Domon and Costello [341], who have also made special recommendations for improved analysis. Isobe et al. [342] have proposed the use of a hexamethylphosphoric acid triamide/glycerol matrix system for negative-ion FAB-MS in examining native gangliosides. The advantages are the higher polarity and potential as an electron pair donor of hexamethylphosphoric acid triamide.

Three glycosylphosphatidylinositolglycolipids, recognized by antibodies from patients with cutaneous leishmaniasis, have been identified by chemical analysis, FAB-MS, and proton NMR [343]. Alkylacyl substitutions in the glycerol backbone showed considerable heterogeneity. These three glycolipids belong to the glycosylphosphatidylinositols, which serve as lipid anchors of membrane-bound proteins. Alkylacyl substitutions in the glycerol backbone have been described in the bovine erythrocyte acetylcholine esterase membrane anchor [344] and proposed in the insulin-sensitive glycophospholipid isolated from H35 hepatoma cells [345]. The phosphatidylinositol glycan anchors of the acetylcholine esterase of human red blood cells also contain an alkylacyl glycerol as the neutral lipid moiety of the PI [346], but the inositol residue is palmitylated in addition [347].

A novel and effective method for structural characterization of glycosphingolipids, based on the glycanase cleavage, has been devised by Hansson et al. [302]. The oligosaccharides and ceramides released by the glycanase are permethylated and analyzed by GC/MS. Nonpolar capillary GLC gave excellent resolution and separated two isomeric 10-sugar oligosaccharides with a molecular mass of 2150 D, differing only by a Gall-3GlcNAc and Gall-4GlcNac linkage. The GLC analyses were performed on 2, 5, or 10 m x 0.25-mm-ID capillary columns, coated with 0.03-0.05 μm of crosslinked PS 264 and with hydrogen or helium (for MS) as the carrier gas. The separations were carried out in the temperature range 70-400°C, programed at 30°C/min.

15.7 DETERMINATION OF TOTAL LIPID PROFILES

Most chromatographic methods are suitable for an examination of total lipid extracts of tissues and body fluids. Because of the nature of the chromatographic process, some of them are better suited than others for obtaining reproducible and quantitatively meaningful lipid profiles. TLC, high-temperature GLC, and HPLC have become well-established for this purpose [1]. Of the three, high-temperature GLC has proved best suited because of the ease of detection and quantitation of the chromatographic peaks by the FID. More recently, effective detection and quantitation of peaks in HPLC has been obtained by means of the light-scattering detector [11], flame-ionization detection scanning of Chromarods [46], and densitometry in HPTLC [115]. Other improved methods of chromatographic profiling of total lipid extracts are based on combinations of chromatographic and mass-spectrometric procedures [10,348]. In all instances, the multicomponent analysis has been facilitated or made possible by automation of the analytical systems.

References on p. B218

15.7.1 Thin-layer chromatography and high-performance thin-layer chromatography

The usefulness of normal-phase TLC for total lipid extract profiling is well recognized and has been the subject of frequent discussions in the literature. The most effective method of quantitating the lipid components resolved by TLC takes advantage of the sintered coating on quartz rods. A detailed examination of the conditions has been reported by Tanaka et al. [46]. During the last few years this system has enjoyed great popularity, and a symposium on the Iatroscan TLC/FID system has been held [111] (see Section 15.4.1.1).

The introduction of HPTLC has greatly enhanced the effectiveness of adsorption TLC for total lipid profiling [45]. Ando and Saito [115] have discussed in great detail the practical aspects of HPTLC separation and quantitation by densitometry. They have also provided excellent illustrations of numerous applications.

15.7.2 High-performance liquid chromatography and liquid chromatography with mass spectrometry

The application of HPLC to total lipid profiling has been exploited only to a limited extent, mainly because lipids lack chromophores that can be detected spectrophotometrically. Nevertheless, UV detection at wavelengths of 190-210 nm, where isolated double bonds absorb, has been used occasionally, but very few solvents are entirely transparent in this absorption range. Although the suitability of the FID in combination with an endless belt has been demonstrated [68], the commercial instruments are costly and not widely used. The use of the light-scattering detector has been extended from monitoring triacyl glycerols to phospholipids in the effluents of HPLC columns [11].

Mass spectrometry coupled to HPLC provides an ideal method for the detection and quantitation of solutes in HPLC effluents. Kuksis et al. [349] have discussed in detail the application of reversed-phase HPLC with CI-MS to the determination of plasma total lipid profiles. Kuksis et al. [280] have employed the direct liquid inlet system for this purpose, but other, more recently developed systems would appear to be better suited for this purpose [350-352].

The application of HPLC and LC/MS to the profiling of the molecular species in total lipid extracts of tissues other than plasma has also been attempted in the past, but the more complex composition of the phospholipid classes prohibits effective assessment of their composition from the mass spectra. Much better results are obtained if the lipid classes are prefractionated before total lipid analyses are attempted.

15.7.3 Gas/liquid chromatography and gas chromatography with mass spectrometry

GLC profiling of plasma neutral lipids has been reviewed by Mares [9], Kuksis and Myher [10] and Kuksis [353]. The major advance in total lipid profiling has been the substitution of capillary [232,354] for packed columns. Lercker [355] and Lercker et al. [356] have employed nonpolar capillary GLC columns for the determination of plasma neutral lipid profiles. Likewise, Mares [9] has used nonpolar capillary GLC for plasma neutral lipid profiling. Kuksis et al. [354] and Myher and Kuksis [232] have described the utilization of nonpolar capillary columns for the determination of human plasma total lipid profiles, following dephosphorylation with phospholipase C.

Wakeham and Frew [245] reported GC/MS of triacyl glycerols and other fatty esters, while Traitler [8] has studied the separation of other neutral lipid mixtures, containing components of a wide molecular-weight range. Limsathayourat and Melchert [357] separated hydrocarbons, fatty acids, mono-, di- and triacyl glycerols, wax esters and cholesteryl esters in a single analysis on a 9-m-long capillary column, while Grob et al. [358] have determined free and esterified sterols in oils and fats by coupled HPLC and GLC. Plasma total lipid profiling by GC/MS with ammonia CI has also been reported [348].

Lohninger et al. [359,360] have described accurate capillary GLC methods for determining the triacyl glycerol profile in serum and other biological sources. Quantitation was obtained by using trinonadecanoylglycerol as an internal standard. After serum extraction, total triacyl glycerol and triacyl glycerol species were determined without any further sample manipulation. Under such conditions, the phospholipids are pyrolyzed, but the cholesteryl esters are resolved as clean peaks, based on carbon number. Tabulation of the triacyl glycerol results gave a CV of 2.08%. A 5 m x 0.9-μm-ID fused-silica capillary column with chemically bonded OV-1 (0.1-μm coating thickness) was used with hydrogen as carrier gas and nitrogen as the makeup gas. The oven temperature was programed from 320°C to 350°C at 4°C/min or from 186°C to 276°C at 8°C/min and from 276°C to 340°C at 6°C/min. Before use, the column was conditioned at 350°C for 3 h, while the injector was temperature-programed. GLC at low temperatures resulted in better separations but enhanced CV values up to ca. 4%. These results confirm the earlier conclusions [9,10,232,353] regarding the suitability of GLC methods for quantitative estimation of plasma triacyl glycerols. The GLC method is not influenced by interfering substances.

Further experiments on the stabilization of polarizable liquid phases in capillary columns have opened up new possibilities for plasma lipid profiling. Kuksis and Myher [361] have noted the suitability of this liquid phase for the separation of plasma triacyl glycerols, and Kuksis and Myher [10] have explored the simultaneous analysis of the triacyl glycerols and other plasma lipid classes. Mares [9] has pointed out overlapping between cholesteryl esters and low-molecular-weight triacyl glycerols, which interferes with the identification and quantitation of critical pairs of cholesteryl esters and triacyl glycerols. Recently, Kuksis et al. [61] have examined in some detail the suitability of two polarizable capillary GLC columns for the resolution of plasma total lipids on the basis of carbon number and molecular weight.

15.8 SUMMARY AND CONCLUSIONS

A review of the chromatography of lipids as practiced since publication of the previous edition of this book reveals many advances in both methodology and applications. HPTLC has now emerged as the most satisfactory method of planar chromatography, although

Chromarod chromatography is also widely used. HPTLC has shown great promise for the accurate and sensitive detection of glycolipids by immunoreactivity with appropriate monoclonal antibodies. It has not been replaced by HPLC, as predicted. Adsorption column chromatography has found new life in the form of adsorbent cartridge columns for rapid sample isolation and purification. In GLC, the flexible quartz capillary columns, containing both polar and nonpolar stationary phases, have now largely replaced packed columns as a separation medium for both low- and high-molecular-weight lipids. There has been a large increase in the use of high-temperature GLC, which reflects the great improvement in the stabilization of the stationary phases. Both polar and nonpolar columns of high temperature stability are now available and are extensively utilized. The availability of high-temperature capillary columns containing the polarizable phenyl-methylsilicone phase promises to revolutionize the GLC of natural triacyl glycerols, although the very highly unsaturated triacyl glycerols may not be stable at the higher temperatures necessary for their elution from the currently available columns. There has also been an enormous increase in the utilization of HPLC columns for the isolation of both lipid classes and molecular species. The lack of a universal detector has been partially overcome by the wider utilization of the light-scattering detector and by more sophisticated application of short-wave UV, RI, and FTIR spectrometry for the detection of the natural products. The preparation of the UV-absorbing derivatives of fatty acids, monoacyl and diacyl glycerols, and ceramides has now established itself as a convenient and accurate method of analyzing molecular species by reversed-phase HPLC. In the stationary phases for HPLC analyses, the most important development has been the introduction of chiral phases, capable of resolution of enantiomeric monoradyl and diradyl glycerols, as well as enantiomeric hydroxy fatty acids and secondary alcohols. There has been a great increase in the utilization of GC/MS and LC/MS for the characterization of the resolved lipid species, and the decreasing cost of these systems heralds their general utilization in lipid analyses. Other developments include the introduction of supercritical-fluid chromatography, which is also suitable for combination with mass spectrometry, multiple column switching, and tandem mass spectrometry.

ACKNOWLEDGEMENTS

The studies by the author and his collaborators were supported by funds from the Medical Research Council of Canada, Ottawa, Ont., and the Heart and Stroke Foundation of Ontario, Toronto, Ont., Canada.

REFERENCES

1 A. Kuksis, in E. Heftmann (Editor), Chromatography, Part B, Journal of Chromatography Library, Vol. 22B, Elsevier, Amsterdam, 1983, p. 75.
2 R.G. Ackman, in H.K. Mangold (Editor), Handbook of Chromatography, Vol. 1, Lipids, CRC Press, Boca Raton, FL, 1984, p. 95.
3 H. Jaeger, Glass Capillary Chromatography in Clinical Medicine and Pharmacology, Marcel Dekker, New York, 1985.

4 R.G. Ackman, in R.J. Hamilton and J.B. Bossell (Editors), *Analysis of Oils and Fats,* Elsevier, London, 1986, p. 137.
5 W.W. Christie, *Gas Chromatography and Lipids,* The Oily Press, Ayr, 1989.
6 C.R. Pace-Asciak, in A. Kuksis (Editor), *Chromatography of Lipids in Biomedical Research and Clinical Diagnosis,* Elsevier, Amsterdam, 1987, p. 107.
7 W.S. Powell, in A. Kuksis (Editor), *Chromatography of Lipids in Biomedical Research and Clinical Diagnosis,* Elsevier, Amsterdam, 1987, p. 76.
8 H. Traitler, *Progr. Lipid Res.,* 26 (1987) 257.
9 P. Mares, *Progr. Lipid Res.,* 27 (1988) 107.
10 A. Kuksis and J.J. Myher, *Adv. Chromatogr.,* 28 (1989) 267.
11 W.W. Christie, *High Performance Liquid Chromatography and Lipids,* Pergamon, Oxford, 1987.
12 V.K.S. Shukla, *Progr. Lipid Res.,* 27 (1988) 5.
13 R.H. McCluer, M.D. Ullman and F.B. Jungalwala, *Adv. Chromatogr.,* 25 (1986) 309.
14 D.J. Hanahan and R. Kumar, *Progr. Lipid Res.,* 26 (1987) 1.
15 S. Hakamori, in S. Hakamori and J. Kanfer (Editors), *Sphingolipid Biochemistry, Handbook of Lipid Research,* Vol. 3, Plenum Press, New York, 1983, p. 1.
16 L.R. Hogge and J.C. Millar, *Adv. Chromatogr.,* 27 (1987) 299.
17 A. Kuksis, L. Marai, J.J. Myher and S. Pind, in A. Kuksis (Editor), *Chromatography of Lipids in Biomedical Research and Clinical Diagnosis,* Elsevier, Amsterdam, 1987, p. 403.
18 K.B. Tomer and C.E. Parker, *J. Chromatogr.,* 492 (1989) 189.
19 A. Kuksis and J.J. Myher, in A.M. Lawson (Editor), *Mass Spectrometry,* Walter de Gruyter, Berlin, 1989, p. 267.
20 J. Folch, M. Lees and G.H. Stanley, *J. Biol. Chem.,* 226 (1957) 497.
21 E.G. Bligh and W.J. Dyer, *Can. J. Biochem.,* 37 (1959) 911.
22 M. Kates, *Techniques in Lipidology,* Elsevier, Amsterdam, 1986.
23 N.S. Radin, in A.A. Boulton, G.B. Baker and L. A. Horrocks (Editors), *Lipids and Related Compounds, Neuromethods, Vol. 7,* Humana Press, Clifton, NJ, 1988, p. 1.
24 A. Kuksis and J.J. Myher, in A. Kuksis (Editor), *Chromatography of Lipids in Biomedical Research and Clinical Diagnosis,* Elsevier, Amsterdam, 1987, p. 1.
25 T. Kyrklund, *Lipids,* 22 (1987) 274.
26 R.D. McDowall, *J. Chromatogr.,* 492 (1989) 3.
27 D.C. Turnell and J.D.H. Cooper, *J. Chromatogr.,* 492 (1989) 59.
28 M. Saito, Y. Tanaka and S. Ando, *Anal. Biochem.,* 132 (1983) 376.
29 G. Freyburger, A. Heape, H. Gin, M. Boisseau and C. Cassagne, *Anal. Biochem.,* 171 (1988) 213.
30 H.G. Rose and M. Oaklander, *J. Lipid Res.,* 6 (1965) 428.
31 T. Nakamura, Y. Hatori, K. Yamada, M. Ikeda and T. Yuzuriha, *Anal. Biochem.,* 179 (1989) 127.
32 N.A. Shaikh, in H.A. Fozzard, E. Haber, R.B. Jennings, A. Katz and H. Morgan (Editors), *The Heart and Cardiovascular System,* Vol. 1, Raven Press, New York, 1986, p. 289.
33 R.J. Hamilton and K. Comai, *Lipids,* 23 (1988) 1146.
34 R.J. Hamilton and K. Comai, *J. Lipid Res.,* 25 (1984) 1142
35 M.A. Kaluzny, L.A. Duncan, M.V. Merritt and D.E. Epps, *J. Lipid Res.,* 26 (1985) 135.
36 D.A. Figlewicz, C.E. Nolon, I.N. Singh and F.B. Jungalwala, *J. Lipid Res.,* 26 (1985) 140.
37 J. Egberts and R. Buiskool, *Clin. Chem.,* 34 (1988) 163.
38 W.S. Powell, *Methods Enzymol.,* 86 (1982) 467.
39 J.D. Eskra, M.J. Pereira and M.J. Ernest, *Anal. Biochem.,* 164 (1986) 332.
40 D. Steinhilber, T. Herrmann and H.J. Roth, *J. Chromatogr.,* 493 (1989) 361.
41 T. Takagi and Y. Itabashi, *Lipids,* 22 (1987) 596.
42 B. Breuer, T. Stuhlfauth and H.P. Fock, *J. Chromatogr. Sci.,* 25 (1987) 302.
43 W.S. Powell, *Methods Enzymol.,* 86 (1982) 530.
44 W.W. Christie, *J. High Resolut. Chromatogr. Chromatogr. Commun.,* 10 (1987) 148.
45 A.M. Siouffi, E. Mincsovics and E. Tyihak, *J. Chromatogr.,* 492 (1989) 471.

46 M. Tanaka, K. Takase, J. Ishii, T. Itoh and H. Kaneko, *J. Chromatogr.*, 284 (1984) 433.
47 T.L. Kaduce, K.C. Norton and A.A. Spector, *J. Lipid Res.*, 24 (1983) 1398.
48 W.W. Christie, *J. Lipid Res.*, 26 (1985) 507.
49 R.J. Hamilton and K. Comai, *Lipids*, 23 (1988) 1150.
50 S. Bahrami, H. Gasser and H. Redl, *J. Lipid Res.*, 28 (1987) 596.
51 J.S. Ellingson and R.L. Zimmerman, *J. Lipid Res.*, 28 (1987) 1016.
52 S.J. Robins and G.M. Patton, *J. Lipid Res.*, 27 (1986) 131.
53 W. Murayama, Y. Kasuge, N. Nakaya, Y. Nanogaki, K. Nunogaki, J. Cazes and H. Nunogaki, *J. Liquid Chromatogr.*, 11 (1988) 283.
54 A. Wilsch and G. Schneider, *Fresenius Z. Anal. Chem.*, 316 (1983) 265.
55 J.W. Hellgeth, J.W. Jordan, L.T. Taylor and M. Ashrat Khorassani, *J. Chromatogr. Sci.*, 24 (1986) 183.
56 H. Kallio, P. Laakso, R. Huopalahti and R.R. Linko, *Anal. Chem.*, 61 (1989) 698.
57 E. Geeraert, in A. Kuksis (Editor), *Chromatography of Lipids in Biomedical Research and Clinical Diagnosis*, Elsevier, Amsterdam, 1987, p. 48.
58 P. Dawes and M. Cumbers, *Am. Lab.*, May (1989) 102.
59 J.J. Myher, A. Kuksis and L.-Y. Yang, *Biochem. Cell Biol.*, 68 (1990) 336.
60 E. Geeraert and P. Sandra, *J. Am. Oil Chem. Soc.*, 64 (1987) 100.
61 A. Kuksis, J.J. Myher and P. Sandra, *J. Chromatogr.*, 500 (1990) 427.
62 A. Hayashi, T. Matsubara, M. Morita, T. Kinoshita and T. Nakamura, *J. Biochem.*, 106 (1989) 264.
63 B. Domon and C.E. Costello, *Biochemistry*, 27 (1988) 1534.
64 H.L. Fredrickson, J.W. de Leeuw, A.C. Tas, J. van der Greef, G.F. LaVos and J.J. Boon, *Biomed. Environ. Mass Spectrom.*, 18 (1989) 96.
65 P.E. Haroldsen and S.J. Gaskell, *Biomed. Environ. Mass Spectrom.*, 18 (1989) 439.
66 J.D. Craske and C.D. Bannon, *J. Am. Oil Chem. Soc.*, 64 (1987) 1413.
67 K. Okumura, H. Hashimoto, T. Ito, K. Ogawa and T. Satake, *Lipids*, 23 (1988) 253.
68 F.C. Phillips, W.L. Erdahl, J.D. Nadenicek, L.J. Nutter, J.A. Schmit and O.S. Privett, *Lipids*, 19 (1984) 142.
69 L.A. Smith, H.A. Norma, S-H. Cho and G.A. Thompson, Jr., *J. Chromatogr.*, 346 (1985) 291.
70 H.A. Norman and J.B. St. John, *J. Lipid Res.*, 27 (1986) 1104.
71 J.F. Lawrence, *J. Chromatogr. Sci.*, 23 (1985) 484.
72 M.L. Blank, M. Robinson, V. Fitzgerald and F. Snyder, *J. Chromatogr.*, 298 (1984) 473.
73 H. Takamura, H. Norita, R. Urade and M. Kito, *Lipids*, 21 (1986) 356.
74 M.D. Ullman and R.H. McCluer, *J. Lipid Res.*, 26 (1985) 501.
75 H. Kadowaki, K.E. Rys-Sikora and R.S. Koff, *J. Lipid Res.*, 30 (1989) 616.
76 M.D. Ullman and R. F. Ventura, *J. Lipid Res.*, 28 (1987) 878.
77 B. Rustow, H. Rabe and D. Kunze, in A. Kuksis (Editor), *Chromatography of Lipids in Biomedical Research and Clinical Diagnosis*, Elsevier, Amsterdam, 1987, p. 191.
78 F.B. Jungalwala, J.E. Evans and R.H. McCluer, *Biochem. J.*, 155 (1976) 55.
79 Y. Nakagawa and L.A. Horrocks, *J. Lipid Res.*, 24 (1983) 1268.
80 K. Ho, A. Suzuki, Y. Kuroki and T. Akino, *Lipids*, 20 (1985) 611.
81 G.M. Patton and S.J. Robins, in A. Kuksis (Editor), *Chromatography of Lipids in Biomedical Research and Clinical Diagnosis*, Elsevier, Amsterdam, 1987, p. 311.
82 J. Terao, I. Asano and S. Matsushita, *Arch. Biochem. Biophys.*, 235 (1984) 326.
83 R.S. Kim and F.S. LaBella, *J. Lipid Res.*, 28 (1987) 1110.
84 G. Sempore and J. Bezard, *J. Chromatogr.*, 366 (1986) 261.
85 A. Stolyhwo, H. Colin, M. Martin and G. Guiochon, *J. Chromatogr.*, 288 (1984) 253.
86 W.W. Christie, *J. Lipid Res.*, 26 (1985) 507.
87 M.A. Grayson, *J. Chromatogr. Sci.*, 24 (1986) 529.
88 J.J. Myher, A. Kuksis, L. Marai and S.K.F. Yeung, *Anal. Chem.*, 50 (1978) 557.
89 J.J. Myher, A. Kuksis, L. Marai and F. Manganaro, *J. Chromatogr.*, 283 (1984) 289.
90 J. Palumbo and F. Zullo, *Lipids*, 22 (1987) 201.
91 T. Nagata, L.L. Poulsen and D.M. Ziegler, *Anal. Biochem.*, 171 (1988) 248.

92 S.D. Fowler, W.J. Brown, J. Warfel and P. Greenspan, *J. Lipid Res.*, 28 (1987) 1225.
93 W.G. Wood, M. Cornwell and L.S. Williamson, *J. Lipid Res.*, 30 (1989) 775.
94 L.J. Macala, R.K. Yu and S. Ando, *J. Lipid Res.*, 24 (1983) 1243.
95 C.F. Poole and S.K. Poole, *J. Chromatogr.*, 492 (1989) 539.
96 L.G. Blomberg, *Adv. Chromatogr.*, 26 (1987) 229.
97 W.W. Christie, *J. Chromatogr.*, 447 (1988) 301.
98 B.D. Beaumelle and H.J. Vial, *J. Chromatogr.*, 356 (1986) 409.
99 B. Lacroix, J.P. Huvenne and M. Deveaux, *J. Chromatogr.*, 492 (1989) 109.
100 M.D. Greenspan, C.Y.L. Lo, D.P. Hauf and J.B. Yudkowitz, *J. Lipid Res.*, 29 (1988) 971.
101 S. Baba, *J. Chromatogr.*, 492 (1989) 137.
102 A. Karmen, G. Malikin and S. Lam, *J. Chromatogr.*, 468 (1989) 279.
103 D.J. Harvey, *Biomed. Mass Spectrom.*, 11 (1984) 340.
104 W.W. Christie, E.Y. Brechany and R.T. Holman, *Lipids*, 22 (1987) 224.
105 H-R. Buser, H. Arn, P. Geurin and S. Rauscher, *Anal. Chem.*, 55 (1983) 818.
106 G. Janssen, G. Parmentier, A. Verhulst and H. Eyssen, *Biomed. Mass Spectrom.*, 12 (1985) 134.
107 J.Y. Zhang, Q.T. Yu, B.N. Liu and Z.H. Huang, *Biomed. Environ. Mass Spectrom.*, 15 (1988) 33.
108 A.J. Yates, in A.A. Boulton, G.B. Baker and L.A. Horrocks (Editors), *Lipids and Related Compounds, Neuromethods, Vol. 7*, Humana Press, Clifton, NJ, 1988, p. 265.
109 P. Strasberg, A. Grey, I. Warren and M.A. Skomorowski, *J. Lipid Res.*, 30 (1989) 121.
110 T.E. Beesley, *J. Chromatogr. Sci.*, 24 (1985) 525.
111 Various authors, *Lipids*, 20 (1985) 501.
112 J.L. Sebedio, Ch. Septier and A. Grandgirard, *J. Am. Oil Chem. Soc.*, 63 (1986) 1541.
113 S. Kim and F.S. La Bella, *J. Lipid Res.*, 28 (1987) 1110.
114 B. Nikolova-Damyanova and B. Amidzhin, *J. Chromatogr.*, 446 (1988) 283.
115 S. Ando and M. Saito, in A. Kuksis (Editor), *Chromatography of Lipids in Biomedical Research and Clinical Diagnosis*, Elsevier, Amsterdam, 1987, p. 266.
116 J.F. Fine and H. Sprecher, *J. Lipid Res.*, 23 (1982) 660.
117 K.T. Mitchell, J.E. Ferrell, Jr. and W.H. Huestis, *Anal. Biochem.*, 158 (1986) 447.
118 A.M. Heape, H. Juguelin, M. Fabre, F. Boiron, B. Garbey, M. Fournier, J. Bonnet and C. Cassagne, *Dev. Brain Res.*, 25 (1986) 173.
119 A.K. Banerjee, W.M.N. Ratnayake and R.G. Ackman, *Lipids*, 20 (1985) 121.
120 R. Homan and H.J. Pownall, *Anal. Biochem.*, 178 (1989) 166.
121 A.G. Andrews, *J. Chromatogr.*, 336 (1984) 139.
122 W. Kuhnz, B. Zimmermann and H. Nau, *J. Chromatogr.*, 334 (1985) 309.
123 T. Heinze, G. Kynast, J.W. Dudenhausen, C. Schmitz and E. Saling, *Chromatographia*, 25 (1988) 497.
124 M. Seewald and H. Eichinger, *J. Chromatogr.*, 469 (1989) 271.
125 R.L. Van Tassell, T. Piccariello, D.G.I. Kingston and T.D. Wilkins, *Lipids*, 24 (1989) 454.
126 M.C. Moschidis and N.K. Andrikopoulos, *J. Chromatogr.*, 403 (1987) 363.
127 B.L. Slomiany, V.L.N. Murty, Y.H. Liau and A. Slomiany, *Progr. Lipid Res.*, 26 (1987) 29.
128 M.C. Byrne, M. Sbaschnig-Agler, D.A. Aquino, J.R. Sclafani and R.W. Ledeen, *Anal. Biochem.*, 148 (1985) 163.
129 W.W. Young, Jr. and C.A. Borgman, *J. Lipid Res.*, 27 (1986) 120.
130 T. Ariga, L.J. Macala, M. Saito, R.K. Margolis, L.A. Greene, R.U. Margolis and R.K. Yu, *Biochemistry*, 27 (1988) 52.
131 M. Saito, N. Kasai and R.K. Yu, *Anal. Biochem.*, 148 (1985) 48.
132 M. Shimamura, T. Hayase, M. Ito, M-L. Rasilo and T. Yamagata, *J. Biol. Chem.*, 263 (1988) 12124.
133 S. Ladisch, C.C. Sweeley, H. Becker and D. Gage, *J. Biol. Chem.*, 264 (1989) 12097.

134 K. Ogawa, Y. Fujiwara, K. Sugamata and T. Abe, *J. Chromatogr.,* 426 (1988) 188.
135 J.G. Alvarez and J.C. Touchstone, *J. Chromatogr.,* 436 (1988) 515.
136 G. Gazotti, S. Sonnino and R. Ghidoni, *J. Chromatogr.,* 348 (1985) 371.
137 G. Gazotti, S. Sonnino, R. Ghidoni, G. Kirschner and G. Tettamanti, *J. Neurosci. Res.,* 12 (1984) 179.
138 E.B. Hoving, J. Prins, H.M. Rutgers and F.A.J. Muskiet, *J. Chromatogr.,* 434 (1988) 411.
139 M.M. Whalen, G.C. Wild, W. Dale Spall and R.J. Sebring, *Lipids,* 21 (1986) 267.
140 K. Watanabe and Y. Tomono, *Anal. Biochem.,* 139 (1984) 367.
141 C. Demandre, A. Tremolieres, A-M. Justin and P. Mazliak, *Phytochemistry,* 24 (1985) 481.
142 N. Satirhos, C-T. Ho and S.S. Chang, *Fette Seifen Anstrichm.,* 88 (1986) 6.
143 D. Marion, G. Gandemer and R. Douillard, in P-A. Siegenthaler and W. Eichenberger (Editors), *Structure, Function and Metabolism of Plant Lipids,* Elsevier, Amsterdam, 1984, p. 139.
144 J.W.M. Heemskerk, G. Bogemann, M.A.M. Scheijen and J.F.G.M. Wintermans, *Anal. Biochem.,* 154 (1986) 85.
145 W.W. Christie and W.R. Morrison, *J. Chromatogr.,* 436 (1988) 510.
146 P. Michelsen, E. Aronsson, G. Odham and B. Akesson, *J. Chromatogr.,* 350 (1985) 417.
147 N. Oi and H. Kitahara, *J. Chromatogr.,* 265 (1983) 117.
148 W.H. Pirkle and T.C. Pochapsky, *Adv. Chromatogr.,* 27 (1987) 73.
149 N. Oi and H. Kitahara, *J. Liquid Chromatogr.,* 9 (1986) 511.
150 W.H. Pirkle and J.E. McCune, *J. Liquid Chromatogr.,* 11 (1988) 2165.
151 W.H. Pirkle and J.H. Hanske, *J. Org. Chem.,* 42 (1977) 2436.
152 Y. Itabashi, T. Takagi and T. Tsuda, *J. Chromatogr.,* 472 (1989) 271.
153 J. Turk, W.T. Stamp, B.A. Wolf, R.A. Easom and M.L. McDaniel, *Anal. Biochem.,* 174 (1988) 580.
154 T. Takagi and Y. Itabashi, *Yakugaku,* 34 (1985) 962.
155 Y. Itabashi and T. Takagi, *Lipids,* 21 (1986) 413.
156 T. Takagi and Y. Itabashi, *J. Chromatogr.,* 366 (1986) 451.
157 Y. Itabashi and T. Takagi, *J. Chromatogr.,* 402 (1987) 257.
158 Y. Itabashi, A. Kuksis and J.J. Myher, *J. Lipid. Res.,* (1990) in press.
159 Y. Itabashi, L. Marai and A. Kuksis, *J. Am. Oil Chem. Soc.,* 66 (1989) 491 (abstract).
160 W.W. Christie and G.H. Breckenridge, *J. Chromatogr.,* 469 (1989) 261.
161 M.A. Quilliam and J.M. Yaraskavitch, *J. Liquid Chromatogr.,* 8 (1985) 449.
162 K. Korte, K.R. Chien and M.L. Casey, *J. Chromatogr.,* 469 (1989) 261.
163 E. Juengling and H. Kammermeier, *Anal. Biochem.,* 171 (1988) 150.
164 G. Lakshminarayana and D.G. Cornwell, *Lipids,* 21 (1986) 175.
165 H.J. Dutton, S.B. Johnson, F.J. Pusch, M.S.F. Lie Ken Jie, F.D. Gunstone and R.T. Holman, *Lipids,* 23 (1988) 481.
166 K. Kawamura and R.B. Gagosian, *J. Chromatogr.,* 438 (1988) 309.
167 B. Schmitz and R.A. Klein, *Chem. Phys. Lipids,* 39 (1986) 285.
168 N.J. Jensen and M.L. Gross, *Mass Spectrom. Rev.,* 6 (1987) 497.
169 D.J. Harvey, *Mass Spectrom. Adv.,* 1985 (1986) 651.
170 D.J. Harvey, *Biomed. Environ. Mass Spectrom.,* 18 (1989) 719.
171 R.A. Klein and B. Schmitz, *Biomed. Environ. Mass Spectrom.,* 13 (1986) 429.
172 W.W. Christie, E.Y. Brechany, S.B. Johnson and R.T. Holman, *Lipids,* 21 (1986) 657.
173 W.W. Christie, E.Y. Brechany and R.T. Holman, *Lipids,* 22 (1987) 224.
174 W.W. Christie, E.Y. Brechany, F.D. Gunstone, M.S.F. Lie Ken Jie and R.T. Holman, *Lipids,* 22 (1987) 664.
175 W.W. Christie, E.Y. Brechany and K. Stefanov, *Chem. Phys. Lipids,* 46 (1988) 127.
176 W.W. Christie, E.Y. Brechany and M.S.F. Lie Ken Jie, *Chem. Phys. Lipids,* 46 (1988) 225.
177 W.W. Christie and K. Stefanov, *J. Chromatogr.,* 392 (1987) 259.
178 J.Y. Zhang, Q.T. Yu, B.N. Liu and Z.H. Huang, *Biomed. Environ. Mass Spectrom.,* 15 (1987) 33.

179 Q.T. Yu, B.N. Liu, J.Y. Zhang and Z.H. Huang, *Lipids*, 24 (1989) 79.
180 Q.T. Yu, B.N. Liu, J.Y. Zhang and Z.H. Huang, *Lipids*, 23 (1988) 804.
181 M. Balazy and A.S. Nies, *Biomed. Environ. Mass Spectrom.*, 18 (1989) 328.
182 D.V. Crabtree, A.J. Adler and G.J. Handelman, *J. Chromatogr.*, 466 (1989) 251.
183 F.O. Gulacar, A. Buchs and A. Susini, *J. Chromatogr.*, 479 (1989) 61.
184 M.I. Aveldano, *J. Biol. Chem.*, 262 (1987) 1172.
185 M.I. Aveldano and H. Sprecher, *J. Biol. Chem.*, 262 (1987) 1180.
186 P. Aubourg, P. Bougneres and F. Rocchiccioli, *J. Lipid Res.*, 26 (1985) 263.
187 G. Janssen, A. Verhulst and G. Parmentier, *Biomed. Environ. Mass Spectrom.*, 15 (1988) 1.
188 G. Janssen, G. Parmentier, A. Verhulst and H. Eyssen, *Biomed. Mass Spectrom.*, 12 (1985) 134.
189 T. Baba, K. Kaneda, E. Kusunose, M. Kusunose and I. Yano, *Lipids*, 23 (1988) 1132.
190 J.L. Le Quere, E. Semon, B. Lanher and J.L. Sebedio, *Lipids*, 24 (1989) 347.
191 H. Frank, D. Thiel and J. Macleod, *Biochem. J.*, 260 (1989) 873.
192 C.N. Christopoulou and E.G. Perkins, *J. Am. Oil Chem. Soc.*, 66 (1989) 1344.
193 C.N. Christopoulou and E.G. Perkins, *J. Am. Oil Chem. Soc.*, 66 (1989) 1360.
194 A. Grandgirard, A. Piconneaux, J.L. Sebedio, S.F. O'Keefe, E. Semon and J.L. Le Quere, *Lipids*, 24 (1989) 799.
195 N.J. Jensen and M.L. Gross, *Lipids*, 21 (1986) 362.
196 K.B. Tomer, N.J. Jensen and M.L. Gross, *Anal. Chem.*, 58 (1986) 2429.
197 D.L. Birkle, H.E.P. Bazan and N.G. Bazan, in A.A. Boulton, G.B. Baker and L.A. Horrocks (Editors), *Lipids and Related Compounds, Neuromethods, Vol. 7*, Humana Press, Clifton, NJ, 1988, p. 227.
198 W.S. Powell, in W.E.M. Lands (Editor), *Biochemistry of Arachidonic Acid Metabolism*, Martinus Nijhoff Publishing, Boston, 1985, p. 375.
199 T. Tonai, K. Yokota, T. Yano, Y. Hayashi, S. Yamamoto, K. Yamashita and H. Miyazaki, *Biochim. Biophys. Acta*, 836 (1985) 335.
200 F. Dray, K. Gerozissis, B. Kouznetzova, S. Mamas, P. Pradelles and G. Trugan, *Adv. Prostaglandin Thromboxane Res.*, 6 (1980) 167.
201 B.C. Beaubien, J.R. Tippins and H.R. Morris, *Biochem. Biophys. Res. Commun.*, 125 (1984) 97.
202 J.W. Cox, R.H. Pullen and M.E. Royer, *Anal. Chem.*, 57 (1985) 2365.
203 W.S. Powell, *Anal. Biochem.*, 128 (1983) 93.
204 V.H. Do, D.G. Ahern, J. Iles, M. Maniscalco and M. Tutujian, *J. Chromatogr.*, 489 (1989) 359.
205 W.S. Powell, in C. Benedetto, R.G. McDonald-Gibson, S. Nigam and T.F. Slater (Editors), *Prostaglandins: A Practical Approach*, IRL Press, Oxford, 1987, p. 75.
206 R. Sugata, H. Iwahashi, T. Ishii and R. Kido, *J. Chromatogr.*, 487 (1989) 9.
207 P. Borgeat and S. Picard, *Anal. Biochem.*, 171 (1988) 283.
208 J.A. Yergey, H.Y. Kim and N. Sulena, *Anal. Chem.*, 58 (1986) 1344.
209 R. Richmond, S .R. Clarke, D. Watson, C.G. Chappell, C.T. Dollery and G.W. Taylor, *Biochim. Biophys. Acta*, 881 (1986) 159.
210 C. Chiabrando, A. Noseda and R. Fanelli, *J. Chromatogr.*, 250 (1982) 100.
211 C. Chiabrando, A. Benigni, A . Piccinelli, C. Carminati, E. Cozzi, G. Remuzzi and R. Fanelli, *Anal. Biochem.*, 163 (1987) 255.
212 C. Chiabrando, V. Pinciroli, A. Campoleoni, A. Benigni, A. Piccinelli and R. Fanelli, *J. Chromatogr.*, 495 (1989) 1.
213 S. Fischer and P.C. Weber, *Biomed. Mass Spectrom.*, 12 (1985) 470.
214 W.S. Powell, *Biochem. Biophys. Res. Commun.*, 145 (1987) 991.
215 W.S. Powell and F. Gravelle, *J. Biol. Chem.*, 264 (1989) 5364.
216 W.S. Powell, *J. Biol. Chem.*, 259 (1984) 3082.
217 C.J. Hartzell and N.H. Andersen, *Biomed. Mass Spectrom.*, 12 (1985) 303.
218 H. Schweer, H.W. Seyberth and C.O. Meese, *Biomed. Environ. Mass Spectrom.*, 15 (1988) 129.
219 H. Schweer, H. Seyberth, C.O. Meese and O. Fuerst, *Biomed. Environ. Mass Spectrom.*, 15 (1988) 139.

220 H. Schweer, H.W. Seyberth, C.O. Meese and O. Fuerst, *Biomed. Environ. Mass Spectrom.*, 15 (1988) 143.
221 J. Gut, J.R. Trudell and G.C. Jamieson, *Biomed. Environ. Mass Spectrom.*, 15 (1988) 509.
222 H. Hughes, C.V. Smith, J.O. Tsokos-Kuhn and J.R. Mitchell, *Anal. Biochem.*, 152 (1986) 107.
223 C. Fischer and C.O. Meese, *Biomed. Mass Spectrom.*, 12 (1985) 399.
224 C.O. Meese, C. Fischer, P. Thalheimer and O. Fuerst, *Biomed. Mass Spectrom.*, 12 (1985) 554.
225 C.R. Pace-Asciak, J.M. Martin and S.P. Lee, *Biochem. Cell Biol.*, 66 (1988) 901.
226 C.R. Pace-Asciak and Z. Domazet, *Biochim. Biophys. Acta*, 796 (1984) 129.
227 C.R. Pace-Asciak and S. Micallef, *Can. J. Biochem. Cell Biol.*, 62 (1984) 709.
228 C.R. Pace-Asciak, *Biochem. Biophys. Res. Commun.*, 151 (1988) 493.
229 E. Malle, H. Gleispach, G.M. Kostner and H.J. Leis, *J. Chromatogr.*, 488 (1989) 283.
230 J.M. Rosenfeld, M. Mureika-Russell and M. Love, *J. Chromatogr.*, 489 (1989) 263.
231 P. Mares and P. Husek, *J. Chromatogr.*, 350 (1985) 87.
232 J.J. Myher and A. Kuksis, *J. Biochem. Biophys. Methods*, 10 (1984) 13.
233 J.J. Myher, A. Kuksis, Y-L. Yang and L. Marai, *Can. J. Biochem. Cell Biol.*, 62 (1984) 301.
234 A. Kuksis, L. Marai, J.J. Myher, J. Cerbulis and H.M. Farrell, Jr., *Lipids*, 21 (1986) 183.
235 J.J. Myher, A. Kuksis, L. Marai and J. Cerbulis, *Lipids*, 21 (1986) 309.
236 E. Defense, *Fett Wiss. Technol.*, 13 (1987) 3.
237 J.J. Myher, A. Kuksis, L. Marai and P. Sandra, *J. Chromatogr.*, 452 (1988) 93.
238 E. Geeraert and P. Sandra, *J. High Resolut. Chromatogr. Chromatogr. Commun.*, 8 (1985) 415.
239 P. Mares, in A. Kuksis (Editor), *Chromatography of Lipids in Biomedical Research and Clinical Diagnosis*, Elsevier, Amsterdam, 1987, p. 128.
240 W.R. Lusby, M.J. Thompson and J. Kochansky, *Lipids*, 19 (1984) 888.
241 R.P. Evershed, V.L. Male and J. Goad, *J. Chromatogr.*, 400 (1987) 187.
242 R.P. Evershed and L.J. Goad, *Biochem. Environ. Mass Spectrom.*, 14 (1987) 131.
243 A. Kuksis, J.J. Myher, L. Marai, J.A. Little, R.G. McArthur and D.A.K. Roncari, *Lipids*, 21 (1986) 371.
244 T. Murata and S. Takahashi, *Anal. Chem.*, 49 (1977) 728.
245 S.G. Wakeham and N.M. Frew, *Lipids*, 17 (1982) 831.
246 T. Rezanka, P. Mares, P. Husek and M. Podojil, *J. Chromatogr.*, 355 (1986) 265.
247 H. Egge, in H.K. Mangold and F. Paltauf (Editors), *Ether Lipids*, Academic Press, London, 1983, p. 17.
248 W.M.N. Ratnayake, A. Timmins, T. Ohshima and R.G. Ackman, *Lipids*, 21 (1986) 518.
249 M.J. Bossant, R. Farinotti, F. De Maack, G. Mahuzier, J. Benveniste and E. Ninio, *Lipids*, 24 (1989) 121.
250 W.D. Nes, *Lipids*, 23 (1988) 9.
251 J.H.P. Tyman, in L.R. Treiber (Editor), *Quantitative TLC and its Industrial Applications*, Dekker, New York, 1987, p. 125.
252 A. Kuksis, J.J. Myher and L. Marai, *J. Am. Oil Chem. Soc.*, 62 (1985) 762.
253 J.J. Myher and A. Kuksis, *Can. J. Biochem.*, 60 (1982) 161.
254 J.J. Myher and A. Kuksis, *Can. J. Biochem. Cell Biol.*, 62 (1984) 352.
255 J.J. Myher and A. Kuksis, *Biochim. Biophys. Acta*, 795 (1984) 85.
256 J.J. Myher, A. Kuksis, S. Pind and E.R.M. Kay, *Lipids*, 23 (1988) 398.
257 J.J. Myher, A. Kuksis and S. Pind, *Lipids*, 24 (1989) 396.
258 J.J. Myher, A. Kuksis and S. Pind, *Lipids*, 24 (1989) 408.
259 L-Y. Yang, A. Kuksis and J.J. Myher, *INFORM*, 1 (1990) 329 (abstract).
260 L-Y. Yang, A. Kuksis and J.J. Myher, *J. Lipid Res.*, 31 (1990) 137.
261 J.J. Myher and A. Kuksis, *J. Chromatogr.*, 471 (1989) 187.
262 J.J. Myher, A. Kuksis and L-Y. Yang, *Biochem. Cell Biol.*, 68 (1990) 336.
263 T. Oshima, H-S. Yoon and C. Koizumi, *Lipids*, 24 (1989) 535.

264 M.J. Wojtusik, P.R. Brown and J.G. Turcotte, *Biochromatography*, 3 (1988) 76.
265 K. Payne-Wahl and R. Kleiman, *J. Am. Oil Chem. Soc.*, 60 (1983) 1011.
266 F.C. Phillips, W.L. Erdahl, J.A. Schmit and O.S. Privett, *Lipids*, 19 (1984) 880.
267 T.A. Foglia, P.D. Vail and T. Iwama, *Lipids*, 22 (1987) 362.
268 A. Stolyhwo, H. Colin and G. Guiochon, *Anal. Chem.*, 57 (1985) 1342.
269 J.L. Robinson, M. Tsimidou and R. Macrae, *J. Chromatogr.*, 324 (1985) 35.
270 A.J. Palmer and F.J. Palmer, *J. Chromatogr.*, 465 (1989) 369.
271 L.J.R. Barron and G. Santa-Maria, *Chromatographia*, 23 (1987) 209.
272 E. Frede, *Chromatographia*, 21 (1985) 29.
273 J.A. Singleton and H.W. Pattee, *J. Am. Oil Chem. Soc.*, 61 (1984) 761.
274 J.A. Singleton and H.W. Pattee, *J. Am. Oil Chem. Soc.*, 62 (1985) 739.
275 J.A. Bezard and M.A. Quedraogo, *J. Chromatogr.*, 196 (1980) 279.
276 E. Frede and H. Thiele, *J. Am. Oil Chem. Soc.*, 64 (1987) 521.
277 I.S. Gilkinson, *Chromatographia*, 26 (1988) 181.
278 V.K.S. Shukla and F. Spener, *J. Chromatogr.*, 348 (1985) 441.
279 Y. Kondoh and S. Takano, *Anal. Chem.*, 58 (1986) 2380.
280 A. Kuksis, L. Marai and J.J. Myher, *J. Chromatogr.*, 273 (1983) 43.
281 L. Marai, J.J. Myher and A. Kuksis, *Can. J. Biochem. Cell Biol.*, 61 (1983) 840.
282 A. Kuksis, L. Marai and J.J. Myher, *J. Am. Oil Chem. Soc.*, 66 (1989) 482 (abstract NN1).
283 A. Kuksis, L. Marai, J.J. Myher and Y. Itabashi, in V.K.S. Shukla and G. Holmer (Editors), *Proceedings, 15th Scandinavian Symposium on Lipids, Rebild Bakker, Denmark*, A Lipidforum Publication, 1989, p. 336.
284 A. Kuksis, L. Marai, J.J. Myher and Y. Itabashi, in E.G. Perkins (Editor), *Analysis of Fats, Oils and Lipoproteins*, AOCS Monograph, American Oil Chemists' Society, Champaign, IL, 1991, in press.
285 Y. Nakagawa, T. Sugiura and K. Waku, *Biochim. Biophys. Acta*, 833 (1985) 323.
286 Y. Nakagawa and K. Waku, *Lipids*, 21 (1986) 155.
287 A. Ojima-Uchiyama, Y. Masuzawa, T. Sugiura, K. Waku, H. Saito, Y. Yiu and H. Tomioka, *Lipids*, 23 (1988) 815.
288 H.G. Choe, R.D. Wiegand and R.E. Anderson, *J. Lipid Res.*, 30 (1989) 454.
289 Y. Nakagawa and K. Waku, in A. Kuksis (Editor), *Chromatography of Lipids in Biomedical Research and Clinical Diagnosis*, Elsevier, Amsterdam, 1987, p. 163.
290 H. Takamura, K. Tanaka, T. Matsuura and M. Kito, *J. Biochem.*, 105 (1989) 168.
291 A. Poulos, P. Sharp, D. Johnson and C. Easton, *Biochem. J.*, 253 (1989) 645.
292 M.V. Bell, *Lipids*, 24 (1989) 585.
293 C.S. Ramesha, W.C. Pickett and D.V.K. Murthy, *J. Chromatogr.*, 491 (1989) 37.
294 H. Rabe, G. Reichman, Y. Nakagawa, B. Rustow and D. Kunze, *J. Chromatogr.*, 493 (1989) 353.
295 L. Binaglia, R. Roberti, A. Vecchini and G. Porcellati, *J. Lipid. Res.*, 23 (1982) 955.
296 S. Pind, A. Kuksis, J.J. Myher and L. Marai, *Can. J. Biochem. Cell Biol.*, 62 (1984) 301.
297 P.E. Haroldsen and R.C. Murphy, *Biomed. Environ. Mass Spectrom.*, 14 (1987) 573.
298 C.S. Ramesha and W.C. Pickett, *Biomed. Environ. Mass Spectrom.*, 13 (1986) 107.
299 J.H. Peters, H.W. Nolen, III, G.R. Gordon, W.W. Bradford, III, J.E. Bupp and E.J. Reist, *J. Chromatogr.*, 488 (1989) 301.
300 L. Marai, A. Kuksis and J.J. Myher, (1989) unpublished results.
301 B. Nilsson and D. Zopf, *Arch. Biochem. Biophys.*, 222 (1983) 628.
302 G.C. Hansson, Y-T. Lee and H. Karlsson, *Biochemistry*, 28 (1989) 6672.
303 A. Kuksis, L. Marai and J.J. Myher, *INFORM*, 1 (1990) 353 (abstract).
304 Y. Nakagawa and K. Waku, *Eur. J. Biochem.*, 152 (1985) 569.
305 Y. Nakagawa, K. Fujishima and K. Waku, *Anal. Biochem.*, 157 (1986) 172.
306 Nakagawa and K. Waku, *J. Chromatogr.*, 487 (1989) 239.
307 R.W. Gross, *Biochemistry*, 24 (1985) 1662.
308 F.H. Chilton and R.C. Murphy, *J. Biol. Chem.*, 261 (1986) 7771.
309 R. Leduc, G.M. Patton, D. Atkinson and S.J. Robins, *J. Biol. Chem.*, 262 (1987) 7680.

310 A.D. Purdon, D. Patelunas and J.B. Smith, *Lipids,* 22 (1987) 116.
311 C.L. Swedsen, J.M. Ellis, F.H. Chilton, J.T. O'Flaherty and R.L. Wyckle, *Biochem. Biophys. Res. Commun.,* 113 (1983) 72.
312 R.K. Satsangi, J.C. Ludwig, S.T. Weintraub and R.N. Pinckard, *J. Lipid Res.,* 30 (1989) 929.
313 F.R. Sugnaux and C. Djerassi, *J. Chromatogr.,* 251 (1982) 189.
314 F.B. Jungalwala, J.E. Evans and R.H. McCluer, *J. Lipid Res.,* 25 (1984) 738.
315 W.L. Erdahl and O.S. Privett, *J. Am. Oil Chem. Soc.,* 62 (1985) 786.
316 M.J. Thomas, M. Samuel, R.L. Wykle, J.R. Surles and C. Piantadosi, *J. Lipid Res.,* 27 (1986) 172.
317 R.W. Gross, *Biochemistry,* 23 (1984) 158.
318 P. Varenne, B.C. Das, J. Polonsky and M. Tence, *Biomed. Mass Spectrom.,* 12 (1985) 6.
319 E. Benfenati and R. Reginato, *Biomed. Mass Spectrom.,* 12 (1985) 643.
320 G.R. Fenwick, J. Eagles and R. Self, *Biomed. Mass Spectrom.,* 10 (1983) 382.
321 A. Poulos, D.W. Johnson, K. Beckman, I.G. White and C. Easton, *Biochem. J.,* 248 (1987) 961.
322 W.R. Sherman, K.E. Ackermann, R.H. Bateman, B.N. Green and I. Lewis, *Biomed. Mass Spectrom.,* 12 (1985) 409.
323 N.J. Jensen, K.B. Tomer and M.L. Gross, *Lipids,* 22 (1987) 480.
324 H. Itabe, Y. Kushi, S. Handa and K. Inoue, *Biochim. Biophys. Acta,* 962 (1988) 8.
325 C. Easton, D.W. Johnson and A. Poulos, *J. Lipid Res.,* 29 (1988) 109.
326 L.M. Smith, H.A. Norma, S.H. Choa and G.A. Thompson, Jr., *J. Chromatogr.,* 346 (1985) 291.
327 T. Rezanka and M. Podojil, *J. Chromatogr.,* 364 (1989) 397.
328 H. Kubo and M. Hoshi, *J. Lipid Res.,* 26 (1985) 638.
329 Y. Hirabayashi, A. Hamaoka, M. Matsumoto and K. Nishimura, *Lipids,* 21 (1986) 710.
330 H. Clausen, S.B. Levery, E. Nudelman, S. Tsuchiya and S. Hakamori, *Proc. Natl. Acad. Sci.,* 82 (1985) 1199.
331 H. Clausen, S.B. Levery, E.D. Nudelman, M. Stroud, M.E.K. Salyan and S. Hakomori, *J. Biol. Chem.,* 262 (1987) 14228.
332 G.K. Ostrander, S.B. Levery, S. Hakomori and E.H. Holmes, *J. Biol. Chem.,* 263 (1988) 3103.
333 W. Fischer, in H.K. Mangold (Editor), *Handbook of Chromatography, Vol. 1, Lipids,* CRC Press, Boca Raton, FL, 1984, p. 555.
334 K. Nishimura, Y. Hirabayashi, A. Hamaoka, M. Matsumoto, A. Nakamura and K. Miseki, *Biochim. Biophys. Acta,* 796 (1984) 269.
335 D.M. Bonafede, L.J. Macala, M. Constantine-Paton and R.K. Yu, *Lipids,* 24 (1989) 680.
336 M. Constantine-Paton, A.S. Blum, R. Mendez-Otero and C.J. Barnstable, *Nature (London),* 324 (1986) 459.
337 W.W. Young and C.A. Borgman, *J. Lipid Res.,* 27 (1986) 120.
338 J. Gottfries, P. Davidson, J.E. Mansson and L. Svennerholm, *J. Chromatogr.,* 490 (1989) 263.
339 S. Degrandis, H. Law, J. Brunton, C. Gyles and C.A. Lingwood, *J. Biol. Chem.,* 264 (1989) 12520.
340 S. Ladisch, in H.S. Oettgen (Editor), *Gangliosides and Cancer,* VCH Verlag, Weinheim, 1988, p. 217.
341 B. Domon and C.E. Costello, *Biochemistry,* 27 (1988) 1534.
342 R. Isobe, Y. Kawano, R. Higuchi and T. Komori, *Anal. Biochem.,* 177 (1989) 296.
343 G. Rosen, P. Pahlsson, M.V. Londner, M.E. Westerman and B. Nilsson, *J. Biol. Chem.,* 264 (1989) 10457.
244 W.L. Roberts, J.J. Myher, A. Kuksis and T.L. Rosenberry, *Biochem. Biophys. Res. Commun.,* 150 (1988) 271.
345 J.M. Mato, K.L. Kelly, A. Abler and L. Jarrett, *J. Biol. Chem.,* 262 (1987) 2131.

346 W.L. Roberts, J.J. Myher, A. Kuksis, M.G. Low and T.L. Rosenberry, *J. Biol. Chem.*, 263 (1988) 18766.
347 W.L. Roberts, S. Santikarn, V.N. Reinhold and T.L. Rosenberry, *J. Biol. Chem.*, 263 (1988) 18776.
348 A. Kuksis, J.J. Myher, L. Marai, K. Geher, A. Angel and D.A.K. Roncari, *Pittsburgh Conference on Analytical Chemistry, New York, March 5-9, 1990* (Abstract No. 686).
349 A. Kuksis, L. Marai and J.J. Myher, *Lipids*, (1990) in press.
350 J. van der Greef, W.M.A. Niessen and U.R. Tjaden, *J. Chromatogr.*, 474 (1989) 5.
351 H.Y. Kim and N. Salem, Jr., *Anal. Chem.*, 59 (1987) 722.
352 F. Hullin, H.Y. Kim and N. Salem, Jr., *J. Lipid Res.*, 30 (1989) 1963.
353 A. Kuksis, in W.D. Nes and J. Parish (Editors), *Analysis of Sterols and Other Biologically Significant Steroids*, Academic Press, New York, 1989, p. 151.
354 A. Kuksis, J.J. Myher, L. Marai, J.A. Little, R.G. McArthur and D.A.K. Roncari, *J. Chromatogr.*, 381 (1986) 1.
355 G. Lercker, *J. Chromatogr.*, 279 (1983) 543.
356 G. Lercker, M. Cocchi, E. Turchetto and S. Savioli, *J. High Resolut. Chromatogr. Chromatogr. Commun.*, 7 (1984) 274.
357 N. Limsathayourat and H-U. Melchert, *Fresenius Z. Anal. Chem.*, 318 (1984) 410.
358 K. Grob, M. Lantranchi and C. Mariani, *J. Chromatogr.*, 471 (1989) 397.
359 A. Lohninger, L. Linhart, M. Landau, D. Glogar, C. Kratochwil and E. Keiser, *Anal. Biochem.*, 171 (1988) 366.
360 A. Lohninger, P. Preis, L. Linhart, S.V. Sommoggy, M. Landau and E. Keiser, *Anal. Biochem.*, 186 (1990) 243.
361 A. Kuksis and J.J. Myher, *J. Chromatogr.*, 379 (1986) 57.

Chapter 16

Carbohydrates

SHIRLEY C. CHURMS

CONTENTS

16.1 INTRODUCTION

Since the chromatography of carbohydrates was last reviewed in this textbook [1] the main area of growth has been HPLC. A survey of the literature of the past 10 years shows that papers describing new methods for HPLC analysis of carbohydrates, improvements to the older techniques or new applications of HPLC far exceed those devoted to any other form of chromatography, even GLC. In addition to development of the modes of chromatography that were designed for HPLC ab initio, those using microparticulate silica as an adsorbent, silica-based packings with bonded phases for partition or reversed-phase chromatography, or microparticulate cation-exchange resins for ligand-exchange or size-exclusion chromatography, much research has been devoted to the adaptation to HPLC of classical methods. Ion-exchange, steric-exclusion, and even affinity chromatography have undergone modification to permit the use of HPLC equipment. The next most important growth point has been improvement in the resolution and sensitivity of GLC methods, due to the advent of highly efficient capillary columns and the ever-increasing use of MS coupled to GLC as the body of mass spectral data for various volatile derivatives of carbohydrates has grown. Some new methods of derivatization have also been used to advantage, especially chiral derivatives, with which resolution of enantiomers is possible. A parallel development aimed at resolution of enantiomers has been the use of chiral stationary phases in GLC. Planar chromatography of carbohydrates has not expanded to the same extent as the column methods, but some improvements have stemmed from the use of more sensitive detection methods and, it TLC, bonded-phase and more efficient HPTLC plates. Like PC, electrophoresis has long been a well-established technique for carbohydrate analysis, and here too there have been few changes, the main trend being increased use of gel electrophoresis, especially PAGE, in studies of glycoproteins and other glycoconjugates.

Some entirely new chromatographic methods have made their appearance during the past decade, notably supercritical-fluid chromatography (see Chapter 8), and others, like ion chromatography, have been revolutionized by improvements in packings and, especially, the detector. The application of these techniques to carbohydrate chromatography is included here for the first time in this textbook.

16.2 LIQUID CHROMATOGRAPHY

The rapid proliferation of HPLC methods in carbohydrate chromatography during the past decade is reflected in a number of reviews devoted entirely or largely to this topic [2-9]. The older, low-pressure methods, with the exception of size-exclusion chromatography on gels (Section 16.2.6) and affinity chromatography (Section 16.2.7), are now used only for preparative purposes. Most of this section is therefore devoted to the application of HPLC systems of various types to chromatographic analysis of a variety of carbohy-drates; HPLC of linear oligosaccharides up to high degree of polymerization (DP), cyclic oligosaccharides, and the complex products of degradation of glycoproteins and glyco-saminoglycuronans is highlighted, since this has been a major success of HPLC in recent years.

The organization of a comprehensive review of HPLC methods is difficult in cases where more than one mechanism can be operative. In carbohydrate chromatography this is true especially of C_{18}-silica packings and microparticulate resins in the calcium and other salt forms; all applications of such columns have been grouped together in the appropriate sections (Sections 16.2.1.4 and 16.2.2, respectively). Packings in which an amino phase is bonded to silica can function by normal-phase partition or ion exchange, but here the distinction is more clear-cut: when an organic solvent, usually acetonitrile, preponderates in the eluent, chromatography is governed largely by the former mecha-nism, but with eluents consisting entirely or mainly of aqueous buffer solutions, the packing acts as weakly basic anion exchanger. HPLC under the former conditions is discussed in Section 16.2.1.3, while applications in which the ion-exchange mechanism is operative are grouped with other ion-exchange methods (Section 16.2.3).

16.2.1 HPLC on silica-based packings

16.2.1.1 Adsorption and partition chromatography on unmodified silica

Unmodified silica packings continue to find application mainly in HPLC of carbo-hydrates that have been derivatized with a view to increasing the sensitivity of detection by introduction of a chromophoric group, thus permitting the use of UV detection rather than the less sensitive differential refractometer. Benzoate and 4-nitrobenzoate derivatives, discussed previously [1, p. 227], have been widely used, but the production of multiple peaks due to resolution of anomers and of pyranose and furanose ring forms, though aiding identification of individual sugars, complicates the chromatograms given by mix-tures [10,11]. As in GLC, this can be obviated by reduction of sugars to alditols prior to derivatization, and good resolution of a mixture of 6-deoxyhexoses, pentoses, and hexoses, as the derived alditol benzoates, on a column (150 x 4.5 mm ID) of 3-μm silica has been reported by Oshima and Kumanotani [12], who used isocratic elution with n-hexane/dioxane/dichloromethane (30:4:1). Amino- and acetamidodeoxyhexoses, similarly derivatized, can be resolved and distinguished from one another, well separated from the neutral sugar derivatives, on the same column with gradient elution (n-hexane/dioxane/di-chloromethane, 22:2:1 to 4:2:1 in 80 min). This simultaneous analysis of neutral, aminodeoxy, and acetamidodeoxy sugars, which can be used also for preparative pur-poses, is a major advantage over GLC as the alditol acetate derivatives, although resolution of the neutral sugar derivatives is inferior to that obtained by capillary GLC of alditol acetates (Section 16.4.1.1).

HPLC of mono- and disaccharides and methyl glycosides as phenyldimethylsilyl de-rivatives on a column (250 x 4.6 mm ID) of 5-μm silica, with n-hexane/ethyl acetate eluents (proportions ranging from 49:1 for glycosides to 99:1 for monosaccharides) has been investigated by White et al. [13,14]. This method was successful when applied to the methyl glycosides, methyl gluco-, galacto-, and xylopyranosides being separated from one another, and with good resolution of α- and β-anomers in each case. Detection of submicrogram quantities (down to ca. 250 ng) was possible with UV detection at 260 nm. However, resolution of sugars by this method was poor and complicated by the production of 2-4 peaks for each sugar, and inferior resolution of perphenyldimethylsilylated alditols (in n-hexane/ethyl acetate, 199:1) indicated that in this case the problem could not be overcome by reduction of the sugars.

Earlier, Honda and Kakehi [15] studied HPLC of aldehydes, produced on periodate oxidation of carbohydrates, which were chromatographed as 2,4-dinitrophenylhydrazones on LiChrosper SI-100, with detection at 352 and 435 nm. This work has been extended recently by Karamanos et al. [16] to the analysis of the neutral sugar components of glycoproteins (fucose, xylose, mannose, galactose, and glucose). With a column (260 x 4.6 mm ID) of 5-μm silica, isocratic elution with chloroform containing methanol (7.6%) and a trace (0.7%) of water, and detection at 352 nm, the five sugars of interest, converted to the 2,4-dinitrophenylhydrazones, were resolved as single peaks. The slight shoulders or minor peaks, believed to be due to the presence of geometric isomers of these derivatives, were sufficiently separated from the major peaks not to interfere with quantitation. Calibration plots were linear over the range 0.05-3.5 nmol for each sugar.

Even greater sensitivity is obtainable by use of 5-dimethylaminonaphthalene-1-sulphonylhydrazine (dansyl- or DNS-hydrazine) as a fluorescent labeling reagent, which permits the use of highly sensitive fluorimetric detection. For example, Takeda et al. [17] have reported good resolution of a mixture of 6-deoxyhexoses, pentoses, hexoses, acetamidodeoxyhexoses, and the disaccharides maltose, cellobiose, gentiobiose, and lactose as their DNS-hydrazone derivatives on a column (250 x 4 mm ID) of LiChrosorb SI-100 in 25 min, by stepwise elution with eluents containing increasing proportions of ethanol (8-20%) in aq. chloroform (0.5-0.6% water). The use of a spectrofluorophotometer to monitor the effluent (excitation 350 nm, emission 500 nm) made possible detection and quantitation of sugars at levels down to 3-20 pmol.

Of particular interest is the use by Oshima et al. [18] of derivatization of neutral aldoses by reductive amination with chiral L-(-)-α-methylbenzylamine, followed by acetylation, to resolve enantiomers by HPLC (cf. Section 16.4.3). On a column (150 x 4.6 mm ID) of 3-μm silica, eluted isocratically with n-hexane/ethanol (19:1) and monitored by UV (230 nm), the enantiomers of all the common neutral aldoses, except 2-deoxy sugars, were well resolved. The configuration at C-2 determines the elution order of the enantiomers of the diastereoisomeric 1-(N-acetyl-α-methylbenzylamino)-1-deoxyalditol acetates.

Unmodified silica is of very limited use for HPLC of sugars without prior derivatization. Iwata et al. [19] succeeded in resolving D-glucose and its α-(1 $\rightarrow$ 4)-linked oligomers (malto-oligosaccharides) to DP 5, as well as the products (nigerosyl- and nigerotriosyl-erythritol) of Smith degradation of lichen polysaccharides, linked (1 $\rightarrow$ 3) and (1 $\rightarrow$ 4), on a column (250 x 4.6 mm ID) of 5-μm silica, eluted with ethyl acetate/methanol/water (7:3:2). Elution with the same solvents in proportions 12:3:1 resolved sucrose, maltose, and lactose. Some resolution of rhamnose, xylose, arabinose, mannose, and glucose was achieved on the same column with ethyl formate/methanol/water (12:3:1). Nikolov and Reilly [20] measured the capacity factors for HPLC of a digitoxose (2,6-dideoxy-D-*ribo*-hexose), 2-deoxy-D-glucose (2-deoxy-D-*arabino*-hexose), D-glucose, sucrose, the 3 isomers of trehalose ($\alpha\beta$, $\alpha\alpha$, and $\beta\beta$), and *myo*-inositol on a column (250 x 4.6 mm ID) of 5-μm silica, with acetonitrile/water eluents of varying acetonitrile content (60-90%). Comparison of these values with those obtained using a similar column, packed with amine-bonded silica (also of 5-μm particle diameter), showed that similar separations required a higher proportion of acetonitrile in the case of the unmodified silica. Correlation of capacity factors with both the proportion of water in the pores of the silica-based packings and the calculated hydration numbers of the carbohydrates indicated that partition, not adsorption, is the predominant mechanism governing HPLC of underivatized carbohydrates on silica and amine-bonded silica.

16.2.1.2 Partition chromatography on amine-modified silica

Partition chromatography of underivatized carbohydrates on silica is best achieved by addition of a polyfunctional amine to the mobile phase, so that the silica is modified in situ to an amino-phase packing for normal partition chromatography. Since the successful application of this method to a few sugars important in the food industry, its use has been extended to a large number of sugars and alditols [21,22]. In these studies, a Radial-Pak cartridge (Waters), 100 x 8 mm ID, packed with 10-μm silica and operated under radial compression (Waters module RCM-100), was modified by equilibration with acetonitrile/water (7:3), containing 0.1% of tetraethylenepentamine, and eluted with acetonitrile/water (81:19 [21] or 3:1 [22]), containing 0.02% of the polyamine. The authors claimed that operation under radial compression minimized the dissolution of the silica packing at the high pH (8.9-9.2) of the eluent, resulting in long column life. Good resolution of many alditols, monosaccharides, and oligosaccharides up to the tetrasaccharide level was obtained. For chromatography of higher oligosaccharides by this method White et al. [23], using ordinary stainless-steel columns, packed with 5-μm silica, tested various amine modifiers. Of these, tetraethylenepentamine, pentaethylenehexamine, commercial "polyamine modifier", and 1,4-diaminobutane gave the best resolution of the first 8 members of the malto-oligosaccharide series on a 100 x 5-mm-ID column, eluted with acetonitrile/water (11:9), containing 0.01% of the amine, after previous equilibration with the same solvent system with 0.1% of the modifier. For resolution of the malto-oligosaccharides up to DP 20, on a 200 x 8-mm-ID column, 1,4-diaminobutane proved to be the modifier of choice. Members of the (1 → 2)-linked D-fructo-oligosaccharide series (inulins) of DP 2-30 have been resolved by Praznik et al. [24], using a similar amine-modified silica system, in this case with the column temperature raised to 35°C, which improved resolution, and with polyethylene glycol 35 000 (0.2%) added to both the equilibrating and eluting solvent system to decrease interaction of the amino groups with the hydroxyl groups of the saccharides.

References on p. B284

Advantages and disadvantages of the use of amine-modified silica, as opposed to amine-bonded silica (Section 16.2.1.3), have been detailed by several authors [3,8,9,22]. In addition to longer life, since the amine adsorbed on the silica is continuously regenerated, the amine-modified silica system is fairly stable over a wide range of pH and solvent composition, has a high capacity for carbohydrate solutes, and is relatively inexpensive. Disadvantages include variations in retention times due to varying amine loading of the silica, fluctuating base lines due to variable delivery of amine into the mobile phase, and the difficulty of using UV detection, owing to the presence of amine in the eluent. Despite claims to the contrary, amine-modified silica shares with other silica-based packings the disadvantage of dissolution of the silica to a small extent in water-rich eluents, especially at high pH.

16.2.1.3 Partition chromatography on amine-bonded silica

Notwithstanding its limited life, due to the formation of glycosylamines by interaction between reducing sugars and the amino groups on the stationary phase, amine-bonded silica is still preferred by many workers as a support for normal-phase partition chromatography of carbohydrates, because of its greater reproducibility. The use of a packing based on 5-μm silica greatly improves the resolution of the common sugars on HPLC in acetonitrile/water (3:1),and detection limits are decreased to 4-12 μg with a UV detector operating at 188 nm [25]. HPLC on two columns, each 300 x 4 mm ID, coupled together, produced a marked improvement in resolution of these sugars on amino phases based on 10-μm silica, with 4:1 acetonitrile/water as eluent [26]. Using bonded amino-phase packings with silica particles of average diameter 5 or 7 μm, Nikolov et al. have determined the capacity factors of 20 disaccharides [27] and 13 trisaccharides [28] on HPLC in acetonitrile/water mixtures containing 69-80% of acetonitrile and have listed the structural features (e.g., position of glycosidic linkage, presence or absence of D-fructofuranosidic linkages) determining the retention times of these oligosaccharides. Differences were such as to permit resolution only of simple mixtures by this method.

Resolution of higher oligosaccharides on amino phases bonded to 5-μm silica requires gradient elution and therefore detection by methods other than the differential refractometer. Mellis and Baenziger [29] resolved the α-(1 → 6)-linked D-gluco-oligosaccharides (isomalto series) to DP 18 within 1h on Micropak AX-5 (Varian) with a reverse linear gradient (65 → 35%) of acetonitrile in water; pre-column labeling of the oligosaccharides by reduction with NaB[^{3}H]$_4$ permitted detection by scintillation counting. The same conditions were applied in resolution of complex hetero-oligosaccharides derived from glycoproteins, which included bi, tri-, and tetraantennary oligosaccharides containing 9-14 sugar residues and those of the high-mannose type with 5-7 D-mannosyl residues.

The development of packings based on 3-μm silica has extended the upper limit of resolution for linear D-gluco-oligosaccharides to DP 30-35. Koizumi et al. [30], using a column (200 x 6 mm ID) designated ERC-NH-1171 (Erma) and a differential refractometer of improved sensitivity, have reported resolution of the malto-oligosaccharide series to DP 30 within 35 min, and of the β-(1 → 2)-linked oligosaccharides of the sophoro series (from partial hydrolysis of cyclosophoraose) to DP 35 within 45 min, on isocratic elution with acetonitrile/water mixtures, containing 57-58% of acetonitrile. Oligosaccharides that were (1 → 6)-linked (α- or β-) were resolved to DP 26 within 40 min with 55-56% acetonitrile, but solubility considerations imposed upper limits of DP 18 and 10, respectively, for those linked β-(1 → 3) and the β-(1 → 4)-linked cellodextrin series.

Koizumi et al. [31] have also tested two novel packings, YMC Pack PA-03 (Yamamura Chemical Co.), in which a polyamine resin is bonded to silica, and TSK gel Amide-80 (Toyo Soda), which has carbamoyl amide groups instead of amino groups in the bonded phase. However, their resolving power for malto-oligosaccharides was inferior to that of ERC-NH-1171, the upper limit of resolution being 25 (in 40 min) with 250 x 4.6-mm-ID columns eluted with acetonitrile/water mixtures, containing 50-53% of acetonitrile. In the same study, these columns and ERC-NH-1171 were applied to HPLC analysis of α-, β-, and γ-cyclodextrins and series in which these cyclodextrins carried branches. Here, the best resolution was obtained with YMC Pack PA-03, eluted with 55% acetonitrile, although the amide column, which is stable at higher temperatures (up to 80°C) than the amino-phase packings, gave improved resolution at 50-70°C, with eluents containing 60-62% of acetonitrile. For cyclosophoraoses, the cyclic β-(1 → 2)-linked D-gluco-oligosaccharides produced by Rhizobium and Agrobacterium strains, the same group [32] achieved resolution of the oligomers of DP 17-40 within 50 min on the ERC-NH-1171 column, eluted with 57% acetonitrile. This performance was superior to that of an amino-phase packing based on 5-μm silica, which was found by Benincasa et al. [33] to resolve the same cyclic oligosaccharides over the range DP 17- 33 (160 x 4.6-mm-ID column eluted with 64% acetonitrile).

A study of the behavior of 55 neutral oligosaccharides, mainly of the type obtained in degradative studies of glycoproteins, by HPLC on a column (250 x 4 mm ID) of LiChrosorb-NH_2 (Merck; 5-μm particles), eluted isocratically for 30 min with a 4:1 mixture of acetonitrile and an aqueous 15 mM potassium phosphate buffer solution (pH 5.2), then with a linear gradient of increasing buffer content (0.5% per min), has enabled the effects of various structural features on retention time to be correlated. Blanken et al. [34] concluded that retention time is decreased by the presence of residues of L-fucose or 2-acetamido-2-deoxy-D-glucose, especially when the latter is at the reducing end, and increased for oligosaccharides having a 1 → 6 linkage. Five complex oligosaccharides of the N-acetyllactosamine type, linear, bi-, tri-, and tetraantennary and containing 5-12 sugar residues, were resolved with a buffer gradient of 0.3% per min, starting with the 4:1 solvent mixture. Detection was by UV (195 nm), phenol/sulfuric acid assay, or liquid scintillation counting of radioactive labeled compounds.

Dua et al. [35] have reported the separation of a series of reduced oligosaccharides, containing up to 7 residues (fucose, galactose, 2-acetamido-2-deoxyglucose, and 2-acetamido-2-deoxygalactitol), that are produced on alkaline borohydride degradation of mucin glycoproteins; columns of Alltech 605 NH and Micropak AX-5 (250 x 4.6 mm and 300 x 4 mm ID, respectively, both 5-μm) were eluted with 3:2 acetonitrile/water or acetonitrile/ 1 mM KH_2PO_4 buffer solution (pH 5.4). Both analytical and preparative separations of lacto-N-hexaose and mono- and difucosylated derivatives, isolated from human milk, have

References on p. B284

been achieved [36] on the Micropak AX-5 column, eluted with 1:1 acetonitrile/buffer. The resolution of oligosaccharides of this type by normal-phase partition chromatography is vastly superior to that given by HPLC on C_{18}-silica packings with water (Section 16.2.1.4).

Brain gangliosides, including highly sialylated members of the series, are well resolved by HPLC on amine-bonded silica (5-7 μm) and gradient elution with phosphate buffers (pH 5.5-5.6) in acetonitrile [37,38].

16.2.1.4 HPLC on octadecylsilica packings

Packings in which a long alkyl chain, usually C_{18}, is bonded to silica, can operate in different modes when applied to HPLC of carbohydrates. The reversed-phase partition mechanism for which they are primarily designed applies only after introduction of nonpolar groups by derivatization of the carbohydrate solutes, a procedure that is employed, as in the case of HPLC on unmodified silica (Section 16.2.1.1), to increase sensitivity of detection by conferring chromophoric or fluorescent properties upon the molecules. Thus, sugars have been analyzed on C_{18}-silica columns as DNS-hydrazones [39-41] and pyridylamino derivatives (prepared by reductive amination with 2-aminopyridine) [42], which permit sensitive fluorimetric detection, and recently as dabsyl-hydrazones (produced on reaction with 4'-N,N-dimethylamino-4-azobenzene-1-sulfonyl hydrazine) [43,44], which are strongly chromophoric (absorbance maximum 460 nm, shifting to 495 nm at low pH). Using a 5-μm C_{18}-silica column (250 x 4.6 mm ID) and eluents containing acetonitrile (21-24%) mixed with aqueous 0.08 M acetic acid, Mopper and Johnson [40] achieved good resolution, within 20 min, of 16 common pentoses, hexoses, deoxy sugars, and reducing disaccharides as DNS-derivatives, fluorimetric detection (excitation 360 nm, emission 540 nm) permitting analysis at levels of 5-15 pmol. Muramoto et al. [43] were able to analyze the neutral sugar compositions of glycoprotein samples as small as 5 μg by HPLC of the hydrolyzates as dabsyl-hydrazones; a short column (50 x 4.6 mm ID), packed with 3-μm C_{18}-silica, was eluted for 15 min at 40°C with a 1:3 mixture of acetone and 0.08 M acetic acid. Under these conditions, all the common monosaccharides, as well as cellobiose, lactose, and gentiobiose, were well resolved. The wavelength for detection was 485 nm at this low pH; with eluents containing acetonitrile and water only (concave gradient), the detection limit for dabsyl-sugars was 10 pmol at 425 nm [44].

Conversion to pyridylamino derivatives has the advantage that simultaneous analysis of neutral and amino sugars in possible: Takemoto et al. [42], using two columns, each 250 x 4.6 mm, connected in series, were able to resolve the common neutral sugars, 2-acetamido-2-deoxymannose, -glucose and -galactose, and N-acetylneuraminic acid as their pyridylamino derivatives on 5-μm C_{18}-silica, eluted with 0.25 M sodium citrate buffer (pH 4.0), containing 1% of acetonitrile. Fluorimetric detection (excitation 320 nm, emission 400 nm) allowed the determination of 0.01-10 nmol of each sugar. Glycoconjugates (glycoproteins and gangliosides) could be analyzed by this method in samples of 100-200 pmol. For simpler mixtures of glycoprotein-derived neutral sugars, Shinomiya et al. [45] have proposed HPLC on C_{18}-silica after labeling of the products of reductive amination with the fluorescent reagent 7-fluoro-4-nitrobenz-2-oxa-1,3-diazole (NBDF), with which detection in the range 40 pmol to 50 nmol is possible.

Perbenzoylation, followed by reversed-phase HPLC, has been proposed as an alternative to GLC of the trimethylsilylated product (Section 16.4.1.4) for the analysis of methanolyzates of glycoproteins, the necessity for use of empirically determined molar response factors, as in GLC, being obviated by the direct relationship between UV absorbance and the number of benzoyl groups in each derivative. Jentoft [46] reported good resolution of the benzoylated methyl glycosides of all the sugar constituents (neutral and amino sugars) of glycoproteins on a column (150 x 4.6 mm ID) of 3-μm C_{18}-silica within 45 min, using stepwise elution with 50% and then 60% acetonitrile. Multiple peaks are obtained by this method, as in HPLC of benzoates and 4-nitrobenzoates on silica (Section 16.2.1.1). The limit of detection under the conditions used by Jentoft was of the order of 1 nmol, but recently Gisch and Pearson [47] have claimed analysis of benzoylated methyl glycosides at picomolar levels by use of a microbore column (150 x 2.1 mm; Supelcosil LC-18) with gradient elution (35 → 90% acetonitrile in 65 min).

For reversed-phase HPLC of higher oligosaccharides the separations achieved by Wells et al. [48,49] of peracetylated malto-oligosaccharides to DP 30 and above on two coupled columns, packed with pellicular C_{18}-silica, by gradient elution (10 → 70% acetonitrile in water) at 65°C, which were mentioned in the previous review [1, p. 231], have never been surpassed. It is surprising that this method has received so little attention. Chromatography of these oligosaccharides as perbenzoylated, reduced derivatives, on C_8-silica at ambient temperature (linear gradient, 80 → 100% acetonitrile in 15 min), gave far inferior resolution, to a maximum DP of 13 [50]. Both of these methods have also been applied to HPLC of high-mannose-containing oligosaccharides from ovalbumin [49] or mannosidosis urine [50].

The use of reversed-phase HPLC to fractionate the mixtures of partially methylated, partially ethylated, reduced oligosaccharides, obtained during sequencing of complex carbohydrates by the procedure developed by Valent et al. [51], which was newly introduced at the time of the last review [1, p. 231], is now a well-established technique. Its value has been greatly enhanced by interfacing of the column (250 x 4.6 mm ID, Zorbax ODS; Du Pont) with the chemical-ionization chamber of a mass spectrometer, the HPLC eluent (aq. acetonitrile) serving as the CI reactant gas [52]. This permits the use of gradient elution (50 → 65% acetonitrile in 45 min [52] or 50 → 70% in 60 min [53]) and direct MS detection, with partial identification of the individual alkylated oligosaccharide-alditols, which are then isolated for further examination by GLC, coupled to EI-MS. This methodology has enabled the Albersheim group [51-54] to sequence complex carbohydrates, using samples of less than 10 mg.

Attempts to use HPLC on C_{18}-silica in analyses of partially methylated sugars without prior reduction have met with only limited success. Owing to the resolution of anomers, double peaks result for most methylated sugars [55,56]. Eluents are mixtures of methanol or ethanol (2-20%) with water or a volatile buffer (1% ammonium acetate (pH 6.9) has been recommended [55], from which isolated methylated sugars are easily recovered. For HPLC of the mixtures of partially methylated sugars obtained in methylation analysis of polysaccharides, reduction to alditols is necessary to eliminate the doublets. The partially methylated alditols are eluted in order of increasing degree of methylation. Resolution of those differing in this respect is good but, with some exceptions when acetonitrile (1-5%, or gradient) replaces the alcohols in the eluent [57], separation of isomers having the same degree of methyl substitution is difficult.

Chromatography of methyl glycosides on C_{18}-silica in water resolves the α- and β-anomers in most cases [58], and this affords a convenient method for studies of relative rates of formation with a view to optimization of conditions for the isolation of a chosen glycoside in maximal yield. The 100 x 8-mm-ID Dextropak cartridge (Waters) used by Cheetham and Sirimanne [58-60] can be loaded with up to 8 mg of sample, and therefore isolation of small amounts of selected glycosides is possible by this method. The mixtures of methyl glycosides, obtained on methanolysis of complex polysaccharides, are readily analyzed by HPLC on C_{18}-silica in water [59,61], which gives a simple alternative, but with much lower sensitivity, to the GLC methods (Sections 16.4.1.4 and 16.4.1.5). Not only neutral sugars but also uronic acids can be analyzed in this way, although the chromatograms given by the acids are complicated by the production of multiple peaks, corresponding to anomers of both the uronic acids themselves and the co-existing lactones [59,62,63].

Separation of underivatized oligosaccharides on C_{18}-silica, eluted with water or aqueous solutions, takes place by a mechanism that is not reversed-phase partition but probably a form of hydrophobic-interaction chromatography, involving interaction of the Van der Waals type between the oligosaccharide chains and the C_{18} on the stationary phase. Early attempts to exploit this effect in chromatography of the malto-oligosaccharides [64,65] were not very successful, owing to the production of very broad or double peaks for each member of the series, due to anomer separation. The broad, flattened peaks became undetectable at DP≥8. Subsequently, various possible methods of sharpening the peaks have been investigated with a view to extending this upper limit. Both retention times and resolution of anomers decrease with increasing temperature [66,67]: Vrátný et al. [66] have demonstrated that the capacity factors for malto- and cellodextrins on stainless-steel columns (250 x 6 mm ID) packed with 10-μm C_{18}-silica decrease nonlinearly over the temperature range 20-80°C, differences between anomers disappearing at 60°C for the maltodextrins, 80°C for the cellodextrins. The maltodextrins could be resolved to DP 10 at 60°C, but the lower members of the series were not well separated. Equilibration between α- and β-anomers is also accelerated in the presence of base. Verhaar et al. [68] found that addition of triethylamine (1 mM) to the eluent in HPLC of the maltodextrins on a column (150 x 4.6 mm ID) of 5-μm C_{18}-silica at ambient temperature improved resolution to DP 6 by elimination of doublets, but there was little effect at higher DP, and the improvement was not sufficient to warrant possible damage to the column by base-catalyzed degradation of the silica. Porsch [69] advocates achieving basic catalysis of anomeric equilibration of the oligosaccharides by use of a packing in which the surface of the C_{18}-silica has been doped with primary amino groups: with such a packing, containing 0.25 mmol/g of primary amino groups, in a high-pressure glass column (150 x 3.2 mm ID) good resolution of glucose and the maltodextrins to DP 8 was achieved within 15 min. The disadvantage of possible glycosylamine formation between the oligosaccharides and the amino groups arises, although it is claimed that this occurs to a much smaller extent on this mixed amino- and C_{18}-silica.

The best resolution of oligosaccharides on C_{18}-silica has been achieved by Cheetham and Teng [70], who used a Dextropak cartridge in which the C_{18}-silica packing had been coated with a nonionic detergent, Triton X-100. This gave good resolution of the members of the malto and isomalto series over the DP ranges 3-15 and 3-8, respectively, each being eluted as a single peak. The effect of the detergent was clearly to reduce the hydrophobicity of the silica surface. Similar improvement through modification of hydrophobic interaction was obtained by addition of tetramethylurea (0.025%) to the eluent. This gave single, well-resolved peaks for oligomers of DP 3-13 in the malto series, 2-7 in the isomalto series.

Cyclic oligosaccharides, which contain no reducing sugars and therefore no anomers, are easily resolved by this method. Koizumi et al. [32] have fractionated the cyclosophoraoses of DP 17-33 on a column (250 × 4 mm ID) of 5-μm C_{18}-silica, eluted with water containing 5.5% methanol; however, the resolution obtained was inferior to that given by normal-phase partition chromatography on a 3-μm NH_2-silica (Section 16.2.1.3). In contrast, the Koizumi group consider C_{18}-silica columns superior to amino-phase columns for analyses of cyclodextrins and branched cyclodextrins, having obtained satisfactory resolution of these oligosaccharides with several C_{18} columns, eluted with water, containing 3-7% methanol [71].

Application of the method to the complex oligosaccharides of human milk and those obtained in degradative studies of glycoproteins has met with limited success. The milk oligosaccharides lacto-N-tetraose and -neotetraose and their fucosylated derivatives are poorly resolved on C_{18}-silica by water, owing to the production of broad or double peaks, which overlap [72,73]. However, for preparative purposes, the technique is useful as a fractionation step prior to purification by methods such as paper chromatography or steric-exclusion chromatography. Good separations are obtained only in the absence of anomeric centers as, for example, in the resolution of glycopeptides with the oligosaccharide chains still linked to asparagine [74]. Reduced oligosaccharides, produced by treatment of glycoproteins with alkaline borohydride, also give single peaks in HPLC on C_{18}-silica but, as has been mentioned, resolution is inferior to that possible with NH_2-silica [35].

An interesting recent development is the use by the Koizumi group of a new column packing (Asahipak ODP-50; Asahi Kasei), in which the C_{18} chains are bonded not to silica but to a vinyl alcohol copolymer gel (5-μm particle diameter). Although this column (150 × 6 mm ID) was less successful than the silica-based packings in HPLC of branched cyclodextrins [71], it gave vastly superior resolution of linear D-gluco-oligosaccharides [75], resolving the oligomers of the malto series to DP 23, the β-(1 → 2)- linked sophoro series to DP 12, and both α- and β-(1 → 6)-linked series to DP 11. Elution by water produced double peaks, but this could be prevented by elution with aq. NaOH (pH 11), which is not usable with silica-based packings. This is a major advantage of the polymer-based packing, which is likely to be much exploited in future work.

References on p. B284

Yet another mode of chromatography in which C_{18}-silica columns have been used to fractionate carbohydrates is ion-pair chromatography. As an alternative to ion-exchange chromatography (Section 16.2.3) ionic carbohydrates can be separated on C_{18}-silica columns with eluents containing an ion-pairing reagent. For anionic solutes, the tetrabutyl-ammonium cation is particularly effective in conferring sufficient hydrophobic character on the ion-pair complex to permit resolution on a reversed-phase column. For example, the normal and unsaturated oligogalacturonic acids produced by degradation of pectic acid with endo-polygalacturonase and endo-pectic acid lyase, respectively, have been re-solved (the normal series to DP 4, the unsaturated acids to DP 7) on a column (250 x 4.6 mm ID) of 10-μm C_{18}-silica by elution at 40°C with eluents containing 0.05 M phosphate buffer (pH 7.0), methanol (10% for normal, 30% for unsaturated acids), and 25 mM tetrabutylammonium bromide [76]. The technique has also been successfully applied to the determination of hyaluronic acid in biological tissues and fluids, by quantitative HPLC of the unsaturated tetra- and hexasaccharide produced on degradation of the hyaluronic acid with *Streptomyces* hyaluronidase. The oligosaccharides were well resolved on a column (250 x 4.6 mm ID) of 5-μm C_{18}-silica by use of an acetonitrile gradient (20 → 22%) in 8 mM H_3PO_4, containing 10 mM tetrabutylammonium hydroxide [77]. The first reported separation of glucose- and fructose-6-phosphate was achieved by ion-pair reversed-phase chromatography on a similar column, eluted at 38°C with 0.02 M formic acid, containing 8-9 mM tetrabutylammonium hydroxide [78].

16.2.1.5 Combination of normal- and reversed-phase HPLC: two-dimensional mapping of oligosaccharides

It should be evident from the preceding sections that normal-phase HPLC and chro-matography on C_{18}-silica columns are complementary, especially when applied to oligosaccharides. This has recently been exploited by two Japanese research groups, who have developed a technique to facilitate identification of glycoprotein-derived oligo-saccharides by reference to a two-dimensional "map", based upon the retention times of standard oligosaccharides on chromatography in each of these two modes [79,80]. In both cases, the oligosaccharides, produced by sequential digestion of glycoproteins with pepsin and *N*-oligosaccharide glycopeptidase [79] or hydrazinolysis followed by *N*-acetylation [80], were derivatized by reductive amination with 2-aminopyridine, allowing fluorimetric detection down to 10 pmol (Section 16.2.1.4). Tomiya et al. [79] used an amide column [31] (cf. Section 16.2.1.3), Hase et al. [80] the amino-silica column Micro-pak AX-5 (Varian) to fractionate the oligosaccharides mainly according to size, whereas several structural factors governed their retention times on the C_{18}-silica columns. The retention times of 113 oligosaccharides, representing the different types (*N*-acetyllactos-amine, oligomannoside, hybrid and xylose-containing) isolated from glycoproteins, were compared by Tomiya et al. [79] with those of the α-(1 → 6)-linked isomalto series, similarly derivatized, on the two columns and plotted as equivalent numbers of glucose units. Those of the 45 oligosaccharides examined by Hase et al. [80] were related to the number of D-mannosyl residues present. These seminal papers have provided an extensive database for identification of unknown oligosaccharides from HPLC data. Recently, Arbatsky et al. [81] have reported studies of the structure and heterogeneity of the carbohydrate chains of several glycoproteins by use of similar oligosaccharide maps. The reduced oligosaccharides, released by treatment of the glycoproteins with alkaline LiBH$_4$ in aqueous t-butanol, were desialylated and fractionated by SEC prior to HPLC on amino- and reversed-phase columns. In this case, the retention times of the underivatized oligosaccharides were plotted on the two-dimensional maps.

16.2.1.6 Novel silica-based HPLC packings

Silica gel, doped with Cu(II), has been used by Guyon and coworkers [82,83] in HPLC of simple mixtures of mono- and disaccharides, by a mechanism probably involving ligand exchange. Ammonia is added to the acetonitrile/water mixtures used as eluents in this system.

In attempts to find a substitute for amino-bonded silica, and thus obviate losses of sugars due to glycosylamine formation, normal-phase partition chromatography on LiChrosorb DIOL (Merck), a diol-modified silica gel, was tested by Brons and Olieman [84]. This sorbent behaved similarly to amino-bonded silica in HPLC of monosaccharides, simple oligosaccharides, and alditols, but a high proportion of acetonitrile (85%) was required in the eluent. A packing in which polyol is bonded to silica gel (Polyol-RSiL; Alltech) is therefore considered preferable [8], since it is capable of similar resolution with eluents containing only 68-70% of acetonitrile [85]. With both diol- and polyol-bonded silica, addition of amine (0.1% of diisopropylethylamine [84] or triethylamine [85]) to the eluent is necessary to avoid peak-broadening due to resolution of anomers, and this must be detrimental to the stability of the silica.

Far more promising results have been achieved in a recent evaluation of new packings in which α- or β-cyclodextrin is bonded to 5-μm silica (Cyclobond; Advanced Separation Technologies). Armstrong and Jin [86] have reported retention times of diverse carbohydrates, including monosaccharides from triose to heptose, alditols, di-, tri-, and tetrasaccharides, and even the cyclodextrins, on columns (250 x 4.6 mm ID) containing these packings, with eluents consisting of 80-85% aq. acetonitrile or 85-90% acetone. Some remarkable separations of complex mixtures, by isocratic or gradient elution, have been demonstrated, indicating that the bonded-cyclodextrin columns are highly efficient and selective. Capacity factors for some important sugars and alditols on these columns are compared in Table 16.1 with those found for other silica-based HPLC packings. Analysis times are generally comparable with those on C$_{18}$-silica and less than those on amino and ion-exchange phases. The stability and reproducibility of the new packings are further advantages.

TABLE 16.1

CAPACITY FACTORS FOR HPLC OF SUGARS AND ALDITOLS ON SILICA AND SILICA-BASED PACKINGS

Compound	Capacity factor (k')				
	Derivatized		Underivatized		
	Unmodified silica [17]	C$_{18}$-silica [40]	Amine-modified silica [22]	Amine-bonded silica [26]	Cyclodextrin-bonded silica [86]
L-Rhamnose	1.44	2.73	1.21	0.93	0.87
L-Fucose	0.93	2.54	1.33	1.40	1.42
D-Ribose	0.75	2.31	1.08	1.00	1.01
D-Xylose	1.80	2.00	1.39	1.36	1.20
L-Arabinose	1.62	2.04	1.77	1.65	1.35
D-Fructose	–	2.03; 2.84[*]	1.80	2.02	1.50
D-Mannose	3.87	1.73	2.04	2.34	1.54
D-Glucose	3.73	1.46	2.22	2.64	1.74
D-Galactose	3.80	1.42	2.62	2.96	1.83
Sucrose	–	–	3.39	4.96	2.40
Cellobiose	6.27	1.24	4.92	6.49	2.64
Maltose	5.76	1.22	4.35	6.53	2.70
Lactose	–	1.13	5.09	7.54	2.97
D-Glucitol	–	–	2.16	2.78	1.86
D-Mannitol	–	–	2.15	2.82	1.88
Galactitol	–	–	2.24	2.89	2.01

* Double peak.

16.2.2 HPLC by ion-moderated partitioning on microparticulate resins

16.2.2.1 Partitioning with aqueous eluents

The use of microparticulate cation-exchange resins in the calcium form, with water as eluent, in HPLC of carbohydrates, particularly alditols and the lower members (to DP 6-8) of homologous series of D-gluco-oligosaccharides, was just becoming recognized as a standard technique when the previous review in this textbook [1, p. 233] was compiled. During the past decade, the special capabilities of resins with certain other counterions to resolve particular classes of carbohydrates have been exploited and such resins are now commercially available in prepacked columns (e.g., the 300 x 7.8-mm-ID columns of the Aminex HPX series, Bio-Rad). The separation mechanism, depending upon the counterion associated with the resin, the degree of crosslinking, and the type of carbohydrate involved, can be ion or size exclusion, ligand exchange, or even hydrophobic interaction with the polystyrene matrix, and therefore this group of methods is best described by the generic term "ion-moderated partitioning" (IMP).

In HPLC involving IMP, columns are usually operated at elevated temperatures (45-85°C) to avoid the broad peaks or doublets that result from anomer resolution at room temperature and to accelerate the slow, diffusion-controlled partitioning process. Water is the eluent, except with resins in the H^+ form, which are eluted with dilute acid solutions. Oligosaccharides are eluted in the order of decreasing molecular weight, the size-exclusion mechanism being predominant here [87].

Resin in the Ca^{2+} form (8% crosslinked) remains the best choice for analysis of alditols [88], owing to the marked differences in strength of complex formation with Ca(II) that arise from differences in stereochemistry among the alditol series. Recently, Sasaki et al. [89] have also achieved the resolution of all eight inositol isomers on such a column (Aminex HPX-87C; 9-μm particles). Chromatography at 85°C with a flowrate of 1 ml/min separated all except *myo*- and *chiro*-inositol, the strongly retarded *cis*-inositol being eluted after 95 min. Further chromatography at 24°C, with a flowrate of 0.65 ml/min, resolved *myo*- and *chiro*-inositol within 14 min. The technique was applied to the analysis of the products resulting from epimerization when *myo*-inositol was heated in solution, with Raney nickel as catalyst, during $^1H/^2H$ exchange in deuterium oxide.

Monosaccharides and small oligosaccharides (to tetrasaccharide level) can be analyzed by HPLC on an 8%-crosslinked Ca^{2+}-form resin, although resolution of some members of the same class (e.g., the pairs maltose/cellobiose, galactose/mannose) is poor. Baust et al. [22], in comparing the performance of this system with that of amine-modified silica (Section 16.2.1.2), have listed the capacity factors of a large number of low-molecular-weight carbohydrates on the Waters Sugar-Pak I column at 90°C. From their data it is evident which separations are feasible by this HPLC method. These authors considered the use of amine-modified silica preferable with respect to column life, capacity and cost, but greater sensitivity is possible when the Ca^{2+} resin is used, especially with UV detection at 190 nm [88], which is possible with the aqueous eluent.

If separation of anomers is desired, this can be achieved by HPLC on 8%-crosslinked Ca^{2+} resin at low temperatures. For example, Baker and Himmel [90] have reported complete separation of the anomers of arabinose, xylose, galactose, mannose, and glucose and their methyl glycosides, and resolution of those of maltose and cellobiose, on the Aminex HPX-87C column at 1.5°C. This offers a simple method for studies of anomeric equilibria as, e.g., in investigations of the action of enzymes on glycosidic bonds. Resin in the Pb^{2+} form was less effective in resolving anomers, and this was ascribed to more rapid mutarotation in the presence of this cation.

The Pb^{2+} form of the resin has proved to be more effective than the Ca^{2+} resin in resolving similar mono- and disaccharides, such as the pairs mentioned above. Pre-packed columns, containing the Pb^{2+} resin (e.g., Aminex HPX-87P; 9 μm, 8% crosslinked), eluted with water at 65-85°C, have found wide application in HPLC analyses of mixtures of pentoses, hexoses, and deoxyhexoses, produced on hydrolysis of plant cell-wall and mucilage polysaccharides [91], plant glycosides [92], and proteoglycans [93]. The products of enzymic degradation of polysaccharides, such as arabinoxylans and -galactans, including not only the monomers but also oligomers of DP 2-4, have been analyzed by this method [94].

Microparticulate resin in the H^+ form (such as Aminex HPX-87H; 9 μm, 8% cross-linked) gives poor resolution of neutral sugars [95], but it is useful in analyses of carbohydrate acids and their lactones. For example, Grün and Loewus [96] developed a method for the determination of L-ascorbic acid in algal cells and growth media, in the presence of galacturonic acid, glucuronic acid and its 6,3-lactone, aldonolactones, and D-erythorbic acid (the C-5 diastereomer of L-ascorbic acid), by HPLC on Aminex HPX-87H, eluted with 0.1 M formic acid at 30°C at a flowrate of 0.7 ml/min. The use of an electrochemical detector distinguished L-ascorbic and D-erythorbic acid from other solutes, which gave no response with this detector. Assay of L-ascorbic acid at nanogram levels is possible by this technique. Hicks et al. [97] have measured the retention times of a variety of aldonic and alduronic acids and their lactones (also including L-ascorbic and D-erythorbic acids) on Aminex HPX-87H, eluted with 4.5 mM sulfuric acid over the temperature range 25-65°C, from which the optimal temperature for each specific separation is apparent. The method affords rapid direct analysis of carbohydrate acids and lactones, and this is useful in studying equilibria between them and in monitoring lactonization or saponification [98].

For HPLC of oligosaccharides of DP > 4, resins of lower crosslinking and therefore larger pores are required. The 4%-crosslinked resin in the Ca^{2+} form first used for this purpose [1,87], which is available commercially as Aminex HPX-42C (25-μm particles) was shown by Schmidt et al. [99] to be capable of resolving the D-gluco-oligosaccharides of the malto and cello series to DP 8 and 7, respectively, at 85°C when three columns (each 300 x 7.5 mm ID) were connected in series. Under the same conditions, the β-(1 → 4)-linked D-xylo-oligosaccharides, obtained on hydrolysis of xylan, were resolved to DP 9, as were the borohydride-reduced oligosaccharide-alditols from the maltodextrin series to DP 10. Brunt [100], using a single column (250 x 9 mm ID) at 90°C, resolved the maltodextrin series to DP 6 and the cyclodextrins, which were eluted in the order $\alpha < \gamma < \beta$, as reported earlier by Hokse [101]. The unexpected early elution of the cyclomaltohexaose before the 7- and 8-membered rings shows that size exclusion is not the only factor operating in this case; specific interaction of the cyclodextrins with the polystyrene matrix of the resin may be involved.

Brunt [100] considered it necessary to add the calcium salt of EDTA (calcium-Titriplex; Merck) in small proportion (50 ppm) to the eluent to protect the Ca^{2+} resin from contamination by traces of metal ion, leached from the stainless-steel column by the hot water. Resin in the Ag^+ form is more stable, and the 25-μm, 4%-crosslinked Ag^+-form resin Aminex HPX-42A has proved very successful in HPLC of higher oligosaccharides to DP 12-14 [9]. With a single column (300 x 7.8 mm ID) at 85°C the cellodextrins have been resolved to DP 8 by Bonn et al. [102], while Hicks and Sondey [103] have separated the maltodextrins to DP 11 on a preparative scale (30 x 2-cm-ID column of Bio-Rad AG,

50W-X4, 20-30 μm, Ag$^+$ form). A two-column combination [104] can resolve the malto-dextrins to DP 14.

Recently, a new 2%-crosslinked resin in the H$^+$ form (Aminex HPX-22H; particle diameter 20-25 μm) has been applied in chromatography of oligosaccharides. Provided that flowrates, and thence backpressures, are low (not more than 0.35 ml/min at 60°C, 0.5 ml/min at 85°C) [105], the softer resin can be used in HPLC. Hicks and Hotchkiss [105] and Derler et al. [106], eluting with 5 mM sulfuric acid at 76-88°C, resolved the maltodextrins to DP 12-14 and the cellodextrins to DP 9 on a single column (300 mm long, 7.8-10 mm ID) of HPX-22H. This H$^+$-form resin has the advantage over those in other ionic forms that it can be used in analyses of ionic oligosaccharides. Thus, oligogalacturonic acids have been resolved to DP 10 and chito-oligosaccharides (the β-(1 → 4)-linked oligomers of 2-acetamido-2-deoxy-D-glucose, obtained on hydrolysis of chitin) to DP 8 at 85°C, though the latter are poorly separated, possibly due to heterogeneity arising form partial de-N-acetylation under these conditions [105]. Lower column temperatures are obviously necessary for chromatography of acid-labile oligosaccharides on this column, which can be used in direct analysis, without work-up procedures, of acid hydrolyzates of polysaccharides.

The properties of resins in different ionic forms can sometimes be combined with advantage. A mixed Ag$^+$- and Pb^{2+}-form resin, obtained by treatment of a 6%-cross-linked, 11-μm Ag$^+$-form resin (Aminex HPX-65A) with a solution containing both AgNO$_3$ and Pb(NO$_3$)$_2$ (mol Pb/mol Ag 1.5%; total concentration 0.1 M) was used by Van Riel and Olieman [107] to optimize resolution of mono-, di-, and trisaccharides by chromatography at 65-85°C. Under these conditions, oligosaccharides such as sucrose, raffinose, melezi-tose, and stachyose decompose on the untreated Ag$^+$ resin, due to acid-catalyzed hydrolysis, resulting from residual sulfonic acid groups in the H$^+$ form. The introduction of Pb^{2+} ions converts these to the Pb^{2+} form, and the resulting column has the selectivity of the Pb^{2+} resin for monosaccharides, combined with the resolving power of a Ag$^+$ resin for oligosaccharides. Further selectivity is achieved by coupling the column of mixed resin to one containing the Pb^{2+} resin Aminex HPX-87P. Such a system is particularly useful in the analysis of the mixtures of sugars occurring in foods. Bonn [108] advocates the use of coupled columns of resins in appropriate ionic forms — Ag$^+$ to resolve oligosaccharides, Pb^{2+} or Ca^{2+} for monosaccharides, and H$^+$ for carbohydrate acids and degradation products, such as furfural and alcohols — to optimize resolution of these components at various stages of the degradation of biomass to ethanol.

16.2.2.2 Partitioning with aqueous organic solvent mixtures

The classical technique of partition chromatography of sugars and alditols on ion-ex-change resins (cation or anion exchangers) with eluents consisting of mixtures of water with a polar organic liquid, usually ethanol [1, p. 234], has also been adapted to HPLC. The sugar constituents of glycoproteins, including the N-acetylated amino sugars, have been analyzed on a 10-μm, highly crosslinked resin, Shodex DC-613 (Showa Denko) in the H$^+$ form, eluted at 30°C with 92% aq. acetonitrile [109]. The neutral components can be separated on Aminex A-6 (15-20 μm, 8% crosslinked), converted to the trimethylam-monium form, by elution at 60°C with 89% aq. ethanol [110]. With detection by sensitive postcolumn derivatization methods both of these techniques allow analysis at nanomolar levels. Glycoprotein-derived sugars can also be analyzed as the derived alditols and acetamidodeoxyalditols, obtained by reduction of the hydrolyzate with $Na[^3B]H_4$, by HPLC on a Pb^{2+}-form resin (Shodex Sugar SP-1010; Showa Denko) at 80°C, with 20% aq. ethanol as the eluent [111]. Analysis of 30-50 μg of glycoproteins was possible by this method with detection by scintillation counting of the labeled alditols. Honda et al. [112] have reported good resolution of anomers of L-rhamnose, L-fucose, and all pentoses and hexoses by chromatography on Shodex DC-613 in the Na^+ or Ca^{2+} forms, eluted with 80% aq. acetonitrile at 4°C. The same resin, in the Na^+ form, has been used in HPLC analysis of the unsaturated disaccharides obtained on digestion of glycosamino-glycuronans with chondroitinase ABC or AC. Using two coupled columns, each 150 x 6 mm ID, eluted at 70°C with a mobile phase containing acetonitrile, methanol, and 0.8 M ammonium formate (pH 4.5) in the ratio 13:3:4, Murata and Yokoyama [113] were able to resolve this mixture of unsaturated oligosaccharides within 20 min.

A macroporous anion-exchange resin, Hitachi 3013 N, in the phosphate form, eluted with 80-83% aq. acetonitrile at 60°C, has been used in HPLC analysis of sugars in urine [114]; and a macroreticular resin (Mitsubishi CDR-10; 5-7 μm), converted to the sulfate form, has proved effective in separating anomers of some mono- and disaccharides when eluted at 20°C with 80% aq. ethanol [115]. More recently, the application of a series of vinylpyridinium polymers, in phosphate and sulfate forms, as column materials for HPLC of sugars, eluted with 80% aq. acetonitrile, has been investigated [116].

It is vital to prevent contamination of any of the ion-exchange resins used in HPLC of the IMP type with inorganic salts, and a guard column, containing resin similar to that in the analysis column, or a mixed-bed resin, is essential. If this precaution is taken [9] and undue pressure on the resin is avoided [8], the lifetime of these columns is satisfactory.

16.2.3 Ion-exchange chromatography

In adaptations of ion-exchange chromatography of carbohydrates to HPLC both silica-based packings with bonded anion exchangers and highly efficient, microparticulate ion-exchange resins have been used. The silica-based anion exchangers, which bear quaternary ammonium (strongly basic) or primary amino (weakly basic) groups, tend to deteriorate rapidly, even when guard columns are used, and therefore their lifetime is much shorter than that of the resins. The latter are relatively stable if protected by suitable guard columns of similar resin and if excess pressure and high temperatures are avoided. However, it is often necessary to operate resin columns at elevated temperature to maximize efficiency, because of the slow diffusion processes involved [9], and this re-duces their lifetime.

16.2.3.1 HPLC on silica-based anion exchangers

The uronic acids are not well resolved on HPLC by strongly basic anion exchangers, bonded to silica, and peaks due to co-existing lactones can cause problems in analysis, unless base (1% of triethylamine [117]) is added to the sample before chromatography. The D-mannuronic and L-guluronic acid present in alginate hydrolyzates can be resolved and determined quantitatively by HPLC on a column (250 x 4.6 mm ID) of Partisil-10 SAX (Whatman), eluted with 0.02 M KH_2PO_4 (pH 4.6), containing 5% of methanol [117]. However, D-glucuronic acid is incompletely separated from D-mannuronic acid under the same conditions. On elution of a similar column with 0.7 M acetic acid at 40°C, better resolution of D-glucuronic and D-mannuronic acid is achieved, but D-galacturonic acid is poorly separated from L-guluronic acid [76]. The order of elution of the acids is reversed with the acid as eluent, indicating a different mechanism (ion exclusion rather than ion exchange).

Voragen et al. [76] resolved the normal oligogalacturonic acids from pectic acid to DP 4, and the unsaturated series to DP 6, by HPLC on silica-based anion exchangers (Nucleosil 10 SB or Zorbax SAX), eluted at 40°C with 0.3 M sodium acetate buffer (pH 5.4). Use of 0.4 M buffer gave resolution of the unsaturated acids to DP 8 within 15 min, whereas ion-pair reversed-phase chromatography achieved resolution to DP 7 in similar time (Section 16.2.1.4). LiChrosorb 10-NH_2, here acting as a weakly basic anion exchanger, resolved the unsaturated oligogalacturonic acids to DP 5 within 15 min by elution with 0.11 M sodium acetate buffer (pH 7.5) at 40°C. Preparative-scale ion-exchange chromatography of oligogalacturonic acids is described in Section 16.2.5.

HPLC on the strongly basic anion exchanger Partisil-10 SAX has been used to resolve the reduced disaccharides produced on nitrous acid deamination of heparin, followed by $NaB[^3H]_4$ reduction [118]; three monosulfated disaccharides were separated by isocratic elution with 40 mM KH_2PO_4 buffer (pH 4.6), a disulfated product being recovered subsequently by gradient elution (40 → 400 mM KH_2PO_4 over 40 min). Separation of the sulfated oligosaccharides (the disaccharide repeating unit and oligomers of this up to the decasaccharide), obtained on hyaluronidase digestion of chondroitin 4- and 6-sulfates, was achieved, after a preliminary fractionation by SEC, and $NaB[^3H]_4$ reduction of the products, by HPLC on Partisil-10 SAX, using stepwise elution with KH_2PO_4 buffers (190 to 400 mM) or gradient elution (250 → 550 mM KH_2PO_4 in 20 min) [119]. Nebinger et al. [120] have reported complete separation of all the saturated oligosaccharides, even- and odd-numbered, derived from hyaluronic acid. In this case, HPLC on a weakly basic anion exchanger, Ultrasil-NH_2 (Beckman), eluted isocratically with 100 mM KH_2PO_4 (pH 4.75), resolved all members of the series, containing 2-8 sugar residues, within 30 min. HPLC on similar weakly basic anion exchangers, eluted with 0.3 M ammonium formate, containing 4% of methanol [121] or 10 mM sodium sulfate containing 1 mM acetic acid [122], gives good resolution of the unsaturated disaccharides, produced on chondroitinase digestion of chondroitin 4- and 6-sulfate and of dermatan sulfate (Section 16.2.2.2). The unsaturated tetra- and hexasaccharides generated by digestion of hyaluronic acid with *Streptomyces* hyaluronidase (Section 16.2.1.4) are also quantitatively resolved, within 12 min, by this method [121]. Recently, a technique has been developed for distinguishing between heparins from various sources by HPLC analysis of the mixtures of unsaturated oligosaccharides produced on digestion with heparin lyase. HPLC on a strong-anion exchanger, bonded to 5-μm silica, with a linear gradient of sodium chloride (0.2 $\rightarrow$ 0.6 M in 50 min) at pH 3.5, gives excellent resolution of the complex mixtures of sulfated di-, tetra-, and hexasaccharides obtained [123].

Neutral (oligomannoside) and sialylated oligosaccharides, (di-, tri-, and tetraantennary and containing 6-18 sugar residues), derived from glycoproteins, have been resolved [124] by HPLC on Micropak AX-10 (isocratic elution with 25 mM KH$_2$PO$_4$, (pH 4.0) for 15 min, followed by a linear gradient of 25 $\rightarrow$ 500 mM KH$_2$PO$_4$ over 30 min). Phosphorylated oligosaccharides can also be fractionated by this method.

16.2.3.2 Ion-exchange chromatography on microparticulate resins

Sialylated oligosaccharides and oligosaccharide alditols, derived from glycoproteins, and those isolated form bovine colostrum, have been resolved by Tsuji et al. [125] by HPLC on a custom-made anion-exchange resin (Hitachi), eluted at 55°C with various concave gradients of sodium chloride in water. For oligosaccharides containing up to 4 sugar residues, isomers differing in sugar sequence and glycosidic linkages could be separated, as well as those of different chainlength or sialic acid content. However, larger, bi- and triantennary oligosaccharides, containing 11 and 13 sugar residues (from porcine thyroglobulin) were poorly resolved. Better resolution of such oligosaccharides is obtained by use of a prepacked column (50 x 5 mm ID) of Mono Q (Pharmacia), a 10-μm strongly basic anion-exchange resin, specially developed for medium-pressure liquid chromatography (Section 16.2.5). Eluting this column with various linear sodium chloride gradients at ambient temperature, and using detection at 214 nm, Van Pelt et al. [126] have analyzed mixtures of mono-, di-, and trisialylated oligosaccharides, containing up to 14 sugar residues, on the microgram scale. Each analysis required only < 10 min.

Aldonic acids, including the 2- and 5-keto and 2,5-diketo acids, have been resolved by anion-exchange chromatography on Aminex resins A-27, A-28, or A-29 (Bio-Rad; average particle diameters, 15, 11, and 5-8 μm, respectively). These strongly basic anion-exchange resins, in the formate or phosphate forms, were eluted at 45 or 60°C with ammonium formate buffers, 0.2 or 0.3 M (pH 3.2 or 3.75) or 0.2 M KH$_2$PO$_4$ (pH 3.35) [127,128]. The order of elution of these acids and their lactones in this chromatographic mode is the reverse of that given by IMP on a cation-exchange resin in the H$^+$ form, which depends upon ion exclusion (Section 16.2.2.1). Aminex A-28 and A-29 (40 x 4.6-mm-ID columns) have also proved effective in analyses of mixtures of N-acetyl- and N-glycolylneuraminic acid, mono-, di-, and triacetylated derivatives, methyl glycosides, and some sialyl oligosaccharides [129,130]. These were well resolved in 15 min at room temperature by elution, at 0.5 ml/min, with 0.75 mM sodium sulfate or, in some cases, 40 mM sodium acetate buffer (pH 5.5). The same method can be used to resolve 3-deoxy-D-manno-octulosonic acid (KDO) and its methyl glycosides, but for KDO disaccharides a higher ionic strength (10 mM sodium sulfate) is necessary to shorten retention times [130].

Chromatographic analysis of aminodeoxyhexoses requires the use of a cation-exchange resin, with sodium citrate buffers as eluents. Adaptation of an amino acid analyzer for this purpose, as described previously [1, p. 238], has remained the method of choice [131,132]. Lohmander [93] achieved good resolution of 2-amino-2-deoxy-D-glucose and -galactose and their alditols on Aminex A-9 (Bio-Rad), a strong-cation-exchange resin (average particle diameter, 11 μm), which was converted to the K$^+$ form and eluted with 88.4 mM KH$_2$PO$_4$ (adjusted to pH 7.00 with KOH), containing 10% acetonitrile and 5% methanol. A rapid and sensitive method for the simultaneous determination of neutral and amino sugars in glycoprotein hydrolyzates involves reductive amination of the neutral sugar components to 1-amino-1-deoxyalditols (glycamines) and analysis of these and the aminodeoxyhexoses with an amino acid analyzer. Elution in this case requires 20 mM sodium tetraborate (pH 8.00) at temperatures varying between 50 and 70°C, depending upon the constituents to be analyzed [133]. Analysis time is 1-2 h. With fluorimetric detection after reaction of the column effluent with o-phthalaldehyde, 0.05-5 nmol of each hexose can be quantitated.

Chromatography of sugars as anionic borate complexes, formerly an important analytical method [1, p. 236], is now seldom used. Honda et al. [134-136], with a specially prepared resin (Hitachi No. 2633, 11-μm particles) at 65°C, reduced the analysis times for sugars and alditols to 65 and 85 min, respectively, by stepwise and gradient elution with 0.25-0.60 M borate buffers (pH 8.2-9.3) of 80 x 8-mm-ID columns at 1 ml/min. The sensitivity was improved by use of fluorimetric detection (postcolumn derivatization by reaction of aldoses with 2-cyanoacetamide, of alditols by periodate oxidation and Hantzsch condensation with pentane-2,4-dione). The method was applied by Shukla et al. [137] to the analysis of N-acetyl- and N-glycolylneuraminic acid and acetylated derivatives, which were eluted within 70 min from a column (65 x 6 mm ID) of the strongly basic anion-exchange resin DA-X8-11 (Durrrum; 11 μm), by isocratic elution at 63°C with 0.8 M borate buffer at pH 8.55 (flowrate, 20 ml/h). Addition of 0.3 M sodium acetate reduced the analysis time to 30 min, but at the expense of resolution. This technique was much inferior in speed and resolution to direct anion-exchange chromatography [129,130]. In this and other applications, anion-exchange chromatography with borate has been superseded by the more efficient HPLC methods now available.

16.2.4 Ion chromatography

The application of the relatively new technique of ion chromatography to carbohydrates was first demonstrated in a seminal paper, published in 1983 by Rocklin and Pohl [138]. Chromatography is performed at high pH ($\geq$12). Under these conditions, most carbohydrates, having pK_a values in the range 12-14, become anionic and can be sorbed by anion exchange on a strongly basic exchanger in the hydroxide form. However, carbohydrates are liable to undergo base-catalyzed reactions – such as isomerization of reducing sugars by the Lowbry de Bruyn/Van Ekenstein transformation or degradation of oligosaccharides by β-elimination — on prolonged exposure to strong alkali. For this reason, and because of the obvious desirability of rapid analysis, packings capable of much faster exchange than the usual exchangers, based on porous resins or silica, are necessary in ion chromatography by this method. High efficiency is also essential, as the carbohydrate anions are very weakly retained, compared to common inorganic anions. Dionex has developed special pellicular packings, with a thin film of strongly basic anion exchanger coated on nonporous latex beads, of 5- or 10-μm particle diameter, for use in ion chromatography of carbohydrates. It was the first of these, designated HPIC-AS6 (10-μm beads), that was introduced by Rocklin and Pohl [138].

Another innovation was the use, in conjunction with the HPIC-AS6 column (250 x 4 mm ID), of a highly sensitive, specific detector, the pulsed amperometric detector (PAD). The carbohydrates eluted from the column are oxidized at the surface of a platinum [139,140] or gold [138,141] working electrode, a selected potential being applied between this and a silver/silver chloride reference electrode; there is also a glassy carbon counter electrode to allow the potentiostat to maintain the selected potential and to prevent excessive current drain on the reference electrode. After a signal from the oxidation of the carbohydrate solutes in the effluent stream has been collected for a short time (100 msec or less), the potential on the working electrode is increased to oxidize fully any material adsorbed on the surface and then it is reversed to a strongly reducing potential to convert the resulting oxide layer back to the noble metal. Although satisfactory results have been obtained with both platinum and gold electrodes [142,143], it is the latter that has become standard in the commercial PAD (Dionex IonChrom system) used in the analysis of carbohydrates. The sensitivity of the detector can be varied over a very wide range, with full-scale deflection at outputs ranging from 1 nA to 100 μA [141].

With the HPIC-AS6 column, eluted at 1 ml/min with 0.15 M sodium hydroxide, a range of alditols, monosaccharides, and simple oligosaccharides, including lactose, sucrose, maltose, raffinose, and stachyose, can be separated within 15 min at 36°C [138], or 25 min at ambient temperature [144]. Capacity factors are reduced in the presence of acetate ions, and this variable is used to improve resolution and to promote elution of higher oligosaccharides. Initially, Rocklin and Pohl [138], eluting isocratically with a solution in which both sodium hydroxide and sodium acetate were 0.2 M, resolved the malto-oligosaccharides to DP 10 within 10 min at 34°C. The potential of gradient elution was demonstrated by excellent resolution of polyols, monosaccharides, and disaccharides on HPIC-AS6 (5-μm particles) with a gradient in which the proportion of a solution containing 50 mM NaOH/1.5 mM acetic acid to pure water was increased from 7% (for 15 min) to 100% (linear gradient for 10 min) [145,146]. Retention data are shown in Table 16.2, in which those from IMP are compared.

In the case just cited, with lower concentrations of sodium hydroxide in the eluent, it was necessary to add 0.3 M NaOH to the eluate emerging from the column to raise the pH to the optimum (13) for the detector. This is required also on elution of aminodeoxy sugars, which is best with eluents containing 10-15 mM NaOH [146]. Use of the dilute barium hydroxide eluents (8 mM $Ba(OH)_2$, containing 0.5 mM acetic acid, or 1 mM $Ba(OH)_2$ with 0.125 mM acetic acid) recommended by Johnson and Polta [142,143] and

TABLE 16.2

RELATIVE RETENTION TIMES FOR ION-MODERATED PARTITIONING AND ION CHRO-
MATOGRAPHY OF SUGARS AND POLYOLS

Compound	Relative retention time[*]		
	Waters Sugar-Pak [22]	Aminex HPX-87P [107]	HPIC-AS6A [145,146]
myo-Inositol	1.17	—	0.14
D-Glucitol	1.26	—	0.21
L-Fucose	1.16	1.25	0.38
L-Rhamnose	1.06	1.19	0.76
L-Arabinose	1.23	1.31	0.69
D-Xylose	1.10	1.09	1.17
D-Galactose	1.08	1.18	0.83
D-Glucose	1.00	1.00	1.00
D-Mannose	1.09	1.35	1.24
D-Fructose	1.11	1.45	1.38
Sucrose	0.79	0.83	0.90
Lactose	0.82	0.91	1.12
Melibiose	0.81	0.93	1.58
Isomaltose	—	—	1.79
Gentiobiose	0.76	0.78	2.13
Cellobiose	0.76	0.82	2.27
Maltose	0.80	0.89	3.10

[*] Relative to retention time for D-glucose.

Edwards et al. [147] for the resolution of monosaccharides also necessitates postcolumn
addition of NaOH. This step, which has the disadvantage that it dilutes the carbohydrates
at the detector, thus adding to baseline noise and drift [145], is unnecessary if the pH of
the eluent is kept constant, and only the concentration of acetate changes during gradient
elution. Such a procedure has proved very effective for the resolution of higher oligosaccharides. With the HPIC-AS6 column, the use of a linear gradient (0 → 600 mM sodium
acetate in 100 mM NaOH over 30 min), followed by isocratic elution at the final concentration for 5 min, at a flowrate of 1 ml/min and ambient temperature, has given excellent
resolution of the oligosaccharides of the maltodextrin series [146]. Detection of oligomers
of DP as high as 43 is possible, if the sensitivity of the PAD is increased 3 times for elution
of those having DP 25-35, 30 times for DP above 35. Ion chromatography is evidently the
most useful method for HPLC of higher oligosaccharides (Table 16.3).

References on p. B284

TABLE 16.3

RESOLUTION OF MALTO-OLIGOSACCHARIDES BY VARIOUS HPLC SYSTEMS

Reproduced from S.C. Churms, *J. Chromatogr.*, 500 (1990) 555.

DP	Relative retention time [*]				
	Silica-based systems		Resin-based systems		
	Amino phase [30]	Reversed phase [70]	HPX-42A [104]	HPX-22H [105]	HPIC-AS6 [146]
1	–	–	1.33	1.23	–
2	0.91	–	1.13	1.11	0.51
3	1.00	1.00	1.00	1.00	1.00
4	1.09	1.16	0.87	0.89	1.34
5	1.18	1.37	0.77	0.82	1.65
6	1.26	1.58	0.71	0.76	1.95
7	1.33	2.00	0.64	0.70	2.25
8	1.39	2.37	0.60	0.65	2.52
9	1.45	2.74	0.56	0.60	2.79
10	1.53	3.26	0.53	0.56	2.99
11	1.62	4.00	0.51	0.53	3.19
12	1.72	5.21	–	0.50	3.39
13	1.82	6.42	–	0.48	3.58
14	1.95	8.37	–	0.46	3.70
15	2.09	10.2	–	–	3.79
20	3.00	–	–	–	4.29
25	4.18	–	–	–	4.73
30	5.90	–	–	–	5.06
35	–	–	–	–	5.29
40	–	–	–	–	5.47
43	–	–	–	–	5.57

* Relative to retention time for maltotriose.

Recently Koizumi et al. [148] have applied ion chromatography on HPIC-AS6 to the analysis of cyclodextrins and branched cyclodextrins, and to cyclosophoraoses of DP 17-25, using isocratic elution with 0.15 M NaOH, containing sodium acetate (0.20 M and 0.14 M for cyclodextrins and cyclosophoraoses, respectively). Hardy et al. [149-151], using the newest Dionex system, comprising a CarboPac PA-1 column (250 x 4.6 mm ID) and a Model PAD 2 detector, have applied the technique extensively to studies of glyco-proteins, the neutral and amino sugars found in hydrolyzates being well resolved on isocratic elution with 22 mM NaOH [149]. With gradient elution, excellent resolution of many of the neutral, sialylated, and phosphorylated glycoprotein-derived oligosaccharides has been achieved [150,151].

Sugar phosphates, which are themselves anionic, have been analyzed by ion chromatography with another system, comprising a Dionex AS4A or AS5 (50 x 4-mm-ID) column, equipped with an anion micromembrane suppressor, and a conductivity detector [152]. In this case, the eluent was 2.4 mM NaHCO$_3$/1.92 mM Na$_2$CO$_3$. The method permitted detection of phosphates in the 20-100 pmol range, and gave good resolution of glucose- and fructose-6-phosphate (Section 16.2.1.4).

16.2.5 Ion-exchange on derivatized gels; medium-pressure liquid chromatography

The application of suitably derivatized dextran and cellulose gels in preparative ion-exchange fractionation of charged oligo- and polysaccharides, reviewed previously [1, p. 239], has remained an important technique in carbohydrate chemistry. During the past decade, several new packings of this type, with the ionic groups attached to more rigid gels or to silica, have been developed for use in automated chromatographic systems. These ion-exchange gels cannot withstand high pressure, but are capable of operating at faster rates of flow than the older gels, and therefore this type of system is often referred to as "medium-pressure liquid chromatography" (MPLC).

In addition to the weakly basic anion exchanger DEAE-Sephadex A-25, the strongly basic QAE-Sephadex A-25 and strongly acidic cation exchanger SP-Sephadex C-25 have found application in the fractionation of low-molecular-weight carbohydrates from plant extracts [153] and in the isolation of specific oligosaccharides believed to be uniquely active in biochemical processes. For example, the saturated oligogalacturonic acids, produced on endo-polygalacturonase digestion of pectic acid, were fractionated (to DP 16) on a column (135 x 2.5 cm ID) of DEAE-Sephadex A-25 by use of a KCl gradient (100 → 300 mM) in 10 mM imidazole solution (pH 7.0) [154]. Oligomers with a 4,5-unsaturated acid at the nonreducing end, produced on digestion of sodium pectate with endo-poly-galacturonic acid lyase, were separated (to DP 13) on high-resolution QAE-Sephadex A-25-120 (45 x 1.7-cm-ID column) with a linear gradient (0.125 → 0.75 M) of imidazole hydrochloride [155]. The isolated oligosaccharides were subsequently tested for specific activity as elicitors of phytoalexins in castor and soybeans.

Fractionation of plant polysaccharides of high molecular weight is exemplified by the use of DEAE-Sepharose CL-6B by Barsett and Paulsen [156] in studies of the mucilage from the inner bark of *Ulmus glabra* Huds. (common elm). Chromatography of a sample (2 g) of the elm extract on a column (60 x 5 cm ID) of this anion exchanger, in the chloride form, eluted first with water then with a sodium chloride gradient (0 → 1 M), gave several polymeric fractions. Subsequent semi-preparative chromatography of a 10-mg sample of the major fraction on the anion-exchange column Mono P (200 x 5 mm ID) in the Fast Protein Liquid Chromatography (FPLC) system (Pharmacia) with the same NaCl gradient in 15 mM phosphate buffer (pH 7) showed the presence of three main components, differing in uronic acid content, and some minor components.

For the fractionation of the even- and odd-numbered oligosaccharides from hyaluronic acid digests (Section 16.2.3.1), Nebinger [157] tested several anion exchangers, including DEAE-Sephacel (Pharmacia) and DEAE-Trisacryl M (LKB), which gave best results when used in the formate forms, eluted with a linear gradient of formic acid (0 → 1 M). The cellulose-based DEAE-Sephacel gave faster flowrates than DEAE-Trisacryl, which swells more in the eluent. Both recovery (99.2%) and resolution of the hyaluronate oligosaccharides (di- to decasaccharide) were excellent with DEAE-Sephacel. Mixtures of undegraded glycosaminoglycuronans (from the aorta) have been fractionated on DEAE-Ultropac TSK 545 (LKB), with a sodium chloride gradient (0.05 → 0.75 M) in 10% aq. methanol [158]. The elution program was in three segments: the first (0.05 → 0.3 M NaCl in 10 min) eluted hyaluronic acid, the second (0.3 → 0.45 M NaCl in 10 min) heparan sulfate, and the third (0.45 → 0.75 M NaCl in 10 min) resulted in the simultaneous elution of chondroitin 4- and 6-sulfates and dermatan sulfate.

There is widespread interest in the application of MPLC to the fractionation of gangliosides, which requires stepwise or gradient elution with ammonium or potassium acetate (usually 0 → 0.5 M) in methanol. Packings used have included DEAE-silica gel [159], Spherosil, coated with DEAE-dextran [160], the resin Mono Q [161], and DEAE-Spheron 1000 [162]. The last-named packing (from Lachema), based on a macroporous, rigid copolymer of the glycol methacrylate type (average particle diameter, 17 μm), is capable of resolving mono- to pentasialogangliosides in only 35 min, with a capacity of 2.4 mg/ml ion exchanger. The paper cited [162] contains a useful comparison of the performance of the various sorbents tested for MPLC of gangliosides. More recently, superior resolution was obtained with a new exchanger, Q-Sepharose (Pharmacia), eluted with a linear gradient (0 → 4 M) of sodium acetate in chloroform/methanol/water (15:30:4), which distinguished some minor components, containing N-glycolylneuraminic acid, as well as C-series polysialogangliosides, among the major gangliosides of bovine brain [163].

16.2.6 Steric-exclusion chromatography

During the past decade, developments in steric (size)-exclusion chromatography (SEC) have concentrated mainly on the adaptation of the technique to HPLC. For this purpose, the packings were generally composed of porous silica, with the adsorption sites deactivated by a chemically bonded, inert (usually polyether or hydroxylic) material, or semi-rigid polymers, which are resistant to high pressure. The properties of many of these have been listed in a review by Barth [164]. The fractionation ranges for SEC of commercially available, characterized dextran fractions (Pharmacia) have been reported for most of the newer packings in that review and subsequent papers dealing specifically with, e.g., the Toyo Soda TSK-gel SW series (silica-based) [165,166], the polymer-based PW series [166-168], and Toyopearl packings [169]. Columns of these types have been widely used in the rapid determination of the average molecular weights and degree of polymolecularity of a variety of polysaccharides, including amylose, amylopectin, and pullulan [170-172], water-soluble derivatives of cellulose [173], guar [174,175], and glycosaminoglycuronans [176-179], as well as the molecular weights of reduced glycopolypeptides [180]. Since the silica-based SEC packings have the disadvantage, already mentioned, of tending to dissolve to some extent in aqueous eluents [181], the polymeric packings are now generally preferred.

The speed of high-performance SEC with these rigid packings is achieved at the expense of the resolution possible at the slower flowrates given by the gels used in the classical, low-pressure SEC, so that distinct peaks, corresponding to various components of polymolecular polysaccharides are seldom obtained. However, the molecular-weight distribution can be determined directly from high-performance SEC, if the carbohydrate emerging from the column is monitored by low-angle laser light scattering, which gives the weight-average molecular weight, $\overline{M}_w$, of each fraction. This method has been success-fully employed, e.g., in analyses of amyloses [171] and guar [175].

For carbohydrates of lower molecular weight, and in SEC of polysaccharides having several, structurally significant components which must be distinguished and isolated, a situation often encountered in degradative studies of polysaccharide structures, conven-tional, low-pressure gel chromatography, with its higher resolving power, remains the method of choice. The tightly crosslinked polyacrylamide gel Bio-Gel P-2 is invaluable in studies of small oligosaccharides. In this case, elution volume is determined not only by molecular size but also by the presence of certain structural features, such as methyl groups or uronic acid residues, both of which cause early elution [182,183]. The position of the glycosidic linkages in various homologous series of oligosaccharides also has a profound effect upon their behavior in SEC on gels such as Bio-Gel P-2 [184,185] or Sephadex G-15 [186].

For resolution of higher oligosaccharides, the upper limit of DP 15, reported in the previous review [1, p. 243] has now been raised considerably by John et al. [187], who used Bio-Gels P-4 and P-6 (>400 mesh), further fractionated by suspension in warm water (ca. 50°C) and decantation of very fine particles, floating at the surface, to give gels having average particle diameters of 40 and 47 μm, respectively. SEC on Bio-Gel P-4 was performed with two water-jacketed columns, each 109 x 2.54 cm ID, connected in series by a zero-deadvolume union; these were operated at 60°C, the degassed water used as eluent flowing at 55 ml/h, upwards through the first column, downwards through the second. By this procedure, it was possible to resolve, not only analytically but also on a preparative scale (200-300 mg), maltodextrins to DP 25 and pullulans (which have α-(1 → 6)-linkages between maltotriose repeating units) to DP 39. The softer Bio-Gel P-6 is less stable to pressure, and therefore this gel was packed in a single column (210 x 1.8 cm ID), eluted in the upward direction only. This effected resolution of the pullulan series up to DP 60. All of these separations were complete in ca. 20 h. Even the cellodextrins, less soluble than the others tested, could be resolved to DP 13 on Bio-Gel P-4 under these conditions [188]. Thus, with this degree of resolution attainable with only water as eluent, low-pressure SEC remains the best method for preparative chromatography of oligosac-charides.

Yamashita et al. [189], using a similar technique, have resolved the oligosaccharides of the α-(1→ 6)-linked isomalto series (from the hydrolysis of dextran) to DP 24. The calibra-tion thus obtained has been applied to interpret the behavior in SEC of diverse oligosaccharides of the type isolated from human milk or released in degradative studies of glycoproteins. Comparison of the elution volumes of these oligosaccharides with those of the D-gluco-oligosaccharides has permitted correlation with structural features, such as the number, position, and glycosidic linkage of residues of L-fucose and 2-acetamido-2-deoxy-D-glucose. This approach to SEC of the complex oligosaccharides derived from glycoproteins was also adopted by Natowicz and Baenziger [190].

Some of the newer gels are more rigid than the polyacrylamide and dextran gels, and therefore higher flowrates are possible, but not at the pressures prevailing in HPLC equipment. However, these gels can be used in MPLC systems, and they probably represent the best compromise between speed and resolution. One example of these gels is Trisacryl GF05 (LKB), a polymer of N-acryloylamino-2-hydroxymethyl-2-propane-1,3-diol, which has proved very useful in preparative fractionation of oligosaccharides, such as those obtained on Smith degradation of polysaccharides [191]. Elution with a volatile buffer is recommended, and 0.1 M pyridinium acetate (pH 5.0) has been found effective. For SEC of polysaccharides, the mixed agarose/polyacrylamide gels of the Ultrogel AcA series (LKB) and the artificially crosslinked agarose gels Sepharose CL-4B and -6B can operate at enhanced flowrates. The resolution of dextran standards on these gels at various flowrates has been determined by Hagel [192]. More recently, highly crosslinked 6% and 12% agarose gels of small average particle diameter (13 and 10 μm, respectively) have become available in 30 x 1-cm-ID prepacked columns, Superose 6 and 12 HR 10/30 (Pharmacia). These columns have given good resolution of dextrans over wide fractionation ranges, at flowrates up to ca. 80 and 140 ml/h, respectively [193]. A preparative grade of Superose 6 (particle diameter 32-36 μm) is also effective in SEC of dextrans [194]. Praznik et al. [195,196] have found the performance of the Superose 6 HR 10/30 column in fractionation of amyloses and pullulans to be superior to that of allyldextran (Sephacryl) gels. A system of coupled columns of Bio-Gel P-6 and Sephacryl S-200, used by Praznik and Beck [197] in SEC of inulins, has now been superseded by the Superose 12 HR 10/30 column [198].

The problem of obtaining suitable calibration standards for the determination of molecular weights of polysaccharide fractions by SEC can be overcome by application of the "universal calibration", in which V_e or K_{av} is plotted against log $[\eta] \cdot M$, where $[\eta]$ is the intrinsic viscosity, and the product $[\eta] \cdot M$ is directly related to the hydrodynamic volume of the polymer molecule in solution. A calibration of this type, obtained with well-characterized dextran standards, has been found to hold satisfactorily for pullulans, amyloses, and amylopectins on TSK gels of the PW series [199,200] and Waters μ-Bondagel columns [201], and for a variety of plant-gum polysaccharides of the arabinogalactan and glucuronomannoglycan types on Sepharose 4B [202]. For glycosaminoglycuronans, charged polymers must be used as standards. For example, sodium polystyrene sulfonate has proved satisfactory in the application of this method to hyaluronic acid [203], on a Shodex OHPak B-806 column (Showa Denko). Linear correlations of SEC elution parameters for polysaccharide fractions with log $[\eta] \cdot \overline{M}_w$, determined by viscometry and light-scattering measurements of each individual fraction, have been reported for other, diverse polysaccharides, including gum arabic [204], carrageenans [205], and highly esterified citrus pectins [206].

16.2.7 Affinity chromatography

The application of affinity chromatography on immobilized lectins to the recognition or isolation of polysaccharides and glycoproteins having specific structural features has assumed great importance in biochemistry, where it has proved invaluable, e.g., in the typing of membrane glycoproteins in cells, such as erythrocytes and lymphocytes [207], and in distinguishing between normal and malignant cells [207-209]. The numerous lectins now known have been classified mainly according to their binding specificity for simple sugars, often with pronounced anomeric specificity [210]. More recent investigations, involving affinity chromatography per se rather than the hapten-inhibition technique originally used, have revealed that, in many instances, binding between a lectin and an oligosaccharide is determined by the presence or absence of a particular sequence of sugar residues in the oligosaccharide chain. For example, the binding of glycopeptides to concanavalin A on agarose has been shown to depend upon the number of α-D-mannopyranosyl residues present as nonreducing end groups or 2-O-substituted chain units, a minimum of two such interacting residues being required for binding to the column [211]. The strength of binding depends upon the residues linked to the 2-O-substituted D-mannosyl residue, and is greatest if there is a terminal 2-acetamido-2-deoxy-D-glucose residue in this position. The binding site on the lectin evidently accommodates the branched structure GlcNAcβ1-2Manα1-6(GlcNAcβ1-2-Man-α1-3)Man.

Similar studies have shown that the presence of unsubstituted Galβ1-4-GlcNAcβ1-6(Galβ1-4GlcNAcβ1-2)Man is a necessary condition for firm binding to the *Datura stramonium* lectin. Oligosaccharides containing unsubstituted Galβ1-4GlcNAcβ1-4(Galβ1-4GlcNAcβ1-2)Man are merely retarded on a column containing this lectin immobilized on agarose [212]. The lectin of elderberry (*Sambucus nigra* L.) bark recognizes the sequence in which *N*-acetylneuraminic acid is linked α-(2 → 6) to D-galactose or 2-acetamido-2-deoxy-D-galactose, rather than the α-(2 → 3)-linked isomer [213]. Lectins that bind galactose or 2-acetamido-2-deoxy-D-galactose residues have been classified by Sueyoshi et al. [214] according to their specificity for the typical mucin-like glycopeptides Galβ1-3GalNAcα1-3-Ser/Thr and GalNAcα1-3Ser/Thr, and by Wu and Sugii [215] according to their binding of various disaccharide determinants, such as the human blood group type-specific disaccharides.

Ligands other than lectins have been successfully used in affinity chromatography of carbohydrates in certain cases. Sutherland [216] fractionated microbial polysaccharides, such as xanthan, on the basis of pyruvate content by chromatography on an affinity matrix, prepared by coupling to CNBr-activated Sepharose 6MB the globulin fraction of antiserum prepared against the pyruvylated polysaccharide of *Rhizobium* strain TA1. The use of antibodies as ligands is currently attracting interest again (see below). Serotonin (5-hydroxytryptamine), immobilized by coupling to Sepharose 4B-CNBr, has proved effective in affinity chromatography of polysaccharides and glycoproteins containing

N-acetylneuraminic acid [217]. This too has been adapted to high-performance chromatography, with the serotonin coupled to LiChrosorb DIOL [218].

With the objective of increasing the speed of affinity chromatography by use of HPLC equipment, some of the lectins that have proved most useful in conventional affinity chromatography have been coupled to microparticulate silica [219,220] or LiChrosorb DIOL [221]. These lectins include concanavalin A [219,221], wheat germ agglutinin [221], and phytohemagglutinin, the lectin from red kidney beans [220]. Use of such sorbents has made possible affinity chromatography of glycoproteins [219,220] and of DNS derivatives of ovalbumin-derived glycopeptides [221] within 20-30 min. An interesting variation is the immobilization of concanavalin A on silica through complexing interaction with Cu(II), chelated with iminodiacetic acid groups coupled to the silica. This results in a column functioning by both affinity and ligand-exchange mechanisms, and thus capable of simultaneous fractionation of both glycoproteins and proteins [222].

A problem encountered in chromatography on packings in which lectins are coupled to silica is that broad peaks are obtained, indicating poor resolution. After a detailed thermodynamic and kinetic study of affinity chromatography on silica-based concanavalin A, Muller and Carr [223-225] concluded that the excessive peak width was due mainly to the slow dissociation of the carbohydrate from its complex with the immobilized lectin. Dissociation constants for glycosides in such complexes are about tenfold smaller than those of the complexes in solution, and those for complexes involving glycoproteins even smaller. The feasibility of using weakly interactive ligands, instead of lectins, for rapid affinity chromatography on silica is therefore under investigation, and some success has been achieved with monoclonal antibodies [226]. In affinity chromatography of the reduced oligosaccharide Glcα1-6Glcα1-4Glcα1-4Glc on a column containing antibody 39.5 (IgG2b) coupled to microparticulate silica, elution with 0.2 M NaCl in 0.02 M sodium phosphate buffer (pH 7.5) gave sharp peaks at temperatures of 37-50°C, and the dissociation constants involved were such that the column efficiency approached that generally associated with HPLC. It seems likely that affinity chromatography of carbohydrates will develop in this direction, especially in applications such as clinical analysis, for which highly specific but also efficient methods are essential.

16.3 SUPERCRITICAL-FLUID CHROMATOGRAPHY

To date, there have been few applications of supercritical-fluid chromatography (SFC) to carbohydrates. As CO_2, the mobile phase most employed in SFC (see Chapter 8) is nonpolar, polar solutes, such as carbohydrates, require derivatization to enhance miscibility, and therefore HPLC by methods not requiring derivatization is generally preferred. However, the use of the flame-ionization detector in SFC gives the technique sensitivity comparable with that of GLC, and the recent interfacing of SFC with MS greatly increases its power as an analytical method, so that its application in chromatography of higher oligosaccharides and glycoconjugates could well be the next area of rapid growth in this field.

Chester and Innis [227] were the first to report the analysis of oligosaccharides by SFC, the malto-oligosaccharides in corn syrup being examined. After trimethylsilylation, they were chromatographed on a fused-silica, microbore capillary column (10 m x 50 μm ID) with bonded methylpolysiloxane as the stationary phase (DB-1; J & W Scientific). Supercritical CO_2 at 89°C with linear pressure program, 115 → 355 atm (3 atm/min) served as the mobile phase. The oligosaccharides were resolved to DP 18 in 75 min, with splitting of each peak due to separation of α- and β-anomers. Pinkston et al. [228] examined the same mixture by SFC, coupled to a 3000-D mass range quadrupole mass spectrometer, in CI mode, with ammonia as the reagent gas. They were able to detect the trimethyl-silylated oligosaccharides only to DP 7 (for which m/z of the $M + NH_4^+$ ion was just below the limit for the instrument). Reinhold et al. [229] used a double-focusing magnetic-sector instrument, operating at 8 kV, also in CI (ammonia) mode, interfaced with a DB-5 micro-bore capillary column operated at 90°C, with the CO_2 pressure programed from 115 to 400 atm at 3 atm/min. They succeeded in matching the FID chromatogram exactly by MS detection. Permethylated maltodextrins were resolved to DP 10 on the same column at 90°C, with CO_2 pressure 100 → 405 atm at 5 atm/min. With selected ion monitoring of the $M + NH_4^+$ ions for individual oligomers, the detection limit of the SFC/MS technique approached 2 pmol.

Some glycosphingolipids have been examined with the DB-5 column operating at 120°C, the CO_2 pressure being held at 110 atm for 5 min and then programed at 5 atm/min to 400 atm. Under these conditions, resolution of the permethylated maltodex-trins up to DP 15 was achieved [230]. With FID detection, multiple peaks were observed for the permethylated glycolipids. This heterogeneity was ascribed, after separate studies by direct-CI and fast-atom bombardment (FAB) MS, to variation in the sphingoid or N-acyl alkane chains, rather than in the carbohydrate moiety. The high resolution and nanogram sensitivity possible by SFC, at relatively low temperatures, should prove invaluable in profiling of glycolipids and glycoproteins.

16.4 GAS/LIQUID CHROMATOGRAPHY

Although gas/liquid chromatography remains an important technique in carbohydrate analysis, fewer reviews dealing with GLC of carbohydrates have been published during the past decade, owing to the rapidly growing interest in HPLC methods. An excellent survey of qualitative and quantitative GLC methods for analysis of mono- and disac-charides, with emphasis on those occurring in biological fluids [231], and an article concerned specifically with GLC/MS of derivatives of amino sugars [232], both appeared in 1980. In 1986 a comprehensive review of chromatographic analysis of mono- and disaccharides, which encompassed both HPLC and GLC methods, was published by Robards and Whitelaw [5], and GLC profiling of complex carbohydrates was covered together with other methods, by Kakehi and Honda [6].

In the present review, a survey of developments since 1980 in GLC by the established methods of derivatization is followed by a critical assessment of several novel derivatiza-tion techniques that have been proposed in recent years. The objective of some of these techniques has been resolution of enantiomers, which is discussed separately, as is the use of MS in conjunction with the various GLC methods.

16.4.1 Developments in established GLC methods

The use of the methyl glycosides in GLC of partially methylated sugars, reviewed previously [1, p. 252], is a valuable technique, especially where the characteristic patterns of multiple peaks can aid in the identification of individual methylated sugars. However, the only recent developments have been in the accumulation of mass spectral data for these derivatives (Section 16.4.4), and therefore this method will not be further discussed here. Improvements in the other GLC methods included in Ref. 1 − those with acetylated alditols or aldononitriles, permethylated oligosaccharide alditols, trimethylsilyl ethers or trifluoroacetate esters as volatile derivatives − have stemmed largely from the introduction of capillary GLC (especially the use of fused-silica columns with bonded phases) or new stationary phases capable of functioning at higher temperatures. Modifications in derivatization procedures have also proved advantageous.

16.4.1.1 Acetylated alditols

Conversion of sugars and methylated sugars to their alditol acetate derivatives for GLC analysis is still a universally accepted method; this applies also to GLC/MS, for which a wealth of mass spectral data for alditol acetates has been accumulated (Section 16.4.4). Recent improvements include changes in the method of derivatization, introduced with a view to greater speed and more quantitative recovery of the products. For rapid conversion of sugars to alditol acetates, Blakeney et al. [233] recommend use of a concentrated (20 mg/ml) solution of sodium borohydride in anhydrous dimethyl sulfoxide and elevated temperature (40°C) to reduce the sugars, dissolved in 1 M ammonia, within 90 min, and addition of 1-methylimidazole, instead of pyridine or sodium acetate, as the catalyst in acetylation with acetic anhydride (complete in 10 min at ambient temperature). Acetylation under these conditions does not demand prior removal of the borate formed on decomposition of excess borohydride by acidification with acetic acid after the reduction step. Thus, the repeated evaporation with methanol, to volatilize borate as methyl borate, that has been accepted as part of the standard derivatization procedure, is rendered unnecessary. In addition to the obvious advantage of speed, elimination of this step prevents selective losses of volatile alditols during evaporation.

For derivatization of methylated sugars Harris et al. [234] advocate reduction of the dry sample with 0.5 M sodium borohydride in 2 M ammonia at 60°C for 1 h (reduction of these sugars being incomplete in dimethyl sulfoxide) and, to permit acetylation without removal of borate, catalysis by acid rather than base, with 70% perchloric acid added to the acetylating mixture of acetic anhydride and ethyl acetate. This reaction, which requires only 5 min at ambient temperature, is followed by addition of 1-methylimidazole (after cooling the reaction mixture in an ice bath) to catalyze conversion of excess acetic anhydride to acetic acid (5 min). Both this new derivatization procedure and that of

Blakeney et al. [233] can be carried out in a single receptacle, which minimizes losses of products.

Further significant improvements in GLC analysis by the alditol acetate method have been achieved by using very efficient capillary columns. A glass capillary (SCOT) column (28.5 m x 0.5 mm ID), coated with the highly polar phase Silar 10C, with the temperature program 190°C (4 min), 190 → 230°C at 4°C/min, 230°C (8 min) gave excellent resolution of the peracetylated alditols [233,235]. Partially methylated alditol acetates were well resolved on a fused-silica column (25 m x 0.22 mm ID) with the polar phase OV-275 bonded (column designated BP-75; SGE), with the temperature program 150 → 250°C at 4°C/min, 250°C (10 min) [236], or a fused-silica column (15 m x 0.24 mm ID) wall-coated with the low-polarity phase SP-2100, with the temperature program 120 → 200°C at 2°C/min [237]. There were some differences in the order of elution of the partially methylated alditol acetates on the two columns of different polarity, which afforded useful evidence to aid in the identification of methylated sugars analyzed on both columns. This multiple-column approach was also recommended by Lomax et al. [238,239], who analyzed partially methylated alditol acetates on three capillary columns differing in polarity: a glass column (20 m x 0.3 mm ID), wall-coated with SP-1000 and operated isothermally at 206°C [238], and two fused-silica columns (50 m x 0.33 mm ID), one wall-coated with CP-Sil 88 (Chrompack) and the other with the low-polarity phase OV-1 bonded (BP-1, SGE), which were operated at 210°C and 195°C, respectively [239]. Cold, on-column injection (at 60°C) was used in all three cases. Using as a database the retention coefficients obtained for a wide range of partially methylated alditol acetates on each of the three columns, these authors programed a computer to process the large body of data produced by GLC of complex samples on these columns, all of the results being merged to assist identification in difficult cases. Confirmation of identification from retention times was obtained from MS in all of these studies, with the BP-75 [236], SP-2100 [237], and SP-1000 [238] columns interfaced with the ion source of a mass spectrometer.

Much information of fundamental importance for GLC and GLC/MS analysis of partially methylated alditol acetates is available from the results of a very comprehensive study by Klok et al. [240]. This group synthesized all possible methyl ethers of the common pentitols, hexitols, and 6-deoxyhexitols, by Haworth methylation of each alditol under carefully controlled conditions, and listed retention times of the acetylated products on GLC on a glass capillary column (25 m x 0.25 mm ID) coated with OV-275, with the temperature program 165 → 215°C at 2°C/min. Both EI and CI mass spectral data were also obtained (Section 16.4.4).

The very long retention times of alditol acetates derived from aminodeoxy- and acetamidodeoxyhexoses [241-243] can be obviated by nitrous acid deamination of the amino sugars [235,244], which yields 2,5-anhydromannose and -talose from 2-amino-2-deoxy-D-glucose and -galactose, respectively, and D-glucose from 2-amino-2-deoxy-D-mannose. An alternative method is selective methylation of the amino groups by reaction of the aminodeoxyalditols with formaldehyde and sodium cyanoborohydride prior to acetylation [245]. This gives alditol acetates eluted between the pentitol and the hexitol acetates in GLC on a packed column (2% EGSS-X on Chromosorb W, 60-80 mesh) at 195°C.

16.4.1.2 Acetylated aldononitriles

The procedure for derivatization of sugars to aldononitrile acetates for GLC analysis is, like the preparation of alditol acetates, improved by the use of 1-methylimidazole instead of pyridine as solvent and catalyst for both the oximation with hydroxylamine hydrochloride and the subsequent simultaneous dehydration to nitrile and acetylation with acetic anhydride. This modification decreases the time required for each of these two steps to 5 min [246], and the reaction is not affected by the presence of traces of water or mineral acid [247].

Resolution of peracetylated aldononitriles in GLC is much improved by use of capillary columns. For example, Guerrant and Moss [248] have reported good resolution of all the common neutral and amino sugars as their peracetylated aldononitriles on a fused-silica column (50 m x 0.2 mm ID), coated with the low-polarity phase OV-1, with the temperature program 175°C (4 min), 175 → 260°C at 4°C/min, 260°C (5 min). The high final temperature was necessary to elute the amino sugar derivatives within 1 h; as in the analysis by the alditol acetate method, deamination of these sugars prior to derivatization is recommended [249].

The earlier work of Seymour et al. on GLC/MS of acetylated aldononitriles, derived from *O*-methyl ethers of D-mannose and D-glucose [1, p. 256], has now been complemented by Tanner and Morrison [250], who have published further data for *O*-acetyl-*O*-methyl-D-glucononitriles on packed columns (5% OV-225 on Chromosorb W, 100-120 mesh, temperature program 160 → 210°C at 1°C/min, and 3% SP-2340 on Supelcoport, 100-120 mesh, temperature program 150 → 225°C at 2°C/min). GLC/MS data for *O*-acetyl-*O*-methyl-D-galactononitriles have been reported by Stortz et al. [251], who also used packed columns (3% ECNSS-M or OV-225 on Gas-Chrom Q, 100-120 mesh, the former isothermally at 180°C, the latter with temperature programs 150 → 245°C at 8°C/min or 185 → 245°C at 4°C/min; in both, the temperature was held at 245°C until elution was complete). Methylation analysis of dextrans by this method is facilitated by the superior resolution of capillary columns. Slodki et al. [252] found that, as in the case of partially methylated alditol acetates, GLC on each of two capillary columns, differing in polarity, gave the best results. Fused-silica columns (25 m x 0.2 mm ID), one coated with the polar phase Carbowax 20M and operated at the temperature program 135°C (5 min), 135 → 165°C at 4°C/min, the other coated with SP-2100 and operated isothermally at 150°C, were used in this work.

The validity of quantitative analysis of sugars by GLC as peracetylated aldononitriles has been questioned by Furneaux [253], after identification of by-products, *N*-hydroxy-D-glycosylamine hexaacetates, mainly in the β-furanose form, in appreciable proportion (33-37%) in addition to the nitriles, when D-glucose and D-galactose were derivatized to peracetylated aldononitriles by the standard method. However, only a trace (3-5%) of by-product was detected by Slodki et al. [252] among the aldononitrile acetates derived from permethylated dextrans, and these authors contend that, since glycofuranosylamine formation should be blocked in methylated sugars, except those derived from residues linked 1 → 4 in the original polysaccharide, this side reaction does not pose a serious problem in methylation analysis of dextrans by the aldononitrile acetate method.

16.4.1.3 Permethylated oligosaccharide alditols

Analysis of oligosaccharides by GLC of their permethylated, reduced derivatives has remained an important technique, especially in degradative structural studies of the carbohydrate portions of glycoproteins. Column packings that are stable at high temperatures are necessary. For example, Dexsil 300, a *meta*-carborane stable at temperatures up to 500°C, was used by Fournet et al. [254] in a GLC packing (1% on Supelcoport, 100-200 mesh) that was capable of resolving permethylated oligosaccharide alditols, derived from products of partial acetolysis of glycopeptides, up to tetrasaccharide level. For lower oligosaccharides the temperature program used was 150 → 320°C at 4°C/min; for tetra-saccharides the starting temperature was 200°C. Sialylated oligosaccharides, viz. the trisaccharides in which *N*-acetylneuraminic acid is linked α-(2 → 3) or α-(2→ 6) to the D-galactosyl residue in a lactose or lactosamine sequence, were examined by Mononen [255] by GLC/MS on SE-30 (2%, in a packed column) at 265°C, after *N*-deacetylation and deamination of the sialic acid residues in the reduced oligosaccharides. Those originally carrying 2-amino-2-deoxy-D-glucose at the reducing end were terminated by 2,5-anhydro-D-mannitol, from previous deamination. For GLC of higher oligosaccharides, containing up to 7 sugar residues, Nilsson and Zopf [256] recommend increasing their volatility by *N*-trifluoroacetylation of 2-acetamido-2-deoxyhexose residues (through transamidation by trifluoroacetolysis under carefully controlled conditions). Using a 10-m fused-silica column, wall-coated with a methyl silicone phase, and the temperature program 200°C (2 min), 200 → 350°C at 4°C/min, these authors have obtained GLC data for the permethylated, *N*-trifluoroacetylated oligosaccharide alditols derived from various glycoconjugates. A further advantage of *N*-trifluoroacetylation is that it allows the use of the ECD, increasing sensitivity more than 100 times (Section 16.4.1.5).

16.4.1.4 Trimethylsilyl ethers

The extensive application of trimethylsilylation, which has long been a well-established technique in GLC of carbohydrates, has been discussed in detail previously [1, p. 257]. The more recent literature has been concerned mainly with improved resolution, resulting from the replacement of packed by capillary columns. A valuable contribution to the study of relationships between molecular structure and GLC behavior has been made recently in a series of papers by García-Raso, Páez, Martínez-Castro, and coworkers [257-259]. These authors have listed retention indices for the TMS ethers of all four of the tautomeric cyclic forms of each of the four aldopentoses and eight aldohexoses, and five forms, including a strongly retained open-chain form, of each of the ketohexoses. Several capillary columns, coated with phases ranging in polarity from low (SE-54), through OV-17, OV-215, and OV-225, to the highly polar Carbowax 20M, were used. GLC was performed isothermally at temperatures in the range 160-180°C. From the large body of data thus

References on p. B284

obtained, mathematical correlations have been found between the retention indices and various molecular features of the sugar derivatives, enabling predictions to be made. By measurement of retention times of the trimethylsilyl ethers of 17 disaccharides at 240°C on a fused-silica capillary column (30 m x 0.26 mm ID), coated with SE-54, Nikolov and Reilly [260] were able to draw some conclusions regarding the effects of glycosidic linkage and constituent sugars on the order of elution of the sugars and of anomeric pairs. The malto-oligosaccharides have been resolved to DP 6 in about 40 min by Traitler et al. [261], by GLC of the trimethylsilylated derivatives on a short (10-m) glass capillary column, coated with OV-1, with cold on-column injection (80°C), followed by the temperature program 80 → 140°C at 5°C/min, 140 → 190°C at 20°C/min, 190 → 340°C at 8°C/min, the column temperature being held isothermal for 1 min after each stage.

Capillary GLC of the trimethylsilylated derivatives of the methyl glycosides produced on methanolysis of the carbohydrate components of glycoproteins is a highly sensitive method, allowing the determination of individual components at levels down to 100 pmol, and analysis of glycoprotein samples of 0.5-10 μg. Chaplin [262], using a fused-silica capillary column (25 m x 0.32 mm ID), coated with CP-Sil 5 (Chrompack), and the temperature program 140°C (2 min), 140 → 260°C at 8°C/min, achieved excellent resolution of the trimethylsilylated methyl glycosides of all the carbohydrate components of glycoproteins. These included the derivatives from 2-acetamido-2-deoxy-D-glucose and -galactose, and N-acetylneuraminic acid, which were well separated from the neutral sugar derivatives but still emerged within 20 min. For GLC/MS, however, Mononen [263] advocates N-deacetylation and deamination of these components, which simplifies the mass spectra. The TMS methyl ester methyl glycosides, derived from D-glucuronic and D-galacturonic acid, are also well resolved and may be analyzed by capillary GLC together with the TMS glycosides from neutral sugars in methanolyzates of acidic polysaccharides, a method that has proved particularly useful in the analysis of gums of industrial importance, such as gums arabic, karaya, and tragacanth [264]. The characteristic pattern of multiple peaks given by each sugar facilitates identification.

The chromatograms are simplified by reduction of sugars to alditols prior to trimethylsilylation, but resolution of TMS pentitols and hexitols remains poor in GLC at 190°C on a 25-m fused-silica capillary column, coated with OV-101, demanding a 50-m column and, consequently, long analysis times [265]. Resolution of TMS deoxyalditols is better, and an interesting result of capillary GLC/MS of human urine by this method (30-m glass capillary, OV-101, temperature program 120 → 260°C at 3°C/min) was the identification of several uncommon deoxyalditols, including 4-deoxyerythritol and -threitol, 5-deoxyxylitol and -arabinitol, and 6-deoxyallitol and -gulitol [266,267]. The cyclitols myo-, chiro-, and scyllo-inositol, also well resolved as their TMS derivatives, were found to be present at increased levels in the urine of uremic patients [266]. The cyclitols, including the mono-O-methyl inositols, often encountered in grain and forage legumes, are easily resolved as their TMS ethers, even on packed columns (3% SE-30 on Gas-Chrom Q, 80-100 mesh, temperature program 130 → 190°C at 2°C/min) [268]. Anhydroalditols can also be analyzed as TMS ethers, by capillary GLC. The method was used by Gerwig et al. [269] to investigate the formation of anhydro derivatives under the conditions normally used in methanolysis of polysaccharides and glycoproteins, a side reaction that affects quantitation in GLC analysis.

Because of the poor resolution of the TMS alditols, capillary GLC of TMS oximes or O-methyloximes is generally used to simplify the chromatograms in GLC analysis of sugars. The products of condensation of formaldehyde, C_2-C_7 carbohydrates, have been analyzed as the TMS oximes [270] on a 60-m SP-2100 glass capillary. Sugars in honey (glucose, fructose, and several di- and trisaccharides) were similarly derivatized and chromatographed on a 25-m fused-silica column, coated with OV-101 (temperature program 180 → 280°C at 3°C/min, 280°C for 4 min, 280 → 290°C at 2°C/min; held at 290°C until elution of trisaccharides) [271]. Good resolution of a variety of sugars (aldoses, ketoses, gluconic and glucuronic acid, and acetamidodeoxyhexoses) as the TMS O-methyloxime derivatives has been achieved by GLC at 180°C with a 50-m fused-silica column, coated with SP-2100 [272]. Most sugars give two peaks, corresponding to geometric isomers, but this does not seriously interfere with their separation and identification by GLC/MS.

16.4.1.5 Trifluoroacetate esters

The use of N-methylbis(trifluoroacetamide) is now generally recommended for trifluoroacetylation of carbohydrates, since this reagent does not affect the GLC column as does trifluoroacetic anhydride. Englmaier [273] has reported satisfactory analysis of D-glucose, D-fructose, sucrose, maltose, raffinose, stachyose, and D-fructo-oligosaccharides (to DP 6), thus derivatized, on a packed column, containing thermally stable Dexsil 410 (3%, on Chromosorb W-HP, 80-100 mesh). On-column injection (100°C) was followed by the temperature program 100 → 310°C at 3.5°C/min for 3.5 min, 6°C/min for 5 min, 15°C/min for 5 min, and 25°C/min for 3.7 min; finally the column temperature was held at 310°C until the hexasaccharide was eluted. Some trifluoroacetylated alditols and cyclitols were also resolved under these conditions. Selosse and Reilly [274], extending an earlier study of GLC of disaccharides as their TMS derivatives [260], have published retention times for TFA derivatives of 13 trisaccharides on a 30-m fused-silica bonded-phase column (DB-5; J & W Scientific), with the temperature program 180°C (5 min), 180 → 200°C at 5°C/min, 200°C to the end. Anomers of reducing oligosaccharides were well resolved, and some conclusions were drawn regarding the effects of structural features on retention time.

The excellent resolution of TFA methyl glycosides in the analysis of methanolyzates by this method, and the high sensitivity made possible by the use of the ECD [1, p. 262] make this GLC technique a powerful analytical method, especially for the sugar constituents of complex polysaccharides. Even with the FID, analyses of bacterial lipopolysaccharide samples as small as 0.1 mg, with components including the dideoxy sugars abequose and tyvelose, heptoses, and KDO, as well as the more common sugars, aminodeoxyhexoses, and N-acetylneuraminic acid, have been achieved by GLC of the trifluoroacetylated methanolyzate [275]. A 25-m fused-silica capillary column, coated with SE-30 (on-column injection at 90°C followed by the temperature program 90°C for 4 min, 90 → 250°C at 8°C/min) was used. A glass capillary, coated with CP-Sil 5, with a similar temperature program, is also effective [276]. Here again, the multiple peaks given by each component can aid identification.

To simplify the chromatograms in analyses of mixtures of many sugars, reduction to the alditols before trifluoroacetylation is suitable, since resolution of the TFA alditols on GLC (especially capillary) columns is satisfactory (cf. Section 16.4.1.4). An advantage of the volatility of these derivatives is that aminodeoxy sugars, thus derivatized, do not have the long retention times found with other derivatives of these sugars and can be analyzed, without deamination, together with neutral sugars. For example, Shinohara [277] resolved the glycoprotein constituents fucose, mannose, galactose, and 2-amino-2-deoxyglucose and -galactose within 10 min on a packed column (2 m x 3 mm ID, 5% OV-101 on Chromosorb W AW DMCS, 60-80 mesh) at 120°C. With a fused-silica capillary column (25 m x 0.25 mm ID) having a cyanopropyl bonded phase (Shimadzu CBP 10-M25-025), operating at 150°C, baseline resolution of the TFA alditols derived from all of the common sugars, from tetroses to hexoses, has been achieved [278].

Decker and Schweer have published chromatograms showing resolution of TFA derivatives of C_3-C_6 alditols on a packed column (4 m x 3 mm ID, 1% OV-225 on Chromosorb W-HP, 80-100 mesh) with the temperature program 100°C (3 min), 100 → 180°C at 5°C/min [279], and of the pentitols and common hexitols on a 50-m glass capillary column, wall-coated with OV-225, with the temperature program 70°C for 2 min, 70 → 180°C at 5°C/min, 180°C for 15 min [280]. The emphasis in these papers was on the use of TFA O-methyl- and O-butyloximes as volatile derivatives for the analysis of sugars by GLC on the OV-225 capillary column with the latter temperature program, which resolved all aldopentoses [279] and the mixtures of 2- and 3-pentuloses and -hexuloses and 2,5-hexodiuloses, produced on bromine oxidation of pentitols and hexitols [280]. Schweer published a series of papers [281-283] reporting GLC and CI-MS data for these and other TFA oximes, including O-methyl-, O-butyl-, and O-2-methyl-2-propyloximes of all aldopentoses [281] and hexoses [282], and the O-butyloximes of the ketoses produced on bromine oxidation of alditols [283], with GLC on the OV-225 capillary column, programed between 120 and 180°C. The method was applied by Decker et al. [284] to the analysis, by GLC and GLC/MS as the TFA O-butyloximes, of the complex mixture of sugars (trioses to hexoses, including pentuloses and hexuloses) formed by autocatalytic condensation of formaldehyde. Owing to the presence of geometric isomers for each oxime, the GLC chromatogram was complicated but most components were resolved by GLC/MS, as in the case of TMS oximes (Section 16.4.1.4). The TFA oximes have the advantage that both preparation and GLC are performed at relatively low temperatures. However, they are far less stable than TMS oximes, requiring the use of all-glass equipment, as they decompose on metal surfaces [279,280].

16.4.2 Novel derivatization methods

In view of the successful use of TMS and TFA oximes as volatile derivatives for GLC and GLC/MS of sugars, there has been some interest in the similar use of acetylated oximes and O-methyloximes. Seymour et al. [285] extended their earlier study of GLC of aldononitrile acetates on packed columns [1, p. 256] by a detailed investigation of GLC/MS of ketoses, pentulose to heptulose, as the derived peracetylated oximes. A packed column (2% OV-17 on Chromosorb W-HP, 80-100 mesh) programed at 130 → 300°C at 20°C/min was used in this work. Mawhinney et al. [286] used various packed columns, containing polar phases, in developing a method for analysis of aminodeoxy sugars as their acetylated O-methyloximes. The neutral sugars, also present in hydroly-zates of glycoproteins, were simultaneously analyzed as aldononitrile acetates, which are eluted much earlier. Mawhinney has based another analytical method, in which the acet-amidodeoxyhexoses are analyzed together with their alditols, on good GLC separation of the O-methyloxime acetates from the alditol acetates [287].

It is in the analysis of amino sugars in hydrolyzates from glycoproteins and bacterial cell wall polysaccharides that this derivatization method has been most used, since the derivatives are well separated from neutral sugar derivatives, at retention times that are not excessively long. For example, Neeser and Schweizer [288] resolved the acetylated O-methyloximes, derived from the common neutral sugars, and those from amino- and acetamidodeoxyhexoses and muramic acid, in a single chromatogram (40 min) on a fused-silica capillary column (25 m x 0.3 mm ID), coated with Carbowax 20M (temperature program 80 → 180°C at 20°C/min, 180 → 210°C at 2°C/min, 210 → 230°C at 10°C/min, isothermal at 230°C until elution of muramic acid). The same method has been used to distinguish the neutral and amino sugars from the alditols likely to be present in hydro-lyzates of carbohydrate chains released from glycoproteins by alkaline borohydride [243]. The capillary column gave resolution much superior to that obtained with packed columns by Mawhinney [287]. Using a 50-m fused-silica capillary column coated with OV-1, under the conditions already described for their separation of aldononitrile acetates (Section 16.4.1.2), Guerrant and Moss [248] resolved the O-methyloxime acetates of the neutral and amino sugars, including muramic acid, KDO, and heptoses that are important con-stituents of bacterial cell wall polysaccharides, in less than 30 min. The double peaks resulting from geometric isomerism do not affect separation of the sugars on an efficient capillary column. Recently, GLC/MS of neutral sugars as acetylated O-pentafluorobenzyl-oximes, on a packed column (3% SP-2340 on Supelcoport) and a capillary column (15-m fused silica, CP-Sil 88) has been described [289], and the method was tested by analyz-ing some glycoproteins of known sugar composition.

Lehrfeld [290,291] developed a method for the GLC determination of aldonic acids, involving lactonization followed by treatment with a 1-alkylamine in pyridine to form the N-alkylaldonamide, which, after acetylation, gives a well-defined single peak on GLC on SP-2340 (3%, on Supelcoport, 100-120 mesh). He then proposed that this technique be used in simultaneous analysis of aldonic acids and aldoses [292]. The latter, in the form of alditol acetates, were eluted much earlier than the N-alkylaldonamide acetates. Of several 1-alkylamines tested [290,291], 1-propylamine was selected as the most suitable for this derivatization. The acetylated N-propylaldonamides derived from ribonic, xylonic, man-nonic, gluconic, and galactonic acids are well separated from one another and from the alditol acetates within 16 min by GLC on the packed SP-2340 column (temperature program 190 → 260°C at 5°C/min) [290,292]. The method can be applied also to alduronic acids [293], if they are first converted to aldonic acids by sodium borohydride reduction of the alduronates, formed by treatment of the uronic acids with 0.5 M sodium carbonate at 30°C for 45 min. This prolonged treatment is essential to hydrolyze any alduronolactone, which would give alditol on borohydride reduction. The inverted aldonic acids are formed. For example, D-glucuronic acid is converted to L-gulonic acid and L-guluronic acid to D-gluconic acid. Lehrfeld [293] improved the resolution of the N-propylaldonamide acetates by GLC on a wide-bore capillary column with the temperature program 200°C (2 min), 200 → 235°C at 3°C/min, 235°C (5 min). Good separations of these derivatives and the alditol acetates were achieved within 20 min. Recently, an adaptation of this method, in which the uronic acids are converted to N-hexylaldonamide acetates and GLC is performed on a fused-silica capillary column (30 m x 0.25 mm ID) having a bonded phase (DB-1701; J & W Scientific), with the temperature program 220 → 270°C at 1°C/min, has been proposed [294].

Derivatization of neutral sugars, uronic acids and the products of deamination of aminodeoxy sugars as TMS diethyl dithioacetals [295,296], which was newly introduced at the time of the previous review [1, p. 263], gives a single, sharp peak for each sugar and permits analysis of μmol quantities on a capillary column (50-m SCOT column, SF-96, 225°C). Pentoses are well separated from hexoses, but resolution within each group is poor. For example, the ribose and xylose derivatives are not resolved, and those from glucose and mannose are separated very little. The method has been applied successfully to the analysis of hydrolyzates of gum arabic and gum tragacanth, which do not contain these pairs but do contain uronic acids, analyzed simultaneously with the neutral sugar constituents by this technique [295]. The TMS diethyl dithioacetals derived from partially methylated sugars [297] give simple, characteristic mass spectra, which can be used in identification of individual methylated sugars in GLC/MS (Section 16.4.4), but resolution of sugars having the same degree of methyl substitution is poor. This derivatization method has been applied in GLC analysis of sialic acids, the mannosamine derivatives released by the action of N-acetylneuraminate lyase being converted to TMS diethyl dithioacetals before GLC on a packed column (2% OV-1 on Chromosorb W AW DMCS, 80-100 mesh) at 190°C [298].

In a novel derivatization method, recently developed by Das Neves et al. [299], sugars are converted to O-methyloximes and the products undergo reductive amination on treatment with sodium cyanoborohydride, forming the deoxy(methoxyamino)alditols. These are resolved by capillary GLC after acetylation or trimethylsilylation, each aldose giving a single peak, each ketose two peaks due to the formation of diastereomers. Resolution of isomeric sugars is better with the acetates (on a glass capillary column, 25 m x 0.25 mm ID, coated with OV-101, temperature program 170°C for 10 min, 170 → 210°C at 2°C/min) than with the TMS ethers (on a glass capillary column, 25 m x 0.18 mm ID, coated with SE-30, temperature program 160°C for 10 min, 160 → 170°C at 0.5°C/min). However, optimal resolution is achieved after permethylation of the deoxy(methoxyamino)alditols [300]. The derivatives from rhamnose, fucose, and all aldopentoses and hexoses are completely separated within 35 min on a glass capillary column (50 m x 0.5 mm ID), coated with Carbowax 20M (temperature program as for acetates on OV-101). With the

FID, the detection limit for each sugar derivative is ca. 4 nmol, but with the NPD samples of 200 pmol can be analyzed. The presence of intense diagnostic ions in the mass spectra (Section 16.4.4) also allows highly sensitive selected-ion monitoring. The use of this method of derivatization has been extended to GLC of reducing disaccharides, which are well resolved as the permethylated deoxy(methylmethoxyamino)alditol glycosides [301] on a glass capillary column (25 m x 0.2 mm ID), coated with Superox 0.1 (Alltech), with the temperature program 240°C (5 min), 240 → 260°C at 1°C/min. Here again, high sensitivity is achieved by use of the NPD or selected-ion monitoring. The excellent resolution and high sensitivity possible with this derivatization method should prove invaluable in analyses of small samples containing complex mixtures of sugars.

16.4.3 Resolution of enantiomers

16.4.3.1 Use of chiral derivatives

Among the newer methods of derivatization used in GLC of sugars, there have been several aimed at resolution of enantiomers by production of chiral derivatives. The successful attainment of this objective by formation of glycosides with chiral alcohols (+)-2-octanol [302] and (−)-2-butanol [303,304], was included in the previous review [1, p. 263]. As always in GLC of glycosides, the production of 3 or 4 peaks by each sugar is a disadvantage if a mixture of several components, including isomeric sugars, is to be analyzed. For this reason, the emphasis in more recent work has been on the formation of chiral derivatives giving only one or two peaks for each sugar. Thus, Little [305] used (+)-1-phenylethanethiol in a mercaptalation reaction that converted pairs of sugar enantiomers into diastereoisomeric, acyclic dithioacetals. These could be resolved as acetylated or TMS derivatives by capillary GLC on SE-30 or SE-54 (fused-silica columns, 30 m x 0.25 mm ID), isothermally at 280°C or with the temperature programs 275 → 300°C at 2°C/min (acetates) or 275 → 290°C at 1°C/min (TMS ethers). In general, the acetylated diastereomers were better separated. The order of elution of the diastereomers depended upon the stereochemistry at C-2 of the parent sugar, those having the S-configuration at C-2 being eluted before those with the R-configuration. For example, the D-arabinose derivative preceded the L-enantiomer, but for galactose the reverse order was observed. By this method the unusual occurrence of galactose as its L-enantiomer in flax seed mucilage was confirmed.

The successful use, by Oshima et al. [18], of HPLC on silica to resolve enantiomers of sugars as the acyclic diastereomers, namely 1-(N-acetyl-α-methylbenzylamino)-1-deoxy-alditol acetates, obtained by reductive amination with chiral L-(−)-α-methylbenzylamine, followed by acetylation, has been mentioned (Section 16.2.1.1). Attempted resolution of enantiomers by GLC of the same derivatives at 250°C on a fused-silica capillary column (50 m x 0.2 mm ID), coated with SE-54, was unsuccessful, as the enantiomers were indistinguishable in most cases [306]. Better results were obtained with the TMS ethers of the methylbenzylamino-deoxyalditols on a 25-m fused-silica column, coated with Carbowax 20M. GLC was performed at 158°C for aldoses, 190°C for acetamidodeoxyhexoses, and 155°C for ketoses, which gave 2 peaks for each enantiomer, due to epimerization at C-2. However, resolution of enantiomeric pairs by GLC was generally much inferior to that obtained by HPLC [18].

Much better resolution of enantiomeric aldoses has recently been reported by Hara et al. [307], who converted the sugars to methyl 2-(polyhydroxyalkyl)-thiazolidine-4(R)-car-boxylates by reaction with L-cysteine methyl ester in pyridine, prior to trimethylsilylation and capillary GLC at 200°C (glass SCOT column, 50 m x 0.3 mm ID, coated with OV-17). A single peak was obtained for each enantiomer, and the pairs were well separated. This method of obtaining chiral derivatives for resolution of enantiomeric sugars by GLC seems the most promising of those reported to date.

Other derivatives tested for this purpose include the trifluoroacetates of (−)-menthyl- or (−)-bornyloximes. They were chromatographed by Schweer [308,309], on a 50-m OV-225 capillary column at 180°C. As noted for other oxime derivatives, the existence of two geometric isomers for each oxime resulted in double peaks, which overlapped in some cases, but for simple mixtures of sugars from different classes enantiomers could be distinguished. For example, analysis of the neutral sugars in hydrolyzates of gum arabic and gum tragacanth as the TFA(−)-bornyloximes [309] confirmed the assignments of D- or L-configuration from other data.

16.4.3.2 Use of chiral stationary phases

An alternative approach to the resolution of enantiomeric sugars by GLC is the use of chiral stationary phases. Leavitt and Sherman [310,311] used a commercially available stationary phase, Chirasil-Val (Applied Science Laboratories; N-t-butyl-L-valinamide in α-amide linkage with an organosiloxane copolymer), coated in glass capillary columns (25 m). They thus resolved the D- and L-enantiomers of fucose, arabinose, mannose, and *chiro*-inositol, derivatized as the per(heptafluorobutyric) esters (temperature program 95°C for 8 min, 95 → 125°C at 3°C/min). However, the enantiomers of glucose were best resolved as the TMS-methaneboronate or bis(methaneboronate) heptafluorobutanoate derivatives, at 130°C and 180°C, respectively. Resolution of the enantiomers of *myo*-inositol 1-phosphate, as penta-O-(TMS)-1-O-dimethyl phosphates, was achieved by GLC at 190°C. Alditols and aldonolactones were not resolved, and it was evident that the potential of this phase in the carbohydrate field was very limited.

König and coworkers [312,313] prepared the first chiral stationary phase to allow separation of enantiomeric pairs of a wide variety of carbohydrates, derivatized as simple trifluoroacetate esters. In this phase L-valine-(S)-α-phenylethylamide was coupled to the methyl(cyanoethyl)silicone phase XE-60 through the carboxylate groups produced on saponification of XE-60 with base. Using a glass capillary column (40 m x 0.2 mm ID) coated with this phase, this research group achieved separations of the enantiomers of fucose, all the aldopentoses, glucose, galactose, and mannose, and their methyl glyco-sides, in several cases in all four possible forms (α- and β-anomers of pyranose and furanose ring forms). GLC was performed isothermally at various temperatures in the range 100-140°C or, for the methyl glycosides, with temperature programing (100 →

145°C or 120 → 180°C at 3°C/min). A similar phase carrying L-valine-(R)-α-phenylethyl-amide, which showed high selectivity for chiral alcohols in general, separated the enantiomers of arabinitol and mannitol when used in a 15-m capillary column at 145°C [314].

Recently, König et al. [315] have produced another, highly successful chiral stationary phase, modifying α-cyclodextrin, in order to lower the melting point and increase thermal stability, by introducing hydrophobic groups. All free hydroxyls were pentylated by reaction of the cyclodextrin with 1-bromopentane and NaOH in dimethyl sulfoxide, followed by further pentylation in the presence of NaH in tetrahydrofuran. This pentylated cyclodextrin has been found to exhibit a high degree of enantioselectivity towards trifluoroacetylated aldoses, methyl glycosides, and alditols, including the 1,4- and 1,5-anhydroalditols, which have assumed importance in studies of polysaccharides by the reductive cleavage method. Baseline resolution of enantiomers is achieved within 5-10 min on glass capillary columns (20 or 40 m), coated with the phase, GLC being performed isothermally at temperatures varying from 80°C to 120°C [316,317]. The thermal stability of the pentylated cyclodextrin is such that no deterioration of column performance has been observed, even after operation at 200°C [316]. Since trifluoroacetylated carbohydrate enantiomers are resolved, but not peralkylated or TMS derivatives, chiral recognition probably depends upon specific dipole/dipole interactions between the TFA groups and the asymmetric centers in the pentylated cyclodextrin, rather than inclusion phenomena involving the cavities in the cyclodextrin matrix [317]. Investigation of these novel phases is continuing, and it is possible that appropriate modification of the molecular geometry of α- and β-cyclodextrin by introduction of alkyl or acyl substituents will eventually produce a range of chiral stationary phases tailored to achieve specific separations.

16.4.4 Use of mass spectrometry combined with GLC

The use of a mass spectrometer in EI mode coupled to the gas chromatograph is now considered essential for the identification of components separated by GLC, especially in the analysis of complex mixtures of closely related compounds, such as the partially methylated sugars in hydrolyzates of methylated polysaccharides. The sensitivity of GLC analysis can be increased considerably if the usual FID is replaced by the mass spectrometer, and the intensities of certain selected, diagnostic ions are used to monitor the emergence from the column of specific components. Deuterium labeling can be advantageous in increasing the specificity of the ions chosen and, hence, the sensitivity of detection. For example, Waeghe et al. [318] have reported methylation analysis of oligosaccharides and polysaccharides using samples of only 1 and 5 μg, respectively, by capillary GLC in conjunction with multiple selected-ion monitoring of the deuterium-labeled, partially methylated alditol acetates. The latter are obtained by sodium borodeuteride reduction of the methylated sugars in the hydrolyzate of the permethylated sample, prior to acetylation. Oligosaccharides are reduced with sodium borodeuteride before methylation, so that the reducing endgroup is specifically labeled. If uronic acid residues are present, the carboxylate groups in the methyl-esterified product of methylation are reduced to 6,6-dideuteriohexosyl residues by treatment with sodium boro-deuteride in a 27:73 mixture of 95% aq. ethanol and oxolane [318], or with lithium aluminium deuteride in oxolane [53]. The diagnostic fragment ions for EI-MS of the deuterium-labeled, partially methylated alditol acetates derived from hexosyl and 6,6-dideuteriohexosyl residues in all possible glycosidic linkages have been listed by Waeghe et al. [318]. Some examples are given in Table 16.4, together with m/z values of the most important ions in the normal mass spectra [319].

CI-MS is a useful complement to EI-MS. The simplicity of the mass spectra obtained by CI, in which the masses $M + 1$, $(M + 1) - 32$, and $(M + 1) - 60$ are predominant [320], facilitates discrimination between partially methylated alditol acetates with different degrees of methyl substitution. Data recorded by Klok et al. [240] from CI-MS (with isobutane reagent gas) of all possible partially methylated alditol acetates derived from pentitols, hexitols, and deoxyhexitols are shown in Table 16.4.

Comprehensive data from EI-MS of partially methylated alditol acetates, derived from aminodeoxyhexoses, are included in the review by Wong et al. [232]. The m/z values of some important ions are given in Table 16.4, with those from the mass spectrum of 2-acetamido-1,3,4,5,6-penta-O-acetyl-2-deoxyhexitol. Reduction with sodium borodeuteride results in an increase of 1 unit in some of these values, and thus deuterium labeling permits distinction between 2-acetamido-2-deoxy-D-galactose occurring in the oligosaccharide chain in glycoproteins (which is present as such in hydrolyzates) and that involved in linkage with serine or threonine, which is released as the alditol on alkaline borohydride treatment of the glycoprotein [321].

In addition to the large body of mass-spectral data on alditol acetates, many of the other derivatives discussed in Sections 16.4.1 and 16.4.2 have been examined by MS with a view to their application in GLC/MS of carbohydrates. Some m/z values of ions important in the identification of these derivatives are listed in Table 16.4.

16.5 PLANAR CHROMATOGRAPHY

16.5.1 Paper chromatography

The techniques used in paper chromatography of carbohydrates have not changed since they were last discussed in this textbook [1, p. 264]. This long-established method has been superseded by the more rapid TLC in many laboratories, but there are separations, notably those of partially methylated sugars in hydrolyzates of methylated polysaccharides, in which the resolution obtainable with PC remains advantageous. Resolution of homologous oligosaccharides by multiple development with 1-butanol/pyridine/water mixtures of increasing water content is much slower than the newer HPLC methods, but still useful as a preparative method: for example, since linear malto- oligosaccharides of DP up to 25 can be separated in this way, the method has proved useful in studies of the average linear chainlength in amylodextrins [330].

Modified papers with bonded ionic groups are used with advantage in specific separations. DEAE-cellulose paper (Whatman DE-81), with the solvent system ethyl

TABLE 16.4

SOME IMPORTANT DIAGNOSTIC IONS IN MASS SPECTRA OF VOLATILE DERIVATIVES
USED IN GLC/MS OF SUGARS AND METHYLATED SUGARS

Reproduced from S.C. Churms, *CRC Handbook of Chromatography: Carbohydrates*,
Vol. III, CRC Press, Boca Raton, FL, 1990, with permission.

Derivative	Diagnostic ions (*m/z*)[*]		Ref.
	EI-MS	CI-MS	
Alditol acetates			
Tetritol-Ac$_4$	73,86,**103**,**115**,128,**145**,217		319
Pentitol-Ac$_5$	73,**85**,**103**,**115**,**127**,**145**,175,**187**,217,289	303,363	240, 319
Hexitol-Ac$_6$	73,**85**,**103**,**115**,**127**,**128**,**139**,**145**,**157**,170, 187,217,259,289,315	375,435	
6-Deoxyhexitol-Ac$_5$	73,86,**99**,**103**,**115**,**128**,145,157,**170**,187, 201,217,231	317,377	
2-Acetamido- 2-deoxyhexitol-Ac$_5$	60,**84**,**102**,114,126,139,**144**,151,156,168, 216,318,360,390		232, 321
From NaBD$_4$ reduction	**85**,**103**,**145**		321
Partially methylated alditol acetates			
2(4)-Me-pentitol-Ac$_4$	**85**,99,103,115,**117**,**127**,139,145,187,217	275,303, 335	240, 319
3-Me-pentitol-Ac$_4$	**87**,**129**,**189**	275,303, 335	
5-Me-pentitol-Ac$_4$	**85**,**87**,**103**,**115**,127,128,**129**,145,187	275,303, 335	
2,3(3,4)-Me$_2$-pentitol- Ac$_3$	71,**87**,99,101,**117**,**129**,189	247,275, 307	
2,4-Me$_2$-pentitol-Ac$_3$	85,**117**,127,159,173,201,233	247,275, 307	
2,5-Me$_2$-pentitol-Ac$_3$	**87**,101,**117**,**129**,161,189,233		**
3,5-Me$_2$-pentitol-Ac$_3$	**87**,101,**129**,161,189	247,275, 307	240, 319

TABLE 16.4 (continued)

Derivative	Diagnostic ions (m/z)[*]		Ref.
	EI-MS	CI-MS	
2,3,4-Me$_3$-pentitol-Ac$_2$	87,**101,117**,161	219,247, 279	
2,3,5-Me$_3$-pentitol-Ac$_2$	71,**87,101,117,129**,161	219,247, 279	
2-Me-hexitol-Ac$_5$	87,**97,117,139**,259,333	347,375, 407	
3(4)-Me-hexitol-Ac$_5$	**85,87**,99,**127,129**,159,**189**,201,261	347,375, 407	
6-Me-hexitol-Ac$_5$	87,97,103,**115,129,139,157**,184	347,375, 407	
2,3-Me$_2$-hexitol-Ac$_4$	**85,87,99,101,117,127**,159,201,261	319,347, 379	
2,4-Me$_2$-hexitol-Ac$_4$	**87,117,129**,159,189,201,233,305	319,347, 379	
2,6-Me$_2$-hexitol-Ac$_4$	**87,117,129**,143,159,185,203,231,305	319,347, 379	
3,4-Me$_2$-hexitol-Ac$_4$	**87,99,129**,189,233	319,347, 379	
3,6-Me$_2$-hexitol-Ac$_4$	**87,99,113,129**,189,233	319,347, 379	
4,6-Me$_2$-hexitol-Ac$_4$	85,87,99,101,127,**129,161**,201,261	319,347, 379	
2,3,4-Me$_3$-hexitol-Ac$_3$	**87,99,101,117,129**,173,189,233	291,319, 351	
2,3,6-Me$_3$-hexitol-Ac$_3$	**87,99,101,113,117**,129,131,161,173,233	291,319, 351	
2,4,6-Me$_3$-hexitol-Ac$_3$	87,99,**101,117,129**,161,201,233	291,319, 351	
3,4,6-Me$_3$-hexitol-Ac$_3$	87,99,101,**129**,145,**161**,189	291,319, 351	
2,3,4,6-Me$_4$-hexitol-Ac$_2$	71,87,**101,117,129,145,161**,205	263,291, 323	
2,3,5,6-Me$_4$-hexitol-Ac$_2$	**89,101,117**,129,161,205	263,291, 323	
2-Me-6-deoxyhexitol-Ac$_4$	57,**87,99,117**,129,141,159,173,201	289,317, 349	
3-Me-6-deoxyhexitol-Ac$_4$	**87,101**,117,**129,143**,159,189,203	289,317, 349	
4-Me-6-deoxyhexitol-Ac$_4$	85,87,**89**,99,127,**131**,159,201,261	289,317, 349	

TABLE 16.4 (continued)

Derivative	Diagnostic ions (m/z)[*]		Ref.
	EI-MS	CI-MS	
2,3-Me$_2$-6-deoxyhexitol-Ac$_3$	**101**,**117**,129,**143**,161,203	261,289, 321	
2,4-Me$_2$-6-deoxyhexitol-Ac$_3$	**89**,**101**,**117**,127,**131**,173,201,233	261,289, 321	
3,4-Me$_2$-6-deoxyhexitol-Ac$_3$	**87**,**89**,99,115,**129**,**131**,189	261,289, 321	
2,3,4-Me$_3$-6-deoxy-hexitol-Ac$_2$	59,71,**89**,**101**,**115**,**117**,**131**,161,175	233,261, 293	
3-O-Me-2-N-Me-hexaminitol-Ac$_4$	74,87,98,116,124,142,158,202,261		232
4-O-Me-2-N-Me-hexaminitol-Ac$_4$	74,87,98,116,129,158,170,189		
6-O-Me-2-N-Me-hexaminitol-Ac$_4$	74,87,98,116,128,142,158,170		
3,4-O-Me-2-N-Me-hexaminitol-Ac$_3$	74,87,98,116,129,142,145,158,161, 189,202		
3,6-O-Me-2-N-Me-hexaminitol-Ac$_3$	74,87,98,116,124,129,142,158,170, 173,202,233		
4,6-O-Me-2-N-Me-hexaminitol-Ac$_3$	74,87,98,116,129,142,158,161,170,230,274		
3,4,6-O-Me-2-N-Me-hexaminitol-Ac$_2$	74,87,98,116,129,142,158,161,202,205		

Deuterium-labelled partially methylated alditol acetates[***]

2-Me-hexitol-Ac$_5$	118,333	318
3-Me-hexitol-Ac$_5$	190,261	
4-Me-hexitol-Ac$_5$	189,262	
6-Me-hexitol-Ac$_5$	129,185	
2,3-Me$_2$-hexitol-Ac$_4$	118,261	
2,4-Me$_2$-hexitol-Ac$_4$	118,189,234,305	
2,6-Me$_2$-hexitol-Ac$_4$	118,305	
3,4-Me$_2$-hexitol-Ac$_4$	189,190	
3,6-Me$_2$-hexitol,Ac$_4$	190,233	
4,6-Me$_2$-hexitol-Ac$_4$	161,262	
2,3,4-Me$_3$-hexitol-Ac$_3$	118,189,233	
2,3,6-Me$_3$-hexitol-Ac$_3$	118,162,233	

References on p. B284

TABLE 16.4 (continued)

Derivative	Diagnostic ions (m/z)[*]		Ref.
	EI-MS	CI-MS	
2,4,6-Me$_3$-hexitol-Ac$_3$	118,161,234		
3,4,6-Me$_3$-hexitol-Ac$_3$	161,190		
2,3,4,6-Me$_4$-hexitol-Ac$_2$	118,161,162,205		
2-Me-6,6-d$_2$-hexitol-Ac$_5$	118,275,335		
3-Me-6,6-d$_2$-hexitol-Ac$_5$	190,263		
4-Me-6,6-d$_2$-hexitol-Ac$_5$	191,262		
2,3-Me$_2$-6,6-d$_2$-hexitol-Ac$_4$	118,203,263		
2,4-Me$_2$-6,6-d$_2$-hexitol-Ac$_4$	118,191,234,307		
3,4-Me$_2$-6,6-d$_2$-hexitol-Ac$_4$	190,191		
2,3,4-Me$_3$-6,6-d$_2$-hexitol-Ac$_3$	118,191,235		

Aldononitrile acetates

Derivative	EI-MS	CI-MS	Ref.
Erythrononitrile-Ac$_3$	73,99,103,141,145,170	184,261	322
2-Deoxy-D-erythro-pentononitrile-Ac$_3$	112,115,125,145,154,167,184	198,215,275	
Arabinononitrile-Ac$_4$	73,103,127,141,145,183,187,200,242,289	256,333	
Rhamnononitrile-Ac$_4$	55,87,99,117,129,141,159,183,200,212,242,272,314	270,347	
2-Deoxy-D-arabino-hexononitrile-Ac$_4$	73,83,103,112,115,125,127,142,145,154,157,175,184,187,214,217,256,289	270,287,347	
Mannononitrile-Ac$_5$	73,103,115,127,141,145,157,170,175,183,187,200,212,217,242,272,289,314	328,345,405	
2-Acetamido-2-deoxy-hexononitrile-Ac$_4$	73,85,98,103,115,127,140,145,164,169,182,187,211,224,289	207,267,327,387,404	

Partially methylated aldononitrile acetates

Derivative	EI-MS	CI-MS	Ref.
2,3-Me$_2$-hexononitrile-Ac$_3$	**85**,87,**99**,115,**127**		323
2,4-Me$_2$-hexononitrile-Ac$_3$	75,**87**,96,99,**112**,127,**129**,**154**,**159**,**186**,189	240,272,349	323,324

TABLE 16.4 (continued)

Derivative	Diagnostic ions (m/z)*		Ref.
	EI-MS	CI-MS	
2,6-Me$_2$-hexononitrile-Ac$_3$	75,**87**,99,115,117,129,**159,169**,184		323
3,4-Me$_2$-hexononitrile-Ac$_3$	**87**,99,126,**129**,142,**189**		
3,6-Me$_2$-hexononitrile-Ac$_3$	75,83,84,85,**87**,98,**99**,113,117,**129,131,142**, 147,184,189,233		
4,6-Me$_2$-hexononitrile-Ac$_3$	71,87,99,**101**,112,**129**,154,161,214		
2,3,4-Me$_3$-hexono-nitrile-Ac$_2$	**71,73,87,90,99**,101,113,**129**,173,**189**		
2,3,6-Me$_3$-hexono-nitrile-Ac$_2$	**87,99,113,129**,131,147,173,189,233		
2,4,6-Me$_3$-hexono-nitrile-Ac$_2$	87,99,**101**,112,**129**,161,186	244,272, 304	323, 324
3,4,6-Me$_3$-hexono-nitrile-Ac$_2$	71,74,**87,101**,119,126,**129**,142,**161**		323
2,3,4,6-Me$_4$-hexono-nitrile-Ac	71,73,85,**87,88**,89,96,**101**,113,114,119,**129**, 130,131,**145**,158,**161**,162,205	216,244, 276	323, 324

Peracetylated keto-oximes (PAKO)

D-*erythro*-2-Pentulose PAKO	81,**99,111,129,141**,145,154,**158,171**,183, 184,213,231,**273**	**138,198**, 256,**258**, 318,376	285
D-Fructose PAKO	99,**103,111,115,123,141,153,158,183**,184, 201,243,345	**150,210**, 268,**270**, 328,**330**, 388,**390**	
D-*manno*-2-Heptulose PAKO	**103,115,123,128**,139,**145,153,165**,187,**195**, 196,200,213,**315**,417		

TFA-O-methyloximes

Pentose		338,450, **563,564**	281
Hexose		449,450, 464,576, **689,690**	282

TABLE 16.4 (continued)

Derivative	Diagnostic ions (m/z)[*]		Ref.
	EI-MS	CI-MS	
TFA-O-butyloximes			
Pentose		380,492, 605,**606**	281
Hexose		492,506, 618,731, 732	282
2-Pentulose		492,**605**, **606**	283
3-Pentulose		**380**,436, 492,**605**, **606**	
2-Hexulose		280,392, 394,505, **506**,617, 618,619, 620,**731**, **732**	
3-Hexulose		280,392, 394,505, **506**,522, 617,618, 731,**732**	
2,5-Hexodiulose		367,478, **479**,591, 592,593, **704**,**705**	
TMS oximes			
Hexose	103,205,218,307,320,409,422,524		325
Hexulose	103,205,307,320,422,524		
TMS O-methyloximes			
Tetrose	103,160,205,262		326
Pentose	103,160,205,262,307,364		
Hexose	103,160,205,262,307,364,409,466		
Heptose	103,160,205,262,319,364,421,466,568		

TABLE 16.4 (continued)

Derivative	Diagnostic ions (m/z)*		Ref.
	EI-MS	CI-MS	

TMS methyl glycosides

Pentoside	103,**133,204**,205,**217**		327
6-Deoxyhexoside	103,**133,204,205,217**,319		
Hexopyranoside	103,**133,204,205,217**,319		
Hexofuranoside	103,133,**204**,205,**217**,319		
2-Acetamidodeoxy-hexoside	103,**131**,133,**147,173**,204,205,217,247		232, 327

TMS partially methylated methyl glycosides

Me 2-Me-(TMS)$_2$-rhamnoside	75,133,**146**,204		328
Me 3-Me-(TMS)$_2$-rhamnoside	75,133,**146,217**		
Me 4-Me-(TMS)$_2$-rhamnoside	75,133,146,**204**		
Me 2,3-Me-(TMS)-rhamnoside	75,**88**,133,146,159		
Me 2,4-Me-(TMS)-rhamnoside	75,133,**146**,159		
Me 3,4-Me-(TMS)-rhamnoside	75,133,**146,159**,		
Me 2-Me-(TMS)$_3$-glucopyranoside	75,133,**146**,204		
Me 3-Me-(TMS)$_3$-glucopyranoside	75,133,**146,217**		
Me 4-Me-(TMS)$_3$-glucopyranoside	**73**,75,133,146,**204**		
Me 6-Me-(TMS)$_3$-glucopyranoside	**73**,133,146,**204**,217		

Partially methylated methyl glycosides

Me pentopyranoside	57,59,**60,61**,71,**73,74**,86		329
Me 2-Me-pento-pyranoside	57,59,60,61,71,73,**74**,85,**87**,88		

References on p. B284

TABLE 16.4 (continued)

Derivative	Diagnostic ions (m/z)[*]		Ref.
	EI-MS	CI-MS	
Me 3-Me-pento-pyranoside	55,57,58,59,60,61,69,71,73,**74,75**,85,86,87,100,114		
Me 4-Me-pento-pyranoside	57,**58,59,60**,69,73,74,75,**87**		
Me 2,3-Me$_2$-pento-pyranoside	57,59,61,69,71,73,74,**75**,85,87,**88**,101,114,129,161		
Me 2,4-Me$_2$-pento-pyranoside	57,58,59,61,71,73,**74**,85,87,88,**101**		
Me 3,4-Me$_2$-pento-pyranoside	57,58,59,69,71,73,**74,75,87**,88,101		
Me 2,3,4-Me$_3$-pento-pyranoside	55,58,71,73,**75,88**,99,**101**,115,176		

TMS diethyl dithioacetals, partially methylated sugars

2,3,4-Me$_3$-glucose	**190,205**,234,249		297
2,3,6-Me$_3$-glucose	**147,190,248**		
2,4,6-Me$_3$-glucose	**147**,190		
3,4,6-Me$_3$-glucose	**147,190**,237		

TMS deoxy(methoxyamino)alditols

Pentose, deoxyhexose	103,205,307,366,454,468		299
Hexose	**73**,102,147,205,**217,276**,291,306,320,**332**,360,421,450,525,539,556	572	

Permethylated deoxy(methylmethoxyamino)alditols

Pentose	**74**,236		300
Deoxyhexose	**74**,250		
Hexose	**74,101,131,145**,174,**189**,221,280		

[*] Ions of higher intensity are in bold print; for alditol acetates m/z 43 is present in all cases, as the base peak.

[**] Author's data.

[***] Reduced with NaBD$_4$; 6,6-d$_2$-hexitol indicates carboxylate-reduced uronic acid residue, converted to 6,6-dideuteriohexosyl unit.

acetate/acetic acid/water (3:1:1), is effective in separating D-glucuronic and D-galacturonic acid from each other and from aldobiouronic acids and neutral sugars [331]. Carboxymethylcellulose paper, converted to Ca^{2+}, Ba^{2+}, or La^{3+} forms, has been successfully used in separations of alditols and some aldoses, with 1-butanol/ethanol/water (10:1:2) as the mobile phase [332]. The latter separation, which depends upon ligand exchange, is slow (requiring 7 days for resolution of hexitols), but sharp spots or bands are obtained, particularly with the paper in the La^{3+} form. This method, too, is applicable to preparative chromatography, in which paper chromatography is still widely used.

16.5.2 Thin-layer chromatography

16.5.2.1 Improvements in TLC separation on cellulose plates

Separations of monosaccharides of all types — amino sugars as well as neutral and acidic monosaccharides — on cellulose plates have been much improved by the use of two-dimensional TLC. Thus, the aminodeoxyhexoses and aminodideoxyhexoses found in hydrolyzates of the lipopolysaccharide and peptidoglycan constituents of bacterial cell walls have been separated from one another, and from the amino acid components, by TLC on cellulose plates (20 x 20 cm), with development in one direction with 2-propanol/90% formic acid/water (20:1:5) and in the orthogonal direction with 65% aq. 2,6-dimethylpyridine (lutidine) [333]. Complete separation of the neutral and acidic sugars occurring in plant cell-wall polysaccharides was reported by Métraux [334], who used a two-dimensional TLC system in which two successive developments in one direction with 1-butanol/2-butanone/formic acid/water (8:6:3:3) were followed by a single development in the orthogonal direction with phenol/water/formic acid (50:49:1; organic phase only). Triple development in unidimensional TLC remains the best method of resolving small oligosaccharides on cellulose plates, but not necessarily with the same solvent each time. For example, Horikoshi et al. [335], in analyzing the products of reaction of sucrose with different sucrases, achieved excellent resolution of D-glucose, D-fructose, and oligosaccharides up to DP 5 (isomaltopentaose) by two developments with 1-butanol/pyridine/acetic acid/water (10:6:1:3), then a third with 2:1 phenol in 1.5% aq. ammonia.

16.5.2.2 Use of HPTLC and bonded-phase plates

While good separations of sugars and alditols, improved by impregnation of the plates with borate [336] or phosphate [337] buffers or the presence of boric acid in the mobile phase [337], have been achieved by use of HPTLC silica plates with various solvent systems, it is in TLC of higher oligosaccharides that the power of these plates has been most strikingly demonstrated. The upper limit of DP for resolution of malto-oligosaccharides by HPTLC, initially 10-12, with multiple development of Silica Gel 60 HPTLC plates [338,339], has been extended by Koizumi et al. [30] to 26, by using macroporous Silica Gel 50,000 plates (Merck), developed 4 times with 1-butanol/pyridine/water (6:5:4). The β-(1 → 2)-linked D-gluco-oligosaccharides of the sophoro series were resolved to DP

References on p. B284

30 under the same conditions. Resolution of the cellodextrins to DP 8, and of the β-(1 → 4)-linked D-xylo-oligosaccharides from xylan to DP 10, was achieved by Doner [340], with the same plates, developed 3 times with 1-butanol/pyridine/water (8:5:4). Oligogalacturonic acids from pectin were resolved to DP 6 on Whatman HPTLC silica plates (type HP-KF), developed twice with 1-butanol/formic acid/water (4:6:1) at ambient temperature [340], but better resolution (to DP 9) was obtained by irrigation at 35°C (single development with 21:29 ethanol/25 mM acetic acid) [341]. Excellent resolution of gangliosides has been reported by Ando et al. [342] on Silica Gel 60 HPTLC plates with the solvent systems acetonitrile/2-propanol/50 mM aq. KCl (10:67:23) or acetonitrile/2-propanol/2.5 mM aq. ammonia (10:65:25). A combination of the two solvent systems in two-dimensional HPTLC was effective in the analysis of complex mixtures of gangliosides, including polysialo-gangliosides. Good resolution of branched cyclodextrins was obtained by Koizumi et al. [343], who used Silica Gel 50,000 HPTLC plates, developed twice with 1-butanol/pyridine/water (6:3:2 then 6:4:3).

The Koizumi group [343] reported another successful resolution of the branched cyclodextrins on HPTLC plates carrying a bonded amino phase, which were developed five times with acetonitrile/water (3:2). Doner et al. [344] found such plates, developed with acetonitrile/water (7:3), effective in resolving disaccharides, and tri- and tetrasaccharides, such as raffinose, melezitose and stachyose. However, among the monosaccharides only the ketoses were separated under these conditions; aldotetroses, -pentoses, and some -hexoses remained at the origin. The covalent interaction between these sugars and the amino groups on the plates (Section 16.2.1.3) can be precluded by impregnation of the amine-bonded plates with NaH_2PO_4. Using the impregnated plates, Doner and Biller [345] have resolved a diversity of carbohydrates: monosaccharides, including aldopentoses and -hexoses, ketoses, deoxy- and aminodeoxyhexoses, disaccharides, malto-oligosaccharides to DP 7, acetals, and methyl glycosides. The separations were achieved in less than 30 min on development with 60-90% aq. acetonitrile. These bonded-phase TLC plates are a useful adjunct to similar HPLC systems.

16.5.3 Sensitive detection reagents for planar chromatography of sugars

Detection reagents that produce fluorescence in reducing sugars, thus permitting fluorimetric scanning of the chromatograms, greatly enhance the sensitivity of PC. Such reagents include malonamide [346] and fluorescamine [347], the latter being specific for aminodeoxy sugars. Aldoses, ketoses, uronic acids, aminodeoxy sugars, and reducing disaccharides can be detected at levels as low as 0.25 nmol with the malonamide spray, while the detection limit for aminodeoxy sugars with fluorescamine is 50 pmol. Klaus and coworkers [337,348] have described a procedure for quantitative TLC which is applicable not only to reducing sugars but also to alditols, aldonic acids, and lactones, since all compounds with vicinal diol groups in the molecule are oxidized in situ on the plates by treatment with lead tetraacetate. The resulting polyaldehydes produce strong fluorescence with 2,7-dichlorofluorescein (0.1-1%), which is present, together with lead tetraacetate (saturated solution in glacial acetic acid), in the toluene solution in which the plates are dipped after development. After activation by heating at 100°C for 3 min, the plates are scanned fluorimetrically for detection at nanogram levels.

Derivatization of sugars with chromogenic or fluorogenic reagents before TLC, like pre-column derivatization in liquid chromatography (Sections 16.2.1.1 and 16.2.1.4), also gives high sensitivity of detection. For example, TLC of aminodeoxy sugars as dansyl derivatives, with fluorimetric detection, permits analysis at levels down to 0.3 nmol [349]. After development with a relatively nonpolar solvent system (necessitated by the nature of the derivatives), such as cyclohexane/ethyl acetate/ethanol (6:4:3), the silica plates are sprayed with triethanolamine/2-propanol (1:4) to stabilize and intensify fluorescence. Derivatization of neutral sugars to dabsylhydrazones (Section 16.2.1.4), followed by TLC on silica plates, gives yellow spots, turning bright pink on exposure to HCl vapor, and this permits visual detection down to 0.1 nmol [44].

16.6 ELECTROPHORESIS

As in paper chromatography, there have been few changes in the standard electro-phoretic methods for carbohydrates since the previous review [1, p. 275]. In paper electrophoresis, metavanadate has been added to the list of complexing anions which can be used to give mobility to neutral sugars and alditols, and electrophoresis in sodium metavanadate solution has proved particularly useful in separations of hexitols and hexuloses, as well as some isomeric disaccharides [350]. An interesting new medium for electrophoresis, which may replace glass-fiber paper in electrophoretic separations of polysaccharides, is silanized silica gel [351]. With the surface coated with 1-octanol, the silica gel is suspended in a borate buffer (0.3 M, pH 10, containing 20 mM Titriplex III) and coated onto thin-layer plates, which, after drying, are connected through strips of glass-fiber paper to the electrolyte vessels, containing the same buffer. This method of electrophoresis has proved useful in the examination of biomass degradation products, which contain mono-, oligo-, and polysaccharides and derivatives [351]. The plates are mechanically more stable than silanized glass-fiber paper, adsorption and endosmosis are low, and chemically aggressive reagents can be used.

Electrophoresis on cellulose acetate membranes has been extensively used in separations of glycosaminoglycuronans [352-355]. The method clearly demonstrates structural heterogeneity in heparin, which gives several bands, the distribution of which varies according to the source of the heparin [354,355]. Barium acetate (0.1-1.0 M) is the electrolyte commonly used in cellulose acetate electrophoresis of glycosaminoglycuro-nans, although other solvents, containing mixtures of electrolytes, have been proposed in attempts to improve resolution [353,354].

The alternative procedure for electrophoretic separation of glycosaminoglycuronans, on slides (7.5 x 5.0 cm, 1 mm thick) of agarose gel [1, p. 276], has been improved by use of a discontinuous method [356]. The sample was subjected to electrophoresis first in 0.04 M barium acetate (pH 5.8), on a slide of 0.5% agarose gel, prepared in this solution, and then, after transfer of the slide to another chamber, in 0.1 M 1,3-diaminopropane acetate buffer (pH 9.0). Heterogeneity of heparin was demonstrated by this technique also, most heparins being fractionated into at least 3 components. These were well resolved from dermatan sulfate and chondroitin 4- and 6-sulfate, but the fastest-moving heparin band overlapped with that of heparan sulfate. These components were resolved by electrophoresis in the 0.06 M barbital buffer (pH 8.6) used previously [1], but dermatan sulfate and the chondroitin sulfates were incompletely separated by this method. Complete separation of all of the glycosaminoglycuronans was achieved only by a two-dimensional method. Electrophoresis in the barbital buffer was followed, after excision of the gel strip containing the fractionated components and insertion in a slot of the same dimensions situated transversely in another gel slide, by discontinuous electrophoresis in the orthogonal direction.

Polyacrylamide gel electrophoresis (PAGE) is an important technique in studies of glycoproteins and other glycoconjugates. The method generally used is discontinuous PAGE in the presence of sodium dodecyl sulfate (SDS), a technique widely applied in electrophoresis of proteins. The most sensitive staining procedure for glycoconjugates is that recommended by Dubray and Bezard [357]. After the separated components have been fixed in the gel by soaking overnight in an aqueous solution containing 25% 2-propanol and 10% acetic acid, the gel slab is washed in 7.5% acetic acid for 30 min and then left for 1 h in 0.2% periodic acid solution at 4°C, washed for 3 h in several changes of distilled water, and finally dipped into an ammoniacal silver nitrate solution. Destaining is effected with a fresh solution containing 0.05% citric acid, 0.019% formaldehyde, and 10% methanol, followed by a photographic fixative. The lower limit of detection for glycoproteins by this stain is 20 ng, which corresponds on average to about 0.4 ng of bound carbohydrate; the procedure is thus 64 times more sensitive than the alternative periodic acid/Schiff's reagent method. SDS/PAGE with periodic acid/silver staining has been successfully applied to a diversity of glycoproteins [357] and to the arabinogalactan proteins present in the gum of *Acacia erioloba* [358]. The technique has proved useful also in studies of the structural heterogeneity of lipopolysaccharides form *E. coli* bacterial strains [357,359]. Recently, replacement of SDS by sodium deoxycholate has been found to improve resolution in PAGE of lipopolysaccharides from *Salmonella* bacteria [360], and this modification is likely to be widely adopted in the future.

REFERENCES

1 S.C. Churms, in E. Heftmann (Editor), *Chromatography*, Journal of Chromatography Library Series, Vol. 22B, Elsevier, Amsterdam, 1983, pp. 223-286.
2 G.D. McGinnis and P. Fang, *Methods Carbohydr. Chem.*, 8 (1980) 33.
3 L.A.Th. Verhaar and B.F.M. Kuster, *J. Chromatogr.*, 220 (1981) 313.
4 S. Honda, *Anal. Biochem.*, 140 (1984) 1.
5 K. Robards and M. Whitelaw, *J. Chromatogr.*, 373 (1986) 81.
6 K. Kakehi and S. Honda, *J. Chromatogr.*, 379 (1986) 27.
7 N.B. Beaty and R.J. Mello, *J. Chromatogr.*, 418 (1987) 187.
8 M. Verzele, G. Simoens and F. Van Damme, *Chromatographia*, 23 (1987) 292.
9 K.B. Hicks, *Adv. Carbohydr. Chem. Biochem.*, 46 (1988) 17.
10 C.A. White, J.F. Kennedy and B.T. Golding, *Carbohydr. Res.*, 76 (1979) 1.

11 M. Petchey and M.J.C. Crabbe, *J. Chromatogr.*, 307 (1984) 180.
12 R. Oshima and J. Kumanotani, *J. Chromatogr.*, 265 (1983) 335.
13 C.A. White, S.W. Vass, J.F. Kennedy and D.G. Large, *Carbohydr. Res.*, 119 (1983) 241.
14 C.A. White, S.W. Vass, J.F. Kennedy and D.G. Large, *J. Chromatogr.*, 264 (1983) 99.
15 S. Honda and K. Kakehi, *J. Chromatogr.*, 152 (1978) 405.
16 N.K. Karamanos, T. Tsegenidis and C.A. Antonopoulos, *J. Chromatogr.*, 405 (1987) 221.
17 M. Takeda, M. Maeda and A. Tsuji, *J. Chromatogr.*, 244 (1982) 347.
18 R. Oshima, Y. Yamauchi and J. Kumanotani, *Carbohydr. Res.*, 107 (1982) 169.
19 S. Iwata, T. Narui, K. Takahashi and S. Shibata, *Carbohydr. Res.*, 133 (1984) 157.
20 Z.L. Nikolov and P.J. Reilly, *J. Chromatogr.*, 325 (1985) 287.
21 D.L. Hendrix, R.E. Lee, Jr., J.G. Baust and H. James, *J. Chromatogr.*, 210 (1981) 45.
22 J.G. Baust, R.E. Lee, Jr., R.R. Rojas, D.L. Hendrix, D. Friday and H. James, *J. Chromatogr.*, 261 (1983) 65.
23 C.A. White, P.H. Corran and J.F. Kennedy, *Carbohydr. Res.*, 87 (1980) 165.
24 W. Praznik, R.H.F. Beck and E. Nitsch, *J. Chromatogr.*, 303 (1984) 417.
25 H. Binder, *J. Chromatogr.*, 189 (1980) 414.
26 M.T. Yang, L.P. Milligan and G.W. Mathison, *J. Chromatogr.*, 209 (1981) 316.
27 Z.L. Nikolov, M.M. Meagher and P.J. Reilly, *J. Chromatogr.*, 319 (1985) 51.
28 Z.L. Nikolov, M.M. Meagher and P.J. Reilly, *J. Chromatogr.*, 321 (1985) 393.
29 S.J. Mellis and J.U. Baenziger, *Anal. Biochem.*, 114 (1981) 276.
30 K. Koizumi, T. Utamura and Y. Okada, *J. Chromatogr.*, 321 (1985) 145.
31 K. Koizumi, T. Utamura, Y. Kubota and S. Hizukuri, *J. Chromatogr.*, 409 (1987) 396.
32 K. Koizumi, Y. Okuda, T. Utamura, M. Hisamatsu and A. Amemura, *J. Chromatogr.*, 299 (1984) 215.
33 M. Benincasa, G.P. Cartoni, F. Coccioli, R. Rizzo and L.P.T.M. Zevenhuizen, *J. Chromatogr.*, 393 (1987) 263.
34 W.M. Blanken, M.L.E. Bergh, P.L. Koppen and D.H. van den Eijnden, *Anal. Biochem.*, 145 (1985) 322.
35 V.K. Dua, V.E. Dube and C.A. Bush, *Biochim. Biophys. Acta,* 802 (1984) 29.
36 V.K. Dua, K. Goso, V.E. Dube and C.A. Bush, *J. Chromatogr.*, 328 (1985) 259.
37 G. Gazzotti, S. Sonnino and R. Ghidoni, *J. Chromatogr.*, 348 (1985) 371.
38 S. Ando, H. Waki and K. Kon, *J. Chromatogr.*, 408 (1987) 285.
39 W.F. Alpenfels, *Anal. Biochem.*, 114 (1981) 153.
40 K. Mopper and L. Johnson, *J. Chromatogr.*, 256 (1983) 27.
41 F.M. Eggert and M. Jones, *J. Chromatogr.*, 333 (1985) 123.
42 H. Takemoto, S. Hase and T. Ikenaka, *Anal. Biochem.*, 145 (1985) 245.
43 K. Muramoto, R. Goto and H. Kamiya, *Anal. Biochem.*, 162 (1987) 435.
44 J.-K. Lin and S.-S. Wu, *Anal. Chem.*, 59 (1987) 1320.
45 K. Shinomiya, H. Toyoda, A . Akahoshi, H. Ochiai and T. Imanari, *J. Chromatogr.*, 387 (1987) 481.
46 N. Jentoft, *Anal. Biochem.*, 148 (1985) 424.
47 D.J. Gisch and J.D. Pearson, *J. Chromatogr.*, 443 (1988) 299.
48 G.B. Wells and R.L. Lester, *Anal. Biochem.*, 97 (1979) 184.
49 G.B. Wells, V. Kontoyiannidou, S.J. Turco and R.L. Lester, *Methods Enzymol.*, 83 (1982) 132.
50 P.F. Daniel, D.F. de Feudis, I.T. Lott and R.H. McCluer, *Carbohydr. Res.*, 97 (1981) 161.
51 B.S. Valent, A.G. Darvill, M. McNeil, B.K. Robertsen and P. Albersheim, *Carbohydr. Res.*, 79 (1980) 165.
52 P. Åman, M. McNeil, L.-E. Franzén, A.G. Darvill and P. Albersheim, *Carbohydr. Res.*, 95 (1981) 263.
53 P. Åman, L.-E. Franzén, J.E. Darvill, M. McNeil, A.G. Darvill and P. Albersheim, *Carbohydr. Res.*, 103 (1982) 77.
54 M. McNeil, A.G. Darvill, P. Åman, L.-E. Franzén and P. Albersheim, *Methods Enzymol.*, 83 (1982) 3.

55 N.W.H. Cheetham and P. Sirimanne, *J. Chromatogr.*, 196 (1980) 171.
56 A. Heyraud and P. Salemis, *Carbohydr. Res.*, 107 (1982) 123.
57 S. Saadat and C.E. Ballou, *Carbohydr. Res.*, 119 (1983) 248.
58 N.W.H. Cheetham and P. Sirimanne, *J. Chromatogr.*, 208 (1981) 100.
59 N.W.H. Cheetham and P. Sirimanne, *Carbohydr. Res.*, 112 (1983) 1.
60 N.W.H. Cheetham and P. Sirimanne, *Carbohydr. Res.*, 96 (1981) 126.
61 A. Hjerpe, B. Engfeldt, T. Tsegenidis and C.A. Antonopoulos, *J. Chromatogr.*, 259 (1983) 334.
62 A. Hjerpe, C.A. Antonopoulos, B. Classon, B. Engfeldt and M. Nurminen, *J. Chromatogr.*, 235 (1982) 221.
63 G. Annison, N.W.H. Cheetham and I. Couperwhite, *J. Chromatogr.*, 264 (1983) 137.
64 A. Heyraud and M. Rinaudo, *J. Liq. Chromatogr.*, 3 (1980) 721.
65 N.W.H. Cheetham, P. Sirimanne and W.R. Day, *J. Chromatogr.*, 207 (1981) 439.
66 P. Vrátný, J. Čoupek, S. Vozka and Z. Hostomská, *J. Chromatogr.*, 254 (1983) 143.
67 E. Rajakylä, *J. Chromatogr.*, 353 (1986) 1.
68 L.A.Th. Verhaar, B.F.M. Kuster and H.A. Claessens, *J. Chromatogr.*, 284 (1984) 1.
69 B. Porsch, *J. Chromatogr.*, 320 (1985) 408.
70 N.W.H. Cheetham and G. Teng, *J. Chromatogr.*, 336 (1984) 161.
71 K. Koizumi, Y. Kubota, Y. Okada, T. Utamura, S. Hizukuri and J.-I. Abe, *J. Chromatogr.*, 437 (1988) 47.
72 N.W.H. Cheetham and V.E. Dube, *J. Chromatogr.*, 262 (1983) 426.
73 V.K. Dua and C.A. Bush, *Anal. Biochem.*, 133 (1983) 1.
74 V.K. Dua and C.A. Bush, *Anal. Biochem.*, 137 (1984) 33.
75 K. Koizumi and T. Utamura, *J. Chromatogr.*, 436 (1988) 328.
76 A.G.J. Voragen, H.A. Schols, J.A. de Vries and W. Pilnik, *J. Chromatogr.*, 244 (1982) 327.
77 L.E. Chun, T.J. Koob and D.R. Eyre, *Anal. Biochem.*, 171 (1988) 197.
78 S.K. Henderson and D.E. Henderson, *J. Chromatogr. Sci.*, 24 (1986) 198.
79 N. Tomiya, J. Awaya, M. Kurono, S. Endo, Y. Arata and N. Takahashi, *Anal. Biochem.*, 171 (1988) 73.
80 S. Hase, K. Ikenaka, K. Mikoshiba and T. Ikenaka, *J. Chromatogr.*, 434 (1988) 51.
81 N.P. Arbatsky, M.D. Martynova, A.O. Zheltova, V.A. Derevitskaya and N.K. Kochetkov, *Carbohydr. Res.*, 187 (1989) 165.
82 J.L. Leonard, F. Guyon and P. Fabiani, *Chromatographia*, 18 (1984) 600.
83 F. Guyon, A. Foucault, M. Caude and R. Rosset, *Carbohydr. Res.*, 140 (1985) 135.
84 C. Brons and C. Olieman, *J. Chromatogr.*, 259 (1983) 79.
85 M. Verzele and F. Van Damme, *J. Chromatogr.*, 362 (1986) 23.
86 D.W. Armstrong and H.L. Jin, *J. Chromatogr.*, 462 (1989) 219.
87 L.E. Fitt, W. Hassler and D.E. Just, *J. Chromatogr.*, 187 (1980) 381.
88 J.A. Owens and J.S. Robinson, *J. Chromatogr.*, 338 (1985) 303.
89 K.Sasaki, K.B. Hicks and G. Nagahashi, *Carbohydr. Res.*, 183 (1988) 1.
90 J.O. Baker and M.E. Himmel, *J. Chromatogr.*, 357 (1986) 161.
91 W. Blaschek, *J. Chromatogr.*, 256 (1983) 157.
92 A. Lenherr, T.J. Mabry and M.R. Gretz, *J. Chromatogr.*, 388 (1987) 455.
93 L.S. Lohmander, *Anal. Biochem.*, 154 (1986) 75.
94 A.G.J. Voragen, H.A. Schols, M.F. Searle-van Leeuwen, G. Beldman and F.M. Rombouts, *J. Chromatogr.*, 370 (1986) 113.
95 R. Pecina, G. Bonn, E. Burtscher and O. Bobleter, *J. Chromatogr.*, 287 (1984) 245.
96 M. Grün and F.A. Loewus, *Anal. Biochem.*, 130 (1983) 191.
97 K.B. Hicks, P.C. Lim and M.J. Haas, *J. Chromatogr.*, 319 (1985) 159.
98 K.B. Hicks, *Carbohydr. Res.*, 145 (1986) 312.
99 J. Schmidt, M. John and C. Wandrey, *J. Chromatogr.*, 213 (1981) 151.
100 K. Brunt, *J. Chromatogr.*, 246 (1982) 145.
101 H. Hokse, *J. Chromatogr.*, 189 (1980) 98.
102 G. Bonn, R. Pecina, E. Burtscher and O. Bobleter, *J. Chromatogr.*, 287 (1984) 215.
103 K.B. Hicks and S.M. Sondey, *J. Chromatogr.*, 389 (1987) 183.
104 *Bio-Rad Catalogue L,* (1986) 92.

105 K.B. Hicks and A.T. Hotchkiss, Jr., *J. Chromatogr.*, 441 (1988) 382.
106 H. Derler, H.F. Hörmeyer and G. Bonn, *J. Chromatogr.*, 440 (1988) 281.
107 J.A.M. van Riel and C. Olieman, *J. Chromatogr.*, 362 (1986) 235.
108 G. Bonn, *J. Chromatogr.*, 322 (1985) 411.
109 S. Honda and S. Suzuki, *Anal. Biochem.*, 142 (1984) 167.
110 H.G. van Eijk, W.L. van Noort, C. Dekker and C. van der Heul, *Clin. Chim. Acta*, 139 (1984) 187.
111 M. Takeuchi, S. Takasaki, N. Inoue and A. Kobata, *J. Chromatogr.*, 400 (1987) 207.
112 S. Honda, S. Suzuki and K. Kakehi, *J. Chromatogr.*, 291 (1984) 317.
113 K. Murata and Y. Yokoyama, *J. Chromatogr.*, 415 (1987) 231.
114 M. D'Amboise, T. Hanai and D. Noël, *Clin. Chem.*, 26 (1980) 1348.
115 R. Oshima, N. Takai and J. Kumanotani, *J. Chromatogr.*, 192 (1980) 452.
116 A. Sugii, K. Harada and Y. Tomita, *J. Chromatogr.*, 366 (1986) 412.
117 G. Gacesa, A. Squire and P.J. Winterburn, *Carbohydr. Res.*, 118 (1983) 1.
118 S.R. Delaney, M. Leger and H.E. Conrad, *Anal. Biochem.*, 106 (1980) 253.
119 S.R. Delaney, H.E. Conrad and J.H. Glaser, *Anal. Biochem.*, 108 (1980) 25.
120 P. Nebinger, M. Koel, A. Franz and E. Werries, *J. Chromatogr.*, 265 (1983) 19.
121 T. Gherezghiher, M.C. Koss, R.E. Nordquist and C.P. Wilkinson, *J. Chromatogr.*, 413 (1987) 9.
122 J. Macek, J. Krajíčková and M. Adam, *J. Chromatogr.*, 414 (1987) 156.
123 R.J. Linhardt, K.G. Rice, Y.S. Kim, D.L. Lohse, H.M. Wang and D. Loganathan, *Biochem. J.*, 254 (1988) 781.
124 J.U. Baenziger and M. Natowicz, *Anal. Biochem.*, 112 (1981) 357.
125 T. Tsuji, K. Yamamoto, Y. Konami, T. Irimura and T. Osawa, *Carbohydr. Res.*, 109 (1982) 259.
126 J. van Pelt, J.B.L. Damm, J.P. Kamerling and J.F.G. Vliegenthart, *Carbohydr. Res.*, 169 (1987) 43.
127 R.A. Lazarus and J.L. Seymour, *Anal. Biochem.*, 157 (1986) 360.
128 J.D. Blake, M.L. Clarke and G.N. Richards, *J. Chromatogr.*, 312 (1984) 211.
129 A.K. Shukla and R. Schauer, *J. Chromatogr.*, 244 (1982) 81.
130 A.K. Shukla, R. Schauer, F. Unger, U. Zähringer, E. Rietschel and H. Brade, *Carbohydr. Res.*, 140 (1985) 1.
131 D.E. Madden, W.F. Alpenfels, R.A. Mathews and A.E. Newsom, *J. Chromatogr.*, 248 (1982) 476.
132 Dj. Josic, R. Hofermaas, Ch. Bauer and W. Reutter, *J. Chromatogr.*, 317 (1984) 35.
133 F. Perini and B.P. Peters, *Anal. Biochem.*, 123 (1982) 357.
134 S. Honda, M. Takahashi, K. Kakehi and S. Ganno, *Anal. Biochem.*, 113 (1981) 130.
135 S. Honda, M. Takahashi, Y. Nishimura, K. Kakehi and S. Ganno, *Anal. Biochem.*, 118 (1981) 162.
136 S. Honda, M. Takahashi, S. Shimada, K. Kakehi and S. Ganno, *Anal. Biochem.*, 128 (1983) 429.
137 A.K. Shukla, N. Scholz, E.H. Reimerdes and R. Schauer, *Anal. Biochem.*, 123 (1982) 78.
138 R.D. Rocklin and C.A. Pohl, *J. Liq. Chromatogr.*, 6 (1983) 1577.
139 S. Hughes and D.C. Johnson, *Anal. Chim. Acta*, 132 (1981) 11.
140 S. Hughes and D.C. Johnson, *J. Agric. Food Chem.*, 30 (1982) 712.
141 P. Edwards and K.K. Haak, *Amer. Lab.*, No. 4 (1983) 78.
142 D.C. Johnson, *Nature (London)*, 321 (1986) 451.
143 D.C. Johnson and T.Z. Polta, *Chromatogr. Forum*, 1 (1986) 37.
144 R.D. Rocklin, *LC, Liq. Chromatogr. HPLC Mag.*, 1 (1983) 504.
145 *Dionex Technical Note*, No. 20 (1987).
146 J.D. Olechno, S.R. Carter, W.T. Edwards and D.S. Gillen, *Amer. Biotechnol. Lab.*, 5, No. 5 (1987) 38.
147 W.T. Edwards, C.A. Pohl and R. Rubin, *Tappi*, 70 (1987) 138.
148 K. Koizumi, Y. Kubota, T. Tanimoto and Y. Okada, *J. Chromatogr.*, 454 (1988) 303.
149 M.R. Hardy, R.R. Townsend and Y.C. Lee, *Anal. Biochem.*, 170 (1988) 54.
150 M.R. Hardy and R.R. Townsend, *Proc. Natl. Acad. Sci. USA*, 85 (1988) 3289.

151 R.R. Townsend, M.R. Hardy, O. Hindsgaul and Y.C. Lee, *Anal. Biochem.,* 174 (1988) 459.
152 R.E. Smith, S. Howell, D. Yourtee, N. Premkumar, T. Pond, G.Y. Sun and R.A. MacQuarrie, *J. Chromatogr.,* 439 (1988) 83.
153 R.J. Redgewell, *Anal. Biochem.,* 107 (1980) 44.
154 D.F. Jin and C.A. West, *Plant Physiol.,* 74 (1984) 989.
155 K.R. Davis, A.G. Darvill, P. Albersheim and A. Dell, *Plant Physiol.,* 80 (1986) 568.
156 H. Barsett and B.S. Paulsen, *J. Chromatogr.,* 329 (1985) 315.
157 P. Nebinger, *J. Chromatogr.,* 320 (1985) 351.
158 Y.J. Lee, B. Radhakrishnamurthy, E.R. Dalferes, Jr. and G.S. Berenson, *J. Chromatogr.,* 419 (1987) 275.
159 S.K. Kundu, S.K. Chakravarty, S.K. Roy and A.K. Roy, *J. Chromatogr.,* 170 (1979) 65.
160 P. Fredman, O. Nilsson, J.-L. Tayot and L. Svennerholm, *Biochim. Biophys. Acta,* 618 (1980) 42.
161 J.-E. Mansson, B. Rosengren and L. Svennerholm, *J. Chromatogr.,* 322 (1985) 465.
162 F. Šmíd, V. Bradová, O. Mikeš and J. Sedláčková, *J. Chromatogr.,* 377 (1986) 69.
163 Y. Hirabayashi, T. Nakao, M. Matsumoto, K. Obata and S. Ando, *J. Chromatogr.,* 445 (1988) 377.
164 H.G. Barth, *J. Chromatogr. Sci.,* 18 (1980) 409.
165 Y. Kato, K. Komiya, H. Sasaki and T. Hashimoto, *J. Chromatogr.,* 190 (1980) 297.
166 Y. Kato, K. Komiya, H. Sasaki and T. Hashimoto, *J. Chromatogr.,* 193 (1980) 311.
167 R.M. Alsop and G.J. Vlachogiannis, *J. Chromatogr.,* 246 (1982) 227.
168 M.P. Cullen, C. Turner and G.B. Haycock, *J. Chromatogr.,* 337 (1985) 29.
169 P.E. Barker, B.W. Hatt and G.J. Vlachogiannis, *J. Chromatogr.,* 208 (1981) 74.
170 Y. Takeda, K. Shirasaka and S. Hizukuri, *Carbohydr. Res.,* 132 (1984) 83.
171 S. Hizukuri and T. Takagi, *Carbohydr. Res.,* 134 (1984) 1.
172 S. Kobayashi, S.J. Schwartz and D.R. Lineback, *J. Chromatogr.,* 319 (1985) 205.
173 H.G. Barth and F.E. Regnier, *J. Chromatogr.,* 192 (1980) 275.
174 H.G. Barth and D.A. Smith, *J. Chromatogr.,* 206 (1981) 410.
175 B.R. Vijayendran and T. Bone, *Carbohydr. Polym.,* 4 (1984) 299.
176 J. Harenberg and J.X. de Vries, *J. Chromatogr.,* 261 (1983) 287.
177 D.M. Hittner and M.K. Cowan, *J. Chromatogr.,* 402 (1987) 149.
178 N. Motohashi and I. Mori, *J. Chromatogr.,* 299 (1984) 508.
179 N. Motohashi, Y. Nakamichi, I. Mori, H. Nishikawa and J. Umemoto, *J. Chromatogr.,* 435 (1988) 335.
180 N. Ui, *J. Chromatogr.,* 215 (1981) 289.
181 P.E. Barker, B.W. Hatt and S.R. Holding, *J. Chromatogr.,* 206 (1981) 27.
182 H. Yamada, Y. Ohshima, K. Tamura and T. Miyazaki, *Carbohydr. Res.,* 83 (1980) 377.
183 S.P. Djordjevic, M. Batley and J.W. Redmond, *J. Chromatogr.,* 354 (1986) 507.
184 A. Heyraud and M. Rinaudo, *J. Chromatogr.,* 166 (1978) 149.
185 W.W. Luchsinger, S.W. Luchsinger and D.W. Luchsinger, *Carbohydr. Res.,* 104 (1982) 153.
186 Å.C. Haglund, N.V.B. Marsden and S.G. Östling, *J. Chromatogr.,* 318 (1985) 57.
187 M. John, J. Schmidt, C. Wandrey and H. Sahm, *J. Chromatogr.,* 247 (1982) 281.
188 K. Hamacher, G. Schmid, H. Sahm and C. Wandrey, *J. Chromatogr.,* 319 (1985) 311.
189 K. Yamashita, T. Mizuochi and A. Kobata, *Methods Enzymol.,* 83 (1982) 105.
190 M. Natowicz and J.U. Baenziger, *Anal. Biochem.,* 105 (1980) 159.
191 S.C. Churms and A.M. Stephen, *Carbohydr. Res.,* 167 (1987) 239.
192 L. Hagel, *J. Chromatogr.,* 160 (1978) 59.
193 T. Andersson, M. Carlsson, L. Hagel, P.-Å. Pernemalm and J.-C. Janson, *J. Chromatogr.,* 326 (1985) 33.
194 L. Hagel and T. Andersson, *J. Chromatogr.,* 285 (1984) 295.
195 W. Praznik, R.H.F. Beck and W.D. Eigner, *J. Chromatogr.,* 387 (1987) 467.
196 W. Praznik, G. Burdicek and R.H.F. Beck, *J. Chromatogr.,* 357 (1986) 216.

197 W. Praznik and R.H.F. Beck, *J. Chromatogr.*, 348 (1985) 187.
198 R.H.F. Beck and W. Praznik, *J. Chromatogr.*, 369 (1986) 208.
199 T. Kato, T. Tokuya and A. Takahashi, *J. Chromatogr.*, 256 (1983) 61.
200 T. Kuge, K. Kobayashi, H. Tanahashi, T. Igushi and S. Kitamura, *Agric. Biol. Chem.*, 48 (1984) 2375.
201 M. Fishman, W.C. Damert, J.G. Phillips and R.A. Barford, *Carbohydr. Res.*, 160 (1987) 215.
202 S.C. Churms and A.M. Stephen, *S. Afr. J. Sci.*, 84 (1988) 855.
203 M. Terbojevich, A. Cosani, M. Palumbo and F. Pregnolato, *Carbohydr. Res.*, 157 (1986) 269.
204 M.-C. Vandevelde and J.-C. Fenyo, *Carbohydr. Polym.*, 5 (1985) 251.
205 G. Sworn, W.M. Marrs and R.J. Hart, *J. Chromatogr.*, 403 (1987) 307.
206 G. Berth, *Carbohydr. Polym.*, 8 (1988) 105.
207 R. Lotan and G.L. Nicolson, *Biochim. Biophys. Acta*, 559 (1979) 329.
208 I. Koyama, M. Miura, H. Matsuzaki, Y. Sakagishi and T. Komoda, *J. Chromatogr.*, 413 (1987) 65.
209 H.-J. Gabius, *Angew. Chem. Int. Ed. Engl.*, 27 (1988) 1267.
210 I.J. Goldstein and C.E. Hayes, *Adv. Carbohydr. Chem. Biochem.*, 35 (1978) 127.
211 S. Narasimhan, J.R. Wilson, E. Martin and H. Schachter, *Can. J. Biochem.*, 57 (1979) 83.
212 K. Yamashita, K. Totani, T. Ohkura, S. Takasaki, I.J. Goldstein and A. Kobata, *J. Biol. Chem.*, 262 (1987) 1602.
213 N. Shibuya, I.J. Goldstein, W.F. Broekaert, M. Nsimba-Lubaki, B. Peeters and W.J. Peumans, *J. Biol. Chem.*, 262 (1987) 1596.
214 S. Sueyoshi, T. Tsuji and T. Osawa, *Carbohydr. Res.*, 178 (1988) 213.
215 A.M. Wu and S. Sugii, *Proc. 14th Intern. Symp. Carbohydr. Chem.*, (1988) C34; see also A.M. Wu and S. Sugii, in A.M. Wu (Editor), *The Molecular Immunology of Complex Carbohydrates*, Plenum Press, New York, 1988.
216 I.W. Sutherland, *J. Chromatogr.*, 213 (1981) 301.
217 R.J. Sturgeon and C.M. Sturgeon, *Carbohydr. Res.*, 103 (1982) 213.
218 R.A. Pask-Hughes, *J. Chromatogr.*, 393 (1987) 273.
219 A. Borchert, P.-O. Larsson and K. Mosbach, *J. Chromatogr.*, 244 (1982) 49.
220 C.A.K. Borrebaeck, J. Soares and B. Mattiasson, *J. Chromatogr.*, 284 (1984) 187.
221 S. Honda, S. Suzuki, T. Nitta and K. Kakehi, *J. Chromatogr.*, 438 (1988) 73.
222 Z. El Rassi, Y. Truei, Y.-F. Maa and Cs. Horváth, *Anal. Biochem.*, 169 (1988) 172.
223 A.J. Muller and P.W. Carr, *J. Chromatogr.*, 284 (1984) 33.
224 A.J. Muller and P.W. Carr, *J. Chromatogr.*, 294 (1984) 235.
225 A.J. Muller and P.W. Carr, *J. Chromatogr.*, 357 (1986) 11.
226 S. Ohlson, A. Lundblad and D. Zopf, *Anal. Biochem.*, 169 (1988) 204.
227 T.L. Chester and D.P. Innis, *J. High Resolut. Chromatogr. Chromatogr. Commun.*, 9 (1986) 209.
228 J.D. Pinkston, G.D. Owens, L.J. Burkes, T.E. Delaney, D.S. Millington and D.A. Maltby, *Anal. Chem.*, 60 (1988) 962.
229 V.N. Reinhold, D.M. Sheeley, J. Kuei and G.-R. Her, *Anal. Chem.*, 60 (1988) 2719.
230 J. Kuei, G.-R. Her and V.N. Reinhold, *Anal. Biochem.*, 172 (1988) 228.
231 M.F. Laker, *J. Chromatogr.*, 184 (1980) 457.
232 C.G. Wong, S.-S.J. Sung and C.C. Sweeley, *Methods Carbohydr. Chem.*, 8 (1980) 55.
233 A.B. Blakeney, P.J. Harris, R.J. Henry and B.A. Stone, *Carbohydr. Res.*, 113 (1983) 291.
234 P.J. Harris, R.J. Henry, A.B. Blakeney and B.A. Stone, *Carbohydr. Res.*, 127 (1984) 59.
235 R.J. Henry, A.B. Blakeney, P.J. Harris and B.A. Stone, *J. Chromatogr.*, 256 (1983) 419.
236 A. Bacic, P.J. Harris, E.H. Hak and A.E. Clarke, *J. Chromatogr.*, 315 (1984) 373.
237 P.J. Harris, A. Bacic and A.E. Clarke, *J. Chromatogr.*, 350 (1985) 304.
238 J.A. Lomax and J. Conchie, *J. Chromatogr.*, 236 (1982) 385.

239 J.A. Lomax, A.H. Gordon and A. Chesson, *Carbohydr. Res.*, 138 (1985) 177.
240 J. Klok, H.C. Cox, J.W. de Leeuw and P.A. Schenck, *J. Chromatogr.*, 253 (1982) 55.
241 R. Oshima, J. Kumanotani and C. Watanabe, *J. Chromatogr.*, 250 (1982) 90.
242 T. Kontrohr and B. Kocsis, *J. Chromatogr.*, 291 (1984) 119.
243 J.-R. Neeser, *Carbohydr. Res.*, 138 (1985) 189.
244 T. Anastassiades, R. Puzic and O. Puzic, *J. Chromatogr.*, 225 (1981) 309.
245 T. Kiho, S. Ukai and C. Hara, *J. Chromatogr.*, 369 (1986) 415.
246 C.C. Chen and G.D. McGinnis, *Carbohydr. Res.*, 90 (1981) 127.
247 G.D. McGinnis, *Carbohydr. Res.*, 108 (1982) 284.
248 G.O. Guerrant and C.W. Moss, *Anal. Chem.*, 56 (1984) 633.
249 S.H. Turner and R. Cherniak, *Carbohydr. Res.*, 95 (1981) 137.
250 G.R. Tanner and I.M. Morrison, *J. Chromatogr.*, 299 (1984) 252.
251 C.A. Stortz, M.C. Matulewicz and A.S. Cerezo, *Carbohydr. Res.*, 111 (1982) 31.
252 M.E. Slodki, R.E. England, R.D. Plattner and W.E. Dick, Jr., *Carbohydr. Res.*, 156 (1986) 199.
253 R.H. Furneaux, *Carbohydr. Res.*, 113 (1983) 241.
254 B. Fournet, J.-M. Dhalluin, G. Strecker, J. Montreuil, C. Bosso and J. Defaye, *Anal. Biochem.*, 108 (1980) 35.
255 I. Mononen, *Carbohydr. Res.*, 104 (1982) 1.
256 B. Nilsson and D. Zopf, *Methods Enzymol.*, 83 (1982) 46.
257 A. García-Raso, I. Martínez-Castro, M.I. Páez, J. Sanz, J. García-Raso and F. Saura-Calixto, *J. Chromatogr.*, 398 (1987) 9.
258 M. Páez, I. Martínez-Castro, J. Sanz, A. Olano, A. García-Raso and F. Saura-Calixto, *Chromatographia*, 23 (1987) 43.
259 I. Martínez-Castro, M.I. Páez, J. Sanz and A. García-Raso, *J. Chromatogr.*, 462 (1989) 49.
260 Z.L. Nikolov and P.J. Reilly, *J. Chromatogr.*, 254 (1983) 157.
261 H. Traitler, S. Del Vedovo and T.F. Schweizer, *J. High Resolut. Chromatogr. Chromatogr. Commun.*, 7 (1984) 558.
262 M.F. Chaplin, *Anal. Biochem.*, 123 (1982) 336.
263 I. Mononen, *Carbohydr. Res.*, 88 (1981) 39.
264 Y.W. Ha and R.L. Thomas, *J. Food Sci.*, 53 (1988) 574.
265 A.G.W. Bradbury, D.J. Halliday and D.G. Medcalf, *J. Chromatogr.*, 213 (1981) 146.
266 T. Niwa, N. Yamamoto, K. Maeda, K. Yamada, T. Ohki and M. Mori, *J. Chromatogr.*, 277 (1983) 25.
267 T. Niwa, K. Yamada, T. Ohki and A. Saito, *J. Chromatogr.*, 336 (1984) 345.
268 C.W. Ford, *J. Chromatogr.*, 333 (1985) 167.
269 G.J. Gerwig, J.P. Kamerling and J.F.G. Vliegenthart, *Carbohydr. Res.*, 129 (1984) 149.
270 D.E. Willis, *J. Chromatogr. Sci.*, 21 (1983) 132.
271 R. Mateo, F. Bosch, A. Pastor and M. Jimenez, *J. Chromatogr.*, 410 (1987) 319.
272 O. Pelletier and S. Cadieux, *J. Chromatogr.*, 231 (1982) 225.
273 P. Englmaier, *Carbohydr. Res.*, 144 (1985) 177.
274 E.J.-M. Selosse and P.J. Reilly, *J. Chromatogr.*, 328 (1985) 253.
275 K. Bryn and E. Jantzen, *J. Chromatogr.*, 240 (1982) 405.
276 I. Brondz and I. Olsen, *J. Chromatogr.*, 310 (1984) 261.
277 T. Shinohara, *J. Chromatogr.*, 207 (1981) 262.
278 H. Haga and T. Nakajima, *Chem. Pharm. Bull.*, 36 (1988) 1562.
279 P. Decker and H. Schweer, *J. Chromatogr.*, 236 (1982) 369.
280 P. Decker and H. Schweer, *Carbohydr. Res.*, 107 (1982) 1.
281 H. Schweer, *J. Chromatogr.*, 236 (1982) 355.
282 H. Schweer, *J. Chromatogr.*, 236 (1982) 361.
283 H. Schweer, *Carbohydr. Res.*, 111 (1982) 1.
284 P. Decker, H. Schweer and R. Pohlmann, *J. Chromatogr.*, 244 (1982) 281.
285 F.R. Seymour, E.C.M. Chen and J.E. Stouffer, *Carbohydr. Res.*, 83 (1980) 201.
286 T.P. Mawhinney, M.S. Feather, G.J. Barbero and J.R. Martinez, *Anal. Biochem.*, 101 (1980) 112.

287 T.P. Mawhinney, *J. Chromatogr.*, 351 (1986) 91.
288 J.-R. Neeser and T.F. Schweizer, *Anal. Biochem.*, 142 (1984) 58.
289 P.A. Biondi, F. Manca, A. Negri, C. Secchi and M. Montana, *J. Chromatogr.*, 411 (1987) 275.
290 J. Lehrfeld, *Anal. Chem.*, 56 (1984) 1803.
291 J. Lehrfeld, *Carbohydr. Res.*, 135 (1985) 179.
292 J. Lehrfeld, *Anal. Chem.*, 57 (1985) 346.
293 J. Lehrfeld, *J. Chromatogr.*, 408 (1987) 245.
294 J.S. Walters and J.I. Hedges, *Anal. Chem.*, 60 (1988) 988.
295 S. Honda, N. Yamauchi and K. Kakehi, *J. Chromatogr.*, 169 (1979) 287.
296 S. Honda, M. Kakehi and K. Okada, *J. Chromatogr.*, 176 (1979) 367.
297 S. Honda, M. Nagata and K. Kakehi, *J. Chromatogr.*, 209 (1981) 299.
298 K. Kakehi, K. Maeda, M. Tetramae, S. Honda and T. Takai, *J. Chromatogr.*, 272 (1983) 1.
299 H.J.C. das Neves, A.M.V. Riscado and H. Frank, *Carbohydr. Res.*, 152 (1986) 1.
300 H.J.C. das Neves, A.M.V. Riscado and H. Frank, *J. High Resolut. Chromatogr. Chromatogr. Commun.*, 9 (1986) 662.
301 H.J.C. das Neves and A.M.V. Riscado, *J. Chromatogr.*, 367 (1986) 135.
302 K. Leontein, B. Lindberg and J. Lönngren, *Carbohydr. Res.*, 62 (1978) 359.
303 G.J. Gerwig, J.P. Kamerling and J.F.G. Vliegenthart, *Carbohydr. Res.*, 62 (1978) 349.
304 G.J. Gerwig, J.P. Kamerling and J.F.G. Vliegenthart, *Carbohydr. Res.*, 77 (1979) 1.
305 M.R. Little, *Carbohydr. Res.*, 105 (1982) 1.
306 R. Oshima, J. Kumanotani and C. Watanabe, *J. Chromatogr.*, 259 (1983) 159.
307 S. Hara, H. Okabe and K. Mihashi, *Chem. Pharm. Bull.*, 35 (1987) 501.
308 H. Schweer, *J. Chromatogr.*, 243 (1982) 149.
309 H. Schweer, *J. Chromatogr.*, 259 (1983) 164.
310 A.L. Leavitt and W.R. Sherman, *Carbohydr. Res.*, 103 (1982) 203.
311 A.L. Leavitt and W.R. Sherman, *Methods Enzymol.*, 89 (1982) 3.
312 W.A. König, I. Benecke and H. Bretting, *Angew. Chem. Int. Ed. Engl.*, 20 (1981) 693.
313 W.A. König, I. Benecke and S. Sievers, *J. Chromatogr.*, 217 (1981) 71.
314 W.A. König and I. Benecke, *J. Chromatogr.*, 269 (1983) 19.
315 W.A. König, S. Lutz and G. Wenz, *Angew. Chem. Int. Ed. Engl.*, 27 (1988) 979.
316 W.A. König, S. Lutz, P. Mischnick-Lübbecke, B. Brassat and G. Wenz, *J. Chromatogr.*, 447 (1988) 193.
317 W.A. König, P. Mischnick-Lübbecke, B. Brassat, S. Lutz and G. Wenz, *Carbohydr. Res.*, 183 (1988) 11.
318 T.J. Waeghe, A.G. Darvill, M. McNeil and P. Albersheim, *Carbohydr. Res.*, 123 (1983) 281.
319 P.E. Jansson, L. Kenne, H. Liedgren, B. Lindberg and J. Lönngren, *Chem. Commun. (Univ. Stockholm)*, No. 8 (1976).
320 M. McNeil and P. Albersheim, *Carbohydr. Res.*, 56 (1977) 239.
321 P.L. Weber and D.M. Carlson, *Anal. Biochem.*, 121 (1982) 140.
322 F.R. Seymour, E.C.M. Chen and S.H. Bishop, *Carbohydr. Res.*, 73 (1979) 19.
323 F.R. Seymour, R.D. Plattner and M.E. Slodki, *Carbohydr. Res.*, 44 (1975) 181.
324 F.R. Seymour, R.D. Knapp, E.C.M. Chen, S.H. Bishop and A. Jeanes, *Carbohydr. Res.*, 74 (1979) 41.
325 G. Petersson, *Carbohydr. Res.*, 33 (1974) 47.
326 R.A. Laine and C.C. Sweeley, *Carbohydr. Res.*, 27 (1973) 199.
327 H.A.S. Aluyi and D.B. Drucker, *J. Chromatogr.*, 178 (1979) 209.
328 M. Rivière, J.-J. Fournié, B. Monsarrat and G. Puzo, *J. Chromatogr.*, 445 (1988) 87.
329 V. Mihálov, V. Kováčik and P. Kováč, *Carbohydr. Res.*, 73 (1979) 267.
330 K. Umeki and K. Kainuma, *Carbohydr. Res.*, 96 (1981) 143.
331 G. Caldes, B. Prescott and P.J. Baker, *J. Chromatogr.*, 234 (1982) 264.
332 L. Bilisics and L. Petrus, *Carbohydr. Res.*, 146 (1986) 141.
333 E.A. Ryan and A.M. Kropinski, *J. Chromatogr.*, 195 (1980) 127.
334 J.P. Métraux, *J. Chromatogr.*, 237 (1982) 525.
335 T. Horikoshi, T. Koga and S. Hamada, *J. Chromatogr.*, 416 (1987) 353.

336 K. Lajunen, S. Purokoski and E. Pitkänen, *J. Chromatogr.*, 187 (1980) 455.
337 R. Klaus and J. Ripphahn, *J. Chromatogr.*, 244 (1982) 99.
338 D. Nurok and A. Zlatkis, *Carbohydr. Res.*, 81 (1980) 167.
339 P. Würsch and Ph. Roulet, *J. Chromatogr.*, 244 (1982) 177.
340 L.W. Doner, *Methods Enzymol.*, 160 (1988) 176.
341 L.W. Doner, P.L. Irwin and M.J. Kurantz, *Carbohydr. Res.*, 172 (1988) 292.
342 S. Ando, H. Waki and K. Kon, *J. Chromatogr.*, 405 (1987) 125.
343 K. Koizumi, T. Utamura, T. Kuroyanagi, S. Hizukuri and J.-I. Abe, *J. Chromatogr.*, 360 (1986) 397.
344 L.W. Doner, C.L. Fogel and L.M. Biller, *Carbohydr. Res.*, 125 (1984) 1.
345 L.W. Doner and L.M. Biller, *J. Chromatogr.*, 287 (1984) 391.
346 S. Honda, Y. Matsuda and K. Kakehi, *J. Chromatogr.*, 176 (1979) 433.
347 M. Jimenez and C.E. Weill, *Carbohydr. Res.*, 101 (1982) 133.
348 R. Klaus and W. Fischer, *Methods Enzymol.*, 160 (1988) 159.
349 D.C. Farwell and A.S. Dion, *Anal. Biochem.*, 95 (1979) 533.
350 F. Searle and H. Weigel, *Carbohydr. Res.*, 85 (1980) 51.
351 G. Bonn, M. Grünwald, H. Scherz and O. Bobleter, *J. Chromatogr.*, 370 (1986) 485.
352 R. Cappelletti, M. del Rosso and V.P. Chiarugi, *Anal. Biochem.*, 99 (1979) 311.
353 E.H. Schuchman and R.J. Desnick, *Anal. Biochem.*, 117 (1981) 419.
354 E.A. Johnson, *J. Chromatogr.*, 233 (1982) 365.
355 P. Oreste and G. Torri, *J. Chromatogr.*, 195 (1980) 398.
356 P. Bianchini, H.B. Nader, H.K. Takahashi, B. Osima, A.H. Straus and C.P. Dietrich, *J. Chromatogr.*, 196 (1980) 455.
357 G. Dubray and G. Bezard, *Anal. Biochem.*, 119 (1982) 325.
358 D.W. Gammon, A.M. Stephen and S.C. Churms, *Carbohydr. Res.*, 158 (1986) 157.
359 S. Pelkonen, J. Hayrinen and J. Finne, *J. Bacteriol.*, 170 (1988) 2646.
360 T. Kumuro and C. Galanos, *J. Chromatogr.*, 450 (1988) 381.

Chapter 17

Nucleic acids, their constituents and analogs

NAN-IN JANG and PHYLLIS R. BROWN

CONTENTS

17.1 INTRODUCTION

The first part of this chapter is divided into sections based on the molecular size of the compounds. One section is devoted to separation methods for bases, nucleosides, and nucleotides, another to oligonucleotides, and a third to the macromolecular nucleic acids. In the latter part of the chapter gel electrophoresis and capillary electrophoresis are discussed, and another section covers the sequencing of nucleic acids, which is of particular importance in regard to the research in progress on deciphering the structure of the human genome [1].

This chapter is not intended to be a comprehensive review. Although no review would be complete without mentioning some of the pioneering work done on the separation of nucleic acids, their constituents and analogs, this chapter covers literature mainly from the past decade, from 1980 through the present. For earlier work on the separation of nucleic acids and their constituents, there are many reviews in the literature [2-14].

17.1.1 Importance

The building blocks of nucleic acids are made up of nucleotides, which are composed of nucleosides (ribosides or deoxyribosides of purine or pyrimidine bases) and a phosphate group. Free nucleotides, nucleosides, and their bases are present in biological matrices as a result of the catabolism of nucleic acids, enzymatic degradation of tissues, dietary intake, and various salvage pathways [15,16].

In the last two decades conditions have been optimized for HPLC separations so that the normal range of concentrations of nucleosides and their bases in physiological fluids and of nucleotides in cells of healthy normal subjects and in patients with various disease states can be readily determined [17-19]. Certain analogs of nucleosides or their bases are antimetabolites or antitumor agents [20,21] that exert their therapeutic effects after in vivo conversion to their respective deoxyribonucleotides, which are then incorporated into deoxyribonucleic acids (DNA) [22].

Oligonucleotides are nucleotide fragments with various chain lengths, prepared synthetically or produced by digestion of polynucleotides with restriction enzymes of DNA or RNA [23,24]. The chemical authenticity of a particular oligonucleotide can be crucial in clinical diagnostics and molecular biology. In addition, oligonucleotides can be used to generate probes for in situ hybridization when the expression of specific members of a multi-gene family is investigated [25].

In order to understand the proper function of cells, tissues, and organs and to open new avenues for therapy, the metabolic products of nucleic acids in physiological fluids must be determined [26,27]. In addition, sensitive and easily reproducible analytical techniques are needed to determine not only levels of naturally occurring nucleotides, nucleosides, their bases and methylated derivatives, but also to determine the activity of enzymes which catalyze their metabolic reactions. Moreover, assays are needed for nucleotides that regulate metabolic reactions and for pharmacokinetic studies of drugs and their metabolites [28].

17.1.2 The types of compounds

Nucleic acids are macromolecules with negative charges at physiological pH. They are composed of long chains of nucleotides, which are linked to each other via a phosphate group. A 3′,5′-phosphodiester bond bridges the sugar moieties of adjacent nucleosides. Nucleic acids are classified in two broad categories: deoxyribonucleic acids (DNA) and ribonucleic acids (RNA). DNA (Fig. 17.1) [29] contains the 2′-deoxy-D-ribose moiety whereas RNA contains the D-ribose moiety. Since nucleic acids contain heterocyclic purine and pyrimidine bases, they strongly absorb UV light in the region of 250-280 nm.

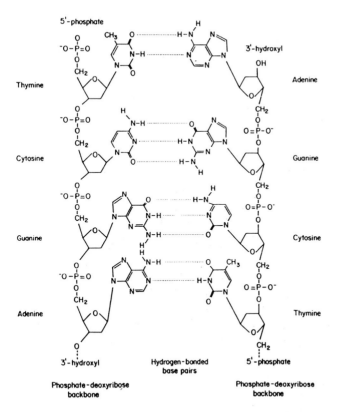

Fig. 17.1. Structure of double-stranded DNA, showing the specific associations between the nucleoside subunits on opposite strands (base pairing). (Reproduced from Ref. 29, with permission.)

TABLE 17.1

COMPOUND ABBREVIATIONS

Nucleotides

Abbreviation	Compound	Abbreviation	Compound
ADP	Adenosine 5'-diphosphate	dTTP	2'-Deoxythymidine 5'-triphosphate
AMP	Adenosine 5'-monophosphate	dUMP	2'-Deoxyuridine 5'-monophosphate
ATP	Adenosine 5'-triphosphate	GDP	Guanosine 5'-diphosphate
cAMP	Adenosine 3',5'-cyclic monophosphate	GMP	Guanosine 5'-monophosphate
cCMP	Cytidine 3',5'-cyclic monophosphate	GTP	Guanosine 5'-triphosphate
cGMP	Guanosine 3',5'-cyclic monophosphate	IDP	Inosine 5'-diphosphate
cIMP	Inosine 3',5'-cyclic monophosphate	IMP	Inosine 5'-monophosphate
CDP	Cytidine 5'-diphosphate	ITP	Inosine 5'-triphosphate
CMP	Cytidine 5'-monophosphate	β-NAD	β-Nicotinamide adenine dinucleotide
CTP	Cytidine 5'-triphosphate	NAD	Nicotinamide adenine dinucleotide
cUMP	Uridine 3',5'-cyclic monophosphate	NADH	Nicotinamide adenine dinucleotide; reduced form
dADP	2'-Deoxyadenosine 5'-diphosphate		
dAMP	2'-Deoxyadenosine 5'-monophosphate	NADP	Nicotinamide adenine dinucleotide 3'-phosphate
dATP	2'-Deoxyadenosine 5'-triphosphate	NADPH	Nicotinamide adenine dinucleotide 3'-phosphate; reduced form
dCMP	2'-Deoxycytidine 5'-monophosphate		
dCTP	2'-Deoxycytidine 5'-triphosphate	TDP	Thymidine 5'-diphosphate
dGMP	2'-Deoxyguanosine 5'-monophosphate	TMP	Thymidine 5'-monophosphate
dGTP	2'-Deoxyguanosine 5'-triphosphate	UDP	Uridine 5'-diphosphate
dm6AMP	2'-Deoxy-6-methyladenosine 5'-monophosphate	UMP	Uridine 5'-monophosphate
dm5CMP	2'-Deoxy-5-methylcytidine 5'-monophosphate	UTP	Uridine 5'-triphosphate
dTMP	2'-Deoxythymidine 5'-monophosphate	XMP	Xanthosine 5'-monophosphate

Nucleosides

Ado	Adenosine
Ctd	Cytidine
dAdo	2'-Deoxyadenosine
dCtd	2'-Deoxycytidine
dGuo	2'-Deoxyguanosine
dIno	2'-Deoxyinosine
dm6Ado	2'-Deoxy-N^6-methyladenosine
dThd	2'-Deoxythymidine
Guo	Guanosine
Ino	Inosine
m6Ado	N^6-Methyladenosine
m5Ctd	5-Methylcytidine
m1Guo	1-Methylguanosine
m2Guo	N^2-Methylguanosine
m7Guo	7-Methylguanosine
m2,2Guo	N^2_2-Dimethylguanosine
m1Ino	1-Methylinosine
m7Ino	7-Methylinosine
m7Xao	7-Methylxanthosine
ψ	Pseudouridine
Thd	Thymidine
Urd	Uridine
Xao	Xanthosine

Bases

Ade	Adenine
Cyt	Cytosine
Gua	Guanine
Hyp	Hypoxanthine
m1Ade	1-Methyladenine
m3Ade	3-Methyladenine
m6Ade	N^6-Methyladenine
m7Ade	7-Methyladenine
m5Cyt	5-Methylcytosine
m1Gua	1-Methylguanine
m2Gua	N^2-Methylguanine
m7Gua	7-Methylguanine
m1Xan	1-Methylxanthine
m3Xan	3-Methylxanthine
m9Xan	9-Methylxanthine
Thy	Thymine
Ura	Uracil
Xan	Xanthine

References on p. B326

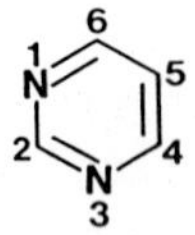

Fig. 17.2. Structure and systematic numbering of the pyrimidine base.

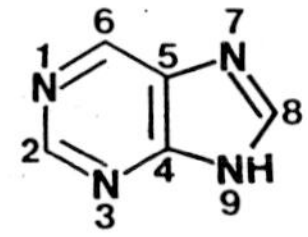

Fig. 17.3. Structure and systematic numbering of the purine base.

For a list of the major naturally occurring purine and pyrimidine bases, their nucleosides and nucleotides, along with the accepted abbreviations, see Table 17.1.

Pyrimidines, which are derivatives of the parent compound pyrimidine, and purines, which consist of a pyrimidine ring fused with an imidazole ring, are weakly basic compounds. They exist in tautomeric forms, and the ring systems show marked aromatic properties. The numbering of substituents on the pyrimidine and purine rings are shown in Figs. 17.2 and 17.3.

Nucleosides consist of a purine or pyrimidine combined with a pentose or a deoxypentose sugar. The nucleoside of adenine is adenosine; of guanine, guanosine; of cytosine, cytidine; of uracil, uridine; and of thymine, thymidine. The major nucleosides obtained from DNA are 2'-deoxyadenosine, 2'-deoxyguanosine, 2'-deoxycytidine, and thymidine. Since the nucleoside of thymine occurs naturally only as the 2'-deoxyribonucleoside, it is always referred to in the literature as thymidine and not 2'-deoxythymidine.

Nucleotides are phosphate esters of the nucleosides (Fig. 17.4). Those derived from ribose nucleosides are referred to as ribonucleotides and those from deoxyribose nucleo-

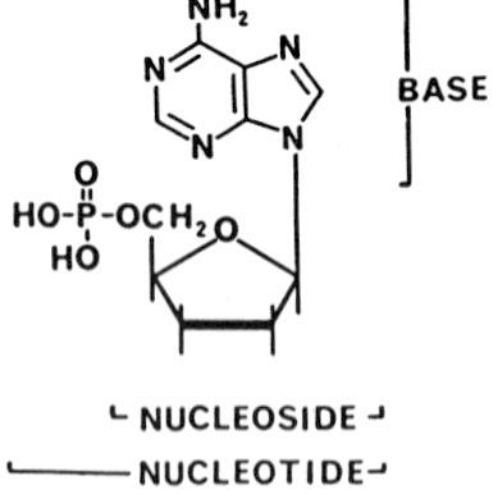

BASE

HO-P-OCH₂

└ NUCLEOSIDE ┘

└──────NUCLEOTIDE┘

Fig. 17.4. Structural relationship of nucleotide, nucleoside, and base.

tides as deoxyribonucleotides. Since the ribonucleosides have three hydroxy groups on the ribose ring, three ribonucleotides can be formed. For example, the monophosphate esters of adenosine are adenosine 5'-monophosphate, adenosine 3'-monophosphate, and adenosine 2'-monophosphate. The free 5'-ribonucleotides and 5'-deoxyribonucleotides in the mono-, di-, or triphosphate form can be present in biological matrices. In addition, nucleotides can be present as the 3'-monophosphates or in the cyclic form, for example, as 3',5'-cyclic adenosine monophosphate (cAMP).

The cyclic nucleotides of cytosine, uracil, and hypoxanthine have been isolated in extremely low levels from biological samples [30-34]. However, the concentrations of cAMP can range from micromolar in urine to nanomolar in cerebrospinal fluid and plasma to picomolar in tissue [35], and the levels of the cyclic guanosine monophosphate (cGMP) in these matrices are one or more orders of magnitude lower than those of the cyclic AMP levels [35].

17.1.3 Biological matrices

Free nucleotides are not normally found in significant quantities in physiological fluids. However, they are present in appreciable quantities in tissues and cells [15,16,27]. Nucleotide profiles of various tissues have been reported.

Since free nucleosides and bases can readily diffuse out of cells, they are present mainly in physiological fluids. In order to understand metabolic disorders and pathological states, there have been many investigations of nucleosides and bases in serum [36], plasma [37-41], urine [40,42,43], and erythrocytes [40].

17.2 PURINE AND PYRIMIDINE BASES, THEIR NUCLEOSIDES AND NUCLEOTIDES

17.2.1 Sample preparation

In preparing biological samples prior to analysis, proteins and other macromolecules are removed to maximize isolation and preservation of the desired components. Enzymes released from cells can cause degradation of the components or cell lysis; thus, the enzymatic activity is halted as soon as possible after the sample is collected. With tissue samples, immediate freezing by the freeze-clamp method [30,44,45] or submerging the sample in liquid nitrogen [46-53] has been used to retard enzyme action prior to deproteinization. For plasma, blood samples are usually collected in heparinized tubes, and defibrination is achieved by shaking the blood sample at room temperature with glass beads [54]. Cells obtained from cultures or blood samples must be kept cold prior to deproteinization. Sodium fluoride can also be added to freshly drawn blood to inhibit phosphatase activity [55]. The sample is then centrifuged to prevent clogging the columns.

Non-enzymatic proteins can interfere with analyses due to irreversible binding of the protein to the analytical column, thus decreasing retention reproducibility and column lifetime. Deproteinization can be achieved by changing the ionic strength, temperature, or pH of a biological matrix, or by the addition of an organic solvent. Proteins also can be removed by filtration, adsorption, or specific complexation [56,57]. A combination of methods may be advisable to obtain very clean samples. However, deproteinization may result in a loss of the solutes of interest due to the adsorption of the analyte on the precipitate.

17.2.1.1 Deproteinization

Insoluble salts are formed when deproteinization is achieved by the addition of chilled strong acid. Perchloric acid (PCA) and trichloroacetic acid (TCA) are widely used to precipitate proteins in biological fluids prior to chromatographic analysis [18,19,30,31, 37,48-62]. The resulting precipitate is removed by centrifugation, and the excess acids or salts in the supernatant are then removed by extraction with water-saturated diethyl ether [46,56,61] or Freon/amine solution [42,47,50], or the acids are neutralized by alkali.

The use of a solution of trioctylamine in Freon to extract the excess TCA has the advantages of simplicity, rapidity, and good recovery of nucleotides [63]. This procedure was found to be less suitable for nucleosides and bases because of the coprecipitation of the purine and pyrimidine compounds with the protein [56]. When PCA is used instead of TCA, the resulting acidic supernatant is usually neutralized with a potassium hydroxide solution and phosphate buffers, and better recovery of nucleosides and bases is achieved [18,30,64]. However, when high sensitivity is needed for the determination of low levels of bases and nucleosides, as in the analysis of cerebrospinal fluid, UV-absorbing compounds can interfere with the analysis [58]. In addition, coprecipitation is a serious problem that can cause loss of analytes [64]. Instead of using PCA or TCA, Dutta and O'Donovan [65] used trifluoroacetic acid (TFA) as the solvent for the extraction of nucleosides and nucleotides from *Escherichia coli*. This acid provided higher sensitivity and reliability than extraction with TCA, PCA, or formic acid.

To remove nicotinamide adenine dinucleotide (NAD^+) and its reduced form (NADH) from human platelets, an alkaline solution was used [66]. The dinucleotides were extracted with ice-cold NaOH, immediately sonicated to ensure maximal extraction, and then neutralized with KH_2PO_4. The yield of total adenine nucleotides was better in the alkaline extract than in an acid extract. In a basic solution, ATP and ADP are more stable, and their degradation to AMP or inosine is slower than in the PCA extract. Moreover, only the alkaline extraction method allows measurements of both NADH and NAD^+, since the reduced form is unstable in an acidic medium.

Most proteins are insoluble in distilled water but dissolve in dilute inorganic salt solutions. The use of high concentrations of salt solutions, commonly called the "salting-out" technique, is a classical method for precipitating proteins from biological samples. In this technique, a solution of 3 M ammonium sulfate is added to biological samples, and the supernatant is filtered and analyzed after centrifugation. One advantage of this procedure is that the process is reversible and proteins can be reconstituted with full recovery of their biological activity. Another advantage of the salting-out technique is that nucleosides and bases are extracted from serum samples with 100% recovery [46]. However, the sample may become contaminated with ammonium sulfate. In addition, sample dilution can cause a significant decrease in the sensitivity of HPLC analysis.

Ultrafiltration has been established as a reliable and convenient method for the preparation of protein-free samples [42,66-72]. Disposable ultrafiltration cones or ultrafilters separate solutes quickly and easily according to molecular size with high recovery. Compounds larger than 25 kD are filtered out of the samples. Either pneumatic or centrifugal force is used to push the compounds smaller than 25 kD through the membranes. The method does not alter the pH of the medium nor does it dilute the samples; however, only the concentrations of free nucleotides, nucleosides, or bases in the filtrate can be determined, since protein-bound components will not penetrate the filtration barrier. Because ultrafiltration membranes rapidly become clogged with cellular debris, this method is not recommended for the treatment of cellular or tissue samples.

17.2.1.2 Solid-phase extraction

In solid-phase extraction, or precolumn concentration, a cartridge of bonded phase is used prior to chromatographic analysis. This cartridge allows most interfering proteins to pass through, while retaining the compounds of interest. By changing to an appropriate solvent, the analytes are subsequently eluted selectively. Because the analytes are concentrated on the column, the sensitivity is increased. Hydrostatic pressure, centrifugal force, or pressure obtained by means of a syringe can be used to push the components through the cartridge.

During the last few years, disposable short columns with reversed-phase packings, such as the SepPak C_{18} Cartridge, have been marketed for this purpose [73,74]. This technique has practical advantages, such as ease of preparation, time saving, and smaller sample requirements [74-79]. In addition, no interfering materials are introduced into the sample. The method is rapid and the sensitivity is increased. However, the acidic solution used to remove proteins from the cartridge may hydrolyze nucleoproteins thus causing errors in the concentrations of free nucleosides or nucleotides.

17.2.1.3 Selective removal of interfering materials

Affinity chromatography has been successfully used in sample preparation for the selective removal of interfering materials. This method involves the formation of complexes between the particular interfering material and the stationary phase. A crude sample is introduced into a cartridge or column, packed with an affinity stationary phase. The components that have selective affinity for the stationary phase are retained, while components that have no affinity are eluted with a buffer. Subsequently, the column is cleaned by changing the eluent to remove the retained interfering materials [73,80-84]. It is also possible to retain only the compounds of interest and to elute all other compounds. The analytes are then stripped from the column by using the appropriate mobile phase.

For instance, an immobilized phenylboronic acid gel has been used as the affinity sorbent to bind selectively the *cis*-diols of ribose in ribonucleosides and/or ribonucleotides under mild and reversible conditions [81-86]. With this gel, the ribonucleosides are readily separated from the deoxyribonucleosides and the purine and pyrimidine bases in physiological fluids. Gehrke et al. [80] developed modified boronate affinity gels to isolate ribonucleosides as boronate complexes from biological samples. This is a rapid procedure, which results in high sensitivity for the analysis of either ribonucleosides [80] or deoxyribonucleosides [82] in physiological fluids. Recently, Echizen et al. [86] used a Bond-Elut column, packed with phenylboronate-bonded silica, to isolate adenosine (Ado) and dopamine (Do) from urine. This affinity adsorbent has been shown to interact selectively with the *cis*-diol group of Ado and Do, particularly under alkaline conditions. Another specific sorbent, a metal chelate affinity sorbent introduced by Hubert and Porath [85], was used for the group separation of various nucleotides by the formation of complexes of copper-loaded packings with purine mononucleotides.

17.2.2 Chromatography

17.2.2.1 Anion-exchange chromatography

Most of the early chromatographic methods for the separation of nucleic acid constituents were based on ion exchange. In 1949 when Cohn [87] separated various nucleotides on polymeric ion exchangers, the analyses were slow and efficiency limited. Later, in 1967 Horváth et al. [88] developed pellicular ion-exchange packings, which were the first high-efficiency stationary phases used in HPLC. Subsequently, microparticulate, chemically bonded, silica-based ion-exchange packings were developed and were applied to the separation of nucleotides [89,90]. With these packings, nucleotide analyses by HPLC were improved [91,92]. However, because there is no ionic moiety in the purine and pyrimidine bases or their nucleosides, it was difficult to separate the nucleosides concomitantly with the bases by ion exchange.

The addition of an organic solvent to the ionic mobile phase changes selectivity in the ion-exchange separation of nucleosides and bases dramatically [93]. For example, Nguyen et al. [94] developed a simple, rapid method of analyzing the major monophosphate nucleotides in the hydrolyzates of RNA by isocratic elution from a Partisil 10 SAX anion-exchange column with a mobile phase of 3% methanol in 8 mM KH_2PO_4 buffer (pH 4.15).

Ion-exchange separations have several disadvantages: columns tend to be less efficient and less stable than RPLC columns; the choice of packing materials is limited; and long equilibration times are required when gradient elution is used.

17.2.2.2 Reversed-phase chromatography

Due to their ionic character, nucleotides (with the exception of cyclic nucleotides) in biological fluids cannot be well resolved on reversed-phase packings, unless ion-suppression or ion-pairing methods are employed. The retention of nucleotides on a stationary phase such as a C_{18} column can be improved by ion suppression. For example, a mobile phase of low pH shifts the chemical equilibrium to the nonionic state of the phosphate groups [95]. Another approach is to use ion pairing; in the presence of low concentrations of a cationic ion-pairing agent, nucleotides can be well resolved on a reversed-phase column [96]. In general, at a suitable pH value the k' values for most nucleotides can be increased 3- or 4-fold by the inclusion of 1-5 mmol/l of an ion-pairing reagent [93].

Perrone and Brown [97] investigated some of the variables that affect the retention behavior of nucleotides in ion-pair chromatography and presented a number of guide-lines. The variables included the concentration of the ion-pairing agents, the chainlength or solvophobicity of the alkyl group of the quaternary amines, and the pH of the mobile phase. The optimum concentration of tetrabutylammonium ion (TBA) was between 10 and 25 μmol/l. The k' value, as expected, was found to be a direct function of solvophobicity of the alkyl groups of the ion-pairing reagent; e.g., the k' value is largest with tetrabutyl and smallest with tetramethyl. TBA phosphate, the pairing agent with the longest chainlength, is most effective for retarding the elution of the majority of the nucleotides, especially the triphosphates. For ion pairing to occur, the pH of the mobile phase must be greater than the acid dissociation constants of the phosphate moiety, especially in those nucleotides containing Ade, Gua, and Cyt. Retention of the ion-paired nucleotides was greater at pH 5.7 than at 3.0, the only two pH values studied. In most cases, the elution order at pH 5.7 was mono- < di- < triphosphates. In addition, ion-pair chromatography of nucleotides is also affected by the concentration of buffer salts in the eluent, since the ions of the buffer will compete with the ion-pairing agents for complexation with the nucleotides [98].

A variation of ion-pair chromatography, called zwitterion chromatography, was intro-duced by Knox and Jurand [99]. The zwitterionic pairing agent, 11-aminoundecanoic acid, forms a quadrupolar ion pair with the nucleotide phosphate group. Separations were optimized by adjusting the pH of the mobile phase and the concentration of the pairing agent. The method does not necessarily require a gradient and has been successfully used for the separation of 10 nucleotides.

Since nucleosides and bases lack the charged phosphates of the nucleotides, they are prime candidates for RPLC. In order to optimize the separation of nucleosides and bases by RPLC, the mobile-phase conditions that affect the retention and selectivity parameters were investigated [100,101]. The conditions include: organic modifier, temperature, pH, ionic strength of the buffer, and the kind and amount of ion-pairing reagents. In addition, structure/retention relationships were studied for a selected group of nucleic acid compo-nents, and rules were formulated for prediction of retention behavior [102]. These rules indicated that the type of substituent and the position of substituents on the ring of the bases affect the k'. In purines, the 6-position is more important in controlling retention than the 2-position. For the pyrimidines, the predicted trend of retention is Ura < Cyt < Thy. The presence of a ribosyl moiety greatly increases the k' of the purine or pyri-midine, and the 2'-deoxyribosyl compounds have longer retention times than the corresponding ribonucleosides. Whether or not a methyl group increases the k' depends on the location of the substituent and whether a charge is generated by the substitution.

References on p. B326

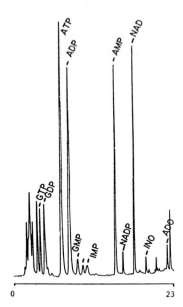

Fig. 17.5. Chromatograms of human kidney biopsy taken during donor nephrectomy. The purine compounds were separated on a 4-μm LiChrosorb RP-18 column (25 x 0.4 cm ID) by a linear gradient from 85% Eluent A to 15% Eluent B at a flowrate of 0.7 ml/min. Eluent A, 2% H_3PO_4 (pH 4.0); Eluent B, acetonitrile/methanol (1:1). (Reproduced from Ref. 46, with permission.)

In most of the reports on applications of RPLC to the separation of nucleosides and bases, either 5- or 10-μm ODS packings and gradient elution were used to separate the large number of purine and pyrimidine compounds in biological fluids. A gradient from 0% to 10% acetonitrile in a phosphate buffer, with TBA hydroxide as the ion-pairing agent, was used to monitor nucleotide pools during smooth-muscle contraction [31]. Levels of 14 nucleotides, including cAMP, in isolated muscle fibers were investigated to study energy metabolism and contractility. Maessen et al. [46] also used RPLC for determining the nucleotides, nucleosides, and oxypurines in renal cortical tissue. The method they used, with gradient elution but without an ion-pairing agent in the mobile phase, provides a rapid, reproducible diagnostic tool for assessing the chemical energy status of human kidneys in renal surgery and transplantation. A standard solution, containing 19 purine compounds, was resolved in 23 min (Fig. 17.5).

With isocratic elution, the system is simpler and more sensitive without undue baseline drift. Since no equilibration is required, there is also a higher throughput of samples. Recently, Lim and Peters [34] used an isocratic RPLC system to separate 20 ribonucleotides, deoxyribonucleotides, cyclic nucleotides, and deoxycyclic nucleotides. The effects of methanol content, pH, and ionic strength of the buffer on the retention and resolution of these nucleotides were studied. Olivares and Verdys [37] described a rapid and efficient isocratic HPLC method for the analysis of all pyrimidine bases and their nucleosides in plasma in 20 min.

In the application of RPLC to analogs of nucleic acid constituents, Brubaker and Prusoff [103] developed a gradient RPLC method for the separation of nitrosourea nucleosides. The detection limits were in the range of 1-2 ng. Blau et al. [104] analyzed and identified a 2′,3′-dideoxyribonucleoside and its metabolites in biological fluids by using a thermospray RPLC/MS system. Recently RPLC methods for the analysis of 3′-azido-2′,3′-dideoxythymidine (AZT) and other dideoxyribonucleosides in various biological matrices were developed [105-110].

17.2.2.3 Fast, microbore, and fast microbore chromatography

In order to use HPLC for the routine determination of trace levels of nucleosides and bases in physiological fluids and for therapeutic drug monitoring, Simpson and Brown [111] investigated HPLC techniques to maintain chromatographic selectivity and resolution while increasing sensitivity and sample throughput. These techniques include fast HPLC, microbore HPLC, and fast microbore HPLC. In their work, they defined fast HPLC as the use of columns, 5-10 cm in length with an internal diameter of 4.6 mm, microbore HPLC as the use of columns, 25-30 cm x 1-2 mm ID, and fast microbore HPLC as the use of columns, 3-5 cm x 1-2 mm ID. At present, columns with an internal diameter of 1-2 mm are referred to as small-bore columns, in contrast to microbore columns, which have an internal diameter of a fraction of a micrometer [112]. The performance of these columns was compared with the performance of columns, 25-30 cm x 4.6 mm ID. Both isocratic and gradient elution were used. The model compounds were purines and pyrimidines. Besides increased sensitivity, efficiency, and linear response range, the practical advantages of the fast microbore technique (columns 5 cm x 2 mm ID) included the reduction of sample size and volume of mobile phase, short analysis time, and low cost per analysis [113-116]. The fast microbore system, which is 40 times more sensitive than conventional HPLC, was the most suitable system for the rapid separation and quantitation of trace amounts of nucleosides and bases in very small volumes of physiological samples.

Another application of RPLC with a small-bore C_{18} column of 2.1 mm ID was developed by Hammer et al. [47]. A mobile phase containing the ion-pairing agent, TBA phosphate, was used to resolve all of the myocardial nucleotides and AMP degradation products in cardiac tissue samples. The gradient elution method described provided excellent resolution, reproducibility, and linearity. In 31 min, which included both separation and re-equilibration time, a large amount of information was obtained from a single chromatogram. Microbore columns, also referred to as capillary columns [117-119], have been applied to the separation of the nucleic acid constituents from the enzymatic hydrolysis products of tRNA-Phe [120].

References on p. B326

17.2.2.4 Column switching

The use of a column switching technique is another way to separate groups of compounds of widely differing properties and to establish a profile of the most important bases, nucleosides, and nucleotides in a single chromatogram. In an on-line two-stage column switching technique, reported by Lang and Rizzi [121], a RP C_{18} column was followed by an anion-exchange column for separating purine nucleic acid constituents in standard mixtures and human erythrocytes. The elution program consisted of two gradients. The bases, nucleosides, and deoxynucleosides were separated on the C_{18} column and, thereafter, the nucleotides and the nucleic acids on the anion exchanger. A column switching technique in the reverse order, i.e. anion exchanger followed by a C_{18} column, described by Halfpenny and Brown [122], was used to separate the ionic nucleotides from the nonionic nucleosides and bases (Fig. 17.6). By increasing the ionic strength, nucleosides and bases were eluted from the void volume of the anion-exchange column into a reversed-phase column. These nonionic compounds were then analyzed on the reversed-phase column by increasing the concentration of organic modifier.

In recent work, Mathes et al. [123] used a column switching technique to determine the concentrations of 2′,3′-dideoxycytidine (DDC) and AZT directly in plasma of cats infected with feline leukemia virus. The sample was first loaded onto a Pinkerton internal-surface reversed-phase column, which was packed with 5-μm silica particles, bonded with a tripeptide consisting of glycine-phenylalanine-phenylalanine groups, and then chromatographed on a Cyclobond I column.

17.2.2.5 Mixed-bed chromatography

Another approach to separating classes of compounds is the use of mixed-bed columns. These columns are prepared by mixing physically an appropriate ion-exchange matrix with a reversed-phase matrix [124] so that the resulting column will have sites for ionic as well for hydrophobic interactions. Hartwick and coworkers [125] separated the nucleotides from nucleosides on a mixed-bed column containing a 1:1 mixture by weight of a SAX and a C_8 packing. Good separation of nucleotides and nucleosides was obtained. When this approach was compared with the mixed-mode method (cf. Section 17.3.2.1), it was found that mixed-bed columns were less suitable for the analysis of high-molecular-weight solutes.

17.2.2.6 Affinity chromatography

Affinity chromatography is based on the interactions of matrix-bound affinity ligands with nucleic acid constituents. Elution of the nucleic acid components can be achieved either by changing the pH or ionic strength of the mobile phase or by addition of a competitive inhibitor to the mobile phase. Hubert and Porath [85] used metal chelate affinity chromatography to separate purine mononucleotides from pyrimidine mononucleotides. The column was packed with a copper chelate stationary phase. Only the

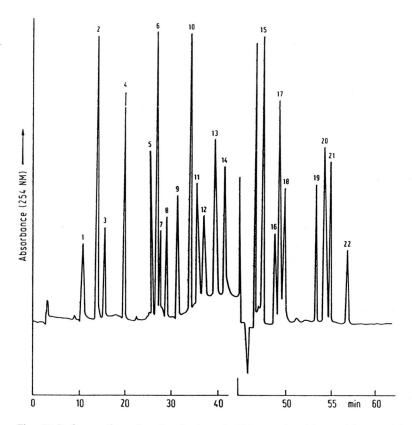

Fig. 17.6. Separation of a standard nucleotide, nucleoside, and base mixture by column switching. The nucleotides were separated on a Partisil 10-SAX column by a nonlinear gradient of buffer from 0.005 M KH$_2$PO$_4$ to 0.5 M KH$_2$PO$_4$ at a flowrate of 2 ml/min. After elution of the XTP peak, the column was switched out of line and a radially compressed C$_{18}$ column switched in line. The nucleosides were then separated by a linear gradient of 0.005 M KH$_2$PO$_4$/methanol (7:3). (Reproduced from Ref. 122, with permission.)

purine mononucleotides were retained on the column by complexation with the copper on the stationary phase. Another type of affinity chromatography, with immobilized boronates as the column packing, has also been applied to the separation of nucleosides in various biological samples, such as urine, and tissue extracts [126-128].

17.2.2.7 Preparative chromatography

Recently, Turcotte et al. [129] successfully isolated the AZT 5'-phosphate from a synthetic mixture on a preparative C$_{18}$ column. This compound was used as a starting material in the synthesis of target-site-directed AZT analogs. The method is also useful for the purification of other antiretroviral dideoxyribonucleotides.

References on p. B326

17.2.3 Peak identification

17.2.3.1 Retention times and mixed chromatograms

Peaks of nucleic acid constituents in chromatograms of biological samples can be identified only tentatively by retention times and the standard addition method. Therefore, other techniques must also be used to characterize these compounds.

The retention behavior of the bases, nucleosides, and nucleotides in RPLC has been found to be a function of pH. Jacobson et al. [130,131] and Assenza and Brown [132] formulated quantitative structure/retention relationships (QSRR) for nucleosides to estimate relative retention times from chemical structure. This information aids in identification of peaks in HPLC chromatograms.

17.2.3.2 Postcolumn ionization

The postcolumn ionization technique can be employed effectively to enhance the fluorescence and UV spectra of purine and pyrimidine compounds [132]. When an acid or base is introduced into the effluent in a postcolumn reactor ahead of the detector, the nucleic acid constituents are ionized by either deprotonation or protonation. The ionized purine and pyrimidine compounds have optical properties different from those of non-ionized ones, and their UV spectra and fluorescence spectra are distinctive. Hence, the effect of pH on UV spectra and fluorescence spectra can be used for the identification of these nucleic acid constituents.

Common postcolumn reagents used for this method include acids, such as sulfuric and phosphoric acid, and bases, such as potassium and sodium hydroxide. The maximum allowable concentration of their ions in the effluent is limited, since the fluorescence of nucleic acid compounds in solution can be quenched by excess phosphate and sulfate ions. However, neither potassium ions nor sodium ions will affect the fluorescence properties. Postcolumn ionization has several advantages. With the use of this method, the identity of two closely eluted compounds can be determined. For example, 7-methyl-adenine and 2-methylguanosine have retention times differing by < 0.5 min in RPLC. However, in basic solutions methylated guanosine fluoresces, while the adenine analog does not. Other advantages of this on-line technique are that the ionization process is rapid, nondestructive, and reversible, permitting collection of chromatographic fractions for further analysis. Additionally, the method is sensitive and specific. High sensitivity for the characterization of methylated guanines and other methylated purines has been reported, with detection limits as low as 1 pmol [132].

17.2.3.3 Selective detection

The high sensitivity and selectivity of fluorescence detection complements the more general detection of purines and pyrimidines by UV [133]. Relatively few nucleic acid constituents have fluorescence quantum efficiencies large enough to make fluorometric detection useful analytically. However, the fluorescence signal of a molecule such as uric acid can often be measured without interference. Moreover, because fluorometry possesses a high degree of "tunable" selectivity, both excitation and emission spectra can be determined. In addition, laser-induced fluorescence can greatly increase the sensitivity [134].

The most important advantage of the fluorescence detector is its ability to aid in the identification of the separated compounds. Peak identification with a fluorescence detector includes the same identification techniques as are commonly used with UV absorbance detectors, such as the determination of retention times compared to standards, chromatography with internal standards, and ratios of the peak heights of eluted components to those of reference standards at two different wavelengths or pH values. Since both excitation and emission spectra can be monitored, two measurements reflect a unique energy level diagram; thus, fluorometry can be a selective method for the identification of chromatographic peaks.

17.2.3.4 Absorbance ratios

The use of absorbance ratios at two wavelengths is a fast and easy method for distinguishing between two closely eluted peaks and for establishing the purity of a peak prior to quantitation. The ratios are usually obtained at 280/254 nm when monitoring nucleic acid constituents in biological samples. If a peak is composed of more than one compound, the absorbance ratio will not be characteristic of either compound.

The effect of pH changes on absorbance ratios is especially helpful in characterizing peaks of nucleotides, nucleosides, and their bases [132]. Since the purines and pyrimidines undergo protonation and deprotonation at various pH values, there are characteristic changes in their absorption spectra as a function of pH. For example, in the RPLC separation of nucleosides and bases, inosine and adenine are two closely eluted peaks having identical absorbance ratios in near-neutral eluents. However, since adenine has a much higher absorbance ratio in a basic environment than inosine, the adenine peak can be distinguished from the inosine peak. UV absorbance ratios for purine and pyrimidine compounds at 280/254 nm, taken at different pH values, are shown in Table 17.2 [132].

17.2.3.5 Enzymatic peak shift

The enzymatic peak-shift technique is a unique method used to identify peaks and establish peak purity in chromatograms of biological compounds, especially of nucleic acid constituents [42,49,135]. Generally, a peak representing the compound is characterized tentatively by retention times prior to using the enzymatic peak-shift technique. One aliquot of the sample is chromatographed and another is incubated with an excess amount of an enzyme that catalyzes a specific reaction of the compound of interest. The enzyme is then deactivated and the sample is rechromatographed; the disappearance of a peak and/or the appearance of a peak corresponding to the retention time of a known

References on p. B326

TABLE 17.2

ULTRAVIOLET ABSORBANCE RATIOS FOR PURINES AND PYRIMIDINES AT 280/254 nm

Compound	pH		
	Acidic (1-2)	Neutral (6-7)	Basic (11-13)
Ade	0.38	0.13	0.61
m1-Ade	0.21	0.82	0.87
m2-Ade	0.56	0.15	0.84
m6-Ade	0.70	0.48	1.20
m6,6-Ade	1.36	1.26	2.63
m7-Ade	1.06	0.67	0.80
Cyt	1.54	0.58	3.15
m5-Cyt	2.62	1.21	3.29
Gua	0.84	1.04	1.20
m1-Gua	0.81	0.93	1.20
m7-Gua	0.79	1.86	1.87
Guo	0.70	0.67	0.97
m1-Guo	0.71	0.53	0.63
m2-Guo	0.56	0.56	0.82
m2,2-Guo	0.56	0.63	0.79
m7-Guo	0.68	1.05	1.52
Ado	0.22	0.10	0.16
m2-Ado	0.40	0.13	0.19
m6-Ado	0.41	0.64	0.68
m6,6-Ado	0.94	1.13	1.66
Hyp	0.04	0.09	0.17
Ino	0.17	0.12	0.17
Xan	0.21	1.10	1.14
Ura	0.25	0.16	1.40
Cyd	2.10	0.93	2.54
m5-Cyd	3.57	1.42	1.59

TABLE 17.3

ENZYMES IN PURINE AND PYRIMIDINE METABOLIC PATHWAYS THAT CAN BE USED FOR THE IDENTIFICATION OF PURINES AND PYRIMIDINES

Substrate(s)	Enzyme(s)	Product(s)
Ade	Adenine phosphoribosyl transferase	AMP
Ado	Adenosine deaminase	Ino
	Adenosine kinase	AMP
Cyd	Cytidine deaminase	Urd
Gua	Guanase	Xan
	Phosphoribosyl transferase	GMP
	Purine nucleoside phosphorylase	Gua
Hyp	Xanthine oxidase	Xan
		Uric acid
Hyp	Hypoxanthine/guanine phosphoribosyl transferase	IMP
Gua		GMP
Ino	Purine nucleoside phosphorylase	Hyp
Uric acid	Uricase	Allantoin
Ura	Uridine phosphorylase	Urd
Urd	Uridine kinase	UMP
Thy	Thymidine phosphorylase	Thd
Thd	Thymidine kinase	TMP

product identify the peak in question. An example of the enzymatic peak-shift technique is the peak shift of adenosine to inosine in a serum sample that has been incubated with the enzyme adenosine deaminase. After the incubation of the sample with this enzyme, the adenosine peak is no longer present in the chromatogram, and the peak representing inosine has formed or increased in size [135]. Some commercially available enzymes that can be used in the enzyme peak-shift technique for nucleotides, nucleosides, or their bases are listed in Table 17.3.

References on p. B326

17.2.3.6 Spectroscopic identification

Spectral techniques – MS, NMR, IR and Raman spectrometry, and inductively coupled plasma atomic emission spectrometry (ICPAES) – provide valuable information for the fingerprinting and structural elucidation of compounds. The usefulness of spectroscopic methods as a tool for positive identification and structural interpretation of nucleic acids and their constituents has been well documented [136-160]. Combination of chromatography with these spectroscopic methods, especially MS, have shown great potential as universal techniques for unambiguous peak identification [136-147].

NMR spectrometry has powerful capability for studies of structure, dynamics, and molecular interactions of biopolymers, like nucleic acids and their constituents [148]. More recently, ^{31}P-NMR has been applied to the investigation of the ATP in tumors in vivo by means of inversion transfer techniques [149]. Since NMR spectra furnish valuable information, such as chemical shifts and coupling constants, LC/NMR can provide sufficient structural information for the classification of unknown compounds [150]. In addition, the use of FTNMR as an on-line detector for HPLC [151,152] can provide information for rapid structural elucidation of components present in complexes. Although the application of HPLC/NMR offers a nondestructive detection method for HPLC, the NMR detector lacks sensitivity for analytical purposes, especially for trace analysis.

Infrared spectra can reveal unique characteristics of the functional groups in a molecule and of the overall configuration of atoms. LC coupled with FTIR greatly aids in peak identification [153-155]. With FTIR, rapid spectrum acquisition and averaging are possible, with greatly improved signal-to-noise ratio in comparison with dispersive IR spectroscopy. FTIR also is a nondestructive detection method [153]. However, interference of the mobile phase is a problem in LC/FTIR, because most common solvents used in HPLC absorb strongly in the IR; thus, the direct coupling of HPLC and FTIR is difficult. A very short pathlength (typically 100 μm) has been used successfully to obtain spectra in an IR-absorbing solvent [154].

Raman spectrometry is a useful analytical method for the elucidation of molecular structure, for locating various functional groups or chemical bonds in molecules, and for the quantitative analysis of major components in mixtures. Recently, surface-enhanced Raman spectrometry (SERS), which offers excellent sensitivity, has been developed as a detection method for LC [156,157]. Freeman et al. [156] have used flow-injection analysis with SERS as a detection method for HPLC, and Sequaris and Koglin [157] used it to identify purine derivatives separated by HPTLC.

ICPAES is a spectrochemical method with versatile capability, low detection limits, and element specificity, suitable for multielement analysis. Interfaced with HPLC, ICPAES has been used as phosphorus-sensitive detector for the quantitative analysis of nucleotides in biological samples [158-160].

17.3 OLIGONUCLEOTIDES

17.3.1 Sample preparation

Oligonucleotides are obtained by chemical synthesis or by the enzymatic degradation of DNA or RNA with restriction endonucleases. With the enzymatic method, oligonucleotides are found together with a complex mixture of proteins and the reagents used in the cleavage reaction. In the oligonucleotides produced by the synthetic method or automatic synthesizer, the functional groups are fully blocked. Therefore, prior to further analyses or applications, effective, rapid, and inexpensive techniques for sample preparation (cf. Section 17.4.1) are needed to eliminate interfering compounds or to release the protecting groups from the oligonucleotides.

Generally, the incubation mixture of proteins and oligonucleotides from enzymatic reactions is precipitated with TCA. Proteins are then removed by extraction with phenol. To remove the phenol from the mixture, dialysis, ethanol precipitation, or extraction with chloroform or ether follows [161]. The oligomers are then gently heated to 37°C to disrupt any reannealed termini of the oligomers. If TCA is not used to precipitate the oligonucleotides, other methods for sample preparation can be used. These methods include extraction with 2-butanol [162], disposable cartridge columns packed with gels, ion exchangers, reversed-phase or hydroxyapatite packings [163-169], or centrifugation in potassium iodide, cesium chloride, or sucrose density gradients [170,171].

When oligonucleotides are prepared synthetically, the first step in sample preparation is deblocking the fully protected oligonucleotides. The deblocking can be achieved simply by treatment with ammonium hydroxide, which is then evaporated. Impurities such as benzamide [172-177] are removed by extraction with organic solvents. The deblocked oligonucleotide mixture can be directly applied to a Sephadex disposable column [175] or a DEAE-cellulose column [176,177], where ammonium hydroxide and other impurities are removed.

17.3.2 Chromatography

17.3.2.1 Mixed-mode chromatography

Most of the original chromatographic work on oligonucleotides was performed on mixed-mode column materials that provide at least two types of sites for interaction with the solutes [178]. The mixed functionalities of the stationary phase permit the separation of nucleic acid fragments according to size as well as nature of the nucleobases. In mixed-mode chromatography, the separation of nucleic acid solutes is based on electrostatic as well as hydrophobic interactions between the stationary phase and the solutes. The elution of oligonucleotides requires complex mobile-phase mixtures in order to optimize each type of interaction. Separation can be achieved with a gradient of organic solvents, salt concentration, or a combination of both.

In 1960, Mandell and Hershey [179] separated nucleic acids with open columns of methylated albumin/silica-based materials. This stationary phase, which contained lysine and arginine side chains, functioned as an anion-exchange matrix with a secondary effect due to the added methyl groups. Gillam et al. [180] extended this idea by coupling benzoyl or naphthoyl residues to DEAE-cellulose in order to introduce sites for secondary hydrophobic interactions on an anion-exchange matrix. The chromatographic material, RPC-5, another type of mixed-mode material, consists of a charged reversed-phase matrix with a solution of Adogen 464 (methyltrioctylammonium chloride) coated on nonporous, spherical polymer materials, such as polychlorotrifluoroethylene beads (Plascon 2300). RPC-5 columns have been effective for the separation of oligonucleotides [181,182]. However, since commercial material of consistently high quality is not available, there has been no widespread application. In addition, because the tetraalkylammonium groups are physically adsorbed on the polymeric support, bleeding of the modified stationary phase occurs at salt concentrations $< 0.2\,M$ [183].

With the advent of microparticulate HPLC-grade silica gel supports, more efficient mixed-mode materials were produced and the optimization of mixed-mode chromatographic parameters was investigated [124]. Crowther et al. [125,184] developed column materials by modifying a silica gel support with a mixture of 3-chloropropyl- and octyl-dimethylmonochlorosilanes. Oligonucleotides up to 20-mers were successfully separated on these materials. Bischoff and McLaughlin [186-189] investigated the separation of oligonucleotides on mixed-mode materials by modifying commercially available aminopropylsilyl (APS) matrices with organic acids, such as phenylalanine or n-alkyl carboxylic acids. Using these materials, the effects of increasing the concentration of the organic modifier at constant pH and ionic strength were studied. At concentrations of organic solvents $> 30\%$, these modified materials functioned almost exclusively by electrostatic interactions. Mixed-mode packings were also used to separate polyuridylic acids, containing up to 100 uridyl residues, on a tetraalkylammonium C_{18} matrix [189,190]. These acids were obtained from homopolymer digests. El Rassi and Horváth [191,192] resolved oligonucleotides on a multifunctional stationary phase with a salt gradient in the presence of 2-propanol or n-decylbetaine. In addition, poly(ethyleneimine)-based silica gel matrices have been prepared for the separation of adenylic acid oligomers ranging from 40- to 60-mers [193].

17.3.2.2 Anion-exchange chromatography

In anion-exchange chromatography, separation of oligonucleotides involves an electrostatic interaction of the negatively charged phosphodiester residue with a positively charged matrix. The oligonucleotides are then eluted as a function of chainlength by using a salt gradient. The effects of ionic strength, pH, and temperature on the retention behavior in anion exchange of oligomers of varying chainlengths were investigated [177]. Whereas early work on anion-exchange chromatography in open columns was carried out with soft gel media, such as DEAE-cellulose [194] or DEAE-Sephadex [196], the availability of HPLC microparticle packings of large poresize has now greatly improved analysis time and resolution [168,177,193,196-199]. However, the procedure requires high concentrations of eluting salts, high pH, and time-consuming gradient elution. Thus, its use has not been widespread. In addition, the oligonucleotides used as primers or probes require an additional desalting step. To avoid the desalting step, Munholland et al. [200] developed a volatile buffer system for the HPLC anion-exchange separation of oligonucleotides. The use of the volatile buffer system, containing triethylamine acetate, now provides a flexible, efficient, and convenient method for the separation of synthetic oligonucleotides.

Two types of ion exchangers are in use: silica-based anion exchangers and polymer-based anion exchangers. Since the polymer-based materials are chemically resistant, the chromatographic separation can be performed at high pH. For example, with a polymer-based weak-anion-exchange resin oligonucleotides can be separated according to chainlength at an operating pressure of < 500 psi [201,202]. This material is used mainly for the separation of short-chain oligonucleotides, up to 30 bases. Since the functional groups of the resin are adsorbed on the solid support by weak hydrophobic interactions, the resin must be handled carefully to maintain both retention times and lifetime of the column [164,165,202-204].

Partisil 10-SAX, a silica-based strong-anion exchanger, has been used to separate short-chain oligonucleotides having the same chainlength and base composition but different sequences [199,204]. However, this column has a shorter lifetime than analogous reversed-phase columns (cf. Section 17.4.2.3) [205].

Recently, large-pore resins were developed that have large surfaces for solute interaction but optimal poresize to prevent size-exclusion effects; thus, oligonucleotides or DNA fragments of high molecular mass can be separated. The polymer-based resins include TSK-DEAE-5PW, having functional groups of diethylaminoethyl [206], and Mono Q with quaternary amino functional groups [207]. The modified silica-based packings are TSK-DEAE-3SW [206] and Nucleogen-DEAE [98] with diethylamino functional groups. These commercially available materials are suitable for the separation of restriction fragments with various chainlengths [98,206-209]. In addition, Mono Q columns have been used for the rapid fractionation and chainlength determination of oligomers [205,206]. The semi-preparative separations of oligonucleotides with less than 20 units were also achieved on a Mono Q column [209].

17.3.2.3 Reversed-phase chromatography

Reversed-phase HPLC of oligonucleotides is based on hydrophobic interactions between the solute and the stationary phase [210]. Retention depends on the base composition of the oligonucleotide. Most RPLC separations of oligonucleotides have been performed on C_{18} columns with a poresize of 125 Å [210,211]. A C_{18} column is well suited for the separation of oligonucleotides containing 2 [212] to 50 base residues [213]. For example, isomeric pyrimidine oligonucleotides of short chainlengths (n = 2-5) were separated on a Nucleosil C_{18} column [214]. The introduction of packings with a larger poresize, e.g., a 300-Å-poresize C_4 column, improved the resolution for the larger oligonucleotides with up to 104 base residues [215]. The use of volatile organic solvents and volatile buffers, such as triethylammonium bicarbonate [216], obviates the desalting step after purification of the oligonucleotides.

17.3.2.4 Ion-pair chromatography

In ion-pair chromatography, with cationic ion-pairing agents added to the mobile phase, the retention of the oligonucleotides on reversed-phase columns is longer than in the absence of the ion-pairing agent [217]. The increase in hydrophobicity of the solutes is proportional to the number of the phosphodiester residues present. Therefore, the resolution of oligonucleotides by ion-pair chromatography is based on the chainlength of the solute. Elution of the solutes can be achieved by adjusting the concentration of an organic modifier in the mobile phase. Ion-pair chromatography is well established for the separation of short oligonucleotides [199,207,217-221]. The resolution of oligonucleotides of isomeric sequence, such as CpU and UpC, CpA and ApC, UpG and GpU, and GpA and ApG, on a Radial-Pak C_{18} column has also been accomplished [221].

17.3.3 Detection and identification

The most widely used detection method for monitoring oligomers is the use of a UV detector at a fixed wavelength, because oligonucleotides contain the purine and pyrimidine chromophores which absorb in the vicinity of 260 nm. However, it is preferable to monitor the effluent at 280 nm or above with a variable-wavelength detector, because the high absorbance of synthetic oligonucleotides can swamp the detector at 254 nm or 260 nm. A combination of a UV detector and a flow-through radioactive detector has been used to identify and quantify radiolabeled oligonucleotides [222]. Another detection and identification method for oligomer sequence analysis is the use of mass spectrometry. Grotjahn et al. [223] used fast-atom bombardment (FAB) mass spectrometry to analyze the sequence of an oligomer with 10 units. This highly sensitive and selective method required only about 1 nmole of sample per analysis.

The positive identification of collected oligonucleotides can be accomplished by using a sequencing technique, such as chemical or enzymatic sequencing (Section 17.6). Crowther et al. [224,225] developed a method in which the enzymatic digestion of the oligonucleotides with phosphodiesterase I or II, followed by ion-pair chromatography of the nucleosides, is used to determine both chainlength and base composition. Hakam and Kun [177] determined the chainlength of poly(adenosine diphosphate ribose), which was separated from a phosphodiesterase digestion mixture on an anion-exchange column. Zon and Thompson [168] reported another method, in which the base composition of a 16-base oligomer was determined by RPLC after enzymatic digestion of the oligomer to nucleoside components. A modification of Zon and Thompson's method has been described by Eadie et al. [24] for the determination of the base composition of oligomers from 18 to 150 bases in length. In this approach, two RP C_{18} columns were connected in tandem, and a shallow-gradient elution was used to increase the resolution. This method provides direct and unambiguous data for a wide range of synthetic oligomers and for routine quality assessment of probes, primers, and gene fragments.

Recently, a method for determining the chainlength of oligo- and polynucleotides by ion-exchange HPLC was developed by Tsygankov et al. [209]. In this approach, two types of gradients were used according to the chainlength of oligomers: a linear salt gradient used for separating short oligomers up to 10-mers and a hyperbolic salt gradient for eluting oligomers up to 30 units. An equation proposed for this experimental model describes the relationship between the eluting salt concentration and the chainlength, and thus chainlengths of oligomers up to 500 units can be determined.

17.4 NUCLEIC ACIDS

17.4.1 Sample preparation

In order to prepare a biological specimen for the chromatographic isolation of a nucleic acid, the association of nucleoproteins with other cellular constituents and of nucleic acids with proteins must be disrupted [15]. The disruption methods include freezing-and-thawing, bursting by osmotic shock, or using lytic chemical agents. Some mechanical methods used are grinding in a mortar with sand, alumina, or fine glass beads, or in a Potter homogenizer. For small amounts of sample, exposure to ultrasonic vibration can be very useful. Cell lysis can also be achieved with a detergent, such as sodium dodecyl sulfate or by using the enzyme lysozyme.

After cell disruption and removal of cell debris by centrifugation, the nucleic acids must be separated from associated proteins. This separation can be achieved by standard procedures of extraction with phenol, chloroform/isoamyl alcohol, or chloroform/octanol [227]. Proteins can also be removed by digestion with proteases, which degrade most proteins to amino acids.

During sample preparation, care must be taken to avoid degradation of nucleic acids by enzymes, either from the experimental materials or the operator's fingers. A purified ribonuclease can be used to remove RNA in the preparation of DNA; a purified deoxyribonuclease can remove DNA in the analysis of RNA [15].

The method of choice depends upon the type of sample and the particular nucleic acid to be isolated. Separation of RNA from DNA can be accomplished in sodium chloride solution, because the corresponding nucleoproteins have very different solubilities in this solution [228]. For example, RNA-protein is much more soluble than DNA-protein in 0.14 M sodium chloride solution. The most common method for the preparation of RNA in good yield is based on treatment of the sample with a mixture of phenol and water, which precipitates protein and DNA [229,230]. The aqueous supernatant obtained by centrifugation contains both RNA and polysaccharides. This supernatant is treated with ethanol. The RNA, which precipitates, is then extracted with a mixture of 2-methoxyethanol and phosphate buffer. The salts are removed by dialysis. A more detailed method of a modified phenol treatment was described by Wisniewski et al. [231].

References on p. B326

Since DNA exists in nature as deoxyribonucleoprotein, the associated protein can be removed by extraction with sodium chloride solution. When the solution is shaken with chloroform/octanol or chloroform/isoamyl alcohol, the protein forms a gel at the chloroform/water interface, while the sodium salt of the DNA remains in the aqueous phase. The salt can be removed by dialysis. DNA can also be precipitated by either polyethylene glycol [232] or 2-propanol [233]. DNA in 10% (w/v) polyethylene glycol on ice is allowed to precipitate for 2 h or longer. DNA is selectively precipitated with 2-propanol. This precipitation is accomplished by first diluting the sample with an equal volume of water, then adding one volume of 2-propanol, and keeping the solution at −20°C for 2 or more hours. The precipitate is collected and washed with 75% ethanol solution. DNA can be redissolved in a low-salt buffer, such as 10 mM Tris-HCl/0.1 mM EDTA (pH 7.0), and stored at −20°C. Small amounts of RNA, contaminating the DNA preparation, can be removed by treatment with alkali [234]. Recently, more rapid and effective methods have been reported for the isolation of DNA from yeast [235] and for the extraction of human genomic DNA from whole blood [179].

Nucleic acids can be isolated from a sample more rapidly by the use of disposable columns. A column packed with methylated albumin on kieselguhr (MAK) was found to bind single-stranded DNA more firmly than double-stranded DNA [161,179,236]. It was also found that hydroxyapatite can retain double-stranded DNA while allowing single-stranded DNA to pass through the column [237]. Recently, with a new disposable cartridge column, Nensorb 20, DNA and RNA were separated from proteins, salts, unincorporated radioactive nucleotides, and other low-molecular-weight materials [165]. For the purification of RNA, some disposable columns with size-exclusion [238,239], ion-exchange [240,241], or affinity packings [242] have been used.

17.4.2 Chromatography

17.4.2.1 Anion-exchange chromatography

Separation of nucleic acids by anion-exchange HPLC depends on the number of phosphodiester residues and is independent of the base composition. An increasing concentration of salt in the mobile phase is used to elute the retained nucleic acids as a function of molecular weight. Therefore, small-molecular-weight fragments are eluted first and large-molecular-weight fragments are eluted last. Investigations have been carried out on the effects of pore size, pH, temperature, eluting salts, bivalent metal ions, denaturing reagents, and flowrate that affect the separation of various nucleic acids on ion-exchange columns [98,207,232,243]. Optimal poresize is required to eliminate size-exclusion effects; a high flowrate and a shallow gradient can decrease the detection limits; steep gradients are required for fast analyses, whereas shallow gradients lead to a better separation for preparative applications.

Nucleogen is a silica-based material, modified with DEAE groups. The silica has different poresizes, ranging from 60 to 4000 Å. The Nucleogen-DEAE column has been employed for the separation of high-molecular-weight nucleic acids, such as natural

RNAs, double-stranded RNA, DNA fragments, and supercoiled plasmids [98]. Fractionation of DNA fragments up to 880 base pairs (bp) has been reported on a Nucleogen-DEAE column with a 500-Å poresize [98]. Resolution of DNA fragments ranging from 25 to 1500 bp has been achieved by using a Nucleogen-DEAE 4000 column [232].

Several other commercially available ion exchangers, developed for the separation of high-molecular-weight nucleic acids include: NACS, TSK-DEAE-5PW, TSK-DEAE-3SW, Mono Q, and Mono P. The fractionation capabilities of some of these high-performance ion-exchange resins for various DNA restriction fragments have been studied [207].

17.4.2.2 Mixed-mode chromatography

High-performance mixed-mode materials, consisting of both ionic and hydrophobic ligands on the same solid support (Section 17.3.2.1), have been successfully employed to separate high-molecular-mass nucleic acids. Using mixed-mode packings, Floyd et al. [185] separated a 267-bp restriction fragment from one of 234 pb with a gradient of KCl and methanol. The mixed-mode material has also been very effective for the isolation of tRNAs [191,244].

Optimization of a mixed-mode chromatographic separation of nucleic acid fragments has been systematically studied by Wells et al. [183]. Varying the pH of the mobile phase can alter the ratio of the hydrophobic to ionic interactions experienced by the solute and hence can improve resolution [187]. Another way of optimizing a separation is to increase the amount of organic solvent in the mobile phase, thus increasing the hydrophobicity of the stationary phase and decreasing the capacity factors.

17.4.2.3 Reversed-phase chromatography

Because volatile solvents and buffers can be used, reversed-phase liquid chromatography is a more convenient method for nucleic acid purification than other modes of chromatography. Some separations on reversed-phase columns have been optimized by the addition of ion-pairing agents. Resolution depends on the molecular mass of the nucleic acids. The separation of DNA restriction fragments, ranging from 10 to 2000 bp, on Prep RPC columns with ion-pairing agents has been established as an effective technique for the separation of nucleic acids [245]. On columns packed with 5-μm, 100-Å C_2/C_{18} materials, separation was achieved within 2-4 h. The load capacity of the column is limited by the poresize. Although fragments larger than 100 bp are excluded from the inner pores, only small changes in performance are seen with a 10-fold load.

Another ion-pair method for resolving nucleic acids depends on the use of an inert, irregularly shaped, nonporous polymer as the stationary phase [162,236]. The use of this polymer has some advantages over silica in nucleic acid research [161,236]: high capacity, high recoveries, absence of size-exclusion effects, and resistance of the packing to chemical degradation. Fragments up to 5000 bp were eluted in 60 min with gradient

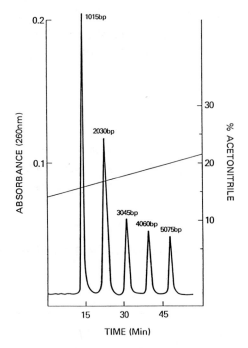

Fig. 17.7. Paired-ion chromatographic system (PICS-HPLC) tracing obtained following injection of 0.2 mg of a mixture of DNA restriction fragments. The DNA sample was loaded on a glass PICS column (1 x 15 cm), previously equilibrated with 10% acetonitrile, buffered to pH 7.0 with 0.1 M triethylammonium acetate. Fractions were collected and detected by monitoring the absorbance at 260 nm. Bound DNA fragments were eluted for 60 min with a linear gradient from 15 to 22% acetonitrile, buffered with 0.1 M triethylammonium acetate (pH 7.0). Aliquots (0.5 μg) of each fraction were lyophilized, analyzed, and identified (number of base pairs) by 0.8 and 1.0% agarose gel electrophoresis. (Reproduced from Ref. 161, with permission.)

elution and triethylammonium acetate as the ion-pairing agent (Fig. 17.7) [161]. In a preparative application, up to 10 mg of DNA have been isolated.

17.4.2.4 Size-exclusion chromatography

With polymeric stationary phases, such as agarose-based gels [246-248] or micropar-ticulate HPLC materials [249-253], which are chemically stable, the separation of nucleic acids can be performed by using either a single column or by connecting columns of various pore sizes in tandem. The nucleic acids do not interact with these polymeric stationary phases, and hence the recovery of the desired component(s) is increased. With a polymer-based packing having pores of 6000 Å, DNA fragments with 10 to 40 x 10^3 bp were separated as discrete peaks. On a silica-based packing (TSK-GEL SW) with an exclusion limits of either 1 x 10^5 or 5 x 10^5 D [253] Schmitter et al. [252] obtained a linear relationship between the logarithm of DNA fragments and the distribution coefficient. The linear portion of the calibration curve indicated that this column is suitable for the separation of DNA fragments ranging from 75 to 10 000 bp. However, compared with RP chromatography, SEC requires an additional step to remove salts and additives after the chromatographic separation.

17.4.2.5 Affinity chromatography

Affinity chromatography has been widely used for the separation of nucleic acids [236,254,255]. Recently, Holstege et al. [256] separated a mixture of DNA and RNA on a histone-silica column, based on the differential affinities of DNA and RNA for histone. This is a rapid and gentle method for the preparative and analytical separation of nucleic acids.

17.4.3 Detection and identification

Gehrke and Kuo [257,258] developed a method for rapid enzymatic hydrolysis, followed by RPLC/UV for the determination of the composition of nucleic acids. In this procedure, nucleic acids are enzymatically digested to the component nucleosides for 2 h. Even the more labile nucleosides were not degraded. The nucleoside mixture from the enzymatic hydrolysis was then injected directly into a RPLC system. A multiple-wavelength detector was used to provide the absorbance ratios at different wavelengths for each nucleoside. Thus, by a combination of retention times and absorbance ratios, each nucleoside was identified. The method, which is accurate and highly selective, has been successfully used to analyze the major and modified nucleotides in RNA, DNA, and mRNA. Ludlum [259] reported a similar method, in which the modified nucleosides are released by the enzymatic digestion of DNA and analyzed by RPLC. After the addition of an internal standard, UV and fluorescence detectors are used to monitor the components in the digestion mixtures. The identity of peaks is established by comparison with an external or internal standard. A radioactive detector can also be used to monitor minor components which may not be separated from the major constituents.

Determination of the complete sequence of large DNA molecules by means of sequencing strategies and computerized data handling have been developed [260-263]. These methods are very rapid and efficient (Section 17.6). Recently, with the development of appropriate computer software, the automated DNA sequencing analyzer has become an important instrument for the determination of the structure of DNAs [264,265].

17.5 GEL ELECTROPHORESIS

For many years electrophoresis has been used in biochemistry (Chapter 11) as a technique for separation and characterization of biopolymers, and now gel electrophoresis is widely used in molecular biology to separate nucleic acids [161,266-271] and for DNA sequencing [271-274].

Polyacrylamide gel electrophoresis (PAGE) is very useful for the isolation and purification of synthetic oligonucleotides [266] and the restriction fragments of nucleic acids [267,268]. In order to maintain molecular sieving properties, the proper poresize of the polyacrylamide gel is produced by controlling the concentration of the acrylamide and the relative proportion of the cross-linking agent, bisacrylamide [269]. Agarose gels have larger pores than acrylamide and thus permit separation of much larger molecules [171]. Schwartz and Cantor [270] demonstrated that whole chromosomes from yeast can be separated by electrophoresis on agarose gels. Gel electrophoresis can separate nucleic acid fragments ranging from 10 to 60 000 bp (60 kb) [161], but it was more recently reported that fragments larger than 20 kb are not well resolved [271]. Although this method is time-consuming, labor-intensive, and gives only poor recoveries from gels, gel electrophoresis has been widely used for the determination of DNA sequences [272-274].

A new electrophoretic technique, pulsed-field gel electrophoresis (PFGE), has shown improved capabilities for the separation and characterization of large DNA molecules [271]. In PFGE, DNA molecules are subjected alternately to two electric fields. A second electric field, which is oriented at a different angle from the first, is applied to the gel; thus, the DNA molecules are realigned differently from the alignment that resulted from the first electric field. The molecules immediately begin to move in the direction of the new field after the field switching. The larger the molecule, the longer it takes to change direction. During each pulse period, the larger molecules have less time to move and hence migrate more slowly than the smaller molecules. Resolution in PFGE is dramatically influenced by the number and the geometry of the electrodes used as well as the angle of the two consecutive electric fields. The use of an obtuse angle (generally ca. 120°) of the electric fields provides superior separation [270,275,276], and various electrodes and field angles can be used. PFGE includes pulsed-field gradient gel electrophoresis (PFGGE) [270], orthogonal-field alternation gel electrophoresis (OFAGE) [277], transverse alternating field electrophoresis (TAFE) [278], contour-clamped homogeneous field electrophoresis (CHFE) [279], rotating field gel electrophoresis (RFGE) [280] and a programable autonomously controlled gel electrophoresis system (PACE) [275]. These methods are potentially very powerful for the separation and detection of DNA molecules. Carle and Olson [281] have presented detailed procedures for using OFAGE to separate yeast chromosomal DNA molecules. Bio-Rad Laboratories developed the CHFE-DR II electrophoresis system [282], which combines the benefits of CHFE and PACE. Using the CHFE-DR II system, DNA fragments ranging from a few hundred bases to about 12 million bases have been resolved into sharp bands [283]. Schwartz and Koval [284] separated the bacteriophage DNA molecules with 700 kb and investigated their dynamic conformation by using a PFGE technique with different field angles.

17.6 CAPILLARY ELECTROPHORESIS

Capillary electrophoresis (CE) is one of the most rapidly expanding areas in analytical biotechnology [285-288]. Since the separations are performed in capillary tubes, CE has several advantages. The use of capillary tubes allows the effective dissipation of Joule heat. Therefore, the separation can be performed with high-potential fields. CE is a fast and highly efficient technique for both trace and micropreparative separations of nucleic acids. In addition, CE is more easily automated than gel electrophoresis, because electroosmosis, which is generated by the application of a tangential potential field in capillary tubes, can cause all solutes to be eluted at one end of the capillary tube. Because the zone volumes are extremely small, on-column detection methods, which maintain the high resolution, can be used.

Electrophoresis in narrow-bore glass tubes was first reported by Virtanen in 1974 [289]. In 1979, Mikkers et al. [290] gave the first report of electrophoresis in narrow-bore Teflon tubes. Although the technique provided high efficiency, the detection of the small amounts of analytes in the capillary was very difficult. Jorgenson and Lukacs [291,292] in 1981 developed an electrophoretic system with open-tubular glass capillaries and an on-column fluorescence detector. Since then, research in instrumentation and applications of the technique, especially in biotechnology, has increased dramatically. Generally, CE produces fast, highly efficient separations of ionic and nonionic compounds. There are many types of analytical systems: capillary zone electrophoresis (CZE), capillary gel electrophoresis (CGE), capillary isotachophoresis, isoelectric focusing, and micellar electrokinetic capillary electrophoresis. The detection methods used are based on absorption, fluorescence, conductivity, electrochemical reactions, refractive index, radioisotope labeling, or mass spectrometry. Presently, there are at least seven instruments for CE commercially available [285].

CE has been found to be very useful in nucleic acid research for the purification of nucleotides, bases, and DNA fragments, as well as for DNA sequencing. Tsuda et al. [293,294] developed a CZE method for the analysis of nucleotides in tissue or serum, and Karger and his coworkers reported the separation and micropreparative purification of oligonucleotides on polyacrylamide-gel-filled capillary columns. They demonstrated a rapid separation of DNA restriction fragments, ranging from 72 to 23 130 bp [295-297]. Novotny and his coworkers [298] reported the use of an on-line laser-induced fluorescence detector for the CE separation of nucleic acid fragments at the femtomole level.

17.7 SEQUENCING OF NUCLEIC ACIDS

In the early Seventies, two groups of researchers independently developed rapid and powerful methods for the determination of DNA sequences. These methods are the enzyme method [299] and the chemical degradation method [300,301]. In both techniques, which are based on similar principles, radioactivity-labeled DNA fragments are prepared with the sequence information encoded by the varying lengths of these fragments. Electrophoresis is used to separate the fragments in a gel according to their lengths. After a fixed period of electrophoresis, autoradiograms are developed, and the relative positions of the fragments are indicated by the bands on the film. Since the

References on p. B326

chemical degradation method is not efficient for data accumulation and is not suitable for automation, it will not be discussed in this chapter.

17.7.1 Sanger method

The Sanger enzymatic sequencing method [299] is the most widely used technique for DNA sequencing. In this method, a single strand of the DNA to be sequenced is used as a primer and template. The complementary strand of the DNA is synthesized when this primer is incubated with DNA polymerase in the presence of a 2′,3′-dideoxyribonucleotide and all four deoxyribonucleotides. One of these four deoxyribonucleotides is labeled with ^{32}P. The synthesis of DNA is terminated by the enzymatic incorporation of a dideoxyribonucleotide, since the latter does not contain a hydroxy group in either the 2′- or 3′-position [302]. For example, if 3′-deoxythymidine triphosphate (dTTP) is used as the terminator, a mixture of DNA fragments, all having the same 5′-end and ddT residues at the 3′-end, is obtained. Four different base-specific reactions are carried out in separate test tubes; each test tube contains a radiolabeled deoxyribonucleotide, the other three deoxyribonucleotides, and a different single terminator (in ca. 1:5 mol ratio) [300]. This mixture results in four sets of nested radiolabeled fragments, each terminated by ddAMP, ddGMP, ddCMP, or ddTMP. Gel electrophoresis is then used to separate the fragments according to their lengths. X-ray film is exposed to the gel after electrophoresis, and the sequence information is recorded on the film.

The techniques used in the Sanger method are tedious and expensive, requiring highly skilled scientists and the use of hazardous chemicals and unstable radioisotopes. The amount of sequence information obtained is limited by the dimensions of the electrophoretic apparatus. The longer DNA fragments migrate more slowly, and the bands associated with these fragments are so close together that they are difficult to resolve. Another limiting factor in obtaining large amounts of sequencing data is the time needed to remove the gel after electrophoresis, to dry it, and then to expose and develop the film. Moreover, this approach is not suitable for automation; to obtain an accurate computerized analysis of the film image is difficult [303]. These considerations have limited large-scale sequencing projects, such as the analysis of the human genome, which is encoded in three billion bases of DNA.

17.7.2 Automation

Over the past decade, the rapidly expanding interest in molecular biology for the analysis of gene structure and function has increased the demand for more rapid and powerful DNA analyses. Particular interest has focused on DNA sequencing methods that can improve the throughput while maintaining high accuracy. A recent innovative approach is based on introducing fluorophores into the DNA sequencing fragments; thus, these fragments can be detected and analyzed during electrophoresis. In this approach, fluorophores are attached either to the chain-terminating base analog itself [304] or to the DNA fragments, which are then converted to a dye/oligonucleotide primer [29,265,305-

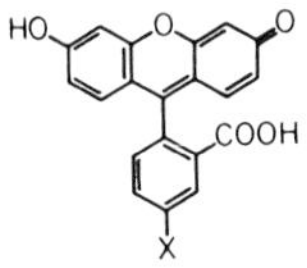

FLUORESCEIN

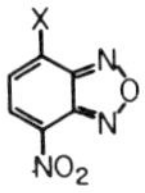

NBD or
7-NITROBENZO-
2-OXA-1,3-DIAZOLE

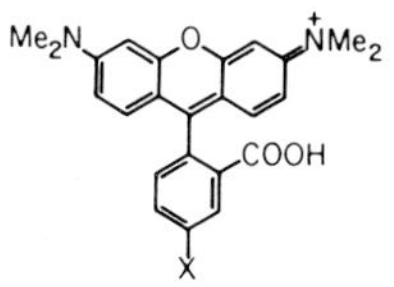

TETRAMETHYLRHODAMINE

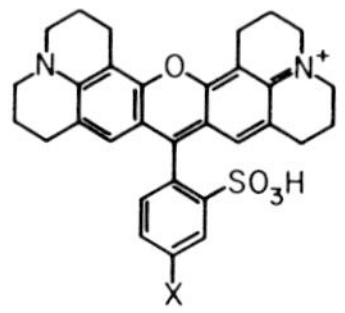

TEXAS RED

Fig. 17.8. Chemical structures of the four fluorophores used in automated sequencing. X is the moiety to which the dye is bound, e.g., an oligonucleotide primer. (Reproduced from Ref. 265, with permission.)

307]. With the use of laser-induced fluorescence detection, electrophoresis of the sequencing reaction mixture in a single lane can be used. By this detection method, the instrument throughput is increased and also the degree of complexity in the data analysis is determined. Automated DNA sequence analysis becomes more feasible for both relatively short or very large DNA fragments.

Four fluorophores (Fig. 17.8) are covalently attached to the DNA fragments by using chemical synthesis to prepare dye/oligonucleotide primers [265]. When these dye primers are used in the Sanger-type sequencing reaction in place of the normal primer, the dye is attached to the newly synthesized DNA fragments. A differently colored fluorophore is used in each of the four base-specific reactions. The four reaction mixtures are then combined and electrophoresed in a single polyacrylamide tube gel. The fragments are detected near the bottom of the gel by a high-sensitivity fluorescence detector. The data can be directly stored and analyzed by a microcomputer to yield the DNA sequence. A commercially available instrument from Applied Biosystems Inc. [308,309] is based on the use of dye primers.

A research group at Du Pont, using the attachment of the dye to the chain terminator itself, developed an automated DNA sequencing instrument for the four-dye fluorescence-based DNA sequencing [310]. In this approach, the dye compound attaches itself with the fluorophore to the end of the DNA molecule. The DNA synthesis is terminated at the same time by incubation with a polymerase enzyme. The Du Pont instrument [307] measures the ratios of the fluorescence emission at two wavelengths and identifies the fluorophore, based on the characteristic spectral fingerprint. However, with this instrument the relative amounts of the four dyes present cannot be determined when peaks of different colors overlap in the detector. In general, there are two advantages of the four-dye fluorescence-based sequence method over the dye-primer method. First, the primers for the fluorescence-based sequence method can be more rapidly and easily synthesized in a commercial DNA analyzer [29]. Second, the four separate incubation reactions can all be performed in one tube.

Instead of using four fluorophores, an alternative method has been developed in which a single fluorophore is used [307,312-314]. With this instrument the DNA molecules detected are those to which only a single fluorophore, fluorescein, in attached. In this case, the four sequencing reaction products are electrophoresed in contiguous lanes of the gel.

Other systems for automating conventional radioisotope-based sequence analyses include real-time systems that detect radioactivity in the gels during electrophoresis [315] and systems that use scanning densitometry to analyze autoradiograms obtained in conventional sequencing [316].

REFERENCES

1 L. Smith and L.E. Hood, *Bio/Technology*, 5 (1987) 933.
2 C.J. Cowling, in E. Heftmann (Editor), *Chromatography*, 4th Edn., Elsevier, Amsterdam, 1983, p. B372.
3 H.A. Scoble and P.R. Brown, in Cs. Horváth (Editor), *High Performance Liquid Chromatography - Advances and Perspectives, Vol. 3,* Academic Press, New York, 1983, p. 1.
4 P.R. Brown, *Cancer Invest.,* 1 (1983) 439.
5 P.R. Brown, *Cancer Invest.,* 1 (1983) 527.
6 R.W. Zumwalt, K.C.T. Kuo, P.F. Agris, M. Mehrlich and C.W. Gehrke, *J. Liq. Chromatogr.,* 5 (1982) 2041.
7 M. Zakaria and P.R. Brown, *Anal. Biochem.,* 120 (1982) 25.
8 M. Zakaria and P.R. Brown, *J. Chromatogr.,* 226 (1981) 267.
9 P.R. Brown, A.M. Krstulovic and R.A. Hartwick, *Adv. Chromatogr.,* 18 (1980) 101.
10 P.R. Brown, R.A. Hartwick and A.M. Krstulovic, in G.L. Hawk (Editor), *Biological/Biomedical Application of Liquid Chromatography II,* Dekker, New York, 1979, p. 307.
11 P.R. Brown and A.M. Krstulovic, *Anal. Biochem.,* 99 (1979) 1.
12 P.R. Brown (Editor), *HPLC in Nucleic Acid Research,* Dekker, New York, 1984.
13 K. Hohse, R. Meyer, W. Lin, I. Clark and R.A. Hartwick, *LC Liq. Chromatogr. HPLC Mag.,* 2 (1984) 266.
14 E. Freise, Z. Olempska-Beer and M. Eisenberg, *J. Chromatogr.,* 284 (1984) 125.
15 E. Harbers, G.F. Domagk and W. Muller, *Introduction to Nucleic Acids,* Chapman-Reinhold, New York, 1968.
16 J.N. Davidson, *The Chemistry of the Nucleic Acids,* Academic Press, New York, 7th Edn., 1972, p. 67.
17 M. Zakaria, in P.R. Brown (Editor), *HPLC in Nucleic Acid Research,* Dekker, New York, 1984, p. 365.
18 G.P.J.M. Gerrits, A.A.M. Haagen, R.A. De Abreu, L.A.H. Monnons, F.J.M. Gabreels, F.J.M. Trijbels, A.L.M. Theeuwes and J.M. van Baal, *Clin. Chem.,* 34 (1988) 1439.
19 A. Werner, W. Siems, G. Gerber, H. Schmidt, S. Gruner and H. Becker, *Chromatographia,* 25 (1988) 237.
20 R. Ellison, J. Holland and M. Wei, *Blood,* 32 (1968) 507.
21 Y.M. Rustum and D.H. Preisler, *Cancer Res.,* 30 (1979) 42.

22 D.M. Tidd and S. Dedhar, *J. Chromatogr.*, 145 (1978) 237.
23 H. Schott and H. Eckstein, *J. Chromatogr.*, 296 (1984) 363.
24 J.S. Eadie, L.J. McBride, J.W. Efcavitch, L.B. Hoff and R. Cathcart, *Anal. Biochem.*, 165 (1987) 442.
25 K. Taneja and R.H. Singer, *Anal. Biochem.*, 166 (1987) 389.
26 A.P. Halfpenny and P.R. Brown, in H. Engelhardt (Editor), *Practice of High Performance Liquid Chromatography*, Springer-Verlag, Heidelberg, 1986, p. 324.
27 R.D. Simpson and P.R. Brown, *J. Chromatogr.*, 379 (1986) 269.
28 P.R. Brown, in J.C. Giddings, E. Grushka, R.A. Keller and J. Cazes (Editors), *Advances in Chromatography*, Vol. 12, Dekker, New York, 1975.
29 L.M. Smith, *Anal. Chem.*, 60 (1988) 381A.
30 F. Grummt, *Proc. Natl. Acad. Sci. USA*, 75 (1978) 371.
31 B. Guattari, *J. Chromatogr.*, 489 (1989) 394.
32 A. Gies, *Comp. Biochem. Physiol.*, 91B (1988) 483.
33 A. Alajoursijarri and E. Nissinen, *Anal. Biochem.*, 165 (1987) 128.
34 C.K. Lim and T.J. Peters, *J. Chromatogr.*, 461 (1989) 259.
35 C. Brooker, J.F. Harper, W.L. Terasaki and R.D. Moylan, *Adv. Cyclic Nucleotides Res.*, 10 (1979) 1.
36 J.R. Wermeling, J.M. Pruemer, F.M. Hassan, A. Warner and A.J. Pesce, *Clin. Chem.*, 35 (1989) 1011.
37 J. Olivares and M. Verdys, *J. Chromatogr.*, 434 (1988) 111.
38 V. Stocchi, L. Cucchiarini, F. Canestrari M.P. Piacentini and G. Fornaini, *Anal. Biochem.*, 167 (1987) 181.
39 G.A. Taypor, P.J. Dady and K.R. Harrap, *J. Chromatogr.*, 183 (1980) 421.
40 R. Boulieu, C. Bory and P. Baltassat, *Anal. Biochem.*, 129 (1983) 394.
41 J. Ontyd and J. Schrader, *J. Chromatogr.*, 307 (1984) 404.
42 P.R. Brown, *J. Chromatogr.*, 52 (1970) 257.
43 B. Assmann and H.J. Hass, *J. Chromatogr.*, 434 (1988) 202.
44 P.D. Reiss, P.F. Zuurendonk and R.L. Veoch, *Anal. Biochem.*, 140 (1984) 162.
45 G. Lazzarino, M. Nuutinen, B. Tavazzi, D. Di Pierro and B. Giardina, *Anal. Biochem.*, 181 (1989) 239.
46 J.G. Maessen, G.J. van der Vusse, M. Vork and G. Kootstra, *Clin. Chem.*, 34 (1988) 1087.
47 D.F. Hammer, D.V. Unverferth, R.E. Kelley, P.A. Harvan and R.A. Altschuld, *Anal. Biochem.*, 169 (1988) 300.
48 M. Andersson, P.I. Christensson, L. Lewan and U. Stenram, *Int. J. Biochem.*, 19 (1987) 641.
49 D.L. Jenkins, H.G. McDaniel, W. Grizzle, S.W. Parrish and H.B. McDaniel, *J. Chromatogr.*, 426 (1988) 249.
50 H.M. Maguire and F.A. Westemeyer, *J. Chromatogr.*, 380 (1986) 55.
51 I. Dutta, G.A. O'Donovan and D.W. Smith, *J. Chromatogr.*, 444 (1988) 183.
52 O.C. Ingebretsen, A.M. Bakken, L. Segadal and M. Farstad, *J. Chromatogr.*, 242 (1982) 119.
53 C.A. Cordis, R.M. Engelman and D.K. Das, *J. Chromatogr.*, 459 (1988) 229.
54 D. de Korte, W.A. Haverkort, A.H. van Gennip and D. Roos, *Anal. Biochem.*, 147 (1985) 197.
55 G.S. Nakai and C.G. Craddock, *Cancer Res.*, 25 (1965) 575.
56 R.A. Hartwick, D. Van Haverbeke, M. McKeag and P.R. Brown, *J. Liq. Chromatogr.*, 2 (1979) 725.
57 M.E. Dwyer and P.R. Brown, *J. High Resolut. Chromatogr. Chromatogr. Commun.*, 11 (1988) 420.
58 R.J. Simmonds and R.A. Harkness, *J. Chromatogr.*, 226 (1981) 369.
59 E.A. Hull-Ryde, R.G. Cummings and J.E. Lowe, *J. Chromatogr.*, 275 (1983) 411.
60 T. Kremmer, M. Boldizsar and L. Holczinger, *J. Chromatogr.*, 415 (1987) 53.
61 R. Boulien, C. Bory and C. Gonnet, *J. Chromatogr.*, 339 (1985) 380.
62 F. Arezzo, *Anal. Biochem.*, 160 (1987) 57.
63 J.X. Khym, *Clin. Chem.*, 21 (1975) 1245.

64 A. Werner, W. Siems, H. Schmidt, I. Rapoport, G. Gerber, R.T. Toguzov, Y.V. Tikhonov and A.M. Pimenov, *J. Chromatogr.*, 421 (1987) 257.
65 P.K. Dutta and G.A. O'Donovan, *J. Chromatogr.*, 385 (1987) 119.
66 G. Leoncini, E. Buzzi, M. Maresca, M. Mazzei and A. Balbi, *Anal. Biochem.*, 165 (1987) 379.
67 H. Mitsuya, K.J. Weinhold, P.A. Furman, M.H. St. Clair, S. Nusinoff Lehrman, R.C. Gallo, D. Bolognesi, D.W. Barry and S. Broder, *Proc. Natl. Acad. Sci. U.S.A.*, 82 (1985) 7096.
68 S.P. Assenza and P.R. Brown, *J. Liq. Chromatogr.*, 181 (1980) 169.
69 S.P. Assenza and P.R. Brown, *Anal. Chim. Acta,* 123 (1981) 33.
70 R.P. Agarwal, P.P. Major and D.W. Kufe, *J. Chromatogr.*, 231 (1982) 418.
71 B. Alick, C. Bridges, T. Cox, Jr., V. Earl and R. Thedford, *J. Chromatogr.*, 430 (1988) 309.
72 T.W. Kurtz, P.M. Kabra, B.E. Booth, H.A. Al-Bander, A.A. Portale, B.G. Serena, H.C. Tsai and R.C. Morris, Jr., *Clin. Chem.*, 32 (1986) 782.
73 W.H. Scouten, *Affinity Chromatography: Bioselective Adsorption on Inert Matrices,* Wiley, New York, 1981.
74 J.F. Lawerence, R. Leduc and J.J. Ryan, *Anal. Biochem.*, 116 (1981) 433.
75 A.P. Woodbridge, in E.H. McKerell and E. Reid (Editors), *Trace Organic Sample Handling,* Ellis Horwood, Chichester, 1980.
76 R.D. McDowall and G.S. Murkitt, *J. Chromatogr.*, 317 (1984) 475.
77 C.W. Gehrke, K.C. Kuo and R.W. Zumwalt, *J. Chromatogr.*, 188 (1980) 128.
78 C.D. Lothrop and M. Uziel, *Anal. Biochem.*, 109 (1980) 160.
79 J. Harmenberg, A. Larsson and C.-E. Hagberg, *J. Liq. Chromatogr.*, 6 (1983) 655.
80 C.W. Gehrke, K.C. Kuo, G.E. Davis, R.D. Suits, P.T. Waalkes and E. Borek, *J. Chromatogr.*, 150 (1978) 445.
81 M. Glad, S. Ohlson, L. Hansson, M.-O. Mánsson and K. Mosbach, *J. Chromatogr.*, 200 (1980) 254.
82 E. Hagemeier, K.-S. Boos, E. Schlimme, K. Lechtenborger and A. Kettrup, *J. Chromatogr.*, 268 (1983) 291.
83 E. Hagemeier, K. Kemper, K.-S. Boos and E. Schlimme, *J. Clin. Chem. Clin. Biochem.*, 22 (1984) 175.
84 E. Hagemeier, K. Kemper, K.-S. Boos and E. Schlimme, *J. Chromatogr.*, 282 (1983) 663.
85 P. Hubert and J. Porath, *J. Chromatogr.*, 206 (1981) 164.
87 W.E. Cohn, *Science,* 109 (1949) 377.
88 C.G. Horváth, B.A. Preiss and S.R. Lipsky, *Anal. Chem.*, 39 (1967) 1422.
89 R.A. Hartwick and P.R. Brown, *J. Chromatogr.*, 112 (1975) 651.
90 I. Halasz, H. Schmidt and P. Voltel, *J. Chromatogr.*, 126 (1975) 19.
91 M. McKeag, P.R. Brown and J.D. Sallis, *Comp. biochem. Physiol.*, 70B (1981) 541.
92 L.W. McLaughlin and E. Romaniuk, *Anal. Biochem.*, 124 (1982) 37.
93 D. Perrett, in C.K. Lim (Editor), *HPLC of Small Molecules - a Practical Approach,* IRL, Oxford, 1986, p. 221.
94 T.T. Nguyen, M.M. Palcic and D. Hadziyev, *Agric. Biol. Chem.*, 52 (1988) 1105.
95 P.A. Perrone, in P.R. Brown (Editor), *HPLC in Nucleic Acid Research,* Dekker, New York, 1984, p. 161.
96 N. Hoffman and J.C. Liao, *Anal. Chem.*, 49 (1977) 2225.
97 P.A. Perrone and P.R. Brown, *J. Chromatogr.*, 317 (1984) 301.
98 M. Colpan and D. Riesner, *J. Chromatogr.*, 296 (1984) 339.
99 J.H. Knox and J. Jurand, *J. Chromatogr.*, 203 (1981) 85.
100 P.A. Perrone and P.R. Brown, *J. Chromatogr.*, 307 (1984) 53.
101 C. Yi, J.L. Fasching and P.R. Brown, *J. Chromatogr.*, 352 (1986) 221.
102 P.R. Brown and E. Grushka, *Anal. Chem.*, 52 (1980) 1210.
103 W.F. Brubaker and W.H. Prusoff, *J. Chromatogr.*, 322 (1985) 455.
104 P.A. Blau, J.W. Hines and R.D. Voyksner, *J. Chromatogr.*, 420 (1987) 1.
105 J.D. Unadkat, S.S. Crosby, J.P. Wang and C.C. Hertel, *J. Chromatogr.*, 430 (1988) 420.

106 S.S. Good, D.J. Reynolds and P. De Miranda, *J. Chromatogr.*, 431 (1988) 123.
107 J. Balzarini, M. Cools and E. De Clercq, *Biochem. Biophys. Res. Commun.*, 158 (1989) 413.
108 H. Irth, R. Tocklu, K. Welten, G.J. de Jong, U.A.Th. Brinkman and R.W. Frei, *J. Chromatogr.*, 491 (1989) 321.
109 R. Kupferschmidt and R.W. Schmid, *Clin. Chem.*, 35 (1989) 1313.
110 M. Magnani, L. Rossi, M. Bianchi, L. Cucchiarini and V. Stocchi, *J. Chromatogr.*, 491 (1989) 215.
111 R.C. Simpson and P.R. Brown, *J. Chromatogr.*, 400 (1987) 297.
112 J.H. Knox, *J. Chromatogr. Sci.*, 18 (1980) 453.
113 R.C. Simpson, P.R. Brown and M.K. Schwartz, *J. Chromatogr. Sci.*, 25 (1985) 89.
114 M.W. Dong and J.R. Gant, *LC, Liq. Chromatogr. HPLC Mag.*, 2 (1984) 294.
115 R.P.W. Scott, *J. Chromatogr. Sci.*, 23 (1985) 233.
116 R.C. Simpson and P.R. Brown, *J. Chromatogr.*, 385 (1987) 41.
117 M. Novotny, *Clin. Chem.*, 26 (1980) 1474.
118 Y. Hirata, M. Novotny, T. Tsuda and D. Ishii, *Anal. Chem.*, 51 (1979) 1807.
119 Y. Hirata and M. Novotny, *J. Chromatogr.*, 186 (1979) 521.
120 J.F. Banks, Jr. and M.V. Novotny, *J. Chromatogr.*, 475 (1989) 13.
121 H.R.M. Lang and A. Rizzi, *J. Chromatogr.*, 356 (1986) 115.
122 A.P. Halfpenny and P.R. Brown, *Chromatographia*, 21 (1986) 317.
123 L. Mathes, G. Muschik, P. Polas, L. Demby, D.W. Melini, H.J. Issaq and R. Sams, *J. Chromatogr.*, 432 (1988) 346.
124 Z. El Rassi and Cs. Horváth, *J. Chromatogr.*, 359 (1986) 255.
125 J.B. Crowther, S.D. Fazio and R.A. Hartwick, *J. Chromatogr.*, 282 (1983) 619.
126 G. Davis, R. Suits, K. Kuo, C. Gehrke, T. Waalkes and E. Borek, *Clin. Chem.*, 23 (1977) 1427.
127 R.A. Gonzales, H.J. Salinas, E.L. Jacobson and M.K. Jacobson, *Anal. Biochem.*, 135 (1983) 69.
128 R.A. Olsson, *J. Chromatogr.*, 176 (1979) 239.
129 J.G. Turcotte, P.E. Pivarnik, S.S. Shirali, H.K. Singh, R.K. Sehgal, D. Macbride, N.-I. Jang and P.R. Brown, *J. Chromatogr.*, 499 (1990) 55.
130 J. Jacobson, Z. El Rassi and Cs. Horváth, in I. Molnar (Editor), *Practical Aspects of Modern HPLC*, Walter de Gruyter, Berlin, New York, 1982, pp. 1-14 and 15-23.
131 J. Jacobson, Z. El Rassi and Cs. Horváth, *J. Chromatogr.*, 253 (1982) 252.
132 S.P. Assenza and P.R. Brown, *J. Chromatogr.*, 289 (1984) 355.
133 S.P. Assenza, in P.R. Brown (Editor), *HPLC in Nucleic Acid Research*, Dekker, New York, 1984, p. 139.
134 E.S. Yeung, *LC/GC*, 7 (1989) 118.
135 M.J. Wojtusik, in P.R. Brown (Editor), *HPLC in Nucleic Acid Research*, Dekker, New York, 1984, p. 81.
136 P.J. Arpino, *J. Chromatogr.*, 323 (1985) 3.
137 E.D. Lee and J.D. Henion, *J. Chromatogr. Sci.*, 23 (1985) 253.
138 T.R. Covey and J.D. Henion, *Anal. Chem.*, 55 (1983) 2275.
139 G.M. Kresbach, T.R. Baker, R.J. Nelson, J. Wronka, B.L. Karger and P. Vouros, *J. Chromatogr.*, 394 (1987) 89.
140 R.S. Anna, G.M. Kresbach, R.W. Giese and P. Vouros, *J. Chromatogr.*, 465 (1989) 285.
141 K.J. Krosat, *Anal. Chem.*, 57 (1985) 763.
142 J. van der Greef, M.C. ten Noever de Brauw, T. Plomp, R.A. Maes, M. Hohn, G. Meyerhoff and U. Rapp, *J. Chromatogr.*, 343 (1984) 397.
143 C.-P. Tsai, A. Sahil, J.M. McGuire, B.L. Karger and P. Vouros, *Anal. Chem.*, 58 (1986) 2.
144 R.D. Voyksner, W.M. Hagler, Jr. and S.P. Swanson, *J. Chromatogr.*, 394 (1987) 183.
145 M.L. Vestal, *Int. J. Mass Spectrom. Ion Phys.*, 46 (1983) 193.
146 J.S. Thompson and R.S. Houk, *Anal. Chem.*, 58 (1986) 2541.
147 S.J. Jiang and R.S. Houk, *Spectrochim. Acta*, 43B (1988) 405.

148 K. Wuthrich, *NMR of Proteins and Nucleic Acids,* Wiley-Interscience, New York, 1986.
149 L.J. Haseler, B.T. Bulliman, P.W. Kuchel, D.M. Doddrel, J.R. Bell and M.G. Irving, *Cancer Invest.,* 6 (1988) 47.
150 E. Bayer, K. Abert, M. Nieder and E. Grom, *J. Chromatogr.,* 186 (1979) 497.
151 D.A. Laude, Jr. and C.L. Wilkins, *Anal. Chem.,* 56 (1984) 2471.
152 H.C. Dorn, *Anal. Chem.,* 56 (1984) 747A.
153 M. Sabo, J. Gross, J.-S. Wang and I.E. Rosenberg, *Anal. Chem.,* 57 (1985) 1822.
154 P.R. Griffiths, S.L. Pentoney, Jr., A. Giorgetti and K.H. Shafer, *Anal. Chem.,* 58 (1986) 1349A.
155 J.W. Hellgeth and L.T. Taylor, *Anal. Chem.,* 59 (1987) 295.
156 R.D. Freeman, R.M. Hammaker, C.E. Meloan and W.G. Fateley, *Appl. Spectrosc.,* 42 (1988) 456.
157 J.-M. Sequaris and E. Koglin, *Anal. Chem.,* 59 (1987) 527.
158 M. Morita and T. Uehiro, *Anal. Chem.,* 53 (1981) 1997.
159 D.R. Heine, M.B. Denton and T.D. Schlabach, *Anal. Chem.,* 54 (1982) 81.
160 K. Yoshda, H. Haraguchi and K. Fuwa, *Anal. Chem.,* 55 (1983) 1009.
161 J.A. Thompson, *BioChromatography,* 2 (1987) 4.
162 D.W. Stafford and D. Bieber, *Biochim. Biophys. Acta,* 378 (1975) 18.
163 D.B. Danner, *Anal. Biochem.,* 125 (1982) 139.
164 J.G. Guinther, *BioTechniques,* 2 (1984) 320.
165 M.T. Johnson, B.A. Read, A.M. Monko, G. Pappas and B.A. Johnson, *BioTechniques,* 4 (1986) 64.
166 J.A. Thompson, R.W. Blakesley, K. Doron, C.J. Hiugh and R.D. Wells, *Methods Enzymol.,* 100 (1983) 368.
167 M. Wende, T. Dorbic and B. Wittig, *Nucl. Acids Res.,* 13 (1985) 9043.
168 G. Zon and J.A. Thompson, *Biochromatography,* 1 (1986) 22.
169 R.E. Wu, E. Jay and R. Roychoudhury, *Methods Can. Res.,* 12 (1976) 87.
170 H.O. Smith, *Methods Enzymol.,* 65 (1980) 371.
171 N. Blin, A.V. Gabain and H. Bujard, *FEBS Lett.,* 53 (1975) 84.
172 R.S. Lloyd, A. Recinos III and S.T. Wright, *BioTechniques,* 4 (1986) 8.
173 M.S. Urdea and T. Horn, *Tetrahedron Lett.,* 27 (1986) 2933.
174 M.A. Urdea and R. Sanchez-Pescador, *BioTechniques,* 5 (9187) 106.
175 K. Jayaraman, *BioTechniques,* 5 (1987) 627.
176 T.G. Lawson, F.E. Regnier and H.L. Weith, *Anal. Biochem.,* 133 (1983) 85.
177 A. Hakam and E. Kun, *J. Chromatogr.,* 330 (1985) 287.
178 L.W. McLaughlin, *Chem. Rev.,* 89 (1989) 309.
179 J.D. Mandell and A.D. Hershey, *Anal. Biochem.,* 1 (1960) 66.
180 I. Gillam, S. Millward, D. Blew, M. von Tigerstrom, E. Wimmer and G.M. Tener, *Biochemistry,* 6 (1967) 3043.
181 J.B. Dogson and R.D. Wells, *Biochemistry,* 16 (1977) 2367.
182 G.C. Walkers, O.C. Uhlenbeck, E. Bedows and R.I. Gumport, *Proc. Natl. Acad. Sci. USA,* 72 (1975) 122.
183 R.D. Wells, S.C. Hardies, G.T. Horn, B. Klein, J.E. Larson, S.K. Heuendorf, N. Panayotatos, R.K. Patient and E. Selsing, *Methods Enzymol.,* 65 (1980) 327.
184 J. Crowther and R.A. Hartwick, *Chromatographia,* 16 (1982) 349.
185 T.R. Floyd, S.E. Cicero, S.D. Fazio, R.V. Raglione, S.H. Hsu, S.A. Winkle and R.A. Hartwick, *Anal. Biochem.,* 154 (1986) 570.
186 R. Bischoff and L.W. McLaughlin, *J. Chromatogr.,* 270 (1983) 117.
187 R. Bischoff and L.W. McLaughlin, *J. Chromatogr.,* 296 (1984) 329.
188 R. Bischoff and L.W. McLaughlin, *J. Chromatogr.,* 317 (1984) 251.
189 R. Bischoff, E. Graeser and L.W. McLaugahlin, *J. Chromatogr.,* 257 (1983) 305.
190 R. Bischoff and L.W. McLaughlin, *Anal. Biochem.,* 151 (1985) 526.
191 Z. El Rassi and Cs. Horváth, *Chromatographia,* 19 (1984) 9.
192 Z. El Rassi and Cs. Horváth, *J. Chromatogr.,* 326 (1985) 79.
193 R. Drager and R. Regnier, *Anal. Biochem.,* 145 (1985) 47.
194 G.M. Tener, *Methods Enzymol.,* 12 (1967) 398.

195 M. Staehelin, *Prog. Nucl. Acid Res.*, 2 (1963) 169.
196 M. Dizdaroglu, W. Hermes and C. Vonsonntag, *J. Chromatogr.*, 169 (1979) 429.
197 M. Dizdaroglu and W. Hermes, *J. Chromatogr.*, 171 (1979) 321.
198 M.J. Gait, H.W.D. Matthes, M. Singh, B.S. Sproat and R.C. Timas, *Nucl. Acids Res.*, 10 (1982) 6243.
199 M. Haupt and A. Pingoud, *J. Chromatogr.*, 260 (1983) 419.
200 J.M. Munholland, K.A. Bright and R.N. Nazar, *Anal. Biochem.*, 178 (1989) 320.
201 J.A. Thompson, in *NACS Application Manual*, Bethesda Research Laboratories, Life Technologies, Inc., 1984.
202 J.A. Thompson, R.W. Blakesley, K. Doran, C.J. Hough and R.D. Wells, *Methods Enzymol.*, 100 (1983) 372.
203 F. Marashi, G. Stein, J. Stein and C. Shubert, *BioTechniques*, 3 (1985) 238.
204 H. Scott and H. Eckstein, *J. Chromatogr.*, 296 (1984) 363.
205 J.B. Crowther, R. Jones and R.A. Hartwick, *J. Chromatogr.*, 210 (1981) 479.
206 Y. Kato, K. Nakamura and T. Hashimoto, *J. Chromatogr.*, 266 (1983) 385.
207 W. Muller, *Eur. J. Biochem.*, 155 (1986) 203.
208 M.V. Cubellis, G. Marino, L. Mayol, G. Piccialli and G. Sannia, *J. Chromatogr.*, 329 (1985) 406.
209 A.Yu. Tsygankov, Yu.A. Motorin, A.D. Wolfson, D.B. Kirpotin and A.F. Orlovsky, *J. Chromatogr.*, 465 (1989) 325.
210 H.-J. Fritz, R. Belagaje, E.L. Brown, R.H. Fritz, R.A. Jones, R.G. Lees and H.R. Khorana, *Biochemistry*, 17 (1978) 1268.
211 R.A. Jones, H.-J. Fritz and H.G. Khorana, *Biochemistry*, 17 (1978) 1268.
212 V.V. Demidov and V.N. Potaman, *J. Chromatogr.*, 285 (1984) 135.
213 G. Zon, in W.S. Hancock (Editor), *High-Performance Liquid Chromatography in Biotechnology*, Wiley, New York, 1986.
214 H. Schott, H.D. Meyer and E. Bayer, *J. Chromatogr.*, 280 (1983) 297.
215 C.R. Becker, J.W. Efcavitch, C.R. Heiner and N.F. Kaiser, *J. Chromatogr.*, 326 (1984) 293.
216 K.-M. Lo, S.S. Jones, N.R. Hackett and H.G. Khorana, *Proc. Natl. Acad. Sci. USA*, 81 (1984) 2285.
217 W. Jost, K. Unger and G. Schill, *Anal. Biochem.*, 119 (1982) 214.
218 M. Kwiatkowski, A. Sandstrom, N. Balgobin and J. Chattopadhyaya, *Acta Chem. Scand.*, B38 (1984) 721.
219 P.N. Ngyen, J.L. Bradley and P.M. McGuire, *J. Chromatogr.*, 236 (1982) 508.
220 M.T.W. Hearn, *Ion Pair Chromatography*, Dekker, New York, 1985.
221 B. Allinquant, C. Musenger and E. Schuller, *J. Chromatogr.*, 326 (1985) 281.
222 H.K. Webster and J.M. Whaun, *J. Chromatogr.*, 209 (1981) 283.
223 L. Grotjahn, R. Frank and H. Blocker, *Nucl. Acids Res.*, 10 (1982) 4671.
224 J.B. Crowther, J.P. Caronia and R.A. Hartwick, *Anal. Biochem.*, 124 (1982) 65.
225 J.P. Caronia, J.B. Crowther and R.A. Hartwick, *J. Liq. Chromatogr.*, 6 (1983) 1673.
226 M. Kawaichi, K. Ueda and O. Hayaishi, *J. Biol. Chem.*, 256 (1981) 9483.
227 T. Maniatis, E.P. Fritsch and J. Sambrook, in *Molecular Cloning: A Laboratory Manual*, Cold Spring Harbor Laboratory, Cold Spring Harbor, NY, 1982.
228 G. Frick, *Biochim. Biophys. Acta*, 13 (1954) 41.
229 K.S. Kirby, *Biochem. J.*, 64 (1956) 405.
230 K.S. Kirby, *Biochem. J.*, 96 (1965) 266.
231 J. Wisniewski, J. Fronk and K. Toczko, *Anal. Biochem.*, 148 (1985) 245.
232 R. Hecker, M. Colpan and D. Riesner, *J. Chromatogr.*, 326 (1985) 251.
233 H. Lehrach and A.M. Frischauf, *EMBL Letter. Manual*, EMBL, Heidelberg, 1982.
234 W. Mann and J. Jeffery, *Anal. Biochem.*, 178 (1989) 82.
235 M.B. Johns, Jr. and J.E. Paulus-Thomas, *Anal. Biochem.*, 180 (1989) 276.
236 J.A. Thompson, *BioChromatography*, 1 (1986) 68.
237 G. Bernardi, *Methods Enzymol.*, 21 (1971) 95.
238 S.Y. Chan and P.J. Simpson, *BioTechniques*, March/April (1985) 115.
239 D.A. Popovic and V. Leskovac, *Anal. Biochem.*, 153 (1986) 139.

240 M.T. Macdonell, S.C. Morris, B.A. Ortiz-Conde, C.J. Pillidge and R.R. Colwell, *J. Chromatogr.*, 363 (1986) 438.
241 K. Rahman, U. Voss, P.J. Nicholls and D.B. Malcolm, *Biochem. Soc. Trans.*, 16 (1988) 368.
242 M. Hirama, A. Takeda and K.J. McKune, *Anal. Biochem.*, 155 (1986) 385.
243 Y. Kato, M. Sasaki, T. Harshimoto, T. Morotsu, S. Fukushige and K. Matsubara, *J. Chromatogr.*, 265 (1983) 342.
244 R. Bischoff and L.W. McLaughlin, *J. Chromatogr.*, 418 (1987) 51.
245 S. Eriksson, G. Glad, P.-A. Pernemalm and E. Westman, *J. Chromatogr.*, 359 (1986) 265.
246 J.A. Thompson, *BioChromatography*, 1 (1986) 16.
247 T. Andersson, M. Carlsson, L. Hagel, P.-A. Pernemalm and J.-C. Jason, *J. Chromatogr.*, 326 (1985) 33.
248 S. Hjerten and Y. Junquan, *J. Chromatogr.*, 205 (1981) 317.
249 R. Dornburg, J. Kruppa and P. Foldi, *Liq. Chromatogr.*, 4 (1986) 22.
250 Y. Kato, M. Sasaki, T. Hashimito, T. Morotsu, S. Fukushige and K. Matsubara, *J. Biochem.*, 95 (1984) 83.
251 Y. Kato, Y. Yamasaki, T. Hashimoto, T. Morotsu, S. Fukushige and K. Matsubara, *J. Chromatogr.*, 320 (1985) 440.
252 J.-M. Schmitter, Y. Mechulam, G. Fayat and M. Anselme, *J. Chromatogr.*, 378 (1986) 462.
253 J. Hirabayashi and K.-I. Kasai, *Anal. Biochem.*, 178 (1989) 336.
254 J.A. Thompson, S. Garfinkel, R.B. Cohen and B. Safer, *BioChromatography*, 2 (1987) 160.
255 J.R. Mazzeo and I.S. Krull, *BioChromatography*, 4 (1989) 124.
256 C.P. Holstege, M.J. Pickaart and L.L. Louters, *J. Chromatogr.*, 455 (1988) 401.
257 C.W. Gehrke and K.C. Kuo, *Bull. Mol. Biol. Med.*, 10 (1985) 119.
258 C.W. Gehrke and K.C. Kuo, *J. Chromatogr.*, 471 (1989) 3.
259 D.B. Ludlum, *Pharm. Ther.*, 34 (1987) 145.
260 R. Staden, *Nucl. Acids Res.*, 14 (1986) 1.
261 C.A. Hutchinson, *Nucl. Acids Res.*, 14 (1986) 1917.
262 H.S. Bilofsky, C. Burks, J.W. Fickett, W.B. Good, F.I. Lewitter, W.P. Rindone, C.D. Swindell and C.S. Tung, *Nucl. Acids Res.*, 14 (1986) 1.
263 G.H. Hamm and G.N. Cameron, *Nucl. Acids Res.*, 14 (1986) 5.
264 D.W. Mount, *BioTechniques*, March/April (1985) 102.
265 L.M. Smith, *Am. Biotechnol. Lab.*, May (1989) 10.
266 V.A. Vorndam and J. Kerschner, *Anal. Biochem.*, 152 (1986) 221.
267 C.W. Chen and C.A. Thomas, *Anal. Biochem.*, 101 (1980) 339.
268 N.C. Stellwagen, *Biochemistry*, 22 (1983) 6180.
269 A.T. Andrews, *Electrophoresis: Theory, Techniques, and biochemical and Clinical Applications*, Clarendon Press, Oxford, 1981, Chaps. 4 and 5.
270 D.C. Schwartz and C.R. Cantor, *Cell*, 37 (1984) 67.
271 S. Boots, *Anal. Chem.*, 61 (1989) 551A.
272 J.K. Elder, D.K. Green and E.M. Southern, *Nucl. Acids Res.*, 14 (1986) 473.
273 W. Ansorge, B. Sproat, J. Stegemann, C. Schwager and M. Zenke, *Nucl. Acids Res.*, 15 (1987) 4593.
274 J.C. Sutherland, B. Lin, D.C. Monteleone, J. Mugavero, B.M. Sutherland and J. Trunk, *Anal. Biochem.*, 163 (1987) 446.
275 S.M. Clark, E. Lai, B.W. Birren and L. Hood, *Science*, 241 (1988) 1203.
276 G. Chou, D. Vollrath and R. Davis, *Science*, 235 (1986) 1582.
277 G.F. Carle and M.V. Olson, *Nucl. Acids Res.*, 12 (1984) 5647.
278 K. Gardiner, W. Lass and D. Patterson, *Somatic Cell Molecular Genetics*, 12 (1986) 185.
279 G. Chu, D. Vollrath and R. Davis, *Science*, 234 (1986) 1582.
280 P. Serwer, *Electrophoresis*, 8 (1987) 301.
281 C.F. Carle and M.V. Olson, *Methods Enzymol.*, 155 (1987) 468.

282 S. Ferris, S. Freeby, P. Zoller, C. Ragsdale and A. Stevens, *Am. Biotechnol. Lab.*, January (1989) 36.
283 M.J. Orbach, D. Vollrath, R.W. Davis and C. Yanofsky, *Mol. Cell Biol.*, 8 (1988) 1468.
284 D.C. Schwartz and M. Koval, *Nature (London)*, 338 (1989) 520.
285 M. Warner, *Anal. Chem.*, 61 (1989) 795A.
286 A.G. Ewing, R.A. Wallingford and T.M. Olefirowicz, *Anal. Chem.*, 61 (1989) 292A.
287 R.A. Wallingford and A.G. Andrew, in J.C. Gidding, E. Grushka and P.R. Brown (Editors), *Advances in Chromatography*, Vol. 29, Dekker, New York, 1989.
288 M. Warner, *Anal. Chem.*, 69 (1988) 1159A.
289 R. Virtanen, *Acta Polytech. Scand.*, 123 (1974) 1.
290 F.M. Mikkers, F.M. Everaerts and Th.P.E.M. Verheggen, *J. Chromatogr.*, 169 (1979) 11.
291 J.W. Jorgenson and K.D. Lukacs, *Anal. Chem.*, 53 (1981) 1298.
292 J.W. Jorgenson and K.D. Lukacs, *Science*, 222 (1983) 266.
293 T. Tsuda, K. Takagi, T. Watanabe and T. Satake, *J. High Resolut. Chromatogr. Chromatogr. Commun.*, 11 (1988) 721.
294 T. Tsuda, G. Nakagawa, M. Sato and K. Yagi, *J. Appl. Biochem.*, 5 (1983) 330.
295 A.S. Cohen, D.R. Majarian, A. Paulua, A. Guttman, J.A. Smith and B.L. Karger, *Proc. Natl. Acad. Sci. USA*, 85 (1988) 9660.
296 A.S. Cohen, S. Terabe, J.A. Smith and B.L. Karger, *Anal. Chem.*, 59 (1987) 1021.
297 A.S. Cohen, D. Majarian, J.A. Smith and B.L. Karger, *J. Chromatogr.*, 458 (1988) 323.
298 M. Novotny, K. Cobb, J.F. Banks, Jr., S. Beale, Y.-Z. Hsieh and D. Wiesler, *J. Chromatogr.*, 480 (1989) 321.
299 F. Sanger, S. Nicklen and A.R. Coulson, *Proc. Natl. Acad. Sci. USA*, 74 (1977) 5463.
300 A.M. Maxam and W. Gilbert, *Proc. Natl. Acad. Sci. USA*, 74 (1977) 560.
301 A.M. Maxam and W. Gilbert, *Methods Enzymol.*, 65 (1980) 449.
302 M.R. Atkinson, M.P. Deutscher, A. Kornberg, A.F. Russell and J.G. Moffatt, *Biochemistry*, 8 (1969) 4897.
303 J.K. Elder, D.K. Green and E.M. Southern, *Nucl. Acids Res.*, 14 (1986) 417.
304 D.A. Amorese and A.M. Hochberg, *Am. Biotechnol. Lab.*, March (1989) 40.
305 L.M. Smith, S. Fung, M.W. Hunkapiller, T.J. Hunkapiller and L.E. Hood, *Nucl. Acids Res.*, 13 (1985) 2399.
306 L.M. Smith, J.Z. Sanders, R.J. Kaiser, P. Hughes, C. Dodd, C.R. Connel, C. Heiner, S.B.H. Kent and L.E. Hood, *Nature (London)*, 321 (1986) 674.
307 L.R. Middendorf, J.A. Brumbaugh, D.L. Grone, C.A. Morgan and J.L. Ruth, *Am. Biotechnol. Lab.*, August (1988) 14.
308 C. Connell, S. Fung, C. Heiner, J. Bridgham, V. Chakerian, E. Heron, B. Jones, S. Menchen, W. Mordan, M. Raff, M. Recknor, L. Smith, J. Springer, S. Woo and M. Hunkapiller, *Biotechniques*, 5 (1987) 342.
309 J. Gocayne, D.A. Robison, M.G. Fitzgerald, F.-Z. Chung, A.R. Kerlavage, K.-V. Lentes, J. Lai, C.-D. Wang, C.M. Fraser and J.C. Venter, *Proc. Natl. Acad. Sci. USA*, 84 (1987) 8296.
310 J.M. Prober, G.L. Trainor, R.J. Dam, F.W. Hobbs, C.W. Robertson, R.J. Zagursky, A.J. Cocuzza, M.J. Jensen and K. Baumeister, *Science*, 238 (1987) 336.
311 D.A. Amorese and A.M. Hochberg, *Am. Biotechnol. Lab.*, March (1989) 38.
312 W. Ansorge, *Nucl. Acids Res.*, 15 (1987) 4593.
313 H. Kambara and T. Nishikawa, *Hayashibara Forum, 1987*, International Workshop on Automatic and High Speed DNA-Base Sequencing, Okayama, Japan, 1987.
314 J.A. Brumbaugh, L.R. Middenforf and J.L. Ruth, 193rd ACS National Meeting, Denver, Colorado, 1987, unpublished.
315 J.E. Bateman, J.F. Connally and R. Stephenson, *Nucl. Instrum. Methods*, A241 (1988) 1847.
316 J. West, *Nucl. Acids Res.*, 16 (1988) 1847.

Chapter 18

Porphyrins

KARL JACOB

CONTENTS

18.1 INTRODUCTION

The porphyrins constitute an exceptionally important class of compounds in nature [1,2]. Thus, the porphyrin/iron complex heme, in particular, forms the prosthetic group of many important hemoproteins, namely hemoglobin, myoglobin, cytochrome, catalases, and peroxidases. They are responsible for the transport and storage of oxygen, for electron transfer, oxidative detoxification, and the decomposition of hydrogen peroxide. The conversion of light energy to chemical energy in photosynthesis is accomplished by the porphyrin/magnesium complex chlorophyll. The common precursor of these porphyrin complexes is uroporphyrinogen III, which also forms the basic structure of the cobalt complex vitamin B_{12} [3], although in this case the porphyrin skeleton is extensively modified.

In the human body free porphyrins are usually found in larger amounts when there is some dysfunction of the enzymes involved in heme biosynthesis. Such abnormalities may be either inborn or acquired, and these conditions are known as porphyrias [4]. Since nonspecific clinical symptoms occur in a number of porphyrias, a biochemical diagnosis or confirmation – as a rule by chromatographic porphyrin analyses – is essential.

Porphyrins are also found in geological samples, e.g., in petroleum, as so-called petroporphyrins [5]. There they occur mainly in the form of vanadyl or nickel complexes. These petroporphyrins find increasing use as markers in petroleum exploration, and this accounts for the steady increase in interest in the analysis of these compounds (Section 21.6.3.1) [6].

18.2 PROPERTIES

The basic porphyrin skeleton is a macrocyclic structure in which four pyrrole nuclei are linked by methyne bridges. The structural formulae of some of the important porphyrins are evident from Fig. 18.1. The conjugated system, having 18 π-electrons, shows aromatic character. The substituents on Positions 1-8 as well as the metals, being complex formers, also materially affect the chemical and physico-chemical properties of these compounds.

Porphyrins are generally soluble in organic solvents, but the highly polar porphyrins, e.g., the uroporphyrins with eight carboxyl groups, are almost insoluble in organic solvents. Such polar porphyrins dissolve well in acids and bases. For estimating the solubility behavior, the so-called hydrochloric acid number is helpful. It is a measure of the extent to which a porphyrin can be extracted from an ether solution by hydrochloric acid [7].

Porphyrins have a relatively high heat stability but are rapidly destroyed in the presence of light and oxygen. This is particularly true of porphyrins that have unsaturated substituents, like protoporphyrin. In handling such porphyrins care must be taken to exclude light as completely as possible to prevent their decomposition.

Because of their relatively high molecular weight the volatility of porphyrins is low. However, it is possible to increase their volatility through the introduction of substituents

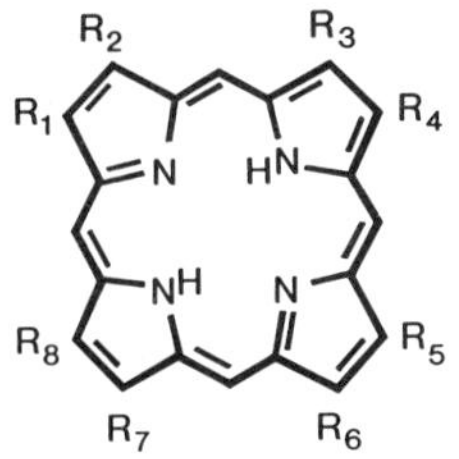

Fig. 18.1. Structures of porphyrins. Abbreviations: Me = methyl; Et = ethyl; V = vinyl; A = -CH2COOH; P = -CH2CH2COOH.

Porphyrin	R_1	R_2	R_3	R_4	R_5	R_6	R_7	R_8
Uroporphyrin I	A	P	A	P	A	P	A	P
Uroporphyrin II	A	P	P	A	A	P	P	A
Uroporphyrin III	A	P	A	P	A	P	P	A
Uroporphyrin IV	A	P	A	P	P	A	P	A
Heptacarboxylic porphyrin I	A	P	A	P	A	P	Me	P
Heptacarboxylic porphyrin III	A	P	A	P	A	P	P	Me
Hexacarboxylic porphyrin I	Me	P	A	P	A	P	Me	P
Hexacarboxylic porphyrin III	Me	P	A	P	A	P	P	Me
Pentacarboxylic porphyrin I	Me	P	Me	P	A	P	Me	P
Pentacarboxylic porphyrin III	Me	P	Me	P	A	P	P	Me
Coproporphyrin I	Me	P	Me	P	Me	P	Me	P
Coproporphyrin II	Me	P	P	Me	Me	P	P	Me
Coproporphyrin III	Me	P	Me	P	Me	P	P	Me
Coproporphyrin IV	Me	P	Me	P	P	Me	P	Me
Isocoproporphyrin	Me	Et	Me	P	A	P	P	Me
Mesoporphyrin IX	Me	Et	Me	Et	Me	P	P	Me
Protoporphyrin IX	Me	V	Me	V	Me	P	P	Me
Deuteroporphyrin IX	Me	H	Me	H	Me	P	P	Me

such as silyl groups so that a number of porphyrins can be analyzed by gas chromatography [8].

The behavior of porphyrins in absorption spectroscopy is of special interest. The macrocycle with its fully conjugated aromatic π-electron system is responsible for the intense color of these compounds. As early as the Thirties, Hans Fischer and his coworkers [9] classified the absorption spectra into a number of spectral types and showed that they were related to the type of substitution. The theoretical treatment of absorption spectra of porphyrins has been reviewed by Gouterman [10]. All porphyrins show a very strong absorption band, the so-called Soret band, in the near-UV at ca. 400 nm. This band is especially useful for the quantitation of porphyrins by spectrophotometry. In the visible range there are several typical absorption bands between 490 and 630 nm, but their extinction is distinctly lower than that of the Soret band.

Due to the rigid structure of the porphyrin skeleton these compounds show a strong fluorescence. As a rule, the maximum of the Soret band is used for excitation, and

References on p. B359

emission is measured at wavelengths above 600 nm. The characteristic red fluorescence allows an exceedingly sensitive and specific determination of porphyrins. Sagen and Romslo [11] have compared the analytical sensitivity by photometric and fluorometric detection. Applied to HPLC, the sensitivity of fluorometric detection was more than ten times higher. However, the intensity of the fluorescence of porphyrins greatly depends on the analytical conditions, e.g., the pH and the solvent composition [12].

18.3 CHROMATOGRAPHIC SEPARATION OF PORPHYRINS

This chapter deals primarily with those chromatographic methods which have been published since the appearance of the 4th edition of this textbook [13] or which were not included there. In the meantime several review articles [14-16] have been published, which describe mainly applications of HPLC to the clinico-chemical analysis of porphyrins.

18.3.1 Gas chromatography

Owing to their low volatility, the porphyrins are usually separated by LC. However, by means of special devices it is possible to separate porphyrins in the gas phase. Thus, Corwin and coworkers [17-19] were first to accomplish the separation of alkyl porphyrins and their metal complexes by supercritical-fluid chromatography. However, peakshapes and chromatographic resolution were less than satisfactory. Boylan and Calvin [20] were able to show that alkyl porphyrins can be separated in the form of their silicon(IV) complexes by gas chromatography in packed columns. The bis-trimethylsilyl ethers, obtained by derivatization of the $Si(OH)_2$/porphyrin complexes, proved to be best for this purpose. The introduction of trimethylsilyl groups markedly increases the volatility of these complexes. It was not until stably coated (OV-1, CPSil-5) glass or quartz capillary columns came into use that Marriott et al. [8,21] were able to prepare satisfactory gas chromatograms of free alkyl porphyrins and of trimethylsiloxy metal complexes (Si(IV), Al(III), Ga(III), Rh(III)) of alkyl porphyrins. Besides the trimethylsiloxy (TMSO) group, the t-butyldimethylsiloxy (TBDMSO) group also proved to be an excellent protective group, the peakshape of these derivatives approaching those of hydrocarbon standards (n-C_{30} to n-C_{40}) [21]. By using this derivatization technique Marriott and Eglinton [22] succeeded for the first time in converting a series of polar porphyrins with ester groups to compounds suitable for GC. Thus, they separated deutero-, meso-, and rhodoporphyrin in the form of $(TMSO)_2Si(IV)$ derivatives on a 20-m CPSil-5 glass capillary column.

Marriott et al. [23] have described the analysis of complex mixtures of alkylated porphyrins in geological samples by the use of these $(TMSO)_2Si(IV)$ and $(TBDMSO)_2Si(IV)$ derivatives. By GC/MS analysis on an OV-1 quartz or a CPSil-5 glass capillary column they were able to classify the petroporphyrins in bitumen and crude oil into various pseudo-homologous series. Because of their more favorable mass spectrometric properties the TBDMSO derivatives were more suitable for GC/MS analysis. Gill et al. [24] then succeeded in determining the age of geological materials by this procedure from the GC/MS data of petroporphyrins.

The analysis of porphyrins by GC is successful only in special instances. It is primarily useful for separating and determining homologous alkyl porphyrins, e.g., the petroporphyrins. The success of these analyses depends on the introduction of protective groups that increase their volatility.

18.3.2 Paper chromatography

The methodology of porphyrin analysis was decisively improved when Nicholas and Rimington [25] introduced paper chromatography in 1949, as that technique made it possible for the first time to work with microgram quantities of porphyrins. Paper chromatography permits the separation of free porphyrin carboxylic acids in aqueous solutions of various lutidines as mobile phases [25-29]. Porphyrin methyl esters can likewise be separated by means of paper chromatography, using kerosene with the addition of other solvents as mobile phase [30,31]. This allowed the resolution of the isomeric coroporphyrin esters I and III [30], but problems arose in the separation of the isomeric uroporphyrin esters I and III [32].

Because the running times in paper chromatography are so long, it is practically no longer used now for porphyrin separations. Paper chromatography has been completely replaced by thin-layer chromatography and later on by HPLC.

18.3.3 Thin-layer chromatography

Thin-layer chromatography is a very versatile technique for separating porphyrins. It requires only a minimum of equipment; moreover, several samples can be analyzed simultaneously on a single thin-layer plate. This technique permits the separation of practically all of the common classes of porphyrins, e.g., the porphyrin carboxylic acids and their esters, alkyl porphyrins, as well as porphyrin/metal complexes. Most of these separations have already been reviewed by Dolphin [13] in the 4th edition of this textbook.

Silica gel [13], talc [33], cellulose [34], and polyamide [35] have been used as sorbents. Of these, primarily silica gel has come to be of practical importance.

The separation by TLC of the diagnostically relevant porphyrins, e.g., uro-, heptacarboxy-, hexacarboxy-, pentacarboxy-, copro-, and protoporphyrin, has proved its worth for the chromatographic analysis of clinical specimens, such as urine, stool, or blood. For this, the methyl esters of the porphyrin carboxylic acids have usually been separated on silica gel plates, but free porphyrin carboxylic acids could also be chromatographed on silica gel. Extensive development work on routine methods for the separation by TLC of porphyrins in human specimens has been carried out by Doss [36,37]. With the solvent system benzene/ethyl acetate/methanol and silica gel the porphyrin methyl esters can be completely separated according to the number of carboxyl groups, the R_F values being inversely proportional to that number. One disadvantage of this solvent system is that benzene is toxic, requiring special safety precautions.

Other solvent systems for the separation of porphyrin methyl esters in clinical specimens have been described, in particular by With [38] and Day et al. [39,40]. These authors used benzene/ethyl acetate/ethanol [38] or carbon tetrachloride/dichloromethane/ethyl acetate/ethyl propionate [39,40].

Free porphyrin carboxylic acids (from uro- to protoporphyrin) were neatly resolved on silica gel with a complex six-component solvent system by Friedmann and Baldwin [41]. The system was composed of N,N-dimethylformamide, methanol, ethylene glycol, glacial acetic acid, 1-chlorobutanol, and chloroform. Again, the R_f values were inversely proportional to the number of carboxyl groups. A simpler system for the separation of free porphyrin carboxylic acids (from uro- to protoporphyrin) was recently published by Henderson [42]. The solvent system consisted of chloroform/methanol/ammonium hydroxide/water and was used for the analysis of urine and stool extracts. A separation of free porphyrin dicarboxylic acids on reversed-phase plates, coated with a C_{18} layer, was described by Garbo et al. [43]. The mobile phase used, methanol/aq. tetrabutylammonium phosphate solution, was suitable for the separation of mixtures of porphyrin dicarboxylic acids present in the so-called hematoporphyrin derivatives. The hematoporphyrin derivatives are used in the photodynamic therapy of tumors [44]. Vever-Bizet et al. [45] also successfully separated the hematoporphyrin derivatives on silica gel with an acetone/ethyl acetate/water system.

For the separation of the position isomers I and III of coproporphyrin Schermuly and Doss [46] used the free porphyrin carboxylic acids. These two isomers were completely resolved on silica gel in an ammonia atmosphere by using 2,6-dimethylpyridine/water as the mobile phase. With the advent of high-performance (HP) TLC a very rapid separation of porphyrins became possible. Thus, the methyl esters of urinary porphyrins could be resolved within 14 minutes with the solvent system toluene/ethyl acetate/methanol by Seubert et al. [47]. The resolution of the free porphyrin carboxylic acids (from uro- to mesoporphyrin) was accomplished by HPTLC by using N-cetyl-N,N,N-trimethylammonium bromide as ion-pair reagent [48], but this required the use of two eluent systems, which were applied in succession.

Not only one-dimensional TLC but also two-dimensional TLC has been employed in porphyrin separations. Using this technique, Elder [49,50] was able to detect isocoproporphyrin in the feces of a patient suffering from porphyria cutanea tarda.

Thus, TLC is characterized by simplicity and general applicability. It is also suitable for work on a preparative scale, but it does have the disadvantage that it is relatively inefficient in the resolution of position isomers, particularly when porphyrin profiles are needed. Moreover, the thin-layer chromatograms are unstable and must be protected against light and air.

18.3.4 High-performance liquid chromatography

The greatest advances in the chromatography of porphyrins have been achieved in recent years by the application of HPLC techniques. While the emphasis was initially on adsorption chromatography, as time went on, reversed-phase chromatography came into ever increasing use [51]. One of the principal advantages of HPLC over planar chromatography is that the light- and oxygen-sensitive porphyrins are completely protected inside the HPLC column. Thus, they can be analyzed by HPLC in the picogram range, being absolutely stable during separation.

18.3.4.1 Normal-phase HPLC

The use of normal-phase HPLC for the separation of porphyrin methyl esters from biological fluids generally furnishes cleaner chromatograms than the direct analysis of these fluids, because esterification acts to some extent as a purification step. Also, the higher capacity of normal-phase columns gives them an advantage over reversed-phase columns in the separation of porphyrins on a semi-preparative scale. On the other hand, a disadvantage is that the esterification reaction takes time and may also lead to the formation of artifacts in the shape of partially esterified products or mixed esters. The equilibration of the column for adsorption chromatography also often takes too much time.

18.3.4.1.1 Isocratic techniques

Table 18.1 lists HPLC systems for the isocratic separation of porphyrins by normal-phase chromatography.

Gray et al. [52] have demonstrated that porphyrin HPLC profiles are exceptionally useful for the diagnosis of the known porphyrias. These authors achieved a complete separation of the methyl esters of porphyrins in the series from proto- to uroporphyrin on μ-Porasil columns. As in TLC, the elution sequence of the porphyrin esters is the reverse of the number of ester groups, i.e. the porphyrin dicarboxylic acid esters (meso- and protoporphyrin) are eluted first and the uroporphyrins with 8 ester groups are eluted last. Similar separations, likewise on μ-Porasil columns, were later reported by Straka et al. [53], while Seubert and Seubert [54] performed the same separations on radial-compression columns (Radial Pak Silica). Using aminopropyl-modified silica columns (Separon Six NH$_2$), Kotal et al. [55,56] prepared profiles of porphyrins in the form of their methyl esters and copper complexes. Although none of these procedures allowed a simultaneous separation of the natural isomers of series I and III, Straka et al. [53] were able to resolve a methyl/ethyl ester mixture of uroporphyrin on μ-Porasil columns.

Isocratic normal-phase HPLC has also been applied to the separation of isomers of individual porphyrins. Thus, originally 5 consecutive chromatographic cycles were required for the difficult separation of the uroporphyrin methyl esters I and III according to the procedure of Bommer et al. [57], but later on, several groups of researchers [54,58-61] accomplished the same separation in a single chromatographic step. Jackson et al. [62] resolved the coproporphyrin isomers I and III as well as isocoproporphyrin in the form of methyl esters by normal-phase HPLC on Hypersil. The same column was also suitable for the separation of several di- and tricarboxyporphyrin methyl esters [63]. Dellinger and Brault [64] were able to chromatograph porphyrin dicarboxylic acids as well as hematoporphyrin derivatives also in unesterified, free form, but optimal chromatographic conditions were very dependent on the amount of water and pH of the mobile phase.

References on p. B359

TABLE 18.1

NORMAL-PHASE HPLC SYSTEMS FOR THE ISOCRATIC SEPARATION OF PORPHYRINS

Column	Eluent	Porphyrins	Ref.
μ-Porasil	Heptane/methyl acetate	Di- to octacarboxylic porphyrin methyl esters	52
μ-Porasil	Benzene/ethyl acetate/methanol	Di- to octacarboxylic porphyrin methyl esters	53
μ-Porasil	n-Heptane/ethyl acetate	Uroporphyrin mixed methyl/ethyl esters	53
Radial Pak Silica	n-Heptane/ethyl acetate/chloroform/methanol	Di- to octacarboxylic porphyrin methyl esters	54
Separon Six NH₂	n-Heptane/ethyl acetate	Di- to octacarboxylic porphyrin methyl esters and Cu complexes	55, 56
μ-Porasil	n-Heptane/acetic acid/acetone/water	Uroporphyrin octamethyl esters I and III	57
μ-Porasil	n-Heptane/acetic acid/acetone	Uroporphyrin octamethyl esters I and III	58
APS-Hypersil	Hexane/ethyl acetate/butanol/acetic acid	Uroporphyrin octamethyl esters I and III	59
Hypersil	Hexane/ethyl acetate	Uroporphyrin octamethyl esters I-IV	60
Adsorbosphere Silica	n-Hexane/ethyl acetate/dichloromethane	Uroporphyrin octamethyl esters I and III	61
Hypersil	Light petroleum/ethyl acetate	Tetracarboxylic porphyrin methyl esters	62
Hypersil	Light petroleum/ethyl acetate	Di- and tricarboxylic porphyrin methyl esters	63
Partisil 5	Acetone/ethyl acetate/water (HCl)	Dicarboxylic porphyrins	64

TABLE 18.2

NORMAL-PHASE HPLC SYSTEMS FOR THE GRADIENT SEPARATION OF PORPHYRINS

Column	Eluent	Porphyrins	Ref.
Silica A	Acetone/acetic acid	Di- to octacarboxylic porphyrins	65
Hypersil	Acetonitrile/water/tetraethylenepentamine	Di- to octacarboxylic porphyrins	66
Silica A	n-Heptane/methyl acetate	Di- to octacarboxylic porphyrin methyl esters	67
Radial Pak Silica	Hexane/ethyl acetate	Di- to octacarboxylic porphyrin methyl esters	68
Spherisorb 3W	Dichloromethane/acetone/hexane/pyridine/acetic acid	Alkylporphyrins	69
Spherisorb 3W	Dichloromethane/acetone/hexane/pyridine/acetic acid	Etioporphyrins, cycloalkano porphyrins	70

TABLE 18.3

REVERSED-PHASE HPLC SYSTEMS FOR THE SEPARATION OF PORPHYRINS

Eluents consisting of organic solvents and acids or bases.

Column	Eluent	Porphyrins	Ref.
μBondapak C_{18}	Methanol/acetic acid/water (gradient mode)	Proto-, meso-, hemato-, and coproporphyrin	71
μBondapak C_{18}	Methanol/acetic acid/water (gradient mode)	Di- to octacarboxylic porphyrins	72
μBondapak C_{18}	Acetone/methanol/formic acid/water (isocratic mode)	Protoporphyrin, Zn protoporphyrin	73
Brownlee RP18	Acetonitrile/methanol/acetone/dimethylpyridine /phosphoric acid/water (gradient mode)	Proto-, meso-, deutero-, copro-, and uroporphyrin	74

References on p. B359

18.3.4.1.2 Gradient techniques

Only a few methods involving normal-phase gradient elution have been described (Table 18.2). These methods were applied to porphyrin carboxylic acids both in the free form and in the form of their methyl esters.

Although Longas and Poh-Fitzpatrick [65] were able to chromatograph picomole quantities of porphyrin carboxylic acids from meso- to uroporphyrin in free form on Silica A columns in ca. 12 min, the separation of uro- from heptacarboxyporphyrin was not satisfactory. Complete separation of these two porphyrins and of coproporphyrin I, III, and isocoproporphyrin was achieved by the method of Lim and Chan [66], in which a Hypersil column, modified with tetraethylenepentamine, was eluted with an acetonitrile/water gradient.

Colombi et al. [67] obtained profiles of porphyrin methyl esters of the series from meso- to uroporphyrin with a Silica A column and a heptane/methyl acetate gradient, but no resolution of isomeric porphyrins was achieved. Profiles of porphyrin methyl esters were also produced with radial-compression columns, using a hexane/ethyl acetate gradient as the mobile phase [68].

Finally, the gradient elution of alkyl-, etio-, and cycloalkanoporphyrins in petroleum and sediments on Spherisorb 3 W columns was also reported [69,70]. Peak resolution was considerably improved by the addition of acetic acid and pyridine to the mobile phase.

18.3.4.2 Reversed-phase HPLC

Porphyrins can be separated on reversed-phase columns by isocratic as well as gradient-elution techniques. The mobile phases essentially belong to one of 4 different eluent systems:
1. organic solvents with additions of either acids or bases;
2. organic solvents in combination with aqueous buffer solutions;
3. organic solvents combined with aqueous ammonium acetate solutions;
4. organic solvents in the presence of ion-pair reagents.

Common to all reversed-phase systems is the fact that, in contrast to normal-phase systems, they elute the more polar porphyrins, having a higher number of carboxyl groups, before the less polar porphyrins. Moreover, the porphyrin carboxylic acids are, as a rule, chromatographed in the free form, whereas the esters are chromatographed only in rare instances.

18.3.4.2.1 Reversed-phase systems with organic solvents and acids or bases

Table 18.3 lists reversed-phase systems containing, in addition to organic solvents, either acids or bases.

Culbreth et al. [71] separated a series of porphyrin dicarboxylic acids and coproporphyrin on μBondapak C_{18} columns with a gradient of methanol/acetic acid/water. This method was used to study the acid-catalyzed hydrolysis of protoporphyrin IX dimethyl ester, and hematoporphyrin was identified as a byproduct. A very similar system was used by Salmi and Tenhunen [72] for the determination of erythrocyte porphyrins. Also on μBondapak, but by isocratic elution with a mobile phase of acetone/methanol/formic acid/water, Bailey and Needham [73] separated protoporphyrin from zinc protoporphyrin. This separation was used for the determination of erythrocyte porphyrins as well as for quality control of commercial porphyrin standards. Beukefeld et al. [74] have studied the fecal porphyrin excretion pattern on RP-18 columns with a complex eluent mixture, containing acetic acid and dimethylpyridine. In this study the emphasis was on the influence of intestinal bacteria on the porphyrin excretion in the stool.

18.3.4.2.2 Reversed-phase systems with organic solvents and aqueous buffers
18.3.4.2.2.1 Isocratic techniques. The already mentioned difficult separation of uroporphyrin isomers was studied by Wayne et al. [75] with the aid of column switching. After the adsorption of uroporphyrins on an ODS precolumn at pH 2.7, they were eluted with acetonitrile/phosphate buffer (pH 6.95) and then completely separated into the isomers I, II, and III on a μBondapak column. However, the resolution of isomers III and IV was only partly successful. Very similar eluent systems (acetonitrile/acetic acid/phosphate buffer) were used by several groups of investigators [76-78] for the isocratic resolution of coproporphyrin isomers I and III (Table 18.4).

Grubina et al. [79] used acetonitrile and an acetate buffer to separate coproporphyrin methyl esters I and III and chromatographed the free isomers with methanol/phosphate buffer. Copro-, meso-, and protoporphyrin as well as zinc protoporphyrin were separated by Scoble et al. [80] on a μBondapak C_{18} column with methanol/phosphate buffer. This isocratic system was applied to the determination of erythrocyte porphyrins. The analysis time was less than 6 min.

Guthrie's research group [81,82] developed isocratic HPLC procedures for the simultaneous determination of porphyrin precursors (5-aminolevulinic acid, porphobilinogen) and porphyrins (from uro- to protoporphyrin, zinc protoporphyrin) with the aid of elution systems consisting of methanol, tetrahydrofuran, and phosphate buffer. Following a suitable extraction step, this procedure allowed the quantitative analysis of the porphyrins including their precursors in whole blood as well as dried blood [82].

Porphyrin c, an addition product of cysteine to the two vinyl groups of protoporphyrin, and the *N,N'*-diacetyl porphyrin c derivatives were chromatographed on Nova Pak C_{18} columns by Scourides et al. [83]. Mixtures of methanol and either ammonium formate or ammonium bicarbonate were used as mobile phases.

18.3.4.2.2.2 Gradient techniques. Numerous methods for the establishment of porphyrin profiles by means of gradient systems consisting of organic solvents and buffer solutions are known (see Table 18.5). A simultaneous resolution of the porphyrin isomers with these systems was achieved only rarely for coproporphyrins I and III. Thus, e.g., Englert et al. [84] and Takayama [85] achieved the simultaneous separation of coproporphyrins I and III besides uro-, hepta-, hexa-, and pentacarboxyporphyrin. The mobile phase was a gradient of acetonitrile and phosphate buffer. Establishment of porphyrin profiles in the series from uro- to meso- and protoporphyrin, but without simultaneous isomer resolution, with a methanol/phosphate buffer gradient system has been described

TABLE 18.4

REVERSED-PHASE HPLC SYSTEMS FOR THE ISOCRATIC SEPARATION OF PORPHYRINS

Eluents consisting of organic modifiers and aqueous buffer solutions.

Column	Eluent	Porphyrins	Ref.
μBondapak C$_{18}$	Acetonitrile/phosphate buffer (pH 6.9)	Uroporphyrins I-IV	75
Hitachi 3053 RP	Acetonitrile/acetic acid/phosphate buffer	Coproporphyrins I and III	76
Finepak Sil C$_{18}$	Acetonitrile/acetic acid/phosphate buffer	Coproporphyrins I and III, Zn coproporphyrins I and III	77
Shim-pack CLC-ODS	Acetonitrile/acetic acid/phosphate buffer	Coproporphyrins I and III	78
Zorbax RP	Acetonitrile/acetate buffer (pH 5.16)	Coproporphyrin methyl esters I and III	79
Zorbax RP	Methanol/phosphate buffer (pH 6.52)	Coproporphyrins I and III	79
μBondapak C$_{18}$	Methanol/phosphate buffer (pH 3.4)	Di- to tetracarboxylic porphyrins, Zn protoporphyrin	80
Partisil-10 ODS	Methanol/tetrahydrofuran/phosphate buffer (pH 5.2)	Di- to octacarboxylic porphyrins, Zn protoporphyrin	81
μBondapak C$_{18}$	Methanol/tetrahydrofuran/phosphate buffer (pH 5.38)	Di- to octacarboxylic porphyrins, Zn protoporphyrin	82
Nova Pak C$_{18}$	Methanol/ammonium formate buffer (pH 5.4)	Porphyrin c	83
Nova Pak C$_{18}$	Methanol/aq. ammonium bicarbonate	N,N-Diacetyl porphyrin c	83

TABLE 18.5

REVERSED-PHASE HPLC SYSTEMS FOR THE GRADIENT SEPARATION OF PORPHYRINS

Eluents consisting of organic modifiers and aqueous buffer solutions.

Column	Eluent	Porphyrins	Ref.
μBondapak C_{18}	Acetonitrile/phosphate buffer (pH 7.5)	Tetra- to octacarboxylic porphyrins	84
Hitachi ODS	Acetonitrile/acetic acid/phosphate buffer	Tetra- to octacarboxylic porphyrins	85
μBondapak C_{18}	Methanol/phosphate buffer (pH 3.5)	Di- to octacarboxylic porphyrins	86, 87
μBondapak C_{18}	Methanol/phosphate buffer (pH 3.0)	Di- to octacarboxylic porphyrins	88
Radial Pak C_{18}	Methanol/phosphate buffer (pH 3.5)	Di- to octacarboxylic porphyrins	89
Perkin-Elmer ODS	Methanol/phosphate buffer (pH 3.5)	Di- to octacarboxylic porphyrins	90
LiChrosorb RP-18	Methanol/lithium citrate (pH 2.5)	Di- to octacarboxylic porphyrins	91
Radial Pak C_{18}	Methanol/lithium citrate (pH 2.5)	Di- to octacarboxylic porphyrins	92
Spherisorb ODS	Methanol/lithium citrate (pH 3.0)	Di- to octacarboxylic porphyrins	93
LiChrosorb RP-18 (coated with tributyl phosphate)	Methanol/phosphate buffer (pH 4.4-6.5)	Tetra- to octacarboxylic porphyrins	94
Vydac ODS	Acetonitrile/acetate buffer (pH 3.0)	Meso-substituted porphyrins and metal complexes	95

References on p. B359

TABLE 18.6

REVERSED-PHASE HPLC SYSTEMS FOR THE ISOCRATIC SEPARATION OF PORPHYRINS

Eluents consisting of organic modifiers and ammonium acetate buffer.

Column	Eluent	Porphyrins	Ref.
ODS-Hypersil	Acetonitrile/ammonium acetate (pH 5.16)	Coproporphyrins I-IV	97
ODS-Hypersil	Acetonitrile/ammonium acetate (pH 5.16)	Uroporphyrins I-IV	98
ODS-Hypersil	Acetonitrile/ammonium acetate (pH 5.16)	Uroporphyrins I and III	99, 100
ODS-Hypersil	Acetonitrile/ammonium acetate (pH 5.16)	Pentacarboxylic porphyrin isomers	101
ODS-Hypersil	Acetonitrile/ammonium acetate (pH 5.16)	Tetra- to octacarboxylic porphyrin isomers	102
SAS-Hypersil (C$_1$)	Methanol/ammonium acetate (pH 5.16)	Deutero-, meso-, proto-, and hematoporphyrin	103
SAS-Hypersil (C$_1$)	Methanol/ammonium acetate (pH 4.6)	Co, Fe, Zn, and Cu complexes of meso- and protoporphyrin	103
MOS-Hypersil (C$_8$)	Methanol/ammonium acetate (pH 4.6)	Hematoporphyrin derivatives	104
ODS-Hypersil	Methanol/ammonium acetate (pH 5.16)	Meso- and protoporphyrin	105

TABLE 18.7

REVERSED-PHASE HPLC SYSTEMS FOR THE GRADIENT SEPARATION OF PORPHYRINS

Eluents consisting of organic modifiers and ammonium acetate buffer.

Column	Eluent	Porphyrins	Ref.
ODS-Hypersil	Acetonitrile/ammonium acetate (pH 5.16)	Tetra- to octacarboxylic porphyrins I and III (simultaneous separation)	109
SAS-Hypersil (C$_1$)	Acetonitrile/methanol/ammonium acetate (pH 5.16)	Dicarboxylic porphyrins and tetra- to octacarboxylic porphyrins I and III (simultaneous separation)	110
ODS-Hypersil	Acetonitrile/ammonium acetate (pH 5.16)	Uroporphyrins I and III	111
Shandon RP-C$_1$	Acetonitrile/methanol/ammonium acetate (pH 5.16)	Meso-, copro-, and pentacarboxylic porphyrin	112

References on p. B359

by several groups of researchers [86-90]. Separation was particularly fast (4 min) when a 3-μm ODS column, only 3 cm in length, was used [90].

Methanol/lithium citrate gradients were also applied to porphyrin profiles (from uro- to protoporphyrin) but did not allow simultaneous isomer resolution [91-93]. Johansson and Niklasson [94] used columns of LiChrosorb RP-18, coated with tributyl phosphate, for the separation of the porphyrins in the uro- to coproporphyrin series. Their eluent system consisted of methanol with a pH gradient of 4.4-6.5, established with a phosphate buffer. The separation of meso-substituted porphyrins and their indium and manganese complexes was carried out by Duff et al. [95] on Vydac 5-μm ODS columns with an acetonitrile/acetate buffer gradient as the mobile phase.

18.3.4.2.3 Reversed-phase systems with organic solvents and ammonium acetate buffer

The general applicability of ammonium acetate as a highly effective buffer system for reversed-phase HPLC has been reported by Lim's research group [96]. In numerous papers Lim et al. [97-110] have amply demonstrated the special usefulness of this buffer for the chromatography of porphyrins and investigated systematically the effects of eluent modifiers, pH, and columns. They were able to separate by isocratic (Table 18.6) and gradient (Table 18.7) techniques not only a large number of porphyrin isomers but also to establish porphyrin profiles showing for the first time a simultaneous separation of the natural isomers I and III in a single chromatogram. Thus, the combination of ammonium acetate with organic solvents, such as acetonitrile and/or methanol in conjunction with RP columns, can presently be regarded as the universal separation method for porphyrins. The retention mechanism is presumably hydrophobic interaction, but ion-pair and ion-exchange mechanisms may also play a role.

18.3.4.2.3.1 Isocratic techniques. The first simultaneous and complete separation of the 4 coproporphyrin isomers I-IV was accomplished by Lim et al. [97] on C_{18} columns with the acetonitrile/ammonium acetate eluent system at pH 5.16. This value has proved to be optimal with very few exceptions for the separation of porphyrins and their isomers. Although the uroporphyrin isomers I, II, and III could be completely separated under similar conditions, this was not the case for isomers III and IV [98-100]. The 4 isomeric pentacarboxyporphyrins III, prepared by partial decarboxylation of uroporphyrin III, and the pentacarboxyporphyrin I as well as all naturally occurring isomers from uro- to coproporphyrin could be resolved on a C_{18} column with appropriate compositions of acetonitrile/ammonium acetate eluents, again at pH 5.16 [101,102].

For the separation of the relatively nonpolar porphyrin dicarboxylic acids, e.g., deutero-, meso-, proto-, and hematoporphyrin, a RP material with shorter chainlength is more suitable, because the retention times would be very long with C_{18} columns due to hydrophobic interactions. C_1 columns gave very good results [103]. In this case, the methanol content of the mobile phase can be kept lower.

The cobalt, iron, zinc, and copper complexes of meso- and protoporphyrin were also chromatographed on C_1 columns by Lim et al. [103]. In contrast to the free porphyrins, the metal complexes exhibited a pH optimum of 4.6 for the ammonium acetate buffer. A series of hematoporphyrin derivatives could also be separated at pH 4.6, but in this case a C_8 column served as the stationary phase [104]. The dicarboxylic acids meso- and protoporphyrin were also chromatographed on C_{18} columns, but this required a methanol concentration of 88% in the mobile phase [105]. In addition to the porphyrins, Lim's research group [106-108] also studied the isocratic separation of a series of isomeric porphyrinogens on C_{18} with the solvent system ammonium acetate/acetonitrile or methanol, using an electrochemical detector. They completely separated the isomeric coproporphyrinogens [106], the isomeric pentacarboxyporphyrinogens [107], and the isomeric heptacarboxyporphyrinogens [108]. Among the uroporphyrinogens, only the isomers I, II, and III were successfully resolved, but not the isomers III and IV [106].

18.3.4.2.3.2 Gradient techniques. The first simultaneous separation of the isomers I and III in the uro- to coproporphyrin series was accomplished by Lim et al. [109] with an acetonitrile/ammonium acetate gradient on ODS-Hypersil columns. By means of a ternary gradient, consisting of acetonitrile, methanol, and ammonium acetate, they were able to separate not only the isomers I and III of uro- to coproporphyrin, but also meso- and protoporphyrin in a single chromatogram on C_1 columns [110]. This system proved to be of particular value for establishing porphyrin profiles in urine and feces. Thus, it is also suitable for the biochemical diagnosis of all known porphyrias.

Tsai et al. [111] have described the separation of uroporphyrins I and III on ODS-Hypersil columns by the use of an acetonitrile/ammonium acetate gradient. In this procedure, the separation was improved by keeping the column temperature constant at 20°C with a thermostat. The separation of pentacarboxyporphyrin, coproporphyrins I/III, and mesoporphyrin was accomplished by McManus et al. [112] on C_1 columns with a ternary gradient of acetonitrile, methanol, and ammonium acetate. This procedure was used for the determination of the uroporphyrinogen decarboxylase activity in erythrocytes.

18.3.4.2.4 Reversed-phase systems with organic solvents and ion-pair reagents

Ion-pair chromatography has proved to be another very widely applicable method for separating porphyrins and their isomers. In this techniques an ion-pair reagent – usually a tetrabutylammonium compound – is added to the mobile phase, which consists of water-miscible solvents and water or buffer solutions. This reagent forms neutral ion pairs, consisting of the lipophilic tetrabutylammonium cation and the porphyrin carboxylate anion. This not only improves the solubility of the porphyrin carboxylic acids in the mobile phase but also increases their affinity for the stationary phase [113]. Isocratic (Table 18.8) as well as gradient (Table 18.9) techniques have been developed for the ion-pair chromatography of porphyrins. In ion-pair chromatography the polar porphyrins with several carboxyl groups are also eluted first, while the relatively less polar porphyrin dicarboxylic acids show the longest retention times.

18.3.4.2.4.1 Isocratic techniques. Ion-pair chromatography of porphyrin carboxylic acids was first described by Bonnett et al. [113]. They were able to produce porphyrin profiles for the series from uro- to coproporphyrin with aqueous methanol in the presence of tetrabutylammonium phosphate under isocratic conditions. However, the separation of position isomers was not achieved. Using acetonitrile/water/tetrabutylammonium phosphate, Gotelli et al. [114] separated copro-, protoporphyrin, and Zn protoporphyrin on

TABLE 18.8

REVERSED-PHASE HPLC SYSTEMS FOR THE ISOCRATIC SEPARATION OF PORPHYRINS

Eluents consisting of organic modifiers and ion-pairing reagents.

Column	Eluent	Porphyrins	Ref.
μBondapak C$_{18}$	Methanol/water/tetrabutylammonium dihydrogen phosphate	Tetra- to octacarboxylic porphyrins, hematoporphyrin derivatives	113
μBondapak C$_{18}$	Acetonitrile/water/tetrabutylammonium phosphate (pH 7.5)	Proto- and coproporphyrin, Zn protoporphyrin	114
LiChrosorb RP-18	Methanol/water/tetrabutylammonium phosphate (pH 7.5)	Meso- and protoporphyrin, Zn protoporphyrin	115
Supelcosil LC-18	Methanol/water/tetrabutylammonium hydrogen sulfate (pH 6.5)	Proto- and deuteroporphyrin, proto- and deuteroheme	116
LiChrosorb RP-18	Methanol/acetonitrile/phosphate buffer (pH 6.8)/ tetrabutylammonium phosphate (pH 7.4)	Coproporphyrins I-IV	117
LiChrosorb RP-18	Methanol/phosphate buffer (pH 5.4)/ tetrabutylammonium phosphate (pH 7.4)	Uroporphyrins I and III	117
LiChrospher RP-18	Methanol/acetonitrile/phosphate buffer (pH 6.6)/ tetrabutylammonium phosphate (pH 7.3)	Coproporphyrins I-IV	118-120
LiChrosorb RP-18	Methanol/phosphate buffer (pH 5.0)/ tetrabutylammonium phosphate (pH 7.4)	Uroporphyrins I-IV	118-120
Capcell Pak C$_{18}$	Acetonitrile/water/tetrabutylammonium phosphate (pH 7.5)	Protoporphyrin, Zn protoporphyrin	121
PRP-1 (polymeric gel)	Acetonitrile/methanol/tetrabutylammonium hydroxide (pH 12)	Sn, Zn, and Mn protoporphyrin	122
Bondapak Phenyl	Methanol/water/EDTA/sodium 1-pentanesulfonate (pH 2.1)	Tetra- to octacarboxylic porphyrins	123

TABLE 18.9

REVERSED-PHASE HPLC SYSTEMS FOR THE GRADIENT SEPARATION OF PORPHYRINS

Eluents consisting of organic modifiers and ion-pairing reagents.

Column	Eluent	Porphyrins	Ref.
μBondapak C$_{18}$	Methanol/water/tetrabutylammonium phosphate (pH 7.5)	Tetra- to octacarboxylic porphyrins	124, 125
LiChrosorb RP-18	Methanol/water/tetrabutylammonium phosphate (pH 7.5)	Di- to octacarboxylic porphyrins	126
HC-ODS Sil-X-1	Methanol/phosphate buffer (pH 6.0)/ tetrabutylammonium phosphate (pH 6.0)	Di- to octacarboxylic porphyrins	127
LiChroCART RP-18	Methanol/phosphate buffer (pH 7.5)/ tetrabutylammonium phosphate (pH 7.5)	Tetra- to octacarboxylic porphyrins	128
LiChrosorb RP-18	Methanol/phosphate buffer (pH 5.4)/ tetrabutylammonium phosphate (pH 7.4)	Dicarboxylic porphyrins and tetra- to octacarboxylic porphyrins I and III (simultaneous separation)	117

References on p. B359

μBondapak C_{18} columns. The separation of meso-, protoporphyrin, and Zn protopor-phyrin was performed by Jacob et al. [115] on LiChrosorb RP-18 columns with methanol/water/tetrabutylammonium phosphate. Tangeras [116] separated proto- and deuteroporphyrin as well as proto- and deuteroheme with a very similar solvent system on Supelcosil LC-18 columns.

The simultaneous and complete resolution of the 4 isomeric coproporphyrins I-IV was achieved by Jacob's research group with an eluent consisting of methanol/acetoni-trile/phosphate buffer (pH 6.8)/tetrabutylammonium phosphate, first on LiChrosorb RP-18 columns [117] and later also on LiChrospher RP-18 columns [118-120]. The elution sequence of the isomers – I, III, IV, and II – was the same as that for the ammonium acetate system of Lim et al. [97]. Jacob et al. were also able to separate the uroporphyrin isomers I and III in the presence of tetrabutylammonium phosphate, but the required pH of the phosphate buffer (pH 5.4) was decidedly lower than that used for the copropor-phyrin isomers [117]. The uroporphyrins I, II, and III could be completely resolved, just as in the case of the ammonium acetate system [98], but resolution of the isomers III and IV was not achieved [118-120].

Ion-pair chromatography on RP materials, surface-coated with silicone polymers (Cap-cell Pak C_{18}), has also been studied [121]. These phases were used in the separation of protoporphyrin from Zn protoporphyrin. For the chromatographic analysis of Sn, Zn, and Mn complexes of protoporphyrin an alkali-resistant polymer gel column (PRP-1) was used at pH 12 [122]. In this case the mobile phase was acetonitrile/methanol/tetrabutylammoni-um hydroxide. Finally, Hill et al. [123] also described an ion-pair system with an anionic counterion for the isocratic separation of porphyrins of the uro- to coproporphyrin series. In this instance, sodium pentanesulfonate at pH 2.1 served as the ion-pair reagent. In contrast to the tetrabutylammonium compounds, it forms ion pairs not with the carboxyl groups but with the basic nitrogen atoms of the porphyrin ring.

18.3.4.2.4.2 Gradient techniques. By means of a methanol/water gradient in the presence of tetrabutylammonium phosphate Jacob et al. [124,125] were able to obtain porphyrin profiles for the series from uro- to coproporphyrin. When the methanol concen-tration was increased, this system was also applicable to porphyrin profiles including porphyrin dicarboxylic acids, enabling the analysis of fecal porphyrins [126].

The gradient systems of Chiba and Sassa [127] contained, in addition to methanol and tetrabutylammonium phosphate, an aqueous phosphate buffer (pH 6.0). Such a system was applied to the analysis of porphyrin profiles in biological fluids. The system containing a phosphate buffer of pH 7.5, developed by Jacob's group [128], gives a similar separa-tion.

All ion-pair gradient systems with constant pH described so far can simultaneously separate only the coproporphyrin isomers I and III, but not the uroporphyrins I and III. On the basis of the results of isocratic ion-pair chromatography of uro- and coproporphyrins Jacob et al. [117] have developed a novel gradient system for the simultaneous separa-tion of the isomers I and III from uro- to coproporphyrin. This required not only a solvent gradient but also a pH gradient, because the separation of uroporphyrin isomers occurs at a considerably lower pH than that of coproporphyrin isomers. This double gradient was accomplished simply by using an aqueous phosphate buffer with a pH of 5.4 and tetra-butylammonium phosphate with a pH of 7.2, dissolved in the methanol phase. This procedure allowed the simultaneous separation of the uro- and coproporphyrin isomers I and III in a single chromatogram by means of ion-pair chromatography for the first time.

18.4 SAMPLE PREPARATION FOR THE HPLC ANALYSIS OF PORPHYRINS

Biological samples, such as urine, blood, and stool, contain, of course, a multitude of various compounds. For best results, the HPLC analysis of porphyrins, which are usually present in these materials in only trace amounts, therefore requires as a rule a preliminary separation and concentration. Only rarely, e.g., in the case of urine, can porphyrins in biological materials be analyzed directly. When normal-phase systems are used for the HPLC analysis, an additional esterification step is required to produce the porphyrin methyl esters, whereas the naturally occurring porphyrin carboxylic acids can be directly chromatographed as the free acids on reversed phases.

18.4.1 Direct analysis of urine

Numerous reversed-phase systems suitable for the direct analysis of the free porphyrin carboxylic acids in urine have been described [15,84,86,110,123,125]. In these procedures it is essential to protect the analytical column by an appropriate guard column. Moreover, precipitates in urine samples should be removed by centrifugation, but it should be noted that urinary porphyrins will also be removed by adsorption on these precipitates if the pH is below 7 [129]. When the pH is raised above 7 no adsorption occurs. Alternatively, the precipitates, which normally consist of calcium phosphate, can be dissolved by adding a small amount of conc. HCl [15,16,84,86].

Since a considerable portion of the porphyrins is excreted in reduced form as porphyrinogens in the urine [130], they must first be converted to the corresponding porphyrins by suitable oxidizing agents. This oxidation has been carried out with atmospheric oxygen [123], hydrogen peroxide [67], or even iodine [131].

18.4.2 Solvent extraction

Extraction of porphyrins from urine with organic solvents has been applied only rarely, as urine is preferably analyzed by direct injection or solid-phase extraction. Diethyl ether [76] and ethyl acetate [85] have been used as extraction solvents. HPLC analysis of erythrocytes usually requires extraction with organic solvents. The procedures published by Culbreth et al. [71] and Salmi and Tenhunen [72] allowed only a determination of total porphyrins, because the Zn protoporphyrin is cleaved during back extraction with hydrochloric acid. Subsequently described procedures did allow a separate determination of protoporphyrin and of Zn protoporphyrin by the use of the following extraction solvents: propanol in the presence of Triton X and tetrabutylammonium phosphate [114], ethyl acetate/acetic acid [80,115], acetone/water [132], acetone/water/formic acid [73], and N,N-dimethylformamide [121].

The porphyrins in plasma are adequately extracted by ethyl acetate/acetic acid [65] or diethyl ether/acetic acid [15], then transferred to aqueous hydrochloric acid, and finally analyzed by HPLC. In the analysis of porphyrins in feces by normal-phase HPLC the reaction with methanol/sulfuric acid to form the necessary methyl esters and the extraction step can be combined in a single operation [52]. For the determination of free porphyrins by reversed-phase HPLC this esterification is unnecessary. Extracts prepared by the procedure of Lockwood et al. [133] with diethyl ether/hydrochloric acid are suitable for HPLC analysis [110]. A high yield of porphyrins was obtained by extracting freeze-dried stool with a methanolic tetrabutylammonium phosphate solution, and the extracts were then analyzed directly by ion-pair chromatography [126].

18.4.3 Adsorption on talc

Talc adsorption of urinary porphyrins at pH 3-4 is an isolation method which has been known for many years and is widely practiced. If analysis of the corresponding methyl esters by normal-phase HPLC is intended, a simultaneous esterification and extraction with methanol/sulfuric acid is recommended [54,134]. For the HPLC separation in reversed-phase systems, on the other hand, the porphyrins can be extracted from talc with acetone/hydrochloric acid [110] or with a solution of tetrabutylammonium phosphate in methanol [117]. However, a disadvantage of talc adsorption is that the porphyrins are not completely recovered and that uroporphyrin in particular is to some extent irreversibly bound by the talc [15,16].

18.4.4 Adsorption on reversed-phase materials

The recently described adsorption in cartridges containing C_2-, C_8-, or C_{18}-coated silica gel has produced nearly quantitative recoveries of porphyrins from biological materials. The porphyrins are completely adsorbed on these reversed-phase materials from acid solutions, and they can be eluted from them with organic solvents, such as methanol, acetone, or acetonitrile. The cartridge technique has been applied to the extraction of porphyrins from urine [41,56,88,135,136], erythrocytes [100], stool [136], and tissues [137-139]. In view of its simplicity the cartridge technique is eminently suitable for automatic sample preparation in the handling of a large number of samples [100,136].

18.5 APPLICATIONS OF PORPHYRIN ANALYSIS BY HPLC

Progress in the field of porphyrin analysis by HPLC has led to a considerable broadening of applications. This progress is primarily due to improved separation efficiency through the development of more effective stationary and mobile phases and to the simplification of the techniques of sample preparation.

18.5.1 Biochemical diagnosis of porphyrias

The most important application of porphyrin analysis is the biochemical diagnosis of porphyrias. They can be diagnosed on the basis of either the determination of porphyrin precursors and the establishment of porphyrin profiles in various clinical materials or on the determination of the activity of enzymes in the pathway of porphyrin biosynthesis.

18.5.1.1 Porphyrin profiles

Porphyrin profiles have been obtained from urine, stool, or blood with normal-phase (Tables 18.1 and 18.2) as well as reversed-phase (Tables 18.3-18.9) systems. The simultaneous separation of the natural isomers I and III of the series uro- to coproporphyrin in the same chromatogram has proved to be especially advantageous for the differential diagnosis of porphyrias. This particular separation has so far been accomplished only with the ammonium acetate system [110] or by ion-pair chromatography [117]. Having, at the same time, a knowledge of the isomer ratio considerably facilitates the diagnosis of the following porphyrias: congenital erythropoietic porphyria (preponderance of isomers of series I), hereditary coproporphyria (high proportion of coproporphyrin III), and porphyria cutanea tarda (occurrence of heptacarboxyporphyrin III and isocoproporphyrin). Acquired, secondary coproporphyrinurias can occur in a number of liver diseases and in lead poisoning. They show characteristic ratios of coproporphyrins I and III, which are helpful in the differentiation of these diseases [140].

18.5.1.2 Determination of enzymes in the heme biosynthetic pathway

Hereditary porphyrias are caused by inborn defects in the enzymes of the heme biosynthetic pathway, whereas acquired porphyrias arise from later damage to these enzymes. Thus, the determination of the pertinent enzyme activities can make a valuable contribution to the diagnosis of their causes.

HPLC methods have been developed for all enzymes in the heme biosynthetic pathway. For the determination of the first two enzymes of the pathway, 5-aminolevulinic acid synthase and porphobilinogen synthase, there is no porphyrin analysis; here, the formation of the precursors, 5-aminolevulinic acid and porphobilinogen, is determined. Porphyrin analyses are used in the determination of the activity of hydroxymethylbilane synthase [99,100], uroporphyrinogen III synthase [99,100,111], uroporphyrinogen decarboxylase [93,100,112,141], coproporphyrinogen oxidase [105,143], protoporphyrinogen oxidase [143], and ferrochelatase [103,116,142,143].

18.5.2 Separation of porphyrin isomers

Numerous methods have been developed for the separation of position isomers of porphyrins (Tables 18.1, 18.4, 18.6-18.9). Isomer separations are used not only for the diagnosis of porphyrias but also in testing the purity of synthetic and natural porphyrins.

References on p. B359

The isomer composition of coproporphyrins, isolated from the urine and feces of porphyria patients, was studied by Jackson et al. [62] by adsorption chromatography in the form of their methyl esters. Complete resolution of the free coproporphyrin isomers I-IV by means of the ammonium acetate system allowed the examination of the uroporphyrin mixture produced by condensation of porphobilinogen, which had subsequently been converted to the corresponding coproporphyrin isomers by partial decarboxylation [97]. The isomeric purity of many synthetic coproporphyrins has been tested by ion-pair chromatography [117].

Isomer separations of uroporphyrins in the form of their methyl esters [54,58,60] and free acids [75,98,118,120] have been described. However, only isomers I, II, and III can be completely separated by these procedures. Only Wayne et al. [75] succeeded in the partial separation of isomers III and IV, but without studying their retention behavior in detail. The other procedures could not be made to resolve isomers III and IV. Hence, it is necessary to convert the uroporphyrins to the corresponding coproporphyrins by partial decarboxylation, as these are the only ones that can be separated completely. Thus, Jackson et al. [60] separated the uroporphyrins isolated from the urine and stool of porphyria patients on an analytical as well as semipreparative scale in the form of methyl esters. Various workers [15,99] have used HPLC methods in the course of the preparation of uroporphyrin standards and purity tests. Additional isomer separations have been reported, e.g., by Lim and Rideout [101], for the elucidation of the pathway of the in vivo decarboxylation reactions of porphyrinogens.

18.5.3 Determination of atypical porphyrin isomers of series II and IV in human urine

Up to now it had been assumed that only the porphyrin isomers of the I and III series occur in nature. However, using ion-pair chromatography, Jacob et al. have been able to show that 2-5% uroporphyrin II and 13-19% uroporphyrin IV may occur in the urine of patients with acute intermittent porphyria, in addition to the usual uroporphyrin isomers I and III [118,120]. However, these atypical isomers are formed from porphobilinogen, which is present in large quantities in these urines, by nonenzymatic selfcondensation [118]. Selfcondensation occurs in the human body as well as during collection and storage of urines [118]. These findings were confirmed by isomer analysis of the so-called Waldenström uroporphyrin, which had been isolated by Waldenström et al. [144] from the urine of a patient with acute porphyria and had originally been regarded as uroporphyrin III. But ion-pair chromatography showed that the Waldenström uroporphyrin also contains, in addition to the isomers I and III, considerable quantities of the atypical isomers II and IV [120]. It was surprising that the atypical isomers II and IV could be found in the coproporphyrin fraction of the urine of not only patients with acute intermittent porphyria [120] but also that of normal persons [145], albeit in considerably lower concentration than in the case of the uroporphyrins, isomer II making 1-2% and isomer IV making 2-5% of the total. Isomerization experiments on coproporphyrinogen III under physiological conditions showed that the atypical coproporphyrins II and IV are formed in the human body by nonenzymatic isomerization at the coproporphyrinogen level [145]. The discovery of isomers II and IV of uroporphyrin and coproporphyrin in human urine indicates that purely chemical, nonenzymatic pathways exist in porphyrin metabolism. These processes can go on in vivo as well as in vitro.

ACKNOWLEDGEMENT

The work of the author and his associates, cited in this chapter, has been carried out with the support of the Hans Fischer Society, Munich, FRG.

REFERENCES

1 K.M. Smith (Editor), *Porphyrins and Metalloporphyrins,* Elsevier, Amsterdam, 1975.
2 D. Dolphin (Editor), *The Porphyrins,* Academic Press, New York, 1978/79.
3 A. Eschenmoser, *Angew. Chem.,* 100 (1988) 5.
4 M.R. Moore, K.E.L. McColl, C. Rimington and A. Goldberg, *Disorders of Porphyrin Metabolism,* Plenum Medical Book Company, New York, 1987.
5 A. Treibs, *Angew. Chem.,* 53 (1940) 202.
6 M.I. Chicarelli and J.R. Maxwell, *Trends Anal. Chem.,* 6 (1987) 158.
7 R. Willstätter and W. Mieg, *Liebigs Ann. Chem.,* 350 (1906) 1.
8 P.J. Marriott, J.P. Gill, R.P. Evershed, G. Eglinton and J.R. Maxwell, *Chromatographia,* 16 (1982) 304.
9 H. Fischer and H. Orth, *Die Chemie des Pyrrols,* Vol. II, Part 1, Akademische Verlagsgesellschaft, Leipzig, 1937, pp. 579-590.
10 M. Gouterman, in D. Dolphin (Editor), *The Porphyrins,* Vol. III, Academic Press, New York, 1978, pp. 1-165.
11 E. Sagen and I. Romslo, *Scand. J. Clin. Lab. Invest.,* 45 (1985) 309.
12 C.F. Polo, A.L. Frisardi, E.R. Resnik, A.E.M. Schoua and A.M. del C. Batlle, *Clin. Chem.,* 34 (1988) 757.
13 D. Dolphin, in E. Heftmann (Editor), *Chromatography,* Elsevier, Amsterdam, 4th Edn., 1983, pp. B377-B406.
14 H.D. Meyer, in A. Henschen, K.P. Hupe, F. Lottspeich and W. Voelter (Editors), *High Performance Liquid Chromatography in Biochemistry,* VCH Verlagsgesellschaft, Weinheim, 1985, pp. 445-479.
15 E. Rossi and D.H. Curnow, in C.K. Lim (Editor), *HPLC of Small Molecules,* IRL Press, Oxford, 1986, pp. 261-303.
16 C.K. Lim, F. Li and T.J. Peters, *J. Chromatogr.,* 429 (1988) 123.
17 E. Klesper, A.H. Corwin and D.A. Turner, *J. Org. Chem.,* 27 (1962) 700.
18 N.M. Karayannis and A.H. Corwin, *Anal. Biochem.,* 26 (1968) 34.
19 N.M. Karayannis and A.H. Corwin, *J. Chromatogr.,* 47 (1970) 247.
20 D.B. Boylan and M. Calvin, *J. Amer. Chem. Soc.,* 89 (1967) 5472.
21 P.J. Marriott, J.P. Gill and G. Eglinton, *J. Chromatogr.,* 249 (1982) 291.
22 P.J. Marriott and G. Eglinton, *J. Chromatogr.,* 249 (1982) 311.
23 P.J. Marriott, J.P. Gill, R.P. Evershed, C.S. Hein and G. Eglinton, *J. Chromatogr.,* 301 (1984) 107.
24 J.P. Gill, R.P. Evershed and G. Eglinton, *J. Chromatogr.,* 369 (1986) 281.
25 R.E.H. Nicholas and C. Rimington, *Scand. J. Clin. Lab. Invest.,* 1 (1949) 12.
26 R.E.H. Nicholas and C. Rimington, *Biochem. J.,* 48 (1951) 306.
27 R. Kehl and W. Stich, *Hoppe-Seyler's Z. Physiol. Chem.,* 289 (1951) 6.
28 J.E. Falk, E.I.B. Dresel, A. Benson and B.C. Knight, *Biochem. J.,* 63 (1956) 87.
29 L. Eriksen, *Scand. J. Clin. Lab. Invest.,* 10 (1958) 319.
30 T.C. Chu, A.A. Green and E.J. Chu, *J. Biol. Chem.,* 190 (1951) 643.

31 T.C. Chu and E.J. Chu, *J. Biol. Chem.*, 227 (1957) 505.
32 L. Bogorad and G.S. Marks, *Biochim. Biophys. Acta*, 41 (1960) 356.
33 T.K. With, *J. Chromatogr.*, 42 (1969) 389.
34 M. Yuan and C.S. Russell, *J. Chromatogr.*, 87 (1973) 562.
35 D.W. Lamson, A.F.W. Coulson and T. Yonetani, *Anal. Chem.*, 45 (1973) 2273.
36 M. Doss, *Z. Klin. Chem. Klin. Biochem.*, 8 (1970) 197.
37 M.O. Doss, in H.C. Curtius and M. Roth (Editors), *Clinical Biochemistry - Principles and Methods*, Walter de Gruyter, Berlin, New York, 1974, pp. 1323-1371.
38 T.K. With, *Clinical Science and Molecular Medicine*, 52 (1977) 463.
39 R.S. Day, N.R. Pimstone and L. Eales, *Int. J. Biochem.*, 9 (1978) 897.
40 R.S. Day and L. Eales, *Nephron*, 26 (1980) 90.
41 H.C. Friedmann and E.T. Baldwin, *Anal. Biochem.*, 137 (1984) 473.
42 M.J. Henderson, *Clin. Chem.*, 35 (1989) 1043.
43 G.M. Garbo, J.B. Kramer, R.W. Keck, S.H. Selman and M. Kreimer-Birnbaum, *Anal. Biochem.*, 151 (1985) 70.
44 T.J. Dougherty, J.E. Kaufman, A. Goldfarb, K.R. Weishaupt, D.G. Boyle and A. Mittleman, *Cancer Res.*, 38 (1978) 2628.
45 C. Vever-Bizet, O. Delgado and D. Brault, *J. Chromatogr.*, 283 (1984) 157.
46 E. Schermuly and M. Doss, *Z. Klin. Chem. Klin. Biochem.*, 13 (1975) 299.
47 A. Seubert, S. Seubert and H. Ippen, *Deut. Med. Wochenschr.*, 104 (1979) 1459.
48 A. Junker-Buchheit and H. Jork, *J. Planar Chromatogr.*, 1 (1988) 214.
49 G.H. Elder, *J. Chromatogr.*, 59 (1971) 234.
50 G.H. Elder, *J. Clin. Pathol.*, 28 (1975) 601.
51 J.E. Francis and A.G. Smith, *Trends Anal. Chem.*, 4 (1985) 80.
52 C.H. Gray, C.K. Lim and D.C. Nicholson, *Clin. Chim. Acta*, 77 (1977) 167.
53 J.G. Straka, J.P. Kushner and B.F. Burnham, *Anal. Biochem.*, 111 (1981) 269.
54 A. Seubert and S. Seubert, *Anal. Biochem.*, 124 (1982) 303.
55 P. Kotal, B. Porsch, M. Jirsa and V. Kordac, *J. Chromatogr.*, 333 (1985) 141.
56 P. Kotal, M. Jirsa, P. Martasek and V. Kordac, *Biomed. Chromatogr.*, 1 (1986) 159.
57 J.C. Bommer, B.F. Burnham, R.E. Carlson and D. Dolphin, *Anal. Biochem.*, 95 (1979) 444.
58 H. Nordlöv, P.M. Jordan, G. Burton and A.I. Scott, *J. Chromatogr.*, 190 (1980) 221.
59 I.C. Walker, M.T. Gilbert and K. Stubbs, *J. Chromatogr.*, 202 (1980) 491.
60 A.H. Jackson, K.R.N. Rao and S.G. Smith, *Biochem. J.*, 203 (1982) 515.
61 M.D. Gonzalez, S.K. Grant, H.J. Williams and A.I. Scott, *J. Chromatogr.*, 437 (1988) 311.
62 A.H. Jackson, K.R.N. Rao and S.G. Smith, *Biochem. J.*, 207 (1982) 599.
63 A.H. Jackson, K.R.N. Rao, S.G. Smith and T.D. Lash, *Biochem. J.*, 227 (1985) 327.
64 M. Dellinger and D. Brault, *J. Chromatogr.*, 422 (1987) 73.
65 M.O. Longas and M.B. Poh-Fitzpatrick, *Anal. Biochem.*, 104 (1980) 268.
66 C.K. Lim and J.Y.Y. Chan, *J. Chromatogr.*, 228 (1982) 305.
67 A. Colombi, M. Maroni, A. Ferioli, C. Valla, G. Coletti and V. Foa, *Amer. J. Industrial Med.*, 4 (1983) 551.
68 N.A. McCarroll, *Clin. Chem.*, 34 (1988) 2390.
69 A.J.G. Barwise, R.P. Evershed, G.A. Wolff, G. Eglinton and J.R. Maxwell, *J. Chromatogr.*, 368 (1986) 1.
70 M.I. Chicarelli, G.A. Wolff and J.R. Maxwell, *J. Chromatogr.*, 368 (1986) 11.
71 P. Culbreth, G. Walter, R. Carter and C. Burtis, *Clin. Chem.*, 25 (1979) 605.
72 M. Salmi and R. Tenhunen, *Clin. Chem.*, 26 (1980) 1832.
73 G.G. Bailey and L.L. Needham, *Clin. Chem.*, 32 (1986) 2137.
74 G.J.J. Beukeveld, B.G. Wolthers, J.J.M. van Saene, T.H.I. de Haan, L.W. de Ruyter-Buitenhuis and R.H.F. van Saene, *Clin. Chem.*, 33 (1987) 2164.
75 A.W. Wayne, R.C. Straight, E.E. Wales and E. Englert, Jr., *J. High Resolut. Chromatogr. Chromatogr. Commun.*, 2 (1979) 621.
76 M. Udagawa, Y. Hayashi and C. Hirayama, *J. Chromatogr.*, 233 (1982) 338.
77 T. Sakai, Y. Niinuma, S. Yanagihara and K. Ushio, *Clin. Chem.*, 29 (1983) 350.
78 K. Tomokuni and Y. Hirai, *Clin. Chem.*, 32 (1986) 872.

79 L.A. Grubina, I.F. Gurinovich, E.P. Demidchik and S.V. Trofimovich, *J. Chromatogr.*, 380 (1986) 232.

80 H.A. Scoble, M. McKeag, P.R. Brown and G.J. Kavarnos, *Clin. Chim. Acta*, 113 (1981) 253.

81 J. Ho, R. Guthrie and H. Tieckelmann, *J. Chromatogr.*, 375 (1986) 57.

82 J. Ho, R. Guthrie and H. Tieckelmann, *J. Chromatogr.*, 417 (1987) 269.

83 P.A. Scourides, C. Morstyn and R.W. Henderson, *J. Liq. Chromatogr.*, 9 (1986) 2879.

84 E. Englert, Jr., A.W. Wayne, E.E. Wales and R.C. Straight, *J. High Resolut. Chromatogr. Chromatogr. Commun.*, 2 (1979) 570.

85 M. Takayama, *J. Chromatogr.*, 423 (1987) 313.

86 R.E. Ford, C.-N. Ou and R.D. Ellefson, *Clin. Chem.*, 27 (1981) 397.

87 M.J. Richard, B. Fovet, M. Fontaine and A. Boersma, *Ann. Biol. Clin.*, 44 (1986) 639.

88 W.E. Schreiber, V.A. Raisys and R.F. Labbe, *Clin. Chem.*, 29 (1983) 527.

89 H.L. Bonkovsky, S.G. Wood, S.K. Howell, P.R. Sinclair, B. Lincoln, J.F. Healey and J.F. Sinclair, *Anal. Biochem.*, 155 (1986) 56.

90 P.M. Johnson, S.L. Perkins and S.W. Kennedy, *Clin. Chem.*, 34 (1988) 103.

91 B. Johansson and B. Nilsson, *J. Chromatogr.*, 229 (1982) 439.

92 F.P. Armbruster, H. Schmidt-Gayk, M. Kohlmeier and H. Gräfinger, *Ärztl. Lab.*, 29 (1983) 379.

93 J.E. Francis and A.G. Smith, *Anal. Biochem.*, 138 (1984) 404.

94 I.M. Johansson and F.A. Niklasson, *J. Chromatogr.*, 275 (1983) 51.

95 G.A. Duff, S.A. Yeager, A.K. Singhal, B.C. Pestel, J.M. Ressner and N. Foster, *J. Chromatogr.*, 416 (1987) 71.

96 C.K. Lim and T.J. Peters, *J. Chromatogr.*, 316 (1984) 397.

97 D.J. Wright, J.M. Rideout and C.K. Lim, *Biochem. J.*, 209 (1983) 553.

98 J.M. Rideout, D.J. Wright and C.K. Lim, *J. Liq. Chromatogr.*, 6 (1983) 383.

99 D.J. Wright and C.K. Lim, *Biochem. J.*, 213 (1983) 85.

100 C.K. Lim, F. Li, J.M. Rideout, D.J. Wright and T.J. Peters, *J. Chromatogr.*, 371 (1986) 293.

101 C.K. Lim and J.M. Rideout, *J. Liq. Chromatogr.*, 6 (1983) 1969.

102 C.K. Lim, J.M. Rideout and D.J. Wright, *J. Chromatogr.*, 282 (1983) 629.

103 C.K. Lim, J.M. Rideout and T.J. Peters, *J. Chromatogr.*, 317 (1984) 333.

104 J.C.M. Meijers, C.K. Lim, A.M. Lawson and T.J. Peters, *J. Chromatogr.*, 352 (1986) 231.

105 R. Guo, C.K. Lim and T.J. Peters, *Clin. Chim. Acta*, 177 (1988) 245.

106 C.K. Lim, F. Li and T.J. Peters, *Biochem. J.*, 234 (1986) 629.

107 F. Li, C.K. Lim and T.J. Peters, *Biochem. J.*, 243 (1987) 621.

108 C.K. Lim, F. Li and T.J. Peters, *Biochem. J.*, 247 (1987) 229.

109 C.K. Lim, J.M. Rideout and D.J. Wright, *Biochem. J.*, 211 (1983) 435.

110 C.K. Lim and T.J. Peters, *Clin. Chim. Acta*, 139 (1984) 55.

111 S.-F. Tsai, D.F. Bishop and R.J. Desnick, *Anal. Biochem.*, 166 (1987) 120.

112 J. McManus, D. Blake and S. Ratnaike, *Clin. Chem.*, 34 (1988) 2355.

113 R. Bonnett, A.A. Charalambides, K. Jones, I.A. Magnus and R.J. Ridge, *Biochem. J.*, 173 (1978) 693.

114 G.R. Gotelli, J.H. Wall, P.M. Kabra and L.J. Marton, *Clin. Chem.*, 26 (1980) 205.

115 H.D. Meyer, K. Jacob and W. Vogt, *Chromatographia*, 16 (1982) 190.

116 A. Tangeras, *J. Chromatogr.*, 310 (1984) 31.

117 K. Jacob, W. Sommer, H.D. Meyer and W. Vogt, *J. Chromatogr.*, 349 (1985) 283.

118 K. Jacob, I. Kossien, E. Egeler and M. Knedel, *J. Chromatogr.*, 441 (1988) 171.

119 K. Jacob, E. Egeler, B. Hennel and D. Neumeier, *Fresenius' Z. Anal. Chem.*, 330 (1988) 386.

120 K. Jacob, E. Egeler, D. Neumeier and M. Knedel, *J. Chromatogr.*, 468 (1989) 329.

121 T. Sakai, Y. Takeuchi, T. Araki and K. Ushio, *J. Chromatogr.*, 433 (1988) 73.

122 J. Bauer, C. Linton and B. Norris, *J. Chromatogr.*, 445 (1988) 429.

123 R.H. Hill, Jr., S.L. Bailey and L.L. Needham, *J. Chromatogr.*, 232 (1982) 251.

124 H.D. Meyer, K. Jacob and W. Vogt, *J. High Resolut. Chromatogr. Chromatogr. Commun., 3 (1980)* 85.
125 H.D. Meyer, K. Jacob, W. Vogt and M. Knedel, *J. Chromatogr.,* 199 (1980) 339.
126 H.D. Meyer, K. Jacob, W. Vogt and M. Knedel, *J. Chromatogr.,* 217 (1981) 473.
127 M. Chiba and S. Sassa, *Anal. Biochem.,* 124 (1982) 279.
128 H.D. Meyer, W. Vogt and K. Jacob, *J. Chromatogr.,* 290 (1984) 207.
129 R.H. Hill, Jr. and S.L. Bailey, *Clin. Chem.,* 32 (1986) 377.
130 C.J. Watson, R. Pimenta de Mello, S. Schwartz, V.E. Hawkinson and I. Bossenmaier, *J. Lab. Clin. Med.,* 37 (1951) 831.
131 P. Martasek, M. Jirsa and V. Kordac, *J. Clin. Chem. Clin. Biochem.,* 20 (1982) 113.
132 D. Hart and S. Piomelli, *Clin. Chem.,* 27 (1981) 220.
133 W.H. Lockwood, V. Poulos, E. Rossi and D.H. Curnow, *Clin. Chem.,* 31 (1985) 1163.
134 V. Miller and L. Malina, *J. Chromatogr.,* 145 (1978) 290.
135 M. Muraca and W. Goossens, *Clin. Chem.,* 30 (1984) 338.
136 F. Li, C.K. Lim and T.J. Peters, *Biomed. Chromatogr.,* 1 (1986) 93.
137 S.W. Kennedy, D.C. Wigfield and G.A. Fox, *Anal. Biochem.,* 157 (1986) 1.
138 J. Sanitrak, J. Krijt, J. Coupek, V. Janousek and I.A. Magnus, *J. Chromatogr.,* 415 (1987) 129.
139 V.R. Reddy, W.R. Christenson and W.N. Piper, *J. Pharmacol. Methods,* 17 (1987) 51.
140 M. Doss, in H. Greiling and A.M. Gressner (Editors), *Lehrbuch der Klinischen Chemie und Pathobiochemie,* Schattauer Verlag, Stuttgart, 1987, pp. 311-339.
141 D.G. Adjarov and G.H. Elder, *Clin. Chim. Acta,* 177 (1988) 123.
142 F. Li, C.K. Lim and T.J. Peters, *Biomed. Chromatogr.,* 2 (1987) 164.
143 F. Li, C.K. Lim, K.J. Simpson and T.J. Peters, *J. Hepatology,* 8 (1989) 86.
144 J. Waldenström, H. Fink and W. Hoerburger, *Hoppe-Seylers Z. Physiol. Chem.,* 233 (1935) 1.
145 K. Jacob, E. Egeler, B. Hennel and P. Luppa, *J. Clin. Chem. Clin. Biochem.,* 27 (1989) 659.

Chapter 19

Phenolic compounds

JEFFREY B. HARBORNE

CONTENTS

19.1 INTRODUCTION

Phenolic compounds continue to command our attention as scientists because of their ubiquity in nature and of the many different functions they perform within the plant or its environment. Salicylic acid, for example, has just been identified as the trigger molecule for heat production in the spadix of the voodoo lily, *Sauromatum guttatum*, which is an essential feature of pollination attraction in this and many other members of the family Araceae [1]. Again, certain phenylpropanoids, such as acetosyringone and hydroxyaceto-syringone are involved in the infective process by which the crown gall disease organism, *Agrobacterium tumefaciens*, is able to enter the roots of most dicotyledonous plants [2]. Phenolics are widely present in food plants and in many popular beverages, so that there is current interest in their fate in and effects on the human body [3]. There has also been a renewed interest in the use of phenolics, and especially flavonoids, as medicinal drugs. The flavanolignan, silybin, is widely used in the treatment of liver diseases [4], and it has a unique role in acting as an antidote to the peptide toxins of the poisonous toadstool, *Amanita phalloides* [5].

References on p. B389

The single most important development in phenolic analysis of the past decade is the application of high-performance liquid chromatography (HPLC) to their separation and detection. In particular, the introduction in 1976 of reversed-phase silica columns and their elution with aqueous acidic methanol solvents [6] has provided an almost ideal system for resolving the complex mixtures of phenolic components that are encountered in most plants. This technique has been widely exploited ever since, and HPLC on reversed-phase columns is regarded today as the method of choice. This does not mean that other chromatographic techniques have been abandoned in favor of HPLC. Paper chromatography and TLC are just as important as they have ever been for the rapid screening of phenolics in crude extracts, while column chromatography on Sephadex gel is essential for preliminary purification or large-scale separation. A range of other techniques, from electrophoresis to countercurrent distribution, are still regularly applied to phenolic analysis in many laboratories.

The term 'phenolic compound' embraces some eight thousand naturally occurring substances [7], the majority of which are of plant rather than animal origin. There are two main groups: simple phenolics and flavonoids. Simple phenolics are phenols, such as catechol and resorcinol; phenolic acids, such as salicylic and protocatechuic acids; and hydroxycinnamic acids (e.g., caffeic acid) and their lactone derivatives, the coumarins. The flavonoids comprise the widely occurring water-soluble plant pigments, the anthocyanins and their colorless co-pigments, the flavones, together with the related isoflavones, catechins, tannins, and biflavonoids. A recent listing of all the plant flavonoids showed that over 4000 structures had been described in nature up to 1986 [8]. Because quinone pigments contain phenolic hydroxyl groups, their separation will be described here, as well as that of the related xanthones. A small but significant number of phenolics also contain a nitrogen function (e.g., adrenaline, mescaline, betanin), but, in general, these compounds will not be considered here.

A new, comprehensive volume on phenolic analysis, including chromatographic techniques, has been published [9], and the reader is referred to it for the more detailed treatment that cannot be provided here. Recent volumes of *Methods in Enzymology* may also be usefully consulted for detailed chromatographic procedures for the purification of particular phenolic substrates. Several general reviews dealing, inter alia, with the HPLC of phenolics have appeared [10-13]. The present review will concentrate on the application of chromatographic techniques to the identification of phenolics in plant tissues, although the methods described are usually equally applicable to animal tissues. Applied aspects of the subject will not be covered in any detail.

19.2 HIGH-PERFORMANCE LIQUID CHROMATOGRAPHY

19.2.1 General

The major advantage of HPLC over other chromatographic methods for phenolic separation is that it provides high resolution and a sensitive quantitative analysis in one and the same operation. Coupled with a diode-array detection system, HPLC may also provide identification through the characteristic UV spectra of the different components as they are eluted [14]. These advantages have to be set against the considerable capital and running costs of such an apparatus. Some authors, such as Van de Casteele et al. [15], have demonstrated that it is possible with HPLC to separate several classes of phenolics in a single analysis, and that up to 55 phenols can be resolved at one time. However, the main application of HPLC to date has been to the separation of compounds of the same structural class, with usually no more than 5 to 10 components being resolved in the same chromatogram. In such cases, it has proved to be a powerful technique, combining rapidity with high sensitivity. Separations are usually reproducible, and relative retention times have become as useful as R_F values for characterization, at least within the same laboratory.

Although many different column packings and solvent systems have been described for separating phenolics by HPLC (see Table 19.5), there are two main operations: the chromatography of nonpolar phenolics on normal-phase silica columns with isocratic elution; and the chromatography of polar phenolics (e.g., the water-soluble flavonoid glycosides) on reversed-phase chemically bonded octadecylsilane-treated silica columns with gradient elution. A typical isocratic separation, e.g., with heptane/2-propanol (3:2) is that of the methylated flavones in citrus peel on a LiChrosorb Si-60 column [16]. A widely used solvent system for gradient elution of polar phenolics, such as the flavonol glycosides in *Chondropetalum*, from a μBondapak C_{18} column is water/acetic acid/methanol, in variable proportions [17]. Some authors combine isocratic and gradient elution to achieve separation of closely similar phenolics on a reversed-phase column [15]. Guard columns are now used routinely, and crude plant extracts are usually cleaned up by Sephadex chromatography before application.

Ultraviolet spectral detection is most widely used for phenols, the detection of the colored anthocyanins being monitored in the visible region at 500-550 nm. Dual-wavelength detection (at, e.g., 265 and 272 nm) is useful when working with crude plant extracts for checking the purity of the components as they separate [18]. The purity of individual peaks can also be checked routinely with a UV diode-array detector [14].

Phenols are electro-oxidizable compounds and are therefore amenable to electrochemical detection. Such detection has proved to be more sensitive and selective when measuring, e.g., the trace amounts of phenolics that are present in beers; measurements are possible in the 3- to 15-pg range [19]. Electrochemical detection has also been employed for analyzing the metal/flavonoid complexes formed during the HPLC of food extracts in the presence of iron(III) or copper(II) [20].

While isoflavonoids are usually detected after HPLC by UV absorbance, some improvements in sensitivity are claimed for amperometric [21] and circular dichroism measurements [22]. Isoflavones in silage have been analyzed, both qualitatively and quantitatively, by use of fluorescence detection [23].

References on p. B389

19.2.2 Anthocyanins

Anthocyanins are separated by HPLC on a reversed-phase column, using acidic solvents, which are necessary to prevent the pigments from fading [24]. However, there is the danger of forming artifacts with acetic or formic acids, if the concentrations used are too high [25]. Anthocyanins can be monitored both in the visible (500-540 nm) and in the ultraviolet (275 nm) region, and measurement of the relative absorbances at two wavelengths provides a useful check on purity. Retention times are closely related to the number and type of substituents present in the anthocyanin nucleus. Glycosylation and hydroxylation both increase mobility, so that a triglycoside of delphinidin (with six hydroxyl groups) will be eluted first and a diglycoside of malvidin (with four hydroxyl groups) will be eluted last of a mixture of common anthocyanins (Table 19.1). By contrast, O-methylation (cf. malvidin and delphinidin glycosides in Table 19.1) and acylation with either aromatic or aliphatic acids will increase elution time. Acylation with malonic acid, e.g., nearly doubles elution time, while acylation with two malonic acid residues increases elution time threefold (Table 19.2) [26].

TABLE 19.1

RETENTION TIMES (RT) OF ANTHOCYANINS ON A LICHROSORB RP-18 COLUMN, ELUTED WITH A WATER/ACETIC ACID/ACETONITRILE GRADIENT

Pigment		RT (min)[**]
3-Rhamnosylglucoside 5-glucoside	Dp[*]	15.4
	Cy	19.1
	Pt	22.0
	Pg	22.2
	Pn	25.8
	Mv	28.4
3,5-Diglucoside	Dp	18.2
	Cy	21.7
	Pt	25.0
	Pn	25.0
	Pg	25.3
	Mv	32.1
3-Rhamnosylglucoside	Dp	27.0
	Cy	31.1
	Pt	33.6
	Pg	34.8
	Pn	39.0
	Mv	42.8

[*] Dp = delphinidin, Cy = cyanidin, Pt = petunidin, Pg = pelargonidin, Pn = peonidin, Mv = malvidin.

[**] Data from Ref. 24.

TABLE 19.2

COMPARISON OF RELATIVE RETENTION TIMES (RRT) OF MALONATED ANTHO-
CYANINS WITH THE CORRESPONDING NONACYLATED PIGMENTS IN HPLC[*]

Pigment	RRT
Pg 3,5-diglucoside	1.00
Pg 3-(6"-malonylglucoside) 5-glucoside	1.87
Pg 3,5-di(malonylglucoside)	2.67
Cy 3-glucoside	1.00[*]
Cy 3-malonylglucoside	2.14
Cy 3-dimalonylglucoside	2.96

[*] On a Spherisorb-hexyl column at 35°C, eluted with 0.6% aq. $HClO_4$/MeOH gradient [26].
For abbreviations see Table 19.1

The high resolving power of C_8 or C_{18} columns means that it is easy to separate pairs of closely related pigments (e.g., the 3-glucosides and 3-galactosides of cyanidin) which would tend to be inseparable by PC. For example, the 15-pigment mixture, based on the 3-glucosides, 3-galactosides, and 3-arabinosides of all the common anthocyanidins, except pelargonidin, present in *Vaccinium uliginosum* fruit, is readily resolved on a Supelcosil LC-18 column (3 μm) after gradient elution with methanol/formic acid/water [27]. The rapidity of HPLC has allowed the separation for the first time of the chalcones formed in solution by ring opening of malvidin 3-glucoside and malvidin 3,5-diglucoside [28]. The sensitivity of HPLC analysis provides the means of detecting and identifying the trace amounts of pigments that are present in rare flower mutants [29] or in tissue cultures [30].

HPLC with the columns and solvents mentioned above is particularly valuable for the quantitative analysis of anthocyanins. It has been applied, e.g., to the quantification of the pigment mixtures in different cultivars of ornamental plants, e.g., *Gladiolus* [24], and for measuring the anthocyanin content of red wines [31]. The ratio of pigments in grape skins, determined by HPLC, has been shown to be characteristic of particular grape cultivars, and it is possible by HPLC to follow the fate of these pigments during port or table wine production [25,32].

19.2.3 Flavonol and flavone glycosides

Flavonol glycosides are best separated on either C_8 or C_{18} reversed-phase columns by using a methanol/acetic acid/water gradient elution program, with UV detection at 335 or 365 nm [17,33]. The main structural feature affecting mobility on such columns is the number of sugar residues attached, since increasing glycosylation generally lowers retention time (Table 19.3). 3-Glucosides (e.g., of quercetin) are eluted ahead of 3-arabinosides, which emerge ahead of 3-rhamnosides. Under these conditions, 3-galactosides and 3-glucuronides are eluted at the same time as 3-glucosides, but 3-glucosides and 3-galactosides can be separated by using instead an aq. acetic acid/acetonitrile elution gradient [34]. 3-Diglycosides with only glucose moieties will be eluted sooner than those containing rhamnose (compare 3-gentiobioside and 3-sophoroside with 2-neohesperidoside and 3-rutinoside in Table 19.3). The effects of glycosylation at other positions in the flavonol nucleus than the 3-hydroxyl are less predictable, but in the quercetin series the 5-glucoside is eluted ahead of the 3-glucoside, while the 4'-glucoside has a longer retention time [17].

The effect of methylation in the flavonol moiety of flavonol glycosides expectedly increases retention on a C_8 column, as in the anthocyanin series (see Table 19.1). This can be seen by comparing the HPLC of the monoglycosides of myricetin, its 3'-methyl ether (larycitrin), and its 3',5'-dimethyl ether (syringetin). Thus, myricetin 3-galactoside is eluted at 9.38 min, larycitrin 3-galactoside at 11.64 min, and syringetin 3-galactoside at 17.65 min. Likewise, the corresponding 3-rhamnosides (which, incidentally, have a methyl sugar) are eluted at 11.06, 13.92, and 16.32 min, respectively [17].

TABLE 19.3

HPLC RETENTION TIMES AND R_F VALUES OF QUERCETIN GLYCOSIDES

Glycoside	RRT Q3G (min)[*]	$h\,R_F$ in[**]	
		BAW	water
Monoglycosides			
5-Glucoside	0.84	22	02
3-Glucoside	1.00	58	08
3-Arabinopyranoside	1.10	70	07
4'-Glucoside	1.12	48	01
3-Rhamnoside	1.17	72	19
Di- and triglycosides			
3-Gentiobioside	0.68	36	46
3,4'-Diglucoside	0.74	35	29
3-Sophoroside	0.77	45	31
3-(2^G-Glucosyl)rutinoside	0.79	37	19
3-Neohesperidoside	0.85	50	43
3-Rutinoside	0.96	45	23

[*] Retention time relative to quercetin 3-glucoside (Q3G) on a Partisil CC3/C_8 column eluted with a gradient. Eluent A, water; Eluent B, methanol/acetic acid/water (90:5:5). The initial composition was 25% B in A. A linear increase of 2% B/min in A was used for 20 min at a flowrate of 1.7 ml/min [17].

[**] Separations in 1-butanol/acetic acid/water (BAW, 4:1:5, top layer) and pure water at room temperature on Whatman No. 1 paper.

Comparison of HPLC retention times with R_F values on paper in two commonly used solvents, 1-butanol/acetic acid/water (BAW, 4:1:5, top layer) and water (Table 19.3) shows that there is little direct relationship between the behavior of a given glycoside in the two systems. Relative retention time on a reversed-phase silica column is therefore a useful, independent criterion for establishing the novelty of a particular flavonol glycoside. Similarly, HPLC is a valuable additional means of comparing an isolated plant glycoside with a standard by mixed chromatography [17].

Flavonols occur naturally in over 20 plant families in conjugation with sulfate anion, and such conjugates tend to overlap with the commonly occurring glycosides when chromatographed under the conditions described above. However, they can be nicely separated from the neutral glycosides by ion pairing with tetrabutylammonium phosphate, when they are consistently eluted after the glycosides (Table 19.4). Different 3-sulfates are also well separated from each other, and HPLC will resolve sulfate mixtures that are essentially inseparable on paper chromatograms [17]. HPLC with an ion-pairing reagent can also be employed to distinguish the anionic glucuronides of flavonols or flavones from other glycosides or from the corresponding aglycones [35].

Different flavone O-glycosides separate well under similar conditions of HPLC, as do C-glycosylflavones. However, Lardy et al. [36] found that isocratic rather than gradient elution of a reversed-phase LiChrosorb RP-18 column with methanol/acetic acid/water (40:2:58) will separate apigenin C-galactoside from the C-glucoside, the C-arabinoside from the C-xyloside, and also the C-arabinofuranoside from the C-arabinopyranoside. Flavone-di-C-glycosides, previously only separable by TLC as the permethyl ethers, can be directly separated by this HPLC method. This also applies, e.g., to resolving a mixture

TABLE 19.4

HPLC RETENTION TIMES OF FLAVONOL SULFATES WITH AND WITHOUT ION PAIRING*

Flavonol derivatives	RT (min)	
	Without ion pairing	With ion pairing
Quercetin 3-sulfate	6.0	18.2
Kaempferol 3-sulfate	7.2	19.3
Isorhamnetin 3-sulfate	8.4	19.7
Quercetin 3'-sulfate	9.2	22.5
Quercetin 3-rutinoside	14.2	14.2
Quercetin 3-rhamnoside	16.2	16.2

* On a C_{18} column, eluted with a linear gradient of Eluent A, water, and Eluent B, methanol/acetic acid/water (90:5:5). Ion pairing in the presence of 0.01 M tetrabutylammonium phosphate; eluent buffered at pH 2.2 [17].

References on p. B389

TABLE 19.5

COLUMN PACKINGS AND SOLVENT SYSTEMS FOR HPLC OF DIFFERENT CLASSES
OF PHENOLIC COMPOUNDS

Phenolic class	Column packing	Solvents[*]	Reference
Acyl phloroglucinols	μBondapak C_{18}	Tetrahydrofuran/ phosphoric acid/water (650:1:350)	40
Anthocyanidins	LiChrosorb RP-18	Water/formic acid/ methanol (gradient)	41
Anthocyanins	LiChrosorb RP-18	Water/acetic acid/ acetonitrile (gradient)	24
Anthraquinones	Spherisorb 5ODS	Water/formic acid/ methanol (32:5:163)	42
Biflavonoids	Hibar LiChrosorb Diol	Hexane/chloroform/ tetrahydrofuran (gradient)	43
	Novapak C_{18}	21% Tetrahydrofuran, 79% propanol/water (10:69)	44
Depsides and depsidones	LiChrosorb RP-8 (5 μm)	Methanol/aq. phosphoric acid (gradient)	45
Dihydrochalcones	Partisil 5CCS C_8	Acetic acid/water (1:20); methanol/ acetic acid/water (18:1:1) (gradient)	46
Flavone glycosides	μBondapak C_{18}	Methanol/acetic acid/ water (300:7:693)	47
Flavones and flavonols	Perkin Elmer C_{18}	Water/formic acid/ acetonitrile (gradient)	48
Flavone and flavonol glycosides	Partisil 5CCS C_8	Methanol/acetic acid/water (90:5:5) (gradient in water)	17
Furanocoumarins	Silica	Hexane/ethyl acetate (4:1)	49
	μBondapak C_{18}	Water/methanol/ acetic acid (40:58:2)	50
Glycosylflavones	LiChrosorb RP-18	Methanol/acetic acid/ water (40:2:58) or acetonitrile/water(13:87)	36
Hydroxycinnamic and esters	LiChrosorb RP-18	Water/acetic acid/ methanol (gradient)	51

TABLE 19.5 (continued)

Phenolic class	Column packing	Solvents[*]	Reference
Isoflavonoids	μPorasil	Dichloromethane/ ethanol/acetic acid (97:3:0.2) (gradient in hexane)	52
	LiChrosorb RP-18	Methanol/formic acid/ water (gradient)	15
Methylated flavones	LiChrosorb Si 60	Heptane/2-propanol (3:2)	16
Naphthoquinones	μBondapak CN (10 μm)	Hexane/acetic acid (94:1)	53
Phenolic acids	Spherisorb C_{18} (5 μm)	Water/acetic acid/ 1-butanol (342:1:14)	54
Phenolic aldehydes	LiChrosorb RP-8 (10 μm)	Water/methanol/tetra- hydrofuran (7:2:1)	55
Phenols	LiChrosorb RP-18 (10 μm)	Formic acid/water methanol (gradient)	15
Stilbenes	Hypersil ODS	Acetonitrile/water (7:13)	56
Tannins, condensed	C_{18}	Methanol/1% acetic acid (1:4)	57
Tannins, hydrolyzable	Develosil 60-5	Hexane/methanol/tetra- hydrofuran/formic acid (60:45:15:1) containing oxalic acid (500 mg/ 1.21 l)	58
Xanthones free	CN-bonded silica	Hexane/chloroform (13:7)	59
Xanthone glycosides	LiChrosorb RP-18	Acetonitrile/water/ acetic acid (20:80:1)	60

[*] Unless otherwise stated, isocratic elution is used. UV detection is normally applied.

of schaftoside (apigenin 6-glucoside-8-α-arabinoside), isoschaftoside (the 6-α-arabinoside-8-glucoside isomer) and neoschaftoside (the 6-glucoside-8-β-arabinoside) [36].

The HPLC of flavonol or flavone glycosides on a C_8 column, coupled to a photodiode-array detector, provides simultaneous detection of each compound through its UV spectrum as it separates. However, most flavonoids in any given series have almost identical neutral spectra, so that the spectrum alone is insufficient for characterization. This problem has been solved by Hostettmann et al. [14], who have introduced a postcolumn derivatization system, in which spectra are additionally measured in the presence of shift reagents (e.g., alkali, $AlCl_3$). This allows more complete characterization of the indi-

References on p. B389

vidual flavonoid peaks. The method works well, e.g., for the HPLC identification of isovitexin (apigenin 6-*C*-glucoside) and iso-orientin (luteolin 6-*C*-glucoside) in *Gentiana* extracts [14].

The concentrations of flavonol and flavone glycosides in direct plant extracts are readily determined by HPLC, as long as authentic markers are available for standardization. Measurements on stem extracts of five species of *Chondropetalum* showed that species with qualitatively the same seven flavonol glycosides are readily and reliably differentiated on quantitative grounds [37]. Analyses were regularly obtained on 0.1-2.0 g of dried stem samples, in which the total flavonoid content varied from 0.5 to 1.0%. Similar HPLC analysis has been applied to map the flavonoid concentrations in 135 plants of the grass *Dactylis glomerata* [38], and to the determination of the flavonoid content in medicinal plants, such as *Combretum micranthum* [39].

19.2.4 Other phenolics

Separation systems for the HPLC of most of the common classes of natural phenolics are given in Table 19.5. The choice of column packing and eluent depends largely on the relative polarity of the compounds to be separated. For example, most of the known hydroxyxanthones are partly *O*-methylated and, hence, moderately lipophilic. Thus, they are appropriately separated on a CN- or NH_2-bonded silica column, by isocratic elution with a nonpolar solvent mixture, such as hexane/chloroform (13:7) [59]. By contrast, the few known xanthone glycosides are water-soluble, so that a reversed-phase C_8 or C_{18} column is necessary, and gradient elution with an aqueous acidic methanol system [60].

When determining the right conditions for separating a particular phenolic mixture by HPLC, it is still best to start with the most appropriate literature procedure and then modify that system to suit the components that are being resolved. One of the methods available for optimizing the mobile phase in isocratic separations is the so-called PRISMA model. This has been applied, e.g., to improving the resolution of furanocoumarins in the roots of *Peucedanum palustre* on a Spherisorb ODS II (3 μm) column [61]. The four-solvent system tetrahydrofuran/acetonitrile/methanol/water in the ratio 2.9:32.5:5.0:59.6 emerged from these calculations, and, in practice, the 9 coumarins were separated with good resolution ($R_S > 1.25$ for all pairs of compounds) in 18 min.

In general, the unconjugated phenolic compounds within any given class separate according to the number of hydroxyl, methoxyl, or other (e.g., isoprenyl, C-methyl) substituents. The ratio of hydroxyl to methoxyl groups, in particular, determines the relative mobility in reversed-phase HPLC. Increasing hydroxylation increases mobility, and increasing methoxylation decreasing it. This is nicely illustrated in an isocratic separation of flavone aglycones on an AC18 reversed-phase column, eluted with a mixture of 23% acetonitrile and 77% water/formic acid (19:1) [62]. Luteolin (4 OH groups) was eluted first, at 2.67 min, followed by apigenin (3 OH) at 4.70 min, and chrysoeriol (3 OH, 1 OMe) at 5.47 min. Cirsiliol (3 OH, 2 OMe) was then eluted at 6.27 min, cirsimaritin (2 OH, 2 OMe) at 11.89 min, and, finally, xanthomicrol (2 OH, 3 OMe) at 15.67 min.

The presence of conjugation, i.e. attachment through a phenolic hydroxyl of a sugar or other organic moiety, has a profound effect on HPLC mobility. Glycosylation, e.g., always increases the mobility of phenolics on reversed-phase columns. The separation of glycosides and of sulfates has already been discussed in earlier sections, but some mention of quinic acid esters is necessary, since hydroxycinnamic acids occur widely in this combination. As expected (Table 19.6), quinic esters are eluted according to the degree of hydroxylation/methoxylation of the cinnamic acid moiety (caffeoyl esters are ahead of feruloyl esters), and disubstitution with quinic acid increases retention time [63]. Each hydroxycinnamic acid ester is present in plant extracts as a mixture of the *E*- and *Z*-forms, and these two forms usually separate during HPLC, the *E*-form being eluted ahead of the *Z*-form [51].

The HPLC conditions required for separating free phenolics are usually different from those required for phenolic glycosides (Table 19.5) and there will be a loss in resolution if both are chromatographed together. This does not seem to apply to the isoflavones, since excellent separations of biochanin A, formononetin, genistein, and their respective 7-*O*-glucosides have been achieved on a RP-18 column, eluted with varying proportions of acetic acid/water/acetonitrile [64]. Likewise, free isoflavones and their *O*- and *C*-glucosides present in *Pueraria* root have been resolved in a single chromatogram on a Phenyl 1252 N column, eluted with a gradient of the same three solvents [65].

TABLE 19.6

RETENTION TIMES OF MONO- AND DI-CINNAMOYL QUINIC ACIDS IN REVERSED-PHASE HPLC[*]

Acids	RT (min)
5-*O*-Caffeoylquinic	16.1
4-*O*-Caffeoylquinic	18.1
3-*O*-Caffeoylquinic	19.0
4-*O*-Feruloylquinic	20.6
3-*O*-Feruloylquinic	22.0
4,5-*O*-Dicaffeoylquinic	22.5
3,5-*O*-Dicaffeoylquinic	22.8
3,4-*O*-Dicaffeoylquinic	24.6
3-*O*-Caffeoyl-4-*O*-feruloylquinic	26.4
3-*O*-Feruloyl-4-*O*-caffeoylquinic	27.0

[*] On a Finepack C_{18} column (250 x 4.6 mm); flowrate, 1 ml/min; gradient elution with 5% Eluent B (methanol) in Eluent A (10 m*M* phosphoric acid) to 50% B in A within 15 min, followed by 70% B in A within 15 min [63].

TABLE 19.7

TLC SOLVENTS FOR ALL THE MAJOR CLASSES OF PHENOLIC COMPOUNDS

For methods of detection, see Section 19.3.1. References are collected in Ref. 67.

Class	Support	Solvent system
Anthocyanidins	cellulose	conc. HCl/acetic acid/water (3:30:10)
Anthocyanins } Aurones	cellulose	1-Butanol/acetic acid/water (4:1:5, top layer)
Biflavonoids	silica gel	Toluene/ethyl formate/formic acid (5:4:1)
Chalcones	cellulose	1-Butanol/acetic acid/water (4:1:5, top layer)
Coumarins		
hydroxy-	cellulose	10% aq. acetic acid
furano-	silica gel	Chloroform
Depsidones	silica gel	n-Hexane/ether/formic acid (5:4:1) (see also Table 19.9)
Dihydrochalcones	cellulose	Water
Flavanones	cellulose	5% aq. acetic acid
Flavones/flavonols		
aglycones	cellulose	Acetic acid/conc. HCl/water (30:3:10)
glycosides	cellulose	1-Butanol/acetic acid/water (4:1:5, top layer)
methylated	polyamide	Toluene/methyl ethyl ketone (9:1)
Glycoflavones	silica gel	Ethyl acetate/pyridine/water/methanol (16:4:2:1)
Hydrolyzable tannins	cellulose	2-Methyl-1- propanol/acetic acid/water (14:1:5)
Hydroxycinnamic acid		
free	cellulose	Toluene/acetic acid/water (6:7:3)
esterified	silica gel	Toluene/ethyl formate/formic acid (2:1:1)
Isoflavones	silica gel	Chloroform/methanol (89:11)
Lignans	silica gel	Ethyl acetate/methanol (19:1)
Phenolic acids	silica gel	Acetic acid/chloroform (1:9)
Phenylpropenes	silica gel	Hexane/chloroform (3:2)
Phenols	cellulose	Benzene/methanol/acetic acid (45:8:4)
Proanthocyanidins	cellulose	2-Methyl-1-propanol/acetic acid/water (14:1:5)

TABLE 19.7 (continued)

Class	Support	Solvent system
Quinones		
anthra-	silica gel	Ethyl acetate/methanol/water (100:17:13)
benzo-	silica gel	Hexane/ethyl acetate (17:3)
isoprenoid	silica gel	Benzene/petroleum ether (br 40-60°C) (2:3)
naphtho-	silica gel	Petroleum ether (br 60-80°C)/ ethyl acetate (7:3)
Stilbenes	cellulose	1-Butanol/acetic acid/water (4:1:5, top layer)
Xanthones	silica gel	Chloroform/acetic acid (4:1)

19.3 THIN-LAYER CHROMATOGRAPHY

19.3.1 General

A major application of TLC to phenolics is to the analytical or preparative separation of the less hydrophilic compounds (methylated flavones, isoflavonoids, and quinonoids) on layers of silica gel. However, water-soluble phenolics are readily separable by TLC if microcrystalline cellulose or polyamide plates are used. Hence, TLC is a procedure which is applicable to all classes of phenolics without exception. Different solvent systems have been devised for separating them according to class (Table 19.7). The advantages of TLC over other chromatographic procedures include the great flexibility in the choice of solvent, the ability to chromatograph samples in several systems simultaneously, the speed of separation, and the diagnostic identification of the separated components, without necessarily using spray reagents, on the TLC plates. The resolving power of TLC can be further improved by using the specially prepared fine particles of silica gel that are used in HPLC columns, in a technique known as HPTLC.

Some phenolics (e.g., anthocyanins, aurones, chalcones) can be visualised immediately on TLC plates by their colors. Many others appear in UV light either as dark, absorbing (e.g., flavones) or fluorescent (e.g., furanocoumarins) spots. Furthermore, most absorbing compounds give bright yellow, green, or brown colors in ultraviolet light when the plate is fumed with ammonia.

Additional information about the class of compound or structural substitution can be obtained by using one or more spray reagents. For flavones, flavonols, and their derivatives, the "Naturstoffreagenz A" (diphenylboric acid β-aminoethyl ester) is particularly

TABLE 19.8

TLC SEPARATION OF ISOMERIC DIHYDROXY- AND TRIHYDROXYBENZOIC ACIDS

Benzoic acid	$h\,R_F$ in solvent[*]		Color with diazotized p-nitroaniline
	1	2	
2,3-Dihydroxy-	53	60	purple
2,4-Dihydroxy-	38	57	brown
2,5-Dihydroxy-	60	51	yellow
2,6-Dihydroxy-	81	28	orange
3,4-Dihydroxy-	19	43	brown
3,5-Dihydroxy-	29	31	orange
2,3,4-Trihydroxy-	22	17	yellow
2,4,6-Trihydroxy-	44	18	orange
3,4,5-Trihydroxy-	17	23	brown

[*] Solvent 1 = 2-propanol/ammonia/water (8:1:1), Solvent 2 = anisole/acetic acid/water (70:24:1) on microcrystalline cellulose [68].

useful, since it gives a range of fluorescent colors, observable in both daylight and in the ultraviolet, which vary according to the structure [66]. A general spray reagent, to which all phenolics respond, is the Folin-Ciocalteu reagent. Catechol or p-quinol derivatives become blue immediately on spraying, while all other phenols five blue or gray colors after the plate has been fumed with ammonia [67]. Diazotized p-nitroaniline is another general reagent, which gives a range of colors with differently substituted phenols. For identifying isoflavonoids, the latter reagent is often used in conjunction with the Gibbs reagent (2% 2,6-dichloroquinone 4-chlorimide in methanol, followed by 10% carbonate) which specifically interacts with those phenols (e.g. phloroglucinol) in which the para-position to the hydroxyl is unsubstituted [67]. Finally, when separating phenolics that are completely substituted or that barely show up under UV light on TLC plates, it is advisable to use fluorescent silica gel plates; they then readily appear as quenching spots.

19.3.2 Phenols and lichen products

For the resolution of closely related isomers of simple phenols or phenolic acids two-dimensional TLC may well be the answer. For phenols, such as catechol, pyrogallol, resorcinol, and the three isomeric orcinols (2-, 4-, and 6-methylresorcinol), silica gel TLC with 10% acetic acid in chloroform, followed by 45% ethyl acetate in benzene (or toluene) has proved to be satisfactory [67]. Mixtures of dihydroxy- and trihydroxybenzoic acids are even harder to resolve, but Peck et al. [68] have discovered that a two-dimensional separation on microcrystalline cellulose (10 x 10-cm plates) with 2-propanol/ammonia in

TABLE 19.9

SOLVENT SYSTEMS FOR SEPARATING LICHEN PHENOLICS ON SILICA GEL PLATES

Solvent system*	Application
1. Toluene/dioxane/acetic acid (180:45:5)	Separates compounds according to number of phenolic groups
2. Hexane/methyl *t*-butyl ether/formic acid (140:72:18)	Separates compounds according to the length of alkyl side chain or number of *C*-methyl groups
3. Toluene/acetic acid (170:30)	Good general solvent
4. Cyclohexane/ethyl acetate (3:1)	Separates methyl esters of depsidones, usnic acids, and xanthones
5. Toluene/ethyl acetate/formic acid (139:83:8)	Separates β-orcinol depsidones

* From Ref. 69.

the first dimension, followed by anisole/acetic acid/water in the second (Table 19.8) can be applied directly to the detection of these phenolic acids in human urine. Indeed, the method proved to be both simpler and more effective than either HPLC or GC for this purpose.

The depsidones and other phenolics found in lichens also occur as closely related mixtures of compounds, which are difficult to resolve without the use of chromatography. They are highly characteristic of these plants, and the detection of their presence or absence plays a very important part in modern lichen taxonomy. Chromatography is the ideal technique for this purpose, and a standardized procedure, TLC on silica gel plates with three or more solvent systems, two internal controls, and several spray reagents, has been in use for a number of years [69]. The types of solvent used and their ability to separate different types of lichen product are listed in Table 19.9. In difficult cases, two-dimensional TLC with a pair of the listed solvents can be used [70].

Most lichen products show up, after development, as fluorescing or quenching spots under short- or long-wave UV. They can be visualised in daylight by spraying the plate with 10% sulfuric acid and heating at 110°C for 15 min [71]. Confirmation of structure can be obtained by spectral measurement on the TLC spots [72], while chromatographic scanners can be employed to quantify the components after TLC separation [73]. Identification of lichen products on the basis of R_F values and colors in TLC has been computerized [74]. However, extracts must be chromatographed in six solvent systems and compared with eight control compounds in order to provide the computer with sufficient information for full identification.

Finally, HPTLC provides a technique that is complementary to other TLC procedures for the resolution of closely related phenols. It has not been used to any great extent yet with lichen products. However, it has been applied successfully to the separation of hydroxycinnamic acid esters, which are chromatographed on reversed-phase silica gel plates and eluted with ethanol/water (11:9) [75] and of furanocoumarins, eluted with dichloromethane/acetonitrile mixtures [76].

19.3.3 Flavonoids

TLC is widely used in the separation and identification of almost all classes of flavonoids. Suitable adsorbents and solvents for routine separations are listed in Table 19.7; further detail is available in Refs. 8 and 9. In this Section, only the more difficult separations will be considered, and the more recent literature in the field quoted. Anthocyanins and anthocyanidins are usually separated on microcrystalline cellulose, but different solvents are normally used for the glycosides, as opposed to the aglycones (Table 19.7). Andersen and Francis [77] have found that it is possible to separate them all in one system, if a particular mixture (24.9:23.7:51.4) of conc. HCl, formic acid, and water is used. The R_F values obtained included those for delphinidin (0.03), cyanidin (0.06), pelargonidin (0.11), cyanidin 3-glucoside (0.26), cyanidin 3-rutinoside (0.48), and pelargonidin 3-sophoroside-5-glucoside (0.95). However, the relative mobilities of acylated anthocyanins in this system were not reported.

The chromatographic separation of flavones and flavonols, which are substituted additionally by hydroxylation in the 6- and/or 8-positions and which are also partly O-methylated, can present a considerable problem. In conventional TLC on cellulose, they tend to overlap with the more commonly occurring compounds, such as kaempferol and luteolin. However, the use of polyamide layers and solvents, such as toluene/methyl ethyl ketone (9:1) or toluene/methyl ethyl ketone/acetic acid (18:5:1) may resolve mixtures of such compounds; in particular, 5,6-dihydroxy-7,8-dimethoxyflavones can be separated from the isomeric 5,8-dihydroxy-6,7-dimethoxy derivatives [78]. The application of the Naturstoff A reagent is especially important (see p. B376), since flavones and flavonols respond differently, according to both the A-ring and B-ring substitution pattern [79]. More highly methylated flavones, which can occur in plant surface extracts in complex mixtures of very closely related isomers, are readily resolved on silica gel plates eluted with toluene/acetic acid (4:1) [80]. In some cases, two-dimensional TLC may be advantageous. Bohm and Banek [81] used Polyamide 6.6 plates and the solvent pair water/1-butanol/methyl ethyl ketone/dioxane (70:15:10:5) and benzene/methanol/butanone/water (55:25:23:2) to separate quercetin and patuletin (6-methoxyquercetin) glycosides present in the plant *Lasthenia burkei* (Compositae).

Naturally occurring glycosylflavone mixtures can be resolved by one-dimensional TLC on silica gel (Table 19.7) or by two-dimensional TLC on cellulose, with 1-butanol/acetic

TABLE 19.10

TLC DATA FOR TEN COMMON BIFLAVONOIDS

Key: 1 = toluene/ethyl formate/formic acid (5:4:1); 2 = toluene/pyridine/formic acid (100:20:7); 3 = toluene/methyl ethyl ketone/methanol (4:3:3); 4 = nitromethane/methanol (4:3); 5 = 1-butanol/2 N ammonium hydroxide (1:1, top layer). Polyamide 11 is from Merck, Microcryst. polyamide is from Schleicher & Schüll, ME = monomethyl ether; DIME = dimethyl ether; TRIME = trimethyl ether. Data from Williams et al. [83].

Biflavonoid	$h\,R_F$ on				
	Silica gel Solvent 1	Silica gel Solvent 2	Polyamide 11 Solvent 3	Microcryst. Polyamide Solvent 4	Cellulose Solvent 5
Amentoflavone	43	02	13	07	36
4''' ME (Podocarpusflavone A)	43	07	17	19	64
7 ME (Sequoiaflavone)	43	07	17	19	74
4' ME (Bilobetin)	42	08	28	40	36
7,4'-DIME (Ginkgetin)	42	23	71	79	49
4',4''' DIME (Isoginkgetin)	43	23	71	79	46
4',7'',4''' TRIME (Kayaflavone)	48	34	73	37	48
7,4',4''' TRIME (Sciadopitysin)	46	34	73	48	46
2,3-Dihydroamentoflavone	39	02	05	16	16
Hinokiflavone	38	06	07	14	16

acid/water and 15% aq. acetic acid. They are readily detected on TLC plates by their UV absorption, which may change from dark to brighter colors in the presence of ammonia. Van Genderen et al. [82] have discovered that it is possible to pick out isovitexin 7-glucoside on TLC plates from many closely related glycosylflavones by the blue color it produces on spraying with iodine/KI reagent.

No one TLC system works very well for separating biflavonoids (flavone dimers), and it is usually necessary to use more than one adsorbent and one or more solvents. Some typical results in the separation of amentoflavone and related methyl ethers are presented in Table 19.10 [83]. For more highly hydroxylated biflavonoids (e.g., luteolin dimers), separation on polyamide in ethyl acetate/methyl ethyl ketone/formic acid/water (5:3:1:1) has been proposed [84].

Isoflavonoids are commonly separated on silica gel plates, although polyamide plates have also been used (Table 19.11) [85]. Separation difficulties emerge with isoprenylated isoflavonoids, where there are very small structural differences between isomeric compounds. The mixtures of three glyceollins, which are isoflavonoid phytoalexins of soybean, are almost impossible to resolve by any conventional silica gel TLC method. However,

TABLE 19.11

h R$_F$ VALUES FOR TLC OF SELECTED ISOFLAVONOIDS

Compounds	TLC systems[*]		
	A	B	C
5-Deoxyisoflavonoids			
Daidzein	41	40	40
7,2',4'-Trihydroxyisoflavone	31	25	33
7,2',4'-Trihydroxyisoflavanone	23	17	38
5-Deoxykievitone	28	23	40
3,9-Dihydroxypterocarpan	67	73	48
Phaseollidin	76	84	47
Phaseollin	87	89	59
6a-Hydroxyphaseollin	70	78	64
7,2',4'-Trihydroxyisoflavan	43	36	37
Phaseollinisoflavan	76	84	44
Coumestrol	54	50	9
5-Hydroxyisoflavonoids			
Genistein	59	57	26
2'-Hydroxygenistein	36	26	25
Dalbergioidin	33	20	29
Kievitone	41	29	30
1'',2''-Dehydrocyclokievitone	59	53	44
Licoisoflavone A	61	58	24
2,3-Dehydrokievitone	54	45	29

[*] Supports and solvents are: A, silica gel, petroleum ether (br 55-65°C)/ether/acetic acid (25:75:1); B, silica gel, petroleum ether (br 55-65°C)/ethyl acetate/methanol (10:10:1), C, polyamide, methanol/water (17:3). (From Woodward [85].)

Komives [86] achieved success with these compounds by impregnating the silica gel plate beforehand with 5% formamide in acetone and then carrying out multiple development (x 4) with diethyl ether/hexane (3:2). The R_F values obtained were 0.50 (glyceollin I), 0.42 (glyceollin II), and 0.35 (glyceollin III).

HPTLC has also been applied to flavonoid separations, but the resolving power of silanized adsorbent does not appear to be that much greater than that of ordinary silica gel. Thus, Heimler [87] has separated some common flavones and flavonols on RP-18 and Sil C$_{18}$-50 plates, using 1 *M* acetic acid in 60% methanol and acetic acid/methanol/water (3:3:4), respectively, as mobile phases. The relatively high cost of such plates has, perhaps, restricted the use of this technique in this particular field.

19.4 PAPER CHROMATOGRAPHY AND ELECTROPHORESIS

19.4.1 Paper chromatography

The main application of paper chromatography is to water-soluble phenolics of all types, but especially to the flavonoid glycosides. Advantages include the relative cheapness of paper, the fact that paper chromatography works well with crude extracts, and the very ready detection of phenolics on paper chromatograms, even in the presence of impurities, such as free sugars, other plant glycosides, etc. Additionally, R_F values in paper chromatography are very reproducible, and identification is often possible on the basis of comparison with reported data for known compounds.

The main disadvantage of paper chromatography is the limited resolving power, compared to TLC and HPLC. For example, the 3-glucoside and 3-galactoside of a given flavonoid, e.g. quercetin, will only separate on paper with considerable difficulty, whereas the pair can be resolved by HPLC within 20 min (Section 19.2.3). Nevertheless, most glycosides can be separated from each other by paper chromatography, especially if several solvent systems are employed in turn. Different anthocyanidin glycosides, e.g., are readily separated on paper, if four solvents are used. The data produced, e.g., for cyanidin derivatives (Table 19.12) can be used to identify unknown glycosides with considerable confidence.

TABLE 19.12

GLYCOSYLATION AND BEHAVIOR OF THE CYANIDIN SERIES IN PAPER CHROMATOGRAPHY

Cyanidin glycosides	$h\,R_F$ values[*]			
	BAW	BuHCl	1% HCl	AAH
Aglycone	68	72	0	3
3-Glucoside	23	23	6	24
5-Glucoside	32	43	6	24
7-Glucoside	34	27	4	20
3'-Glucoside	–	44	2	15
3,5-Diglucoside	14	4	13	40
3,7-Diglucoside	11	3	20	49
3,3'-Diglucoside	16	7	20	49
3-Sophoroside 5-glucoside	15	6	53	75
3,7,3'-Triglucoside	8	1	47	67

[*] Solvent key: BAW = 1-butanol/acetic acid/water (4:1:5, top layer); BuHCl = 1-butanol/2 N HCl (1:1, top layer); 1% HCl = water/conc. HCl (97:3); AAH = acetic acid/conc. HCl/water (15:3:82). (From J.B. Harborne, unpublished data.)

References on p. B389

Two-dimensional chromatography on Whatman No. 1 paper provides an excellent general procedure for screening crude plant extracts for their phenolic constituents. The compounds are readily detectable after development by their colors in UV light and characterized by the color changes produced in the ultraviolet upon fuming the paper with ammonia. Suitable solvent pairs for two-dimensional separation include: 1-butanol/acetic acid/water (4:1:5, top layer) and 15% aq. acetic acid; and t-butanol/acetic acid/water (3:1:1) and 15% acetic acid. R_F data for 55 different C-glycosylflavonoids in the latter two solvents have been collected by Markham and Wilson [88].

Phenolics are so regular in their behavior on paper that it is often possible to discern relationships between chromatographic mobility and chemical structure. R_M values can be calculated for known substances and used to predict the behavior of new compounds. The number of hydroxyl, methyl, and glycosyl substituents are the most important factors in determining mobility, but the planarity or lack of planarity of the molecule will also be important. Planar flavonoids (flavonol aglycones), for example, can be separated from nonplanar derivatives (e.g., flavonol 3-glycosides) by their relative immobility in aqueous solvents. The relationship between hydroxylation of a flavonoid and R_F in benzene/acetic acid/water (125:72:3) is so regular that these measurements can be used to estimate the number of free phenolic groups in an unknown flavonoid [89].

Other solvents used for paper chromatography, in addition to those mentioned above, include 1-butanol/2 N ammonia (1:1) (good for cinnamic acids and biflavonoids), phenol/water (4:1) (for separating flavonoid methyl ethers), and 2-propanol/ammonia/water (8:1:1) (for phenolic acids). Colorless phenolics can be visualized on chromatograms, in most cases, by examination in UV light with or without ammonia vapor. A more permanent record is obtained by using one or other of the spray reagents that have been mentioned under TLC (Section 19.2.1). For further details about the paper chromatography of phenolics, see Ref. 67.

19.4.2 Paper electrophoresis

Paper electrophoresis is mainly employed with acidic buffers to distinguish, in plant surveys, those phenolics which are cationic (anthocyanins), anionic (O-sulfates or O-glucuronides) or zwitterionic (anthocyanin malonates) from neutral compounds. It is equally valuable for separating and purifying such compounds, when they occur as mixtures, although HPLC (Section 19.2.3) or column chromatography on an ion-exchange resin is also applicable. The extent of electrophoretic mobility also provides structural information regarding the number and position of attachment of sulfate or glucuronide moieties in a conjugated phenolic.

Anthocyanins are generally mobile in an acetate/formate buffer (pH 2.2), but they only separate from each other with difficulty, unless the time of electrophoresis is extended beyond 1-2 h. Thus, Markakis [90] could only separate cyanidin 3-rutinoside from the 3-sophoroside by electrophoresis for 5 h at 7 V/cm. However, such an acid buffer is excellent for distinguishing the positively charged anthocyanins, as a class of purple pigment, from the betacyanins (one positive nitrogen atom, three ionizable carboxyl

TABLE 19.13

ELECTROPHORETIC MOBILITIES OF SELECTED MALONATED ANTHOCYANINS

Pigment	Mobility (cm)[*]
Pelargonidin glycosides	
3-Glucoside	0.0
3-Malylglucoside	1.2
3-Malonylglucoside	2.1
3-Malonylglucoside 5-glucoside	3.6
3,5-Di(malonylglucoside)	7.1
Cyanidin glycosides	
3-Malylglucoside	0.6
3-Malonylglucoside	1.3
3-Dimalonylglucoside	3.4
3-Malonylglucuronosylglucoside	4.5
3-Glucuronosylglucoside	1.6
3-Malonylglucoside 5-glucoside	3.7
3,5-Di(malonylglucoside)	7.0
Delphinidin glycosides	
3-Malonylglucoside 5-glucoside	3.5
3,5-Di(malonylglucoside)	7.0

[*] On Whatman No. 3 paper at pH 4.4, 40 V/cm, 1 mA/cm for 1.5 h at 25°C [91].

groups), since the two classes of pigment clearly move in different directions during 1 h of electrophoresis [67].

In a less acidic buffer, such as acetate (pH 4.4), ordinary anthocyanins are normally immobile, whereas those which are zwitterionic, due to the acylation of a sugar moiety by a dibasic organic acid, such as malonic, move towards the cathode. Electrophoresis on Whatman No. 3 paper is therefore an excellent technique for distinguishing malonated anthocyanins from unsubstituted pigments, and for separating one malonated pigment from another [91]. Typical mobilities for malonated anthocyanins are given in Table 19.13, which clearly shows that dimalonated pigments usually move twice as far as monomalonated derivatives and that the degree of glycosylation of the anthocyanin also affects mobility. Electrophoresis at pH 4.4 also distinguishes anthocyanins acylated with other dibasic acids besides malonic; malylated pigments, for example, have about half the relative mobility of the corresponding malonates (Table 19.13) [92]. Anthocyanidin glucuronides are also zwitterionic in behavior and are mobile under these conditions. However, there is no difficulty in distinguishing anthocyanins acylated with dibasic acids, due to the lability of the linkage, from the glucuronides, which are exceptionally stable to acid hydrolysis [91].

Zwitterionic anthocyanins have now been recorded in about a third of all angiosperm species surveyed for their presence [92], so that electrophoretic screening of crude plant extracts at pH 4.4 has an important place in modern anthocyanin research. Flavones and flavonols can also occur naturally with malonic acid substitution, but more frequently they are found in association with sulfate groups, which may be attached directly to one or more phenolic hydroxyl groups. Paper electrophoresis in formate/acetate buffer (pH 2.2) can be employed to screening plant extracts for such sulfate salts, and also for their separation and purification [93]. Having an overall negative charge, the sulfate derivative migrates to the anode, whereas other flavones and flavonols are immobile at this pH. Disulfates migrate further than monosulfates, although the position of attachment of the sulfate has a modifying effect on relative mobility (Table 19.14) [94]. Two other classes of flavones – glucuronides and malonates – are also anionic, but they do not move significantly in an electric field at pH 2.2. They are, however, mobile at pH 4.0, so that an acetate buffer of this pH can be used to screen plant extracts for the presence of malonate and glucuronide conjugates [67].

Other applications of paper electrophoresis in the phenolic field include the separation of simple phenols from phenolic amines (e.g., tyramine) or phenolic acids (e.g., p-hydroxybenzoic). This can be achieved in either acetate (pH 5.2) or phosphate (pH 7.2) buffer, where the neutral phenols are immobile and the amines or acids move [95]. Complexing

TABLE 19.14

ELECTROPHORETIC MOBILITIES OF SELECTED FLAVONE SULFATES

Flavone structures	Electrophoretic mobility[*]
5,7,3',4'-Tetrahydroxy-6-methoxyflavone (nepetin)	0.0
5,7,4'-Trihydroxy-6,3'-dimethoxyflavone (jaceosidin)	0.0
5,7,4'-Trihydroxy-6-methoxyflavone (hispidulin)	0.0
Hispidulin 7-sulfate	1.8
Hispidulin 4'-sulfate	1.7
Nepetin 7-sulfate	1.0
Jaceosidin 7-sulfate	1.8
6-Hydroxyluteolin 7-sulfate	1.1
6-Hydroxyluteolin 6-sulfate	0.8
Nodifloretin 7-sulfate[**]	1.2
6-Hydroxyluteolin 6,7-disulfate	3.1
Nodifloretin 6,7-disulfate	4.4
Nepetin 3',4'-disulfate	4.4
Jaceosidin 7,4'-disulfate	3.5
Hispidulin 7,4'-disulfate	4.4

[*] Mobilities are relative to quercetin 3'-sulfate, at pH 2.2, 1 h, 40 V/cm [94].

[**] Nodifloretin = 5,6,7,4'- tetrahydroxy-3'-methoxyflavone.

electrolytes have been applied to the separation of catechol derivatives, which form metal complexes, from related monophenols. Alkaline borate (pH 8.8) has been utilized, but catechol derivatives tend to oxidize or decompose in alkaline medium. Better separations are obtained with sodium molybdate at pH 5.2. This buffer has the advantage that the catechols form visible brown molybdenum complexes in it [95].

19.5 GAS/LIQUID CHROMATOGRAPHY

Whenever complex mixtures of relatively volatile phenols are encountered in plant or animal tissues, GC is probably the method of choice for achieving separation [96-102]. Capillary GC is conducted with silica glass columns of between 10 and 100 meters in length and of 0.20 to 0.32 mm diameter, coated with an appropriate silicone oil (Table 19.15). In order to increase volatility, phenols are generally chromatographed after derivatization as their trimethylsilyl (TMS) ethers. Otherwise, rather high temperatures, e.g. 250-300°C, have to be reached to achieve the separations [96]. Flame ionization and electrochemical detection are commonly employed. For quantitative analysis of phenolic acid methyl esters, FID will serve for detection in the range 0.01-10 μg, while ECD will measure 0.05-1 ng of phenolics [103].

GC/MS has found favor in a number of laboratories for phenolic analysis. This technique has been applied successfully to the identification of the phenolics present in bee propolis and in the bud exudates of poplar and other trees from which the propolis is

TABLE 19.15

CAPILLARY GLC OF PHENOLS

Phenols	Derivatives	Column dimensions and coating	Column temp., °C	Ref.
Long-chain phenols	None	12.5 m x 0.20 mm, methylsilicone	250-300	96
Monoalkylphenols	Acetates	100 m x 0.25 mm, phenylmethylsilicone	185	97
Phenolic acids	TMS	10 m x 0.22 mm, CP-Sil 5CB	140	98
Phenolic glycosides	TMS	25 m x 0.32 mm, silicone	190-295	99
Phenols	2,4-Dinitro-phenyl ethers	50 m x 0.30 mm, OV-210	60-220	100
Phenylacetic acids	TMS	25 m x 0.30 mm, SE-54	80-280	101
Phenylpropanoids	TMS	50 m x 0.30 mm, OV-1	–	102

References on p. B389

derived [102]. Likewise, GC/MS of urinary phenolic acids has a place in clinical diagnosis in hospitals [104].

While GC or GC/MS is less applicable to the analysis of the relatively involatile flavonoids, the choice between GC and HPLC for separating simple plant phenolics is finely balanced. For example, Van de Casteele et al. [105] were able to quantify accurately the phloroglucinol and phenolic acids present in *Medinilla magnifica* extracts by GC and, similarly, Guyot et al. [106] the phenolics from smoky-tasting cocoa beans. Indeed, Ford and Hartley [98] found capillary GC better than HPLC for the analysis of phenolics (e.g., ferulic and *p*-coumaric acids) present in plant cell walls.

19.6 LARGE-SCALE CHROMATOGRAPHIC SEPARATIONS

Some ten systems in common use for the preparative separation of phenolic mixtures are shown in Table 19.16, and a general indication of the applicability of each system is listed. In practice, more than one of these procedures are likely to be used in a sequential operation. Much further detail for the preparative separation of phenolics can be found in the book of Hostettmann et al. [107]. Preliminary purification, to separate other classes of natural product from the phenolic fraction, is nearly always necessary with most plant extracts. While Sephadex LH-20 is widely used, Hiermann [108] has claimed rather better results with Fractogel PGM 2000, a gel specially developed by Merck for this purpose. Alternatively, various ion-exchange resins have been employed, especially for obtaining flavonoids free from other contaminants. Rosler and Goodwin [109] recommend a non-ionic resin, Amberlite XAD-2, for such separations. After the flavonoids are adsorbed from an aqueous extract and washed with water, elution is achieved with 50% aq. methanol and eventually with 100% methanol.

Anthocyanins are among the easiest of flavonoids to separate preparatively, since it is possible to monitor their separation by eye. For their purification, it is advisable to use a preliminary fractionation on an ion-exchange resin; for malonated anthocyanins, Sephadex LH-20 should be used with methanol/acetic acid/water (9:1:1) as eluent. The pigments are then separated by paper chromatography on thick paper [91] or by column chromatography on a Perlon-type polyamide [110]. Final purification is then achieved on Sephadex LH-20. Droplet countercurrent chromatography (DCCC) has limited use for anthocyanin preparation; the upper layer of butanol/acetic acid/water (4:1:5) has been employed to isolate blackberry and raspberry pigments, starting with 300-mg amounts of crude anthocyanin [111]. Finally, preparative HPLC has been used to separate the anthocyanins of blackberry and cranberry on the mg scale and above [112].

Flavone and flavonol aglycones are best separated on silica (Merck Kieselgel 60), though it is important to prewash the column with conc. HCl to remove traces of iron, which would otherwise hold hydroxylated aglycones onto the column. Eluents range from benzene/chloroform (1:1), through pure chloroform to pure ethyl acetate. The corresponding glycosides are preferably fractionated on a polyamide column, eluted with water containing increasing amounts of methanol or with chloroform/methanol (1:1) of increas-

TABLE 19.16

PREPARATIVE SEPARATIONS OF PHENOLICS

Separation system	Comments
Thick-paper or thick-layer chromatography	Convenient for 5- to 10-mg samples, but otherwise laborious
Centrifugal TLC	Excellent for flavonoid or xanthone aglycones and lipophilic phenols on 50- to 100-mg scale
Droplet countercurrent chromatography	Only a few solvent systems applicable; useful for hydrolyzable tannins, anthraquinones, and anthocyanins
Open-column chromatography Polyamide	Excellent capacity, simple aq. alcoholic solvents used, up to gram scale
Cellulose	Time-consuming and lacking in reproducibility
Sephadex G-50	System of choice for proanthocyanidins, eluted with acetone/water mixtures
Sephadex LH-20	Useful for cleanup and for simple separations
Silica	For nonpolar phenolics
Amberlite XAD-2 resin	General cleanup of crude extracts to give a total flavonoid fraction
Pressure column chromatography Low pressure	200- to 300-mg samples can be separated in 1-2 h
HPLC on bonded silica	Separations in 10-30 min, wide range of applications

ing methanol content [113]. For the DCCC of flavone and flavonol glycosides, the solvent system chloroform/1-butanol/methanol/water (10:1:10:6) is generally applicable. Aglycones may be fractionated by using the bottom (chloroform) layer, while the glycosides will separate in the top layer; starting with 2 g of crude flavonoid, DCCC will provide up to 100-mg amounts of the separated products, since no material is lost in the process [113].

Isoflavonoids are usually separated on silica gel columns, eluted with chloroform, containing increasing amounts of methanol. More elaborate procedures were used by Tahara et al. [114] to separate the prenylated isoflavonoids of *Lupinus albus*, because of the complex mixture of closely related substances present. Major fractionation was achieved on a Wako-gel C-200 column, with benzene, containing increasing amounts of ethyl acetate, as eluent. Final purification was obtained by preparative TLC on thick silica gel plates.

References on p. B389

Condensed tannins (or flavolans) can be obtained in quantity by adsorbing crude 50% aq. methanol plant extracts on Sephadex LH-20 columns and then washing with the same solvent to remove other phenolics. The tannins are recovered by eluting with acetone/water (1:1). Dimeric tannins can then be purified by rechromatographing them on Sephadex LH-20 and eluting with ethanol/water/acetone mixtures. Higher oligomers are best dealt with on either a high-porosity polystyrene gel (MC1 gel CHP-20P), eluted with aq. methanol, or a TSK gel (HW-405), eluted with methanol [115]. Condensed tannins can also be obtained in milligram amounts, following preparative HPLC on a nitrile column [116].

Hydrolyzable tannins (gallic and ellagic acid derivatives) can be purified by any of the methods mentioned above for condensed tannins. They are also separable by means of DCCC or a recent modification of DCCC, centrifugal partition chromatography (CPC) [117]. While DCCC normally takes 2 days or more for separating a tannin mixture, DCCC performed in a centrifuge cuts down this time to a few hours. The shortening of developing time also reduces the amount of solvent used, making it easier to concentrate the final extracts. CPC has been applied, e.g., to the purification of the ellagitannins in *Lythrum anceps*, using the solvent system 1-butanol/propanol/water (4:1:5) with entirely beneficial results [118].

An advantage of using preparative HPLC for tannin fractionation is that it is normally possible to determine the molecular weights of the purified components at the same time. Hydrolyzable tannins can be separated on a gel HPLC column, as their methyl or acetyl derivatives. However, more conveniently, both hydrolyzable and condensed tannins can be purified on a normal-phase HPLC column of Develosil 60-5 (15 cm x 4 mm), eluted with hexane/methanol/tetrahydrofuran/formic acid (60:45:15:1) containing oxalic acid (500 mg/1.2 l). In this system, monomers are eluted after 1-2 min, dimers after 3-4 min, trimers at 7 min, and tetramers at 13 min [117].

19.7 CONCLUSION

Chromatography of phenolics has developed considerably since the early days when chromatographic separation on filter paper was the main procedure available [119]. The sophistication of modern apparatus, such as HPLC, is considerable, but it can involve large capital expenditures. For quantitative estimation of the less volatile phenolics, such equipment is almost obligatory. For determining estrogenic isoflavone levels in silage or tannin levels in beer, measurement of phenolic content is undoubtedly best achieved by analytical HPLC. By using a pulsed amperometric detector with HPLC, it is possible to measure phenols, such as those which commonly pollute drinking water, down to the levels of ten parts per trillion [120]. HPLC is thus competitive with GC for determining trace phenols, and time-saving in that TMS derivatization is not needed.

For the qualitative analysis of phenolics, the simple and inexpensive techniques of PC still offer a very satisfactory methodology. The great advantage of PC over HPLC is that one can recognize the different phenolics as they separate by their colors in ordinary or

UV light, or after separation, by the colors produced with appropriate spray reagents. The application of TLC to the identification of medicinal plants, such as the aloes, by their phenolic profiles is well illustrated in the color guide of Wagner et al. [121]. A particular merit of TLC is the wide choice of adsorbents and solvent systems available for phenolics. This means that the direct comparison of an unknown with a standard can be repeated several times to confirm an identification [17], a point not always recognized but one that is essential in order to avoid analytical errors.

For the preparative chromatography of phenolics, a wide range of procedures are available but none is of universal application. For the lipophilic phenolics, as found in the bud exudates of trees or the leaf washings of herbs, GC/MS provides a technique capable of resolving and ultimately identifying the 100 or more components present. However, much effort must be devoted to building up the necessary reference library of known compounds beforehand. For the water-soluble phenolics of plant tissues, it is possible to identify the majority of components with HPLC, a diode-array detector, and postcolumn derivatization [14]. However, a reference library is again needed, and new structures, when they appear, will still require separate analysis.

For most workers in the phenolic field, preparative separations on thick paper or thick silica plates or on columns of cellulose, silica gel, or Sephadex are still the order of the day. This is followed by appropriate cleanup and standard spectral (UV, IR, NMR, MS) identifications of the separated components. Newer techniques, such as DCCC, centrifugal TLC, and low-pressure column chromatography are coming into regular use. However, there is still room for further technical development in the preparative chromatography of the highly complex phenolic mixtures encountered in living matter.

REFERENCES

1 I. Raskin, I.M. Turner and W.R. Melander, *Proc. Natl. Acad. Sci. U.S.A.*, 86 (1989) 2214.
2 P.A. Spencer and G.H.N. Towers, *Phytochemistry*, 27 (1988) 2781.
3 W.S. Pierpoint, in V. Cody, E. Middleton and J.B. Harborne (Editors), *Plant Flavonoids in Biology and Medicine,* Vol. 1, Liss, New York, 1986, p. 125.
4 V. Cody, E. Middleton, J.B. Harborne and A. Beretz (Editors), *Plant Flavonoids in Biology and Medicine,* Vol. 2, Liss, New York, 1988.
5 E.M. Williamson and F.J. Evans, *Potter's New Encyclopedia of Botanical Drugs and Preparations,* C.W. Daniel, Saffron Walden, 1988.
6 L.W. Wulf and C.W. Nagel, *J. Chromatogr.*, 116 (1976) 271.
7 J.B. Harborne, in E.A. Bell and B.V. Charlwood (Editors), *Encyclopedia of Plant Physiology, New Series,* Vol. 8, Springer, Berlin, 1980, p. 329.
8 J.B. Harborne (Editor), *The Flavonoids: Advances in Research since 1980,* Chapman and Hall, London, 1988.
9 J.B. Harborne (Editor), *Methods in Plant Biochemistry, Volume 1, Plant Phenolics,* Academic Press, London, 1989.
10 J.B. Harborne, *Adv. Med. Plant. Res., Plenary Lect. Int. Cong.*, 32 (1985) 135.
11 D.J. Daigle and E.J. Conkerton, *J. Liq. Chromatogr.*, 11 (1988) 309.
12 R.N. Strange, in H.F. Linskens and J.F. Jackson (Editors), *High Performance Liquid Chromatography in Plant Science,* Springer, Berlin, 1987, p. 121.
13 K. Hostettmann and A. Marston, *Prog. Clin. Biol. Res.*, 213 (1986) 43.

14 K. Hostettmann, B. Domon, D. Schaufelberger and M. Hostettmann, *J. Chromatogr.,* 283 (1984) 137.
15 K. Van de Casteele, H. Geiger and C.F. Van Sumere, *J. Chromatogr.,* 258 (1983) 111.
16 J.P. Bianchini and E.M. Gaydou, *J. Chromatogr.,* 259 (1983) 150.
17 J.B. Harborne and M. Boardley, *J. Chromatogr.,* 299 (1984) 377.
18 A. Bilyk, K.B. Hicks, D.D. Bills and G.M. Sapers, *J. Liq. Chromatogr.,* 11 (1988) 2829.
19 P.J. Hayes, M.R. Smyth and I. McMurrough, *Analyst (London),* 112 (1987) 1197.
20 G. Weber, *Chromatographia,* 26 (1988) 133.
21 Y. Kitada, M. Mizabuchi, Y. Ueda and H. Nakazawa, *J. Chromatogr.,* 347 (1985) 438.
22 S.A. Westwood, D.E. Games and L. Sheen, *J. Chromatogr.,* 204 (1981) 103.
23 E. Farmakalidis and P.A. Murphy, *J. Agric. Food Chem.,* 33 (1985) 385.
24 N. Akavia, D. Strack and A. Cohen, *Z. Naturforsch.,* 36C (1981) 378.
25 J. Bakker and C.F. Timberlake, *J. Sci. Food Agric.,* 36 (1985) 1315.
26 K. Takeda, J.B. Harborne and R. Self, *Phytochemistry,* 25 (1986) 1337.
27 O.M. Andersen, *J. Food Sci.,* 52 (1987) 665.
28 N.W. Preston and C.F. Timberlake, *J. Chromatogr.,* 214 (1981) 222.
29 A.W. Schram, L.M.V. Jonsson and P. de Vlaming, *Z. Naturforsch.,* 38C (1983) 342.
30 C.M. Colijn, L.M.V. Jonsson, A.W. Schram and A.J. Kool, *Protoplasma,* 107 (1981) 63.
31 J. Bakker, N.W. Preston and C.F. Timberlake, *Am. J. Enol. Vitic.,* 37 (1986) 121.
32 J. Bakker and C.F. Timberlake, *J. Sci. Food Agric.,* 36 (1985) 1325.
33 K. Van de Casteele, H. Geiger and C.F. Van Sumere, *J. Chromatogr.,* 284 (1984) 269.
34 V. Cheynier and J. Rigaud, *Am. J. Enol. Vitic.,* 37 (1986) 248.
35 K. Sagara, Y. Ito, T. Oshima, T. Misaki and H. Murayama, *J. Chromatogr.,* 328 (1985) 289.
36 C. Lardy, M.L. Bouillant and J. Chopin, *J. Chromatogr.,* 291 (1984) 307.
37 J.B. Harborne, M. Boardley and H.P. Linder, *Phytochemistry,* 24 (1985) 273.
38 M. Jay, D. Plenet, P. Ardouin, R. Lumaret and P. Jacquand, *Biochem. System. Ecol.,* 12 (1984) 193.
39 E. Bassene, A. Laurence, D. Olschwang and J.L. Pousset, *J. Chromatogr.,* 346 (1985) 428.
40 C.J. Widen, H. Pyysalo and P. Salovaara, *J. Chromatogr.,* 188 (1980) 213.
41 K. Van de Casteele, H. Geiger, R. De Loose and C.F. Van Sumere, *J. Chromatogr.,* 259 (1983) 291.
42 A.S.J. van den Berg and R.P. Labadie, *J. Chromatogr.,* 329 (1985) 311.
43 F. Briancon-Scheid, A. Lobstein-Guth and R. Anton, *Planta Med.,* 49 (1983) 204.
44 M. Pietta, P. Muri and A. Rava, *J. Chromatogr.,* 437 (1988) 453.
45 K. Huovinen, R. Hiltunen and M. von Schantz, *Acta Pharm. Fenn.,* 94 (1985) 99.
46 R. Grayer, in J.B. Harborne (Editor), *Methods in Plant Biochemistry. Volume 1. Plant Phenolics,* Academic Press, London, 1989, p. 283.
47 D.J. Daigle and E.J. Conkerton, *J. Chromatogr.,* 240 (1982) 202.
48 F.A.T. Barberan, F. Tomas, L. Hernandez and F.J. Ferreres, *J. Chromatogr.,* 347 (1985) 443.
49 H.J. Thompson and S.A. Brown, *J. Chromatogr.,* 314 (1984) 803.
50 G.F. Spencer, L.W. Tjarks and R.G. Powell, *J. Agric. Food Chem.,* 35 (1987) 803.
51 W. Brandl and K. Herrmann, *J. Chromatogr.,* 260 (1983) 447.
52 R.E. Carlson and D.H. Dolphin, *Phytochemistry,* 20 (1981) 2281.
53 F. Gafner, J.C. Chapnis, J.D. Msonthu and K. Hostettmann, *Phytochemistry,* 26 (1987) 2501.
54 R.D. Hartley, in H.F. Linskens and J.F. Jackson (Editors), *HPLC in Plant Sciences,* Springer, Heidelberg, 1987, p. 92.
55 E. Roggendorf and R. Spatz, *J. Chromatogr.,* 204 (1981) 263.
56 P. Langcake, C.A. Cornford and R.J. Pryce, *Phytochemistry,* 18 (1979) 1025.
57 A.G.H. Lea, *J. Chromatogr.,* 238 (1982) 253.
58 T. Okuda, T. Yoshida and T. Hatano, *J. Nat. Prod.,* 52 (1989) 1.

59 K. Hostettmann and H.M. McNair, *J. Chromatogr.*, *116 (1976) 201.*
60 T. Hayashi and T. Yamagishi, *Phytochemistry*, 27 (1988) 3696.
61 H. Vuorela, K. Dallenbach-Tolke, S. Nyiredy, R. Hiltunen and O. Sticher, *Planta Med.*, 55 (1989) 181.
62 F.A.T. Barberan, J.M. Nunez and F. Tomas, *Phytochemistry*, 24 (1985) 1285.
63 H. Morishita, H. Iwahashi, N. Osaka and R. Kido, *J. Chromatogr.*, 315 (1984) 253.
64 J. Koster, D. Strack and W. Barz, *Planta Med.*, 48 (1983) 131.
65 Y. Ohshima, T. Okuyama, K. Takahashi, T. Takizawa and S. Shibata, *Planta Med.*, 54 (1988) 250.
66 H. Homberg and H. Geiger, *Phytochemistry*, 19 (1980) 2443.
67 J.B. Harborne, *Phytochemical Methods*, Chapman and Hall, London, 2nd Edn., 1984.
68 H. Peck, A.W. Stott and J.B. Turner, *J. Chromatogr.*, 367 (1986) 289.
69 C.F. Culberson and A. Johnson, *J. Chromatogr.*, 238 (1982) 483.
70 C.F. Culberson and A. Johnson, *J. Chromatogr.*, 128 (1976) 253.
71 F.J. White and P.W. James, *Br. Lichen Soc. Bull.*, No. 57 (Suppl) (1985) 1.
72 C. Leuckert and H. Mayrhofer, *Herzogia*, 7 (1985) 99.
73 D. Fahselt, *Lichenologist*, 13 (1981) 253.
74 J.A. Elix, J. Johnston and J.L. Parker, *Mycotaxon*, 31 (1988) 89.
75 M. Vanhaelen and R. Vanhaelen-Fastre, *J. Chromatogr.*, 187 (1980) 255.
76 M.L. Bieganowska and K. Glowniak, *Chromatographia*, 25 (1988) 111.
77 O.M. Andersen and G.W. Francis, *J. Chromatogr.*, 318 (1985) 450.
78 F.A.T. Barberan, F. Ferreres and F. Tomas, *Tetrahedron*, 41 (1985) 5733.
79 E. Wollenweber, *Supplement Chromatographia*, (1982) 50.
80 L.M. Hernandez, F.A.T. Barberan and F.T. Lorente, *Biochem. System. Ecol.*, 15 (1987) 61.
81 B.A. Bohm and H.M. Banek, *Biochem. System. Ecol.*, 15 (1987) 57.
82 H.H. van Genderen, J. van Brederode and G.J. Niemann, *J. Chromatogr.*, 256 (1983) 151.
83 C.A. Williams, J.B. Harborne and F.A.T. Barberan, *Phytochemistry*, 26 (1987) 2553.
84 R. Becker, R. Mues, H.D. Zinsmeister and H. Geiger, *Z. Naturforsch.*, 41C (1986) 507.
85 M.D. Woodward, *Phytochemistry*, 19 (1980) 921.
86 T. Komives, *J. Chromatogr.*, 261 (1983) 423.
87 D. Heimler, *J. Chromatogr.*, 366 (1986) 407.
88 K.R. Markham and R.D. Wilson, *Phytochem. Bull.*, 21 (1989) 2.
89 K. Yoshitama, *Bot. Mag. (Tokyo)*, 91 (1978) 207.
90 P. Markakis, *Nature (London)*, 187 (1960) 1092.
91 K. Takeda, J.B. Harborne and R. Self, *Phytochemistry*, 25 (1986) 1337.
92 J.B. Harborne, *Phytochemistry*, 25 (1986) 1887.
93 D. Barron, L. Varin, R.K. Ibrahim, J.B. Harborne and C.A. Williams, *Phytochemistry*, 27 (1988) 2375.
94 F.A.T. Barberan, J.B. Harborne and R. Self, *Phytochemistry*, 26 (1987) 2281.
95 J.B. Pridham, in J.B. Pridham (Editor), *Methods in Polyphenol Chemistry*, Pergamon, Oxford, 1964, p. 111.
96 Y. Du and R. Oshima, *J. Chromatogr.*, 318 (1985) 378.
97 V. Raverdino and P. Sassetti, *J. Chromatogr.*, 153 (1978) 181.
98 C.W. Ford and R.D. Hartley, *J. Chromatogr.*, 436 (1988) 484.
99 R. Julkunen-Tiitto, *J. Chromatogr.*, 324 (1985) 129.
100 M. Lehtonen, *J. Chromatogr.*, 202 (1980) 413.
101 M.E. Snook, P.F. Mason, R.F. Arrendale and O.T. Chortyk, *J. Chromatogr.*, 324 (1985) 141.
102 W. Greenaway, T. Scaysbrook and F.R. Whatley, *Phytochemistry*, 27 (1988) 3513.
103 M. Lehtonen and M. Ketola, *J. Chromatogr.*, 370 (1986) 465.
104 T. Niwa, *J. Chromatogr.*, 379 (1986) 313.
105 K.L. Van de Casteele, M.I.D.V. Keymeulen, P.C. Debergh, L.J. Maene, M.C. Flamee and C.F. Van Sumere, *Phytochemistry*, 20 (1981) 1105.

106 B. Guyot, D. Gueule, I. Morcrette and J.C. Vincent, *Cafe, Cocoa, The,* 30 (1986) 113.
107 K. Hostettmann, M. Hostettmann and A. Marston, *Preparative Chromatography Techniques,* Springer, Berlin, 1986.
108 A. Hiermann, *J. Chromatogr.,* 362 (1986) 152.
109 K.H. Rosler and R.S. Goodwin, *J. Nat. Prod.,* 47 (1983) 188.
110 D. Strack, E. Busch, V. Wray, L. Grotjahn and E. Klein, *Z. Naturforsch.,* 41C (1986) 707.
111 G.W. Francis and O.M. Andersen, *J. Chromatogr.,* 283 (1984) 445.
112 K.B. Hicks, S.M. Sondey, D. Hargrave, G.M. Sapers and A. Bilyk, *LC Mag.,* 3 (1985) 981.
113 K.R. Markham, *Techniques of Flavonoid Identification,* Academic Press, London, 1982.
114 S. Tahara, S. Orihara, J.L. Ingham and T. Mizutani, *Phytochemistry,* 28 (1989) 901.
115 L.J. Porter, in J.B. Harborne (Editor), *The Flavonoids: Advances in Research since 1980,* Chapman and Hall, London, 1988, p. 21.
116 R.W. Hemingway, L.Y. Foo and L.J. Porter, *J. Chem. Soc. Perkin I,* (1982) 1209.
117 T. Okuda, T. Yoshida and T. Hatano, *J. Nat. Prod.,* 52 (1989) 1.
118 T. Okuda, T. Yoshida, T. Hatano, K. Yazaki, R. Kiro and Y. Ikeda, *J. Chromatogr.,* 362 (1986) 375.
119 J.B. Harborne, *Chromatogr. Rev.,* 1 (1958) 209.
120 D.A. Baldwin and J.K. Debowski, *Chromatographia,* 26 (1988) 186.
121 H. Wagner, S. Bladt and E.M. Zgainski, *Drug Analysis, the TLC Analysis of Drug Plants,* Springer, Berlin, 1984, p. 102.

Chapter 20

Drugs

K. MACEK and J. MACEK

CONTENTS

20.1 INTRODUCTION

Chemically, drugs represent perhaps the most heterogeneous group of compounds in this book. In this category, low-molecular-weight compounds with a nitrogen-containing functional group prevail. Besides, there are compounds which are devoid of nitrogen, both organic and inorganic compounds, and some biopolymers. In this chapter, we shall focus on pharmacologically active compounds and we shall omit drug excipients and diagnostics.

The tasks facing an analyst in the area of drugs are quite diverse. One group of problems are those that have to be solved in the pharmaceutical industry and research. This concerns the analysis of crude materials, analysis of synthetic intermediates and, more recently, a special branch of analytical chemistry, the analysis of biotechnology products. The final products must be identified, and the active compounds and impurities in the drug form must be quantitated. A number of these procedures are legally required by the Pharmacopoeias of various countries. Another important task is the estimation of the stability of the drug form, calling for the analysis of decomposition products, including the determination of their structure. In all these situations, stringent requirements exist for the speed, accuracy, and selectivity of the assay, and there may be a need for analyzing a large number of samples. Therefore, particularly in industry, some of these procedures are automated or carried out with the help of robots [1,2].

Another area of actual interest is the analysis of drugs in biological materials. This includes monitoring drug levels, pharmacokinetic studies, estimation of drugs of abuse, and various toxicological and forensic applications. Analyses in this category demand high sensitivity (in the range of picograms to micrograms per gram of biological material),

TABLE 20.1

SURVEY OF PAPERS, PUBLISHED IN 1988, DEALING WITH DRUG CHROMA-
TOGRAPHY

	Total	Percent
Total number of papers on chromatography	9831	
Number of papers devoted to drug analysis	1549	15.9
Total number of HPLC papers	5716	
HPLC papers on drug analysis	1008	17.6
Total number of GC papers	2698	
GC papers on drug analysis	270	10.0
Total number of PC papers	1417	
PC papers on drug analysis	271	19.1

the possibility of assaying a series of structurally related compounds (including enantiomers and metabolites), great selectivity and speed, and, of course, the capability of
assaying a large number of samples. Consequently, it is not surprising that chromatographic methods are preferred for this purpose, though for routine analysis most clinicians
would prefer simpler tests, e.g., immunochemical assays. The advantage of the latter is
speed, sensitivity, simplicity, and ease of automation. However, there are also drawbacks,
such as lower selectivity, impossibility of estimating optical isomers, and relatively high
cost per analysis.

The bibliography of chromatographic papers published in 1988 shows that chromatography of drugs occupies the first place among all applications of chromatographic
procedures. Table 20.1 was compiled from the data published in Ref. 3. It is evident from
this survey that since the last edition of this monograph [4] over 10 000 papers were
published concerning chromatographic drug analysis. Thus, the present chapter can give
only a general picture of the status of chromatography in drug analysis and the advances
achieved within the last five years. For a more detailed study, bibliographic data [5],
monographs, and reviews [6-10] should be consulted.

20.2 GENERAL PART

Although the chromatographic techniques used in drug analysis are basically identical
with those described in Part A of this monograph, certain aspects that are characteristic of
these assays will be dealt with in this section.

20.2.1 Chromatographic techniques

20.2.1.1 High-performance liquid chromatography

HPLC is now the most frequently used technique in drug analysis. The reason is that HPLC methods fullfil most of the requirements for assaying drugs, both in biological materials and in pharmaceuticals. Although drug analysis by HPLC is usually carried out with conventional equipment, an increased interest in miniaturization is apparent [11,12]. Column switching is frequently applied in the analysis of biological materials. For analyzing a large number of samples at least some steps can be conveniently automated. Simple tasks in the quantitative analysis of drugs or drug forms can be carried out by means of a UV spectrophotometer with a variable wavelength, but the disadvantage is that not all impurities in the sample can be detected. To a certain extent, this can be achieved with a multichannel photodiode-array detector. Since reference impurities are frequently unavailable and since the most likely impurities or metabolites are structurally related to the drug itself, quantitation is often based on the assumption that the detector response is identical for all compounds analyzed. For identification by Pharmacopoeia standards or for the assay of biological material selective detectors are preferred, e.g., fluorescence detectors (sometimes with laser excitation for increased sensitivity), chemiluminescence detectors, and electrochemical detectors. Compounds that do not exhibit intrinsic fluorescence or are not electrochemically active can be appropriately derivatized (cf. also p. B401). For unambiguous identification a combination of HPLC/MS with a thermospray interface is used increasingly [13].

In situations where none of the detectors is sufficiently sensitive it is possible to use a combination with nonchromatographic, mostly immunochemical procedures, like RIA [14]. Radioactively labeled compounds are most useful for drug metabolism studies. In modern radio-HPLC procedures on-line detection is mainly used, both in heterogeneous and homogeneous systems [15].

Most drug separations are carried out on columns packed with a C_{18}-bonded phase, having a particle size of 3-10 μm. Unmodified silica gel, C_8 phase, nitrile-bonded sorbents, polymeric beads, and ion exchangers are used less frequently. With reversed-phase columns, residual silanol groups are mainly responsible for peak tailing of basic drugs. Therefore, these groups must be deactivated by the end-capping. Microbore columns, which speed up the analysis and decrease sample size and solvent consumption, are increasing in popularity [11,12]. Direct injection of biological samples into HPLC columns is enabled by a new type of sorbents in which the protein ballast materials are separated from the analyzed drugs by gel permeation and the drugs are simultaneously chromatographed in a reversed-phase mode. In this category, the internal-surface reversed-phase (ISRP) silica introduced by Pinkerton et al. [16] and the shielded hydrophobic phase (SHP) [17], which can be useful in drug analysis [18], should be at least mentioned.

Aqueous methanol or acetonitrile is generally used as mobile phase in reversed-phase chromatography. For acidic drugs the pH of the mobile phase must be decreased by using buffers in order to repress dissociation. For basic drugs a small amount of a strongly polar compound is added to the mobile phase. The drug forms an ion pair with that compounds, acquiring more favorable retention properties than the free base [19]. Moreover, its adsorption on the free silanol groups of the sorbent is thus prevented. In analyzing drug metabolites alongside the parent drug, gradient elution must often be applied because of considerable differences in polarity.

De Smet and Massart [20-22] published a series of papers devoted to automated method selection (expert systems) for the chromatography of drugs. Their articles include the selection of the mode of detection, stationary phase, mobile-phase solvents, ion suppression or ion enhancement, and the use of buffers.

20.2.1.2 Gas chromatography

Gas chromatography has been used frequently in drug analysis, but since the Seventies it is gradually being replaced by HPLC. Generally speaking, most drugs exhibit properties unfavorable for gas chromatography: They are mostly polar compounds of low volatility, which are often thermolabile. Thus, prior to the gas-chromatographic separation, they must be converted to volatile derivatives. This is time-consuming and laborious, and the reaction is not necessarily stoichiometric. Moreover, derivatization does not always guarantee thermal stability. Still, there are several groups of drugs for which gas chromatography continues to be used alongside with HPLC. A number of studies in which the suitability of both approaches to drug analysis was compared have demonstrated that approximately identical results can be obtained by both methods. Therefore, the choice of a particular chromatographic procedure reflects frequently the availability of equipment, laboratory practice, and experience. For the purpose of forensic analysis and for the structure analysis of decomposition products or metabolites in biological materials it is necessary to couple GC with MS [23] or FTIR [24]. In such cases, gas chromatography is the method of choice [25].

In the area of column techniques there is a strong tendency towards miniaturization, at least in the analysis of biological material. Fused-silica capillary columns with bonded stationary phases [26] are increasing in popularity. The choice of the detector depends on the purpose of the analysis. In situations where estimating the purity of a particular drug is not the primary objective, selective detectors are preferred: nitrogen detectors, electron-capture detectors, or mass spectrometers with deuterated analogs of the drug as internal standard are most frequently applied.

20.2.1.3 Thin-layer chromatography

In the Fifties and Sixties planar chromatography used to be the leading technique in drug analysis. Paper chromatography was first used, but beginning with the Sixties, it was gradually replaced by TLC. Because of its simplicity and flexibility TLC continues to be the most popular technique in laboratories involved in drug synthesis, drug isolation from plant material, and toxicology [27]. The obvious advantages of TLC are low cost, simplicity, and the possibility of chromatographing a large number of samples in parallel. All these features allow TLC to be used for field applications. There are hundreds of selective and specific detection reagents that enable the characterization of compounds.

Regarding the sorbents, considerable interest has been shown recently in chemically bonded phases [28], which have replaced the impregnated layers formerly used in pharmaceutical analysis. However, most of the attention was focused on instrumentation [29]: In addition to devices for automated and precise sample application and devices for densitometry [30], instruments have appeared for forced-flow planar chromatography (PC). In combination with sorbents of small particle size (so-called HPTLC plates), which require less developing time than conventional plates, forced flow is obtained by applying either increased pressure (overpressured-layer PC [31]) or centrifugal force [32]. These instruments find application mainly in the pharmaceutical industry. With these modern instrumental techniques PC can compete with column techniques.

20.2.1.4 Additional techniques

Currently, three other techniques are of considerable interest in drug analysis: supercritical-fluid chromatography (SFC) [33], capillary zone electrophoresis [34,35], and electrokinetic chromatography [36,37]. SFC, particularly in capillary columns, finds application mainly in industrial laboratories, where it complements HPLC. Its advantage is seen in the possibility of more universal detection, important for the purity estimation of final products. Capillary zone electrophoresis and electrokinetic chromatography are promis-

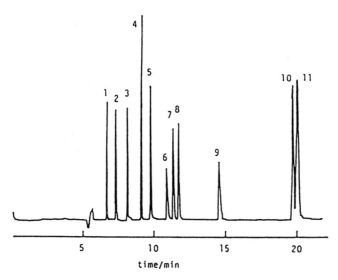

Fig. 20.1. Micellar electrokinetic chromatogram of a cold medicine. Capillary, 0.05 mm ID. 650 mm overall length, 500 mm to the detector; applied voltage, 20 kV; detection wavelength 220 nm. 1 = Acetaminophen; 2 = caffeine; 3 = sulpyrin; 4 = naproxen; 5 = guaiphenesin; 6 = impurity; 7 = phenacetin; 8 = ethenzamide; 9 = 4-isopropylantipyrine; 10 = noscapine; 11 = chlorpheniramine and tipepidine. (Reprinted from Ref. 36 with permission.)

ing methods for the analysis of biological material. Their advantages are the high number of theoretical plates (cf. Fig. 20.1) and the possibility of separating structurally related compounds with minimal sample size. The problem here is the lack of highly sensitive detectors.

20.2.2 Sample preparation

One of the areas of particular interest in drug analysis by chromatography is sample preparation. While no problems are encountered in the analysis of drugs in pure form, in pharmaceutical preparations and plants, in the analysis of biological materials, like blood, urine or tissues, considerable difficulties are encountered. This is due to the sample complexity, low drug concentration, and binding to proteins. The field of sample preparation is covered by a number of reviews [38,39], so that we can limit ourselves to the principles most frequently used in HPLC.

As long as the drug concentration is sufficiently high, e.g., in urine or saliva, it is possible to introduce the sample directly into the column. With plasma or serum direct sample application without any pretreatment is an exception. One of the options offered is the use of micellar mobile phases. In this case, the micelles prevent proteins from precipitating [40], but the chromatographic efficiency of this procedure is usually low. Another possibility is the use of special types of sorbents, as discussed on p. B396. Serum may be injected in some cases on reversed-phase columns packed with stationary phases generally less hydrophobic than C_{18} (PRP1 or C_1) when the content of organic modifier in the mobile phase is low [41,42] and, consequently, protein precipitation does not occur. Otherwise, deproteination of serum or plasma must be performed, either by precipitation, which, however, involves the danger of losses due to drug adsorption to the precipitate, or by ultrafiltration.

Most of the current methods require some special sample preparation before chromatography. The classical extraction with organic solvents is now used infrequently. Supercritical solvents have proved suitable for the extraction of plant material [43]. Solid-phase extraction of biological samples [44] is carried out in small columns or cartridges, packed with a suitable sorbent (silica gel, bonded silica, ion exchangers, polymers, immunosorbents [45], etc.), which enables a selective cleanup of complex matrices. These columns serve not only for the removal of ballasts (cleanup) and for concentrating the drugs but also, last but not least, for improving the selectivity of the assay. The drug in question is sorbed on the column, while the ballasts are eluted. In the next step, the sorbed drug is eluted by a suitable solvent. In order to prevent clogging of the column one can use an ionic surfactant, which solubilizes plasma proteins so that they are flushed out while the drugs are retained on the column [46]. An important factor in this type of sample preparation is the selection of a suitable internal standard.

There are basically three approaches in solid-phase extraction: (1) The simplest one is the off-line procedure, which is used when the number of samples is small or when then samples are prepared for GC or TLC. (2) For a large number of samples of similar type, to be analyzed by HPLC, the most suitable treatment is that which exploits semiautomated

References on p. B433

handling, e.g., the AASP (Advanced Automated Sample Processor, Varian [2]). In this equipment a cassette containing 10 cartridges (miniature extraction columns packed with a solid phase) is used for extraction. The cartridges, containing the samples, are inserted into an autoinjector, and up to 100 samples can thus be loaded on the column. The whole procedure is controlled by a microprocessor. (3) The third type of approach is repre-sented by on-line coupling of the extraction column (or precolumn) with the analytical column. The principle of most of the described procedures in this category is identical [38,47] and can well be illustrated by Fig. 20.2.

In the simpler, manual arrangement the sample loop is replaced by a precolumn, containing the same sorbent as the analytical column, but more coarsely grained (30 to 50 μm). When the biological fluid is applied to this precolumn by a syringe, the drug is retained on the column, while interfering compounds are eluted. By switching the injection valve, the drug can be transferred from the precolumn to the analytical column by means of the mobile phase [48]. The advantage of on-line techniques is the higher speed of analysis, higher accuracy, and minimum requirement of technical skill. A certain disad-

1) preconcentration step

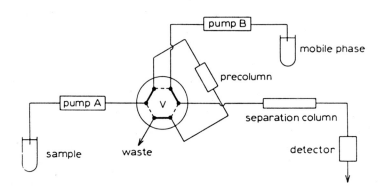

2) separation step

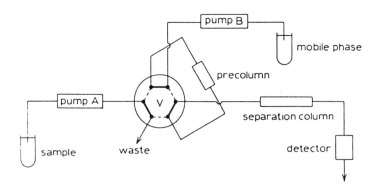

Fig. 20.2. Scheme of the column switching system for sample preparation. (Reprinted from Ref. 38 with permission.)

vantage is the occurrence of the so-called system peaks, which are due to disturbances of the system equilibrium.

Another interesting method of on-line sample preparation is the combination of zone electrophoresis with HPLC (so-called ZEST), in which the proteins of serum or plasma are quantitatively removed by on-line electrophoresis [49]. The method can also be automated and is therefore suitable for routine drug analysis [50].

One of the important steps in the routine drug analysis of biological materials is the preservation of samples for analysis. Because the samples in solution are frequently subject to decomposition, it is more convenient to use AASP cartridges. The samples sorbed in these cartridges can be stored without the danger of decomposition. For field samples intended for the determination of optimum drug levels dry blood spots can be used as well [51].

In analyzing drugs, particularly in urine, another problem is that, besides the parent drugs and their metabolites, their conjugates are also present. In most cases these conjugates are much more hydrophilic, and for their analysis it is necessary to use either gradient elution or to perform a parallel analysis of the free drug in the sample and the drug released by acidic or enzymic hydrolysis. Because of the problems accompanying precolumn hydrolysis, some authors prefer analysis of conjugates and utilize, e.g., a postcolumn reactor with immobilized enzymes, e.g., β-glucuronidase [52], for the determination. An additional complication is drug binding to serum proteins [53], which allows only the monitoring of free (unbound) drugs.

Derivatization is a large topic [54,55]. While in GC derivatization serves mainly to prepare volatile derivatives, in liquid chromatography (both planar and column) derivatization is applied to increase assay sensitivity and selectivity, to decrease sample complexity, and to improve separating conditions, particularly for enantiomer separation. Derivatization reactions can be performed before chromatography in order to prepare derivatives with more favorable chromatographic properties, or after chromatography if the aim is to achieve simpler or more sensitive detection [38,56]. Precolumn derivatization is usually carried out off-line. A promising new development in this regard is micelle-mediated precolumn derivatization, based on phase-transfer catalysis [57]. The main advantage of such procedures is that no extraction or solvent evaporation steps are required. In postcolumn reactors frequently UV irradiation [58,59] or a change in the pH of the eluate [60] is sufficient. Quite interesting, though of little practical use, is the application of biological detectors, represented, e.g., by isolated organ preparations [61].

20.2.3 Enantiomer separations

Next to sample preparation, the greatest attention in drug analysis nowadays has been given to chromatography of optically active drug forms. Although about one-half of the most frequently prescribed drugs contains a chiral center, most of them are still administered as racemates [62]. This contrasts with the fact that individual isomers can differ considerably in their pharmacological activity [63] and that drug metabolism is strictly stereoselective. A classical example is the case of thalidomide, the administration of which

References on p. B433

in its racemic form led to catastrophic consequences. The $S(-)$-isomer of this drug is responsible for the teratogenic effect, while the sleep-inducing effect is due only to the $R(+)$-form. Unfortunately, until recently no simple analytical methods capable of assaying the individual optically active forms were available.

For the purpose of chiral separation either HPLC, GC, or TLC may be used [64,65]. In the indirect mode enantiomers are converted to diastereomers by a reaction with a suitable optically pure chiral reagent before chromatography. Such diastereomers exhibit different physicochemical properties and can be chromatographically separated on non-chiral columns and without application of chiral mobile phases. Such approaches were particularly popular at the beginning of enantiomer separation, because they are relatively fast, they do not require special sorbents, and offer relatively good separation. A certain disadvantage is seen in the fact that during derivatization and subsequent chromatography racemization may occur to some extent (mainly in GC at higher temperatures) or that individual isomers may react at different speed. For the unequivocal identification of separated peaks the application of mass spectrometry is frequently inevitable.

For these reasons, direct methods have been introduced into drug chromatography more recently. In the direct mode separations are achieved mainly by using chiral stationary phases in HPLC, GC, or TLC. Of the numerous phases developed for HPLC [66] the attention is focused on the dinitrobenzoyl phases introduced by Pirkle et al. [67], on the phases bonded with proteins (most frequently α-acid glycoprotein [68]) or cyclodextrins [69], and on phases based on synthetic polymers. Procedures in which a chiral reagent is added to the mobile phase [70] are used much less. Since chiral separations are more difficult than nonchiral procedures, suitable internal standards are important. In HPLC, all conditions (temperature, pH, ionic strength, mobile-phase composition, etc.) must be carefully controlled.

20.3 INDIVIDUAL GROUPS OF DRUGS

Because in the past few years attention has focused upon drug analysis in biological materials, this section will deal mainly with these types of applications. Drugs are categorized according to their pharmacological effect, regardless of chemical differences between members of a category. The fundamental papers on the determination of the majority of these compounds were published during the late Seventies and early Eighties and have been reviewed in the last edition of this monograph [4]. In this chapter preference will be given to improvements in methodology, e.g., sample preparation, separation of enantiomers, and increased assay sensitivity. It would be beyond the scope of this chapter to include such groups as alkaloids, most steroids and vitamins, as well as industrial, toxicological, or forensic applications.

The chromatography of compounds with central nervous system activity, antibiotics, and autonomous and cardiovascular drugs has been of great interest recently. The individual categories of drugs chromatographed most frequently in 1988 are presented in Table 20.2, which is based on data published in the *J. Chromatogr.* Bibliography Section

TABLE 20.2

THE MOST FREQUENTLY CHROMATOGRAPHED DRUG CATEGORIES, PUBLISHED IN 1988

Drug categories	Total number of papers	Technique used		
		HPLC	GC	TLC
Central nervous system	423	282	62	79
Autonomic and cardiovascular	268	181	43	44
Antibiotics	239	197	4	38
Cytostatics	108	82	14	12
Other chemotherapeutics	174	109	45	20
Antirheumatics and anti- inflammatory drugs	71	57	6	8
Plant extracts	121	59	13	49
Toxicological applications	145	41	83	21

[3]. From this table the contribution of individual chromatographic techniques in the analysis of a particular category of compounds is obvious. With the exception of toxicological applications the most frequently used technique is HPLC.

20.3.1 Psychotropic drugs

Psychotropic drugs represent a large group of compounds acting on the central nervous system. Classification of these drugs is difficult, because many of them may also have other pharmacological properties. The great number of drugs introduced into the market in recent decades is reflected by an increasingly large number of papers devoted to their chromatographic analysis (Table 20.2). Precise and accurate methods are required, because psychotropic drugs are being administered for long periods, and their concentrations should stay within the therapeutic range. Moreover, many of these compounds are important from a toxicological point of view. Plasma levels of psychotropic drugs are usually in the ng/ml range, dictating relatively complex sample preparation techniques and the application of sensitive detection. GC methods enjoy the advantage of selective detectors (NPD, ECD, and MS) but have the drawback of requiring an extensive cleanup of plasma samples and often also derivatization. Sample preparation for HPLC may be simpler, and the UV detector is sufficiently sensitive for most applications. TLC techniques are also widely used, especially in toxicological applications.

20.3.1.1 Antidepressants

The monitoring of tricyclic and tetracyclic antidepressants has been reviewed recently [71-74].

References on p. B433

TABLE 20.3

METHODS FOR THE DETERMINATION OF ANTIDEPRESSANTS IN BIOLOGICAL FLUIDS

Drug	Sample and preparation*	Column	Detection (nm)	Detection limit	Ref.
(a) HPLC					
Amitriptyline, +3 metabolites	P. Column switching	CN	UV 215	5 ng/ml	92
Amitriptyline Nortriptyline	P. Column switching	C_{18}	UV 238	2.5 nmol/l	93
Clomipramine +4 metabolites	P.U. Back-extraction	Ultrasphere IP	Coulometric	0.2 ng/ml	75
Clomipramine Imipramine +metabolites	AASP	CN	UV 252	10 ng/ml	96
Doxepin	U. Back-extraction	Hexyl	UV 205	20 ng/ml	81
Nordoxepin	S. SPE	CN	UV 214	1 ng/ml	94
Imipramine Metapramine Trimipramine	P. Extraction	C_{18}	UV 254	<5 ng/ml	83
Imipramine Desipramine +metabolites	T. Extraction	C_{18}	UV 254	10 ng/g	95
Mianserin	P.T. Back extraction	C_{18}	UV 215	12.5 ng/ml	84
Trazodone	P.T. Back extraction	C_3	UV 214	5 ng/ml	85
(b) GC Amitriptyline Nortriptyline Imipramine Desipramine Doxepin	P (S). SPE	SP-2250	NPD	0.1 μmol/l	97
Clomipramine +metabolite	B. Back extraction	Methyl silicone	GC/MS	2 nmol/l	76
Desipramine Imipramine	P. Extraction	CBP1	GC/MS	50(100) pg	82
Maprotiline +metabolite	P.U. Extraction	Phenylmethyl-silicone	NPD	<10 ng/ml	77

TABLE 20.3 (continued)

Drug	Sample and preparation[*]	Column	Detection (nm)	Detection limit	Ref.
Nomifensine	P.T. Extraction	OV-17	GC/MS	1 ng/ml	78
Trimipramine +2 metabolites	P. Extraction	SE-30	GC/MS	1 ng/ml	79
Trimipramine +9 metabolites	U. Extraction	SE-54	GC/MS	?	80

[*] Abbreviations: B = blood, U = urine, S = serum, P = plasma, T = tissue, SPE = solid-phase extraction, AASP = Advanced Automated Sample Processor, PP = protein precipitation, DI = direct injection, F = fluorescence detection (excitation nm/emission nm), UV = spectrophotometric detection (nm).

Liquid/liquid extraction still remains the most popular sample preparation technique [75-85]. Typically, alkalinized plasma or urine is extracted with n-hexane, dichloromethane, or ethyl acetate and then either back-extracted with dilute acid or evaporated and processed further, as desired. This technique is gradually being replaced by solid-phase extraction [86-91] and column switching techniques [92,93].

In HPLC reversed-phase columns are most frequently used. In order to improve peak shape ion-pair reagents, mostly triethylamine [84,85,93], are often added to the mobile phase. Detection is accomplished with a UV detector in the range of 210-254 nm. Eight tricyclic antidepressants were quantitated simultaneously on a Supelcosil LC-PCN column by isocratic elution and UV detection at 254 and 280 nm [91]. The limit of detection was 25 ng/ml for all compounds studied.

For GC derivatization acetic [77], trifluoroacetic [78-80], or pentafluoropropionic anhydride [76] and N-trifluoroacetylimidazole [82] have been applied. For sensitive detection NPD [77] and ECD are used. The GC/MS combination is frequently used for this class of compounds [76,78-80,82]; deuterium-labeled compounds are most often employed as internal standards [76,78,82].

A selection of methods for the determination of individual antidepressants can be found in Table 20.3.

20.3.1.2 Benzodiazepines

Comprehensive reviews on this subject have been published by Boulton et al. [74] and by de Silva [98].

By virtue of the presence of halogen in the 7-position of the molecule and/or in the 2'-position of the 5-phenyl ring an ECD can be used in GC [99-103]. The NPD is also

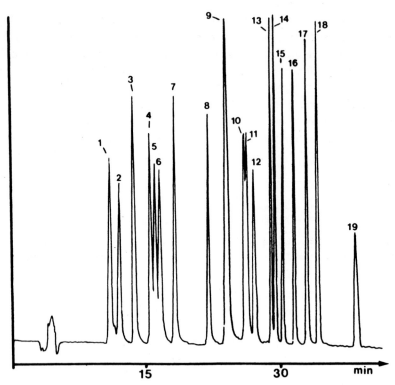

Fig. 20.3. Chromatogram of a plasma extract containing 1 μg/ml of each drug. Chromatographic conditions: μBondapak 5-μm reversed-phase column (300 x 4.6 mm ID); mobile-phase components, acetonitrile (A) and phosphate buffer (pH 5.40) (B); gradient elution, 0-15 min: A/B, 38:62, 15-22 min: gradient from A/B 38:62 to 70:30, 22-40 min: A/B 70:30; flowrate 0.7 ml/min. Peaks: 1 = bromazepam; 2 = oxazepam; 3 = estazolam; 4 = nitrazepam; 5 = chlordiazepoxide; 6 = triazolam; 7 = clonazepam; 8 = chlorazepate; 9 = desmethyldiazepam; 10 = tofizopam; 11 = clobazam; 12 = flunitrazepam; 13 = diazepam; 14 = loflazepate; 15 = desalkylprazepam; 16 = clotiazepam; 17 = tetrazepam; 18 = prazepam; 19 = medazepam. (Reprinted from Ref. 108 with permission.)

applicable [104]. When screening and positive identification of individual benzodiazepines is required, the combination of capillary GC with MS is the method of choice [105,106]. Twenty-nine 1,4- and 1,5-benzodiazepines and their metabolites in urine were identified after acetylation by means of their mass spectra and retention indices on OV-101 [106]. Seven benzodiazepines and four metabolites in plasma were determined in underivatized form (only hydroxy compounds were silylated) and quantified by SIM [105] in the 10- to 5000-ng/ml range.

The retention data of 16 benzodiazepines were determined by HPLC on silica, cyano, and amino phases [107]. Linear relationships between log k' and the mobile-phase composition were found to depend on solute and stationary-phase characteristics. The benzodiazepine molecule acted as a proton donor towards the adsorbent sites of the stationary phases investigated. The majority of the applicable methods employed reversed-phase (C_{18} or C_8) columns. Using a gradient of acetonitrile and a C_{18} column, Mura et al. [108] were able to separate 19 benzodiazepines in plasma and urine in 50 min (Fig. 20.3). The identity of individual compounds was confirmed with a diode-array detector; the sensitivity was 3-5 ng/ml. Another approach was chosen by De Giovanni and Chiarotti [109]. Samples of plasma and urine were hydrolyzed with β-glucuronidase and, after solid-phase extraction, the benzodiazepines were hydrolyzed in 6 M HCl at 100°C for 60 min to benzophenones, which were extracted and cyclized to 9-acridones with dimethylsulfoxide or sodium hydroxide. The column eluate was monitored with a fluorescence detector, and the derivatives were separated by isocratic elution with acetonitrile/1 mM acetate buffer (pH 4.0) (1:1, v/v).

20.3.1.3 Neuroleptics

Screening for phenothiazines and analogous neuroleptics is of importance in analytical toxicology. A method for the identification of 29 phenothiazines and analogous neuroleptics and their metabolites in urine after acid hydrolysis has been described by Maurer and Pfleger [110]. The acetylated extract was analyzed by computerized GC/MS, using SIM. For the interpretation of the mass fragmentograms, the individual mass spectra and retention indices were used.

Eight phenothiazines plus several thioxanthenes, dibenzazepines, and butyrophenones were separated by LC on a cyano column [111]. The drugs, which had been extracted from plasma, were detected with either a UV detector at 254 nm in concentrations down to 10 ng/ml or with an electrochemical detector, which was more sensitive. A HPLC method for the determination of chlorpromazine and 13 metabolites, based on isocratic elution, has been described by Smith et al. [112]. For the quantitation of chlorpromazine, whole blood was directly injected into a TSK HW-65F precolumn and, after removal of proteins, chlorpromazine was backflushed into an ODS column [113]. By this technique not only the amount of the drug bound to proteins was determined, but also that bound to cytomembranes. Enantiomers of thioridazine have been resolved on Spherisorb packing with bonded R,N-(α)-phenethyl-N-propylurea [114]. The method described by Hoffman et al. [115] was very sensitive for the determination of fluphenazine and its four metabolites. The limit of detection was 300 fmol for an electrochemical detector.

Haloperidol appears to have a therapeutic window between 6 and 20 ng/ml. Thus, detection limits < 1 ng/ml are required for pharmacokinetic studies. Using a 10-cm-long column with 3-μm packing to minimize dilution, Nilsson [116] achieved a sensitivity of 0.2 ng/ml. The GC/NPD method of Abernethy et al. [117] also offers such a sensitivity (0.5 ng/ml).

20.3.1.4 Lysergic acid diethylamide

Lysergic acid diethylamide (LSD) is an extremely potent hallucinogen: oral doses as low as 25 μg can cause central nervous system disturbances [118]. Therefore, very low concentrations of LSD and its metabolites in blood and urine must be analyzed. HPLC with fluorescence detection has been used to determine LSD in blood and urine [119,120], but the method is subject to interferences. A GC/resonance electron-capture ionization mass-spectrometric assay for LSD and N-demethyl-LSD in urine has been developed [121], in which the drug and its metabolite are converted to their N-trifluoro-acetyl derivatives prior to GC/MS analysis. Calibration curves for LSD and its metabolite are linear from 0.05 to 5 ng/ml and from 0.03 to 5 ng/ml, respectively.

20.3.2 Hypnotics and sedatives

The most important class in this category is that of the barbiturates, which are still prescribed in large amounts as sedatives and hypnotics. Their determination is carried out either for toxicological purposes or for monitoring of therapeutic concentrations. The determination of another class, the benzodiazepines, is discussed in Section 20.3.1.2.

20.3.2.1 Barbiturates

The chromatography of barbiturates has been extensively reviewed recently [122-124]. Various approaches for sample preparation have been proposed, including liquid/liquid extraction [125-138], solid-phase extraction [139], column switching [140,141], and pro-tein precipitation with perchloric acid [142] or acetonitrile [143,144].

All modes of chromatography are applicable to the separation of barbiturates. Using GC, barbiturates can be separated either as free acids [128-132] or in the form of alkyl derivatives [133-135,145,146]. For a long time separation of free acids was hampered by their strong adsorption to the column packing. The situation improved substantially with the advent of fused-silica columns with chemically bonded stationary phases [131]. The NPD is the detector of choice for the determination of barbiturates [128,131,132,146], although the FID provides acceptable sensitivity [129,130,133-135,145].

In recent years, a large number of HPLC method has been published, based mainly on reversed-phase systems with C_{18} columns. Detection is the main problem, as barbiturates in nonionized form at pH < 9 show extremely weak absorbance at wavelengths > 220 nm. In ionized form they show an intense band in the 240 nm region. The following solutions to the detection problem have been proposed: detection at low wavelengths [127,136,137,140-142,144], postcolumn addition of a base [125,147], and separation at alkaline pH on a non-silica-based column [138].

Because barbiturates have long been of concern to toxicologists, methods suitable for screening of barbiturates have been developed [124,142]. A diode-array detector collects the spectra of barbiturates before and after addition of alkali and thus produces spectra of the ionized and nonionized forms for the additional identification of an individual barbitu-rate [125]. An isocratic separation system is able to separate 18 barbiturates. Another isocratic system, developed for clinical pharmacology and forensic toxicology, separates 19 barbiturates [142]; detection is performed at 230 nm for thiobarbiturates (thiopental, thialbarbital) and at 198 nm for other barbiturates. The detection limit is within the 0.2- to 0.4-μg/ml range.

20.3.2.2 Other hypnotics

Into this group fall various compounds prescribed to a much lesser extent than the barbiturates. These drugs include glutethimide [148], meprobamate [149], methaqualone [150], methyprylon [151], bromisoval and carbromal [152,153], and chlormethiazole [154,155]. The need for determining these drugs in biological materials is limited.

20.3.3 Antiparkinsonics and skeletal-muscle relaxants

20.3.3.1 Antiparkinsonics

Antiparkinsonics include procycliden, orphenadrine, diphenhydramine, levodopa, carbidopa, and deprenyl. Phenothiazines are treated in the chapter on psychotropic drugs (Section 20.3.1.3). A screening method for the identification of antiparkinsonics and their metabolites was described by Maurer and Pfleger [156]. The acetylated extract was analyzed by computerized GC/MS, using SIM at m/z 86, 98, 136, 150, 165, 196, 197, 208. The individual compounds were characterized by mass fragmentograms and individual mass spectra and by retention indices on OV-101.

Procycliden has been determined by GC on packed columns with FID [157], NPD [158], and MS detection [159]. Capillary GC methods developed for the quantitation of orphenadrin in plasma or serum exhibit limits of detection of 5 ng/ml [160] and 2 ng/ml [161], respectively. For the determination of diphenhydramine the method of Lutz et al. [161] as well as that of Yoo and Axelson [162] are applicable. Using capillary GC and NPD, Juvance et al. [163] were able to quantitate deprenyl down to a level of 3 ng/ml. HPLC with electrochemical detection is the method of choice for the determination of levodopa [162,164-167] and carbidopa [164,167]; ion-pair reversed-phase chromatographic systems are preferably employed.

20.3.3.2 Skeletal-muscle relaxants

A rapid HPLC method for the determination of dantrolene and its metabolites 5-hydroxydantrolene, aminodantrolene, and acetamidodantrolene was published by Lalande et al. [168]. The sample is chromatographed after protein precipitation, and quantitation can be performed in the range of 1-10 μg/ml. Better sensitivity (20 ng/ml) can be achieved by extraction of plasma with a chloroform/1-butanol mixture [169]. Baclofen and its γ-hydroxy metabolite were measured by HPLC in urine [170], and baclofen alone in plasma [171] and in cerebrospinal fluid [172]. For the latter method two procedures are available: (1) direct injection and UV detection at 220 nm (detectable concentrations down to 50-100 ng/ml) and (2) extraction, derivatization with phenylisothiocyanate, and detection at 254 nm (limit of detection, 5-10 ng/ml). Baclofen enantiomers can be separated

References on p. B433

after derivatization with o-phthalaldehyde in the presence of N-acetyl-L-cysteine by reversed-phase chromatography [173].

20.3.4 Antiepileptics

Antiepileptics can be grouped into several classes, including barbiturates (Section 20.3.2.1), primidone, hydantoins (phenytoin, mephenytoin), succinimides (ethosuximide, methsuximide, phensuximide), carbamazepine, valproic acid, benzodiazepines (clonazepam), and progabide. They exhibit narrow therapeutic ranges, and this makes drug monitoring essential [174,175]. Their therapeutic plasma levels are in the range of 1-100 μg/ml, except for clonazepam with levels in the ng/ml range.

Very often a combination of two or more antiepileptics is administered simultaneously. Therefore, the determination of these agents in a single chromatographic system is highly desirable for either HPLC [176-196] or GC methods [197-199]. Antiepileptics can be chromatographed by GC either in underivatized form [198-201] or as methyl, ethyl [197,200,202,203], TBDMS [204], or BSTFA [205] derivatives.

HPLC methods employ almost exclusively C_{18} stationary phases [176,179,180-183,206-209] and less frequently C_1 [178], C_8 [177,183,210], and CN [184] columns. Detection is performed mostly in the 195- to 210-nm range.

20.3.4.1 Valproic acid

Valproic acid is a short-chain volatile carboxylic acid, and thus it can be determined by GC either in the free state or following derivatization. Up to 12 metabolites of valproic acid have been assayed as TBDMS derivatives [204] by capillary GC/MS. Semmes and Shen [205] published an assay for the determination of $E\text{-}\Delta^2$-valproic acid, the parent drug, and eight other metabolites after derivatization with BSTFA, by capillary GC with a FID.

Valproate exhibits a weak UV absorbance, but HPLC methods based on UV detection at 210 nm have nevertheless been developed [207,211]. Their sensitivity is moderate (3 μg/ml), but sample preparation is simple (protein precipitation only). Better sensitivity can be obtained after attachment of a suitable chromophore or fluorophore to the carboxylic function, e.g. with 4-bromomethyl-7-methoxycoumarin [211a].

20.3.4.2 Carbamazepine

Two methods allowing direct injection of blood and serum into the LC column have been reported. Tamai et al. [208] injected whole blood into a precolumn, where hemolysis occurred. The precolumn was washed and the carbamazepine retained was transferred to an analytical column by column switching. The other approach, by Shihabi and Dyer [212], was simpler. They injected serum directly into an analytical column packed with a C_1 stationary phase of 30-nm pore size. Because of the large pores and low hydrophobicity of this column, protein precipitation does not occur. For the determination of carbamazepine metabolites Refs. 210, 213, and 214 should be consulted.

20.3.4.3 Phenytoin

Three procedures for the determination of phenytoin in urine were compared by Juergens et al. [215]: GC after derivatization, a HPLC extraction procedure, and HPLC with column switching. The last one proved to be the simplest and most accurate method. Lum et al. [206] separated phenytoin and its five metabolites on a C_{18} column by gradient elution. Electrochemical immunoassay, combined with HPLC, was developed by Sayo et al. [216]. An interesting approach to sample preparation was chosen by Johansson [209]: phenytoin-specific polyclonal immunoglobulin was attached to a silica packing in a precolumn, serum was injected, and, after a precolumn wash, the retained phenytoin was transported to the analytical column.

20.3.4.4 Other anticonvulsants

Ascalone et al. [217] described an assay for measuring progabide and its acid metabolite by HPLC on a silica column with UV detection. The limit of quantitation was 50 ng/ml. More sensitive (but also more complex) is the method by Padovani et al. [218], where a limit of quantitation of 1 ng/ml is achieved by using electrochemical detection. Separation of primidone and its metabolite, phenylethylmalonamide, can be achieved by reversed-phase HPLC [219]: the other metabolite, phenobarbital, is removed during the extraction procedure.

20.3.5 Cardiovascular drugs

Diseases of the cardiovascular system are the most frequent cause of death. Consequently, research in this field has attracted a great deal of attention. In the last 25 years a tremendous variety of new cardiovascular drugs has been developed and introduced into the market. All these new drugs have required analytical methods for pharmaceutical applications, pharmacokinetic studies, and therapeutic drug monitoring. For some recent reviews Refs. 220 and 221 should be consulted.

20.3.5.1 Cardiac glycosides

Among the substances used for the treatment of congestive heart failure and certain anomalies of the cardiac rhythm the cardiac glycosides (especially digoxin) are still prominent. These drugs are highly efficient at plasma levels within the range of 0.5-2 ng/ml. Overdoses may be fatal, and thus, monitoring them is of utmost importance. In the past these substances were usually determined by immunoassays, as HPLC with UV detection was not sensitive enough. However, because the immunoassays lack specificity, later on, HPLC was used to separate drugs from their metabolites, and quantitation was performed off-line by immunoassays [222,223]. More recently, a postcolumn reaction system has been developed. It converts digoxin to fluorescent derivatives [224], which can be detected with acceptable sensitivity. As sample cleanup plays an important role at these low plasma levels, some automated procedures have been proposed. In a method described by Reh [45] serum was injected into an anti-digoxin immunoadsorptive cleanup cartridge, connected to a reversed-phase column via a switching valve. The effluent was mixed with hydrochloric acid prior to fluorescence detection; the limit of detection was claimed to be 300 pg/ml.

20.3.5.2 Antiarrhythmics

In this group the most frequently chromatographed compounds are amiodarone, disopyramide, flecainide, mexiletine, procainamide, and quinidine. Except for immunoassays for some antiarrhythmics, like procainamide and quinidine, HPLC is currently the technique of choice for these drugs. Systems suitable for the simultaneous determination of various antiarrhythmics were elaborated by several authors [225,226].

Amiodarone can be chromatographed on C_{18} columns with methanol as the mobile phase with additions of aqueous ammonia [227,228]. Fluorescence detection (ex: 310 nm, em: 380 nm) allows flecainide to be determined in the range of 5-50 ng/ml [229]. A comparison of HPLC with fluorescence detection with a fluorescence polarization immunoassay, performed by Woollard [230], demonstrated good agreement between the two independent technologies.

Derivatization with trifluoroacetic anhydride and capillary GC with a NPD was the method used for determining disopyramide and its mono-dealkylated metabolite [231]. A simpler HPLC procedure [232] is capable of determining the parent drug alone. Enantiomers of disopyramide and its metabolite can be separated according to Le Corre et al. [233]. Two columns are connected in series: a reversed-phase C_8 column separates the racemate of the drug from that of the metabolite and an α_1-acid glycoprotein column separates the individual enantiomers (Fig. 20.4).

For the determination of mexiletine in plasma [234] solid-phase extraction on Clin-Elut silica was used. Mexiletine and its two major hydroxylated metabolites were determined in human liver microsomes [235] and in serum [236]. Enantiomers of mexiletine can be separated on a Pirkle-type 1A column. Plasma and urine are deproteinized, and mexiletine is derivatized to the fluorescent 2-naphthoyl derivative [237].

Procainamide and its N-acetyl metabolite were assayed in whole blood by using a column switching system [238]. Blood was injected into a Butyl Toyopearl 650-M column, which retained proteins, hydrophobic compounds, and blood cytomembranes. Hydrophilic compounds, such as procainamide and its metabolite, were eluted into an analytical column and were determined after column switching. Yamaji et al. [239] reported a determination of these compounds by packed GC/NPD with a limit of detection of 0.5 μg/ml. Quinidine can be assayed by several HPLC procedures [240,241]. Good correlation was obtained between a fluorescence polarization immunoassay and HPLC [240].

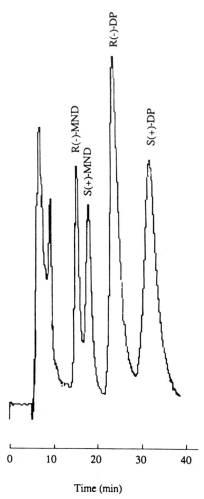

Time (min)

Fig. 20.4. Chromatogram of a plasma extract, containing 1 μg/ml of each R(–)-disopy-ramide (DP) and S(+)-DP, and 0.5 μg/ml of each R(–)-mono-N-desisopropyldisopyramide (MND) and S(+)-MND. Chromatographic conditions: Two columns, coupled in series: Supelcosil LC-8-DB (50 x 4.6 mm ID) and LKB EnantioPac (100 x 4 mm ID); mobile phase, 8 mM phosphate buffer (pH 6.20)/2-propanol (92:8); flowrate, 0.3 ml/min; detection, 220 nm. (Reprinted from Ref. 233 with permission.)

20.3.5.3 β-Adrenoreceptor antagonists

β-Adrenoreceptor antagonists are widely used in the treatment of cardiac arrhythmias, hypertension, angina pectoris, and thyrotoxicosis, but they are also sometimes used as doping agents in sports. Assays sensitive down to 1-20 ng/ml may be required for their determination, since plasma levels of these drugs are usually low. Relatively complex GC/ECD methods have been mostly replaced by simpler HPLC methods. Most β-blockers

TABLE 20.4

HPLC METHODS FOR THE DETERMINATION OF β-BLOCKERS IN BIOLOGICAL FLUIDS

For abbreviations, see footnote to Table 20.3.

Drug	Sample and preparation	Column	Detection (nm)	Detection limit	Ref.
Propranolol	U. SPE	C18	UV 260 diode-array	2 ng	247
Pindolol Oxprenolol				2 ng 13 ng	
Propranolol enantiomers	S. Extraction + derivatization	C8	F 220/ > 300	100 pg	248
Propranolol enantiomers	P. Extraction + derivatization	C18	F 290/335	1 ng/ml	249
Propranolol + 5 metabolites	S. Back-extraction	C18	F 300/375	2.5 ng/ml	250
Propranolol + 1 metabolite	P. Extraction		F 238/ > 360	2 ng/ml	251
Sotalol	S. SPE	C18	UV 235	20 ng/ml	252
Sotalol	P.U. SPE	C18	F 240/310	10 ng/ml	253
Atenolol	P. Column switching	C18	F ex.197	15 ng/ml	254
Atenolol enantiomers	Chiral derivatization	silica or C18	UV 280	?	255
Alprenolol Metoprolol	P. SPE	C18	F 195/320	5 ng/ml	256
Alprenolol	S. Extraction	C18	F 200/ > 300	330 pg/ml	257
Metoprolol + 1 metabolite	U. DI	Phenyl	F 220/318	20 ng/ml	258
Metoprolol enantiomers	P. Extraction + chiral derivatization	C18	UV 222	10 ng/ml	259
Metoprolol enantiomers	P.U. Extraction + chiral derivatization	C18	F 265/313	2 ng/ml	260
Betaxolol enantiomers		Chiracel OD	UV 273 F 270/310		261

TABLE 20.4 (continued)

Drug	Sample and preparation	Column	Detection (nm)	Detection limit	Ref.
Betaxolol	P.S. Extraction	Silica	F ex.195	5 ng/ml	262
Nadolol	S. PP	C_{18}	F 227/299	5 ng/ml	263

contain a phenoxy or aryloxy group, which exhibits a high fluorescence intensity. Thus, fluorescence detection is the most popular detection technique for this category of compounds. Most β-blockers are racemic mixtures and, therefore, there is a need for separating the individual enantiomers. A book devoted to the determination of β-blockers in biological materials has been published recently [242].

Capillary GC/MS has been used in doping tests for screening urine samples for the presence of β-blockers. After hydrolysis of conjugates and extraction, the β-blockers were converted to trimethylsilyl/trifluoroacetyl [243], pentafluoropropionyl [244], or trifluoroacetyl [245] derivatives. In the latter paper, liquid/liquid extraction was compared with solid-phase extraction on different extraction columns; Extrelut-1 proved to be the most suitable one. TLC with Fast Black K salt detection [246] was employed for the determination of eleven β-blockers in urine and liver extracts.

In recent years, many HPLC methods have been applied to pharmacokinetic studies and to the therapeutic drug monitoring of propranolol, metoprolol, sotalol, atenolol, betaxolol, nadolol, and other β-blockers. Some data are presented in Table 20.4.

20.3.5.4 Vasodilators

Nitroglycerine and isosorbide dinitrate must be determined in very low concentrations (< 1 ng/ml), and therefore, very sensitive methods for their determination are required. GC/ECD is mostly used for this purpose. Detection limits of 50 pg/ml [264,265] and 25 pg/ml [266] have been reported. Two of the procedures mentioned [264,266] are also applicable to metabolites of nitroglycerine in biological fluids.

20.3.5.5 Calcium antagonists

Verapamil and norverapamil can be determined in plasma by HPLC with fluorescence detection in the range of 1-1000 ng/ml [267]. Bremseth et al. [268] separated verapamil and its four metabolites on an alkyl-phenyl column, using a mobile phase of 0.0005% sulfuric acid in methanol. Another approach was proposed by Pieper and Rutledge [269]. Verapamil and its three metabolites were separated on a silica column by ion-pair adsorption with an inorganic counterion, NaBr, in the mobile phase.

Nifedipine was separated by HPLC in a normal-phase system on a silica column. Therefore, after plasma extraction into an organic solvent, sample evaporation was not needed [270]. Light had to be excluded throughout the analysis, as nifedipine is sensitive to photodecomposition. Nifedipine and its three metabolites can also be determined by capillary GC with on-column injection [271]. This obviates oxidative degradation by metal surfaces.

Diltiazem and its four metabolites can be determined by HPLC [272]. An alternative procedure is GC/ECD [273], which separates diltiazem and its desacetyl metabolite with a detection limit of 3 ng/ml. After solid-phase extraction [274], flunarizine can be determined by HPLC with detection at 250 nm at a detection limit of 13 ng/ml.

20.3.5.6 Antihypertensives

A robotic system that tests the content uniformity of antihypertensive tablet formulations has been described recently [275]. The system consists of a robot, integrated with programable HPLC equipment and a laboratory computer; it allows the automation of sample preparation, HPLC analysis, and report generation. It also selects HPLC columns, mobile phases, and pump and detector parameters.

Some antihypertensives are administered in very low doses, making their determination difficult; e.g. clonidine is given in amounts as low as 100 μg. GC/MS with SIM was used for its sensitive determination [276,277]. By converting clonidine to fluorinated derivatives, a detection limit of 5 pg/ml plasma was obtained [276]. A deuterated analog was used as an internal standard. Arrendale et al. [277] chromatographed clonidine as TBDMS derivative.

Debrisoquine is now used mainly as a model compound for the evaluation of genetic polymorphism in oxidative metabolism. Moncrieff [278] determined debrisoquine and its 4-hydroxy metabolite by HPLC on a CN column. Urine was directly injected into a C_{18} precolumn. Chan [279] compared GC and HPLC methods for the determination of these compounds and found GC to be more sensitive than HPLC; but HPLC was simpler, and its sensitivity was adequate.

HPLC with electrochemical detection can be used for the determination of guanethidine [280] and methyldopa [164,281]. For labetalol, several HPLC procedures were described. A detection limit of 1 ng/ml blood was achieved by Ostrovská et al. [282], who also discussed general guidelines for optimization of fluorescence detection. Similar sensitivity was achieved by Luke et al. [283] with a PRP1 column and fluorescence detection. Problems of determining very low plasma levels are also associated with prazosin. Using HPLC, limits of fluorescence detection of 0.2 ng/ml [284] and 1 ng/ml [285] were achieved. GC/NPD [286] as well as HPLC with fluorescence detection [287] were proposed as assay methods for dihydralazine in biological materials.

20.3.5.7 Diuretics

Several reports describe working conditions for the chromatography of diuretics. A screening procedure for 12 diuretics in urine was published by Fullinfaw et al. [288], who used gradient reversed-phase HPLC with detection at 271 nm. The effect of chainlength of

chemically bonded phases on the retention behavior of 20 diuretics was examined by De
Croo et al. [289]. For the screening of 17 diuretics HPTLC on silica gel plates was used
with neutral, basic, and acidic mobile phases [290]. Several spray reagents were compared for the detection of diuretics on thin-layer plates [291].

A selection of methods for the determination of diuretics is presented in Table 20.5.

TABLE 20.5

HPLC METHODS FOR THE DETERMINATION OF DIURETICS IN BIOLOGICAL FLUIDS

For abbreviations see footnote to Table 20.3.

Drug	Sample and preparation	Column	Detection (nm)	Detection limit	Ref.
Amiloride	P. Extraction	C_{18}	F 368/415	0.5 ng/ml	294
	P.U. SPE	C_{18}	UV 365	<1 ng/ml	298
Chlorthalidone	B. Extraction	C_{18}	UV 214	62 ng/ml	292
Chlorthalidone	P. Extraction	C_8	UV 235	50 ng/ml	293
Furosemide	P. Extraction U. PP	C_{18}	F 235/389	10 ng/ml	299
Hydrochloro-thiazide	P. Extraction	C_{18}	UV 271	10 ng/ml	294
	S. Extraction	C_{18}	UV 271	10 ng/ml	295
Spironolactone +3 metabolites	P. Extraction	C_{18} Gradient elution	UV 254	12.5 ng/ml	296
	S. Extraction	Silica	UV 240	50 ng/ml	297

20.3.6 Antiasthmatic drugs

A comprehensive review on the chromatography of antiasthmatic drugs was published
by Kucharczyk and Segelman [300]. Antiasthmatic drugs can be classified according to
their mode of action as bronchodilators, anti-inflammatory steroids, and asthma prophy-
lactics. In this section the emphasis is on the most important class, the bronchodilators.

20.3.6.1 β-Adrenergic stimulants

The synthetic catecholamines (albuterol, terbutaline, fenoterol) belong to this group.
Their therapeutic levels fall in the range of 0.5-20 ng/ml, and consequently, sensitive
detection methods are required for their determination in plasma. HPLC methods with
electrochemical detection are widely used for this purpose [301-304]; they take advantage

of the sensitivity provided by the oxidizable phenolic hydroxyl group in the catecholamine molecule. For the extraction of biological samples solid-phase extraction columns with C_8 [302] and C_{18} [301,303-305] packings are used almost exclusively.

Emm et al. [301] applied HPLC with an electrochemical detector, equipped with two detector cells, to the determination of albuterol in serum. The potential of the first cell was set to + 0.5 V, so that it selectively oxidized endogeneous substances without affecting the albuterol molecule: the potential of the second cell was set to + 0.8 V. With this arrangement albuterol may be quantitated down to 1 ng/ml.

A relatively complicated column switching system, comprising three columns, was used for the determination of racemic terbutaline in plasma [304]. Even with this arrangement a manual step, entailing solid-phase extraction, was required to eliminate interfering substances and achieve a limit of detection of ca. 0.9 ng/ml. Nevertheless, the method is less costly than the GC/MS procedures usually employed. For instance, in the method of Maes [305], TMS derivatives of terbutaline, albuterol, and fenoterol were separated by capillary GC/MS on CP-Sil 8 as the stationary phase, and deuterium analogs were used as internal standards. The limits of detection were 200, 10, and 500 pg/ml, respectively. Consequently, GC/MS is the method of choice in cases where sensitivity is of prime importance, e.g., in pharmacokinetic studies.

Many β-adrenergic agents are used clinically as a racemic mixture, but the drug activity resides mainly in the $R(-)$-enantiomers. The enantiomers of albuterol were separated on an EnantioPac column by Tan and Soldin [302] and determined in human urine. Enantiomers of terbutaline were separated by coupled-column HPLC, where column switching was employed to pump different eluents through the phenyl and β-cyclodextrin columns [303]. Individual enantiomers were detected either by LC/MS with deuterium-labeled internal standards or by electrochemical detection.

20.3.6.2 Phosphodiesterase inhibitors

The maximal bronchodilating effect of theophylline is observed with plasma concentrations in the range of 10-20 μg/ml, while a range of side-effects is seen with plasma levels > 20 μg/ml. This importance of therapeutic drug monitoring may be deduced from this narrow therapeutic range. GC procedures for the determination of theophylline are laborious, involving multiple solvent extractions or chemical derivatization, and are no longer very popular. The GC/MS combination, which may by useful for some special tasks in pharmacokinetic studies, is an exception [306]. Reversed-phase HPLC with UV detection in the 280-nm region now seems to be the most prevalent method. A common problem in the HPLC assay of theophylline is the potential for interference from concomitantly administered drugs and from metabolites, especially 1,7-dimethylxanthine, the major metabolite of caffeine. Chiou et al. [307] solved this problem by addition of N,N-dimethylformamide to the mobile phase; baseline resolution of theophylline and 1,7-dimethylxanthine was achieved on a C_{18} column. Other methods for the determination of theophylline and its metabolites include an automated determination of theophylline and nine purine derivatives in human serum by column switching HPLC [308]. The analysis was performed on a

C_{18} column by stepwise gradient elution with 10 and 18% methanol in 0.1 M NaH$_2$PO$_4$. All these purines can be determined simultaneously within 40 min. An automated HPLC assay has been developed for the determination of theophylline and its three metabolites in urine [309]. The method involves direct injection of urine into a reversed-phase column, followed by gradient elution and detection at 280 nm at a limit of detection of 1 μg/ml. Theophylline and diphylline can be separated within 6 min on a C_{18} column [310]. Sample preparation is simple – precipitation of serum proteins by trichloroacetic acid.

20.3.7 Anesthetics

Anesthetics include a variety of drugs differing in chemical structure. According to the route of administration they can be divided into inhalational anesthetics (e.g., halothane), intravenous anesthetics (e.g., barbiturates, benzodiazepines, morphine), and local anesthetics (e.g., lidocaine, bupivacaine, articaine). The quantitative analysis of common anesthetic agents has been reviewed extensively [311]. In this section only recent papers on halothane, articaine, lidocaine, and bupivacaine are discussed. For barbiturates and benzodiazepines, see Sections 20.3.2.1 and 20.3.1.2.

GC headspace analysis is a convenient method for the determination of inhalational anesthetics. Koupil et al. [312] compared two quantitation methods for the determination of halothane in blood by this technique: absolute calibration and standard addition. Only the last one gave reliable results: the absolute calibration was inaccurate due to variable amounts of lipids and other compounds in blood.

The most recent methods for the determination of the local anesthetic agents, lidocaine, bupivacaine, and articaine are based on HPLC, although many GC methods were described in the past [311]. Tam et al. [313] described an isocratic HPLC method for determination of lidocaine and its nine metabolites in plasma and urine. The metabolites were acetylated and extracted from an alkaline solution with ethyl acetate. The detection limit of the method was 0.1 μg/ml. Other methods have been published for articaine and its 2-carboxy metabolite [314], bupivacaine and two metabolites [315], and a column switching method for lidocaine and two dealkylated metabolites [316]. Enantiomers of bupivacaine can be resolved on the α_1-acid glycoprotein column Enantiopac [317].

20.3.8 Antibiotics

Rapid development of new antibiotics necessitates their quantitation during isolation from fermentation broths and in studies of toxicology, pharmacokinetics, and bioavailability. Monitoring their plasma levels is of crucial importance, as many antibiotics have a narrow therapeutic window, generally in the concentration range of 1-100 μg/ml of plasma. TLC is now used mainly as a screening method. HPLC procedures are enlisted increasingly for the determination of antibiotics, being more rapid and specific than traditional microbiological techniques. The detection and quantitation by HPLC of various metabolic degradation products of antibiotics is also seen as an advantage. Chromatographic procedures have been reviewed on several occasions [318-321]. Cephalosporins,

TABLE 20.6

HPLC OF SOME CEPHALOSPORINS

Compound	Sample and preparation*	Stationary phase	Detection (nm)	Detection limit	Ref.
Cefpiramide	P.U. Column switching	C_{18} Ion pair	UV 270	0.25 μg/ml	325
Cefpiramide	S. PP (MeOH) U. DI	C_{18}	UV 254	0.92 μg/ml	326
Cefodizime	P.S. PP (MeOH) Bile, feces, tissue	C_{18} Ion pair	UV 264	0.1 μg/ml	327
Ceftetrame	P.U. SPE	Phenyl	UV 225	0.48 μg/ml	328
Cefixime	S. PP (TCA) U. PP (TCA)	C_{18}	UV 280 UV 313	0.05 μg/ml	329
Cefotaxime + metabolite Ceftriaxone	P. PP (MeOH)	C_8	UV 254	0.25 μg/ml	330
Cefatamet	P. PP (HClO$_4$)	C_{18}	UV265	0.2 μg/ml	331
Cefotaxime + metabolite	S. PP (MeCN)	C_{18} Ion pair	UV 254	0.25 μg/ml	332
Cephalexin	S. PP (HClO$_4$) U. Dilution	C_{18}	UV 254	3 ng	333
Cephalexin	Skin. Extraction	Phenyl Ion pair	UV 260 Diode array	100 ng/g	334
Cephalexin	P. PP (HClO$_4$)	C_{18}	UV 254	5 ng/ml	335
Cefadroxil	S. PP (HClO$_4$)	Nucleosil SA	UV 240	1 μg/ml	336
Cefaclor Cephadrine	S. PP (MeCN)	C_{18} Ion pair	UV 265	1μg/ml	337
Cefmenoxime	S. PP (MeCN)	C_{18} Ion pair	UV 254	0.625 μg/ml	338
Cefonicid	P.S. PP (MeCN) U. DI	Phenyl Ion pair	UV 254	0.5 μg/ml	339
Cefpimizole	T. SPE	C_{18} Ion pair	UV 254	?	340

TABLE 20.6 (continued)

Compound	Sample and preparation*	Stationary phase	Detection (nm)	Detection limit	Ref.
Cefoperazone	P.U. PP (MeCN)	C_{18} Ion pair	UV 254	0.5 μg/ml	341
Cefamandole	S. PP (MeCN) U. Dialysis fluid DI	C_{18} Ion pair	UV 254	0.5 μg/ml	342
Ceftriaxone	P.U. PP (MeCN) + extraction	C_{18} Ion pair	UV 270	0.5 μg/ml	343
Cefmetazole	S.U. SPE	C_{18}	UV 280	0.2 μg/ml	344
Cefazolin	S.T. Extraction	Phenyl Ion pair	UV 230	10 ng/ml	345
Cefaloglycine Cefroxadine	P. PP (EtOH) U. DI	TLC	Deriv. F 365/460	8 ng	346

*TCA = trichloroacetic acid, for other abbreviations see footnote to Table 20.3.

penicillins, cyclosporin, and anticancer antibiotics (see Section 20.3.9) are among the compounds of great current interest.

20.3.8.1 Cephalosporins

All HPLC methods for the determination of cephalosporins in biological material are based on similar principles (Table 20.6). Being highly polar, they cannot be extracted by organic solvents and, therefore, sample preparation consists mainly of protein precipitation with various reagents (acetonitrile, methanol, perchloric acid) or, less frequently, solid-phase extraction. Cephalosporins are chromatographed almost exclusively on reversed-phase columns; sometimes ion-pair reagents are added to the mobile phase. UV detection is performed at 220-280 nm, depending on the absorption maximum of the cephalosporin investigated and/or the absorption of interfering endogenous compounds. Several methods have been developed for the simultaneous determination of some cephalosporins [322,323]. Thirteen cephalosporins were resolved on a C_{18} column [322] by isocratic elution with methanol/water/acetic acid (300:700:1). The compounds were detected at 254 nm.

A postcolumn fluorescamine derivatization procedure and HPLC were used for the determination of cephalosporins [324] having an α-primary amino group in their side chain (cefaclor, cephalexin, cephradine, cefroxadine, cefaloglycine, and cefadroxil) in blood and urine. Fluorescence detection was twice as sensitive as UV, and the selectivity was also greater.

20.3.8.2 Penicillins

Due to the amphoteric character of these compounds it is difficult to extract them into organic solvents and, thus, sample preparation is similar to that described for cephalosporins. Another problem from the analytical point of view is the lack of specific chromophores. Penicillins are therefore either monitored at low wavelengths or they are converted to UV-absorbing or fluorescent derivatives.

Postcolumn derivatization was reported for the determination of amoxicillin and its three metabolites in human urine [347]. They were separated from the endogenous compounds on a C_{18} column by using sodium heptylsulfonate as an ion-pairing agent and methanol as a mobile-phase modifier. The derivatization reagent (1.5 *M* sodium hydroxide plus 0.02% sodium hypochlorite) was mixed with the eluent at ambient temperature. The degradation product(s) of each compound were detected at 270 nm. Derivatization with fluorescamine was used by Carlqvist and Westerlund [348]; the detection limits for amoxicillin were 10 and 25 ng/ml in plasma and urine, respectively. The sample preparation could be automated by column switching.

Some limitations of postcolumn derivatization (bandbroadening, use of another pump) were eliminated by Haginaka and Wakai [349]. They described a HPLC method based on a hollow-fiber membrane reactor for the determination of penicillins. The method involves the separation of penicillins on a C_{18} column, postcolumn reaction with NaOH and $HgCl_2$, introduced into the eluent stream in a sulfonated hollow-fiber membrane reactor immersed in each solution. The method was applied to assays of ampicillin and its metabolites in human serum and urine. It had a the detection limit of 1-5 ng.

A sensitive method for determination of benzylpenicillin in milk was reported by Wiese and Martin [350]. Milk was extracted and concentrated, and benzylpenicillin was derivatized to the mercuric mercaptide of benzylpenicillenic acid. HPLC with detection at 325 nm and elimination of interfering peaks by digital subtraction resulted in an analysis with a detection limit of 0.2 ng/ml. The same principles were applied also to the analysis of plasma and lymph [351].

Penicillin derivatives in pharmaceutical preparations [352] can be separated from their degradation products and additives on HPTLC Silica Gel G plates. The plates were developed in a linear chamber, dried, and exposed to iodine vapor. The penicillin spots were measured on a spectrodensitometer at 290 nm. Procaine and procaine penicillin were measured at 360 nm.

The separation and determination of trace amounts of 6-aminopenicillanic acid and penicillins G, V, X, K, and penicillin G penicilloic acid in fermentation broths can be achieved by HPLC following precolumn derivatization with 1-hydroxybenzothiazole [353].

The effect of column temperature on retention and resolution of six penicillins in HPLC was studied by Martín et al. [354]; the effects of pH and acetonitrile concentration of the eluents on the retention and resolution of three isoxazolyl penicillins were investigated by Hung et al. [355]. Several methods were described for ampicillin [356,357], azlocillin [358], and for four acylaminopenicillins, three quinolones, imipenem, and cefixime [359].

20.3.8.3 Cyclosporin A

Cyclosporin A (CsA) is a potent immunosuppressive agent, which is effective for the prevention of acute allograft rejection of transplanted organs. Because of its severe side-effects and the need for maintaining exact blood levels of CsA over a long period of time, therapeutic drug monitoring is required in most medical facilities.

Many of the methods reported lately involve similar separating conditions for the determination of CsA alone [360-363] or in the presence of its metabolites [364-367]: reversed-phase HPLC at elevated temperatures (50-75°C) and UV detection at low wavelengths (205-241 nm). Their detection limits are usually in the 10- to 50-ng/ml range. A summary of these methods can be found in the paper by Wallemacq and Lesne [368]. The effect of temperature on peak shape was studied by Sawchuk and Cartier [360]. Only small changes in retention time were observed as temperature was increased, but peak shapes improved significantly. Different approaches to sample preparation have been reported: liquid/liquid extraction [361,365], solid-phase extraction [363,364,366,368], and protein precipitation, followed by column switching [362,367].

Oka et al. [369] described a determination of CsA in serum by normal-phase HPLC on silica gel. The mobile phase was 3.3 M ammonium hydroxide/ethanol/n-hexane (0.31:10.69:89). The extraction of CsA from serum was achieved by rapid-flow fractionation on a short diatomaceous earth column, eluted with diethyl ether/n-hexane (1:1). The detection limit of the method was 10 ng/ml.

CsA was often measured by RIA methods. The older RIA methods with polyclonal antibodies gave higher results than the HPLC methods because of metabolite interferences [361,370]. However, comparing cyclosporin concentrations in whole blood, as measured by HPLC and by RIA with a monoclonal antibody specific for cyclosporin and with [3]H- or ^{125}I-labeled cyclosporin ligand, Wolf et al. [371] found that concentrations measured with the RIA kit correlated well with HPLC over a wide range of concentrations.

20.3.9 Anticancer drugs

In clinical practice a wide range of anticancer drugs in routinely used, and very high doses are often administered. These drugs are administered for a long time, have serious side-effects, and are given in combination with other drugs. Therefore, monitoring of their plasma levels is essential. Due to biotransformations of anticancer drugs, either intrahepatically or within the tumor cells, a multitude of metabolites with various degrees of cytostatic activity is often formed.

Up to now, about 800 000 substances have been tested for cytostatic activity and about 50 are in clinical use. For their determination, HPLC and GC (often GC/MS) methods are widely used (cf. Table 20.2).

20.3.9.1 Alkylating agents

The alkylating anticancer drugs are susceptible to decomposition, and their stability during analysis must always be checked. Chlorambucil can be determined by GC/MS in the form of alkyl [372] or silyl [373] derivatives and by HPLC. A method involving protein precipitation combined with solid-phase extraction as a sample preparation technique is sensitive down to 10 ng/ml [374]. Direct injection of plasma into the guard column was used by Zakaria and Brown [375] to determine chlorambucil and its metabolite, phenyl-acetic acid.

HPLC methods with UV detection at low wavelengths (190-200 nm) have been developed for the determination of cyclophosphamide at a sensitivity of 0.3 μg/ml [376] and 1 μg/ml [377]. The analysis of cyclophosphamide and its four metabolites was accomplished in a TLC system by reaction with 4-(4-nitrobenzyl)pyridine and densitometry [378].

20.3.9.2 Antimetabolites

Methotrexate is an antifolate drug, which is given in doses up to 500 mg/kg. Its main metabolite is 7-hydroxymethotrexate. Using capillary electrophoresis in either 75- or 200-μm capillaries, Roach et al. [35] were able to achieve detection limits for methotrexate (after oxidation to a fluorescent 2,4-diaminopteridine-6-carboxylic acid) of $3 \cdot 10^{-9}$ M. Detection was performed by laser fluorimeter with excitation at 325 nm and emission at 450 nm. Formation of fluorescent products was also utilized by Šalamoun et al. [379]. On irradiation under short-wavelength light in the presence of hydrogen peroxide, methotrexate and its metabolites were cleaved into highly fluorescent products, which were separated by HPLC. More conventionally, methotrexate can be determined in plasma, urine, and blood cells by reversed-phase HPLC with UV detection at 305 nm [380] down to a detection limit of $6.6 \cdot 10^{-8}$ M.

For the determination of 5-fluorouracil, ion-pair chromatography with a mobile phase containing tetrabutylammonium hydroxide at pH 11 and a PRP-1 stationary phase has been utilized by Rustum and Hoffman [381]. These authors have also reported a similar method for the quantitation of 5-azacytidine in plasma [382].

20.3.9.3 Purine antagonists

Both important agents in this category, 6-thioguanine and 6-mercaptopurine, must be metabolized before they exert their antitumoral effect. Several HPLC methods have been reported for the quantitation of 6-mercaptopurine [383,384] and 6-thioguanine [384] and its metabolite [385]. Active metabolites of 6-mercaptopurine may be determined in human red blood cells with a method developed by Lennard [386]. The assays is based on the specific extraction, via phenyl mercury adduct formation, of the thiopurine released upon acid hydrolysis of the thionucleotide metabolite. The metabolites are separated in a reversed-phase system with detection at 342 or 322 nm.

20.3.9.4 Antibiotics

The anthraquinone glycosides, adriamycin and daunorubicin, are extensively metabolized, and therefore, it is very important to measure also the levels of their metabolites. As cells are the target of these drugs, cellular concentrations are of great importance. Extraction studies [387] showed that various factors affect the extraction recovery from human hematopoietic cells. Assays for adriamycin [387-389], daunomycin [387], bleomycin [390], and doxorubicin [391,392] have been described. Fluorimetric detection is the most commonly used technique for the quantitation of anthracyclines in column eluates.

20.3.9.5 Antihormones

The biological activity of tamoxifen, a nonsteroidal antiestrogen, has been attributed to both the parent compound and its metabolites. A column switching method for the determination of tamoxifen and its four metabolites has been described [393]. Postcolumn irradiation converts these compounds to fluorescent products, monitored by a fluorescence detector (ex: 251 nm, em: 360 nm) with a limit of detection of 1 ng/ml. Other methods make use of GC/MS [394], TLC [395], and HPLC with pre- [396] and postcolumn derivatization [397,398] and fluorescence activation; cis- and trans-isomers of tamoxifen were separated on a β-cyclodextrin column, together with two metabolites of the trans-isomer [399].

20.3.9.6 Vinca alkaloids

Preceded by methods for the determination of Vinca alkaloids in crude materials [400,401], in preformulation studies [402], and in neoplastic tissues [403], procedures for therapeutic drug monitoring of Vinca alkaloids have appeared only recently. De Smet et al. separated vinblastine, vincristine, and vindesine on a CN column after ion-pair extraction from plasma and urine [404] and of vinblastine in mouse fibrosarcoma cells [405]. A detection limit of 6 ng/ml was achieved at 220 nm. The same alkaloids and a desacetyl metabolite of vinblastine were separated on a C_{18} column after solid-phase extraction on a CN column [406].

20.3.9.7 Cisplatin

cis-Diamminedichloroplatinum(II) (cisplatin) has proved to be effective in the treatment of several tumors, but its clinical use is limited, owing to its nephrotoxic potential. Several HPLC methods have been developed for studying the metabolism and the pharmacokinetics of cisplatin [407-414]. Many detection techniques have been devised in order to improve sensitivity (shown in parentheses): detection at 208 nm [407] (150 ng/ml), precolumn derivatization with diethyldithiocarbamate [408] (25 ng/ml), postcolumn reaction with bisulfite in the presence of potassium dichromate [409] (40 ng/ml), electrochemical detection [410] (10-100 ng/ml), quenched phosphorescence detection [411] (90 ng/ml), inductively coupled plasma atomic emission spectrometry [412] (35 ng/ml), off-line atomic absorption spectrometry [413] (10-40 ng/ml) and on-line radioactivity detection [414] (10 ng/ml). None of these techniques appears to be ideal; it is necessary to select methods according to available equipment, sample matrix, and desired sensitivity.

20.3.9.8 Podophyllotoxin derivatives

Etoposide and teniposide, semisynthetic derivatives of podophyllotoxin, are promising antineoplastic agents. Recent methods for their determination in plasma are mainly based on HPLC with electrochemical detection [415-418] and enable quantitation of these drugs with acceptable precision down to 10 ng/ml. Usually, teniposide is used as internal standard in the determination of etoposide and vice versa. A fully automated system with column switching and fluorescence detection has also been described [419].

20.3.10 Antiviral agents

Antivirals are a group that is now of intense interest. Many compounds are being tested for the treatment of AIDS patients, and this trend is expected to continue. Antivirals may be classified into two classes: synthetic nucleoside analogs [acyclovir, ribavirin, dideoxycytidine (DDC), azidodeoxycytidine (AZT)], and others with different chemical structures (foscarnet, suramin).

HPLC is the method of choice for the determination of these compounds in biological fluids. Due to their high polarity, either ion-pair systems [420-423] or reversed-phase systems with a content of organic modifier not higher than 5% must be used [424-426]. An interesting separation method for the determination of either AZT or DDC was developed by Mathes et al. [427]. Filtered plasma samples were directly injected into a precolumn, packed with Pinkerton ISRP material, where drugs were separated from plasma proteins. From this precolumn AZT and DDC were eluted into an analytical column, packed with β-cyclodextrin-bonded silica, where the separation of drugs from endogeneous interfering compounds occurs.

Acyclovir may be detected in the 250-nm region [420,424] or, more sensitively, with a fluorescence detector at low pH [428], which is necessary for the fluorescence detection of purines. A mildly hydrophobic stationary phase, Spheron Micro 300, was used with 0.1 M phosphoric acid/0.1 M sodium sulfate (pH 1.8) as a mobile phase. Plasma was injected directly into the columns, as proteins were not denatured (and precipitated on the stationary phase) under these conditions.

Ribavirin may be detected in biological samples at low wavelengths (207 nm) after sample preparation, involving phenyl boronate affinity chromatography [426] or chromatography on Dowex 1X4-100 [425].

Methods for the determination of suramin [421], foscarnet [422], and carbovir [423] by HPLC and for the determination of DDC by GC/MS [429] and HPLC [427] have recently been published.

20.3.11 Nonsteroidal antiinflammatory drugs, analgesics, and antipyretics

The agents covered in this section exhibit analgetic, antipyretic, and/or antiinflammatory effects. They are among the most commonly prescribed drugs. The measurement of nonsteroidal antiinflammatory drug (NSAID) concentrations in biological fluids is required in pharmacokinetic studies as well as for therapeutic monitoring, in spite of the fact that the relationship between dose and drug action is not so pronounced as in other drug categories. Since NSAIDs are often administered for many years, a knowledge and adjustment of their plasma levels may help to minimize some adverse effects. Clinical plasma levels of these drugs are in most cases in the range 1-100 μg/ml.

The NSAIDs have diverse chemical structures and may accordingly be divided into several categories: salicylates, arylalkanoic acids, oxicams, 3,5-pyrazolidinediones, fenamic acids, etc. Because sample preparation is simpler in HPLC, it is the predominant method for their determination, with the exception of diclofenac, where GC with ECD still plays a significant role on account of its sensitivity [430].

As the plasma levels of NSAIDs are relatively high, fast and simple sample preparation techniques are readily applied. Protein precipitation with either organic solvents (acetonitrile or methanol) alone [431-437] or in combination with strong acids [438-440] has been selected by many authors. Karnes et al. [441] have described the automated solid-phase extraction (SPE) of ibuprofen with the Varian AASP; Jonkman et al. [442] preferred the less expensive off-line SPE.

The classical liquid/liquid extractions are still valued for highly sensitive assays. Satterwhite and Boudinot [443] developed a method for the determination of ketoprofen and naproxen in plasma with detection limits of 10 and 5 ng/ml, respectively. Oka et al. achieved a sensitivity of 2 ng/ml for ketoprofen, but only with a tedious two-column method [444]. A detection limit of 10 ng/ml was reported for diclofenac and its monohydroxylated metabolites in HPLC with detection at 282 nm [445]. In situations where the number of interfering peaks must be minimized (e.g., determination of metabolites) liquid/liquid extraction is often the method of choice. Miyagi et al. [446] reported the determination of aminopyrine and its eight metabolites in rat plasma (linearity range 20 ng/ml to 200 μg/ml) after extraction with chloroform. D'Souza et al. [432] used protein precipitation for quantitation of antipyrine metabolites, but only four metabolites were quantitated, and the linearity range was 0.2-10 μg/ml.

Several papers describe the simultaneous determination of several NSAIDs. Giachetti et al. [447] separated ten major NSAIDs by GC as methyl esters on a SE-52 capillary column. Other authors used HPLC for this purpose: the retention behavior of eighteen NSAIDs on a C_{18} column was studied by Battista et al. [448], and their separation and determination in urine was achieved with gradient elution. Owen et al. [438] developed a rapid and simple isocratic assay for the determination of seven commonly used NSAIDs in plasma. Lapicque et al. [449] described an isocratic HPLC procedure for the screening of plasma samples for the presence of sixteen NSAIDs (Fig. 20.5). Detection was performed at two wavelengths (254 and 370 nm) and the purity of individual peaks was tested by means of absorbance ratios.

References on p. B433

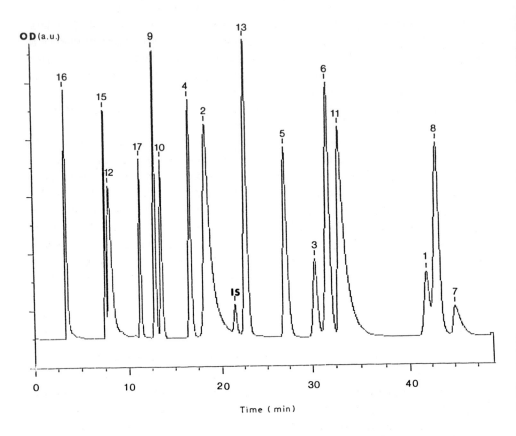

Fig. 20.5. Chromatogram of nonsteroidal anti-inflammatory drugs. Chromatographic conditions: Waters Nova Pak octadecyl reversed-phase column (300 x 3.9 mm ID); mobile phase, acetonitrile/0.3% acetic acid/tetrahydrofuran (36:63.1:0.9); flowrate, 1 ml/min; detection, 254 nm. Peaks: 1 = diclofenac; 2 = diflunisal; 3 = etodolac; 4 = fenbufen; 5 = fenoprofen; 6 = flurbiprofen; 7 = ibuprofen; 8 = indomethacin; 9 = ketoprofen; 10 = naproxen; 11 = niflumic acid; 12 = piroxicam; 13 = pirprofen; 14 = salicylate; 15 = sulindac; 16 = tenoxicam; 17 = tiaprofenic acid. (Reprinted from Ref. 449 with permission.)

Many NSAIDs possess a carboxy group with an acidic proton. Therefore, reversed-phase HPLC with ion suppression is widely used [431,439,450,451]. Usually, isocratic elution is used for a single drug, whereas gradient elution is applied for the separation of parent drugs from their metabolites. A change in selectivity may help to eliminate the need for gradient elution. For example, on a C_{18} column gradient elution was required for the separation of sulindac metabolites [431], but using a phenyl stationary phase, Grgurinovich was able to separate these metabolites under isocratic conditions [450]. A step gradient, which may be produced with an inexpensive isocratic pump, was used by Chai et al. [452] to separate the metabolites of ibuprofen from its parent drug. The C_{18} stationary phase is used predominantly. A CN-bonded stationary phase was employed for the determination of piroxicam in plasma and urine [437,453]. A different approach to chromatography was chosen in the work of Pinkerton et al. [454]. Plasma containing phenylbutazone was injected directly into an ISRP column, the main advantage of this approach being that no sample preparation is required.

Enantiomers of NSAIDs were separated after their derivatization to diastereomers with various agents. Hutt et al. [455] derivatized five NSAIDs with S-(–)-1-(naphthen-1-yl)-ethyl-amine; the diastereomers were separated in a normal-phase system. In the paper by Spahn [456], derivatization of eight NSAIDs with L-leucinamide and subsequent reversed-phase chromatography was described. Determination of individual enantiomers of NSAIDs in body fluids was described for ibuprofen [457,458] and ketoprofen [459]. Recent papers have also described the resolution of ibuprofen enantiomers without chiral derivatization. Nicoll-Griffith et al. [460] have separated ibuprofen enantiomers in the form of 4-methoxyanilide derivatives on a Pirkle-type column; Geisslinger et al. [461] resolved free ibuprofen enantiomers on a β-cyclodextrin (Cyclobond I) stationary phase.

A selection of methods for the determination of individual NSAIDs is presented in Table 20.7.

20.3.12 Drugs of abuse

Drug abuse is one of civilization's major problems. In the United States, e.g., more than 18 million people are current users of marijuana and almost six million are cocaine addicts [463]. Testing for drugs of abuse in body fluids, especially in urine, is thus an important means of identifying drug users in various settings. Usually screening tests are performed by simple techniques at first (TLC, immunoassays, HPLC, GC), and this is followed by confirmatory tests (GC/MS) [464]. Also, seized substances must often be analyzed in order to quantitate the drug of abuse, establish its origin, and determine how it was synthesized.

Morphine, Cannabis products, cocaine, and amphetamines are dealt with in the following sections, while information about barbiturates, benzodiazepines, and LSD can be found in Section 20.3.1.

20.3.12.1 Morphine

HPLC methods for the determination of morphine in biological samples have been reviewed extensively by Tagliaro et al. [465]. HPLC with electrochemical detection is among the most convenient methods of toxicology. Amperometric detectors with single glassy carbon electrodes, operated at potentials ranging from + 0.6 V to 1 V vs. Ag/AgCl electrode are commonly used [466-468], although coulometric detection can serve just as well [469]. An alternative detection method is fluorimetry. Morphine is weakly fluorescent, but under alkaline conditions it can be oxidized to a highly fluorescent dimer, pseudomor-

TABLE 20.7

HPLC OF SOME NONSTEROIDAL ANTI-INFLAMMATORY DRUGS

For abbreviations see footnote to Table 20.3.

Compound	Sample and preparation	Stationary phase	Detection (nm)	Detection limit	Ref.
Acetylsalicylic acid +3 metabolites	P. PP U. Hydrolysis	C_8	P-UV 235 U-UV 313	0.2 μg/ml	451
Aminopyrine +3 metabolites	P. PP	C_{18}	UV 254	0.2 μg/ml	432
Aminopyrine +8 metabolites	P. Extraction	C_{18}	UV 260	20 ng/ml	446
Antipyrine +3 metabolites	U. On-line SPE	C_{18}	UV 244	0.25 μg/ml	48
Fenoprofen	P. PP	C_{18}	UV 240	2.5 μg/ml	424
Flurbiprofen +metabolites	U.P. PP	C_{18}	F 260/320	0.1 μg/ml	435
Ibuprofen	P. On-line SPE P. PP	C_{18} C_{18}	F 253/300 UV 196	1.3 μg/ml 0.5 μg/ml	441 433
Ibuprofen +2 metabolites	U. Extraction	C_{18}	UV 214	2.5 μg/ml	452
Indomethacin Apyramide	P. Extraction	C_{18}	UV 250	0.2 μg/ml	439
Indomethacin	P.U. PP Postcolumn base hydrolysis	C_{18}	F 295/372	10 ng/ml	436
Ketoprofen Naproxen	P. Extraction	C_{18}	UV 229 UV 258	5 ng/ml	443
Phenylbutazone	P. DI	ISRP	UV 241	?	454
Piroxicam	P. PP	CN	UV 360	0.15 μg/ml	437
Piroxicam +metabolite	P.U. Extraction	CN	UV 365	50 ng/ml	453
Sulindac	P. PP U. Hydrolysis	C_{18}	UV 340	0.1 μg/ml	431
Sulindac +2 metabolites	P. Extraction	Phenyl	UV 315	15 ng/ml	450

phine. This reaction can be carried out in a postcolumn reactor [470]. Another possibility is the derivatization with dansyl chloride [471]. A coulometric and a fluorimetric detector, connected in series, were employed for the simultaneous determination of morphine and its principal glucuronide metabolites [472]. Morphine 3-glucuronide, which is not electrochemically active, was detected by fluorimetry, and morphine, normorphine, and morphine 6-glucuronide by coulometric detection.

Heroin is metabolized in vivo to 6-monoacetylmorphine, which is then converted to morphine. These two compounds, together with codeine, can be determined by a procedure reported by Schuberth and Schuberth [473]. It comprises sample preparation by solid-phase extraction, followed by capillary GC/MS with deuterium-labeled analogs as internal standards.

20.3.12.2 Cannabinoids

Δ^9-Tetrahydrocannabinol (THC), the major psychoactive ingredient of marijuana, is metabolized to a large number of compounds; 11-nor-Δ^9-tetrahydrocannabinol-9-carboxylic acid (THC-COOH) is the major one [474]. Immunoassays for cannabinoids are usually targeted for this metabolite, but they also crossreact with other cannabinoids present. Analytical procedures for the identification and quantitation of cannabinoids have recently been reviewed [475- 477].

Lemm et al. [478] extracted urine on columns packed with THC-antibody-bonded Sepharose, which binds THC-COOH as well. In this way, GC/MS enabled quantitation of THC-COOH down to 0.5 ng/ml. For HPLC of six metabolites of THC [479] or three major cannabinoids [480] reversed-phase columns were coupled with electrochemical detectors.

20.3.12.3 Cocaine

Analytical methods should measure cocaine plasma concentrations as low as 50 ng/ml and for pharmacokinetic studies even < 10 ng/ml [481]. Schwartz and David [482] have utilized HPLC with oxidative electrochemical detection for the determination of cocaine and its two metabolites, benzoylecgonine and ecgonine. Cocaine could be detected down to the 2-ng level. These compounds were also determined in urine by a rapid HPTLC procedure [483]. The silica gel plates were sprayed with Ludy-Tenger reagent to visualize individual bands. Using GC/NPD and a rather complex extraction procedure, Jacob et al. [484] were able to determine 3 ng of cocaine per ml plasma.

20.3.12.4 Amphetamines

Both packed and capillary GC columns were used by Christophersen et al. [485] for the determination of amphetamine in whole blood with a NPD. Routine samples were analyzed on the rugged packed column system, while the capillary column system was used for samples presenting special separation problems. Three precolumn derivatization

References on p. B433

reagents, namely o-phthalaldehyde, 4-chloro-7-nitrobenz-1,2,3-oxadiazole and sodium naphthoquinone 4-sulfonate (SNS) have been investigated for their suitability for analysis of urine and plasma samples containing amphetamines [486]. Derivatization with SNS was found to be selective and sufficiently sensitive for the routine determination of amphetamine and methylamphetamine at the ng/ml level.

3,4-Methylenedioxyamphetamine (MDA) and 3,4-methylenedioxymethamphetamine (MDMA) are ring-substituted amphetamines, which have become popular drugs of abuse. Their enantiomers exhibit different toxicities and behavioral effects. Fitzgerald et al. [487] described a method for the determination of the enantiomeric content of MDA and MDMA in whole blood. The method involves liquid/liquid extraction and derivatization with the chiral reagent N-trifluoroacetyl-L-prolyl chloride, followed by separation, identification, and quantitation of the diastereomeric derivatives by GC/MS. The analytical range of the assay is from 0.12 to 48 ng, injected on-column.

20.3.13 Anabolic steroids

Anabolic steroids are compounds that stimulate protein synthesis and thus increase muscle size. In addition to their legitimate medical uses and administration to food-producing animals (not permitted in some countries), they are illegally administered to race horses and abused by athletes. Anabolics are determined in biological samples along with endogeneous hormones long after administration. Evidently, very sensitive and selective techniques are required for the determination of these compounds.

20.3.13.1 Monitoring the urine of athletes

The increasing abuse of anabolic steroids in sports has prompted several investigators to develop methods for the reliable identification and quantitation of trace amounts of these agents in body fluids, especially urine. The detection of these substances was originally performed by radioimmunoassays [488], but such procedures lack selectivity and were therefore replaced by GC/MS with fused-silica capillaries [489-491]. Briefly, the urine sample is purified by solid-phase extraction, the conjugated steroids are hydrolyzed, extracted with diethyl ether, silylated, and then injected into the temperature-programed gas chromatograph. The mass spectrometer is operated in the selected-ion monitoring mode.

HPLC with diode-array detection may also be useful, although the results must be confirmed by an independent procedure. A simple systematic approach to the optimization of the HPLC separation of anabolics by isocratic elution with a multi-component mobile phase was reported by Jansen et al. [492]. Experimental retention data with three quaternary solvent mixtures are required to calculate the optimum solvent composition by the use of a geometric model.

20.3.13.2 Determination in biological materials from cattle and in cattle-breeding

Another application of anabolic agents involves their administration to calves and cattle. Silyl derivatives of anabolics were determined in veal by GC on a capillary SE-54 column and identified by MS [493]. A system for routine detection of various anabolics in bovine urine samples by HPTLC with fluorescence detection was described by De Brabander et al. [494]. Van Ginkel et al. [495] developed a method for the detection and identification of nortestosterone (NT) and one of its major metabolites in bovine urine and bile. The method is based on sample cleanup by immunoaffinity chromatography and determination by HPLC with UV detection at 254 nm (sensitivity 1 ng/ml) and/or by GC/MS (sensitivity 0.1 and 0.5 ng/ml for urine and bile, respectively). The same approach was chosen for the determination of NT in beef [496], where this compound can be quantitated down to 0.1 ng/g by GC/MS. Haasnoot et al. [497] used on-line immunoaffinity sample pretreatment for the determination of NT in urine, bile, and tissue samples. A reversed-phase HPLC/diode-array spectrometric method for the identification of more than 60 different steroidal compounds, anabolics as well as endogenous steroids, in cattle was developed for breeding work [498]. It is based on a gradient elution system with methanol as modifier and it can distinguish between esterified and nonesterified steroids. For those steroids that cannot be reliably identified because of their chromatographic and spectroscopic similarity, other separation systems with the same solvent strength, but with different organic modifiers are utilized.

Image analysis techniques have been applied to the quantitation of anabolics on TLC plates [499], but they cannot compete with GC/MS in terms of selectivity and sensitivity.

REFERENCES

1 H.G. Fouda, *J. Chromatogr.*, 492 (1989) 85.
2 D.C. Turnell and J.D.H. Cooper, *J. Chromatogr.*, 492 (1989) 59.
3 Z. Deyl, J. Janák, V. Schwarz and K. Macek, *J. Chromatogr.*, 460 (1989) No. 1-6.
4 L. Fishbein, in E. Heftmann (Editor), *Chromatography,* Elsevier, Amsterdam, 1983, pp. B287- B330.
5 Bibliography Section of *J. Chromatogr.*
6 Z. Deyl and J.A.F. de Silva (Editors), Drug Level Monitoring, *J. Chromatogr.*, 340 (1985) 1-495.
7 G. Piemonte, F. Tagliaro, M. Marigo and A. Frigerio (Editors), *Developments in Analytical Methods in Pharmaceutical, Biomedical and Forensic Sciences,* Plenum Press, New York, 1987.
8 K. Macek, *Cs. Farm.*, 37 (1988) 263.
9 R.K. Gilpin and L.A. Pachla, *Anal. Chem.*, 59 (1987) 174R.
10 International Symposium of Pharmaceutical and Biomedical Analysis, *J. Pharm. Biomed. Anal.*, 6 (1988) No. 6-8.
11 P. Kucera (Editor), *Microcolumn High-Performance Liquid Chromatography,* Elsevier, Amsterdam, 1984.
12 M. Novotny, *Anal. Chem.*, 60 (1988) 500A.
13 K.B. Tomer and C.E. Parker, *J. Chromatogr.*, 492 (1989) 189.
14 J. Plum and T. Daldrup, *J. Chromatogr.*, 377 (1986) 221.
15 R.F. Roberts and M.J. Fields, *J. Chromatogr.*, 342 (1985) 25.
16 T.C. Pinkerton, J.A. Perry and J.D. Rateike, *J. Chromatogr.*, 367 (1986) 412.

17 D.J. Gisch, B.T. Hunter and B. Feibush, *J. Chromatogr.*, 433 (1988) 264.
18 T. Nakagawa and A. Shibukawa, *J. Chromatogr.*, 420 (1987) 297.
19 R.K. Gilpin, S.S. Yang and G. Werner, *J. Chromatogr. Sci.*, 26 (1988) 388.
20 G. Musch and D.L. Massart, *J. Chromatogr.*, 370 (1986) 1.
21 M. De Smet and D.L. Massart, *J. Chromatogr.*, 410 (1987) 77.
22 M. De Smet and D.L. Massart, *Trends Anal. Chem.*, 6 (1987) 266.
23 F.W. Karasek and R.E. Clement, *Basic Gas Chromatography-Mass Spectrometry; Principles and Techniques,* Elsevier, Amsterdam, 1988.
24 B. Lacroix, J.P. Huvenne and M. Deveaux, *J. Chromatogr.*, 492 (1989) 109.
25 M. Warner, Anal. Chem., 59 (1987) 855A.
26 A.C. Mehta, *J. Chromatogr.*, 494 (1989) 1.
27 A.M. Siouffi, E. Mincsovics and E. Tyihak, *J. Chromatogr.*, 492 (1989) 471.
28 J. Dingenen and A. Pluym, *J. Chromatogr.*, 475 (1989) 95.
29 R.E. Kaiser (Editor), *Instrumental Thin-layer Chromatography/Planar Chromatography,* Institute for Chromatography, Bad Dürkheim, 1989.
30 C.F. Poole and S.K. Poole, *J. Chromatogr.*, 492 (1989) 539.
31 E. Tyihak, *J. Pharm. Biomed. Anal.*, 3 (1987) 191.
32 Sz. Nyiredy, K. Dallenbach-Toelke and O. Sticher, in F.A.A. Dallas, H. Read, R.J. Ruane and I. Wilson (Editors), *Recent Advances in Thin-layer Chromatography,* Plenum Press, London, 1988, pp. 45-54.
33 W.M.A. Niessen, U.R. Tjaden and J. van der Greef, *J. Chromatogr., 492 (1989) 167.*
34 B.L. Karger, A.S. Cohen and A. Guttman, *J. Chromatogr.*, 492 (1989) 585.
35 M.C. Roach, P. Gozel and R.N. Zare, *J. Chromatogr.*, 426 (1988) 129.
36 S. Terabe, *Trends Anal. Chem.*, 8 (1989) 129.
37 D.F. Swaile, D.E. Burton, A.T. Balchunias and M.J. Sepaniak, *J. Chromatogr. Sci.*, 26 (1988) 406.
38 R.W. Frei and K. Zech (Editors), *Selective Sample Handling and Detection in High-Performance Liquid Chromatography, Part A,* Elsevier, Amsterdam, 1988.
39 R.D. McDowall, *J. Chromatogr.*, 492 (1989) 3.
40 L.J. Cline Love, S. Zibas, J. Noroski and M. Arunyanart, *J. Pharm. Biomed. Anal.*, 3 (1985) 511.
41 Z.K. Shibabi, R.D. Dyer and J. Scaro, *J. Liquid Chromatogr.*, 10 (1987) 663.
42 Z.K. Shibabi and R.D. Dyer, *J. Liquid Chromatogr.*, 10 (1987) 2383.
43 G.M. Schneider, E. Stahl and G. Wilke (Editors), *Extraction with Supercritical Gases,* Verlag Chemie, Weinheim, 1980.
44 R.D. McDowall, J.C. Pearce and G.S. Murkitt, *J. Pharm. Biomed. Anal.*, 4 (1986) 3.
45 E. Reh, *J. Chromatogr.*, 433 (1988) 119.
46 M.J. Koenigbauer and M.A. Curtis, *J. Chromatogr.*, 427 (1988) 277.
47 B. Karger, R.W. Giese and L.R. Snyder, *Trends Anal. Chem.*, 2 (1983) 106.
48 J. Moncrieff, *J. Chromatogr.*, 383 (1986) 425.
49 W.Th. Kok, K.P. Hupe and R.W. Frei, *J. Chromatogr.*, 436 (1988) 421.
50 A.J.J. Debets, R.W. Frei, K.P. Hupe and W.Th. Kok, *J. Chromatogr.*, 465 (1989) 315.
51 R. Tawa, S. Hirose, T. Fujimoto, *J. Chromatogr.*, 490 (1989) 125.
52 V.K. Boppana, K.L.L. Fong, J.A. Zemniak, R.K. Lynn, *J. Chromatogr.*, 353 (1986) 231.
53 A.C. Mehta, *Trends Anal. Chem.*, 8 (1989) 107.
54 R.W. Frei and J.F. Lawrence (Editors), *Chemical Derivatization in Analytical Chemistry,* Plenum Press, New York, 1981.
55 E.L. Johnson, D.L. Reynolds, D.S. Wright, L.A. Pachla, *J. Chromatogr. Sci.*, 26 (1988) 372.
56 U.A.Th. Brinkman, R.W. Frei and H. Lingeman, *J. Chromatogr., 492 (1989) 251.*
57 F.A.L. van der Horst, G.G. Eikelboom and J.J.M. Holthuis, *J. Chromatogr.*, 456 (1988) 191.
58 J. Šalamoun, M. Smrz, F. Kiss and A. Šalamounová, *J. Chromatogr.*, 419 (1987) 213.
59 M. Nieder and H. Jaeger, *J. Chromatogr.*, 413 (1987) 207.
60 H. Jansen, C.J.M. Vermunt, U.A.Th. Brinkman and R.W. Frei, *J. Chromatogr.*, 366 (1986) 135.

61 M. Idei, A. Pajor, J. Guoth and J. Menyhart, *J. Chromatogr.*, 490 (1989) 247.
62 A.M. Krstulovic, *J. Chromatogr.*, 488 (1989) 53.
63 I.W. Wainer and D.E. Drayer (Editors), *Drug Stereochemistry. Analytical Methods and Pharmacology*, Marcel Dekker, New York, 1988.
64 D.W. Armstrong, *Anal. Chem.*, 59 (1987) 84A.
65 A.C. Mehta, *J. Chromatogr.*, 426 (1988) 1.
66 A.M. Krstulovic, *J. Pharm. Biomed. Anal.*, 6 (1988) 641.
67 W.H. Pirkle, M.H. Hyun and B. Bank, *J. Chromatogr.*, 316 (1984) 585.
68 J. Hermansson, *J. Chromatogr.*, 269 (1983) 71.
69 D.W. Armstrong, X. Yang, S.M. Han and R.A. Menges, *Anal. Chem.*, 59 (1987) 2594.
70 G. Szepesi and M. Gazdag, *J. Pharm. Biomed. Anal.*, 6 (1988) 623.
71 S.H.Y. Wong, *Clin. Chem.*, 34 (1988) 848.
72 A. Fazio, E. Spina and F. Pisani, *J. Liquid Chromatogr.*, 10 (1987) 223.
73 T.R. Norman and K.P. Maguire, *J. Chromatogr.*, 340 (1985) 173.
74 A.A. Boulton, G.B. Baker and R.T. Coutts (Editors), *Neuromethods*, Vol. 10, *Analysis of Psychiatric Drugs*, Humana Press, Clifton, NJ, 1988, 547 pp.
75 O. Spreux-Varoquaux, D. Morin, C. Advenier and M. Pays, *J. Chromatogr.*, 416 (1987) 311.
76 A. Sioufi, F. Pommier and J.P. Dubois, *J. Chromatogr.*, 428 (1988) 71.
77 R. Drebit, G.B. Baker and W.G. Dewhurst, *J. Chromatogr.*, 432 (1988) 334.
78 S.P. Bagchi, T. Lutz and S.P. Jindal, *J. Chromatogr.*, 344 (1985) 362.
79 A.M. Bougerolle, J.L. Chabard, M. Jbilou, H. Barguoux, J. Petit and J.A. Berger, *J. Chromatogr.*, 434 (1988) 232.
80 C. Köppel and J. Tenczer, *J. Chromatogr.*, 431 (1988) 197.
81 Y.H. Park, C. Goshorn and O.N. Hinsvark, *J. Chromatogr.*, 375 (1986) 202.
82 Y. Sasaki and S. Baba, *J. Chromatogr.*, 426 (1988) 93.
83 P.P. Rop, T. Couquy, F. Gonezo, A. Viala and F. Grimaldi, *J. Chromatogr.*, 375 (1986) 339.
84 K. Kurata, M. Kurachi and Y. Tanii, *J. Chromatogr.*, 434 (1988) 278.
85 R.L. Miller and C.L. Devane, *J. Chromatogr.*, 374 (1986) 388.
86 G.L. Lensmeyer and M.A. Evenson, *Clin. Chem.*, 30 (1984) 1774.
87 F.A. Beierle and R.W. Hubbard, *Ther. Drug Monit.*, 5 (1983) 279.
88 F.A. Beierle and R.W. Hubbard, *Ther. Drug Monit.*, 5 (1983) 293.
89 J.J. Tasset and F.M. Hassan, *Clin. Chem.*, 28 (1982) 2154.
90 J.J. Tasset and A.J. Pesce, *J. Anal. Toxicol.*, 8 (1984) 124.
91 W. Lin and P.D. Frade, *Ther. Drug Monit.*, 9 (1987) 448.
92 D. Dadgar and A. Power, *J. Chromatogr.*, 416 (1987) 99.
93 C. Svensson, G. Nyberg and E. Martensson, *J. Chromatogr.*, 432 (1988) 363.
94 T. Emm, L.J. Lesko and M.B. Perkal, *J. Chromatogr.*, 419 (1987) 445.
95 S. Sugita, A. Kobayashi, S. Suzuki, T. Yoshida and K. Nakazawa, *J. Chromatogr.*, 421 (1987) 412.
96 P. Ni, F. Guyon, M. Caude and R. Rosset, *J. Liquid Chromatogr.*, 11 (1988) 1087.
97 G.M. Roberts and C.S. Hann, *Biomed. Chromatogr.*, 1 (1986) 49.
98 J.A.F. de Silva, *J. Chromatogr.*, 278 (1983) 19.
99 N.R. Badcock and G.D. Zoanetti, *J. Chromatogr.*, 421 (1987) 147.
100 R. Jochemsen and D.D. Breimer, *J. Chromatogr.*, 227 (1982) 199.
101 R.G. Lister, D.R. Abernethy, D.J. Greenblatt and S.E. File, *J. Chromatogr.*, 277 (1983) 201.
102 H.R. Ha, B. Funk, P.O. Maitre, A.M. Zbinden, F. Follath and D.A. Thomson, *Clin. Chem.*, 34 (1988) 676.
103 G. Baktir, J. Bircher, H.-U. Fisch and G. Karlaganis, *J. Chromatogr.*, 339 (1985) 192.
104 H.T. Karnes, L.A. Beightol, R.J. Serafin and D. Farthing, *J. Chromatogr.*, 424 (1988) 398.
105 C. Drouet-Coassolo, C. Aubert, P. Coassolo and J.-P. Cano, *J. Chromatogr.*, 487 (1989) 295.
106 H. Maurer and K. Pfleger, *J. Chromatogr.*, 422 (1987) 85.

107 M.C. Pietrogrande, F. Dondi, G. Blo, P.A. Borea and C. Bighi, *J. Liquid Chromatogr.*, 11 (1988) 1313.
108 P. Mura, A. Piriou, P. Fraillon, Y. Papet and D. Reiss, *J. Chromatogr.*, 416 (1987) 303.
109 N. De Giovanni and M. Chiarotti, *J. Chromatogr.*, 428 (1988) 321.
110 H. Maurer and K. Pfleger, *J. Chromatogr.*, 306 (1984) 125.
111 S.H. Curry, E.A. Brown, O.Y.-P. Hu and J.H. Perrin, *J. Chromatogr.*, 231 (1982) 361.
112 C.S. Smith, S.L. Morgan, S.V. Greene and R.K. Abramson, *J. Chromatogr.*, 423 (1987) 207.
113 G. Tamai, H. Yoshida and K. Imai, *J. Chromatogr.*, 423 (1987) 163.
114 R. Whelpton, G. Jonas and D.G. Buckley, *J. Chromatogr.*, 426 (1988) 223.
115 D.W. Hoffman, R.D. Edkins, S.D. Shillcutt and A. Salama, *J. Chromatogr.*, 414 (1987) 504.
116 L.B. Nilsson, *J. Chromatogr.*, 431 (1988) 113.
117 D.R. Abernethy, D.J. Greenblatt, H.R. Ochs, C.R. Willis, D.D. Miller and R.I. Shader, *J. Chromatogr.*, 307 (1984) 194.
118 J.H. Jaffe, *The Pharmacological Basis of Therapeutics*, MacMillan, New York, 7th ed., 1985, Chap. 23.
119 J.P. Twitchet, S.M. Fletcher, A.T. Sullivan and A.C. Moffat, *J. Chromatogr.*, 150 (1978) 73.
120 J. Christie, M.W. White and J.M. Wiles, *J. Chromatogr.*, 120 (1976) 496.
121 H.K. Lim, D. Andrenyak, P. Francom, R.L. Foltz and R.T. Jones, *Anal. Chem.*, 60 (1988) 1420.
122 R.N. Gupta, *J. Chromatogr.*, 340 (1985) 139.
123 H. Heusler, *J. Chromatogr.*, 340 (1985) 273.
124 J.T. Burke and J.P. Thénot, *J. Chromatogr.*, 340 (1985) 199.
125 I. Minder, R. Schaubhut and D.J. Vonderschmitt, *J. Chromatogr.*, 428 (1988) 369.
126 M. van der Graaff, N.P.E. Vermeulen, P.H. Hofman and D.D. Breimer, *J. Chromatogr.*, 375 (1986) 411.
127 P.J. Soine and W.H. Soine, *J. Chromatogr.*, 422 (1987) 309.
128 B.H. Dvorchik, *J. Chromatogr.*, 105 (1975) 49.
129 R.J. Flanagan and D.J. Berry, *J. Chromatogr.*, 131 (1977) 131.
130 M. Marigo, S.D. Ferrara and L. Tedeschi, *Arch. Toxicol.*, 37 (1977) 107.
131 M.J. Dunphy and M.K. Pandya, *J. High Resolut. Chromatogr. Chromatogr. Commun.*, 6 (1983) 317.
132 T. Villén and I. Petters, *J. Chromatogr.*, 258 (1983) 267.
133 J. De Graeve and J. Vanroy, *J. Chromatogr.*, 129 (1976) 171.
134 R.H. Greeley, *Clin. Chem.*, 20 (1974) 192.
135 F. Vincent, C. Feuerstein, M. Gavend and J. Faure, *Clin. Chim. Acta*, 93 (1979) 391.
136 U.R. Tjaden, J.C. Kraak and J.F.K. Huber, *J. Chromatogr.*, 143 (1977) 183.
137 S. Björkman and J. Idvall, *J. Chromatogr.*, 307 (1984) 481.
138 R.N. Gupta, P.T. Smith and F. Eng, *Clin. Chem.*, 28 (1982) 1772.
139 M.J. Avram and T.C. Krejcie, *J. Chromatogr.*, 414 (1986) 484.
140 J. de Jong, M.W.F. Nielen, R.W. Frei and U.A.T. Brinkman, *J. Chromatogr.*, 381 (1986) 431.
141 M.J. Koenigbauer and M.A. Curtis, *J. Chromatogr.*, 427 (1988) 277.
142 P.P. Rop, J. Spinazola, A. Zahra, M. Bresson, J. Quicke and A. Viala, *J. Chromatogr.*, 427 (1988) 172.
143 H. Hosotsubo, K. Takeda, K. Hosotsubo and I. Yoshiya, *J. Chromatogr.*, 487 (1989) 204.
144 P.M. Kabra, B.E. Stafford and L.J. Marton, *J. Anal. Toxicol.*, 5 (1981) 177.
145 M. Garle and I. Petters, *J. Chromatogr.*, 140 (1977) 165.
146 B. Kinberger, A. Holmen and P. Wahrgren, *J. Chromatogr.*, 224 (1981) 449.
147 C.R. Clark and J.-L. Chan, *Anal. Chem.*, 50 (1978) 635.
148 I.W. Wainer, M.C. Alembik and L.J. Fischer, *J. Pharm. Biomed. Anal.*, 5 (1987) 735.
149 B. Johansson, *J. Chromatogr.*, 341 (1985) 462.
150 R.A. Hux, H.Y. Mohammed and F.F. Cantwell, *Anal. Chem.*, 54 (1982) 113.

151 M. van Boven and I. Sunshine, *J. Anal. Toxicol.*, 3 (1979) 174.
152 W. Butte, *Fresenius' Z. Anal. Chem.*, 290 (1978) 156.
153 M. Eichelbaum, B. Sonntag and G. von Unruh, *Arch. Toxicol.*, 41 (1978) 187.
154 R. Hartley, M. Lucock and M. Becker, *J. Chromatogr.*, 415 (1987) 357.
155 R. Heipertz and Ch. Reimer, *Clin. Chim. Acta*, 110 (1981) 131.
156 H. Maurer and K. Pfleger, *Fresenius' Z. Anal. Chem.*, 321 (1985) 363.
157 R. Varma, *J. Chromatogr.*, 155 (1978) 182.
158 K. Dean, G. Land and A. Bye, *J. Chromatogr.*, 221 (1980) 408.
159 G. Paeme, W. Sonck, D. Tourwé, R. Grimée and A. Vercruysse, *Drug Metab. Disp.*, 8 (1980) 115.
160 M. Contin, R. Riva, F. Albani and A. Baruzzi, *Biomed. Chromatogr.*, 2 (1987) 193.
161 D. Lutz, W. Gielsdorf and H. Laeger, *J. Clin. Chem. Clin. Biochem.*, 21 (1983) 595.
162 S.D. Yoo and J.E. Axelson, *J. Chromatogr.*, 378 (1986) 385.
163 Z. Juvance, I. Rátonyi, A. Tóth and M. Vajda, *J. Chromatogr.*, 286 (1984) 363.
164 Y. Michotte, M. Moors, D. Deleu, P. Herregodts and G. Ebinger, *J. Pharm. Biomed. Anal.*, 5 (1987) 659.
165 T. Ishimitsu and S. Hirose, *Anal. Biochem.*, 150 (1985) 300.
166 E. Nissinen and J. Taskinen, *J. Chromatogr.*, 231 (1982) 459.
167 R.C. Causon, M.J. Brown, K.L. Leenders and L. Wolfson, *J. Chromatogr.*, 277 (1983) 115.
168 M. Lalande, P. Mills and R.G. Peterson, *J. Chromatogr.*, 430 (1988) 187.
169 E.W. Wuis, A.C.L.M. Grutters, T.B. Vree and E. van der Kleyn, *J. Chromatogr.*, 231 (1982) 401.
170 E.W. Wuis, L.E.C. van Beijsterveldt, R.J.M. Dirks, T.B. Vree and E. van der Kleyn, *J. Chromatogr.*, 420 (1987) 212.
171 A.M. Rustum, *J. Chromatogr.*, 487 (1989) 107.
172 B. Sallerin-Caute, M. Monsarrat, Y. Lazorthes, J. Cros and R. Bastide, *J. Liquid Chromatogr.*, 11 (1988) 1753.
173 E.W. Wuis, E.W.J. Beneken Kolmer, L.E.C. Van Beijsterveldt, R.C.M. Burgers, T.B. Vree and E. van der Kleyn, *J. Chromatogr.*, 415 (1987) 419.
174 J.T. Burke and J.P. Thénot, *J. Chromatogr.*, 340 (1985) 199.
175 U. Juergens, *J. Liquid Chromatogr.*, 10 (1987) 507.
176 K. Kushida, K. Chiba and T. Ishizaki, *Ther. Drug Monit.*, 5 (1983) 127.
177 N. Wad, *J. Chromatogr.*, 305 (1984) 127.
178 J.A. Christofides and D.E. Fry, *Clin. Chem.*, 26 (1980) 499.
179 M. Pesh-Iman, D.W. Fretthold, I. Sunshine, S. Kummar, S. Tenentine and C.E. Willis, *Ther. Drug Monit.*, 1 (1979) 289.
180 P.M. Kabra, M.A. Nelson and L.J. Marton, *Clin. Chem.*, 29 (1983) 473.
181 S.J. Soldin and J.G. Hill, *Clin Chem.*, 22 (1976) 856.
182 H.M. Neels, J.A. Tollé, R.M. Verkerk, A.J. Vlietinck and S.L. Scharpé, *J. Clin. Chem. Clin. Biochem.*, 21 (1983) 295.
183 J.W. Dolan, S. van der Wall, S.J. Bannister and L.R. Snyder, *Clin. Chem.*, 26 (1980) 871.
184 Y. Haroon and D.A. Keith, *J. Chromatogr.*, 276 (1983) 445.
185 K. Matsumoto, H. Kiruchi, S. Kano, H. Iri, H. Takahashi and M. Umino, *Clin. Chem.*, 34 (1988) 141.
186 R. Meatherall and D. Ford, *Ther. Drug Monit.*, 10 (1988) 101.
187 S.A. Mira, Y.M. El-Sayed and S.I. Islam, *Analyst (London)*, 112 (1987) 57.
188 U. Juergens, *J. Chromatogr.*, 371 (1986) 307.
189 U. Juergens, *J. Chromatogr.*, 385 (1987) 233.
190 C.P. Pang, K.C. Ho and C.K. Keung, *Biomed. Chromatogr.*, 1 (1986) 173.
191 M. Okamoto and K. Jinno, *J. Chromatogr.*, 395 (1987) 171.
192 A. Aasberg and F. Haffner, *Scand. J. Clin. Lab. Invest.*, 47 (1987) 389.
193 J.D. Berg and B.M. Buckley, *Ann. Clin. Biochem.*, 24 (1987) 488.
194 M. Okamoto, K. Jinno, K. Nobuhara and K. Fukushima, *Chromatographia*, 23 (1987) 373.
195 U. Juergens and B. Rambeck, *J. Liquid Chromatogr.*, 10 (1987) 1847.

196 D.E. Fry, J. Iqbal and J.A. Christofides, *Anal. Clin. Biochem.*, 23 (1986) 559.
197 N. Inotsume, A. Higashi, E. Kinoshita, T. Matsuoka and M. Nakano, *J. Chromatogr.*, 383 (1986) 166.
198 J.J. Thoma, T. Ewald and M. McCoy, *J. Anal. Toxicol.*, 2 (1978) 219.
199 B. Rambeck and J.W.A. Meijer, *Arzneim.-Forsch.*, 29 (1979) 99.
200 R. Nishioka, M. Takeuchi, S. Kawai, M. Nakamura and K. Kondo, *J. Chromatogr.*, 342 (1985) 89.
201 M. Werner, J. Mohrbacher and C.T. Riendeau, *Clin. Chem.*, 25 (1979) 2020.
202 C.A. Cramers, E.A. Vermeer, L.G. van Kuik, J.A. Hulsman and C.A. Meijers, *Clin. Chim. Acta*, 73 (1976) 97.
203 J. De Graeve and J. Vanroy, *J. Chromatogr.*, 129 (1976) 171.
204 F.S. Abbott, J. Kassam, A. Acheampong, S. Ferguson, S. Panesar, R. Burton, K. Farrell and J. Orr, *J. Chromatogr.*, 375 (1986) 285.
205 R.L.O. Semmes and D.D. Shen, *J. Chromatogr.*, 432 (1988) 185.
206 J.T. Lum, N.A. Vassanji and P.G. Wells, *J. Chromatogr.*, 338 (1985) 242.
207 K. Kushida and T. Ishizaki, *J. Chromatogr.*, 338 (1985) 131.
208 G. Tamai, H. Yoshida and H. Imai, *J. Chromatogr.*, 423 (1987) 147.
209 B. Johansson, *J. Chromatogr.*, 381 (1986) 107.
210 G.P. Menge, J.P. Dubois and G. Bauer, *J. Chromatogr.*, 414 (1987) 477.
211 L.J. Lovett, G.A. Nygard, G.R. Erdmann, C.Z. Burley and S.K. Wahba Khalil, *J. Liquid Chromatogr.*, 10 (1987) 687.
211a F.A.L. van der Horst, G.G. Eikelboom and J.J.M. Holthuis, *J. Chromatogr.*, 456 (1988) 191.
212 Z.K. Shihabi and R.D. Dyer, *J. Liquid Chromatogr.*, 10 (1987) 2383.
213 R. Hartley, M. Lucock, W.I. Forsythe and R.W. Smithells, *J. Liquid Chromatogr.*, 10 (1987) 2393.
214 D.K. Robbins, S.-L. Chang, R.J. Baumann and P.J. Wedlund, *J. Chromatogr.*, 415 (1987) 208.
215 U. Juergens, T. May and B. Rambeck, *J. Liquid Chromatogr.*, 9 (1986) 2921.
216 H. Sayo, H. Hatsumura, M. Hosokawa and T. Michida, *J. Chromatogr.*, 417 (1987) 129.
217 V. Ascalone, P. Catalani and L. Dal Bo, *J. Chromatogr.*, 344 (1985) 231.
218 P. Padovani, C. Deves, G. Bianchetti, J.P. Thénot and P.L. Morselli, *J. Chromatogr.*, 308 (1984) 229.
219 J. Sato, Y. Sekizawa, E. Owada, K. Ito, N. Sakuta, M. Yoshihara, T. Goto and Y. Kobayashi, *Chem. Pharm. Bull.*, 34 (1986) 3049.
220 M. Ahnoff, M. Ervik, P.-O. Lagerström, B.-A. Persson and J. Vessman, *J. Chromatogr.*, 340 (1985) 73.
221 R.J. Flanagan, R.K. Bhamra, S. Walker, S.C. Monkman and D.W. Holt, *J. Liquid Chromatogr.*, 11 (1988) 1015.
222 J.A. Stone and S.J. Soldin, *Clin. Chem.*, 34 (1988) 2547.
223 H. Nakashima, K. Tsutsumi, M. Hashiguchi, Y. Kumagai and A. Ebihara, *J. Chromatogr.*, 489 (1989) 425.
224 L. Embree and K.M. McErlane, *J. Chromatogr.*, 496 (1989) 321.
225 H.F. Proelss and T.B. Townsend, *Clin. Chem.*, 32 (1986) 1311.
226 M. Erdweg, M. Hoffman and H.A. Weinand, *Med. Welt*, 39 (1988) 1082.
227 M. Petrarulo, S. Pellegrino and E. Mentasti, *Chromatographia*, 25 (1988) 593.
228 M.J. Moor, P.A. Wyss and M.H. Bickel, *J. Chromatogr.*, 431 (1988) 455.
229 T.M. Welscher, R.L. McQuinn and S.F. Chang, *J. Chromatogr.*, 431 (1988) 438.
230 G.A. Woollard, *J. Chromatogr.*, 487 (1989) 409.
231 R.P. Kapil, F.S. Abbott, C.R. Kerr, D.J. Edwards, D. Lalka and J.E. Axelson, *J. Chromatogr.*, 307 (1984) 305.
232 M. Radwan, J. Price and R. Tackett, *Anal. Lett.*, 20 (1987) 1125.
233 P. Le Corre, D. Gibassier, P. Sado and R. Le Verge, *J. Chromatogr.*, 450 (1988) 211.
234 F. Susanto, S. Humfeld and H. Reinauer, *Chromatographia*, 21 (1986) 41.
235 F. Broly, C. Libersa and M. Lhermitte, *J. Chromatogr.*, 431 (1988) 369
236 F. Filipek, D. Paczkowski and J. Podlesny, *J. Chromatogr.*, 430 (1988) 406.

237 K.M. McErlane, L. Igwemezie and C.R. Kerr, *J. Chromatogr.*, 415 (1987) 335.
238 G. Tamai, H. Yoshida and H. Imai, *J. Chromatogr.*, 423 (1987) 155.
239 A. Yamaji, K. Kataoka, M. Oishi, N. Kanamori, E. Hiraoka and M. Mishima, *J. Chromatogr.*, 415 (1987) 143.
240 B.H. Chen, E.H. Taylor, E. Kennedy, B. Ackerman, K. Olsen and A.A. Pappas, *Clin. Chim. Acta*, 175 (1988) 107.
241 L.N. Ace and B. Chaudhuri, *Anal. Lett.*, 20 (1987) 1483.
242 V. Marko (Editor), *Determination of beta-Blockers in Biological Material*, Elsevier, Amsterdam, 1989.
243 M.S. Leloux, E.G. de Jong and R.A.A. Maes, *J. Chromatogr.*, 488 (1989) 357.
244 G.P. Cartoni, M. Ciardi, A. Giarrusso and F. Rosati, *J. High Resolut. Chromatogr. Chromatogr. Commun.*, 11 (1988) 528.
245 F.T. Delbeke, M. Debackere, N. Desmet and F. Maertens, *J. Pharm. Biomed. Anal.*, 6 (1988) 827.
246 I. Ojanpera and A. Ruohonen, *J. Anal. Toxicol.*, 12 (1988) 108.
247 Y. Li and X. Zhang, *Anal. Chim. Acta*, 196 (1987) 255.
248 S. Laganiere, E. Kwong and D.D. Shen, *J. Chromatogr.*, 488 (1989) 407.
249 W. Lindner, M. Rath, K. Stoschitzky and G. Uray, *J. Chromatogr.*, 487 (1989) 375.
250 S.A. Quereshi and H.S. Buttar, *J. Chromatogr.*, 431 (1988) 465.
251 R.P. Koshakji and A.J.J. Wood, *J. Chromatogr.*, 422 (1987) 294.
252 G.L. Hoyer, *J. Chromatogr.*, 427 (1988) 181.
253 M.J. Bartek, M. Vekshteyn, M.P. Boarman and D.G. Gallo, *J. Chromatogr.*, 421 (1987) 309.
254 M. Johansson and H. Forsmo-Bruce, *J. Chromatogr.*, 432 (1988) 265.
255 M.J. Wilson, K.D. Ballard and T. Walle, *J. Chromatogr.*, 431 (1988) 222.
256 K. Padmalatha Devi, K.V. Ranga Rao, S.K. Baveja, T. Leemann and P. Dayer, *J. Chromatogr.*, 434 (1988) 265.
257 G.S.M.J.E. Duchateau, W.M. Albers and H.H. van Rooij, *J. Chromatogr.*, 383 (1986) 212.
258 J. Moncrieff and D. Simpson, *J. Chromatogr.*, 488 (1989) 498.
259 D. Schuster, M. Woodruff Modi, D. Lalka and F.M. Gengo, *J. Chromatogr.*, 433 (1988) 318.
260 G. Pflugmann, H. Spahn and E. Mutschler, *J. Chromatogr.*, 421 (1987) 161.
261 A.M. Krstulovic, M.H. Fouchet, J.T. Burke, G. Gillet and A. Durand, *J. Chromatogr.*, 452 (1988) 477.
262 R.K. Bhamra, A.E. Ward and D.W. Holt, *J. Chromatogr.*, 417 (1987) 229.
263 J. Moncrieff, *J. Chromatogr.*, 342 (1985) 206.
264 X. Svobodová, D. Kovácová, V. Ostrovská, A. Pechová, O.Poláčiková, S. Kusala and M. Svoboda, *J. Chromatogr.*, 425 (1988) 391.
265 A. Sioufi and F. Pommier, *J. Chromatogr.*, 339 (1985) 117.
266 F.W. Lee, N. Watari, J. Rigod and L.Z. Benet, *J. Chromatogr.*, 426 (1988) 259.
267 P.A. Hynning, P. Anderson, U. Bondesson and L.O. Borens, *Clin. Chem.*, 34 (1988) 2502.
268 D.L. Bremseth, J.J. Lima and J.J. MacKichan, *J. Liquid Chromatogr.*, 11 (1988) 2731.
269 J.A. Pieper and D.R. Rutledge, *J. Chromatogr. Sci.*, 26 (1988) 473.
270 H. Mascher and H. Vergin, *Chromatographia*, 25 (1988) 919.
271 B.J. Schmid, H.E. Perry and J.R. Idle, *J. Chromatogr.*, 425 (1988) 107.
272 L.M. Dubé, N. Mousseau and I.J. McGilveray, *J. Chromatogr.*, 430 (1988) 103.
273 O. Grech-Belanger, E. Leboeuf and S. Langlois, *J. Chromatogr.*, 417 (1987) 89.
274 X. Aparicio, J. Gras, A. Campos, E. Fernandez and E. Gelpi, *J. Pharm. Biomed. Anal.*, 6 (1988) 167.
275 B.L. Cohen, *J. Chromatogr. Sci.*, 25 (1987) 202.
276 J. Girault and J.B. Fourtillan, *Biomed. Environ. Mass Spectrom.*, 17 (1988) 443.
277 R.F. Arrendale, J.T. Stewart and R.L. Tackett, *J. Chromatogr.*, 432 (1988) 165.
278 J. Moncrieff, *J. Chromatogr.*, 428 (1988) 178.
279 K. Chan, *J. Chromatogr.*, 425 (1988) 311.

280 J.T. Stewart and S.S. Clark, *J. Pharm. Sci.,* 75 (1986) 413.
281 C. Dilger, Z. Salama and H. Jaeger, *Arzneim.- Forsch.,* 37 (1987) 1399.
282 V. Ostrovská, X. Svobodová, A. Pechová, S. Kusala and M. Svoboda, *J. Chromatogr.,* 446 (1988) 323.
283 D.R. Luke, G.R. Matzke, J.T. Clarkson and W.M. Awni, *Clin. Chem.,* 33 (1987) 1450.
284 D.D. Shen and A.B. Pitterman, in I. Sunshine (Editor), *Methodology in Analytical Toxicology,* Vol. 3, CRC Press, Boca Raton, FL, 1985, p. 159.
285 R.K. Bhamra, R.J. Flanagan and D.W. Holt, *J. Chromatogr.,* 380 (1986) 216.
286 W. Siegmund, M. Zschiesche, R. Kallwellis, G. Franke, T. Schneider, U. Sill, A. Scherber and H. Hueller, *Pharmazie,* 40 (1985) 779.
287 M.C. Rouan and J. Campestrini, *J. Pharm. Sci.,* 74 (1985) 1270.
288 R.O. Fullinfaw, R.W. Bury and R.F.W. Moulds, *J. Chromatogr.,* 415 (1987) 347.
289 F. De Croo, W. Van den Bossche and P. De Moerloose, *J. Chromatogr.,* 349 (1985) 301.
290 B.M.J. De Spiegeleer and P. De Moerloose, *J. Planar Chromatogr.,* 1 (1988) 61.
291 S.P. Agarwal and J. Nwaiwu, *J. Chromatogr.,* 351 (1986) 383.
292 D.C. Muirhead and R.B. Christie, *J. Chromatogr.,* 416 (1987) 420.
293 D. Dadgar and M.T. Kelly, *Analyst (London),* 113 (1988) 1223.
294 M.J. Van der Meer and L.W. Brown, *J. Chromatogr.,* 423 (1987) 351.
295 V. Palosi-Szantho, M. Kurcz and S. Fritsch, *Symp. Biol. Hung.,* 37 (1988) 321.
296 J.H. Sherry, J.P. O'Donell and H.D. Colby, *J. Chromatogr.,* 374 (1986) 183.
297 J.W.P.M. Overdick, W.A.J.J. Hermens and F.W.H.M. Merkus, *J. Chromatogr.,* 341 (1985) 279.
298 G. Forrest, G.T. Mc Innes, A.P. Fairhead, G.G. Thompson and M.J. Brodie, *J. Chromatogr.,* 428 (1988) 123.
299 L.J. Lovett, G. Nygard, P. Dura and S.K.W. Khalil, *J. Liquid Chromatogr.,* 8 (1985) 1611.
300 M. Kucharczyk and F.H. Segelman, *J. Chromatogr.,* 340 (1985) 243.
301 T. Emm, L.J. Lesko, J. Leslie and M.B. Perkal, *J. Chromatogr.,* 427 (1988) 188.
302 Y.K. Tan and S.J. Soldin, *J. Chromatogr.,* 422 (1987) 187.
303 L.-E. Edholm C. Lindberg, J. Paulson and A. Walhagen *J. Chromatogr.,* 424 (1988) 61.
304 B.-M. Kennedy, A. Blomgren, L.-E. Edholm and C. Roos, *Chromatographia,* 24 (1987) 895.
305 R. Maes, in I. Sunshine (Editor), *Methodology in Analytical Toxicology,* Vol. 3, CRC Press, Boca Raton, FL, 1985, p. 175.
306 E. Bailey, P.B. Farmer, J.A. Peal, S.A. Hotchkiss and J. Caldwell, *J. Chromatogr.,* 416 (1987) 81.
307 R. Chiou, R.J. Stubbs and W.F. Bayne, *J. Chromatogr.,* 422 (1987) 281.
308 K. Matsumoto, H. Kikuchi, H. Iri, H. Takahasi and M. Umino, *J. Chromatogr.,* 425 (1988) 323.
309 S.A. Hotchkiss and J. Caldwell, *J. Chromatogr.,* 423 (1987) 179.
310 M.B. Kester, C.L. Saccar and H.C. Mansmann, Jr., *J. Chromatogr.,* 416 (1987) 91.
311 H. Heusler, *J. Chromatogr.,* 340 (1985) 273.
312 P. Koupil, J. Novák and J. Drozd, *J. Chromatogr.,* 425 (1988) 99.
313 Y.K. Tam, S.R. Tawfik, R. Ke, R.T. Coutts, M.R. Gray and D.G. Wyse, *J. Chromatogr.,* 423 (1987) 199.
314 T.B. Vree, A.M. Baars, G.E.C.J.M. van Oss and L.H.D.J. Booij, *J. Chromatogr.,* 424 (1988) 440.
315 R.L.P. Lindberg, J.H. Kanto and K.K. Pihlajamäki, *J. Chromatogr.,* 383 (1986) 357.
316 N. Daoud, T. Arvidsson and K.-G. Wahlund, *J. Pharm. Biomed. Anal.,* 5 (1987) 533.
317 E.J.D. Lee, S.B. Ang and T.L. Lee, *J. Chromatogr.,* 420 (1987) 203.
318 G.H. Wagman and M.J. Weinstein, *Chromatography of Antibiotics,* Elsevier, Amsterdam, 2nd ed., 1984.
319 R.K. Gilpin and L.A. Pachla, *Anal. Chem.,* 59 (1987) 176R.
320 M.C. Rouan, *J. Chromatogr.,* 340 (1985) 361.
321 P.A. Ristuccia, *J. Liquid Chromatogr.,* 10 (1987) 241.

322 S. Ting, *J. Assoc. Offic. Anal. Chem.*, 71 (1988) 1123.
323 J.A. McAteer, M.F. Hiltke, B.M. Silber and R.D. Faulkner, *Clin. Chem.*, 33 (1987) 1788.
324 M.D. Blanchin, H. Fabre and B. Mandron, *J. Liquid Chromatogr.*, 11 (1988) 2993.
325 F. Demotes-Mainard, G. Vincon, C. Jarry, J. Necciari and H. Albin, *J. Chromatogr.*, 419 (1987) 388.
326 J.E. Conte, Jr. and E. Zurlinden, *J. Chromatogr.*, 417 (1987) 452.
327 T. Marunaka, E. Matsushima and M. Maniwa, *J. Chromatogr.*, 420 (1987) 329.
328 N. Oldfield, D. Chang, W. Garland and C. Town, *J. Chromatogr.*, 422 (1987) 135.
329 A.J. Falkowski, L.M. Look, H. Noguchi and B.M. Silber, *J. Chromatogr.*, 422 (1987) 145.
330 L. Hakin, D.W.A. Bourne and E.J. Triggs, *J. Chromatogr.*, 424 (1988) 111.
331 R. Wyss and F. Bucheli, *J. Chromatogr.*, 430 (1988) 81.
332 L. Hary and M. Andrejak, *J. Chromatogr.*, 419 (1987) 396.
333 T.A. Emm, J. Leslie, M. Chai, L.J. Lesko and M.B. Perkal, *J. Chromatogr.*, 427 (1988) 162.
334 K. Tyczkowska and A.L. Aronson, *J. Chromatogr.*, 427 (1988) 103.
335 M.C. Rouan, *J. Chromatogr.*, 426 (1988) 335.
336 K. Lindgren, *J. Chromatogr.*, 413 (1987) 347.
337 K. Lindgren, *J. Chromatogr.*, 413 (1987) 351.
338 I.L. Smith, D.J. Swanson, L.S. Welage, C. De Angelis, S.A. Boudinot and J.J. Schentag, *Anal. Lett.*, 18 (1985) 1077.
339 R. Phelps, E. Zurlinden, J.E. Conte, Jr. and E. Lin, *J. Chromatogr.*, 375 (1986) 111.
340 J.M. Friis and D.B. Lakings, *J. Chromatogr.*, 382 (1986) 399.
341 G. La Follette, S. Kaubisch, J.G. Gambertoglio and E.T. Lin, *J. Liquid Chromatogr.*, 11 (1988) 683.
342 M. Bliss and M. Mayersohn, *Clin. Chem.*, 32 (1986) 197.
343 F.M. Denotes-Mainard, G.A. Vincon, C.H. Jarry and H.C. Albin, *J. Pharm. Biomed. Anal.*, 6 (1988) 407.
344 J. Martín and R. Mendéz, *J. Liquid Chromatogr.*, 11 (1988) 1729.
345 K. Tyczkowska, D.P. Aucoin, D.C. Richardson and A.L. Aronson, *J. Liquid Chromatogr.*, 10 (1987) 2613.
346 M.D. Blanchin and M.L. Rondot-Dudragne, *J. Chromatogr.*, 432 (1988) 407.
347 J. Haginaka and J. Wakai, *J. Chromatogr.*, 413 (1987) 219.
348 J. Carlqvist and D. Westerlund, *J. Chromatogr.*, 344 (1985) 285.
349 J. Haginaka and J. Wakai, *Anal. Biochem.*, 168 (1988) 132.
350 B. Wiese and K. Martin, *J. Pharm. Biomed. Anal.*, 7 (1989) 95.
351 B. Wiese and K. Martin, *J. Pharm. Biomed. Anal.*, 7 (1989) 107.
352 N. Omar, G. Saleh, M. Neugebauer and G. Rücker, *Anal. Lett.*, 21 (1988) 1337.
353 A.J. Shah, M.W. Adlard and G. Holt, *Analyst (London)*, 113 (1988) 1197.
354 J. Martín, R. Mendéz and A. Negro, *J. Liquid Chromatogr.*, 11 (1988) 1707.
355 C.T. Hung, J.K.C. Lim, A.R. Zoest and F.C. Lam, *J. Chromatogr.*, 425 (1988) 331.
356 E.-S.A. Ibrahim, M.E. Abdel-Hamid, M.A. Abuirjeie and A.M. Hurani, *Anal. Lett.*, 21 (1988) 423.
357 A.H. Hikal and A.B. Jones, *J. Liquid Chromatogr.*, 8 (1985) 1455.
358 T. Valenza and P. Roselli, *Chromatographia*, 24 (1987) 862.
359 J. Knöller, W. König, W. Schönfeld, K.D. Bremm and M. Köller, *J. Chromatogr.*, 427 (1988) 257.
360 R.J. Sawchuk and L.C. Cartier, *Clin. Chem.*, 27 (1981) 1368.
361 R.G. Buice, F.B. Stentz and B.J. Gurley, *J. Liquid Chromatogr.*, 10 (1987) 421.
362 H. Hosotsubo, J. Takezawa, N. Taenaka, K. Hosotsubo and Y. Yoshiya, *J. Chromatogr.*, 383 (1986) 349.
363 P.M. Kabra, J.H. Wall and N. Blanckaert, *Clin. Chem.*, 31 (1985) 1717.
364 W.M. Awni and J.A. Maloney, *J. Chromatogr.*, 425 (1988) 233.
365 A.K. Shah and R.J. Sawchuk, *Clin. Chem.*, 34 (1988) 1467.
366 G.L. Lensmeyer, D.A. Wiebe and I.H. Carlson, *Clin. Chem.*, 33 (1987) 1841.
367 D.J. Gmur, P. Meier and G.C. Yee, *J. Chromatogr.*, 425 (1988) 343.

368 P.E. Wallemacq and M. Lesne, *J. Chromatogr.*, 413 (1987) 131.
369 K. Oka, K. Hosoda, T. Hirano, K. Sakurai and M. Kozaki, *J. Chromatogr.*, 490 (1989) 145.
370 G.L. Lensmeyer, D.A. Wiebe and I.H. Carlson, *Clin. Chem.*, 33 (1987) 1851.
371 B.A. Wolf, M.C. Daft, J.W. Koenig, M.W. Flye, J.W. Turk and M.G. Scott, *Clin. Chem.*, 35 (1989) 120.
372 H. Ehrsson, S. Eksborg, I. Wallin, Y. Mårde and B. Joansson, *J. Pharm. Sci.*, 69 (1980) 710.
373 S.Y. Chang, B.J. Larcom, D.S. Alberts, B. Larsen, P.D. Walson and I.G. Sipes, *J. Pharm. Sci.*, 69 (1980) 80.
374 C.G. Adair, D.T. Burns, A.D. Crockard and M. Harriott, *J. Chromatogr.*, 342 (1985) 447.
375 M. Zakaria and P.R. Brown, *J. Chromatogr.*, 230 (1982) 381.
376 A.M. Rustum and N.E. Hoffman, *J. Chromatogr.*, 422 (1987) 125.
377 L.C. Burton and C.A. James, *J. Chromatogr.*, 431 (1988) 450.
378 A.-H.F.A. Hadidi and J.R. Idle, *J. Chromatogr.*, 427 (1988) 121.
379 J. Šalamoun, M. Smrž, F. Kiss and A. Šalamounová, *J. Chromatogr.*, 419 (1987) 213.
380 P.A. Brimmell and D.J. Sams, *J. Chromatogr.*, 413 (1987) 320.
381 A.M. Rustum and N.E. Hoffman, *J. Chromatogr.*, 426 (1988) 121.
382 A.M. Rustum and N.E. Hoffman, *J. Chromatogr.*, 421 (1987) 387.
383 S. Azeemuddin, Z.H. Israili and F.M. Bharmal, *J. Chromatogr.*, 430 (1988) 163.
384 J.M. van Baal, M.B. van Leeuwen, T.J. Schouten and R.A. De Abreu, *J. Chromatogr.*, 336 (1984) 422.
385 T. Dooley and J.L. Maddocks, *J. Chromatogr.*, 337 (1985) 321.
386 L. Lennard, *J. Chromatogr.*, 423 (1987) 169.
387 P.A.J. Speth, P.C.M. Linssen, J.B.M. Boezeman, J.M.C. Wessels and C. Haanen, *J. Chromatogr.*, 377 (1986) 415.
388 R. Mahdadi, N. Pommery, J. Pommery and M. Lhermitte, *Biomed. Chromatogr.*, 2 (1987) 91.
389 R. Mahdadi, M. Lhermitte and J.J. Lafitte, *Biomed. Chromatogr.*, 2 (1987) 38.
390 R.P. Klett and J.P. Chovan, *J. Chromatogr.*, 387 (1985) 182.
391 L.M. Rose, K.F. Tillery, S.M. El Dareer and D.L. Hill, *J. Chromatogr.*, 425 (1988) 419.
392 P.A. Maessen, H.M. Pinedo, K.B. Mross and W.J.F. van der Vijgh, *J. Chromatogr.*, 424 (1988) 103.
393 E.A. Lien, P.M. Ueland, E. Solheim and S. Kvinnsland, *Clin. Chem.*, 33 (1987) 1608.
394 C.P. Daniel, S.J. Gaskel, K. Bishop and R.J. Nicholson, *J. Endocrinol.*, 83 (1979) 401.
395 H.K. Adam, M.A. Gay and R.H. More, *J. Endocrinol.*, 84 (1980) 35.
396 Y. Golander and L.A. Sternson, *J. Chromatogr.*, 181 (1980) 41.
397 R.R. Brown, R. Bain, V.C. Jordan, *J. Chromatogr.*, 272 (1983) 351.
398 M. Nieder and H. Jaeger, *J. Chromatogr.*, 413 (1987) 207.
399 R.D. Armstrong, T.J. Ward, N. Pattabiraman, C. Benz and D.W. Armstrong, *J. Chromatogr.*, 414 (1987) 192.
400 S. Görög, B. Herenyi and K. Jovanovics, *J. Chromatogr.*, 139 (1977) 203.
401 M. Verzele, L. De Taeye, J. Van Dyck, G. De Decker and C. De Pauw, *J. Chromatogr.*, 244 (1981) 95.
402 J.E. Bodnar, J.R. Chen, W.H. Johns, E.P. Mariani and E.C. Shinal, *J. Pharm. Sci.*, 72 (1983) 535.
403 J.A. Houghton, P.M. Torrance and P.J. Houghton, *Anal. Biochem.*, 134 (1983) 450.
404 M. De Smet, S.J.P. Van Belle, G.A. Storme and D.L. Massart, *J. Chromatogr.*, 345 (1985) 309.
405 M. De Smet, S.J.P. Van Belle, V. Seneca, G.A. Storme and D.L. Massart, *J. Chromatogr.*, 416 (1987) 375.
406 D.E.M.M. Vendrig, J. Teeuwsen and J.J.M. Holthuis, *J. Chromatogr.*, 424 (1988) 83.
407 R. Kizu, S. Higashi and M. Miyazaki, *Chem. Pharm. Bull.*, 33 (1985) 4614.
408 S.J. Bannister, L.A. Sternson and A.J. Repta, *J. Chromatogr.*, 173 (1979) 333.
409 K.C. Marsh, L.A. Sternson and A.J. Repta, *Anal. Chem.*, 56 (1984) 491.

410 I.S. Krull, X.D. Ding, S. Braverman, C. Selavka, F. Hochberg and L.A. Sternson, *J. Chromatogr. Sci.*, 21 (1983) 166.
411 C. Gooyer, A.C. Veltkamp, R.A. Baumann, N.H. Velthorst, R.W. Frei and W.J.F. van der Vijgh, *J. Chromatogr.*, 312 (1984) 337.
412 W.A.J. de Waal, F.J.M.J. Maessen and J.C. Kraak, *J. Chromatogr.*, 407 (1987) 253.
413 P.T. Daley-Yates and D.C.H. McBrien, *Biochem. Pharmacol.*, 33 (1984) 3063.
414 G.S. Baldew, K.J. Volkers, J.J.M. de Goeij and N.P.E. Vermeulen, *J. Chromatogr.*, 491 (1989) 163.
415 P. Canal, C. Michel, R. Bugat, G. Soula and M. Carton, *J. Chromatogr.*, 375 (1986) 451.
416 M.A.J. van Opstal, P. Krabbenborg, J.J.M. Holthuis, W.P. van Bennekom and A. Bult, *J. Chromatogr.*, 432 (1988) 395.
417 G.F. Duncan, R.H. Farmen, H.S. Movahhed and K.A. Pittman, *J. Chromatogr.*, 380 (1986) 357.
418 A. El-Yazigi and C.R. Martin, *Clin. Chem.*, 33 (1987) 803.
419 C.E. Werkhoven-Goewie, U.A.T. Brinkman, R.W. Frei, C. de Ruiter and J. de Vries, *J. Chromatogr.*, 276 (1983) 349.
420 R.L. Smith and D.W. Walker, *J. Chromatogr.*, 343 (1985) 203.
421 O. Teirlynck, M.G. Bogaert, P. Demedts and H. Taelman, *J. Pharm. Biomed. Anal.*, 7 (1989) 123.
422 K.-J. Pettersson, T. Nordgren and D. Westerlund, *J. Chromatogr.*, 488 (1989) 447.
423 R.P. Remmel, Y.-H. Yeom, M. Hua, R. Vince and C.L. Zimmerman, *J. Chromatogr.*, 489 (1989) 323.
424 J. Cronqvist and I. Nilsson-Ehle, *J. Liquid Chromatogr.*, 11 (1988) 2593.
425 R. Paroni, C.R. Sirtori, C. Borghi and M.G. Kienle, *J. Chromatogr.*, 420 (1987) 189.
426 R.H.A. Smith and B.E. Gilbert, *J. Chromatogr.*, 414 (1987) 202.
427 L.E. Mathes, G. Muschik, L. Demby, P. Polas, D.W. Mellini, H.J. Isaaq and R. Sams, *J. Chromatogr.*, 432 (1988) 346.
428 J. Šalamoun, V. Šprta, T. Sládek and M. Smrž, *J. Chromatogr.*, 420 (1987) 197.
429 F.R. Rubio, T. Crews, W.A. Garland and E.K. Fukuda, *Biomed. Environ. Mass Spectrom.*, 17 (1988) 399.
430 C. Giachetti, P. Poletti and G. Zanolo, *J. High Resolut. Chromatogr. Chromatogr. Commun.*, 10 (1987) 469.
431 R.J. Stubbs, L.L. Ng, L.A. Entwistle and W.F. Bayne, *J. Chromatogr.*, 413 (1987) 171.
432 M.J. D'Souza, M.A. Zemaitis, G.J. Burckart and R. Venkataramanan, *J. Chromatogr.*, 421 (1987) 198.
433 A. Shah and D. Jung, *J. Chromatogr.*, 344 (1985) 408.
434 Y. Katogi, T. Ohmura and M. Adachi, *J. Chromatogr.*, 278 (1983) 475.
435 W.J. Adams, B.E. Bothwell, W.B. Bothwell, G.J. Van-Giessen and D.G. Kaiser, *Anal. Chem.*, 59 (1987) 1504.
436 R.J. Stubbs, M.S. Schwartz, R. Chiou, L.A. Entwistle and W.F. Bayne, *J. Chromatogr.*, 383 (1986) 432.
437 J. Macek and J. Vácha, *J. Chromatogr.*, 420 (1987) 445.
438 S.G. Owen, M.S. Roberts and W.T. Friesen, *J. Chromatogr.*, 416 (1987) 293.
439 D. Sauvaire, M. Cociglio and R. Alric, *J. Chromatogr.*, 375 (1986) 101.
440 V.A. Johnson and J.T. Wilson, *J. Chromatogr.*, 382 (1986) 367.
441 H.T. Karnes, K. Rajasekharaiah, R.E. Small and D. Farthing, *J. Liquid Chromatogr.*, 11 (1988) 489.
442 J.H.G. Jonkman, R. Schoenmakers, A.H. Holtkamp and J. Hempenius, *J. Pharm. Biomed. Anal.*, 3 (1985) 433.
443 J.H. Satterwhite and F.D. Boudinot, *J. Chromatogr.*, 431 (1988) 444.
444 K. Oka, S. Aoshima and M. Noguchi, *J. Chromatogr.*, 382 (1986) 367.
445 J. Godbillon, S. Gauron and J.J. Metayer, *J. Chromatogr.*, 338 (1985) 151.
446 N. Miyagi, N. Hikichi and H. Niwa, *J. Chromatogr.*, 375 (1986) 91.
447 C. Giachetti, S. Canali and G. Zanolo, *J. Chromatogr.*, 279 (1983) 587.
448 H.J. Battista, G. Wehinger and R. Henn, *J. Chromatogr.*, 345 (1985) 77.

449 F. Lapicque, P. Netter, B. Bannwarth, P. Trechot, P. Gillet, H. Lambert and R.J. Royer, *J. Chromatogr.*, 496 (1989) 301.
450 N. Grgurinovich, *J. Chromatogr.*, 414 (1987) 211.
451 R.J. O'Kruk, M.A. Adams, R.B. Philp, *J. Chromatogr.*, 310 (1984) 343.
452 B. Chai, P.E. Minkler and C.L. Hoppel, *J. Chromatogr.*, 430 (1988) 93.
453 C.J. Richardson, S.G. Ross, K.L. Blocka and R.K. Verbeeck, *J. Chromatogr.*, 382 (1986) 382.
454 T.C. Pinkerton, J.A. Perry and J.D. Rateike, *J. Chromatogr.*, 367 (1986) 412.
455 A.J. Hutt, S. Fournel and J. Caldwell, *J. Chromatogr.*, 378 (1986) 409.
456 H. Spahn, *J. Chromatogr.*, 423 (1987) 334.
457 A. Avgerinos and A.J. Hutt, *J. Chromatogr.*, 415 (1987) 75.
458 R. Mehvar, F. Jamali and F.M. Pasutto, *Clin. Chem.*, 34 (1988) 493.
459 R.T. Foster and F. Jamali, *J. Chromatogr.*, 416 (1987) 388.
460 D.A. Nicoll-Griffith, T. Inaba, B.K. Tang and W. Kalow, *J. Chromatogr.*, 428 (1988) 103.
461 G. Geisslinger, K. Dietzel, D. Loew, O. Schuster, G. Rau, G. Lachmann and K. Brune, *J. Chromatogr.*, 491 (1989) 139.
462 D. de Zeeuw, J.F. Leinfelder and D.C. Brater, *J. Chromatogr.*, 380 (1986) 157.
463 C.R. Schuster, *Clin. Chem.*, 33 (1987) 7B.
464 C.M. Selavka and I.S. Krull, *J. Liquid Chromatogr.*, 10 (1987) 345.
465 F. Tagliaro, D. Franchi, R. Dorizzi and M. Marigo, *J. Chromatogr.*, 488 (1989) 215.
466 C. Kim, M.P. Speisky and H. Kalant, *J. Chromatogr.*, 370 (1986) 303.
467 B.K. Logan, J.S. Oliver and H. Smith, *Forensic Sci. Int.*, 35 (1987) 189.
468 J. Zoer, P. Virgili and J.A. Henry, *J. Chromatogr.*, 382 (1986) 189.
469 J.-O. Svensson, *J. Chromatogr.*, 375 (1986) 174.
470 P.E. Nelson, *J. Chromatogr.*, 298 (1984) 59.
471 F. Tagliaro, A. Frigerio, R. Dorrizi, G. Lubli and M. Marigo, *J. Chromatogr.*, 330 (1985) 323.
472 S.P. Joel, R.J. Osborne and M.L. Slevin, *J. Chromatogr.*, 430 (1988) 394.
473 J. Schuberth and J. Schuberth, *J. Chromatogr.*, 490 (1989) 444.
474 R.T. Jones, *Clin. Chem.*, 33 (1987) 72B.
475 L. Vollner, D. Bieniek and F. Korte, *Regul. Toxicol. Pharmacol.*, 6 (1986) 348.
476 R.L. Hawks and C.N. Chiang (Editors), *Urine Testing for Drugs of Abuse*, NIDA Research Monograph 73, National Institute of Drug Abuse, Rockville, MD, 1986.
477 T.P. Moyer, M.A. Palmen, P. Johnson, J.R. Charlson and P.J. Ellefson, *Mayo Clin. Proc.*, 62 (1987) 413.
478 U. Lemm, J. Tenczer, H. Baudisch and W. Krause, *J. Chromatogr.*, 342 (1985) 393.
479 Y. Nakahara and C.E. Cook, *J. Chromatogr.*, 434 (1988) 247.
480 A.N. Masoud and D.W. Wingard, *J. High Resolut. Chromatogr. Chromatogr. Commun.*, 2 (1979) 118.
481 P.I. Jatlow, *Clin. Chem.*, 33 (1987) 66B.
482 R.S. Schwartz and K.O. David, *Anal. Chem.*, 57 (1985) 1362.
483 M.J. Kogan, D.J. Pierson, M.M. Durkin and N.J. Willson, *J. Chromatogr.*, 490 (1989) 236.
484 P. Jacob, III, B.A. Elias-Baker, R.T. Jones and N.L. Benowitz, *J. Chromatogr.*, 306 (1984) 173.
485 A.S. Christophersen, E. Dahlin and G. Pettersen, *J. Chromatogr.*, 432 (1988) 290.
486 B.M. Farrell and T.M. Jefferies, *J. Chromatogr.*, 272 (1983) 111.
487 R.L. Fitzgerald, R.V. Blanke, R.A. Glennon, M.Y. Yousif, J.A. Rosecrans and A. Poklis, *J. Chromatogr.*, 490 (1989) 59.
488 R.V. Brooks, R.G. Firth and N.A. Sumner, *Br. J. Sports. Med.*, 9 (1975) 89.
489 R. Massé, C. Ayotte and R. Dugal, in G. Piemonte, F. Tagliaro, M. Marigo and A. Frigerio (Editors), *Developments in Analytical Methods in Pharmaceutical, Biomedical, and Forensic Sciences*, Plenum Press, New York, 1987, pp. 183-190.
490 R. Massé, C. Ayotte and R. Dugal, *J. Chromatogr.*, 489 (1989) 23.
491 D.H. Catlin, R.C. Kammerer, C.K. Hatton, M.H. Sekera and J.L. Merdink, *Clin. Chem.*, 33 (1987) 319.

492 E.H.J.M. Jansen, R. Both-Miedema and R.H.. van den Berg, *J. Chromatogr.*, 489 (1989) 57.

493 B. Bergner-Lang and M. Kaechele, *Dtsch. Lebensm.- Rundsch.*, 83 (1987) 349.

494 H.F. De Brabander, P. Vanhee, S. Van Hoye and R. Verbeke, *J. Planar Chromatogr.*, 2 (1989) 33.

495 L.A. van Ginkel, R.W. Stephany, H.J. van Rossum, H. van Blitterswijk, P.W. Zoontjes, R.C.M. Hooijschuur and J. Zuydendorp, *J. Chromatogr.*, 489 (1989) 95.

496 L.A. van Ginkel, R.W. Stephany, H.J. van Rossum, H.M. Steinbuch, G. Zomer, E. van de Heeft and A.P.J.M. de Jong, *J. Chromatogr.*, 489 (1989) 111.

497 W. Haasnoot, R. Schilt, A.R.M. Hamers, F.A. Huf, A. Farjam, R.W. Frei and U.A.Th. Brinkman, *J. Chromatogr.*, 489 (1989) 157.

498 J.O. de Beer, *J. Chromatogr.*, 489 (1989) 139.

499 E.H.J.M. Jansen, D. Van den Bosch, R.W. Stephany, L.J. Van Look and C. Van Peteghem, *J. Chromatogr.*, 489 (1989) 205.

Chapter 21

Fossil fuels

R.P. PHILP and F. XAVIER DE LAS HERAS

CONTENTS

21.1 INTRODUCTION

The major objective of this chapter is to highlight applications of various chromato-graphic techniques, such as gas chromatography (GC), high-performance liquid chromatography (HPLC), supercritical-fluid chromatography (SFC) and size-exclusion chromatography (SEC), along with related techniques, principally mass spectrometry (MS), to the analysis and characterization of various types of fossil fuels, such as crude oils, coals, and oil shales. Furthermore, the significance of results obtained from these studies will be discussed. In a chapter of this nature it is impossible to provide a compre-hensive review of all aspects of such a topic. However, it should be recognized that in the previous edition of this book there were two chapters that are of direct relevance to the current topic, namely those concerned with porphyrins [1, Chap. 19] and hydrocarbons [1, Chap. 24]. Since then, additional applications of chromatography to fossil fuel charac-terization have been described in several review articles [2-13]. The annual reviews of gas chromatography, mass spectrometry, petroleum, solid and gaseous fuels, published in *Analytical Chemistry* [2-8], provide excellent comprehensive reviews of these topics and their application to fossil fuel characterization. Another excellent source of reference material on the topic is The Proceedings of the International Organic Geochemistry meetings [10-12] along with the Proceedings of the Danube Chromatography Conference [9]. The latter review is particularly useful for obtaining an insight into developments that have occurred in chromatography in various Eastern European countries.

The methods used in the characterization of fossil fuels will vary greatly according to the nature of the fossil fuel being analyzed and the type of compounds present in the sample. In the part of this chapter dealing with soluble organic matter we will discuss the use of GC, then GC/MS, HPLC, SFC, and SEC. In addition, the use of pyrolysis techniques combined with GC and GC/MS for the characterization of the insoluble organic fractions of fossil fuels, such as kerogens and asphaltenes will be discussed. In most applications of chromatography to fossil fuel analyses, the actual chromatographic techniques have become fairly routine and advances tend to parallel major developments in the associated technology, such as new injector designs, high-temperature phases or liquid-crystal phases for capillary column. A major breakthrough in the application of chromatography to exploration for crude oil came as a result of the development of the so-called biomarker concept. A biomarker can be best thought of as an organic compound in a geological sample which is structurally related to its precursor molecule, which in turn occurs as a natural product in higher plants, algae, bacteria, or any other potential source material. Although the idea was initially proposed in the Thirties, it was not widely exploited until the major analytical advances in GC, MS, and GC/MS occurred in the Sixties and Seventies. These developments included coupling of gas chromatographs and mass spectrometer systems, development of glass capillary columns and, more recently, fused-silica columns, bonded phases, and phases stable at high temperatures. The ultimate result of these analytical developments, for geochemical purposes, has been the ability to determine trace amounts of components, i.e. biomarkers, in very complex organic extracts derived from geological samples. The distribution of biomarkers in such samples provides information on the source or maturity of organic matter in a sample and, in the case of oils, on the extent of biodegradation and migration. The nature of the depositional environments will also play an important role in determining the amount and rate at which organic matter will accumulate in a basin [14]. The ability to recognize changes in depositional environments through different stratigraphic horizons plays a major role in the evaluation of a basin for petroleum exploration purposes [15]. If a specific set of biomarker parameters can be assigned to a clearly defined depositional environment, it will permit recognition of this type of environment in previously unexplored regions. In many cases the presence of a particular biomarker can be attributed to a specific organism that will grow only under the conditions peculiar to a particular depositional environment, e.g., gammacerane (hypersaline). Alternatively, an unusual biomarker fingerprint, such as the predominance of even-numbered n-alkanes over odd-numbered ones or the predominance of C_{35} hopanes over C_{34} hopanes, may be associated with a highly reducing environment. It is interesting to note that in the Seventies much of the biomarker work was directed at studies of recent sediments [16,17]. The results and information from those studies have subsequently been invaluable for the application of the biomarker concept to petroleum studies. More recent advances in analytical technology, the continued development of HPLC systems, SFC, and a new generation of mass spectrometers (namely triple-stage quadrupole and hybrid quadrupole/magnetic instruments) are now opening the way for alternative and more sophisticated methods of biomarker determinations.

The three major fractions, commonly associated with virtually all fossil fuels and analyzed by various chromatographic techniques, are the organic-solvent-soluble extracts, the pentane-insoluble asphaltenes, and the insoluble kerogens. The characteristics of these three fractions vary as a result of differences in polarity, molecular weight, solubility, and heteroatom content. Characterization of the solvent-extractable material generally requires further fractionation into saturates, aromatics, and a fraction containing nitrogen, sulfur, and oxygen compounds (NSO) by column liquid or thin-layer chromatography (CLC or TLC), or more recently, HPLC and supercritical-fluid extraction (SFE) prior to any analysis by GC, GC/MS, or GC/MS/MS.

Asphaltenes and kerogens are commonly subjected to pyrolysis in order to break down their geopolymer structure prior to GC analyses and to determine the nature and characteristics of the various pyrolysis products. The insoluble organic portions of coals have also been extensively characterized by various pyrolysis techniques, combined with GC and GC/MS [18].

Whereas crude oils, coal, or oil shale extracts are typically analyzed as part of developing exploration strategies, the major aim of analyzing synthetic fossil fuels is to characterize the various components present in the fuel for purposes of improving their industrial utilization (avoiding catalyst poisoning) and of anticipating environmental problems that may result from their combustion due to the presence of various nitrogen and sulfur compounds. Many of the chromatographic techniques used for the characterization of synthetic fuels are identical to those used in the analyses of natural fossil fuels.

21.2 GAS CHROMATOGRAPHY

For many years gas chromatography, originally with packed columns and now with high-resolution fused-silica capillary columns with bonded phases, has been the method of choice for the analysis of the extremely complex mixtures of organic compounds present in fossil fuels. As a result of the complexity of the mixtures, a number of pre-analysis steps are generally required to fractionate the extracts from these samples. The most common method of extraction involves crushing the source rocks, coals, or shales and then using organic solvent mixtures, such as toluene/chloroform or dichloromethane/methanol to obtain a solvent-soluble extract. Extraction systems, such as Soxhlet extractors, are commonly used to recycle solvents and to obtain a concentrated extract. As an alternative to the Soxhlet extraction process, SFE and the thermal desorption technique, described in Section 21.2.2, are used in certain applications. The higher-molecular-weight, typically more polar, and partially insoluble fraction, referred to as asphaltenes, was, and still is to a large extent, isolated from the solvent extracts by

References on p. B465

pentane precipitation. Kerogen concentrates are isolated from rock samples by removal of carbonates and silicates by HCl and HF respectively. The kerogen fraction is typically the most abundant form of organic matter in sedimentary rocks.

After solvent extraction, the extracts require an intermediate separation of fractionation step to reduce the complexity of the extract and to isolate groups or families of compounds of interest. The fractionation process is accomplished by various chromatographic techniques, such as TLC, CLC, or medium-high-pressure liquid chromatography. These techniques will provide saturate fractions, comprising linear, branched, and cyclic hydrocarbons, aromatic fractions, consisting of mono-, di-, tri-, and polyaromatic compounds, and polar fractions, containing nitrogen, sulfur, and/or oxygen compounds. The fractionation process simplifies the extract for the next step of the analysis, namely GC. Recent innovations which have been incorporated into the separation processes to achieve the best results include sequential chromatographic steps to obtain a narrower range of compounds [19,20]. Introduction of preliminary fractionation involving CLC of crude oils replaces the previous deasphalting steps [21]. Takayama et al. [22] have recently described the use of high-temperature columns to expand the molecular weight range of compounds in fossil fuels that can be characterized by GC. Unequivocal identification of unknown compounds in any of these fractions requires an additional analytical method or associated technique, such as MS or tandem mass spectrometry (MS/MS), which will be discussed below.

21.2.1 Combined gas chromatography/mass spectrometry

Combined GC/MS has been utilized for many years for the characterization of all types of fossil fuels, and it is the first choice of chromatographic techniques for most organic geochemistry laboratories. There are masses of GC/MS data from fossil fuel analyses recorded in the literature, and many reviews have been published which include these data [2-12]. Bench-top mass spectrometers are now capable of providing results similar to those obtained with more expensive, low-resolution mass spectrometers for the routine analysis of fossil fuel samples [23]. Most of the GC/MS reports in the literature are concerned with MS in single-ion monitoring (SIM) or multiple-ion detection (MID) modes, which provide fingerprints of specific classes of biomarkers. The complexity of fossil fuel extracts makes it exceedingly difficult to resolve all components chromatographically. Hence, MID analyses have been invaluable in the detection of trace amounts of specific biomarkers or other compounds of interest. More recently, MS/MS has provided an additional dimension to aid in simplifiyng these complex mixtures. For most applications in fossil fuel analyses based on GC/MS, ionization of the molecules is typically performed in electron-ionization (EI) mode at an electron energy of -70 eV. Alternative methods of ionization, such as chemical ionization (CI), have been used in a few cases to improve the intensity of the molecular ions and to assist in the identification of unknown components. Various authors [24,25] have described the use of ammonia as a reagent gas for the determination of aromatic nitrogen compounds in fossil fuels by GC/MS. Philp and Johnston [26] used methane as a reagent gas for the characterization of saturated hydrocarbon fractions from various crude oils in order to obtain information of biomarker distributions by the direct insertion probe. Schmitter et al. [27] characterized carboxylic acids of crude oils by GC/MS and LC/MS. Synthetic fuels, such as liquid coals, because of their high-molecular-weight range and polarity, are more commonly studied in the positive-ion mode [28-30] than are the naturally occurring fossil fuels. Chemical-ionization and negative-ion mass spectrometry have also been described by various researchers [31,32]. Relatively few examples of fast-atom bombardment (FAB) or chemical desorption have been reported in the characterization of fossil fuels.

In recent years, tandem mass spectrometry (MS/MS) has started to play an increasingly important role in the analysis of fossil fuels with or without GC. Tandem mass spectrometers can be used in several ways. One common system includes a magnetic sector, for high-resolution studies. In such a system, metastable-ion monitoring (MIM) detects the spontaneous fragmentation of the parent ion occurring in the first field-free region of a double-focusing mass spectrometer. These ions can be separately observed by using a programable power supply to vary the accelerating voltage while holding magnetic and electrostatic fields at constant values. One of the first examples of a MS/MS application to fossil fuel analysis was the work of Gallegos [33], who used MIM to analyze terpanes and steranes from crude oils. MS/MS with MIM has also been used to characterize various steranes in fossil fuels [34-36]. Tricyclic terpanes from C_{19} up to C_{45} have been identified by Moldowan et al. [37] in various crude oils by the MS/MS approach. More recently, Summons et al. [38,39] used this approach to characterize and identify a number of branched and isoprenoid alkanes in various Proterozoic oils and source rock samples. Warburton and Zumberge [35] described the use of metastable-peak monitoring to resolve sterane mixtures into individual C_{27}, C_{28}, and C_{29} components. Brooks and others have used high-resolution selected MIM to determine the distribution of various biomarkers in oils and source rocks [40,41].

The second major type of tandem mass spectrometer consists of three quadrupoles, coupled in tandem. The middle quadrupole is a collision cell, operated only in a radio frequency mode. Ions actually formed in the ion source of the mass spectrometer are separated in the first quadrupole, and selected ions pass into the collision cell. The resultant ions, formed as a result of the collisions with the inert gas, such as argon, in the collision cell pass into the third quadrupole for subsequent separation and analysis. The system can be operated in three different scan modes (parent, daughter, neutral), and all three of these have proved to be extremely helpful for the analysis of the complex fossil fuel-type samples and the determination of compounds, such as steranes, hopanes, and bicyclic terpanes in oils and source rock extracts [42-47].

In the daughter-scan mode, two stages of analysis are performed. In the first stage, ions or parent ions, formed in the ion source, enter Q1, which is referred to as the parent mass analyzer. Parent ions, selected by Q1, enter Q2, which is the collision cell. In the second stage of analysis, ions in the collision cell can fragment further to produce daughter ions by a process referred to as collision-activated dissociation (CAD). Ions formed in the collision cell enter Q3 (the daughter-mass analyzer) for the second stage of mass analysis. The Q3 is scanned to obtain a mass spectrum of the daughter ions,

References on p. B465

produced from the fragmentation of the selected parent ion. A mass spectrum, obtained in the daughter-scan mode, is the mass spectrum of fragments of a selected parent ion. Experiments in which the daughter-scan mode is employed can be used to survey complex organic mixtures very rapidly for specific compounds.

The parent-scan mode also employs two stages of analysis. In the first stage, ions formed in the ion source are introduced into the parent-mass analyzer. The mass analyzer is scanned to transmit parent ions sequentially into the collision cell. In the second stage of analysis, in the collision cell, parent ions undergo CAD to produce daughter ions. Ions formed in the collision cell enter the daughter-mass analyzer, which transmits a selected daughter ion. The resultant spectrum shows all the parent ions that have fragmented to produce the selected daughter ion. For a mass spectrum obtained in the parent-scan mode, data for the mass-to-charge ratio axis are obtained from Q1 (the parent ions), whereas data for the ion intensity axis are obtained from Q3 (the daughter ion being monitored). Experiments in which the parent-scan mode is employed are useful for the rapid detection of a series of structural homologs that have a common fragment ion (e.g., m/z 191 for the triterpanes or m/z 217 for the steranes).

In the neutral-loss-scan mode, the two mass analyzers (Q1 and Q3) are linked together in such a way that they are scanned at the same rate over mass ranges of the same width. However, the respective mass ranges are offset by a selected mass, such that the daughter-mass analyzer scans a selected number of mass units lower than the parent mass analyzer. Thus, in the neutral-loss-scan mode, there are two stages of mass analysis: Ions formed in the ion source are separated in the parent-mass analyzer and introduced sequentially into the collision cell, where they undergo fragmentation by CAD. These daughter ions are then separated by mass-to-charge ratio by the daughter-mass analyzer. For an ion to be detected, between the time the ion leaves Q1 and enters Q3, it must lose a neutral moiety having a mass equal to the difference in the mass ranges being scanned by the two mass analyzers. Thus, the spectrum obtained shows all the parent ions that lose a neutral species of selected mass. Experiments in which the neutral-loss-scan mode is used are useful when a large number of compounds is being surveyed for common functionality. Neutral moieties are frequently lost from substituent functional groups, e.g., CO_2 from carboxylic acids, CO from aldehydes, HS from sulfides, and H_2O from alcohols. For a neutral-loss mass spectrum, as for a parent mass spectrum, data for the mass-to-charge ratio axis are obtained from Q1 (the parent ion), whereas data for the ion intensity axis are obtained from Q3 (the daughter ion being monitored).

Selected-reaction monitoring (SRM) can also be used in the MS/MS scan modes and is analogous to SIM in the mass spectrometer scan modes. In SRM, a particular reaction or set of reactions, such as the fragmentation of an ion or the loss of a neutral moiety, is monitored. SRM can be employed in any of the commonly used MS/MS scan modes. Like SIM, SRM allows very rapid analysis of trace components in complex mixtures. However, because two sets of ions are being selected, the specificity obtained in SRM can be much greater than that obtained in SIM. Any interfering compound would not only have to form a parent ion of the same mass-to-charge ratio as the selected parent ion from the target compound, but that parent ion would also have to fragment to form a daughter ion of the same mass-to-charge ratio as the selected daughter ion from the target compound.

Many different variables can be utilized in the analysis, such as collision gas pressure and collison energy. Changing any of these parameters will affect the extent of ionization resulting from the collision process. As an additional variable, the system can be operated in the chemical-ionization mode. The results from several different experiments with methane as an ionizer gas have been described in the literature [26].

21.2.2 Pyrolysis/gas chromatography

Analytical pyrolysis systems coupled to a GC or GC/MS system provide a useful combination for typing or characterizing the most abundant forms of organic carbon in the sedimentary record, namely kerogen and asphaltenes [18]. Pyrolysis can also be used without chromatographic separation by passing pyrolysis products directly from the pyrolyzer into a FID. This approach is suitable for rapid screening of large volumes of samples (Rock-Eval technique). For detailed fingerprinting of the organic material, flash pyrolysis methods are used in conjunction with GC or GC/MS. A variety of pyrolysis methods have been described, which can generally be classified as either anhydrous or hydrous pyrolysis [48]. Anhydrous pyrolysis is useful for characterizing or fingerprinting kerogens [49], whereas hydrous pyrolysis is valuable for simulating the processes of petroleum generation, or maturation, on a relatively short-term scale in the laboratory [50]. A combination of hydrous and anhydrous pyrolysis techniques can provide an extremely powerful laboratory method for monitoring changes resulting from maturation of kerogens [51].

Most of the geochemical literature on pyrolysis deals with the production and identification of hydrocarbon pyrolysis products from asphaltenes or kerogens [52]. In view of the interest in the role of sulfur in the generation of oil and gas, current emphasis is being placed on the production and identification of organosulfur compounds from sulfur-rich source rocks [53]. The use of these compounds as indicators to distinguish depositional environments has also been discussed in some detail [54,55]. Philp and Bakel [56] reported that pyrolyzates from asphaltenes, isolated from limestone-sourced oils, contained high concentrations of benzothiophenes and dibenzothiophenes relative to thiophenes, while asphaltenes from shale-sourced oils produced more thiophenes than benzo- and dibenzothiophenes upon pyrolysis. Pyrolysis of kerogens has been shown to yield primarily thiophenes, with smaller amounts of benzothiophenes, dibenzothiophenes, and thiolanes [56].

There are even fewer papers in the literature concerned with the production of nitrogen-containing compounds by pyrolysis of the various asphaltenes, coals, and kerogens. Such analyses may help to elucidate the nature of the nitrogen molecules present in kerogen structures.

References on p. B465

21.2.2.1 Bulk-flow pyrolysis

Bulk-flow, or Rock-Eval-type, pyrolysis is well suited to the needs of the explorationist for fossil fuels, since it provides a rapid technique capable of characterizing the potential of various kerogen types to produce oil or gas and their maturity level without prior extraction or fractionation of the extracts. The quantity of hydrocarbons released from a source rock at 250°C by thermal desorption is comparable to the quantity of total organic solvent extract. The quantity of hydrocarbons produced from pyrolysis of the kerogen at higher temperatures will give an indication of the petroleum potential of the source rock and maturity level [57,58]. The organic oxygen content can also be measured by using a thermal conductivity detector to measure the CO_2 produced from the kerogen.

21.2.2.2 Flash-flow pyrolysis methods

The combination of pyrolysis with chromatography (Py/GC) and mass spectrometry (Py/GC/MS) enables a detailed analysis of kerogen or asphaltene degradation products. Pyrolysis systems commonly used include microfurnaces, pyroprobes, or Curie-point systems. Several reports in the literature also describe intermediate trapping prior to GC analysis [59-62]. Py/GC at pyrolysis temperatures around 350°C is being used as a thermal desorption technique and as an alternative to solvent extraction of volatile components [63-67]. The organic matter in a sedimentary rock consists of a solvent-soluble and solvent-insoluble fraction. Extraction of the rock, followed by fractionation, prior to GC and GC/MS analyses is a lengthy process. As an alternative, the solvent-soluble fraction can be thermally desorbed directly onto the GC column, and then the GC analyses are performed in the normal manner. Comparison of chromatograms obtained by the two methods show that the results are comparable. At pyrolysis temperatures of 350°C no appreciable breakdown of the kerogen occurs, hence all of the components observed in the chromatograms are those normally in the solvent-soluble fraction.

Hydrocarbons are typically the most abundant components in the pyrolyzates and are generally used to characterize different types of kerogen. Several authors have described applications of Py/GC to the characterization of various source rock and asphaltene samples [68-73]. Pyrolytic characterization of asphaltenes, isolated from both rock extracts and oils, is useful for the purpose of undertaking oil/oil and oil/source rock correlations [74]. This approach is particularly useful in the case of biodegraded oils, where many of the commonly used biomarkers have been removed and are not available for the more traditional biomarker correlation studies [75]. A number of examples demonstrating the use of Py/GC for coal characterization and coal-related liquids can be found in the literature [76,77].

One of the most recently developed fields of pyrolysis is the characterization of kerogens with high sulfur content. These are of interest because of their capability for early oil generation, and also because of problems associated with poisoning of catalysts during synthetic fuel generation and combustion and the environmental impact of these compounds. An updated compilation of papers concerned with this topic can be found in the recent Proceedings of the American Chemical Society Dallas Congress [13]. Many of these papers describe some parameters related with sulfur compounds that can be used to assess crude oil maturation. Eglinton et al. [51] used Py/GC to study various samples that had been artificially matured by pyrolysis. The sulfur compounds produced in this way appeared to show some systematic variations with maturity and nature of source material. Philp and Bakel [56] noticed a number of variations in both the organosulfur and organonitrogen compounds, produced by pyrolysis, with respect to depositional environment of the source material and maturity.

Hydrous pyrolysis has been widely used in the petroleum industry to obtain information on the petroleum potential of immature source rocks and for predicting what would have happened to these rocks if they had been buried deeply and hence become more mature [50]. Hydrous pyrolysis has also become widely accepted in recent years as a method for revealing the molecular structure of kerogen and for kerogen/crude oil correlation studies [78-80]. A study by Telnaes et al. [81] showed that hydrous pyrolysis of the asphaltic fraction of biodegraded and nondegraded oils resulted in the formation of hydrocarbons which may be similar in composition to the hydrocarbons in the original oil and at the same time prevented the formation of unsaturated compounds.

21.3 HIGH-PERFORMANCE LIQUID CHROMATOGRAPHY

TLC is routinely used by many laboratories involved in fossil fuel research, and innovations are relatively rare [82-84]. Similarly, few modifications to liquid adsorption chromatographic techniques have been made in studies of fossil-fuel-related materials [85-89].

New trends include LC coupled with GC. An excellent review on this application to fossil fuel has been written by Davies et al. [90], and recent papers on coal-derived materials have been published [91-93]. LC/GC/MS has been used to determine aromatic and sulfur compounds from solvent-refined coals [94]. Coupling of LC with proton magnetic resonance spectroscopy has been applied [95], along with a novel cleanup technique for HPLC [96].

The use of HPLC to fractionate fossil fuels into their various aromatic fractions has become commonplace. In order to reduce the time-consuming column-chromatographic processes, conventional methods of hydrocarbon group type separation have gradually been replaced by HPLC. For monitoring cut points between the various compound types, different detectors are used, including UV, differential refractometer, and fluorescence detectors. The most commonly used stationary phases for normal- and reversed-phase chromatography are modified alumina or silica. Although alumina is still used extensively as a stationary phase for the separation of PAHs, chemically bonded phases containing functional groups, such as amines, nitriles, pyrrolidone, or dinitroanilinopropyl, are starting to be used for normal-phase chromatography, whereas octadecylsilyl is one of the most commonly used bonded phases for silica in reversed-phase chromatography. In the normal-phase mode, the solvent system is typically hexane with varying amounts of more

References on p. B465

polar solvents for separating aromatics in the order of the number of aromatic rings. In reversed-phase LC (RPLC) polar solvents are used, such as mixtures of acetonitrile or methanol and water. RPLC is typically used to subdivide fractions on the basis of their degree of methylation. Radke et al. [97] described an automated medium-pressure liquid chromatograph, for the fractionation of aromatic hydrocarbons according to their degree of aromatization. This technique has been widely used to isolate aromatic fractions.

A combination of normal- and RPLC has been used by Garrigues et al. [98-104] to discriminate between different aromatic ring systems and degrees of methylation in order to characterize maturity. RPLC has been used to separate nitrogen aromatic compounds in crude oils [105]. Specific LC separation schemes for sulfur compounds have not been reported in any detail, although Radke and Willsch [106] were able to utilize their auto-mated system and methodology to separate such compounds. Synthetic fuels have been the subject of several detailed studies by HPLC [107,108] as have solvent-refined coals by microcolumn LC [109].

Combined LC/MS technology has not been applied to fossil fuel problems as exten-sively as GC/MS. Dark et al. [110] were the first to use HPLC to separate hexane-soluble oil from coal on the basis of aromatic ring numbers. Further separation of individual fractions was performed by HPLC, combined directly with a mass spectrometer operating in the chemical-ionization mode. In recent studies [111-113] CI was also used for the characterization of various fractions from coal-derived materials. Until, recently, porphyrins [114-117] were not widely used as biological markers because their derivatives were not sufficiently volatile for analysis by GC/MS. However, various groups [118-120] have shown that it is possible to analyze porphyrins by GC on fused-silica capillary columns, following the preparation of bis(trimethylsiloxy)-Si(IV) derivatives. An extension of this work was the investigation of the TMS derivatives of Si(IV), Al(III), Ga(III), and Rh(III) alkyl-substituted porphyrins and their analysis. Porphyrins with ester side chains have also been analyzed as their silicon(IV) derivatives by GC/MS [121]. The continued development of GC and GC/MS methods [122] for porphyrin analysis complements the HPLC/MS methods de-veloped for petroporphyrins. Similarly, HPLC/MS was used to study demetallated porphyrins in shales and oils from the Shengli oilfield in China as part of an attempt to correlate the oils with suspected source rocks [123]. Porphyrins are widely distributed in coals, as shown by Palmer et al. [124] in a study of 42 U.S. coals ranging from lignite to anthracite. The major porphyrins were found to be of the etio series, and are thought to have been formed during early stages of coal formation.

21.4 SUPERCRITICAL-FLUID CHROMATOGRAPHY

SFC has now been added as an analytical method for extracting and separating thermally labile compounds and high-molecular-weight oligomers under mild chromato-graphic conditions with powerful solvent efficiency. Even though the utility of this extraction process has been known for a long time, only recently has its utility for the characterization of complex mixtures after separation been recognized. Most of the work performed to date in this area of fossil fuels has been concerned with the optimization of the apparatus and peripheral equipment. A few problems have been solved satisfactorily, and a few applications in the field of fossil fuels have been reported.

Monin et al. [125] coupled the extraction capacity of SFE and SFC for a few samples of crude oils, source-rock extracts, and oil shales in order to compare the efficiency of this method with the classical organic solvent extraction and GC. When extracts were ana-lyzed on-line by GC, results comparable with those for a chloroform extract of a saturated fraction were obtained, although recovery yields for the aromatic and polar fractions were relatively low. Fuhr et al. [126] chromatographed heavy oils, dissolved in CS_2, from several Canadian tar sands at a temperature of 90°C with a pressure program from 1150 to 5600 psi, using a column of polysiloxane-bonded phase, and a FID, which detected hydrocarbons up to C_{90}.

Synthetic fuels, such as coal tar [127], tar pitch [128], and coal liquids [83,129,130] have been analyzed by SFC. Entire oil shales [131] and different fractions [132] have also been studied, and gasoline fractions isolated from crude oils have been examined, using a number of different detectors [133]. SFC, coupled with mass spectrometry, has been applied by Wright et al. [29] to the analysis of nitrogen compounds in heavy-oil distillates by chemical-ionization MS.

21.5 SIZE-EXCLUSION CHROMATOGRAPHY

SEC was introduced when it became feasible to use microgels and HPLC equipment. The technique allows characterization of molecular size and, with some precautions, make average-molecular-weight determinations. The technique, being based on the principle of separation according to size and not according to functionality, plays a role in the charac-terization of fossil fuels that is very different from the other techniques described so far.

A few reports in the current literature describe applications of this technique in the characterization of fossil fuels. Using different polymers as stationary phases, various authors have measured molecular weights of coals and crude-oil fractions [134-142]. The detectors normally used are UV and differential RI detectors and FID. An evaporative index detector has been used in the analysis of heavy oils [143]. Various authors [144-151] have used GPC on-line with an inductively couplex plasma (ICP) mass spectrometer to characterize metals in heavy oils. SEC has also been used to separate sulfur- [152] and nitrogen-containing [153] as well as aromatic [154,155] fractions in synthetic fuels.

21.6 BIOMARKERS

21.6.1 Saturated hydrocarbons

GC has been used for almost three decades for the analysis of saturated hydrocar-bons in fossil fuels, particularly crude oils [156,157]. The distribution of the *n*-alkanes in the chromatograms is easily discerned due to the recognizable presence of the C_{19} and

References on p. B465

C_{20} isoprenoids, pristane and phytane, which have retention times similar to those of the n-C_{17} and n-C_{18} alkanes, respectively. Although n-alkanes are readily determined by GC alone, branched hydrocarbons, including isoprenoids, and cyclic compounds usually require determination by GC/MS due to the fact that they are typically present in relatively low concentrations and are generally hidden in the baselines of most chromatograms.

Bicyclic sesquiterpenoids, for example, were first described to occur in crude oils by Bendoraitis [158]. Several years later, similar types of compounds were discovered in Australian crude oils and extensively characterized by GC, GC/MS, and comparison with chromatographic and spectrometric data of synthetic standards [159,160]. On the basis of this characterization it was established that the sesquiterpane eudesmane was indicative of a higher plant contribution to sedimentary organic matter and is derived from eudesmanol, whereas drimane, whose precursor is thought to be drimanol, is derived from a bacterial source.

The precursors of tri- and tetracyclic diterpenoids are widely distributed in higher plants, in particular the resins of such plants. Their occurrence in Tertiary lacustrine oil shales, coal, and crude oils, derived predominantly from terrestrial source material, has been described in a number of papers [161-165]. The majority of diterpenoid hydrocarbons present in fossil fuels have structures based on the abietane, pimarane, or phyllocladane skeletons. Various other families of bicyclic terpenoids have also been characterized in fossil fuel samples [166-168]. Tetracyclic diterpenoids of the kaurene/phyllocladene type are particularly abundant in leaf resins of conifers. In recent years, a number of reports have appeared concerning their occurrence in oils, source rocks, and brown coals from the Gippsland Basin area in Australia, where they have been used successfully for the purposes of classifying oils into families and making oil/source-rock correlations [160,162,167,168]. Alexander et al. [165] and Noble et al. [167] studied the effect of thermal maturation upon the relative proportions of epimeric diterpanes by comparing the distribution of compounds in sediments and coals of differing maturity. The two structures studied most extensively in this work were phyllocladane and kaurane, both of which exist as the 16α(H)- and 16β(H)-stereoisomers. It was found that in samples of increasing maturity the proportion of the 16α(H)-isomers relative to the 16β(H)-isomers decreased.

Tricyclic terpanes with extended sidechains have been found in a wide range of crude oils, source-rock extracts and oil shale samples [169-171]. The precise origin of these compounds has not been unequivocally established, although the evidence currently suggests that they are biomarkers indicative of a bacterial contribution of organic matter to the sediments. The tricyclic terpane fingerprints have been used extensively for oil/oil and oil/source-rock correlations, as well as indicators of depositional environments and relative maturity determinations. Tricyclic terpanes were first reported to be present in the Green River Shale by Anders and Robinson [172], but it was not until more recently that synthetic structural proof for their occurrence was put forward by Aquino Neto et al. [173] and Heissler et al. [174]. Since the early studies were generally concerned with tricyclic terpanes in crude oils and not shales, only one major stereochemical series was observed, namely 13β(H),14α(H). However, more recently Aquino Neto et al. [169,171]

studied the tricyclic terpanes in a series of immature shales and found that they occurred as a mixture of stereoisomers with the 13β(H),14α(H)-stereoisomers predominating and the 13α(H),14α(H)-stereoisomers present in lower concentration.

Tetracyclic terpanes present in many crude oils and source rocks are thought to be derived from the microbial degradation of pentacyclic terpenoids of the hopanoid structure [175]. The two main types of tetracyclic terpanes found in fossil fuels are derived from degradation of the Ring E of the pentacyclic hopanes, leading to 17,21-secohopanes, or from Ring C degradation, leading to 8,14-secohopanes. Tetracyclic terpanes, based on the lupane structure with a degraded A ring, have also been found in a number of oils derived from terrestrial sources [176].

Pentacyclic triterpanes of the hopanoid type, due to their ubiquitous distribution in the geological record as well as their diversified structural aspects, have been widely used for assessments of source, maturity, migration, biodegradation, and characterization of depositional environments of fossil fuels and precursors [177,178]. So diversified is the literature on pentacyclic hopanes in fossil fuels that no attempt will be made to review it in detail in this chapter. The interested reader is referred to a number of comprehensive review articles cited herein [156,157].

A number of source parameters based on biomarker distributions have been developed, and one of the most widely used of these parameters is based on sterane distributions. The original proposal by Huang and Meinschein [179] was based on sterol distributions but subsequently extrapolated to steranes; they suggested that C_{29} steranes were generally associated with a terrestrial input and that C_{27} steranes were associated with a marine input of organic matter to source rocks and their associated oils. However, in the past two or three years significant concentrations of C_{29} steranes have been observed in oil and source rocks thought to be of predominantly marine origin [180,181]. In a further development of the relationship between sterane distributions in geological samples and source materials, Moldowan et al. [182,183] described the presence of C_{30} steranes as an indicator of oils derived from a marine depositional environment. This was based on the observation that C_{30} steranes were only present in oils known to be derived from marine source rocks. Although the precursor of the C_{30} steranes remains unclear, C_{30} sterols are known to be present in marine organisms. The development of maturity parameters based on the relative proportions of different biomarker stereoisomers has been well documented in several previous publications [156,157,184]. Anomalous values for hopane and sterane maturity indicators have been reported in some samples, particularly those from hypersaline environments [185]. In particular, the premature formation of the 5α(H),14β(H),17β(H) (20R and 20S)-steranes and the complete isomerization of the 17α(H),21β(H)-hopanes at the C-22 position were observed. Ten Haven et al. [185] suggested a possible mechanism for the formation of 14β(H),17β(H)-steranes from Δ^7-sterenes, which are known to occur in hypersaline environments along with spirosterenes [186]. The Δ^7-sterenes will isomerize to Δ^7-, $\Delta^{8(14)}$-, and Δ^{14}-sterenes, and hydrogenation of these intermediates will produce the most thermodynamically stable configuration, i.e., 14β(H),17β(H) and not 14α(H),17α(H) [187]. A detailed study of acid-

References on p. B467

catalyzed rearrangements of Δ^4- and Δ^5-steranes has been recently published by Peak-man and Maxwell [188].

21.6.2 Aromatic hydrocarbons

Aromatic hydrocarbons in fossil fuels, like their saturated counterparts, are present as extremely complex mixtures and typically require some type of fractionation step prior to chromatographic analysis [189]. The fractionation steps permit separation of the aromatics into subfractions, based on the number of rings in the aromatic structure to produce mono-, di-, tri-, and polyaromatic fractions. Fractionation of aromatics into ring classes by HPLC gives subfractions of adequate resolution for satisfactory separation on 25-m fused-silica columns, coated with apolar phases, such as SE-52 or SE-54 methylsilicones or the equivalent chemically bonded stationary phases, e.g., CP Sil-8CB (Chrompack) and Ultra 2 (Hewlett-Packard).

Alkyl-substituted aromatics with either linear, branched, or isoprenoidal sidechains are abundant in coal extracts [190-193], coal-derived oils [191,194-196], and coal pyrolysis products [184,197,198]. They include derivatives of benzene, phenol, acenaphthene, tetralin, naphthalene, phenanthrene and/or anthracene, diphenylene oxide, fluorene, etc. Given and his co-workers [194,195,199] identified several homologous series of long-alkyl-chain aromatics in coal extracts and liquefaction products. Long-chain alkylphenols have been identified in the polar fraction of liquefaction oil [200]. The aromatic fractions in the hydrogenation oil of various ranks of coals have been shown to be composed mainly of aromatics with long alkyl chains [201-203]. Stepwise mild hydrogenation of subbituminous coal [335-375°C] produced polar fractions in the hexane-soluble products that were composed of homologous series, such as C_1-C_{23} alkylphenols and C_1-C_4 alkylhydroxy-polycyclic aromatics [91,204-207]. It was proposed that these long-chain alkylated phenols are important structural units, which may play a dominant role in the crosslinked structure of coals.

Liquefaction products of Texas lignite, as well as products from model experiments [194], produced alkyltetralins or naphthalenes due to reactions of long-chain aliphatic carboxylic acids in the coal with tetralin [208]. However, a detailed analysis of the products prepared by mild liquefaction with benzene as solvent [205] showed that no solvent-derived compounds were found.

The monoaromatic fractions of oils and source-rock extracts typically include some specific biomarkers, such as isoprenoidal alkylbenzenes, characteristic of green sulfur bacteria. Alkylbenzenes have been described by various authors [209-211] and are possibly degradation products of various carotenoids and terpenoids [212,213]. A combination of biomarker and isotopic signatures led Summons and Powell [214] to suggest the presence of microbial communities containing a green sulfur bacterium in ancient restricted seas. Other monoaromatic compounds have been shown to form in the initial stages of aromatization of steranes [215]. Pentacyclic terpenoids and Ring A-degraded terpenoids in the first stage of aromatization give monoaromatic compounds and finally picene and chrysene, which have been identified in Messel oil shale [216] and coals

[217]. Hexacyclic monoaromatic benzohopanes have been detected and shown to be a characteristic feature of carbonate environments [218]. Initial stages of diterpenoid aromatization in coals [219] and crude oils and source rocks [162] also lead to the formation of monoaromatic compounds in crude oils.

Due to their utility in assessing maturity, diaromatic compounds are more widely used in organic geochemistry and fossil fuel exploration than the monoaromatic compounds. Common maturity parameters derived from the aromatic fraction are based on ratios of various naphthalene isomers [220]. In coals Püttmann and Villar [221] have recognized specific sources which can be related to different naphthalene isomers, and similar attempts have been made with crude oils [222]. Utilization of these parameters to calculate temperature ranges, where the reaction occurs under natural conditions, has shown that the reaction occurs in sediments under noncatalytic conditions and the extent of reaction increases with maturity. Triaromatic compounds have also found widespread use in the assessment of maturity, particularly the methylphenathrene index MPI [220]. Triaromatic steroid hydrocarbons, formed as a result of increasing aromatization of monoaromatic steroid hydrocarbons, have also been used to assess the degree of maturation of crude oils [223]. Polyaromatic hydrocarbons have been used to determine the maturity of crude oils on the basis of methylpyrene and methylchrysene isomer distributions [98-104]. Numerous aromatic compounds with differing degrees of aromatization have also been found in coals [224-228].

21.6.3 Polar fraction

Although compounds with mixed functionalities have been identified in fossil fuels, for the sake of convenience, they have been divided according to their principal heteroatom: i.e. nitrogen, sulfur, or oxygen.

21.6.3.1 Nitrogen compounds

The majority of nitrogen-containing compounds in fossil fuels are heterocyclic and occur in the neutral and basic fractions. Basic fractions consist mainly of pyridines and their derivatives, such as quinolines, benzoquinolines, and acridines. The neutral fraction is dominated by pyrroles and their derivatives, such as indoles and carbazoles, as well as amines, nitriles, and lactams. Some analytical work has been done on the identification of these compounds in crude oils [105,229-238], but the geochemical implications of their distributions are far from being completely understood at this time. On the contrary, many studies in synthetic oils have been directed at identification of nitrogen compounds. Nitrogen compounds in liquids derived from oil shales have been analyzed and characterized [239-243]. Different coal-derived materials from various processes, such as liquefaction [244-248], gasification [249], coker gas oil [250,251], solvent-refined coal [91,153,252-254], and coal tars [255-258] have also been analyzed by CLC and HPLC.

References on p. B465

21.6.3.2 Sulfur compounds

In recent studies of petroleum genesis the role of sulfur in the formation of immature oils has received renewed attention for a number of reasons. One is their role as possible indicators of specific depositional environments and another is the proposal that organic matter in sediments can act as a sink for sedimentary sulfur. A wide range of organosulfur compounds, which commonly occur in the aromatic fraction, have been identified in crude oils [259,260], source-rock extracts [55], synthetic fuels [261], and also asphaltenes [56] and kerogen pyrolyzates [76] by pyrolysis techniques (Section 21.2.2). The distributions of these organic sulfur compounds have been shown to be affected by maturity [262] and depositional environment of the original organic matter [56,263]. The distribution of organic sulfur compounds in the pyrolyzates of oil asphaltenes has been shown to resemble that found in the aromatic fraction of the oil from which it was isolated [56], and it also varies according to the depositional environment of the original organic material.

Aromatic benzo- and dibenzothiophene isomers have been described with different degrees of alkylation. The ratios between various dibenzothiophene isomers have been used to characterize relative maturities of crude oils derived from carbonate source rocks. Other aromatic compounds have been characterized by GC with a flame-photometric detector (FPD) as a selective detector for sulfur compounds [264-266] as well as a microwave-induced helium plasma detector [267]. The latter type of detector provides a more linear response compared to the nonlinear, compound-dependent, power-function response of the FPD and is not subject to hydrocarbon quenching. Ligand-exchange chromatography has been used as an alternative to adsorption liquid chromatography [268] for the isolation of sulfur-containing components. A great deal of work has been undertaken on the characterization and identification of sulfur-containing compounds in various coal-derived liquids, including hydrogenates [269], coal tar, and coal-liquid vacuum residues [270]. These systems have been studied in order to understand the potential pollution problems created, if the liquids are to be used as fuels. Organosulfur compounds identified in coal products include: benzothiophene, alkylbenzothiophenes, dibenzothiophenes, alkyldibenzothiophenes, naphthothiophenes, benzonaphthothiophenes, triphenylenothiophenes, benzohexahydrothiophenes, and dinaphthothiophenes. Despite the extensive studies described above, relatively little has been done concerning the presence of organic sulfur compounds in whole coals and the factors which are important in determining their distribution.

One of the earliest studies of sulfur compounds in oils was reported by Ho et al. [271]. It was observed at that time, and more recently by Radke et al. [220], that a variety of changes occur in the distribution of benzothiophenes and methyldibenzothiophenes as a result of thermal maturation. An extension of this work was published by Hughes [259], who compared the distribution of thiophenic organosulfur compounds and was able to distinguish between oils derived from carbonate sources and those from siliciclastic sources. A further extension of this approach was published by Philp and Bakel [56], who showed that pyrolysis of asphaltene fractions could also be used to differentiate oils derived from source rocks of different lithologies.

Sulfur compounds are also well-known constituents of industrial products from petroleum processing. The sulfur content of crude oils is typically 0.5-3% in most samples [272], but it can be up to 8% in the vacuum residue of heavy crude oils [273]. The sulfur is distributed over a wide range of molecular structures: aliphatic mono- and disulfides are sometimes present [274], but a large amount of sulfur occurs in aromatic structures, especially as alkylated thiophene benzologs. Sulfur-containing polynuclear aromatic hydrocarbons (SPAHs), like nitrogen-containing polynuclear aromatic hydrocarbons (NPAHs), are generally associated with adverse effects on the quality of petroleum products, such as catalyst poisoning, corrosion, or pollution. The determination of specific toxic or potentially carcinogenic compounds is important for environmental impact studies [275]. SPAHs are frequently isolated by purification of the PAH fractions obtained from a given starting mixture by narrow-cut short-path distillation [273], and/or column liquid chromatography on silica or alumina adsorbents [276,277], or HPLC with various systems of solvents and sorbents [278]. Alternative preconcentration methods applied to SPAH analyses are based on reversible complexing reactions with aromatic rings [278-280].

Application of these various extraction procedures, applied to crude oils, coal tars, and refined petroleum fractions [246,281-284], produce sulfur-containing fractions which are still contaminated by PAHs. This complicates the characterization of individual SPAHs. This difficulty can be circumvented by selective detection and identification methods, such as capillary GC with FPD [285,286], high-resolution MS with a low ionization voltage [287], chemical ionization with reagent gases (e.g., ammonia) that do not react with PAHs but combine with SPAHs [288], GC/MS with SIM [283], Shpol'skii spectrophotometry [289], photoelectron spectroscopy [290], or MS/MS in the neutral-loss mode. Comparisons and applications of these various processes have been summarized by Arpino et al. [291].

Sulfur and nitrogen compounds are also important components of many coal structures and thus are present in many coal-derived liquids. If coal is burned in power stations, these compounds are released from the coal and may escape unaltered into the atmosphere. Furthermore, sulfur- and nitrogen-containing compounds are often responsible for poisoning catalysts in refining processes. Hence, it is important to understand their distributions in coals prior to selecting a particular coal for liquefaction processes. Many of these compounds are also mutagenic and carcinogenic. The distributions of sulfur and nitrogen compounds in coal, coal precursors, and coal-derived liquids have not been as widely studied as other classes of compounds, such as the hydrocarbons. The heterocyclic compounds have the potential of providing useful information on source materials, depositional environments, and the nature of products that may cause problems with respect to the poisoning of catalysts in large-scale liquefaction reactions and in hydrotreating [292]. Burchill et al. [258] have reported on nitrogen compounds, sulfur compounds [283] and mixed sulfur/nitrogen compounds [248] in coal tars. These papers also include GC methods with nitrogen- and sulfur-selective detectors and MS. Buchanan et al. [293] have used GC/MS to identify a number of nitrogen compounds in coal-derived liquids, some of which (neutral azarenes) are responsible for the mutagenicity of the neutral fraction. Grigsby et al. [294] have described the use of FAB-MS to identify nitrogen-containing compounds in fossil fuels. A LC method for the isolation of thiophenic compounds from shale oil was described by Joyce and Uden [85]. Nishioka et al. [284] described a newly synthesized biphenylpolysiloxane stationary phase that could be used for the separation of sulfur heterocyclic compounds by GC. It was observed that the biphenyl groups in this stationary phase polymer were polarized by the slightly polar SPAH solutes, giving rise to a unique selectivity for specific sulfur compounds.

21.6.3.3 Oxygen compounds

The distribution of oxygen-containing compounds, which generally occur in the polar or acidic fractions, has not been described in great detail in the fossil fuel literature. The acidic fractions consist of hydroxy and carboxylic acids analyzed as their methyl esters. Neutral fractions consist of linear, branched, and cyclic aldehydes, ketones, and alcohols. The alcohols are converted to acetyl or trimethylsilyl derivatives prior to analysis by GC.

Aliphatic monocarboxylic acids are major lipid components of living organisms and have been detected in various fossil fuels, such as crude oils [27,295,296], along with naphthenic acids [297] and polycyclic acids [298,299]. Jaffé et al. [298,299] have discussed the use of hopanoic acid methyl esters as indicators of oil migration. The presence of carboxylic acids in coals has received similar attention [300-302]. Hydroxy acids have been detected as methyl esters and trimethylsilyl ethers [303]. A detailed compilation of cyclic carboxylic acids in coals has been given by Chaffee et al. [300]. The presence of linear alcohols in brown-coals has been described by Chaffee and Johns [304] and that of cyclic functionalized triterpenoids, including acids and alcohols, in peats by Quirk et al. [305]. Cyclic and linear aldehydes have been found as their acetal derivatives in oil shales [306]. Scattered reports of linear and cyclic ketones have been compiled by Chaffee et al. [300]. Aromatic oxygen-containing heterocyclic compounds in coals have been observed [307], but no biological origin for these compounds has been discussed. Chromans have been described and used in the characterization of hypersaline environments [308]. Other aromatic compounds, such as phenols, dibenzofuran, and their alkylated derivatives have been isolated from different crude oils [309], oil shales, and coal-derived liquids [91,310], and polycarboxylic acids have been found in a shale oil fraction [311].

21. 7 SUMMARY

In the preceding sections, an attempt has been made to summarize some of the important areas of fossil fuel research where various forms of chromatography have played an important role in the analysis of these materials. From such analyses has come important information concerning the type of source material in the original sample when it was deposited millions of years ago. Furthermore, it has become possible to predict the type of depositional environment, maturity, extent of migration and biodegradation from the distributions of these compounds. Chromatography also plays a very important role in identifying the types of combustion products derived from the coals. This is particularly useful for predicting the presence of possible environmental pollutants.

In the next few years the developments that are currently taking place in basic chromatographic techniques will lead to an increase in the number of applications designed to monitor compounds with carbon numbers above C_{40}. This is still a relatively new area of fossil fuel research but one that holds great promise for the future.

REFERENCES

1 E. Heftmann (Editor), *Chromatography,* Elsevier, Amsterdam, 4th Edn., 1983.
2 H. Shultz, A.W. Wells, E.A. Frommelt and P.B. Flenory, *Anal. Chem.,* 59 (1987) 103R.
3 H. Shultz, A.W. Wells, E.A. Frommelt and R.M. Hough, *Anal. Chem.,* 61 (1989) 84R.
4 F.C. Trusell, *Anal. Chem.,* 59 (1987) 252R.
5 T.R. McManus, *Anal. Chem.,* 61 (1989) 165R.
6 R.E. Clement, F.I. Onuska, G.A. Eiceman and H.H. Hill, *Anal. Chem.,* 60 (198) 279R.
7 A.L. Burlingame, D. Maltby, D.H. Rusell and P.T. Holland, *Anal. Chem.,* 60 (1988) 294R.
8 L.G. Barth, W.E. Barber, C.H. Lochmüller, R.E. Majors and F.E. Regnier, *Anal. Chem.,* 60 (198) 387R.
9 *J. Chromatogr.,* 446 (1988).
10 *Org. Geochem.,* 6 (1984).
11 *Org. Geochem.,* 10 (1986).
12 *Org. Geochem.,* 13 (1988).
13 American Chemical Society Annual Meeting, Dallas, April 1989, Abstracts only.
14 G.J. Demaison and G.T. Moore, *Org. Geochem.,* 2 (1980) 9.
15 M.R. Mello, P.C. Gaglianone, S.C. Brassell and J.R. Maxwell, *Marine Petrol. Geol.,* 5 (1988) 205.
16 R.P. Philp, J.R. Maxwell and G. Eglinton, *Sci. Prog.,* 63 (1976) 521.
17 S.C. Brassell, G. Eglinton, J.R. Maxwell and R.P. Philp, in O. Hutzinger, I.H. van Lelyveld and B.C.J. Zoeteman (Editors), *Aquatic Pollutants - Transformation and Biological Effects,* Pergamon, Oxford, 1978, p. 69.
18 S.R. Larter, in K. Voorhees (Editor), *Analytical Pyrolysis - Methods and Applications,* Butterworths, London, 1984, p. 212.
19 R.G. Schaefer and J. Höltkemeier, *Chromatographia,* 26 (1988) 311.
20 M.L. Sazanov, M.K. Lunskii, L.I. Zhiltsova, I.L. Paizanskaya and N.I. Chuvilyaeva, *J. Chromatogr.,* 364 (1986) 267.
21 A.H.H. Taneesh and M.H. Hanna, *J. Chromatogr.,* 363 (1986) 303.
22 Y. Takayama, T. Takeichi and S. Kawai, *J. High Resolut. Chromatogr. Chromatogr. Commun.,* 11 (1988) 732.
23 R.P. Philp, C.A. Lewis, C. Campbell and E. Johnson, *Org. Geochem.,* 14 (1989) 183.
24 I. Dzidic, M.D. Balicki and H.V. Hart, *Fuel,* 67 (1988) 1155.
25 M.V. Buchanan, *Anal. Chem.,* 54 (1982) 570.
26 R.P. Philp and M. Johnston, in T.F. Yen and J.M. Moldowan (Editors), *Geochemical Markers,* Harwood Academic Publishers, New York, 1988, p. 253.
27 J.M. Schmitter, P. Arpino and G. Guiochon, *J. Chromatogr.,* 167 (1978) 149.
28 D.J. Miller and S.B. Hawthorne, *Fuel,* 68 (1989) 105.
29 B.W. Wright, H.R. Udseth, E.K. Chess and R.D. Smith, *J. Chromatogr. Sci.,* 26 (1988) 228.
30 K.V. Wood, L.F. Albright, J.S. Brodbelt and R.G. Cooks, *Anal. Chim. Acta,* 173 (1985) 117.
31 M.V. Buchanan and M.B. Wise, *Fuel,* 66 (1987) 954.
32 L.R. Hilpert, G.D. Byrd and C.R. Vogt, *Anal. Chem.,* 56 (1984) 1842.
33 E.J. Gallegos, *Anal. Chem.,* 48 (1976) 1348.

34 W. Windig, T. Chakravarty, J.M. Richards and H.L.C. Meuzelaar, *Anal. Chim. Acta*, 191 (1986) 205.
35 G.A. Warburton and J.E. Zumberge, *Anal. Chem.*, 55 (1982) 123.
36 J.E. Zumberge, *Org. Geochem.*, 11 (1987) 479.
37 J.M. Moldowan, W.K. Seifert and E.J. Gallegos, *Geochim. Cosmochim. Acta*, 47 (1983) 1531.
38 R.E. Summons, J.K. Volkman and C.J. Boreham, *Geochim. Cosmochim. Acta*, 52 (1988) 1747.
39 R.E. Summons, *Org. Geochem.*, 11 (1987) 287.
40 P.W. Brooks, T. Meyer and O.H.J. Christie, *Org. Geochem.*, 6 (1984) 813.
41 N. Telnaes and B. Dahl, *Org. Geochem.*, 10 (1986) 425.
42 R.P. Philp, J.N. Oung and C.A. Lewis, *J. Chromatogr.*, 446 (1988) 13.
43 R.P. Philp and J.N. Oung, *Proc. 1st South American Org. Geochem. Conf. (1988)*, in press.
44 R.P. Philp and J.N. Oung, *Anal. Chem.*, 60 (9188) 887A.
45 T.G. Wang, B.R.T. Simoneit, R.P. Philp and C.P. Yu, *Energy Fuels*, 4 (1990) 177.
46 S. Lewis, Siu Teng and R.P. Philp, *Finnigan Application Notes*, 1989.
47 R.P. Philp, *Chem. Eng. News*, 64 (1986) 28.
48 R.P. Philp and T.D. Gilbert, *J. Anal. Appl. Pyrolysis*, 11 (1987) 93.
49 B. Horsfield, in J. Brooks and D. Welte (Editors), *Advances in Petroleum Geochemistry I*, Academic Press, London, 1984, p. 247.
50 M.D. Lewan, *Phil. Trans. R. Soc. London A*, 315 (1985) 315.
51 T.I. Eglinton, R.P. Philp and S.J. Rowland, *Org. Geochem.*, 12 (1988) 33.
52 F. Behar and M. Vandenbroucke, *Org. Geochem.*, 13 (1988) 927.
53 R.P. Philp and A.J. Bakel, *Abstracts of Papers, Amer. Chem. Soc.*, Paper 16, Geochemistry, 192nd ACS Natl. Meet., Anaheim, CA, September 7-12, 1986.
54 J.S. Sinninghe Damsté, T.I. Eglinton, J.W. de Leeuw and P.A. Schenck, *Geochim. Cosmochim. Acta*, 53 (1989) 873.
55 J.S. Sinninghe Damsté, H. ten Haven, J.W. de Leeuw and P.A. Schenck, *Org. Geochem.*, 10 (1986) 791.
56 R.P. Philp and A.J. Bakel, *Energy Fuels*, 2 (1988) 59.
57 J. Espitalié, J.L. Laporte, M. Madec, E. Marquis, P. Leplat, J. Paulet and A. Boutefer, *Rev. Inst. Fr. Pet.*, 32 (1977) 23.
58 K.E. Peters, *Am. Assoc. Pet. Geol. Bull.*, 70 (1986) 318.
59 H. Dembicki, B. Horsfield and T.Y. Ho, *Bull. Am. Assoc. Pet. Geol. Bull.*, 67 (1983) 1093.
60 B.J. Huizinga, Z.A. Aizenshtat and K.E. Peters, *Energy Fuels*, 2 (1988) 74.
61 B. Horsfield and K.L. Yordy, *Org. Geochem.*, 13 (1988) 121.
62 B. Horsfield and J.C. Crelling, *Geochim. Cosmochim. Acta*, 53 (1989) 891.
63 P.T. Crisp, J. Ellis, J.W. de Leeuw and P.A. Schenck, *Anal. Chem.*, 58 (1986) 258.
64 W. Püttmann, C.B. Eckardt and R.G. Schaefer, *Chromatographia*, 249 (1988) 279.
65 R.G. Schaefer, *J. High Resolut. Chromatogr. Chromatogr. Commun.*, 8 (1985) 267.
66 R.G. Schaefer and W. Püttmann, *J. Chromatogr.*, 395 (1987) 203.
67 W. Püttmann, *Chromatographia*, 26 (1986) 171.
68 M. Bjorøy, P.B. Hall, R. Lølery, J.A. McDermott and N. Mills, *Org. Geochem.*, 13 (1988) 221.
69 J.T. Senftle and S.R. Larter, *Org. Geochem.*, 11 (1987) 407.
70 J.R. Gormly and P.K. Mukhopdhyay, in M. Bjorøy et al. (Editors), *Advances in Organic Geochemistry 1981*, Wiley, Chichester, 1983, p. 597.
71 M. Vandenbroucke, F. Behar and J. Espitalié, *Energy Fuels*, 1 (1987) 452.
72 A.G. Barwise, *Org. Geochem.*, 6 (1984) 343.
73 H. Solli, G. van Graas, P. Leplat and J. Krare, *Org. Geochem.*, 6 (1984) 351.
74 F. Behar, R. Pelet and J. Roucaché, *Org. Geochem.*, 6 (1984) 587.
75 L.H. Lin, G.E. Michael, R.P. Philp, G. Kovachev, H. Zhu and C.A. Lewis, *Org. Geochem.*, 14 (1989) 511.
76 J.S. Sinninghe Damsté, A.C. Kock-van Dalen, J.W. de Leeuw and P.A. Schenck, *J. Chromatogr.*, 435 (1988) 435.

77 R.P. Philp, T.D. Gilbert and J. Friedrich, in T.F. Yen, F.K. Kawahara and R. Hertzberg (Editors), *Chemical and Geochemical Aspects of Fossil Energy Extraction,* Ann Arbor Science Publ., Ann Arbor, MI, 1983, p. 63.
78 C. Vallejos, S. Talukdar and R.P. Philp, *Energy Fuels,* 3 (1989) 366.
79 T.I. Eglinton and A.G. Douglas, *Energy Fuels,* 2 (1988) 81.
80 M. Monthioux and P. Landais, *Chem. Geol.,* 75 (1989) 209.
81 N. Telnaes, G.C. Speers, A. Steen and A.G. Douglas, in B.M. Thomas, A.G. Done, S.E. Eggen, P.C. Home and R.M. Larsen (Editors), *Petroleum Geochemistry in Exploration of the Norwegian Shelf, 1984,* Graham and Trotman, London, 1985, pp. 287-292.
82 M.L. Selucky, P. Hafermann, A. Iacchelli and T. Manske, *Liq. Fuels Technol.,* 1 (1985) 15.
83 B.J. Fuhr, L.R. Holloway, C. Reichert and S.K. Barva, *J. Chromatogr. Sci.,* 26 (1988) 55.
84 J.W. Haas III, M.V. Buchanan and M.B. Wise, *J. Chromatogr. Sci.,* 26 (1988) 250.
85 W.F. Joyce and P.C. Uden, *Anal. Chem.,* 55 (1983) 540.
86 M.M. Boduszynski, R.J. Hurtubise, T.W. Allen and H.F. Silver, *Anal. Chem.,* 55 (1983) 225.
87 C. Liu, G. Que, Y. Chen and W. Liang, *Fuel Sci. Technol. Int.,* 6 (1988) 449.
88 W.K. Robbins and F.C. McElroy, *Liq. Fuels Technol.,* 2 (1984) 113.
89 F.M. Lanas, E. Carrilho, G.H.N. Deane and M.C.F. Camilo, *J. High Resolut. Chromatogr. Chromatogr. Commun.,* 12 (1989) 368.
90 I.L. Davies, K.E. Markides, M.L. Lee, M.W. Raynor and K.D. Bartle, *J. High Resolut. Chromatogr. Chromatogr. Commun.,* 12 (1989) 193.
91 T. Katoh and K. Ouchi, *Fuel,* 66 (1987) 1588.
92 P.C. Hayes and S.D. Anderson, *J. Chromatogr. Sci.,* 26 (1988) 250.
93 I.L. Davies, M.W. Raynor, D.J. Urwin, K.D. Bartle, M. Tolay, E.E. Kinci and H.E. Schwartz, *J. High Resolut. Chromatogr. Chromatogr. Commun.,* 11 (1988) 792.
94 T.V. Raglione, J.A. Troskosky and R.A. Hardwick, *J. Chromatogr.,* 409 (1987) 213.
95 K.A. Caswell, T.E. Glass, M. Swann and H.C. Dorm, *Anal. Chem.,* 61 (1989) 206.
96 P. Garrigues and J. Bellocq, *J. High Resolut. Chromatogr. Chromatogr. Commun.,* 12 (1989) 400.
97 M. Radke, H.W. Willsch and D.H. Welte, *Anal. Chem.,* 52 (1980) 406.
98 P. Garrigues and M. Ewald, *Anal. Chem.,* 55 (1983) 138.
99 P. Garrigues, E. Parlanti, M. Radke, J. Bellocq, H. Willsch and M. Ewald, *J. Chromatogr.,* 395 (1987) 217.
100 P. Garrigues, R. De Sury, M.L. Angelin, M. Ewald, J.L. Oudin and J. Connan, *Org. Geochem.,* 6 (9184) 829.
101 P. Garrigues, J. Connan, E. Parlanti, J. Bellocq, J.L. Oudin and M. Ewald, *Org. Geochem.,* 13 (1988) 1115.
102 P. Garrigues, G. Bourgeois, A. Vegres, J. Rime, M. Lamotte and M. Ewald, *Anal. Chem.,* 57 (1985) 1068.
103 P. Garrigues, R. De Sury, M.L. Angelin, J. Bellocq, J.L. Oudin and M. Ewald, *Geochim. Cosmochim. Acta,* 52 (1988) 375.
104 P. Garrigues, M. Radke, O. Druez, H. Willsch and J. Bellocq, *J. Chromatogr.,* 473 (1989) 207.
105 M. Dorbon, J.M. Schmitter, P. Garrigues, I. Ignatiadis, M. Ewald, P. Arpino and G. Guiochon, *Org. Geochem.,* 7 (1984) 111.
106 M. Radke and H.W. Willsch, in M. Bjorøy et al. (Editors), *Advances in Organic Geochemistry 1981,* Wiley, Chichester, 1983, p. 504.
107 M.R. Khan, K.S. Seshadri and T.E. Kowalski, *Energy Fuels,* 3 (1989) 412.
108 J.F. McKay, M.M. Boduszynski and D.R. Latham, *Liq. Fuels Technol.,* 11 (1983) 35.
109 F. Andreolini, C. Borra, D. Wiesler and M. Novotny, *J. Chromatogr.,* 406 (1987) 375.
110 W.A. Dark, W.H. McFadden and D.L. Bradford, *J. Chromatogr. Sci.,* 15 (1977) 454.
111 M. Novotny, A. Hirose and D. Wiesler, *Anal. Chem.,* 56 (1984) 1243.
112 J. De Guzman, *Fuel,* 66 (1987) 890.

113 A.A. Herod, W.R. Ladner, B.J. Stokes, A.J. Berry, D.E. Games and M. Hohn, *Fuel,* 66 (1987) 935.
114 A. Treibs, *Ann. Chem.,* 509 (1934) 103.
115 A. Treibs, *Angew. Chem.,* 49 (1936) 682.
116 S.K. Hajibrahim, J.M.E. Quirke and G. Eglinton, *Chem. Geol.,* 32 (1981) 173.
117 J.P. Gill, R.P. Evershed, M.I. Chicarelli, G.A. Wolff, J.R. Maxwell and G. Eglinton, *J. Chromatogr.,* 350 (1985) 37.
118 R. Alexander, G. Eglinton, J.P. Gill and J.K. Volkman, *J. High Resolut. Chromatogr. Chromatogr. Commun.,* 3 (1980) 521.
119 P.J. Mariott, J.P. Gill, R.P. Evershed, C.S. Hein and G. Eglinton, *J. Chromatogr.,* 301 (1984) 107.
120 J.P. Gill, R.P. Evershed and G. Eglinton, *J. Chromatogr.,* 369 (1986) 281.
121 P.J. Marriott and G. Eglinton, *J. Chromatogr.,* 249 (1982) 311.
122 C.B. Eckardt, L. Dyas, P.W. Yendle and G. Eglinton, *Org. Geochem.,* 13 (1988) 573.
123 S. Ji-Yang, A.S. Mackenzie, R. Alexander, G. Eglinton, A.P. Gowar, G.A. Wolff and J.R. Maxwell, *Chem. Geol.,* 35 (1982) 1.
124 S.E. Palmer, E.W. Baker, L.S. Charney and J.W. Louda, *Geochim. Cosmochim. Acta,* 46 (1982) 1233.
125 J.C. Monin, D. Barth, M. Perrot, M. Espitalié and B. Durand, *Org. Geochem.,* 13 (1988) 1079.
126 B.J. Fuhr, L.R. Holloway and C. Reichert, *Fuel Sci. Technol. Int.,* 7 (1989) 643.
127 B.W. Wright, C.W. Wright and J.S. Fruchter, *Energy Fuels,* 3 (1989) 474.
128 M.W. Raynor, I.L. Davies, K.D. Bartle, A.A. Clifford, M. Williams, J.M. Chalmer and S.W. Coock, *J. High Resolut. Chromatogr. Chromatogr. Commun.,* 11 (1988) 766.
129 H.C.K. Chang, K.E. Markides, J.S. Bradshaw and M.L. Lee, *J. Chromatogr. Sci.,* 26 (1988) 280.
130 M. Nishioka, H.C.K. Chang and M.L. Lee, *Environ. Sci. Technol.,* 20 (1986) 1023.
131 T. Funazukuri and N. Wakao, *Fuel,* 67 (1988) 875.
132 S.L. Chong and J.F. McKay, *Fuel Sci. Technol. Int.,* 5 (1987) 513.
133 R.M. Campbell, N.M. Djordjevic, K.E. Markides and M.L. Lee, *Anal. Chem.,* 60 (1988) 356.
134 K.D. Bartle, O.G. Mills, H.J. Mulligan, I.O. Amaechina and N. Taylor, *Anal. Chem.,* 58 (1986) 2403.
135 A. Majid, J. Bornais and R.A. Hutchison, *Fuel Sci. Technol. Int.,* 7 (1989) 507.
136 C.N.P. Philip and R.G. Anthony, *Fuel,* 61 (1982) 357.
137 K.D. Bartle, M.J. Mulligan, N. Taylor, T.G. Martin and C.E. Snape, *Fuels,* 63 (1984) 1556.
138 D.H. Buchanan, L.W. Warfel, S. Bailey and D. Lucas, *Energy Fuels,* 2 (1988) 32.
139 J.W. Larsen, M. Mohammadi, I. Yigimsu and J. Kovac, *Geochim. Cosmochim. Acta,* 48 (1984) 135.
140 J.W. Larsen and Y.C. Wei, *Energy Fuels,* 2 (1988) 344.
141 M. Nishioka and J.W. Larsen, *Energy Fuels,* 2 (1988) 351.
142 N. Evans, T.M. Haley, M.J. Mulligan and K.M. Thomas, *Fuel,* 65 (1985) 694.
143 S. Coulombe, *J. Chromatogr. Sci.,* 26 (1988) 1.
144 A. Izquierdo, L. Carbogniani, V. Leon and A. Parisi, *Fuel Sci. Technol. Int.,* 7 (1989) 561.
145 W.R. Biggs, R.J. Brown and J.C. Fetzer, *Energy Fuels,* 1 (1987) 257.
146 W.R. Biggs, R.J. Brown, J.C. Fetzer and J.G. Reynolds, *Liq. Fuels Technol.,* 3 (1985) 397.
147 J. Cerny, J. Mitera and P. Vavrecka, *Fuel,* 68 (1989) 596.
148 J.G. Reynolds and W.R. Biggs, *Fuel Sci. Technol. Int.,* 6 (1988) 329.
149 J.G. Reynolds and W.R. Biggs, *Fuel Sci. Technol. Int.,* 6 (1988) 249.
150 J.G. Reynolds and W.R. Biggs, *Fuel Sci. Technol. Int.,* 6 (1988) 779.
151 M.J. Mulligan, K.M. Thomas and A.P. Tytko, *Fuel,* 66 (1987) 1472.
152 J.G. Reynolds, *Fuel Sci. Technol. Int.,* 5 (1987) 1593.
153 S. Wallace, M.J. Crook, K.D. Bartle and A.J. Pappin, *Fuel,* 65 (1986) 138.
154 M.J. Wornat, A.F. Sarofim and J.P. Longwell, *Energy Fuels,* 1 (1982) 431.

155 A.L. LaFleur and Y. Nakagawa, *Fuel,* 68 (1989) 741.
156 A.S Mackenzie, in J. Brooks and D. Welte (Editors), *Advances in Petroleum Geochemistry I,* Academic Press, London, 1984, p. 115.
157 R.P. Philp, *Mass Spectrom. Rev.,* 4 (1985) 1.
158 J.G. Bendoraitis, in B. Tissot and F. Beinner (Editors), *Advances in Organic Geochemistry 1973,* Editions Technip, Paris, 1974, p. 209.
159 R. Alexander, R.I. Kagi and R.A. Noble, *J. Chem. Soc. Chem. Commun.,* (1983) 226.
160 R.P. Philp, T.D. Gilbert and J. Friedrich, *Geochim. Cosmochim. Acta,* 45 (1981) 1173.
161 L.R. Snowdon, in A.D. Miall (Editor), *Facts and Principles of World Petroleum Occurrence, Can. Soc. Petrol. Geol. Memoir,* 6 (1980) 509.
162 R.P. Philp, B.R.T. Simoneit and T.D. Gilbert, in M. Bjorøy et al. (Editors), *Advances in Organic Geochemistry 1981,* Wiley, Chichester, 1983, p. 698.
163 J.S. Richardson and D.E. Miller, *Anal. Chem.,* 54 (1982) 765.
164 Z. Czochanska, C.M. Sheppard, R.J. Weston, T.A. Wood, A.D. Woolhouse, R.P. Philp and T.D. Gilbert, *Org. Geochem.,* 12 (1988) 123.
165 G. Alexander, I. Hazai, J.O. Grimalt and J. Albaige's, *Geochim. Cosmochim. Acta,* 51 (1987) 2065.
166 J. Zhusheng, R.P. Philp and C.A. Lewis, *Org. Geochem.,* 13 (1988) 561.
167 R.A. Noble, R. Alexander, R.I. Kagi and J. Knox, *Org. Geochem.,* 10 (1986) 825.
168 R.A. Noble, R. Alexander, R.I. Kagi and J. Knox, *Org. Geochem.,* 11 (1987) 151.
169 F.R. Aquino Neto, J.N. Cardoso, R. Rodrigues and A.F. Trindade, *Geochim. Cosmochim. Acta,* 50 (1986) 2069.
170 J.M. Modowan, W.K. Seifert and E.J. Gallegos, *Geochim. Cosmochim. Acta,* 47 (1983) 1531.
171 M.I. Chicarelli, F.R. Aquino Neto and P. Albrecht, *Geochim. Cosmochim. Acta,* 52 (1988) 1955.
172 D.E. Anders and W.E. Robinson, *Geochim. Cosmochim. Acta,* 35 (1971) 661.
173 F.R. Aquino Neto, J.M. Trendel, A. Restlé, P. Albrecht and J. Connan, in M. Bjorøy et al. (Editors), *Advances in Organic Geochemistry 1981,* Wiley, Chichester, 1983, p. 659.
174 G. Heissler, R. Ocampo, P. Albrecht, J.J. Riehl and G. Ourisson, *J. Chem. Soc. Chem. Commun.,* (1984) 496.
175 J.M. Schmitter, W. Sucrow and P.J. Arpino, *Geochim. Cosmochim. Acta,* 46 (1982) 2345.
176 J. Rullkötter, D. Leythaeuser and D. Wendisch, *Geochim. Cosmochim. Acta,* 46 (1982) 2501.
177 G. Ourisson, P. Albrecht and M. Rohmer, *Pure Appl. Chem.,* 51 (1979) 709.
178 G. Ourisson, P. Albrecht and M. Rohmer, *Trends Biochem. Sci.,* 7 (1982) 236.
179 W.-Y. Huang and W.G. Meinschein, *Geochim. Cosmochim. Acta,* 43 (1979) 739.
180 C.C. Walters and M.R. Cassa, *Gulf Coast Assoc. Geol. Soc. Trans.,* 35 (1985) 277.
181 J.G. Palacas, D.E. Anders and J.D. King, *AAPG Stud. Geol. No. 18,* (1984) 71.
182 J.M. Moldowan, W.K. Seifert and E.J. Gallegos, *Am. Assoc. Petrol. Geol. Bull.,* 69 (1985) 1255.
183 J.M. Moldowan, F.J. Fago, C.Y. Lee, S.R. Jacobson, D.S. Watt, N.-E. Slougui, A. Jeganathan and D.C. Young, *Science,* 247 (1990) 309.
184 R.P. Philp and C.A. Lewis, *Ann. Rev. Earth Planet. Sci.,* 15 (1987) 363.
185 H.L. ten Haven, J.W. de Leeuw, T.M. Peakman and J.R. Maxwell, *Geochim. Cosmochim. Acta,* 50 (1986) 853.
186 T.M. Peakman, N.A. Lamb and J.R. Maxwell, *Tetrahedron Lett.,* 24 (1984) 369.
187 S.C. Brassell, C.A Lewis, J.W. de Leeuw, F. de Lange and J.S. Sinninghe Damsté, *Nature (London),* 320 (1986) 160.
188 T.M. Peakman and J.R. Maxwell, *Org. Geochem.,* 13 (1988) 583.
189 M. Radke, *Adv. Pet. Geochem.,* 2 (1987) 149.
190 C.E. Snape, B.J. Stokes and K.D. Bartle, *Fuel,* 60 (1981) 903.
191 K.E. Singleton, R.G. Cooks, K.V. Wood, A. Rabinovich and P.H. Given, *Fuel,* 66 (1987) 74.

192 P.M. Shaw, S.C. Brassell, K.J. Assinder and G. Eglinton, *Fuel,* 67 (1988) 557.
193 A.H. Baset, R.J. Pancirov and T.R. Ashe, in A.G. Douglas and J.R. Maxwell (Editors), *Advances in Organic Geochemistry 1979,* Pergamon Press, Oxford, 1980, p. 619.
194 Z. Mudamburi and P.H. Given, *Org. Geochem.,* 8 (1985) 221.
195 K.V. Wood, R.G. Cooks, Z. Mudamburi and P.H. Given, *Org. Geochem.,* 7 (1984) 169.
196 A.J. Koplick and P.C. Wailes, *Fuel,* 62 (1983) 1161.
197 J. Allan, M. Bjorøy and A.G. Douglas, in A.G. Douglas and J.R. Maxwell (Editors), *Advances in Organic Geochemistry 1979,* Pergamon Press, Oxford, 1980, p. 599.
198 E.J. Gallegos, *Anal. Chem.,* 56 (1984) 701.
199 J.S. Youtcheff, P.H. Given, Z. Baset and M. Sundaram, *Org. Geochem.,* 5 (1985) 157.
200 S. Sugimoto, Y. Miki, S. Yamadaya and M. Oba, *J. Chem. Soc. Jpn.,* (1984) 755.
201 s. Yokoyama, N. Tsuzuki, T. Katoh and Y. Sanada, *Nenryo Kyokaishi (J. Fuel Soc. Jpn.),* 62 (1983) 106.
202 S. Yokoyama, N. Tsuzuki, T. Katoh, Y. Sanada, D.M. Bodily and W.H. Wiser, in M.L. Gorbaty and K. Ouchi (Editors), *Advances in Chemistry Series,* Vol. 192, American Chemical Society, Washington, DC, 1981, p. 257.
203 H. Uchino, S. Yokoyama, T. Katoh and Y. Sanada, *J. Chem. Soc. Jpn.,* (1983) 94.
204 J.Z. Dong, T. Katoh, H. Itoh and K. Ouchi, *Fuel,* 65 (1986) 1073.
205 J.Z. Dong and K. Ouchi, *Fuel,* 67 (1988) 541.
206 T. Katoh and K. Ouchi, *Fuel,* 64 (1985) 1260.
207 T. Katoh and K. Ouchi, *Fuel,* 66 (1987) 58.
208 B.M. Benjamin, E.C. Douglas and E.W. Hagaman, *Energy Fuels,* 1 (1987) 187.
209 H. Solli, S.R. Larter and A.G. Douglas, in A.G. Douglas and J.R. Maxwell (Editors), *Advances in Organic Geochemistry 1979,* Pergamon Press, Oxford, 1980, p. 591.
210 R.P. Philp, N.J. Russell, T.D. Gilbert and J.M. Friedrich, *J. Anal. Appl. Pyrolysis,* 4 (1982) 143.
211 S.R. Larter and A.G. Douglas, in W.E. Krumbein (Editor), *Environmental Biogeochemistry and Geomicrobiology,* Ann Arbor Science Publishers, Ann Arbor, MI, 1978, p. 373.
212 E.J. Gallegos, *Anal. Chem.,* 45 (1973) 1399.
213 E.J. Gallegos, *J. Chromatogr. Sci.,* 19 (1981) 177.
214 R.E. Summons and T.G. Powell, *Geochim. Cosmochim. Acta,* 51 (1987) 557.
215 A.S. MacKenzie, C.F. Hoffmann and J.R. Maxwell, *Geochim. Cosmochim. Acta,* 45 (1981) 1345.
216 A.C. Greiner, C. Spyckerelle and P. Albrecht, *Tetrahedron,* 32 (1976) 257.
217 I. Hazai, G. Alexander, T. Székely, B. Essiger and D. Radek, *J. Chromatogr.,* 367 (1986) 117.
218 J. Connan, J. Bouroullec, D. Dessort and P. Albrecht, *Org. Geochem.,* 10 (1986) 29.
219 B.R.T. Simoneit, in R.B. Johns (Editor), *Biological Markers in the Sedimentary Record,* Elsevier, Amsterdam, 1986, p. 43.
220 M. Radke, D.H. Welte and H. Willsch, *Geochim. Cosmochim. Acta,* 46 (1982) 1.
221 W.I. Püttmann and H. Villar, *Geochim. Cosmochim. Acta,* 51 (1987) 3023.
222 M.G. Strachan, R. Alexander and R.I. Kagi, *Geochim. Cosmochim. Acta,* 52 (1988) 1255.
223 A.S. Mackenzie, C.F. Hoffman and J.R. Maxwell, *Geochim. Cosmochim. Acta,* 45 (1981) 1345.
224 A.L. Chaffee and R.B. Johns, *Geochim. Cosmochim. Acta,* 47 (1983) 2141.
225 A.L. Chaffee and C.J.R. Fookes, *Org. Geochem.,* 12 (1988) 261.
226 A.L. Chaffee, M. Strachan and R.B. Johns, *Geochim. Cosmochim. Acta,* 48 (1984) 2037.
227 H.K. Chang, N. Masaharu, K.D. Bartle, S.A. Wise, J.M. Bayona, K.E. Markides and M.L. Lee, *Fuel,* 67 (1988) 45.
228 R. Hayatsu, R.E. Botto, R.G. Scott, R.L. McBeth and R.E. Winans, *Org. Geochem.,* 11 (1987) 245.
229 J.F. McKay, J.H. Weber and D.R. Latham, *Anal. Chem.,* 48 (1976) 48.

230 I. Dzidic, M.D. Balicki, I.A.L. Rhodes and H.V. Hart, *J. Chromatogr. Sci.*, 26 (1988) 236.

231 P. Burchill, A.A. Herod, J.P. Mahon and E. Pritchard, *J. Chromatogr.*, 281 (1983) 109.

232 J.M. Schmitter and R. Arpino, in M. Bjorøy et al. (Editors), *Advances in Organic Geochemistry 1981*, Wiley, Chichester, 1983, p. 808.

233 J.M. Schmitter, H. Colin, J.L. Excoffier, P. Arpino and G. Guiochon, *Anal. Chem.*, 54 (1982) 769.

234 J.M. Schmitter, Z. Vajta and P. Arpino, in A.G. Douglas and J.R. Maxwell (Editors), *Advances in Organic Geochemistry 1979*, Pergamon Press, Oxford, 1980, p. 67.

235 J.M. Schmitter, I. Ignatiadis, P. Arpino and G. Guiochon, *Anal. Chem.*, 55 (1983) 1685.

236 J.M. Schmitter, I. Ignatiadis and P. Arpino, *Geochim. Cosmochim. Acta*, 47 (1983) 1975.

237 J.M. Schmitter, I. Ignatiadis and G. Guiochon, *J. Chromatogr.*, 248 (1982) 203.

238 C.D. Ford, S.A. Holmes, L.F. Thompson and D.R. Latham, *Anal. Chem.*, 53 (1981) 831.

239 L. Chan, J. Ellis and P.T. Crisp, *J. Chromatogr.*, 292 (1984) 355.

240 G.W. Mushrush, J.V. Cooney, E.J. Beal and R.N. Hazlett, *Fuel Sci. Technol. Int.*, 4 (1986) 103.

241 P.C. Uden, A.P. Carpenter, H.M. Hackett, D.E. Henderson and S. Siggia, *Anal. Chem.*, 51 (1979) 38.

242 F.F. Shue and T.F. Yen, *Anal. Chem.*, 53 (1981) 2081.

243 G.W. Mushrush, J.M. Watkins, E.J. Beal, R.E. Morris, J.V. Cooney and R.N. Hazlett, *Fuel Sci. Technol. Int.*, 7 (1989) 931.

244 R.H. Hardy and B.H. Davis, *Fuel Sci. Technol. Int.*, 7 (1989) 399.

245 M. Novotny, R. Kump, F. Merli and L.J. Todd, *Anal. Chem.*, 52 (1980) 401.

246 D.W. Later, M.L. Lee, K.D. Bartle, R.C. Kong and D.L. Vassilaros, *Anal. Chem.*, 51 (1981) 1612.

247 P. Burchill, A.A. Herod and E. Pritchard, *J. Chromatogr.*, 246 (1982) 271.

248 P. Burchill, A.A. Herod and E. Pritchard, *J. Chromatogr.*, 242 (1982) 65.

249 M. Novotny, J.W. Strand, S.L. Smith, D. Wiesler and F.J. Schwende, *Fuel*, 60 (1981) 213.

250 J.M. Schmitter, I. Ignatiadis, M. Dorbon, P. Arpino, G. Guiochon, H. Toulhoat and A. Huc, *Fuel*, 63 (1984) 557.

251 M. Dorbon, I. Ignatiadis, J.M. Schmitter, P. Arpino, G. Guiochon, H. Toulhoat and A. Huc, *Fuel*, 63 (1984) 563.

252 D.W. Later, R.B. Lucke, E.K. Chess and J.A. Frank, *Fuel*, 66 (1987) 1347.

253 C.Y. Ma, C.H. Ho, J.E. Caton, W.H. Griest and M.R. Guerin, *Fuel*, 66 (1987) 612.

254 C.E. Östman and A.L. Colmsjö, *Fuel*, 67 (1988) 396.

255 J.C. Lauer, D.H. Valles Hernandez and D. Cagniant, *Fuel*, 67 (1988) 1273.

256 J.C. Lauer, D.H. Valles Hernandez and D. Cagniant, *Fuel*, 67 (1988) 1446.

257 P. Burchill, A.A. Herod and E. Pritchard, *Fuel*, 62 (1983) 11.

258 P. Burchill, A.A. Herod and E. Pritchard, *Fuel*, 62 (1983) 20.

259 W. Hughes, *AAPG Stud. Geol. No. 18*, 18 (1984) 181.

260 J.S. Sinninghe Damsté, W.I.C. Rijpstra, J.W. de Leeuw and P.A. Schenck, *Geochim. Cosmochim. Acta*, 53 (1989) 1323.

261 A.O. Bender, T.M. Sarkissian and A.M. Allawi, *Fuel*, 68 (1989) 607.

262 T.I. Eglinton, R.P. Philp and S.J. Rowland, *Org. Geochem.*, 12 (1988) 33.

263 J.S. Sinninghe Damsté, J.W. de Leeuw, A.C. Kock-van Dalen, M.A. de Zeeuw, F. de Lange, W.I.C. Rijpstra and P.A. Schenck, *Geochim. Cosmochim. Acta*, 51 (1987) 2369.

264 M. Nishioka, *Energy Fuels*, 2 (1988) 214.

265 C.M. White, L.J. Douglas, M.B. Perry and C.E. Schmidt, *Energy Fuels*, 1 (1987) 222.

266 R.C. Kong, M.C. Lee, M. Iwao, Y. Tominaga, R. Pratap, R.D. Thompson and R.N. Castle, *Fuel*, 63 (1984) 702.

267 D.S. Sklarew, K.B. Olsen and J.C. Evans, *Chromatographia*, 27 (1989) 44.

268 M. Nishioka, R.M. Campbell, M.L. Lee and R.N. Castle, *Fuel,* 65 (1986) 270.
269 C. Braekman-Danheux, *J. Anal. Appl. Pyrolysis,* 7 (1985) 315.
270 M. Nishioka, M.L. Lee and R. Castle, *Fuel,* 65 (1986) 390.
271 T. Ho, M. Rogers, H. Durshel and C. Koons, *AAPG Bull.,* 58 (1974) 2338.
272 B.P. Tissot and D.H. Welte, *Petroleum Formation and Occurrence,* Springer-Verlag, Heidelberg, 1984.
273 D. Severin and O. Glinzer, *Characterization of Heavy Crude Oils and Petroleum Residues,* Technip, Paris, 1984, p. 19.
274 B. Zygmunt, W. Wardencki and R. Staszewski, *J. Chromatogr.,* 265 (1983) 136.
275 F. Berthou, Y. Dreano and P. Sandra, *J. High Resolut. Chromatogr. Chromatogr. Commun.,* 7 (1984) 679.
276 H.V. Drushel and A.L. Sommers, *Anal. Chem.,* 39 (1967) 1819.
277 H. Castex, J. Roucaché and R. Boulet, *Rev. Inst. Fr. Pet.,* 29 (1974) 3.
278 J.C. Escalier, J.P. Massoué and M. Marichy, *Analusis,* 7 (1979) 55.
279 D.F.S. Natusch and B.A. Tomkins, *Anal. Chem.,* 50 (1978) 1429.
280 F.P. Richter, A.L. Williams and S.L. Meisel, *J. Am. Chem. Soc.,* 78 (1956) 2166.
281 M.A. Poirier and G.T. Smiley, *J. Chromatogr. Sci.,* 22 (1984) 304.
282 R.C. Kong, M.L. Lee, Y. Tominaga, R. Pratap, M. Iwao and R.N. Castle, *J. Chromatogr. Sci.,* 80 (1982) 502.
283 P. Burchill, A.A. Herod and E. Pritchard, *J. Chromatogr.,* 242 (1982) 51.
284 M. Nishioka, J.S. Bradshaw, M.L. Lee, Y. Tominaga, M. Tedjamulia and R.N. Castle, *Anal. Chem.,* 57 (1985) 309.
285 M.L. Lee and B.W. Wright, *J. Chromatogr. Sci.,* 18 (1980) 345.
286 B. Wenzel and R.L. Aiken, *J. Chromatogr. Sci.,* 17 (1979) 503.
287 H.E. Lumpkin and T. Aczel, *Am. Chem. Soc. Div. Pet. Chem. Prepr.,* 70 (1978) 261.
288 J. Guieze, G. Devant and D. Loyaux, *Int. J. Mass Spectrom. Ion Phys.,* 46 (1983) 313.
289 A.L. Colmsjö, Y. Zebühr and C.E. Ostman, *Anal. Chem.,* 54 (1982) 1673.
290 J.M. Ruiz, B.M. Carden, L.J. Lena, E.J. Vincent and J.C. Escalier, *Anal. Chem.,* 54 (1982) 688.
291 P.J. Arpino, I. Ignatiadis and G. de Ryche, *J. Chromatogr.,* 390 (1987) 329.
292 A.E. George, B.B. Pruden and H. Sawatzky, *CANMET Report 78-16,* 1978.
293 M.V. Buchanan, G.L. Kao, B.D. Barkenbus, C.H. Ho and M.R. Guerin, *Fuel,* 62 (1983) 1177.
294 R.D. Grigsby, S.E. Scheppele, Q.G. Grindstaff, G.P. Sturm, L.C.E. Taylor, H. Tudge, C. Wakefield and S. Evans, *Anal. Chem.,* 54 (1982) 1108.
295 F.H. Behar and P. Albrecht, *Org. Geochem.,* 6 (1984) 597.
296 A.S. Mackenzie, G.A. Wolff and J.R. Maxwell, in M. Bjorøy et al. (Editors), *Advances in Organic Geochemistry 1981,* Wiley, Chichester, 1983, p. 637.
297 I. Dzidic, A.C. Somerville, J.C. Raia and H.V. Hart, *Anal. Chem.,* 60 (1988) 1318.
298 R. Jaffé, P. Albrecht and J.L. Oudin, *Org. Geochem.,* 13 (1988) 483.
299 R. Jaffé, P. Albrecht and J.L. Oudin, *Geochim. Cosmochim. Acta,* 52 (1988) 2599.
300 A.L. Chaffee, D.S. Hoover, R.B. Johns and F.K. Scheweighardt, in R.B. Johns (Editor), *Biological Markers in the Sedimentary Record,* Elsevier, Amsterdam, 1986, p. 311.
301 J.F. Branthauer, K.P. Thomas, E.R. Logan and R.E. Barden, *Fuel Sci. Technol. Int.,* 6 (1988) 525.
302 A.L. Chaffee, G.J. Perry, R.B. Johns and A.M. George, in M.L. Gorbaty and K. Ouchi (Editors), *American Chemical Society Symposium Series,* No. 92, American Chemical Society, Washington, DC, 1981, p. 113.
303 D.H. Hunneman and G. Eglinton, in P.A. Schenck and I. Havenaar (Editors), *Advances in Organic Geochemistry 1968,* Pergamon Press, Oxford, 1969, p. 157.
304 A.S. Chaffee and R.B. Johns, *Org. Geochem.,* 8 (1985) 349.
305 M.M. Quirk, A.M.K. Wardroper, R.E. Weatley and J.R. Maxwell, *Chem. Geol.,* 42 (1984) 25.
306 M. Dastillung, P. Albrecht and G. Ourisson, *J. Chem. Res., Synop.,* (1980) 166.
307 C.M. White and M.L. Lee, *Geochim. Cosmochim. Acta,* 44 (1980) 1825.

308 J.S. Sinninghe Damsté, A.C. Kock-van Dalen, J.W. de Leeuw, P.A. Schenck and S. Guoying, *Geochim. Cosmochim. Acta,* 51 (1987) 2393.
309 Y. Sato and T. Yamakawa, *Liq. Fuel Technol.,* 1 (1983) 285.
310 P.H. Neill, L.J. Shadle and P.H. Given, *Fuel,* 67 (1988) 1459.
311 C. Bradley and J.W. Carnahan, *Anal. Chem.,* 60 (1988) 858.

Chapter 22

Synthetic polymers

THOMAS H. MOUREY and TIMOTHY C. SCHUNK

CONTENTS

22.1 INTRODUCTION

Synthetic polymers present unique separations challenges because, unlike many small organic molecules and biopolymers, they always consist of a distribution of structurally different chains. For example, each macromolecule differs in length and may have different end groups. Vinyl monomers can combine in head-to-head, head-to-tail, or tail-to-tail configurations as well as in stereochemically different conformations to give a distribution of homopolymer structural isomers. Homopolymers may also be randomly or nonrandomly branched or possess unique architectures, such as combs, stars, ladders, or macrocycles. Copolymers can be even more complicated, since, along with any of the above repeat unit combinations, each macromolecule can also contain different ratios of comonomers, joined in alternating, random, blocked, or blocky sequence distributions. In addition, many modern polymer-based materials are filled with non-polymeric material and loaded with small organic molecules. Given the inherent complexity of polymeric materials, it is no surprise that nearly every mainstream chromatographic method, along with several that are specific to polymers, has been applied to their fractionation.

There are problems, however, in the characterization of synthetic polymers by chromatography that set them apart from small molecules: (a) they generally cannot be volatilized without decomposition and are frequently characterized in solution by liquid chromatography; (b) there is usually a more limited selection of solvents suitable for chromatographic fractionation of synthetic polymers than for small organic molecules; (c)

References on p. B502

there are thermodynamic considerations (typically entropic), critical to the thermodynamic description of polymer solutions, that are often negligible in solutions of small organic molecules; (d) polymer diffusion constants are smaller than those of small molecules; (e) many commercially important polymers do not contain suitable chromophores for UV/visible detection in liquid chromatography, which in some instances has limited the application of gradient elution liquid chromatography. In most instances, modifications to traditional chromatographic techniques must be made in view of these characteristics.

Frequently, synthetic polymers are described by single, unique values, such as average molecular weight, average chemical composition, average length between branch sites, average comonomer sequence of triads or pentads, or average block length. For many purposes, these averages suffice. However, it is often a shape or characteristic of a structural distribution that is not easily described by an average value that dictates polymer properties. In the case of complex polymers, the information can be obtained only after the superimposed structural distributions are deconvolved. Elucidating these structural distributions is where liquid chromatographic methods become invaluable to the characterization of synthetic polymers and this is the focus of this chapter.

An overview of important applications of liquid chromatography to the solution characterization of synthetic polymers since 1980 is presented. A few important references to earlier literature have been added, since a chapter on synthetic polymers was not included in the Fourth Edition of this book. The chapter is organized by polymer properties, rather than by chromatographic techniques, in an attempt to facilitate solving polymer fractionation problems by use of multiple separation techniques. The characterization of polymers by pyrolysis/gas chromatography, analysis of low-molecular-weight additives, and the measurement of polymer thermodynamic properties, such as diffusivity and solvent/polymer interaction, are extensive and equally important fields in the chromatography of synthetic polymers. None will be reviewed, except for isolated examples that conform to the emphasis on polymer solution characterization and distribution analysis by liquid chromatography in this chapter.

22.2 MOLECULAR WEIGHT

Size-exclusion chromatography (SEC) remains the principal chromatographic method for determining molecular weight distributions of synthetic polymers. Several advances in the last decade have enhanced the applicability and importance of this technique to polymer characterization [1-8] (cf. also Chapter 6). In particular, microparticulate packings and better column packing procedures have reduced the need for resolution correction, although with an accompanying reduction in available pore volume [4, Chapter 3]. The development and commercialization of low-angle laser light scattering (LALLS) and, more recently, multi-angle laser light scattering [9,10] have provided absolute molecular weight distributions directly from size-exclusion experiments. Likewise, on-line viscometric detection [11-18] measures the intrinsic viscosity of the whole polymer directly from the chromatogram and can be used to calculate absolute molecular weight distributions via universal calibration. Coupled with the improvements in solvent delivery and data acquisition systems in the past decade, we are now able to characterize absolute molecular weight distributions in a variety of polymers routinely. Many applications of SEC to the characterization of synthetic polymers have been reviewed recently [19-23]. A selective compilation of size-exclusion applications to some of the more important synthetic polymers, with an emphasis on chromatographic methodology, is given in Table 22.1.

There still remain a number of polymers that are anything but routinely characterized by SEC. These polymers have become the focus of much research in SEC. It is well known that a unique relationship exists between molecular weight and hydrodynamic volume, V_h, through universal calibration

$$[\eta]_1 M_1 = [\eta]_2 M_2 \tag{22.1}$$

where $[\eta]_1$ and M_1 are the intrinsic viscosity and molecular weight, respectively, of one of two chemically different polymers, 1 and 2. However, some polymers have more than one broad distribution and have been defined as "complex" macromolecules. The complexity arises from the fact that polymer chains of equivalent hydrodynamic volume, but different molecular weights, can reside in the SEC detector cell at the same time, even in the case of perfect resolution. Certain copolymers and branched polymers are examples of complex macromolecules.

In the case of copolymers, V_h is a function of molecular weight as well as chemical composition and comonomer sequence length. In some cases, such as statistical and block copolymers, the K and a parameters of the Mark-Houwink relationship

$$[\eta] = KM^a \tag{22.2}$$

relating intrinsic viscosity and molecular weight, may be estimated from a knowledge of the composition of the copolymers [165,198]. In the case of compositionally heterogeneous copolymers, the parameters K and a may not be constant across the molecular size distribution. In addition, the refractive index increment dn/dc of each eluted fraction is a function of chemical composition, and variation across the SEC chromatogram will invalidate the calculation of weight-average molecular weights by LALLS without making suitable corrections to dn/dc at each retention volume. Reviews have appeared recently that discuss the SEC of complex polymers [199,200]. For copolymers in particular, multidetector schemes are useful in elucidating both molecular weight and chemical composition across the size-exclusion chromatogram (see Section 22.5). Evaporative light scattering [120] and densimetry [201] detection have been shown to be less sensitive to copolymer composition than refractive index detection, thus enabling better quantitation of molecular weight distributions. Alternative approaches that define a new molecular weight average have also been proposed [202].

SEC has not completely replaced other chromatographic methods for the characterization of molecular weight distributions. Polymers continue to be fractionated according to molecular weight by column elution [203]. The polymer sample is deposited,

TABLE 22.1

SELECTED SIZE-EXCLUSION CHROMATOGRAPHY APPLICATIONS

Abbreviations: DMAC = dimethylacetamide; DMF = dimethylformamide; HFIP = hexa-fluoro-2-propanol; IR = infrared; MEK = methyl ethyl ketone; MH = Mark-Houwink; MW = molecular weight; NMP = N-methylpyrrolidone; PDMS = poly(dimethyl siloxane); PEI = poly(ethylene imine); PEO = poly(ethylene oxide); PET = poly(ethylene terephthalate); PPO = poly(propylene oxide); PS = polystyrene; SDS = sodium dodecylsulfate; TCB = 1,2,4-trichlorobenzene; THF = tetrahydrofuran; TSK = Toyo Sodo Manufacturing Co., Ltd.

Polymer	Refs.	Comments
Aqueous exclusion (general)	24	Ultrahigh MW, on-line viscometry on Sepha-cryl columns.
	25-28	Various polymers examined on TSK-PW, TSK-SW columns.
	29	Alkaline eluent on Fractogel TSK gels.
	30-32	General articles/reviews.
	33	Poly(vinyl alcohol) columns.
	14	On-line viscometry.
	34	Anionic and neutral polymers on glycero-propyl-bonded silica.
Aqueous exclusion (cationic)	35	Quaternized silica, 0.05 M ammonium acetate.
	36	Poly(2-trimethylammonium ethyl methacryl-ate chloride) examined on quaternized silica, 0.1 M HNO$_3$/0.1 M NaNO$_3$.
	37,38	Polymeric aminopropyl-derivatized silica.
	39	Comparison of neutral and cationic columns for poly(2-vinylpyridine) and PEI.
	40-43	Various cationics, including PEI and poly-(vinylamine) on TSK-PW. Evidence for adsorption in 0.1-0.2 M NaCl.
Cellulose derivatives		
cellulose nitrate	44-46	Broad standard calibration, THF on PS resin columns, LALLS.
cellulose triacetate	47	Dichloromethane eluent on PS resin col-umns.
cellulose tricarbanilate	48-51	Phenylisocyanate derivates of cellulose on PS resin columns in THF.
Epoxies	52-57	
Phenolic resins	58-60	THF
	61	DMF
Polyacrylamides	62-64	Aq. sodium sulfate on controlled-pore glass. LALLS in Refs. 63, 64.
	28	0.3 M NaCl/0.1 M phosphate at pH 7.0 on TSK-PW resin column.

TABLE 22.1 (continued)

Polymer	Refs.	Comments
	65	Formamide/water (1:5) on amine-bonded silica.
	66	1 M NaCl on Sephacryl S1000.
	67	0.1 M phosphate on CPG/200 Glycophase.
	68	Phosphate buffer with 0.025-0.1% SDS and 0.025% formaldehyde on TSK-PW columns.
Poly(acrylic acid)	69	Aq. on silica, 0.1 N NaNO$_3$.
	70	Aq. on PW columns, 0.3 N salt.
	71	Diazomethylated derivative in THF.
Polyacrylonitrile	72, 73	DMF/0.01 M LiBr on PS resin column. MH constants given.
Polyamic acid	74	DMAC with H$_3$PO$_4$/LiBr/1% THF on silica.
	75	DMF/LiBr and DMF/H$_3$PO$_4$.
Polyamides	76, 77	Benzyl alcohol at 130°C on PS resin column, benzyl alcohol or phenol/1-propanol on silanized silica at 100°C.
	78	2,2,2-Trifluoroethanol/0.05 M LiBr on PS resin, LALLS.
	79-81	N-Trifluoroacetyl derivatization.
	82	HFIP/salt on porous silica, LALLS, Nylon 6.
	83	2/8 HFIP/toluene on PS resin columns, Nylon 12.
	84	2,2,2-Trifluoroethanol on silanized silica, Nylon 6.
Poly(aminostyrene)s	85	NMP added to THF to suppress adsorption on PS resin columns.
Polyaniline	86	THF, DMF, and NMP examined as eluents.
Poly(aryl-ether-ether-ketone)	87	Phenol/TCB (1:1) at 115°C on PS resin column.
Polybutadiene	88	High-vinyl-content samples.
	89	Anionically polymerized, narrow distributions.
Poly(butyl methacrylate)	90	Unperturbed dimensions by SEC.
Polycarbonates	91-95, 44	MH constants given in Refs. 91, 92. On-line viscometry detection in Refs. 93, 94. Aggregates in THF observed by LALLS in Ref. 95.
Polyesters PET	96	Chloroform/HFIP (98:2) eluent, PS resin columns, ambient.
	97, 98	Dichloromethane/HFIP (70:30) silanized silica, ambient.
	99, 100	o-Chlorophenol/chloroform, PS resin columns. Samples must be heated to dissolve.
	101	Chloroform/HFIP (90:10) semi-micro PS resin columns.

(Continued on p. B480)

TABLE 22.1 (continued)

Polymer	Refs.	Comments
	102	HFIP and pentafluorophenol on derivatized silica.
	103	HFIP on PS resin columns.
PET copolymers	104	PET-ethylene isophthalate copolymer.
aliphatic	105, 106	Branched, azelaic acid-based, poly(capro-lactone).
Poly(ethyl methacrylate)	107, 90	MH constants in THF, unperturbed dimensions.
Polyethylene	108-111	SEC LALLS, Ref. 111, multi-angle light scattering.
	112, 113	Methods and calibration.
	114-117	SEC/on-line viscometry.
copolymers	118-120	Ethylene/propylene.
	115	Ethylene/vinyl acetate.
	121	Ethylene/norbornene.
Poly(ethylene oxide)	122	Semi-prep of dye-labeled PEO, aq. on TSK-PW columns.
	123	0.2 M NaCl/MeOH (50:50) on bonded silicas.
PEO/PPO block copolymer	124	Examined in THF and aqueous. Aq. elution depends on composition.
Polyisobutylene	125, 94	LALLS, on-line viscometry.
Poly(d,l-lactic acid)	126	THF, PS column. Universal calibration, MH constants given.
Poly(methyl methacrylate)	127, 128	Broad standard calibration methods.
	129, 130	Star microgels and tetrafunctionally branched materials.
	93, 94, 131	On-line viscometry, MH constants.
	132	Concentration effects in 2-ethoxyethanol on silica.
	133	Elution from silica with various solvents.
copolymers	134	Determined percent gel in latex copolymers, THF on controlled pore glass.
Poly(α-methylstyrene)	133	Elution from silica with various solvents.
Poly(p-methylstyrene)	135	Calibration with 2 broad standards.
Poly(phenylene sulfide)	136-138	1-Chloronaphthalene at 120°C. On-line viscometry and MH constants in Ref. 136.
Polypropylene	139-142	Dissolution and calibration in trichloro-benzene at 145°C.
	143, 144	Cyclohexane at 70°C as eluent.
	145	Preparation and use of narrow polypropylene standards.
Polyquinolines	146	SEC/LALLS in N-methylpyrrolidone.
Polysiloxanes	94	On-line viscometry, MH constants.

TABLE 22.1 (continued)

Polymer	Refs.	Comments
	147	Dioxane, THF, dichloromethane, tetrachloroethane eluents for IR detection of functional groups in PDMS.
	148	Poly(dimethylvinyl siloxane), toluene eluent.
Polystyrene	149, 150	Effects of flowrate on shear degradation.
	151-153, 131	Stars, combs, branches, rings.
	154	Hydrodynamic volumes in different solvents by SEC.
copolymers	155	Poly(styrene-co-butyl acrylate) by SEC/LALLS.
	156-158	Poly(styrene-co-methyl methacrylate) method to estimate MH constants, densimetry detection, IR detection for heterogeneity.
	158, 159	Poly(styrene-co-acrylonitrile).
	160	Ion-exclusion effects of poly(styrene-co-methacrylic acid) in DMF.
	161-163	Poly(styrene-co-isoprene), poly(styrene-co-butadiene).
	164	Poly(styrene-co-acrylic acid).
	165, 166	Poly(styrene-co-ethylene oxide).
Poly(styrene sulfonate)	167-170	Controlled-pore glass, aq. salt/buffer eluents.
	171	Sepharose columns, NaOH/NaCl eluent. Universal calibration does not hold.
	69	Silica columns, aq. salt eluent.
	172	μ-Bondagel, 0.025 M sodium perchlorate/0.1% lauryl sulfate/THF (pH 7.2).
	173, 174	I-250 protein column, aq.
	175	Secondary effects on various stationary phases.
	176	TSK-GMPW columns.
Polyurethanes	177	Multidetector methods for block copolymers.
	178	Starting materials, prepolymers, allophanate formation studied in THF on PS resin columns.
Poly(vinyl acetate)	179	Branching
	131	On-line viscometry, MH constants given.
	180	Interlaboratory comparison, THF on PS resin columns.
Poly(vinyl alcohol)	181, 182	TSK-PW columns, 0.05 M sodium nitrate, LALLS/on-line viscometry.
	183	Examined aggregate formation in water with aging.
Poly(vinyl butyral)	184	PS resin column, THF, MH constants given.

References on p. B502   *(Continued on p. B482)*

TABLE 22.1 (continued)

Polymer	Refs.	Comments
Poly(vinyl carbazole)s	185, 186	N-Vinyl(-3,6-dibromocarbazole) in THF on PS resin. MH constants given.
	187	Epoxypropyl oligomers in DMF on Sepha-dex.
Poly(vinyl chloride)	188	Branching by LALLS.
	94, 131	On-line viscometry, MH constants.
Poly(2-vinylpyridine)	189	Polymerized amine column (CATSEC), 0.2 M NaCl/0.1% TFA.
	190, 191	Pyridine, NMP, DMF eluents on μ-Bondagel.
	192	Controlled-pore glass, MEK, acetone, DMF eluents.
Poly(N-vinylpyrrolidone)	193	Aq., agarose gel column.
	72	DMF
	194, 195	50% aq. MeOH containing 0.1 M LiNO$_3$. LALLS in Ref. 195.
Urea-formaldehyde resins	196, 197	DMF and DMAC eluents.

usually by precipitation with a solvent/nonsolvent pair, at the head of a column containing an inert solid packing (frequently glass beads). Polymer fractions of increasing molecular weight are eluted by gradient programing of solvent composition and/or temperature. The principal advantage of this technique is that it is better suited than SEC to preparative fractionations. By this method, Nylon 12 has been fractionated at 170°C with benzyl alcohol/decalin as the solvent/nonsolvent pair [204]. Fractions obtained by the column elution method had narrower molecular weight distributions than fractions isolated by preparative SEC, clearly demonstrating the advantages of the technique for large-scale separation. Programing both the solvent composition and temperature has been used to fractionate polystyrene [205]. The solvent/nonsolvent pairs were 2-butanone/ethanol and cyclohexanone/1-propanol. Poly(vinylbutyral) has been fractionated using stainless steel as the inert support (this polymer is nearly irreversibly adsorbed on glass) with chloroform as solvent and heptane as nonsolvent [184]. The fractionation of linear, low-density polyethylene by column elution has also been achieved using solvent/nonsolvent mixtures of xylene/ethyl cellosolve at 120°C [206].

The application of adsorption and reversed-phase high-performance liquid chromatography (HPLC) to the separation of polymers, based upon molecular weight and molecular weight distribution, has received much attention, particularly with respect to the development of theoretical models for the elucidation and prediction of retention. Belenkii, Gankina, and coworkers have published extensive articles on the use of thin-layer chromatography (TLC) for polymer characterization [207-209]. Theoretical modeling of retention behavior was based on the energy of interaction of the polymer during adsorp-

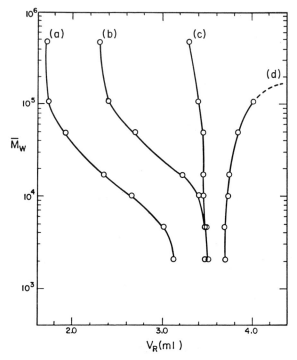

Fig. 22.1. Isocratic retention behavior of polystyrene standards on LiChrospher Si100 with n-hexane/dichloromethane (v/v) ratios of: (a) 0:100; (b) 30:70; (c) 41:59; (d) 42:58.

tion in pores near the "critical condition" region, where a transition from exclusion to adsorption behavior is observed. This type of behavior on silica is shown in Fig. 22.1. Results of TLC experiments were published for polystyrene, poly(methyl methacrylate), poly(dimethyl siloxane), polypropylene diol, polybutadienes, poly(styrene-co-acrylonitrile), and poly(styrene-co-methyl methacrylate) [209]. This model was extended by Gorshkov et al. [210-212] in a series of attempts to relate the energetics of retention of adsorbed polymer molecules under "near-critical conditions" to Snyder's solvent strength equations for adsorption LC.

Gorbunov and Skvortsov et al. [213-220] have taken a somewhat different approach in modeling the observed retention behavior of polymer molecules near "critical conditions". Retention was modeled for flexible Gaussian chains in slit-like or cylindrical pores, a distinction being made between large-pore ($R < d$) and narrow-pore ($R > d$) adsorbents, where d is the pore radius. This model is based upon the ratio of the defined parameter H, correlation length, to the average radius of the polymer chain in solution, R. Under adsorption conditions, H is theoretically related to the thickness of the polymer layer on the adsorbent surface. Experimental retention data have been fit to this model for polyethylene glycol, dextran [216], and polystyrene [219]. In addition to linear molecules, this model has been extended to include the retention behavior of ring macromolecules [214,217]. With consideration to the "critical condition" approach, Klein and Leidigkeit [221] published a separation of polystyrenes by column adsorption LC with gradient elution, spanning the range from 2×10^3 to 2×10^6 D. They argued that the greater resolving power of adsorption LC relative to SEC should make it the preferred technique for molecular size separations.

Independent, and sometimes conflicting, models of chromatographic retention of homopolymers have also been developed by two other groups of researchers over the past decade [222-230]. Boehm et al. have taken a first-principles approach, employing statistical thermodynamics in attempting to derive a retention model that can provide a priori prediction of chromatographic behavior. Snyder et al. [231-235] have developed an empirical model for polymer retention, based upon an extension of the linear solvent strength (LSS) model of reversed-phase retention of small molecules. Both models seek to provide detailed quantitative theories of homopolymer retention in column LC, TLC, and gradient elution LC. Fundamental differences still exist between these two approaches, primarily in the areas of the dependence on solute molecular weight of the rate of change of retention, log k', with eluent solvent strength, Φ ($S = \mathrm{d}\log k'/\mathrm{d}\Phi$), and the effect of polymer solute concentration on retention.

While the work of both Snyder et al. and Boehm et al. has focused on separations in the dilute solution regime, where polymer/polymer interactions and precipitation/redissolution phenomena are minimized, Glöckner [236] has given attention to the exploitation of precipitation phenomena in LC separations. As described by Glöckner, high-performance precipitation LC (HPPLC) is performed in a gradient elution format with intentional choice of solvent and nonsolvent pairs for the polymer solutes. When the point of solubility is reached for a given solute molecule in the solvent gradient, migration and elution occur without further retentive interactions. In order to explain the resolution encountered, a possible mechanism for multiple precipitation/redissolution steps was proposed [237]. Theoretically, the solubilized polymeric solutes advance ahead of the dissolving solvent because of the retardation of the strong solvent front due to the accessibility of the porous packing structure to the small solvent molecules and not the polymer. A comparison of molecular weight separation of polystyrenes by HPPLC, nonaqueous reversed-phase HPLC, and adsorption HPLC has been published [236].

Although many of the literature references describing the chromatographic separation of oligomeric compounds may be considered as fitting into the category of molecular size separations, oligomers are considered outside the scope and definition of this review. For discussions of these applications the reader is directed to Refs. 19-23.

An alternative to conventional chromatographic separation of macromolecular solutes is the techniques of field-flow fractionation (FFF), as described in Chapter 10. FFF avoids many of the difficulties of LC techniques, since it does not employ a stationary phase to support retention of the solute molecules as do other chromatographic methods [238]. Thus far, synthetic polymers have been fractionated by the use of an inertial field generated by a centrifuge in sedimentation FFF (SFFF) [239], a hydraulic field generated by a cross solvent flow in flow FFF (FFFF) [240], and a thermal field generated between a hot and a cold channel wall in thermal FFF (ThFFF) [241]. Theoretical considerations and experimental evidence have been presented for the problems associated with polymer sample overloading in FFF [242]. When the resolving power of FFF was compared with that of SEC, as applied to polymer separations, it was concluded that FFF is superior, owing to its greater selectivity [241,243].

Many applications of synthetic polymer molecular weight separations by FFF have been published over the last decade. Molecular weight distribution separation of a 2.9 x 10^7-D polyacrylamide sample required a force field of 100 000 x g in a SFFF device used by Kirkland et al. [239], indicating that this FFF variant is somewhat limited in polymer separations. Flow FFF in various configurations has been applied to the molecular weight separation of poly(styrene sulfonate) and poly(sodium acrylate) in an aqueous carrier [240] and polystyrene in ethylbenzene [244]. In view of the large number of publications addressing molecular size separation of synthetic polymers by ThFFF, these references are presented in tabular form (Table 22.2).

TABLE 22.2

SELECTED ThFFF APPLICATIONS

Polymer	Refs.	Comments
Polystyrene	245	High-speed separations
	246	High resolution in a 4-pass hairpin channel
	247	Exponential temperature programing
	248	Ultrahigh MW
	249	Thermal diffusion measurements
	241	Comparison with SEC resolution
	250	Supercritical-fluid ThFFF
	251	Linear, star, and comb polymers
	252	Measurement of polydispersity of narrow standards
	253	Hyperlayer mechanism
Polyethylene and polypropylene	254	High-temperature, pressurized-channel ThFFF in tetrachloroethylene
Poly(methyl methacrylate)	247	Exponential temperature programing
Poly(ethylene oxide), poly(ethylene glycol), sodium poly(styrene sulfonate), polyacrylamide, poly(ethylene imine), and poly(vinyl pyrrolidone)	255	Aq., ThFFF

References on p. B502

22.3 BRANCHING

The characterization of branching distribution in synthetic polymers usually involves the measurement of both molecular weight and polymer dimensions in solution. SEC is uniquely suited; separation is based on molecular size (hydrodynamic volume), and molecular weight may be obtained by molecular-weight sensitive detectors such as LALLS and/or on-line viscometers. Excellent reviews describing the use of SEC along with other techniques for the characterization of branching have appeared recently [199,256,257].

A branched molecule will have smaller dimensions in solution than a linear molecule, and it will have a correspondingly lower intrinsic viscosity than the linear analog. A branching parameter, g, defined by Zimm and Stockmayer [258], is the ratio of the mean-square radii of gyration, $<R^2>$, of branched (denoted by subscript b) and linear (subscript l) molecules of equivalent molecular weight

$$g = <R^2>_b/<R^2>_l \qquad (22.3)$$

The branching parameter, g, is not usually measured directly by SEC, although the recent introduction of multi-angle light scattering detection [9] does promise, in theory, direct measurement. Most frequently, intrinsic viscosities and weight-average molecular weights are measured. The relationship between intrinsic viscosity, $[\eta]$, and molecular weight, valid for linear as well as branched molecules, can be generalized as

$$[\eta] = K_v R_h^3 M^{-1} \qquad (22.4)$$

in which K_v is a constant and R_h is the effective hydrodynamic radius. The reduction in intrinsic viscosity resulting from branching provides an alternative parameter, g'

$$g' = [\eta]_b/[\eta]_l \qquad (22.5)$$

Depending on the theoretical assumptions used in describing polymer solution dimensions, g' is related to the Zimm-Stockmayer branching parameter g in a generalized form

$$g' = g^e \qquad (22.6)$$

the exponent e being a value which normally ranges between 0.5 and 1.5, depending upon the type of branching, excluded volume, and other effects. SEC with viscosity detection provides a direct measure of the branching parameter, g', across the molecular size distribution and an estimate of g, provided a suitable value of the exponent e is known. Frequently, this is not the case; however, relative comparisons of samples are still of considerable value.

Alternatively, a measure of the degree of branching can be obtained by the measurement or calculation of molecular weight at each eluted slice of a distribution. The most common procedure is to calibrate a size-exclusion column set with a series of narrow-molecular-weight linear standards of the same composition as the branched samples of interest. Branched polymers will be eluted from the column with an equivalent linear polymer molecular weight, M_l, which is lower than the true molecular weight, M_b. The ratio M_l/M_b is thus an index of branching. It can be shown through application of the universal calibration principle

$$[\eta]_l M_l = [\eta]_b M_b \tag{22.7}$$

that

$$g' = \{M_l/M_b\}^{a+1} \tag{22.8}$$

$$g = \{M_l/M_b\}^{(a+1)/e} \tag{22.9}$$

where a is the exponent of the Mark-Houwink equation (Eqn. 22.2). If an absolute molecular weight detector, such as LALLS, is used, the weight-average molecular weights of linear and branched polymers can be obtained directly without the need for narrow standards, and Eqns. 22.8 and 22.9 apply. A value of g or g' can be obtained at each eluted slice, regardless of the method chosen; however, it is assumed in the derivation of these equations that the sample eluted at each slice is monodisperse, e.g., that there is not a mixture of molecules eluted with equivalent size but different degrees of branching and correspondingly different molecular weights. Hamielec and Ouano [259] have shown that in this case the true molecular weight of an eluted slice, obtained by universal calibration principles, is the number-average molecular weight, M_n. This, of course, may affect the validity of light scattering detection on heterogeneous samples, since in this case the weight-average molecular weight is obtained. In theory, the use of both light scattering and viscosity detection can completely specify the branching distribution. Light scattering provides $\overline{M}_w(v)$, and viscosity detection measures $[\eta](v)$ and permits calculation of $\overline{M}_n(v)$ via universal calibration. The ratio $\overline{M}_w(v)/\overline{M}_n(v)$ is a measure of branching heterogeneity. To date, little work has been done on branched polymers with the detectors in tandem, although an application of both detectors for the characterization of water-soluble polymers has been presented [260,261]. The importance of axial dispersion and the methods of correction have been discussed in detail [262,263] and must be carefully respected if quantitative data are to be obtained. Branching frequency (number of branch units per molecule divided by molecular weight) and the number of branch units per molecule may then be calculated, assuming a suitable model, from the g factors obtained by the size-exclusion experiment.

The characterization of branching distributions in polymers and copolymers can be categorized according to the type of branching.

References on p. B502

22.3.1 Long-chain branching

Most interest has centered around the characterization of high-pressure, low-density polyethylene (LDPE). Rudin et al. [264-267] have described the use of LALLS for polyethylenes containing comonomers with 2-16 carbon branches, and presented evidence that all species in the detector cell have the same constitution (degree of branching) and that Eqns. 22.7 and 22.9 apply. They also indicate that the minimum length of a linear branch for long-chain branching measurements by size exclusion lies between 6 and 12 carbons. The long-chain branching frequency in the samples examined is generally highest at low molecular weights. The weight-average number of long branches per molecule was calculated by the Zimm-Stockmayer relation, assuming randomly branched macromolecules with trifunctional branch sites, although caution was raised that this model applies to polymer in a theta solvent, and that the assumption of branch site functionality may not be universally appropriate. In addition, the exponent e in Eqn. 22.6 may vary from polymer to polymer, and long-chain branch concentrations by SEC may be valid only when comparing polyethylene made by a single vendor and process.

Long-chain branching in poly(vinyl alcohol) [265] and poly(vinyl acetate) have also been studied quantitatively by SEC after establishing a relationship between molecular weight and intrinsic viscosity of isolated fractions [268], and by SEC/LALLS [269], conventional elastic light scattering and viscometry [270], and SEC with viscometry detection [131].

The sol/gel transition in radiation-induced, randomly branched polystyrene has been examined by SEC/LALLS [271]. The experimental results were consistent with scaling concepts [272] for the sol/gel transition. Randomly branched and star-shaped polystyrenes were examined by SEC with viscometric detection [131], and SEC methods were compared with sedimentation methods for model comb- and star-branched polystyrenes [273]. In the latter case, sedimentation was found to be more sensitive than SEC to branching variations; however, the sedimentation method cannot be readily merged with molecular-weight-sensitive detection methods and is considerably less accessible to practitioners.

SEC/LALLS has been used to evaluate the scaling properties of branched polyesters [274], and the critical exponents were compared with percolation and Flory-Stockmayer predictions. Branching in soluble aliphatic polyesters [275] as well as the molecular weight of sol fractions in polyesters of 1,3,5-benzenetriacetic acid with decamethylene glycol [276] have also been reported. The latter were determined by SEC/LALLS. SEC and capillary viscometry methods have been used to characterize long-chain branching in polycarbonates [277], and SEC and light scattering have been used to examine branched epoxy polymers [278].

Poly(methyl methacrylate) with random tetrafunctional branching has been examined by SEC and off-line viscosity measurements [279]. Poly(methyl methacrylate) star microgels have been compared with linear analogs by SEC/LALLS [280]. The reduction in size depends on the number of arms and not on their length. Although the work was not on synthetic polymers (polysaccharides), the reader is also referred to articles by Yu and Rollings on the application of SEC/LALLS to branching in macromolecules [281-283].

22.3.2 Short-chain branching

Most interest is in linear low-density polyethylene (LLDPE), which has short-chain branching from α-olefin comonomers, such as 1-butene, 1-hexene, 1-octene, or 4-methyl-1-pentene. Scholte et al. [284] have proposed a modified Mark-Houwink expression as a function of the mass fraction of side chains, S

$$[\eta] = (1 - S)^{a+1}K_{PE}M^a \tag{22.10}$$

where K_{PE} is the Mark-Houwink coefficient for linear polyethylene. The relationship between linear polyethylene and short-chain branched polyethylene is a simple function of the fraction of short branches

$$M_l = (1 - S)M_b \tag{22.11}$$

The relationships were verified with linear polypropylene and ethylene/propylene copolymers. Chemical heterogeneity in LLDPE has been examined by combined SEC/FTIR, in which methyl group absorption is used as a measure of short-chain branching across the molecular weight distribution [285]. A comparison of SEC with on-line viscosity, LALLS, and off-line light-scattering measurements, temperature rising elution fractionation (TREF), and liquid/liquid extraction was conducted among five laboratories [286]. It was found that SEC methods are more suited for long-chain branching, while TREF separates according to short-chain branching. This technique is the topic of further discussion in the next section.

22.4 CRYSTALLINITY

Crystallinity influences physical properties of synthetic polymers through control of solid-state morphology. It also affects solubility and rate of dissolution; crystalline and semicrystalline materials generally are less soluble than amorphous polymers, and frequently form aggregates in common solvents. The presence and degree of crystallinity is directly related to the molecular architecture of polymer chains and the processing and thermal history of the sample. Generally, regular repeat units encourage crystallinity, and the addition of comonomers or branching agents generates more amorphous materials. Most interest in determining the degree of crystallinity by liquid chromatography has centered on polyethylene. Linear, high-density polyethylene (HDPE) can be highly crystalline and stiff; LDPE contains long-chain branching and is amorphous and pliable; LLDPE contains short-chain branching, and this results in intermediate degrees of crystallinity and properties, depending on the type and degree of branching.

References on p. B502

Characterization of the degree of crystallinity in polyethylenes, which is directly related to the distribution of short-chain branches in these polymers, has been examined by TREF. The polymer is dissolved in a solvent such as xylene or 1,2,4-trichlorobenzene at an elevated temperature (typically 100-140°C) and pumped into the head of a column filled with inert packing material, such as Chromosorb P. The column is cooled at a controlled rate, typically 1.5°C/h over several days. Polymer chains with high degrees of crystallinity deposit on the column material first, followed by chains with decreasing crystallinity. The column is then eluted with solvent, and the temperature is gradually raised. Amorphous material, being more soluble at lower temperatures, is eluted first, followed by progressively more crystalline polymer chains. The effluent is frequently monitored with differential refractive index (DRI) and/or infrared (IR) detection, the latter being capable of monitoring branch content directly through methyl group absorption. A profile of mass vs. elution temperature is obtained. The temperature axis can be converted to short-chain branching by suitable calibration, typically through the use of narrow standards of known short-chain branching content. The experiment can be performed on either analytical or preparative scales, the latter being particularly advantageous when used with secondary analyses of fractions by NMR, x-ray diffraction, differential scanning calorimetry, SEC, and other conventional polymer characterization techniques. Recent articles by Wild et al. [287], Nakano and Goto [28], and Bergström and Avela [289] describe this technique. The methodology is not new; several references are contained in the papers cited above to studies spanning the last three decades.

TREF fractionates polyethylenes primarily according to the degree of short-chain branching (SCB) (comonomer content). It is less sensitive to long-chain branching [286] and relatively insensitive to molecular weight [287,290]. HDPE, LDPE, and LLDPE show distinctly different TREF elution profiles (Figs. 22.2-22.4) [291]. The distribution of SCB is apparently narrow in HDPE, broad in LDPE, and multimodal and broad in LLDPE. This is essentially a measure of copolymer compositional heterogeneity in these materials (Section 22.5). By analysis of preparative TREF fractions of LLDPE, it has been shown that

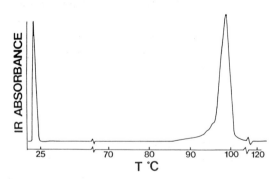

Fig. 22.2. Temperature rising elution fractogram of HDPE. (Reproduced from Ref. 291 with permission.)

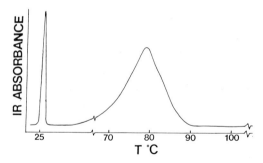

Fig. 22.3. Temperature rising elution fractogram of LDPE. (Reproduced from Ref. 291 with permission.)

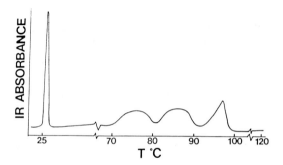

Fig. 22.4. Temperature rising elution fractogram of LLDPE. (Reproduced from Ref. 291 with permission.)

molecular weight decreases with increasing amounts of SCB [291,292]. It is postulated that the multimodal SCB distributions in LLDPE are a result of different active sites on heterogeneous Ziegler catalysts [292]. Kelusky et al. [293] have used the resolving power of TREF for SCB to characterize blends of high-pressure LDPE and LLDPE. The comonomers vinyl acetate, ethyl acrylate, isobutylene, propylene, and hexadiene also reduce crystallinity in linear polyethylenes, and TREF profiles are indicative of the degree of copolymer compositional heterogeneity in these materials [287,293].

22.5 COPOLYMER COMPOSITIONAL HETEROGENEITY

Copolymers are similar to homopolymers in that properties are influenced by molecular weight; however, the ratio of comonomers and the distribution of chemical compositions is frequently of greater importance. Chemical heterogeneity can take several forms, depending on the type of copolymer. Addition polymers from free-radical polymeri-

References on p. B502

zation have a distribution of chemical compositions as a result of monomer drift during the polymerization; a comonomer may prefer to polymerize with itself more than with another monomer. As the polymerization progresses, the relative concentration of one of the monomers in the reaction vessel increases, resulting in the formation of copolymer with a progressively higher concentration of one of the monomers. A range of copolymer chemical compositions of comonomers A and B may form, which in the most severe case may result in a mixture of pure homopolymer A and pure homopolymer B, and in the most favorable cases of alternating or randomly distributed comonomers with equivalent ratios in each polymer chain. This distribution is superimposed on the molecular weight distribution of the copolymer, and may be represented by a contour plot; such as Fig. 22.5. In the case of block copolymers, the ratio and number of blocks may vary from chain to chain, and depending on the method of synthesis and nature of comonomers, it is not uncommon to find a distribution of block structures. The potential for heterogeneity in triblock copolymers is obviously greater. Graft copolymers may show a distribution of graft lengths and frequently contain an ungrafted fraction. Step-growth (condensation) polymers may also contain a distribution of chemical compositions if there are more than two monomers, or if there are prepolymer segments, such as are commonly found in polyurethanes. In fact, most polymers may be compositionally heterogeneous, since endgroups usually have a composition different from that of the main chain.

A principal problem in the characterization of copolymers is the need to deconvolve the molecular weight and chemical composition distributions. Ideally, one would like to fractionate copolymers by molecular weight, independent of chemical composition, and then measure the chemical composition of each molecular weight fraction. Conversely, the copolymer could be separated by composition, independent of molecular weight, and

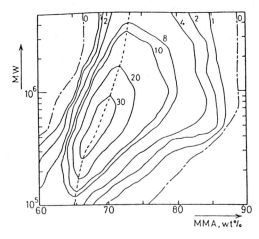

Fig. 22.5. Contour map of the three-dimensional distribution of molecular weight and chemical composition of high-conversion poly(styrene-co-methyl methacrylate). Numerical designations indicate relative abundance. (Reproduced from Ref. 329 with permission.)

the molecular weight of each compositional slice could then be determined along with average sequence length or other molecular properties. The closest approximation to the first case is to fractionate copolymers by SEC. In this case, the primary fractionation is not completely independent of composition, since SEC separates by size rather than molecular weight. An excellent overview of the characterization of copolymers by chromatographic means prior to 1983 has been presented [294].

Selective detection of each comonomer across the size-exclusion distribution provides compositional information directly. Applications prior to 1982 have been reviewed [295]. The most common examples are the characterization of styrene/acrylonitrile (SAN) copolymers by UV absorption spectrophotometry of styrenic units. Garcia-Rubio et al. [296,297] have discussed the use of multidetectors, particularly UV, and the sources of hypochromic effects and band shifts that result in deviations from Beer's law. The extinction coefficient of styrene units is a function of the sequence length of the polymer, and is influenced by the dielectric constant of the solvent as well as the stereochemistry of the polymer. The average styrene concentration, as well as the average sequence length of each fraction, can be estimated by measuring the copolymer absorption at two or more wavelengths. The use of DRI and UV detection for the quantitation of chemical composition in SAN has been compared with analysis of collected SEC fractions by pyrolysis/GC for chemical composition [298]. Agreement between the two methods is fair. In contrast, the multidetection SEC experiment inaccurately estimates chemical composition of styrene/methyl methacrylate copolymers, and better results are obtained when SEC fractions are characterized by pyrolysis/GC [299,300] or when the polymer is separated by adsorption LC [301] (see below). A principal shortcoming is the error introduced, particularly at the wings of the molecular size distribution, when calculating the styrene concentration from the DRI and UV detector signals.

Oxidative aging of poly(styrene/isoprene/styrene) (Kraton) has also been examined semi-quantitatively by SEC with DRI and UV detection [302,303]. In this case, the formation of α, β-unsaturated carbonyls, at the isoprene chain end after scission, can be monitored across the molecular size distribution by the characteristic UV absorption of the endgroups. Compositional heterogeneity in low-molecular-weight styrene/acrylic acid copolymers has been examined by DRI/photodiode-array UV detection, and used to compare reactivity of the monomers at high and low polymerization temperatures [304]. Crosslinking, grafting, and homopolymer formation of polyacrylate/benzyl acrylate polymeric reactions, induced by low-energy electron beams were investigated by SEC with UV and DRI detection [305]. A DRI/UV/LALLS multidetector scheme has been applied to the characterization of compositional heterogeneity in block copolymers of polystyrene and poly(dimethylsiloxane) [306], and a heterogeneity parameter was calculated for correction of RI increments needed for LALLS [307]. A similar multidetector scheme has been used to determine molecular weights and to account for different RI increments in the hard and soft segments of polyurethanes [308].

IR has also been used with SEC for selective characterization of comonomer units across the molecular size distribution. This detection is somewhat more limited than UV spectrophotometry because of eluent absorption, although interfaces that eliminate the solvent have been recently described [309]. Mori et al. [310] have presented results for styrene/acrylonitrile and styrene/methyl methacrylate copolymers obtained with a modified dispersive instrument.

Other combinations of selective detection for SEC are certainly possible, depending on the chemical constitution of the polymer. For example, Parks et al. [311-313] have described the application of element-specific graphite furnace atomic absorption, combined with DRI and/or UV detection, for the characterization of tin-bearing organometallic polymers. The distribution of tin in high- and low-molecular-weight fractions can be estimated form the element-specific and mass responses. Broersen et al. [314] have described a method for derivatizing polymers containing primary amino groups with o-phthalaldehyde in THF. The amine functionality can then be selectively monitored by SEC with fluorescence detection. The example given was low-molecular-weight polyamino acids based on diethylenetriamine or triethylenetetraamine and dimerized diacids.

Many of the difficulties associated with detector response in copolymer characterization by SEC have encouraged the development of cross-fractionation methods, which place more emphasis on fractionating the multiple distributions of these complex polymers, followed by selective detection of the more well-defined fractions. One example demonstrated by Balke is 'orthogonal chromatography' [6,315-319]. In the example given, compositionally heterogenous mixtures of polystyrene, poly(n-butyl methacrylate) and poly(styrene-co-n-butyl methacrylate) are fractionated by SEC in THF on polystyrene gel columns in Chromatograph 1. Chromatograph 2 is coupled to Chromatograph 1 through an injection port, and slices from the size separation are sampled. Chromatograph 2 is equipped with polar size-exclusion columns (μ-Bondagel) and an eluent of THF containing more than 50% n-heptane, which is a nonsolvent for polystyrene and a solvent for poly(n-butyl methacrylate). Polystyrene-rich fractions contract in the poor solvent mixture, resulting in smaller sizes than poly(n-butyl methacrylate)-rich fractions. Polystyrene molecules also adsorptively interact with the size-exclusion resin columns, resulting in further retardation of polystyrene-rich chains. The result is a fractionation in Chromatograph 2 according to chemical composition: poly(n-butyl methacrylate) is eluted first, the copolymer second, and polystyrene last. When coupled with selective UV detection for the styryl components, a two-dimensional profile of molecular size and chemical composition may be generated. In a more recent study on styrene/n-butyl methacrylate copolymers, Chromatograph 2 was equipped with either polystyrene or polyacrylamide gel columns and either THF/heptane or THF/2-propanol eluents [320]. Separation of the copolymer according to chemical composition was obtained, except when polyacrylamide gel and heptane/2-propanol were used in Chromatograph 2.

Cross fractionation, first by preparative SEC, followed by analysis of the chemical composition distribution (CCD) of fractions by TLC, has been performed by Termachi et al. [321] on high-conversion styrene/methyl acrylate (SMA) copolymers. They demonstrated a compositional drift of the high-molecular-weight components. SEC/TLC cross fractionation has similarly been applied to styrene/ethyl methacrylate copolymers [322]. However, because of the difficulties, both in reproducing solvent gradients and in quantitating TLC, many researchers over the past decade have found that cross fractionation for

CCD determination is more effectively accomplished with HPLC techniques. Most of the research has focused on designed mixtures of copolymers, rather than practical samples, for method development and evaluation of CCD separations.

Cross fractionation of SAN was studied by Glöckner in several publications [237,323-327]. HPPLC, as described in Section 22.2, was used for CCD analysis after SEC fractionation of the copolymer sample. By comparing the solvent composition needed for polymer elution in LC with solubility results from turbidimetric titration [324,326,327], the mechanism of polymer separation was determined to be precipitation/redissolution in alkane/THF gradients on alkyl-bonded silica sorbents. The effects of temperature and sample concentration on the HPPLC separation have also been investigated [324]. One of the difficulties with HPPLC is that the effect of molecular weight on solubility is convolved with that of chemical composition in the elution profile, even when prior size separation has been performed. This molecular weight dependence has been evaluated for SAN [324], and an empirical correction factor for polymer retention volumes was calculated [325]. A comparison of the effect of the sorbent phase between octadecyl- and cyanopro-

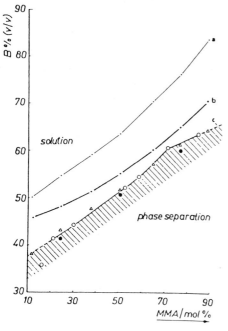

Fig. 22.6. Elution characteristics and solubility boundary vs. composition of poly(styrene-co-methyl methacrylate) in alkane/THF systems. Curves: (a) peak elution solvent concentration in separation on silica with isooctane/THF (10% methanol) gradient; (b) peak elution solvent concentration in separation on silica with isooctane/THF gradient; (c) solubility boundary according to turbidimetric titration with n-hexane of copolymer solutions in THF. (Reproduced from Ref. 327 with permission.)

pyl-bonded silicas, as well as bare silica, was performed for SAN and poly(styrene-co-methyl methacrylate) (SMMA) with the resulting observation that HPPLC separations of SAN are nearly independent of packing type [327]. However, HPPLC separations of SMMA demonstrate an additional contribution to retention on silica beyond solubility, as shown by the data in Fig. 22.6, but not on bonded-phase sorbents.

The technique of cross fractionation for copolymer characterization has also been applied by Mori et al. [300,328-330] to SMMA and by Glöckner et al. [331,332] to poly(styrene-co-ethyl methacrylate) (SEMA). Mori et al. employed the reverse scheme of CCD fractionation on silica with step gradients of 1,2-dichloroethane/chloroform (1% ethanol), followed by SEC [328]. Although they found little molecular weight dependence in the CCD fractionation, they were unable to elute copolymers with higher than 50% MMA content. By the use of very shallow linear gradients of ethanol in chloroform, Mori was able to overcome this limitation and generated detailed maps of CCD vs. molecular weight for both high- [329,330] and low-conversion [330] SMMA samples by SEC, followed by HPLC (Fig. 22.5). He found that, in some cases of low-conversion SMMA, bimodal composition distributions were present in monomodal molecular size distributions. SEC fractions of SEMA were separated by CCD by Glöckner et al. [331] on silica and cyanopropyl-bonded silica with isooctane/THF/methanol gradients. Results similar to those observed previously for SMMA were obtained; higher retention was found on bare silica, indicating that adsorption, in addition to precipitation/redissolution, contributes to retention. Improved CCD resolution was found for SEC fractions of SEMA by nonaqueous reversed-phase HPPLC on octadecyl-bonded silica with methanol/THF gradients [332].

In addition to the many publications describing specific copolymer characterization experiments that have appeared recently, there have been several extensive review articles on the subject. The application of TLC to the CCD separation of copolymers has been reviewed by Inagaki [333], Belenkii and Gankina [334], Belenkii [335], and Inagaki and Tanaka [336]. HPLC and HPPLC for the characterization of copolymers have been reviewed by Glöckner [337].

Unlike the experimental studies described above, in which chromatographic cross fractionation was used to map the molecular weight and chemical composition distributions of copolymers, many more investigations have focused on CCD determination without SEC. TLC was employed to perform CCD separations of SMMA [338] and SAN [339] in studies of models of copolymerization chemistry. An unusual preparative separation scheme was developed by Inagaki et al. [340] in order to perform CCD separations of diblock SMMA copolymers. They used a rising solvent front gradient of ethyl acetate and benzene on a bed of silica, packed into a glass cylinder with filter paper partitions. This simulated a TLC separation on a bulk scale and allowed comparison of CCD separation with light-scattering CCD determination and study of anionic synthesis of diblock copolymers.

As with cross fractionation experiments, researchers studying CCD separations have concentrated on the use of HPLC techniques in preference to TLC. More than a decade ago, SMA copolymers were compositionally separated on a HPLC silica column with a tetrachloromethane/methyl acetate gradient by Teramachi et al. [341]. Danielewicz et al.

performed CCD HPLC separations of SMMA and SEMA on silica with 1,2-dichloroethane (DCE)/THF gradients [342] and of SAN on cyanopropyl-bonded silica with gradients of DCE/n-heptane to DCE/acetonitrile [343]. The theoretical basis of copolymer CCD separations by HPLC techniques, and in particular HPPLC, has been investigated in detail by Glöckner and coworkers. A discussion of the comparison between HPPLC and the older, more tedious technique of Baker-Williams fractionation was published, emphasizing the level of improvement possible with HPPLC [344]. In addition to the correlation between the eluent composition in HPPLC and turbidimetric titration solubility data mentioned previously [324,326,327], further evidence of the precipitation mechanism of CCD separation was presented for mixed SAN copolymers on bare silica and silica bonded with cyanopropyl, octadecyl, or propanediol stationary phases, which indicated that elution was nearly independent of the sorbent type [345]. However, adsorption interactions affected the separation of the more polar copolymer, SMMA, when it was chromatographed on silica, but not on octyl-bonded silica [346]. The control of sample retention and complete elution of both SAN and SEMA during HPPLC CCD separation was addressed in terms of sorbent activity, solvent gradient, and sample volume in a later study [347]. It was also shown that reversal of the elution order of SEMA copolymers is possible when conditions are altered from a normal-phase isooctane/THF/methanol gradient on silica to a reversed-phase methanol/THF gradient on octadecyl-bonded silica [348].

Mori and Uno [349-351] have concentrated their efforts on elucidating the CCD separation of SMMA copolymers with very shallow or step gradients of chloroform/ethanol or DCE/ethanol on silica. By studying the effects of solvent and temperature, they have proposed a retention mechanism of adsorption, rather than precipitation, which is based upon hydrogen-bonding interactions between the carbonyl of MMA and the silanol hydroxyls of the silica surface. The ability of ethanol to displace the copolymer from these hydrogen-bonding interactions is proposed to be the main contribution to selective separation of the copolymers.

Not all copolymer CCD separations have been performed on conventional HPLC silica or bonded silica. Glöckner and Van den Berg [352] performed HPPLC CCD separations of mixtures of SAN copolymers on open-tubular stainless-steel and nonporous glass-bead columns for further confirmation of the lack of influence of the sorbent surface in HPPLC. In addition, they observed that better separations could be performed on short (55-mm) packed silica columns as opposed to longer (150-mm) columns. This was attributable to bandbroadening due to eluent flowpaths after copolymer dissolution, when elution occurs without sample retention. Comparison of different types of sorbent column packings was also made by Sato and coworkers [353-355]. They found that CCD separations of SMMA could be accomplished by an adsorption mechanism on polar sorbents, such as silica and crosslinked acrylonitrile gel beads in a molecular-weight-independent mode. However, the styrene/divinylbenzene beads or glass beads required precipitation conditions for separations, which were molecular-weight dependent [353]. CCD separation of poly(styrene-co-butadiene) was accomplished by Sato on a HPLC column, packed with crosslinked acrylonitrile copolymer beads, with gradients of chloroform and either ethanol, acetone, or hexane [354]. Nonaqueous reversed-phase chromatography on crosslinked polystyrene gel beads with a dichloromethane/acetonitrile gradient was used to separate SMMA by CCD in a molecular-weight-dependent mode [355].

One of the limitations of CCD separations of copolymers has always been the inability to detect the elution of some types of copolymers after separations. Mass detection by DRI, as conventionally applied to SEC, is inapplicable to the solvent gradients used in HPLC of copolymers. In addition, many of the copolymers that require characterization do not possess suitable UV-absorbing chromophores, and many desirable solvents absorb in the UV, limiting the use of UV detection. One of the possible solutions to this detection dilemma was studied by Mourey [356]. He used evaporative light scattering for detection in compositional separations of mixtures of poly(alkyl acrylate) and poly(alkyl methacrylate) homopolymers and in the CCD separation of SMA and poly(methyl methacrylate-co-methyl acrylate) on silica with toluene/2-butanone gradients.

As with the characterization of molecular weight distribution in homopolymers, the technique of FFF has been applied to polymer CCD separations with the use of a thermal gradient as the retentive field (ThFFF). Because the thermal diffusion coefficient is dependent upon solute chemical composition and is integral to the retention process in ThFFF, CCD separations can be performed. This has been demonstrated for homopolymers of polystyrene and poly(methyl methacrylate) of similar molecular weight [357], as well as poly(α-methylstyrene) and polyisoprene [358]. Difficulties encountered in this technique are similar to those in HPPLC, in that molecular weight separation is convolved with chemical composition separation in the elution profile. Since, unlike HPPLC, at present no procedure is known to suppress the molecular weight contribution to retention, the application of ThFFF to the analysis of CCD may remain limited for the near future.

22.6 ANALYSIS OF ENDGROUPS AND FUNCTIONALITY

The analysis of endgroups in synthetic polymers may be viewed as another form of characterization of copolymer compositional heterogeneity. The endgroups have a chemical composition that is different from that of the repeating units of the polymer chain, regardless of whether the polymer is a homopolymer or copolymer. Endgroup analysis is a classical means of determining the number-average molecular weight of polymers, and is of importance in determining the functionality of prepolymers frequently used in block copolymers, such as segmented polyurethanes. The analysis of endgroups is complicated as the molecular weight of the polymer chain increases, for obvious reasons: the sensitivity of the characterization technique must increase proportionally with the length of the polymer chain, and it frequently requires detection and quantitation of one endgroup out of more than 1000 repeat units.

A common approach, as in copolymer compositional analysis, is to separate the polymer by size exclusion and selectively monitor the endgroup absorption spectrophotometrically. In some instances, a suitable chromophore exists for endgroup analysis. Garcia-Rubio et al. [359] demonstrated that benzoyl peroxide initiator fragments could be quantitated by UV spectrophotometry of the whole polymer, indicating that UV detection for SEC could be applicable to endgroup analysis across the molecular weight distribution. More commonly, endgroups do not have distinctive UV absorbance and must be derivatized with a suitable reagent. Manolova et al. [360] distinguished linear from cyclic caprolactone oligomers by SEC with UV detection after derivatizing the hydroxyl endgroups of the linear molecules with phenylisocyanate. Quirk et al. [361] have described a general method of coupling fluorescent aromatic groups to living polymer chains, made by anionic polymerization. This method may also be useful for highly sensitive endgroup analysis by SEC with fluorescence detection. Mori [362] quantitated endgroups in aliphatic polyesters as a function of molecular weight by SEC with IR detection after derivatizing hydroxyl endgroups with 3,5-dinitrobenzoyl chloride and carboxyl endgroups with o-(p-nitrobenzyl)-N,N'-diisopropylisourea. Recently, the feasibility of coupled SEC/500-MHz proton-NMR in CDCl$_3$ for the analysis of t-butyl endgroups and number-average molecular weights of low-molecular-weight, anionically polymerized poly(methyl methacrylate)s was demonstrated [363]. Provided that the currently excessive extra-column bandbroadening can be reduced, the technique holds obvious promise for analysis not only of endgroups, but also of chemical [364], configurational, and conformational heterogeneities in numerous polymers.

Adsorption chromatography can be highly sensitive to endgroup chemical structure if the terminus is more polar than the chain repeat units. Carboxylic acid-terminated polystyrenes can be separated from polystyrenes that are not acid-terminated by HPLC on silica [365], as can hydroxyl-terminated poly(methyl methacrylates) [366]. Winnik et al. [367] described the use of small-pore-diameter silica in a gravity-fed column for the fractionation of labeled from unlabeled polystyrene chains. In all of the above cases, polymer chains that contain polar endgroups are retained more strongly than chains terminated with nonpolar groups. In HPLC on silica [365], the normal elution order for adsorption chromatography of low to high molecular weight was reversed. Gorshkov et al. [368] separated epoxy resins according to endgroup functionality by liquid chromatography in the "critical" mode, which suppresses molecular size separation and fractionates according to chemical composition. No experimental details were presented. Mourey et al. [369] fractionated spiropyran-endlabeled polystyrenes on 6-nm- and 50-nm-pore-diameter silicas by gradient elution, and demonstrated that selectivity for endgroups could be explained by solvent localization theory. Strongly localizing solvents, such as THF, provided little or no selectivity between labeled and unlabeled chains, while weakly localizing solvents, such as dichloromethane, provided differentiation between polar endgroup-labeled and -unlabeled chains as large as 600 repeat units long.

22.7 STEREOCHEMISTRY AND SEQUENCE DISTRIBUTION

As discussed in Section 22.1, the compositional distribution present in synthetic polymers extends beyond simple molecular weight and, for copolymers, chemical composition. Vinyl monomers possess various stereochemistries, also known as tacticity. Separation of polymers by chromatographic means to extract stereochemical information is extremely difficult, due to the subtle ways in which these differences express themselves in a separation system. Tacticity is more readily evaluated for oligomeric compounds, both because of the proportionately high influence of stereochemical orientation on molecular activity for these low-molecular-weight compounds in comparison to high polymers and because of the geometrically increasing complexity of stereochemical mixtures with increasing molecular weight when formed randomly. Lewis et al. [370,371] have published tacticity separations of polystyrene oligomers on various bonded-phase columns. Mourey et al. [372] used shallow, normal-phase gradients on silica with multiple columns to achieve higher resolution than obtained by Elgert et al. [373] and separated up to 11 of 20 possible geometric isomers of polystyrene heptamer, as well as lower oligomers. Semi-preparative separations of polystyrene trimer, tetramer, and pentamer were used to isolate syndiotactic, isotactic, and heterotactic fractions, with subsequent stereochemical identification by NMR.

Mixtures of isotactic and syndiotactic poly(methyl methacrylate) in the 10^5-D molecular-weight range were separated by Inagaki et al. [374] by TLC on silica. The percent syndiotactic content of poly(methyl methacrylate), containing both syndiotactic and atactic chains, was estimated from TLC separations [375]. However, no solvent combination could be found to obtain clean separation, based purely on tacticity without molecular-weight contributions. This behavior was attributed to a mixed-retention mechanism, in which well-retained solutes were separated on the basis of adsorption interactions, and poorly retained solutes by precipitation/redissolution.

For copolymers, the arrangement of monomer units along the polymer backbone may also express itself in terms of block structure with random or perfectly alternating sequences at one extreme and diblock structure at the opposite. Characterization of block and graft copolymers by TLC to elucidate differences in chain architecture has been reviewed by Inagaki [333]. Most investigations in this area have focused on the resolution of block or graft copolymers from residual prepolymer homopolymers after synthesis [376]. However, Inagaki et al. [377] have published a study of the TLC separation of styrene/butadiene (SB) block copolymers, based on chain architecture. Diblock and tri-block SB copolymers were resolved on silica with a cyclohexane/chloroform gradient at elevated temperature. This technique was then applied to five commercially produced SB block copolymers and successfully distinguished their architectural differences. An adsorption mechanism was proposed to account for the separation selectivity for block sequence in these experiments on silica. However, the authors were unable to reproduce their separation by TLC on alumina, and this was attributed to their difficulty in controlling the moisture content of the alumina surface.

22.8 PHYSIOCHEMICAL MEASUREMENTS

Any number of physiochemical measurements can be made on polymer systems by chromatographic means. Most measurements are corroborated with classical polymer characterization methods, e.g., light scattering, viscometry, or sedimentation, and are frequently parts of fundamental studies in polymer physics which are outside the scope of this review. Again, discussion will be limited to a few examples of investigations of polymer phenomena in solution, with emphasis on characterization by LC techniques.

Micelle formation and measurement of hydrodynamic parameters of micelles formed by poly(oxyethylene) *n*-alkyl ether oligomers has been investigated by aqueous SEC on Sepharose columns [378,379]. The surfactants were injected above their critical micelle concentration to avoid complications arising from equilibrium with free surfactant molecules. The size of the micelles could be related to the lengths of the oxyethylene and oxymethylene blocks. Polymer surfactant complexation has been investigated by the reduction in size-exclusion elution volumes of polyethylene oxide in the presence of anionic and neutral surfactants, and by the appearance of vacancy peaks [380]. The vacancy peaks correspond to the deficit of surfactant in the elution zone of the polymer, and can be used to estimate the magnitude of polymer/surfactant complexation. Vacancy peaks have also been used to study selective solvation of poly(alkyl methacrylate)s in various binary solvents [381].

In dilute solutions, block copolymers may form aggregates or micelles, particularly if the solvent is a poor one for one of the blocks. There is an equilibrium between unimer and micelle which is determined by solution thermodynamics, the investigation of which can be undertaken through SEC. A theoretical model for the SEC of associating systems, applicable to block copolymers, has been presented [382].

Micelle formation of polystyrene-block-polyisoprene has been studied by SEC between 26°C and 110°C in *N,N*-dimethylacetamide (DMAC) [383]. DMAC is a poor solvent for polyisoprene blocks, and micelles form which presumably contain polyisoprene-rich cores and polystyrene-rich shells. SEC showed that micelles revert to unimer with increasing temperature, despite some complications from adsorption of the micelles on the column packing. Similar studies by SEC have also been reported for polystyrene-block-hydrogenated polybutadiene-block-polystyrene in mixed dioxane/heptane eluents [384,385]. The unimer is distinctly separated from micelles in the experiment, and the ratio of the two is dependent upon sample concentration and eluent composition. It was shown, by varying the flowrate, that the rate of micelle formation is not negligibly small compared to the duration of the experiment. By making up samples in a solvent that encourages formation of only unimer or micelle, and injecting them into a mixed-eluent SEC system which encourages formation of only micelle/unimer, the rate of micelle formation could be estimated on the basis of the amount of trapped (not eluted) micelles and simple diffusion arguments [385]. SEC has also been used to study the formation of radial polystyrene-block-polybutadiene copolymers in THF/ethanol mixtures [386]. SEC is capable of separating unimer from micelle, thus permitting investigation of the concentration of each species as a function of eluent composition.

The alternative investigation of micelle formation in polystyrene-block-polybutadiene by SEC — varying the temperature while keeping the solvent composition constant (pure 1,4-dioxane) — has also been reported [387]. Variable-temperature studies have also been demonstrated to be useful in the investigation of hydrodynamics in poly(alkyl-block-alkyl methacrylate)s [388] and of transition phenomena of polymers in dilute solution

[389]. In the latter case, a distinct increase in the SEC distribution coefficient, which correlates with a decrease in dilute solution viscosity, is observed for polystyrene and triblock poly(methyl methacrylate-styrene-methyl methacrylate) in benzene/methanol. This discontinuity is attributed to a conformational transition phenomenon, which in the case of the triblock copolymer is likely to be a transition from segregated to nonsegregated conformations. Other types of association in solution may also be studied by SEC. For example, stereoassociation of isotactic and syndiotactic poly(methyl methacrylate) has been examined in THF [390,391].

Phase diagrams of ternary polymer systems of polystyrene/polyisoprene/toluene at 45°C [392] and polystyrene/polybutadiene/THF at 23°C [393] have been generated by analysis of the composition of the separated phases by quantitative SEC with DRI and UV detection. With this method, polymer/polymer interaction parameters, the influence of molecular weight on polymer/solvent interaction parameters, the shape of the binodal curve, and location of the critical point can be evaluated conveniently.

Another method of note is the measurement of the molecular-weight dependence of intramolecular excimer formation in polymer solutions [394]. Poly(2-vinylnaphthalene) was fractionated according to molecular size in toluene, and the fluorescence intensity was measured directly by on-line detection. Combined with a suitable method of molecular weight calibration and a measure of polymer concentration at each eluted slice (obtained by DRI detection), a measure of excimer and monomer fluorescence was obtained across the molecular weight distribution. This example again emphasizes the growing importance and usefulness of multidetection methods in SEC.

REFERENCES

1 P.L. Dubin (Editor), *Aqueous Size-Exclusion Chromatography*, Journal of Chromatography Library, Vol. 40, Elsevier, Amsterdam, 1988.
2 T. Provder (Editor), *Detection and Data Analysis in Size-Exclusion Chromatography*, ACS Symposium Series, Vol. 352, American Chemical Society, Washington, DC, 1987.
3 G. Glöckner, *Polymer Characterization by Liquid Chromatography*, Journal of Chromatography Library, Vol. 34, Elsevier, Amsterdam, 1987.
4 J. Janča, *Steric Exclusion Liquid Chromatography of Polymers*, Chromatographic Science Series, Vol. 25, Dekker, New York, 1984.
5 T. Provder (Editor), *Size Exclusion Chromatography*, ACS Symposium Series, Vol. 245, American Chemical Society, Washington, DC, 1984.
6 S.T. Balke, *Quantitative Column Liquid Chromatography*, Journal of Chromatography Library, Vol. 29, Elsevier, Amsterdam, 1984.
7 B.G. Belenkii and I.Z. Vilenchik, *Modern Liquid Chromatography of Macromolecules*, Journal of Chromatography Library, Vol. 25, Elsevier, Amsterdam, 1984.
8 A. Hamielec and M. Styring, *Pure Appl. Chem.*, 57 (1985) 955.
9 P.J. Wyatt, D.L. Hicks, C. Jackson and G.K. Wyatt, *Am. Lab. (Fairfield, CN)*, 20 (1988) 108, 112.
10 C. Jackson, L.M. Nilsson and P.J. Wyatt, *J. Appl. Polym. Sci., Appl. Polym. Symp.*, 43 (1989) 99.
11 L. Letot, J. Lesec and C. Quivoron, *J. Liq. Chromatogr.*, 3 (1980) 427.

12 F.B. Malihi, C. Kuo, M.E. Koehler, T. Provder and A.F. Kah, in T. Provder (Editor), *Size Exclusion Chromatography,* ACS Symposium Series, Vol. 245, American Chemical Society, Washington, DC, 1984, p. 281.
13 P.K. Dasgupta, *J. Liq. Chromatogr.,* 7 (1984) 2367.
14 C.E. Lundy and R.D.Hester, *J. Liq. Chromatogr.,* 7 (1984) 1911.
15 M.A. Haney, *J. Appl. Polym. Sci.,* 30 (1985) 3037.
16 W.W. Yau, S.D. Abbot, G.A. Smith and M.Y. Keating, in T. Provder (Editor), *Detection and Data Analysis in Size- Exclusion Chromatography,* ACS Symposium Series, Vol. 352, American Chemical Society, Washington, DC, 1987, p. 80.
17 T.A. Chamberlin and H.E. Tuinstra, *J. Appl. Polym. Sci.,* 35 (1988) 1667.
18 J. Lesec, D. Lecacheux and G. Marot, *J. Liq. Chromatogr.,* 11 (1988) 2571.
19 H.G. Barth, W.E. Barber, C.H. Lochmüller, R.E. Majors and F.E. Regnier, *Anal. Chem.,* 60 (1988) 387R.
20 C.G. Smith, R.A. Nyquist, N.H. Mahle, P.B. Smith, S.J. Martin and A.J. Pasztor, Jr., *Anal. Chem.,* 59 (1987) 119R.
21 H.G. Barth, W.E. Barber, C.H. Lochmüller, R.E. Majors and F.E. Regnier, *Anal. Chem.,* 58 (1986) 211R.
22 C.G. Smith, N.H. Mahle, W.R.R. Park, P.B. Smith and S.J. Martin, *Anal. Chem.,* 57 (1985) 254R.
23 R.E. Majors, H.G. Barth and C.H. Lochmüller, *Anal. Chem.,* 56 (1984) 300R.
24 C.E. Lundy and R.D. Hester, *J. Polym. Sci., Polym. Chem. Ed.,* 24 (1986) 1829.
25 Y. Kato, T. Matsuda and T. Hashimoto, *J. Chromatogr.,* 332 (1985) 39.
26 T. Kato, T. Tokuya and A. Takahashi, *J. Chromatogr.,* 256 (1983) 61.
27 T.V. Alfredson, C.T. Wehr, L. Tallman and F. Klink, *J. Liq. Chromatogr.,* 5 (1982) 489.
28 F.C. Lin and G.D. Getman, *GPC Symposium '87,* Waters Division of Millipore, Milford, MA, 1987, p. 225.
29 G. Callec, A.W. Anderson, G.T. Tsao and J.E. Rollings, *J. Polym. Sci., Polym. Chem. Ed.,* 22 (1984) 287.
30 A.R. Cooper, in J.V. Dawkins (Editor), *Developments in Polymer Characterization,* Elsevier, New York, 1986, Chap. 4.
31 J.E. Rollings, A. Bose, J.M. Caruthers, G.T. Tsao and M.R. Okos, in C.D. Craver (Editor), *Polymer Characterization,* Advances in Chemistry Series, Vol. 203, American Chemical Society, Washington, DC, 1983, p. 345.
32 H.C. Barth, in J.E. Glass (Editor), *Water Soluble Polymers, Beauty with Performance,* Adv. Chem. Series, Vol. 213, American Chemical Society, Washington, DC, 1986, Chap. 2.
33 H. Wada, K. Makino, T. Takeuchi, H. Hatano and K. Noguchi, *J. Chromatogr.,* 320 (1985) 369.
34 S. Mori, *J. Chromatogr.,* 471 (1989) 367.
35 A. Domard and M. Rinaudo, *Polym. Commun.,* 25 (1984) 55.
36 M. Stickler and F. Eisenbeiss, *Eur. Polym. J.,* 20 (1984) 849.
37 D.M. Wonnacott and E.V. Patton, *J. Chromatogr.,* 389 (1987) 103.
38 E.V. Patton and D.M. Wonnacott, *J. Chromatogr.,* 389 (1987) 115.
39 G.B. Guise and G.C. Smith, *J. Chromatogr.,* 235 (1982) 365.
40 P.L. Dubin, I.J. Levy and R. Oteri, *J. Chromatogr. Sci.,* 22 (1984) 432.
41 P.L. Dubin and I.J. Levy, *J. Chromatogr.,* 235 (1982) 377.
42 I.J. Levy and P.L. Dubin, *Ind. Eng. Chem. Prod. Res. Dev.,* 21 (1982) 59.
43 Y. Kato and T. Hashimoto, *J. Chromatogr.,* 235 (1982) 539.
44 H. Goetz, H. Elgass and L. Huber, *J. Chromatogr.,* 349 (1985) 357.
45 A. Wirsen, *Makromol. Chem.,* 189 (1988) 833.
46 J.B.F. Lloyd, *Anal. Chem.,* 56 (1984) 1907.
47 F. Mahmud and E. Catterall, in T. Provder (Editor), *Size Exclusion Chromatography,* ACS Symposium Series, Vol. 245, American Chemical Society, Washington, DC, 1984, p. 365.
48 J.J. Cael, D.J. Cietek and F.J. Kolpak, *J. Appl. Polym. Sci., Appl. Polym. Symp.,* 37 (1983) 509.
49 B.F. Wood, A.H. Conner and C.G. Hill, Jr., *J. Appl. Polym. Sci.,* 32 (1986) 3703.

50 L.L. Lloyd, C.A. White, A.P. Brookes, J.F. Kennedy and F.P. Warner, *Brit. Polym. J.*, 19 (1987) 313.
51 R. Evans, R.H. Wearne and A.F.A. Wallis, *J. Appl. Polym. Sci.*, 37 (1989) 3291.
52 E. Dan and M.B. Roller, *J. Polym. Sci., Polym Lett. Ed.*, 21 (1983) 875.
53 G.L. Hagnauer and P.J. Pearce, in T. Provder (Editor), *Size Exclusion Chromatography*, ACS Symposium Series, Vol. 245, American Chemical Society, Washington, DC, 1984 p. 333.
54 F.N. Larsen and D.A. Spieker, *GPC Symposium '87*, Waters Division of Millipore, Milford, MA, 1987, p. 388.
55 D. Noël, K.C. Cole, J.-J. Hechler, A. Chouliotis and K.C. Overbury, *GPC Symposium '87*, Waters Division of Millipore, Milford, MA, 1987, p. 569.
56 D. Noël, K.C. Cole, J.-J. Hechler, A. Chouliotis and K. C. Overbury, *J. Chromatogr.*, 408 (1987) 129.
57 D.J. Russell, *J. Liq. Chromatogr.*, 11 (1988) 383.
58 A. Rudin, C.A. Fyfe and S.M. Vines, *J. Appl. Polym. Sci.*, 28 (1983) 2611.
59 D.R. Bain and J.D. Wagner, *Polymer*, 25 (1984) 403.
60 P. Laurent and Z. Gallot, *J. Chromatogr.*, 236 (1982) 212.
61 A.R. Walsh and A.G. Cambell, *J. Appl. Polym. Sci.*, 32 (1986) 4291.
62 J. Klein and A. Westerkamp, *J. Polym. Sci., Polym. Chem. Ed.*, 19 (1981) 707.
63 W.-M. Kulicke and N. Böse, *Colloid Polym. Sci.*, 262 (1984) 197.
64 C.J. Kim, A.E. Hamielec and A. Benedek, *J. Liq. Chromatogr.*, 5 (1982) 1277.
65 R. Biran and J.V. Dawkins, *Eur. Polym. J.*, 20 (1984) 129.
66 G. Muller and C. Yonnet, *Makromol. Chem., Rapid Commun.*, 5 (1984) 197.
67 S.G. Gharfeh and A. Moradi-Araghi, *J. Chromatogr.*, 366 (1986) 343.
68 G.R. McGowan and G.W. Hawkins, *GPC Symposium '87*, Waters Division of Millipore, Milford, MA, 1987, p. 273.
69 J. Desbrieres, J. Mazet and M. Rinaudo, *Eur. Polym. J.*, 18 (1982) 269.
70 T. Kato, T. Tokuya, T. Nozaki and A. Takahashi, *Polymer*, 25 (1984) 218.
71 D.C. Sherrington and P. Bonner, *Polym. Commun.*, 25 (1984) 71.
72 S. Mori, *Anal. Chem.*, 55 (1983) 2414.
73 C. Azuma, M.L. Dias and E.B. Mano, *Makromol. Chem., Makromol. Symp.*, 2 (1986) 169.
74 C.C. Walker, *J. Polym. Sci., Polym. Chem. Ed.*, 26 (1988) 1649.
75 Y. Mukoyama and H. Sugitani, *Anal. Sci.*, 1 (1985) 299.
76 G. Marot and J. Lesec, *GPC Symposium '87*, Waters Division of Millipore, Milford, MA, 1987, p. 113.
77 G. Pastuska, U. Just and H. August, *Angew. Makromol. Chem.*, 107 (1982) 173.
78 P.J. Wang and R.J. Rivard, *J. Liq. Chromatogr.*, 10 (1987) 3059.
79 T. Ogawa and M. Sakai, *J. Polym. Sci., Polym. Chem. Ed.*, 26 (1988) 3141.
80 W.A. Dark, *GPC Symposium '87*, Waters Division of Millipore, Milford, MA, 1987, p. 18.
81 E. Biagini, G. Costa, E. Gattiglia, A. Imperato and S. Russo, *Polymer*, 28 (1987) 114.
82 H. Schorn, R. Kosfeld and M. Hess, *J. Chromatogr.*, 282 (1983) 579.
83 T. Ogawa, M. Sakai and W. Ishitobi, *J. Polym. Sci., Polym. Chem. Ed.*, 24 (1985) 109.
84 C.A. Veith and R.E. Cohen, *Polymer*, 30 (1989) 942.
85 K. Se, M. Kijima and T. Fujimoto, *Polym J. (Jpn.)*, 20 (1988) 791.
86 Xun Tang, Yan Sun and Yen Wei, *Makromol. Chem., Rapid Commun.*, 9 (1988) 829.
87 J. Devaux, D. Delimoy, D. Daoust, R. Legras, J.P. Mercier, C. Strazielle and E. Nield, *Polymer*, 26 (1985) 1994.
88 Shu-Qin Bo and Rong-Shi Cheng, *J. Liq. Chromatogr.*, 5 (1982) 1405.
89 Wen-Yu Chen, Wen-Hsung Ho and Yie-Shun Chiu, *GPC Symposium '87*, Waters Division of Millipore, Milford, MA, 1987, p. 447.
90 V. Deniz and O. Güven, *J. Appl. Polym. Sci.*, 29 (1984) 433.
91 C. Bailly, D. Daoust, R. Legras, J.P. Mercier, C. Strazielle and A. Lapp, *Polymer*, 27 (1986) 1410.
92 Z. Dobkowski, *J. Appl. Polym. Sci.*, 29 (1984) 2683.

93 J. Lesec, D. Lecacheux and G. Marot, *GPC Symposium '87*, Waters Division of Millipore, Milford, MA, 1987, p. 89.
94 M.A. Haney and J.E. Armonas, *GPC Symposium '87*, Waters Division of Millipore, Milford, MA, 1987, p. 523.
95 H. Schorn, R. Kosfeld and M. Hess, *J. Chromatogr.*, 353 (1986) 273.
96 K. Weisskopf, *J. Polym. Sci., Polym. Chem. Ed.*, 26 (1988) 1919.
97 B.L. Neff and J.R. Overton, *J. Liq. Chromatogr.*, 7 (1984) 1537.
98 J.R. Overton and H.L. Browning, Jr., in T. Provder (Editor), *Size Exclusion Chromatography*, ACS Symposium Series, Vol. 245, American Chemical Society, Washington, DC, 1984, p. 219.
99 S.A. Jabarin and D.C. Balduff, *J. Liq. Chromatogr.*, 5 (1982) 1825.
100 Ming-Min Sang, Nan-Ni Jin and Er-Fang Jiang, *J. Liq. Chromatogr.*, 5 (1982) 1665.
101 K. Hibi, A. Wada and S. Mori, *Chromatographia*, 21 (1986) 635.
102 S. Berkowitz, *J. Appl. Polym. Sci.*, 29 (1984) 4353.
103 S. Mori, *Anal. Chem.*, 61 (1989) 1321.
104 C.V. Uglea, S. Aizicovici and A. Mihaescu, *Eur. Polym. J.*, 21 (1985) 677.
105 P. Helias, D. Durand, J.P. Busnel and C.M. Bruneau, *Eur. Polym. J.*, 18 (1982) 647.
106 N.E. Manolova, I. Gitsov, R.S. Velichkova and I.B. Rashkov, *Polym. Bull.*, 13 (1985) 285.
107 V. Deniz and O. Güven, *Brit. Polym. J.*, 18 (1986) 112.
108 V. Grinshpun, K.F. O'Driscoll and A. Rudin, *J. Appl. Polym. Sci.*, 29 (1984) 1071.
109 V. Grinshpun and A. Rudin, *Makromol. Chem., Rapid Commun.*, 6 (1985) 219.
110 V. Grinshpun, K.F. O'Driscoll and A. Rudin, in T. Provder (Editor), *Size Exclusion Chromatography*, ACS Symposium Series, Vol. 245, American Chemical Society, Washington, DC, p. 273.
111 T. Housaki and K. Satoh, *Makromol. Chem., Rapid Commun.*, 9 (1988) 257.
112 L.A. Utracki and M.M. Dumoulin, in T. Provder (Editor), *Size Exclusion Chromatography*, ACS Symposium Series, Vol. 245, American Chemical Society, Washington, DC, 1984, p. 97.
113 A.C. de Kok and A.C. Oomens, *J. Liq. Chromatogr.*, 5 (1982) 807.
114 D. Lecacheux, J. Lesec and C. Quivoron, *J. Appl. Polym. Sci.*, 27 (1982) 4867.
115 D. Lecacheux, J. Lesec, C. Quivoron, R. Prechner, R. Panaras and H. Benoit, *J. Appl. Polym. Sci.*, 29 (1984) 1569.
116 D. Moldovan and S.C. Polemenakos, *GPC Symposium '87*, Waters Division of Millipore, Milford, MA, 1987, p. 129.
117 F.M. Mirabella, Jr. and L. Wild, *Polym. Mat. Sci. Eng.*, 59 (1988) 7.
118 V. Grinshpun and A. Rudin, *Polym. Mat. Sci. Eng.*, 54 (1986) 174.
119 Ke-Qiang Wang, Shi-Yu Zhang, Jia Xu and Yang Li, *J. Liq. Chromatogr.*, 5 (1982) 1899.
120 V. Grinshpun and A. Rudin, *J. Appl. Polym. Sci.*, 32 (1986) 4303.
121 E. Brauer, H. Wiegleb and M. Helmstedt, *Polym. Bull.*, 15 (1986) 551.
122 S.J. Mumby, P.M. Cotts and T.P. Russell, *J. Chromatogr.*, 319 (1985) 241.
123 J.-F. Fauvarque and J. Malinge, *Makromol. Chem., Rapid Commun.*, 4 (1983) 343.
124 Yun-Zhu Luo, N.K. Reddy, F. Heatley, C. Booth, E.J. Goodwin and D. Jackson, *Eur. Polym. J.*, 24 (1988) 607.
125 K. Lederer, G. Imrich-Schwarz and M. Dunky, *J. Appl. Polym. Sci.*, 32 (1986) 4751.
126 J.A.P.P. van Dijk, J.A.M. Smit, F.E. Kohn and J. Feijen, *J. Polym. Sci., Polym. Chem. Ed.*, 21 (1983) 197.
127 M. Kubin, *J. Appl. Polym. Sci.*, 27 (1982) 2943.
128 H.Kh. Mahabadi, *J. Appl. Polym. Sci.*, 30 (1985) 1535.
129 P. Lang and W. Burchard, *Makromol. Chem., Rapid Commun.*, 8 (1987) 451.
130 L. Mrkvičková and J. Janča, *J. Liq. Chromatogr.*, 9 (1986) 1217.
131 Cheng-Yih Kuo, T. Provder, M. Koehler and A.F. Kah, in T. Provder (Editor), *Detection and Data Analysis in Size- Exclusion Chromatography*, ACS Symposium Series, Vol. 352, American Chemical Society, Washington, DC, 1987, p. 130.
132 O. Chiantore and M. Guaita, *J. Liq. Chromatogr.*, 8 (1985) 1413.
133 O. Chiantore, *J. Liq. Chromatogr.*, 7 (1984) 1.

134 F.B. Malihi, Cheng-Yih Kuo and T. Provder, *J. Liq. Chromatogr.*, 6 (1983) 667.
135 O. Chiantore and A.E. Hamielec, *J. Liq. Chromatogr.*, 7 (1984) 1753.
136 C.J. Stacy, *J. Appl. Polym. Sci.*, 32 (1986) 3959.
137 T. Housaki and K. Satoh, *Polym. J. (Jpn.)*, 20 (1988) 1163.
138 A. Kijugawa, *Kobunshi Ronbunshu*, 44 (1987) 139; *C.A.*, 106 (1987) 139008j.
139 V. Grinshpun and A. Rudin, *J. Appl. Polym. Sci.*, 30 (1985) 2413.
140 R. Lew, D. Suwanda and S.T. Balke, *J. Appl. Polym. Sci.*, 35 (1988) 1049.
141 R. Lew, P. Cheung, D. Suwanda and S.T. Balke, *J. Appl. Polym. Sci.*, 35 (1988) 1065.
142 R. Lew, P. Cheung, D. Suwanda and S.T. Balke, *GPC Symposium '87*, Waters Division of Millipore, Milford, MA, 1987, p. 282.
143 Qicong Ying, Ping Xie and Meiling Ye, *Makromol. Chem., Rapid Commun.*, 6 (1988) 105.
144 Ying Qicong, Xie Ping, Liu Yong and Quian Renyuan, *J. Liq. Chromatogr.*, 9 (1986) 1233.
145 J.W. Mays, S.S. Huang, M.J. Washall and H.G. Barth, *GPC Symposium '87*, Waters Division of Millipore, Milford, MA, 1987, p. 430.
146 P.M. Cotts, *J. Polym. Sci., Polym. Phys.*, 24 (1986) 1493.
147 E. Kohn and M.E. Chisum, in T. Provder (Editor), *Detection and Data Analysis in Size-Exclusion Chromatography*, ACS Symposium Series, Vol. 352, American Chemical Society, Washington, DC, 1987, p. 169.
148 Lu-Zai Min, *J. Liq. Chromatogr.*, 5 (1982) 2241.
149 Jau-Yi Chuang, J.F. Johnson and A.R. Cooper, *J. Appl. Polym. Sci.*, 28 (1983) 473.
150 W.G. Rand and A.K. Mukherji, *J. Polym. Sci., Polym. Lett. Ed.*, 20 (1982) 501.
151 L.H. Tung and A.L. Gatzke, *J. Polym. Sci., Polym. Phys. Ed.*, 21 (1983) 1839.
152 F. Schosseler, H. Benoit, Z. Grubisic-Gallot, Cl. Strazielle and L. Liebler, *Macromolecules*, 22 (1989) 400.
153 He Zhiduan, Yuan Meina, Zhang Xuanqi, Wang Xiaobing, Jin Xiaobing, Huang Jiaxian, Li Chunrong and Wang Linfu, *Eur. Polym. J.*, 22 (1986) 597.
154 S. Mori and M. Suzuki, *J. Liq. Chromatogr.*, 7 (1984) 1841.
155 F.B. Malihi, Cheng-Yih Kuo and T. Provder, *J. Appl. Polym. Sci.*, 29 (1984) 925.
156 J.M. Goldwasser and A. Rudin, *J. Liq. Chromatogr.*, 6 (1983) 2433.
157 W.L. Elsdon, J.M. Goldwasser and A. Rudin, *J. Polym. Sci., Polym. Chem. Ed.*, 20 (1982) 3271.
158 S. Mori, A. Wada, F. Kaneuchi, A. Ikeda, M. Watanabe and K. Mochizuki, *J. Chromatogr.*, 246 (1982) 215.
159 R.A. Mendelson, in T. Provder (Editor), *Detection and Data Analysis in Size-Exclusion Chromatography*, ACS Symposium Series, Vol. 352, American Chemical Society, Washington, DC, 1987, p. 263.
160 D.R. Scheuing, *J. Appl. Polym. Sci.*, 29 (1984) 2819.
161 R.C. Jordan, S.F. Silver, R.D. Sehon and R.J. Rivard, in T. Provder (Editor), *Size Exclusion Chromatography*, ACS Symposium Series, Vol. 245, American Chemical Society, Washington, DC, 1984, p. 295.
162 R.C. Jordan, S.F. Silver, R.D. Sehon and R.J. Rivard, *Org. Coat. Appl. Polym. Sci. Proc.*, 48 (1983) 755.
163 D.J.P. Harrison, W.R. Yates and J.F. Johnson, *J. Liq. Chromatogr.*, 6 (1983) 2723.
164 T. Spychaj and A.E. Hamielec, *GPC Symposium '87*, Waters Division of Millipore, Milford, MA, 1987, p. 599.
165 J.V. Dawkins, M.J. Guest and G.M.F. Jeffs, *J. Liq. Chromatogr.*, 7 (1984) 1739.
166 I.V. Berlinova, N.G. Vladimirov and I.M. Panayotov, *Makromol. Chem., Rapid Commun.*, 10 (1989) 163.
167 P.L. Dubin, C.M. Speck and J.I. Kaplan, *Anal. Chem.*, 60 (1988) 895.
168 P.L. Dubin and M.M. Teklenburg, *Anal. Chem.*, 57 (1985) 275.
169 S.N.E. Omorodion, A.E. Hamielec and J.L. Brash, *J. Liq. Chromatogr.*, 4 (1981) 1903.
170 M.G. Styring, H.H. Teo, C. Price and C. Booth, *Eur. Polym. J.*, 24 (1988) 333.

171 A. Bose, J.E. Rollings, J.M. Caruthers, M.R. Okos and G.T. Tsao, *J. Appl. Polym. Sci.*, 27 (1982) 795.
172 D. Freeman and Xun Liang, in T. Provder (Editor), *Size Exclusion Chromatography*, ACS Symposium Series, Vol. 245, American Chemical Society, Washington, DC, 1984, p. 355.
173 C. Abad, L. Braco, V. Soria, R. Garcia and A. Campos, *Brit. Polym. J.*, 19 (1987) 489.
174 C. Abad, L. Braco, V. Soria, R. Garcia and A. Campos, *Brit. Polym. J.*, 19 (1987) 501.
175 S. Mori, *Anal. Chem.*, 61 (1989) 530.
176 S. Mori, *J. Liq. Chromatogr.*, 12 (1989) 785.
177 Day-Chuan Lee, T.A. Speckhard, A.D. Sorensen and S.L. Cooper, *Macromolecules*, 19 (1986) 2383.
178 M. Furukawa and T. Yokoyama, *J. Polym. Sci., Polym. Chem. Ed.*, 24 (1986) 3291.
179 T.A. Coleman and J.V. Dawkins, *J. Liq. Chromatogr.*, 9 (1986) 1191.
180 J. Bouwma, R. Dietz, L.J. Maisey and B. Wrights, *Eur. Polym. J.*, 20 (1984) 471.
181 D.J. Nagy, *J. Polym. Sci., Polym. Lett. Ed.*, 24 (1986) 87.
182 D.J. Nagy, *GPC Symposium '87*, Waters Division of Millipore, Milford, MA, 1987, p. 250.
183 L. Mrkvičková, E. Prokopová and O. Qudrat, *Colloid Polym. Sci.*, 265 (1987) 978.
184 L. Mrkvičková, J. Daňhelka and S. Pokorný, *J. Appl. Plym. Sci.*, 29 (1984) 803.
185 A. Horta, E. Sáiz, J.M. Barrales-Rienda and P.A. Galera Gómez, *Polymer*, 27 (1986) 139.
186 J.M. Barrales-Rienda, P.A. Galera Gómez, A. Horta and E. Sáiz, *Macromolecules*, 18 (1985) 2572.
187 J.V. Gražulevičius, N. Duobinis and R. Kavaliúnas, *J. Liq. Chromatogr.*, 7 (1984) 1823.
188 T. Hjertberg, L.I. Kulin and E. Soervik, *Polym. Tes.*, 3 (1983) 267.
189 D.L. Gooding, M.N. Schmuck and K.M. Gooding, *J. Liq. Chromatogr.*, 5 (1982) 2259.
190 W.G. Rand and A.K. Mukherji, *J. Chromatogr. Sci.*, 20 (1982) 182.
191 H.J. Mencer and Z. Grubisic-Gallot, *J. Chromatogr. Sci.*, 241 (1982) 213.
192 H.J. Mencer and Z. Vajnaht, *J. Chromatogr.*, 241 (1982) 205.
193 G.E. Fleig and F. Rodriquez, *Chem. Eng. Commun.*, 13 (1982) 219.
194 E.G. Malawer, J.K. DeVasto, S.P. Frankowski and A.J. Montana, *J. Liq. Chromatogr.*, 7 (1984) 441.
195 L. Senak, C.S. Wu and E.G. Malawer, *J. Liq. Chromatogr.*, 10 (1987) 1127.
196 P.R. Ludlam and J.G. King, *J. Appl. Polym. Sci.*, 29 (1984) 3863.
197 T. Hlaing, A. Gilbert and C. Booth, *Brit. Polym. J.*, 18 (1986) 345.
198 J.M. Goldwasser and A. Rudin, *J. Liq. Chromatogr.*, 6 (1983) 2433.
199 A. Hamielec, *Pure Appl. Chem.*, 54 (1982) 293.
200 L.H. Garcia-Rubio, J.F. MacGregor and A.E. Hamielec, in C.D. Craver (Editor), *Polymer Characterization*, Advances in Chemistry Series, Vol. 203, American Chemical Society, Washington, DC, 1983, p. 311.
201 W.L. Elsdon, J.M. Goldwasser and A. Rudin, *J. Polym. Sci., Polym. Chem. Ed.*, 20 (1982) 3271.
202 J.M. Goldwasser, A. Rudin and W.L. Elsdon, *J. Liq. Chromatogr.*, 5 (1982) 2253.
203 L.H. Tung (Editor), *Fractionation of Synthetic Polymers*, Dekker, New York, 1977.
204 T. Ogawa and M. Sakai, *J. Polym. Sci., Polym. Chem. Ed.*, 23 (1985) 1109.
205 H.J. Mencer and Z. Gomzi, *Angew. Makromol. Chem.*, 162 (1988) 163.
206 S. Hosoda, *Polym. J. (Jpn.)*, 20 (1988) 383.
207 B.G. Belenkii and E.S. Gankina, *J. Chromatogr.*, 141 (1977) 13.
208 B.G. Belenky, E.S. Gankina, M.B. Tennikov and L.Z. Vilenchik, *J. Chromatogr.*, 147 (1978) 99.
209 B.G. Belenkii, *Pure Appl. Chem.*, 51 (1979) 1519.
210 A.V. Gorshkov, V.V. Evreinov and S.G. Entelis, *Russ. J. Phys. Chem. Trans. Ed.*, 59 (1985) 552.
211 A.V. Gorshkov, V.V. Evreinov and S.G. Entelis, *Russ. J. Phys. Chem. Trans. Ed.*, 59 (1985) 869.

212 A.V. Gorshkov, V.V. Evreinov and S.G. Entelis, *Russ. J. Phys. Chem. Trans. Ed.*, 59 (1985) 1702.
213 A.A. Gorbunov, E.B. Zhulina and A.M. Skvortsov, *Polymer*, 23 (1982) 1133.
214 A.A. Gorbunov and A.M. Skvortsov, *Polym. Sci. USSR Trans. Ed.*, 26 (1984) 2305.
215 A.M. Skvortsov and A.A. Gorbunov, *J. Chromatogr.*, 358 (1986) 77.
216 A.A. Gorbunov and A.M. Skvortsov, *Polym. Sci. USSR Trans. Ed.*, 28 (1986) 2412.
217 A.M. Skvortsov and A.A. Gorbunov, *Polym. Sci. USSR Trans. Ed.*, 28 (1986) 1878.
218 A.A. Gorbunov, L.Ya. Solov'eva, V.A. Pasechnik and A.Ye. Luk'yanov, *Polym. Sci. USSR Trans. Ed.*, 28 (1986) 2067.
219 A.A. Gorbunov and A.M. Skvortsov, *Dokl. Akad. Nauk. SSSR Phys. Chem. Sect.*, 294 (1987) 396.
220 A.A. Gorbunov and A.M. Skvortsov, *Polym. Sci. USSR Trans. Ed.*, 30 (1988) 1.
221 J. Klein and G. Leidigkeit, *Makromol. Chem.*, 180 (1979) 2753.
222 D.W. Armstrong and K.H. Bui, *Anal. Chem.*, 54 (1982) 706.
223 R.E. Boehm, D.E. Martire, D.W. Armstrong and K.H. Bui, *Macromolecules*, 16 (1983) 466.
224 D.W. Armstrong, K.H. Bui and R.E. Boehm, *J. Liq. Chromatogr.*, 6 (1983) 1.
225 R.E. Boehm, D.E. Martire, D.W. Armstrong and K.H. Bui, *Macromolecules*, 17 (1984) 400.
226 D.W. Armstrong and R.E. Boehm, *J. Chromatogr. Sci.*, 22 (1984) 378.
227 K.H. Bui, D.W. Armstrong and R.E. Boehm, *J. Chromatogr.*, 288 (1984) 15.
228 K.H. Bui and D.W. Armstrong, *J. Liq. Chromatogr.*, 7 (1984) 29.
229 K.H. Bui and D.W. Armstrong, *J. Liq. Chromatogr.*, 7 (1984) 45.
230 R.E. Boehm and D.E. Martire, *Anal. Chem.*, 61 (1989) 471.
231 J.P. Larmann, J.J. DeStefano, A.P. Goldberg, R.W. Stout, L.R. Snyder and M.A. Stadalius, *J. Chromatogr.*, 255 (1983) 163.
232 L.R. Snyder, M.A. Stadalius and M.A. Quarry, *Anal. Chem.*, 55 (1983) 1412A.
233 M.A. Quarry, M.A. Stadalius, T.H. Mourey and L.R. Snyder, *J. Chromatogr.*, 358 (1986) 1.
234 M.A. Quarry, R.L. Grob and L.R. Snyder, *Anal. Chem.*, 58 (1986) 907.
235 M.A. Stadalius, M.A. Quarry, T.H. Mourey and L.R. Snyder, *J. Chromatogr.*, 358 (1986) 17.
236 G. Glöckner, *Chromatographia*, 25 (1988) 854.
237 G. Glöckner, J.H.M. van den Berg, N.L.J. Meijerink, T.G. Scholte and R. Koningsveld, *Macromolecules*, 17 (1984) 962.
238 K.D. Caldwell, *Anal. Chem.*, 60 (1988) 959A.
239 J.J. Kirkland, C.H. Dilks and W.W. Yau, *J. Chromatogr.*, 255 (1983) 255.
240 J.C. Giddings, G.-C. Lin and M.N. Myers, *J. Liq. Chromatogr.*, 1 (1978) 1.
241 J.J. Gunderson and J.C. Giddings, *Anal. Chim. Acta*, 189 (1986) 1.
242 K.D. Caldwell, S.L. Brimhall, Y. Gao and J.C. Giddings, *J. Appl. Polym. Sci.*, 36 (1988) 703.
243 J.C. Giddings, *Pure Appl. Chem.*, 51 (1979) 1459.
244 S.L. Brimhall, M.N. Myers, K.D. Caldwell and J.C. Giddings, *J. Polym. Sci., Polym. Lett. Ed.*, 22 (1984) 339.
245 J.C. Giddings, M. Martin and M.N. Myers, *J. Chromatogr.*, 158 (1978) 419.
246 J.C. Giddings, M. Martin and M.N. Myers, *J. Polym. Sci., Polym. Phys. Ed.*, 19 (1981) 815.
247 J.J. Kirkland and W.W. Yau, *Macromolecules*, 18 (1985) 2305.
248 Y.S. Gao, K.D. Caldwell, M.N. Myers and J.C. Giddings, *Macromolecules*, 18 (1985) 1272.
249 S.L. Brimhall, M.N. Myers, K.D. Caldwell and J.C. Giddings, *J. Polym. Sci., Polym. Phys. Ed.*, 23 (1985) 2443.
250 J.J. Gunderson, M.N. Myers and J.C. Giddings, *Anal. Chem.*, 59 (1987) 23.
251 M.E. Schimpf and J.C. Giddings, *Macromolecules*, 20 (1987) 1561.
252 M.E. Schimpf, M.N. Myers and J.C. Giddings, *J. Appl. Polym. Sci.*, 33 (1987) 117.
253 J.C. Giddings, S. Li, P.S. Williams and M.E. Schimpf, *Makromol. Chem., Rapid Commun.*, 9 (1988) 817.

254 S.L. Brimhall, M.N. Myers, K.D. Caldwell and J.C. Giddings, *Sep. Sci. Technol.*, 16 (1981) 671.
255 J.J. Kirkland and W.W. Yau, *J. Chromatogr.*, 353 (1986) 95.
256 Th.G. Scholte, in J.V. Dawkins (Editor), *Developments in Polymer Chararacterization*, Vol. 4, Elsevier, Amsterdam, 1983, p. 1.
257 A.E. Hamielec and H. Meyer, in J.V. Dawkins (Editor), *Developments in Polymer Chararacterization*, Vol. 5, Elsevier, Amsterdam, 1986, p. 95.
258 B.H. Zimm and W.H. Stockmayer, *J. Chem. Phys.*, 17 (1949) 1301.
259 A.E. Hamielec and A.C. Ouano, *J. Liq. Chromatogr.*, 1 (1978) 111.
260 B. Tinland, J. Mazet and M. Rinaudo, *Makromol. Chem., Rapid Commun.*, 9 (1988) 69.
261 K. Sommermeyer, F. Cech and E. Pfitzer, *Chromatographia*, 25 (1988) 167.
262 A.E. Hamielec, *J. Liq. Chromatogr.*, 4 (1981) 1697.
263 A.E. Hamielec, in J. Janča (Editor), *Steric Liquid Exclusion Chromatogroaphy of Polymers*, Dekker, New York, 1984, Chap. 3.
264 A. Rudin, V. Grinshpun and K.F. O'Driscoll, *J. Liq. Chromatogr.*, 7 (1984) 1809.
265 V. Grinshpun, A. Rudin and D. Potter, *Polym. Bull.*, 13 (1985) 71.
266 V. Grinshpun, A. Rudin, K.E. Russell and M.V. Scammell, *J. Polym. Sci., Polym. Phys. Ed.*, 24 (1986) 1171.
267 D.C. Bugada and A. Rudin, *Eur. Polym. J.*, 23 (1987) 847.
268 T.A. Coleman and J.V. Dawkins, *J. Liq. Chromatogr.*, 9 (1986) 1191.
269 G.N. Foster, T.B MacRury and A.E. Hamielec, in J. Cazes and X. Delamare (Editors), *Liquid Chromatography of Polymers and Related Materials II*, Dekker, New York, 1980, p. 143.
270 Q.-W. Wang, I.H. Park and B. Chu, in T. Provder (Editor), *Detection and Data Analysis in Size-Exclusion Chromatography*, ACS Symposium Series, Vol. 352, American Chemical Society, Washington, DC, 1987, p. 240.
271 F. Schosseler, H. Benoit, Z. Grubisic-Gallot, Cl. Strazielle and L. Liebler, *Macromolecules*, 22 (1989) 400.
272 P.G. de Gennes, *Scaling Concepts in Polymer Physics*, Cornell University Press, Ithaca, NY, 1979.
273 L.H. Tung and A.L. Gatzke, *J. Polym. Sci., Polym. Phys. Ed.*, 21 (1983) 1839.
274 E.V. Patton, J.A. Wesson, M. Rubenstein, J.C. Wilson and L.E. Oppenheimer, *Macromolecules*, 22 (1989) 1946.
275 P. Helias, D. Durand, J.B. Busnel and C.M. Bruneau, *Eur. Polym. J.*, 18 (1982) 647.
276 D.S. Argyropoulos, R.M. Berry and H.I. Bolker, *J. Polym. Sci., Polym. Phys. Ed.*, 25 (1987) 1191.
277 Z. Dobkowski and J. Brzezinski, *Eur. Polym. J.*, 17 (1981) 537.
278 C. Wu, J. Zuo, B. Chu, *Macromolecules*, 22 (1989) 633.
279 L. Mrkvičková and J. Janča, *J. Liq. Chromatogr.*, 9 (1986) 1217.
280 P. Lang and W. Burchard, *Makromol. Chem., Rapid Commun.*, 8 (1987) 451.
281 Li-Ping Yu and J.E. Rollings, *J. Appl. Polym. Sci.*, 35 (1988) 1085.
282 Li-Ping Yu and J.E. Rollings, *J. Appl. Polym. Sci.*, 33 (1987) 1909.
283 A. Corona and J.E. Rollings, *Sep. Sci. Technol.*, 23 (1988) 855.
284 Th.G. Scholte, N.L.J. Meijerink, H.M. Schoffeleers and A.M.G. Brands, *J. Appl. Polym. Sci.*, 29 (1984) 3763.
285 T. Housaki, K. Satoh, K. Nishikida and M. Morimoto, *Makromol. Chem., Rapid Commun.*, 9 (1988) 525.
286 L.I. Kulin, N.L. Meijerink and P. Starck, *Pure Appl. Chem.*, 60 (1988) 1403.
287 L. Wild, T.R. Ryle, D.C. Knobeloch adn I.R. Peat, *J. Polym. Sci., Polym. Phys. Ed.*, 20 (1982) 441.
288 S. Nakano and Y. Goto, *J. Appl. Polym. Sci.*, 26 (1981) 4217.
289 C. Bergström and E. Avela, *J. Appl. Polym. Sci.*, 23 (1979) 163.
290 F.M. Mirabella, Jr., *GPC Symposium '87*, Waters Division of Millipore, Milford, MA, 1987, p. 180.
291 F.M. Mirabella, Jr. and E.A. Ford, *J. Polym. Sci., Polym. Phys. Ed.*, 25 (1987) 777.
292 T. Usami, Y. Gotoh and S. Takayama, *Macromolecules*, 19 (1986) 2722.

293 E.C. Kelusky, C.T. Elston and R.E. Murray, *Polym. Eng. Sci.*, 27 (1987) 1562.
294 S. Mori, in J.C. Giddings, E. Grushka, J. Cazes and P.R. Brown (Editors), *Advances in Chromatography*, Vol. 21, Dekker, New York, 1983, p. 187.
295 H. Inagaki and T. Tanaka, *Pure Appl. Chem.*, 54 (1982) 309.
296 L.H. Garcia-Rubio, in T. Provder (Editor), *Detection and Data Analysis in Size-Exclusion Chromatography*, ACS Symposium Series, Vol. 352, American Chemical Society, Washington, DC, 1987, p. 220.
297 L.H. Garcia-Rubio, A.E. Hamielec and J.F. MacGregor, in T. Provder (Editor), *Computer Applications in Applied Polymer Science*, ACS Symposium Series, Vol. 197, American Chemical Society, Washington, DC, 1982, p. 151.
298 S. Mori, *J. Chromatogr.*, 194 (1980) 163.
299 S. Mori and T. Suzuki, *J. Liq. Chromatogr.*, 4 (1981) 1685.
300 H.J. Cortes, G.L. Jewett, C.D. Pfeiffer, S. Martin and C. Smith, *Anal. Chem.*, 61 (1989) 961.
301 S. Mori, *J. Chromatogr.*, 411 (1987) 355.
302 D.J. Harrison, W.R. Yates and J.F. Johnson, *J. Appl. Polym. Sci.*, 31 (1986) 1393.
303 D.J.P. Harrison, W.R. Yates and J.F. Johnson, *J. Liq. Chromatogr.*, 6 (1983) 2723.
304 T. Spychaj and A.E. Hamielec, *GPC Symposium '87*, Waters Division of Millipore, Milford, MA, 1987, p. 599.
305 R. Takiguchi and T. Uryu, *J. Appl. Polym. Sci.*, 30 (1985) 3961.
306 T. Dumelow, *J. Makromol. Sci.-Chem.*, A26 (1989) 125.
307 T. Dumelow, S.R. Holding, L.J. Maisey and J.V. Dawkins, *Polymer*, 27 (1986) 1170.
308 D. Lee, T.A. Speckhard, A.D. Sorensen and S.L. Cooper, *Macromolecules*, 19 (1986) 2383.
309 A.H. Dekmezian and T. Morioka, *Anal. Chem.*, 61 (1989) 458.
310 S. Mori, A. Wada, F. Kaneuchi, A. Ikeda, M. Watanabe and K. Mochizuki, *J. Chromatogr.*, 246 (1982) 215.
311 E.J. Parks, R.B. Johannesen and F.E. Brinckman, *J. Chromatogr.*, 255 (1983) 439.
312 E.J. Parks, W.F. Manders, R.B. Johannesen and F.E. Brinckman, *J. Chromatogr.*, 351 (1986) 475.
313 E.J. Parks, F.E. Brinckman and L.B. Kool, *J. Chromatogr.*, 370 (1986) 206.
314 J.J. Broersen, H. Jansen, C. de Ruiter, U.A.Th. Brinkman, R.W. Frei, F.A. Buijtenhuijs and F.P.B. van der Maeden, *J. Chromatogr.*, 436 (1988) 39.
315 S.T Balke, in T. Provder (Editor), *Detection and Data Analysis in Size-Exclusion Chromatography*, ACS Symposium Series, Vol. 352, American Chemical Society, Wahington, DC, 1987, p. 59.
316 S.T. Balke and R.D. Patel, in C.D. Craver (Editor), *Polymer Characterization*, Advances in Chemistry Series, Vol. 203, American Chemical Society, Washington, DC, 1983, p. 281.
317 S.T. Balke and R.D. Patel, in T. Provder (Editor), *Size-Exclusion Chromatography*, ACS Symposium Series, Vol. 138, American Chemical Society, Washington, DC, 1980, p. 149.
318 S.T. Balke, *Sep. Purif. Methods*, 11 (1982) 1.
319 S.T. Balke and R.D. Patel, *J. Polym. Sci., Polym. Lett. Ed.*, 18 (1980) 453.
320 J.V. Dawkins and A.M.C. Montenegro, *Brit. Polym. J.*, 21 (1989) 31.
321 S. Teramachi, A. Hasegawa and S. Yoshida, *Macromolecules*, 16 (1983) 542.
322 J.C.J.F. Tacx and A.L. German, *Polymer*, 30 (1989) 918.
323 G. Glöckner and R. Koningsveld, *Makromol. Chem., Rapid Commun.*, 4 (1983) 529.
324 G. Glöckner, *Pure Appl. Chem.*, 55 (1983) 1553.
325 G. Glöckner, J.H.M. van den Berg, N.L.J. Meijerink, T.G. Scholte and R. Koningsveld, *J. Chromatogr.*, 317 (1984) 615.
326 G. Glöckner, V. Albrecht, F. Francuskiewicz and D. Ilchmann, *Angew. Makromol. Chem.*, 130 (1985) 41.
327 G. Glöckner and J.H.M. van den Berg, *J. Chromatogr.*, 384 (1987) 135.
328 S. Mori, Y. Uno and M. Suzuki, *Anal. Chem.*, 58 (1986) 303.
329 S. Mori, *Anal. Chem.*, 60 (1988) 1125.
330 S. Mori, *Anal. Sci.*, 4 (1988) 365.

331 G. Glöckner, M. Stickler and W. Wunderlich, *Fresenius Z. Anal. Chem.*, 328 (1987) 76.
332 G. Glöckner, M. Stickler and W. Wunderlich, *Fresenius Z. Anal. Chem.*, 330 (1988) 46.
333 H. Inagaki, *Polym. Sci.*, 24 (1977) 190.
334 B.G. Belenkii and E.S. Gankina, *J. Chromatogr.*, 141 (1977) 13.
335 B.G. Belenkii, *Pure Appl. Chem.*, 51 (1979) 1519.
336 H. Inagaki and T. Tanaka, *Pure Appl. Chem.*, 54 (1982) 309.
337 G. Glöckner, in *Advances in Polymer Science, Biopolymers, Non-Exclusion HPLC*, Vol. 79, Springer-Verlag, Berlin, 1986, p. 159.
338 S. Teramachi, A. Hasegawa and N. Uchiyama, *J. Polym. Sci., Polym. Lett.*, 22 (1984) 71.
339 T. Ogawa and W. Ishitobi, *J. Polym. Sci., Polym. Chem. Ed.*, 21 (1983) 781.
340 T. Tanaka, M. Omoto, N. Donkai and H. Inagaki, *J. Macromol. Sci. Phys.*, B17 (1980) 211.
341 S. Teramachi, A. Hasegawa, Y. Shima, M. Akatsuka and M. Nakajima, *Macromolecules*, 12 (1979) 992.
342 M. Danielewicz and M. Kubin, *J. Appl. Polym. Sci.*, 26 (1981) 951.
343 M. Danielewicz, M. Kubin and S. Vozka, *J. Appl. Polym. Sci.*, 27 (1982) 3629.
344 G. Glöckner, *Trends Anal. Chem.*, 4 (1985) 214.
345 G. Glöckner and J.H.M. van den Berg, *Chromatographia*, 19 (1984) 55.
346 G. Glöckner and J.H.M. van den Berg, *J. Chromatogr.*, 352 (1986) 511.
347 G. Glöckner, *Chromatographia*, 23 (1987) 517.
348 G. Glöckner, *J. Chromatogr.*, 403 (1987) 280.
349 S. Mori and Y. Uno, *Anal. Chem.*, 59 (1987) 90.
350 S. Mori and Y. Uno, *J. Appl. Polym. Sci.*, 34 (1987) 2689.
351 S. Mori, *J. Liq. Chromatogr.*, 12 (1989) 323.
352 G. Glöckner and J.H.M. van den Berg, *Chromatographia*, 24 (1987) 233.
353 H. Sato, H. Takeuchi and Y. Tanaka, *Macromolecules*, 19 (1986) 2613.
354 H. Sato, H. Takeuchi, S. Suzuki and Y. Tanaka, *Makromol. Chem., Rapid Commun.*, 5 (1984) 719.
355 H. Sato, K. Mitsutani, I. Shimizu and Y. Tanaka, *J. Chromatogr.*, 447 (1988) 387.
356 T.H. Mourey, *J. Chromatogr.*, 357 (1986) 101.
357 J.J. Gunderson and J.C. Giddings, *Macromolecules*, 19 (1986) 2618.
358 M.E. Schimpf and J.C. Giddings, *ACS Div. Polym. Chem., Polym. Prepr.*, 27 (1986) 158.
359 L.H. Garcia-Rubio, N. Ro and R.D. Patel, *Macromolecules*, 17 (1984) 1998.
360 N.E. Manolova, I. Gitsov, R.S. Velichkova and I.B. Rashkov, *Polym. Bull.*, 13 (1985) 285.
361 R.P. Quirk, S. Perry, F. Mendicuti and W.L. Mattice, *Macromolecules*, 21 (1988) 2294.
362 S. Mori, *Anal. Chim. Acta*, 189 (1986) 17.
363 K. Hatada, K. Ute, Y. Okamato, M. Imanari and N. Fujii, *Polym. Bull.*, 20 (1988) 317.
364 K. Hatada, K. Ute, T. Kitayama, M. Yamamoto, T. Nishimura, *Polym. Bull.*, 21 (1989) 489.
365 P. Mansson, *J. Polym. Sci., PartA-1*, 18 (1980) 1945.
366 G.D. Andrews and A. Vatvars, *Macromolecules*, 14 (1981) 1603.
367 M.A. Winnik, K. Paton, J. Danhelka and A.E.C. Redpath, *J. Chromatogr.*, 242 (1982) 97.
368 A.V. Gorshkov, S.S. Verenich, V.V. Eureinov and S.G. Entelis, *Chromatographia*, 26 (1988) 338.
369 T.H. Mourey, I. Noh and H. Yu, *J. Chromatogr.*, 303 (1984) 361.
370 J.J. Lewis, L.B. Rogers and R.E. Pauls, *J. Chromatogr.*, 264 (1983) 339.
371 J.J. Lewis, L.B. Rogers and R.E. Pauls, *J. Chromatogr.*, 299 (1984) 331.
372 T.H. Mourey, G.A. Smith and L.R. Snyder, *Anal. Chem.*, 56 (1984) 1773.
373 K.F. Elgert, R. Henschel, H. Schorn and R. Kosfeld, *Polym. Bull.*, 4 (1981) 105.

374 H. Inagaki, T. Miyamotao and F. Kamiyama, *J. Polym. Sci., Polym. Lett.,* 7 (1969) 329.
375 H. Inagaki and F. Kamiyama, *Macromolecules,* 6 (1973) 107.
376 F. Horii and Y. Ikada, *J. Polym. Sci., Polym. Lett.,* 12 (1974) 29.
377 H. Inagaki, T. Kotaka and T-I. Min, *Pure Appl. Chem.,* 46 (1976) 61.
378 H.H. Teo, M.G. Styring, S.G. Yeates, C. Price and C. Booth, *J. Colloid Interface Sci.,* 114 (1986) 416.
379 M.G. Styring, H.H. Teo, C. Price and C. Booth, *J. Chromatogr.,* 388 (1987) 421.
380 V. Szmereková, P. Králik and D. Berek, *J. Chromatogr.,* 285 (1984) 188.
381 C. Cesteros, I. Katime and C. Strazielle, *Makromol. Chem., Rapid Commun.,* 4 (1983) 193.
382 K. Procházka, B. Bednář, Z. Tuzar and M. Kočiřík, *J. Liq. Chromatogr.,* 11 (1988) 2221.
383 C. Price, *Pure Appl. Chem.,* 55 (1983) 1563.
384 P. Špaček and M. Kubín, *J. Appl. Polym. Sci.,* 30 (1985) 143.
385 P. Špaček, *J. Appl. Polym. Sci.,* 32 (1986) 4281.
386 K. Procházka, G. Glöckner, M. Hoff and Z. Tuzor, *Makromol. Chem.,* 185 (1984) 1187.
387 P.K. Das, J. Hoover, R.J. Dodson, T.C. Ward and J.E. McGrath, *Polym. Prepr.,* 25 (1984) 96.
388 P.K. Das, R.D. Allen, T.C. Ward, J.E. McGrath and R.J. Dodson, *Polym. Prepr.,* 25 (1984) 185.
389 P.C. Tsitsilianis and A. Dondos, *Macromolecules,* 20 (1987) 658.
390 I.A. Katime and J.R. Quintana, *Eur. Polym. J.,* 24 (1988) 775.
391 I.A. Katime and J.R. Quintana, *Makromol. Chem.,* 189 (1988) 1373.
392 H.-S. Tseng and D.R. Lloyd, *J. Polym. Sci., Polym. Phys.,* 25 (1987) 325.
393 D.R. Lloyd, V. Narasimhan and C.M. Burns, *J. Liq. Chromatogr.,* 3 (1980) 1111.
394 H. Itagaki, J.E. Guillet, K. Sienicki and M.A. Winnik, *J. Polym. Sci., Polym. Lett.,* 27 (1989) 21.

Chapter 23

Pesticides

JOSEPH SHERMA

CONTENTS

23.1 INTRODUCTION

The chapter on pesticides in Chromatography, 3rd Edn., by Sherma and Zweig [1], covered the literature of chromatographic analysis up to 1970, and the chapter in the 4th Edn. by Fishbein [2] updated coverage to 1980. The objective of this chapter is to summarize the current status of the field and to extend literature references up to the date of writing, late 1988.

Pesticides are a very diverse group of compounds that can be classified according to their intended use as insecticides, fungicides, herbicides and plant growth regulators, rodenticides, miticides, and nematocides. They can be classified additionally according to their chemical structures: organochlorine (OC) insecticides; organophosphorus (OP) in-secticides; pyrethroid insecticides; substituted chlorophenoxy acid or ester herbicides; carbamate insecticides, herbicides, and fungicides; urea herbicides; triazine herbicides; and other miscellaneous types including uracils, chlorinated phenols, organo-mercury and -tin compounds, etc.

Pesticide analysis is broadly divided into formulation and residue analysis. The former involves determination of macro levels of pesticides and lower levels of impurities in commercial products, usually by spectrometry, GC, or HPLC. This chapter will focus on residue analysis. Readers interested in formulation analysis should consult the review by Hill [3].

Residue analysis involves the determination of parts-per-million (ppm), -billion (ppb), or -trillion (ppt) of pesticides and/or their metabolites and degradation products in a wide variety of samples, including plants, foods, human and animal tissues, and environmental samples, such as water, air, soil, and sludges. Virtually all residue analyses involve the following common sequence of steps: sampling, sample preparation (extraction, cleanup, derivatization), determination, and confirmation. The exact nature and complexity of each stage will depend on the pesticide and the sample matrix.

Chromatography is utilized for the cleanup, determination, and confirmation of pes-ticide residues. The primary determinative method is packed-column GC, but capillary-column GC and HPLC are now being used to an increasing degree. Residue identity is usually confirmed by GC/MS or HPLC/MS. Classical gravity-flow or low-pressure column adsorption chromatography and size-exclusion chromatography (SEC) are among the methods most widely used for sample cleanup. TLC is used primarily as a fast, simple, and inexpensive screening method, but it is also excellent for sample cleanup and quantification. Supercritical-fluid chromatography (SFC) and extraction (SFE) are just be-ginning to be used for pesticide analysis and will also be covered.

In addition to the chapter by Fishbein, the literature of pesticide residue analysis up to about 1980 has been reviewed by Roseboom [4]. Chau and Lee [5] have written a very informative chapter on basic principles and practices of pesticide analysis, which also covers the literature up to 1980. The most important literature of the 1980s on the procedures and applications of residue analysis has been reviewed comprehensively by Sherma and Zweig [6,7] and by Sherma [8] in four biennial reviews in the journal *Analytical Chemistry*, which contain a total of more than 1200 references. Because of the availability of these detailed reviews, this chapter will give only a selection of the most important recent references to illustrate the trends in chromatographic methodology and some single-residue and multiresidue methods that have been developed with the various chromatographic techniques. The 17-volume book series *Analytical Methods for Pesticides and Plant Growth Regulators*, edited by Zweig and Sherma [9], contains chapters on techniques and methods for formulation and residue analysis. The Chemical Rubber Company *Handbook of Chromatography: Pesticides* [10] lists GC, HPLC, and TLC data for all classes of pesticides. Additional books [11-20] and reviews [21-27] contain information on pesticide analysis by chromatography.

In most cases, generic (common) names of pesticides have been used rather than trade names. Chemical names and formulae, trade names, properties, uses, and other information are contained in the Pesticide Dictionary of the *Farm Chemicals Handbook* [28].

23.2 SAMPLE PREPARATION

The term "sample preparation" is sometimes used to indicate the steps required to prepare the sample for extraction, e.g., chopping, grinding, blending, drying, weighing, etc. In this chapter, sample preparation denotes all processes that ready the sample for determination, including extraction and cleanup.

23.2.1 Sampling

Sampling techniques for pesticides analysis have been reviewed by Kratochvil and coauthors [29,30]. Types of samples and sampling processes, sampling theory, and strategy and procedures for collecting reliable samples of different varieties are covered in these reviews.

23.2.2 Multiresidue methods involving extraction and chromatographic cleanup

Most analyses involve solvent extraction to isolate the analyte(s) from the sample matrix, followed by removal of co-extracted interfering material from the analyte (cleanup) prior to the determinative step. The commonly used cleanup techniques include liquid/liquid partitioning, LSC (usually on alumina, silica gel, Florisil, magnesia, Celite, or carbon), TLC, chemical reactions, sweep codistillation, low-temperature precipitation, and SEC. The amount of cleanup required depends upon the particular pesticide and matrix and the

References on p. B540

selectivity of the detection method. For example, use of the selective nitrogen/phosphorus detector (NPD) or flame-photometric detector (FPD) may largely eliminate the need for cleanup of OP pesticide extracts prior to GC. Multiresidue extraction and cleanup methods for pesticide residue analysis in general [31-33] and for food samples [34] have been reviewed, and the cost-effectiveness of some methods has been compared [35].

The most widely used multiresidue screening method for plant substrates for more than 20 years is the Mills-Olney-Gaither (MOG) method, which utilizes a Florisil column eluted with diethyl ether/petroleum ether mixtures, or alternatively, mixtures of dichloromethane, hexane, and acetonitrile to obtain cleaner eluates of fatty samples, and GC with an electron-capture detector (ECD). The MOG method is described in detail in the FDA *Pesticide Analytical Manual* (PAM) [13] and the AOAC book of official methods [17]. The PAM contains a compilation of recovery data for numerous pesticides in different sample matrices, which is continually updated in each new edition. Sawyer [36] modified the MOG method by using high-speed sample milling and homogenization with petroleum ether, ethyl ether/petroleum ether, and ethanol to improve recoveries of pesticides from high-fat, low-moisture plant materials.

M.A. Luke et al. expanded the applicability of the MOG method to include polar pesticides in fruits and vegetables by replacing acetonitrile with acetone for extraction and eliminating the use of the Florisil column for cleanup by using specific thermionic, flame-photometric, and electrolytic conductivity GC detectors [37,38]. Sawyer conducted a successful collaborative study of this method [39], which led to its adoption by the AOAC [17]. Ambrus et al. reported a comprehensive scheme for screening multipesticide residues, metabolites, and conjugates in soil, water, and plants, involving extraction, solvent partitioning, cleanup on alumina and silica gel columns, and determination by GC and TLC [40]. A multiresidue method was described [41] for OC, OP, and pyrethroid pesticides and carbaryl in grain, involving acetone/methanol extraction, partitioning into dichloromethane, and determination of carbaryl by HPLC with ultraviolet detection (HPLC/UV) and OP pesticides by GC/FPD without further cleanup. OCs and pyrethroids were cleaned up on an acidic aluminum oxide column and determined by GC/ECD and HPLC/UV, respectively.

SEC on columns packed with Bio-Beads SX-2 or -3 has proven to be the most generally applicable cleanup method for nearly all types of pesticides in plants, fats, and oils. The eluent dichloromethane/acetone (3:7) was applied for rapid cleanup of 85 pesticides and industrial chemicals from fats and oils [42], and cyclohexane/acetone (1:1) was found to be suitable for cleanup of more than 300 pesticides [43]. The availability of automated instrumentation for GPC [42] is a great advantage. GPC has been used prior to GC for determinations of 33 halogenated compounds in human adipose tissue [44]. Steinwandter has reviewed [45] multiresidue sample preparation procedures and described in detail his recent research on universal extraction and cleanup methods [46].

Simultaneous extraction and cleanup has been carried out in a chromatographic column. OC pesticides and polychlorinated biphenyls (PCBs) were isolated from milk, milk powder, cocoa, and eggs without preliminary fat extraction by mixing aqueous samples with silica gel and placing the powder on top of a 10% water-deactivated silica gel column.

Pesticides and PCBs were eluted with dichloromethane/petroleum ether (2:8) [47]. Animal fats and vegetable oils were blended with deactivated alumina and aqueous acetonitrile to remove lipid prior to petroleum ether partitioning and Florisil cleanup by the official AOAC method [48]. Whole milk and oilseeds were treated the same way to extract pesticides directly from the samples [49]. Silver nitrate-coated alumina has been shown to remove effectively sulfur-containing substances from extracts of samples, such as kale and onions, prior to GC/ECD [50].

23.2.3 Sweep codistillation

The introduction by B.G. Luke et al. in 1984 [51] of the commercial Unitrex apparatus with a redesigned fractionation tube for cleanup of beef fat for OC residue determination caused renewed interest in sweep codistillation, which was originally described in 1965 by Storherr and Watts. The isolation of OP residues in animal fat [52] and cleanup of pesticides from human adipose tissue [53-55] by this unit were subsequently described. Recent developments and applications of sweep codistillation have been reviewed [56].

23.2.4 Solid-phase extraction

One of the clearest trends in residue analysis is the growing application of solid-phase extraction (SPE) on porous polymers and chemically bonded silica gel phases. XAD-4 resin was used to trap and concentrate several classes of nonvolatile pesticides from air. The pesticides were eluted with ethyl acetate, fractionated on a silica gel HPLC column, and determined by capillary GC [57]. Chromosorb 102 sampling tubes [58] and C-18 cartridges [59] were also effective in collecting a variety of pesticides from air. XAD resins [60-62], Tenax-GC cartridges [63,64], and commercial C-18 bonded silica gel cartridges [65,66] have proven superior to conventional extraction in preconcentrating pesticides from water. OC compounds were determined in seafood by SPE followed by GC with the ECD or Hall electroconductivity detector (HECD) [67].

23.2.5 Derivatization

Derivatization reactions are included in many residue analytical methods in order to improve the chromatographic properties, detection selectivity, and/or detection sensitivity of the analyte(s). Derivatization is carried out prior to separation in GC, either before or after chromatography in HPLC, and usually after separation by application of a visualization reagent to the plate in TLC. Chemical derivatization in pesticide analysis has been reviewed [68,69].

23.2.6 Automation

Automated systems for sample preparation were described in the following papers: HPLC three-column switching [70], continuous-flow cleanup for screening pesticides in

References on p. B540

foods [71], SEC cleanup and evaporation for determining residues in fatty samples [72], and robotic sampling, extraction, and cleanup [73].

23.2.7 Supercritical-fluid extraction

SFE has been combined with SFC for the extraction, separation, and identification of sulfonylurea [74] and triazine [75] herbicides from complex matrices. Supercritical carbon dioxide, containing small amounts of polar modifiers (e.g., ethanol and methanol), was the extracting agent. SFE has also been used to extract bound residues from soil and plants [76].

23.3 GAS CHROMATOGRAPHY

23.3.1 Packed columns

The most widely used technique for quantification and preliminary identification of pesticide residues continues to be packed-column GC with sensitive and selective detectors. The liquid phases used most often, in order of increasing polarity, are methyl silicones (e.g., OV-101, OV-1, DC-200, SE-30, SP-2100), phenyl silicones (OV-17, SP-2550, DC-710), fluoropropyl silicones (OV-210, QF-1), cyanopropyl silicones (OV-225), polyethers (Carbowax 210), and polyesters (DEGS, NPGS). Mixed-phase columns that have been used include 10% DC-200 + 15% QF-1 (1:1) [13], 1.95% SP-2401 + 1.5% SP-2250 [40], 1.5% OV-17 + 1.95% QF-1 [77], and 3% OV-225 + 5% OV-101 (1:1) [78]. Columns are typically 0.5-1.8 m x 2-4 mm ID, with 2-10% liquid phase, coated on a 60- to 100-mesh support.

Practical aspects of GC with packed columns, including column selection, preparation and maintenance, and retention data for numerous pesticides were presented by Froberg and Doose [79].

23.3.2 Capillary columns

Pesticide analysis by capillary-column GC was reviewed [80].

Although packed columns continue to be used for routine monitoring of residues by the standard multiresidue methods, capillary-column (high-resolution) GC is being applied to an increasing degree for separations of complex mixtures and GC/MS confirmations. Wall-coated open-tubular (WCOT) borosilicate glass or fused-silica columns, 10-50 m long with 0.05- to 1-μm liquid-phase film thickness, have been generally used for residue analysis [32,81]. Although capillary columns provide advantages in resolution, sensitivity, and analysis time [82], a drawback is the limited capacity compared to packed columns [81]. The injection of involatile coextractives causes greater problems with quantification and peak shape for capillary columns than for packed columns, in which the top glass-wool plug or column material can trap these interfering materials [83]. Improved sensitivity has been achieved by using "wide-bore" (0.5-mm) capillary columns in place of 0.2- to

0.32-mm "small- or normal-bore" columns [32,83-85]. Capillary columns can be made and coated in the laboratory, but commercial fused-silica columns with immobilized stationary phases are most widely used.

Manual and automatic injection has been used in split, splitless, and on-column modes. Split ratios reported in the literature have ranged from 2:1 to 75:1. Splitless and on-column injection offer advantages for sensitivity, but they require more operator experience and they cannot be used under isothermal conditions [86]. Automated and manual hot- and cold-splitless injection with programed-temperature vaporization, and on-column injection were compared for 23 OP pesticides [87]. A modified inlet was designed for injecting a bonded-phase wide-bore capillary column with OC pesticides in a lipid extract [88]. The reliability of on-column and splitless injection have been compared for pesticide residue analysis [89].

The following capillary-column stationary phases are examples of those that have been reported: OV-17 for OC pesticides and PCBs [90]; SE-54 for methylcarbamate insecticides [91]; Carbowax 20M for triazine herbicides [92] and ethylenethiourea (ETU) [93]; SE-52 for triazine herbicides [94], phenylurea herbicides [95], and methylcarbamate insecticides [96]; OV-101 for pyrethroids [97]; SE-30 for fungicides [98]; SE-30 for phenoxy herbicides [99]; and OV-1 for OPs [100].

Multiresidues of OC and OP pesticides have been separated and determined in foods by automated "two-dimensional" capillary GC, which involves the use of two columns of different polarity and multiple detectors [101]. The apparatus, methods, and applications have been reviewed [102]. "Tandem capillary GC" was described for the determination of sub-ppm levels of OC pesticides in grains, fats, and vegetables after SEC cleanup by splitless injection of 100-μl samples [103].

The degree of sample cleanup, the volume injected, and the stationary-phase film thickness must be chosen so that peaks are detected but are not distorted due to column overloading [32].

23.3.3 Detectors

The major pesticide GC detectors for packed and capillary columns are the ECD, nitrogen/phosphorus (NPD), alkali-flame ionization (AFID), thermionic (TID), flame-photometric (FPD), and Hall electrolytic conductivity (HECD) detectors. The mass spectrometer, which is the most selective detector, is used for residue confirmation. The ECD is the primary detector for halogenated pesticides, to which it responds with high sensitivity. It is the least selective pesticide detector and requires adequate sample clean-up. The NPD is an alkali-salt-modified FID that has enhanced response for compounds containing nitrogen and phosphorus. It is the most widely used detector for organo-nitrogen (ON) pesticides. The FPD is selective for compounds containing P or S and is the most widely used detector for OP pesticides. Interference from S in the P mode is more likely than the reverse situation [38]. The variability of response to the P content of pesticides was studied [104]. The HECD can be operated under conditions that render it selective to Cl, N, or S. Its sensitivity is not as good as that of the other detectors, and it is

References on p. B540

relatively difficult to maintain and operate optimally. The microcoulometric detector is another Cl-selective detector that has been used for pesticide analysis. Its sensitivity is lower than that of the ECD and HECD, and it is also difficult to operate and maintain.

A multidetection capillary GC system was described [105] featuring an ECD and NPD in series and a FPD(S) in parallel. By appropriate effluent splitting, pesticides containing N, P, and S were detected simultaneously. Other detectors have been suggested for the analysis of pesticides [6-9], but none has been used to any significant degree for practical analysis.

The detectors mentioned above are reviewed in Refs. 34 and 79, and Refs. 13, 106, and 107 contain considerable practical information on principles of detection, proper operating parameters, sensitivity levels, and interpretation of results.

23.3.4 GC data

Retention data for numerous pesticides on packed GC columns are given in Refs. 10, 13, 79, 106, and 107. The following additional compilations of retention data have been published: 108 pesticides on four methyl silicone capillary columns, with comparison to published values for packed columns [85]; 194 OC, OP, and ON pesticides on a SE-30 capillary column [83]; 50 OP pesticides on nine packed columns with the FPD [108]; 42 OP and 28 OC pesticides on nine packed columns [109]; 600 pesticides and industrial chemicals on methyl silicone columns with temperature programing [110]; and 78 pesticides on four Ultra-Bond columns [111].

23.3.5 GC/MS

The approach to FDA regulatory pesticide analysis by MS and some selected case histories were described [112]. Practical techniques for MS determination of pesticides and their degradation products were reviewed [113], including instrumentation, experimental techniques, and applications of electron-impact (EI), chemical-ionization (CI), negative-ion (NI), field desorption, fast ion bombardment, and atmospheric pressure ionization mass spectrometry. Advances in quantitative and qualitative determination of chlorodibenzo-p-dioxins and chlorodibenzofurans in the environment by GC/MS were reviewed [114]. Compilations of spectra were published for hexachlorocyclopentadiene derivatives [115], 20 carbamates [116], and 90 OP sulfides, sulfoxides, and sulfones [117].

Reports of NI/CI/MS have become increasingly frequent, especially in combination with capillary GC. Examples include Refs. 118-121 on OC insecticides, Ref. 122 on dioxins, Ref. 123 on OP pesticides, Ref. 124 on dinitroaniline and dinitrophenol herbicides, Ref. 125 on triclopyr herbicide, and Ref. 126 on trifluralin herbicide. MS/MS was used to determine the herbicides pyridate [127], and paraquat, diquat, and dibenzoquat [128].

23.3.6 Miscellaneous studies

Fehringer and Walters [86] demonstrated that capillary GC with split injection can be used reliably for regulatory determination of pesticides in foods and feeds.

A book chapter reviewed pesticide determination by GC/Fourier transform (FT) IR [129]. GC/FTIR reference spectra of 47 environmental pollutants were measured by the EPA [130]. Performance of a GC/FTIR system was evaluated [131], and a GC/FTIR spectrometer was linked to a mass-selective detector [132]. Computer programs were described for the interpretation and evaluation of results from GC systems with multiple columns and detectors [133] and for the identification of PCB isomers [134]. Applications of computers to the evaluation of GC data have been reviewed [135].

23.3.7 Organochlorine pesticides and related industrial chemicals

Reviews of many hundreds of general and specific GC methods for OC insecticides and related industrial chemicals, such as PCBs, chlorodibenzofurans, and chlorodibenzo-p-dioxins published in the 1980s may be found in Refs. 6-9. The following are a selection of these references. Most determinations of OC insecticides have involved the traditional pattern of solvent extraction, partition, and column chromatographic cleanup, GC/ECD, and GC/MS confirmation (Chapter 24).

Mixtures of OCs were passed through a mixed adsorbent column for simultaneous fractionation and cleanup prior to capillary GC [136]. Methane Cl/selective ion monitoring (SIM) MS was used to distinguish 19 OC pesticides from PCBs at ppt levels in environmental water extracts with minimal cleanup [137]. An analytical method was reported for automated interpretation of mass spectra of PCBs and 21 OC pesticides [138]. Multi-residues of OC insecticides were determined by capillary GC in moss [118], mixtures with PCBs in water [139], and in human milk [140], feeds [141], and soils [142]. HPLC trace enrichment was combined with capillary GC for determining OC pesticides and PCBs in water [143].

Multi-laboratory validation studies were carried out for GC determinations of OC compounds in swine and beef fats [144]; eel fat [145]; vegetable fat [146]; miscellaneous animal fats [147]; water and wastewater (EPA Method 608) [148]; water, soil, and sediment [149]; and soil and sediment [2,3,7,8-tetrachlorodibenzo-p-dioxin (TCDD) only] [150]. A standard ASTM method for water was described [151]. Volatile halogenated hydrocarbons in sediment were determined at ppb levels, using Porapak N cartridges for analyte collection and an ECD or PID [152]. EPA methods for GC analyses of water [153] and solid wastes [154,155] and UK Department of the Environment methods for sludge and water [156] were outlined.

Capillary GC was applied to the analysis of breast milk for toxaphene [120], milk for heptachlor and octachlor epoxides [157], sediment for PCBs [158], and human serum [159], fish [160], and soil and water [161] for 2,3,7,8-TCDD. The separation of 16 priority pollutant pesticides by capillary-column GC is shown in Fig. 23.1.

References on p. B540

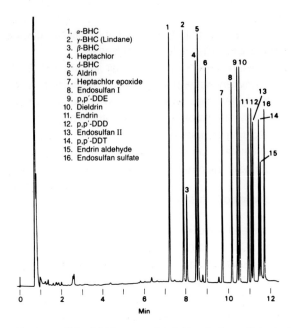

1. *a*-BHC
2. *γ*-BHC (Lindane)
3. *β*-BHC
4. Heptachlor
5. *δ*-BHC
6. Aldrin
7. Heptachlor epoxide
8. Endosulfan I
9. p,p'-DDE
10. Dieldrin
11. Endrin
12. p,p'-DDD
13. Endosulfan II
14. p,p'-DDT
15. Endrin aldehyde
16. Endosulfan sulfate

Fig. 23.1. GC of 200 pg each of 16 priority pollutant chlorinated insecticides on a Supelco SPB-608 polar, fused-silica wide-bore capillary column, 15 m x 0.53 mm ID, 0.5-μm film. Column temperature program: 4 minutes at 150°C, then to 280°C at 16°C/min. Helium carrier gas flowrate, 3.0 ml/min; ECD; sample volume, 0.6 μl in *n*-decane. (Figure supplied by Supelco.)

23.3.8 Organophosphorus pesticides

Most residue determinations of OP pesticides have been carried out by GC with packed columns and the ECD, FPD, or NPD. The latter two specific detectors minimize the amount of cleanup and usually eliminate the need for the higher resolution provided by a capillary column, unless a multiresidue method is required. Liquid phases that have been used include 2% DEGS and 6% DC-200 [162]; 4% SE-30 + 6% QF-1 [163]; 3 or 5% OV-17 with 0.02% Epikote 1001 [164]; and 1.3% Apiezon L [164]. OP residue determinations by GC published up to 1980 and 1986, respectively, have been reviewed comprehensively in Refs. 165 and 166.

The major multiresidue method used by the Canadian Department of Health and Welfare includes capillary GC on DB-17 and -210 columns with a FPD or TID, sample cleanup by SEC, and confirmation of identity by derivatization with pentafluorobenzyl bromide or trifluoroacetic anhydride [167].

Improved sample cleanup by solvent partitioning [168], silica gel columns [169], and solid-phase partitioning [170] for OP determination was reported. Recent multiresidue GC determinations reported include the following: in fruits and vegetables, on DC-200,

DC-200 + QF-1, and OV-17 + QF-1 columns with the rubidium sulfate NPD [171]; in unpolished rice, on DC-200 with the FPD [172]; in rice and rice straw on a fused-silica open-tubular (FSOT) capillary column with the NPD [173]; in grain, on OV-17 with the NPD [174]; and in grain, on an OV-101 capillary column with the FPD [41].

OP residues in soils and sediments were determined at 10 ppb by capillary GC with a NPD after extraction, solvent partitioning, and adsorption column cleanup [175]. A GC method for analysis of 41 foods from the FDA total diet study gave recoveries of 80-118% for 17 OP pesticides [176]. OPs were determined in apples and water by GC/ECD after Florisil/Celite/charcoal column cleanup [177]. Coumaphos and coroxon in milk were determined by blending the sample with Celite, which was poured into a column and eluted with dichloromethane. The eluate was further fractionated by SEC, and analyzed by capillary GC/ECD [178]. Standard methods were described for GC determinations of OP pesticides in waters and sewage sludge [179,180].

23.3.9 Carbamate insecticides

The selective thermionic NPD is the most widely used detector for N-methylcarbamate insecticide determinations, although derivatization reactions to improve electron-capture detection continue to be reported [181-183]. GC determination has been plagued by the thermal instability of these compounds when using packed columns, and the use of Carbowax 20M on modified supports has been suggested to overcome this problem [184]. Capillary-column GC with splitless [83] and split [96] injection has been used successfully for direct determinations. For example, aldicarb and its oxidation product were determined in leaves on a FSOT column, coated with 0.33-μm cross-linked 5% phenylmethylsilicone, by thermionic detection [185]. Carbaryl has been determined without decomposition by SFC [186].

The GC determination of carbamate insecticides in grain and grain products was reviewed [166].

23.3.10 Herbicides

The herbicides comprise a large and diverse group of compounds for which the greatest number of new analytical methods are reported each year. Most herbicide determinations involve packed-column GC. Space does not permit a comprehensive review of these new methods, and interested readers are referred to Refs. 6-9.

Ten acid herbicides were determined in natural waters as pentafluorobenzyl (PFB) esters by capillary GC/ECD [187]. The extraction of neutral and acidic herbicides from waters, derivatization of the acids with diazomethane, and analysis with EC, thermionic, or halogen-mode Hall detectors was described [188]. Eleven triazine herbicides were determined in natural waters by extraction with dichloromethane, cleanup on a deactivated Florisil column, and GC/NPD [189]. Triazines were determined by capillary GC/NPD in drinking water without cleanup of dichloromethane extracts [190] and in butter and milk after alumina column cleanup [191].

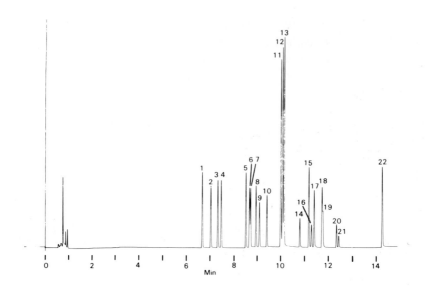

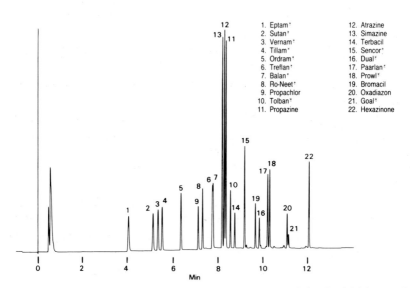

Fig. 23.2. Separation of 5 ng each of 22 nitrogen-containing herbicides on Supelco polar Sup-Herb (top) and nonpolar SPB-5 (bottom) wide-bore capillary columns, 15 m x 0.53 mm ID, 0.5-μm film. Column temperature program: 60°C for 1 min, then to 290°C at 16°C/min and hold for 5 min. Helium carrier gas flowrate, 5 ml/min; NPD; sample volume, 1 μl in ethyl acetate; direct injection. (Figure supplied by Supelco.)

Phenylurea [192] and uracil [193] herbicides were analyzed as methylated derivatives formed by injection into a gas chromatograph together with trimethylanilinium hydroxide. The GC/MS of some thermally labile phenylureas was studied [194].

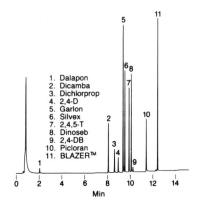

Fig. 23.3. Separation of 11 chlorophenoxy acid herbicide methyl esters on a Supelco SPB-608 polar, fused-silica wide-bore capillary column, 15 m x 0.53 mm ID, 0.5-μm film. Column temperature program: 60°C for 1 min, then to 280°C at 16°C/min and held for 5 min. Helium carrier gas flowrate, 5 ml/min; ECD. (Figure supplied by Supelco.)

Fig. 23.2 illustrates the separation of 22 nitrogen-containing herbicides on two capillary columns with differing selectivities. The use of two columns that provide different elution orders is advantageous for confirming residue identity.

Chlorophenoxy acid herbicides were determined as PFB derivatives by packed-column GC/ECD in grain [195] and soil and water [196], and by capillary GC/ECD in water, plants, and soil [197]. Capillary GC of methyl derivatives was used for determinations in cereal, mushrooms, and berries [198] and water [199]. Different derivatization methods for GC/ECD were compared [200]. ASTM published a standard test method for chloro-phenoxy acid herbicides in water [201]. Fig. 23.3 illustrates the separation of 11 chlorophenoxy acid herbicides by capillary-column GC.

Packed-column GC methods of the EPA for thiocarbamates in wastewaters were presented [202].

Capillary GC with a NPD or ECD was applied to the following determinations: the plant growth regulator indol-3-ylacetic acid in shoots at pg levels [203]; prometryn and its degradation product in parsley [204]; terbutryn and its degradation product in grain with EI/MS and CI/MS confirmation [205]; amitrole in water [206]; chloropropham and two metabolites in potatoes [207]; chlorsulfuron in agricultural runoff water [208]; and nitrofen in fish [209].

23.3.11 Fungicides and ethylenethiourea

The GC determination of dithiocarbamate fungicides [210] and methods for determining viticulture fungicides in grapes, must, and wine [211] were reviewed.

Dithiocarbamates in wastewater were determined by acid hydrolysis to CS_2, extraction with hexane, and GC/FPD(S) [212]. CS_2 formed by acid degradation can be measured instead by headspace GC [213].

References on p. B540

The neutral, nitrogen-containing fungicides fenarimol, MGK 264 and 326, and pronamide were determined in wastewater by dichloromethane extraction, deactivated Florisil column cleanup, and packed-column GC/NPD [214]. Captan, folpet, captafol, vinclozolin, and iprodione were determined in apples and pears by extraction, cleanup on a C-18 Sep-Pak cartridge, and capillary-column GC/ECD [215]. Fenpropimorph, thiabendazole, imazilil, propiconazole, and prochloraz in citrus fruit were determined by SE-54 capillary-column GC of hexane/ethyl acetate extracts [216]. Applicability of GC/ECD and C-18 HPLC in the analysis of dicloran, vinclozolin, meclozolin, iprodione, carbendazim, and thiabendazole were compared [217]. A GC method for 15 fungicides in water involved extraction, cleanup by SEC, and GC with FPD, thermionic, and EC detectors [218]. The AOAC official GC method for ethylenethiourea (ETU) residues in potatoes, spinach, apple sauce, and milk was shown to be applicable to canned fruits and vegetables [219].

Triadimenol, biteranol, paclobutrazol, and diclobutrazol enantiomers were separated on a Chirasil Val FSOT capillary column [220]. Enantiomers of fungicidal triazole alcohols were separated by achiral SE-52 capillary-column GC after chiral derivatization [221]. Capillary GC was applied to the determination of ETU in grapes and wine [222], carbendazim after alkylation with diazomethane [223], and di-n-butyltin and tri-n-butyltin hydrides formed inside the injection port [224].

23.3.12 Pyrethroid insecticides

Most determinations of pyrethroid insecticides have been made by packed-column GC/ECD, and these methods have been reviewed [225-227]. Extracts are usually cleaned up by liquid/liquid partitioning, followed by column chromatography. Derivatization reactions, e.g., to produce PFB derivatives [228], are carried out on non-halogenated pyrethroids to enhance detection. The FID is often used for the analysis of standards and formulations, and it is also occasionally applied to residue analysis [229,230].

The advantages of capillary GC in terms of improved resolution and shorter analysis time have been demonstrated by Bottomley and Baker [41], who used a temperature-programed OV-101 column to determine pyrethroids in grain. A SE-54 capillary column was used to determine cypermethrin in crops [231]. Packed columns that have been used include SE-30 [232,233], SILAR-10-C [229], XE-60 [234], OV-101 [235,236], Ultra-Bond 20M [237], 2% XE-60 + 2% OV-17 [238], OV-225 [239], and OV-210 [240].

Pyrethroid determinations are complicated by the fact that the synthetic compounds consist of a mixture of stereoisomers. Isomers can sometimes be separated on a packed column [239], but a capillary column is usually more successful [241]. Enantiomers of permethrin and cypermethrin were resolved on a non-chiral phase after conversion to diastereoisomers [242]. Positive- and negative-ion CIMS data were given for pyrethroids to aid residue confirmation [243].

23.3.13 Fumigants

Most fumigants are volatile, halogenated industrial solvents that are amenable to GC with the ECD or HECD. Other selective detectors are used as appropriate, e.g., the S-mode FPD for carbon disulfide and the P-mode FPD for phosphine. Multiresidue and specific methods have been reviewed by Scudamore [244].

A variety of methods have been used to prepare samples for GC, including co-distillation [245,246], purge and trap [247], steam distillation [248], vacuum distillation [249], cold solvent extraction [250], trapping from air on charcoal [251], and headspace evolution [252-254]. Packed columns, such as OV-101 and OV-225 + OV-17 [255,256], and capillary columns, such as DB-5 [245,246,248], have been applied.

The most general multiresidue method can determine 20 fumigants in wheat and maize. It is based on GC after cold extraction with mixtures of acetone or acetonitrile and water [257]. A number of GC methods have been collaboratively studied [258] and certified as official by the AOAC [259,260].

23.3.14 Rodenticides

Difficulties arising from thermal instability of some coumarin-based rodenticides preclude the use of GC for routine multiresidue analysis, although capillary GC has been applied for confirmational purposes by identification of degradation products. Available GC methods, along with the more widely used HPLC methods, have been reviewed [261].

23.4 SUPERCRITICAL-FLUID CHROMATOGRAPHY

SFC in packed and capillary columns [262,263] has been growing in popularity for pesticide residue analysis. The properties of supercritical fluids place the method somewhere between GC and HPLC, and both GC and HPLC detectors have been used. SFC can analyze compounds that are thermally labile and have a low volatility (high molecular weight), and resolution and analysis time are superior to those in HPLC [264]. Instrumentation, procedures, and applications of SFC have been reviewed [265].

The following applications of SFC to pesticides have been published. Eight acid and carbamate pesticides were determined in about 2 min by using carbon dioxide as the mobile phase, a 25-μm-ID capillary column with a fast pressure program, and a FID with moving-belt interface [266]. NH_3 and CH_4 CI mass spectra were obtained for these pesticides after direct supercritical-fluid injection [267]. 2-Propanol was used as polar modifier in the SFC/MS of OP pesticides [268]. The herbicide terbacil and its metabolites were separated and identified by directly coupled SFC/MS, the mobile phase acting as the CI reagent gas [269]. Carbamate pesticides were determined with a thermospray interface for SFC/MS [270].

Pesticides were determined by capillary SFC with dual FPDs [271]. Packed and capillary SFC and HPLC were compared for moderate-polarity and moderate-molecular-weight pesticides with different functional groups, and retention characteristics were reported

References on p. B540

[272]. Capillary SFC/ECD was used to determine a triazole fungicide metabolite [273]. Polycyclic aromatic hydrocarbons (PAHs) and PCBs were determined in environmental samples in less than 1 h by coupled SPE/SFC [274]. A FT mass spectrometer was used to detect and identify low-ng levels of pesticides, separated by SFC [275], and SFC/FTIR was applied to the analysis of pyrethrins [276].

23.5 HIGH-PERFORMANCE LIQUID CHROMATOGRAPHY

The application of HPLC to pesticide residue analysis has increased dramatically in the 1980s, particularly for methylcarbamate insecticides; phenoxyacetic acid, phenylurea, and other classes of herbicides; fungicides; pyrethroids; and pesticide metabolites [6-9]. This is because many of these compounds are nonvolatile, polar, and/or thermally labile and thus are not directly amenable to GC determination. In addition, many examples of HPLC analyses that require simpler cleanup or are faster than GC have been reported. The literature of individual and multiresidue pesticide HPLC analysis up to 1980 was reviewed in a book by Lawrence [277]. Muszkat and Aharonson [278] reviewed the practice and applications of HPLC for pesticide analysis through 1984.

23.5.1 Sample preparation for and by HPLC

The usual extraction and cleanup methods (Section 23.2) are most often used to prepare samples for HPLC analysis. In addition, several unique techniques have been employed. Off-line minicolumns and cartridges, e.g., Sep-Pak, have been used to isolate pesticides from water prior to HPLC analysis [279]. Trace enrichment directly on a RP-HPLC column has been used to collect nonpolar compounds from water prior to analysis by application of a solvent gradient. This method was demonstrated for carbofuran and metabolites [280] and difenzoquat [281].

On-line precolumn concentration and cleanup has also been applied, with a variety of sorbents, such as metal-loaded materials, selective ion exchangers, XAD porous polymers, SEC packings, C-18 bonded phases, etc. Concentrations by a factor of 100 to 1000 have been achieved, allowing detection in the pg range [282]. Examples of determinations in water analysis include phenylurea herbicides [282-284]; bromacil, diuron, and 3,4-dichloroaniline [285]; chlorophenoxy acid herbicides [286]; nitroaromatics [287]; and aldicarb species [288]. On-line, precolumn trace enrichment was also applied to N-methylcarbamate insecticide residues in total-diet-sample homogenates [289]. On-line cleanup was reported for the determination of carbamates in plants [290] and carbaryl in wheat [291]. A general description of precolumn technology for environmental analysis has been published [292].

Either pre- or postcolumn reactions can be used to form products that are more sensitively detected than the parent pesticides, mostly fluorescent derivatives. Postcolumn methods, which are usually preferred, are covered in Section 23.5.4.2. Precolumn derivatization has been reported for chlorophenoxy acids [293] and for glyphosate and its major metabolite, aminomethylphosphonic acid (AMPA) [294].

HPLC is frequently used for isolation and/or fractionation prior to GC analysis. Examples are the silica gel cleanup of a glyphosate derivative [295] and OC insecticides [296] prior to GC/ECD. On-line HPLC/GC/ECD was used to determine folpet residues in hops [297] and on-line HPLC/capillary GC for OC pesticides in an aqueous solution [298].

23.5.2 Columns

Although all HPLC modes, including normal phase, reversed phase, ion suppression, ion interaction, ion exchange, and size exclusion, have been used for pesticide analysis, C-18 (ODS) bonded silica gel reversed-phase chromatography has been preferred. Most analysts have used 25-30 cm x 4.6-mm-ID commercially prepacked columns, containing 10-μm, 5-μm, or 3-μm [299] microparticulates, and short guard columns with <20-μm pellicular or porous packing. Shorter (3-15 cm) x 4.6-mm-ID columns have become increasingly popular, because they provide higher analysis speed, efficiency, mass sensitivity, and solvent economy. The separation of 15 carbamate and urea pesticides on a C-8 bonded-phase column is shown in Fig. 23.4.

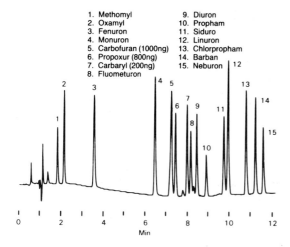

1. Methomyl
2. Oxamyl
3. Fenuron
4. Monuron
5. Carbofuran (1000ng)
6. Propoxur (800ng)
7. Carbaryl (200ng)
8. Fluometuron
9. Diuron
10. Propham
11. Siduro
12. Linuron
13. Chlorpropham
14. Barban
15. Neburon

Fig. 23.4. Separation of 80 ng each (unless otherwise indicated) of 15 carbamate and urea pesticides on a Supelcosil LC-8 HPLC column. Mobile-phase gradient, acetonitrile/water 18:82 v/v to 65:35 v/v in 9 min and held for 3 min; flowrate, 2.0 ml/min; temperature, 35°C; detector at 240 nm, 0.20 AUFS. (Figure supplied by Supelco.)

Another trend is toward the use of higher-efficiency open-tubular [300] and packed-microbore (0.2- to 1-mm-ID) columns. For example, carbamate and phenylurea pesticides were determined in cucumber with 10-fold improved sensitivity, compared to a conventional column, by using a 15 cm x 1-mm, 3-μm C-18 column with acetonitrile/methanol as the mobile phase [301]. Microcolumn HPLC was also used to separate OP [302] and OC [303] pesticides.

The ion-interaction HPLC of pesticides has been reviewed [304]. Chiral-phase columns [305] have enabled the separation of pesticide geometric and optical isomers [306,307].

23.5.3 Retention data

Extensive collections of data for HPLC have not been published as frequently as for GC. However, several reviews have listed the columns, mobile phases, retention data, and optimum UV wavelengths for the HPLC determination of hundreds of pesticides [308-310]. HPLC data are available in the *Handbook of Chromatography: Pesticides* [10]. Retention data were collected for 52 pesticides and other compounds of environmental interest by means of HPLC/thermospray MS [311].

23.5.4 Detectors

The main limitation of HPLC in pesticide residue analysis is the lack of sensitive and selective detectors comparable to those available for GC. The fixed- and continuously variable-wavelength UV absorbance and fluorescence detectors have been used most often, but some use has also been made of electrochemical, FPD, ECD, and of atomic absorption, IR [312], enzyme inhibition [313], and radioactivity detectors.

23.5.4.1 Ultraviolet detectors

Most analyses with the UV detector have been carried out at the two wavelengths available in fixed-wavelength models, 254 nm (e.g., for diuron and its metabolite, 3,4-dichloroaniline, in asparagus [314]) and 280 nm (e.g., for asulam in wheat flour [315] and phosmet [316]). Increased selectivity and/or sensitivity of the variable-wavelength UV detector was demonstrated in the determination of metribuzin (295 nm) [317], azinphosmethyl and its oxon in foods (224 nm) [318], and of simazine and atrazine in water (220 nm) [319]. OP pesticides were screened by simultaneous determination of different compounds at several wavelengths by means of a photodiode-array UV detector [320].

23.5.4.2 Fluorescence detectors and postcolumn derivatization

Fixed- or variable-wavelength fluorescence detectors are suitable for the determination of naturally fluorescent pesticides, or non-fluorescent compounds labeled with a fluorophore. Derivatization reactions can be carried out prior to chromatography, but postcolumn reaction and extraction systems are most commonly used. Postcolumn fluorometric labeling is widely applicable in pesticide analysis by choosing one of two approaches: either direct reaction of the reagent with the analyte, or preliminary hydrolysis or oxidation followed by reaction of the product with the reagent. A solid-phase reactor for postcolumn derivatization was applied to *N*-methylcarbamates in water, using the *o*-phthalaldehyde (OPA) fluorometric reagent after hydrolysis to form methylamine [321]. Post-column fluorometric labeling was also used to determine carbamate insecticides in foods [322,323] and rodenticides [324].

A recent review of postcolumn reaction detectors indicates the variety of reaction types applicable to fluorescence detection [325]. On-line precolumn sample preparation and postcolumn fluorogenic labeling have been combined to determine glyphosate in cereals and vegetables [326], illustrating possibilities for total automation. Postcolumn photolysis was used to produce fluorophores for the determination of nitrogen-containing pesticides [327,328].

23.5.4.3 Postcolumn extraction systems

On-line postcolumn extraction systems have been used to transfer hydrophobic pes-ticides from aqueous eluents, containing nonvolatile ion pairing reagents, acids, or salts, to an appropriate organic phase prior to HPLC detection. This method was applied to phenylurea herbicides [329,330] and chlorophenoxy acid herbicides [331].

23.5.4.4 Electrochemical detectors

Use of an electrochemical detector has been reported for OP pesticides [332,333], carbamates [334-337], triazines [338], and phenylureas [283,337]. In general, com-pounds with electroactive groups will be detected more selectively and sensitively with the electrochemical detector compared to the UV detector, but the latter is more stable and easier to use and maintain [339]. Electrochemical detection has been shown to be com-patible with microbore columns [287,340].

23.5.4.5 Photoconductivity detectors

A Tracor 965 photoconductivity detector with Hg (254 nm) and Zn (214 nm) lamps was evaluated for determining 22 polar or thermally labile pesticides in food. A CN column was eluted with an aqueous methanol mobile phase. Most of the work was done with stand-ards, but the technique was also applied to the determination of captan in strawberries and corn [341].

23.5.4.6 Miscellaneous detectors

The FPD has been adapted to HPLC detection of compounds containing S and P atoms [302,342]. A HPLC/thermionic detection system has been used for analysis of OP pesticides [343,344]. HPLC, coupled with the ECD, was useful for the determination of OC pesticides and PCBs [345], chloroxuron in strawberries [346], and phenylurea herbicides in water [303]. A postcolumn extraction module was used for on-line coupling of a C-18 column with an ECD [347]. Ion-exchange HPLC with an atomic absorption spectrometry (AAS) detector was applied to the analysis of alkyl arsenic herbicides and their metabolites [348]. HPLC analysis of pesticides with a radioactivity detector has been described [349,350]. Carbamate herbicides were separated by HPLC/NPD and identified by on-line coupling with a FTIR spectrometer [351].

23.5.5 HPLC/MS

Applications of the mass spectrometer coupled with HPLC to the quantification and confirmation of pesticide residues are increasing rapidly. Either direct introduction with splitting of effluent, total direct introduction of effluent from a microbore column into the mass spectrometer, or a HPLC/MS interface must be used. Among the different types of commercially available interfaces, the thermospray and moving belt have become the most popular. Use of the thermospray has been reported for OP pesticides [352,353], linuron [354,355], carbamates [354], triazine [352,354-357] and chlorophenoxy acid herbicides [358], and chlorsulfuron [359]. Direct liquid introduction has been used for phenylureas [282,360], chlorophenoxy acid herbicides [361], and chlorsulfuron [359], and a moving-belt interface for phenylurea [362], carbamate [362], and OP [363] pesticides and chlorsulfuron [359]. Direct liquid introduction via a fused-silica capillary interface was reported for HPLC/MS determination of pyrethroids [364].

For HPLC/MS, either EI or CI techniques can be used. However, operation under CI conditions is most common because of compatibility with the high source pressures encountered in HPLC/MS interfacing systems. This leads to problems due to lack of fragmentation of some compounds. CI [365] in both negative-ion [366] and positive-ion [367] modes has been widely reported for different pesticides classes. SIM [365] is used when detection selectivity rather than universality is required. Positive- and negative-ion modes in thermospray HPLC were compared in the determination of chlorophenols and herbicides [355] and of OP pesticides [368]. HPLC/MS/MS has been used for triazine herbicide analysis [369].

Methods and applications of LC/MS have been reviewed in two recent book chapters [113,370] and a review paper [339].

23.5.6 Multiclass, multiresidue procedures

Although a number of HPLC systems have been reported for the determination of multiresidues of a single pesticide class, only a few allow determination of several classes. An analytical procedure was described for the simultaneous determination of 22 nitrogen-containing pesticides in ground and surface water by C-18 HPLC and diode-array detection [371]. Grain was analyzed for OC and OP compounds by GC, and for pyrethroids and carbamates by HPLC [41]. A method was devised to determine 42 pesticides, including OC, OP, carbamate, dinitro, triazine, phenylurea, and phenoxyalkanoic compounds in fruits and vegetables by gradient elution HPLC with a C-18 column and a variable-wavelength UV detector [372]. A method for determining 20 insecticides, herbicides, and fungicides involved extraction with dichloromethane and methanol/dichloromethane, cleanup on Extrelut and Florisil SPE columns, and C-18 HPLC with detection at 240 and

280 nm [373]. Carbamates, OPs, and piperonyl butoxide were determined in fruit and vegetables with a C-8 column, an acetonitrile/water gradient, and postcolumn fluorometric labeling [374]. Carbaryl and OPs were determined in fruit and vegetables on a C-18 column with detection at 220 and 247 nm [375]. Additional determinations include OCs and other pesticides used in the wine industry [376] and OP and triazine pesticides [367].

23.5.7 Organochlorine and organophosphorus pesticides

Although HPLC has been evaluated for some OC and OP pesticides, these compounds are mainly determined by packed- and capillary-column GC with selective detectors. Separation of OC pesticides and PCBs from a high concentration of lipid was accomplished on a polymeric HPLC column [377]. Pentachlorophenol was determined in wood by anion-exchange chromatography and C-18 HPLC at 230 nm [378] and by UV, amperometric, and electron-capture [379] detection. Green wheat forage was analyzed for chlorsulfuron and a metabolite by silica gel HPLC with a photoconductivity detector [380]. Aldrin was determined by C-18 HPLC with UV photolysis and electrochemical detection [381]. Dieldrin was determined in chicken eggs by C-18 HPLC/UV (234 nm) [382].

The following OP pesticide determinations on bonded reversed-phase columns were reported: in fish with thermospray MS confirmation [383]; in green vegetables with UV (270 nm) and electrochemical detectors [384]; and in cabbage and tomatoes with UV (254 nm) [344].

23.5.8 Carbamate pesticides

Aldicarb, barban, carbaryl, chlorbufam, chloropropham, dioxacarb, methomyl, and propoxur were determined in tomatoes on a 10-μm Lichrosorb column with di-oxane/isooctane as the mobile phase after extraction and cleanup by liquid/liquid extraction and Florisil LSC [385].

Seven *N*-methylcarbamate insecticides and 4 metabolites were determined in fruits, vegetables, and grains following postcolumn derivatization. Residues were extracted with methanol, and cleanup was achieved by solvent partition and charcoal/silanized Celite column chromatography. The carbamate residues were separated on a RP Zorbax column with an acetonitrile/water gradient, hydrolyzed to methylamine, and made to react with OPA and 2-mercaptoethanol to form a fluorescent derivative [386,387]. Similar RP-HPLC/fluorometric methods, reported for 7 pesticides and 3 metabolites, incorporated SEC and an on-line Nuchar/Celite cleanup [388]. For 21 pesticides and 10 metabolites, SPE cleanup on NH$_2$-bonded silica was used [389]. Six *N*-methylcarbamates were determined in 4 crops following C-8 separation with an acetonitrile/water gradient and on-line postcolumn basic hydrolysis and electrochemical detection [390]. Silica gel HPLC with postcolumn hydrolysis and fluorometric detection was used to determine carbaryl, carbo-furan, and propoxur in total-diet extracts [289].

References on p. B540

The following additional determinations were carried out by C-18 HPLC: carbofuran and its metabolites with methanol/water gradient elution and UV detection at 200 or 203 nm [391] and in rape plants at 280 nm [392]; aldicarb and its degradation products by isocratic elution with acetonitrile/buffer (pH 7.6) and detection at 220 and 247 nm [393]; carbaryl in wheat on a silica gel column with hexane/ethanol as the mobile phase and detection at 280 nm [291]; aldicarb in the stomach contents of birds of prey with UV (247 nm) detection [394]; and benfuracarb and carbofuran in soil and water with UV(280 nm) detection [395].

23.5.9 Herbicides

Chlorophenoxy acid herbicides are usually determined on C-18 or amino-bonded silica columns [396], either as intact acids in the ion-suppression or ion-interaction mode or as a derivative, such as the PBB or naphthacyl ester [397] with UV detection at 254 or 288 nm [396].

Triazine herbicides are determined on conventional [398] or microbore [399] bonded reversed-phase columns and detected with an electrochemical [400] or UV(220 nm) [401] detector. Triazines are efficiently collected by an on-line C-18 precolumn for their determination in water [402].

Phenylurea herbicides are usually determined on C-18 or ion-exchange [403] columns with either an electrochemical detector [283,404] or a fluorescence detector after derivatization with OPA [405,406]. Column-switching sample preparation technology was described for analysis of river water [407] and berries [408].

Abscisic acid, a widely used antitranspirant plant hormone, has been determined in plants by HPLC at 254 nm and by fluorescence detection [409]; enantiomers were separated on a chiral column [410].

Additional specific HPLC herbicide determinations, in which C-18 columns and UV detection were used unless otherwise noted, include amitrole in tissues in the ion-interaction mode with electrochemical detection [411]; asulam in spinach [412] and peaches [413]; dichlorprop and its conjugates in biological samples on a diol column [414]; diquat in water [415]; ethyl 2-naphthoxyacetate and its metabolite in tomatoes with fluorescence detection [416]; diuron in crops on silica gel with photoconductivity detection [417]; 42 gibberellins (plant growth regulators) and 20 glucosides in the ion-suppression mode [418]; glyphosate in water [419], crops [420], and cereals and vegetables [421] after fluorogenic labeling [419]; indol-3-ylacetic acid in radish plants with fluorometric detection [422]; napropamide in rape [423]; paraquat in liver and blood in the ion-interaction mode [424] and in agricultural products on an NH_2 column [425]; sulfometuron-methyl in soil, water, fish, and plants with a C-2 column and a photoconductivity detector [426]; triallate in soil [427]; chlorimuron-ethyl in crops with a silica gel column and photoconductivity detector [428]; dicamba in water with SPE on an amino-bonded ion-exchange column and determination by ion-interaction HPLC [429], and in asparagus by RP-HPLC [430]; dichlobenil in crops and soil [431]; bromacil in water [285]; fluridone in meat, milk, eggs, and crops [432]; fluazifop and fluazifop-butyl in soil and water [433,434]; metribuzin in soybeans (thermospray HPLC/MS) [435]; and naptalam as 1-naphthylamine in foods with oxidative electrochemical detection [436].

23.5.10 Fungicides and ethylenethiourea

Fungicide analyses are carried out on C-8 [437], C-18 [438], or SCX [439] columns with UV [440], fluorescence [441], or photoconductivity [442] detectors after appropriate extraction and cleanup steps. Postcolumn derivatization systems have also been reported [443].

A number of multiresidue determinations by C-18 HPLC have been described, e.g., dichlozolinate, iprodione, procymidone, and vinclozolin in wine with UV(210 nm) detection [444] and biphenyl, o-phenylphenol, thiabendazole, and diphenylamine in fruits with UV, fluorometric, and electrochemical detectors [445].

Ethylenethiourea (ETU), a degradation product of ethylenebisdithiocarbamate fungicides, has been determined in apples, tomatoes, grapes, and wine by C-8 HPLC/UV(240 nm) [446], in beer by means of an elaborate column-switching system for sample preparation and analysis [447], and in crops by C-8 HPLC with electrochemical detection [448].

Specific fungicide determinations on C-18 columns with UV detection include the analysis of fenpropimorph in fruit [449], of metham sodium and methyl isothiocyanate in aqueous solution [450], and of triflumizole in crops [451].

23.5.11 Pyrethroid insecticides

HPLC has been applied to the analysis of all established pyrethroids [227], but it is mainly used for the analysis of industrial and formulated pyrethroid products and isomer separations rather than residue determinations. Pyrethroids show strong UV absorbance at 235 nm or below and can be as sensitively detected with a variable-wavelength UV detector as by gas chromatography with the ECD.

C-18 HPLC/UV (235 nm) was used to determine pyrethroids in wheat [452]. A multiresidue procedure for 9 pyrethroids in fruits and vegetables at 0.05 ppm involved hexane/acetone extraction, silica gel column cleanup, C-18 HPLC with aqueous methanol, and detection at 206 nm [97]. An identical system was applied to the determination of 10 pyrethroids in grain as part of a large multiclass, multiresidue screening procedure [41]. Silica gel HPLC with an isooctane/2-propanol mobile phase and an IR detector allowed the determination of permethrin residues in lettuce at the 0.2 ppm level [312], and C-18 HPLC/UV(214 nm) was used for the analysis of permethrin and fluvinate in tobacco [453].

Pyrethroid optical isomers have been separated on columns containing chiral packing, (R)-N-(3,5-dinitrobenzoyl)phenylglycine, bonded to silica gel [307,454,455]. This column has been applied to the analysis of fenvalerate isomers in milk [456] and soil [457].

References on p. B540

23.5.12 Rodenticides

HPLC methods are included in a review of rodenticide formulation and residue analyses [261].

23.6 THIN-LAYER CHROMATOGRAPHY

The major use of TLC in pesticide analysis is in qualitative screening of residues and as an aid in the identification of GC or HPLC peaks. It has been demonstrated that by proper application of TLC methods, reproducible results can be obtained at concentrations as low as 0.05-0.1 ppm. Many analyses based on in situ quantification have been published, and many more are anticipated as the excellent precision and accuracy of high-performance TLC (HPTLC) with densitometric scanning become more widely recognized.

The IUPAC Commission on Pesticide Chemistry concluded that TLC is useful in multiresidue analysis of the most important classes of pesticides [458]. The optimization of TLC and GC determinations and cleanup parameters for multiresidue procedures was discussed [459]. General TLC methods were described for determining pesticide residues in plant, soil, and water samples [460,461]. TLC was used to analyze 45 compounds representing aromatic amines, chlorophenols, phenols, and herbicides in water; after extractive enrichment and preliminary group separations, compounds within each group were resolved by combination of several solvents and layers and detected by selective reagents or under UV light [462]. The selectivity and sensitivity of three important detection systems, silver nitrate/2-phenoxyethanol, 4-(4′-nitrobenzyl)pyridine, and 2,6-dibromo-benzoquinone-N-chloroimine, for the detection of OC, OP, and thiocarbamate pesticides were examined [463]. TLC and GC procedures for determining OP pesticides in sewage sludge and drinking and river water were reported, including retention times for 42 pesticides on 6 GC columns and R_F values of pesticides with three mobile phases [464].

The TLC of pesticides is reviewed up to the early 1980s in five book chapters [465-469]. Applications of two-dimensional TLC to the analysis of environmental pollutants and pesticides were reviewed through early 1982 [470]. Table 23.1 updates the coverage of these earlier sources.

23.7 CONCLUSION

This chapter contains a selection of recent references that summarize important techniques and applications of chromatography to the determination of pesticides residues. In addition, a small number of papers on the analysis of plant growth regulators and industrial chemicals related to pesticides and on the analysis of pesticide formulations have been included.

At the present time, GC is the major chromatographic method used for pesticide residue analysis, but applications of HPLC, HPTLC, and SFC have been increasing steadily. Areas in which significant recent advances and promise for future development have

TABLE 23.1

TLC PROCEDURES FOR THE SEPARATION, DETECTION, AND QUANTIFICATION OF
PESTICIDES

Pesticide(s)	Sample; comments	Reference
Abate	Water; RP-TLC + densitometry	471
Abscisic acid	Plants; TLC cleanup for RIA	472
Aminocarb	Separation from metabolites	473
Aphos	Air, water, soil	474
Asulam	Water, spinach; HPTLC and densitometry	475
Asulam	Soil; TLC + colorimetry	476
Asulam, sulfanilamide, sulfanilic acid	Soil	477
Atrazine metabolites	Soil; HPTLC	478
Atrazine, simazine	Water; densitometry	479
Biphenyl	Citrus fruit; densitometry	480
Bolstar	Soil, plants, water	481
Bromophos	Peanut crops; dansyl derivative	482
Captan, folpet, captafol	Water, lettuce; densitometry	483
Carbamates	Separation on zinc acetate-impregnated silica layer	484
Carbamates	Bovine rumen content; RP-TLC	485
Carbamates, thiocarbamates, ureas	Air	486
Carbamate, urea, and anilide herbicides	Soil water; densitometry	487
Carbaryl	Standard; two specific spray reagents	488
Carbaryl, fungicides, herbicides	Tobacco	489
Carbofuran, 2 metabolites	Water, soil, plants	490
Carbofuran, quinalphos	Air	491
Chlormequat chloride	Foods; densitometry	492
Chlormequat chloride	Environmental samples	493
Chlorocholine chloride	Grain, grain products	494
Chloropropham	Onions	495
Chloropropham, meto-bromuron, chlorbromuron	Drugs	496
Chlorpyrifos, TPC	Water, banana; densitometry	497
Curacron, Selecron	Environmental samples	498
DDT	1,4-Dihydroxybenzene detection reagent	499
Dieldrin derivatives	Identification by TLC	500
Dimethoate	Human blood; cresol-impregnated silica layer	501
Dinoseb, DNOC	Water	502
Disulfoton, fenthion, and phorate O-analogs	Detection with m-chloroperbenzoic acid	503
Diquat	Sunflower seeds	504
Disulfoton, monocrotophos, quinalphos	Detection with cupric acetate and KI	505

(Continued on p. B538)

TABLE 23.1 (continued)

Pesticide(s)	Sample; comments	Reference
Endrin	Detection with tin dichloride/fuchsin dye reagent	506
Fenitrothion	Water; enzyme inhibition detection	507
Flucythrinate	Standard	508
Fungicides	Detection with *C. cucumerinium* spores	509
Gibberellins	Fermentation broths; HPTLC + densitometry	510
Herbicides	Detection with chloroplasts + 2,6-dichloroindophenol	511
Herbicides	Standards; detection by inhibition of Hill reaction	512
Herbicides	Standards; 2-D TLC on calcium sulfate layers	513
Herbicides	Sugar beet and sugar	514
Hexazinone metabolites	Rat liver microsomes, peanut seedlings, sugar cane; TLC + MS	515
Imidan and degradation products	Standards; 2-phase layer	516
Isouron	*H. saturnus*; densitometry	517
Lenacil, pyrazon	Water, soil, sugar beets	518
Mecarbam	Water, crops	519
Metalaxyl	Maize	520
Methamidophos	Potato tubers; enzyme inhibition detection	521
Methamidophos	Enzymatic detection	522
Methoxychlor	Water	523
Methoxychlor and methidathion	Clinical samples; NP- and RP-TLC	524
Metoxuron, breakdown product	Determined as fluorescent dansyl derivatives	525
Monocrotophos	Detection with iodine/iodide reagent	526
OC pesticides	Standards; HPTLC	527
OC pesticides	Animal tissue	528
OC pesticides	Water; SPE + densitometry	529
OP pesticides	Standards; metal salt or phenol impregnated layer	530
OP pesticides	Analysis after reduction to amino derivatives	531
OP pesticides	Direct TLC/MS on polyamide layers	532
OP pesticides	Enzyme inhibition detection + reflectance densitometry	533
OP pesticides	Vegetables	534
OP pesticides	Toxicological samples	535
OP pesticides	Standards	536
OP pesticides	Standards; separation with temperature gradient	537
OP pesticides	Standards; concurrent use of 2 silica layers and 2 mobile phases	538

TABLE 23.1 (continued)

Pesticide(s)	Sample; comments	Reference
OP pesticides	Vegetables; cresol-impregnated silica layer	539
OP pesticides	Water; HPTLC	540
OP pesticides containing S	Standards; detection with KI	541
OP pesticides, carbamates	Fruits and vegetables	542
Organotin fungicides	HPTLC	543
Organotin fungicides	Plants, soil; biological detection	544
Organotin fungicides	Standards; HPTLC + densitometry	545
Paraquat	Toxicological samples	546
Paraquat	Plasma, urine; TLC with FID	547
Parathion	Crops; TLC + colorimetry	548
Parathion, paraoxon	Enzyme inhibition detection	549, 550, 551
PCBs	Field test + densitometry	552
Pentachlorophenol	Benzoate derivative separated and detected	553
Phenoxyacetic acid herbicides	Separation on silica and alumina	554
Phenoxyacetic acid herbicides, chlorophenols	Standards; PFB derivatives	555
N-Phenylcarbamates	Carrots, potatoes, wine	556, 557
Phenylureas	Partition TLC on impregnated layers	558
Phorate	Root crops	559
Phosmethylan	Soil, water, crops, milk, meat	560
Propamocarb	Peppers	561
Pyrazophos	Plant products; spectrometry after elution from plate	562
Pyrethroids	Separation of isomers	229
Pyrethroids	Crops; dual-wavelength densitometry	563
Pyrethroids	Tomato, fruit, soil; silver nitrate-impregnated alumina layer	564
Rodenticides	Standards; preadsorbent silica gel layer, phosphomolybdic acid detection reagent	565
Tetrachlorvinphos	Apples; densitometry	566
Thiabendazole	Fruit peel; densitometry	567
Thionophosphoric esters	Detection with iodine/azide reagent	568
Toxaphene	Honey	569
Toxaphene	Surface waters	570
Triazines	Standards; re-development with the same mobile phase	571
Triazines	HPLC behavior of 16 compounds predicted by results of TLC	572
Triazine and phenoxyacetic acid herbicides	Water; SPE and densitometry	573
Triazines, phenylureas, phenylcarbamates, uracils, anilides	Detection with chloroplasts and 2,6-dichloroindophenol	574
Trichlorometaphos-3	Grain	575
Zineb	Foliage; TLC + spectrometry	576

References on p. B540

been noted include capillary-column GC; capillary and microbore HPLC; increased application of HPLC detectors other than UV; sample preparation employing SPE and SFE; coupled methods such as GC/MS/MS, HPLC/MS, and GC/FTIR; and automation of the analysis system. The greatest progress in the next decade may be made in a non-chromatographic area, namely the development of immunoassay techniques for pesticide analysis.

Most available residue methods have been devised for the analysis of only a single pesticide or pesticide class. Public health concerns assure the future prominence of research aimed at extending the available multiclass, multiresidue methods, and finding new ones, to include many more of the pesticides being used world-wide, as well as their polar and conjugated metabolites.

REFERENCES

1 J. Sherma and G. Zweig, in E. Heftmann (Editor), *Chromatography*, Van Nostrand Reinhold Co., New York, 3rd Edn., 1975, p. 781.
2 L. Fishbein, in E. Heftmann (Editor), *Chromatography*, Elsevier, Amsterdam, 4th Edn., 1983, p. B435.
3 D.F. Hill, *Anal. Methods Pestic. Plant Growth Regul.*, 15 (1986) 161.
4 H. Roseboom, *Food Sci. Technol.*, 11 (1984) 489.
5 A.S.Y. Chau and H.B. Lee, *Anal. Pestic. Water*, 1 (1982) 25.
6 J. Sherma and G. Zweig, *Anal. Chem.*, 55 (1983) 57R.
7 J. Sherma and G. Zweig, *Anal. Chem.*, 57 (1985) 1R.
8 J. Sherma, *Anal. Chem.*, 59 (1987) 18R and 61 (1989) 153R.
9 G. Zweig and J. Sherma (Editors), *Analytical Methods for Pesticides and Plant Growth Regulators,* Academic Press, New York, 17 volumes published between 1963 and 1989.
10 J. Follweiler and J. Sherma, *Handbook of Chromatography: Pesticides,* Vol. I, CRC Press Boca Raton, FL, 1984.
11 K.G. Das (Editor), *Pesticide Analysis,* Dekker, New York, 1981.
12 J. Sherma, *Manual of Analytical Quality Control for Pesticides and Related Compounds in Human and Environmental Samples,* EPA-600/2-81-059, NTIS PB81-222721, 1981.
13 J.R. Wessel, Food and Drug Administration (U.S.), *Pesticides Analytical Manual,* Vol. I, FDA, Washington, DC, revised April, 1989.
14 S.T. Preston, Jr. and R. Pankratz, *A Guide to the Analysis of Pesticides by Gas Chromatography,* Preston, Niles, IL, 3rd Edn., 1981.
15 J.R. Wessel, Food and Drug Administration (U.S.), *Pesticide Analytical Manual,* Vol. II, FDA, Washington, DC, revised September, 1989.
16 A.S.Y. Chau and B.K. Afghan (Editors), *Analysis of Pesticides in Water,* Vols. 1/3, CRC Press, Boca Raton, FL, 1982.
17 S. Williams (Editor), *Official Methods of Analysis of the AOAC,* Association of Official Analytical Chemists, Arlington, VA, 1984.
18 H.P. Thier and H. Zeumer (Editors), *Manual of Pesticide Residue Analysis,* Vol. 1, VCH Verlagsgesellschaft, Weinheim, 1987.
19 *The Analysis of Agricultural Materials,* Ministry of Agriculture, Fisheries and Food, London, 3rd Edn., 1986.
20 R. Macrae (Editor), *HPLC in Food Analysis,* Academic Press, New York, 1988.
21 D.E. Bradway, E.M. Lores and T.R. Edgerton, *Residue Rev.,* 75 (1980) 51.
22 F.A. Gunther, *Residue Rev.,* 75 (1980) 113.
23 J. Hylin, *Residue Rev.,* 76 (1980) 203.
24 IUPAC, *Pure Appl. Chem.,* 53 (1981) 1039.

25 S. Khan, *Residue Rev.*, 84 (1982) 1.
26 R.W. Young, *Dev. Food Anal. Tech.*, 3 (1984) 145.
27 V.D. Adams, R.J. Watts and M.E. Pitts, *J. Water Pollut. Control Fed.*, 58 (1987) 449.
28 R.T. Meister (Editor), *Farm Chemicals Handbook*, Meister Publishing Co., Willoughby, OH, 1987.
29 B. Kratochvil, D. Wallace and J.K. Taylor, *Anal. Chem.*, 56 (1984) 113R.
30 B. Kratochvil and J. Peak, *Anal. Methods Pestic. Plant Growth Regul.*, 17 (1988) 1.
31 S.M. Walters, *Anal. Methods Pestic. Plant Growth Regul.*, 15 (1986) 67.
32 A. Ambrus and H.P. Thier, *Pure Appl. Chem.*, 58 (1986) 1035.
33 S. Forbes, *Anal. Chim. Acta*, 196 (1987) 75.
34 M.A. Luke and H.T. Masumoto, *Anal. Methods Pestic. Plant Growth Regul.*, 15 (1986) 161.
35 R.J. Hemmingway, *Pure Appl. Chem.*, 56 (1984) 1131.
36 L.D. Sawyer, *J. Assoc. Off. Anal. Chem.*, 65 (1982) 1122.
37 M.A. Luke, J.E. Froberg and H.T. Masumoto, *J. Assoc. Off. Anal. Chem.*, 58 (1975) 1020.
38 M.A. Luke, J.E. Froberg, G.M. Doose and H.T. Masumoto, *J. Assoc. Off. Anal. Chem.*, 64 (1981) 1187.
36 L.D. Sawyer, *J. Assoc. Off. Anal. Chem.*, 68 (1985) 64.
40 A. Ambrus, J. Lantos, E. Visi, I. Csatlos and L. Sarvari, *J. Assoc. Off. Anal. Chem.*, 64 (1981) 733.
41 P. Bottomley and P.G. Baker, *Analyst (London)*, 109 (1984) 85.
42 M.L. Hopper, *J. Assoc. Off. Anal. Chem.*, 64 (1981) 720.
43 W. Specht, Method S-19 and Method XII-6, *Deutsche Forschungsgemeinschaft (DFG), Ruckstandsanalytik von Pflanzenschutzmitteln*, 1.-8, Lieferung, Verlag Chemie, Weinheim, 1985.
44 G.L. LeBel and D.T. Williams, *J. Assoc. Off. Anal. Chem.*, 69 (1986) 451.
45 H. Steinwandter, *Anal. Methods Pestic. Plant Growth Regul.*, 17 (1989) 35.
46 H. Steinwandter, *Fresenius' Z. Anal. Chem.*, 313 (1982) 536; 314 (1983) 129; 316 (1983) 493; 322 (1985) 752.
47 H. Steinwandter, *Fresenius' Z. Anal. Chem.*, 312 (1982) 342.
48 A.M. Gillespie and S.M. Walters, *J. Assoc. Off. Anal. Chem.*, 67 (1984) 290.
49 M.A. Luke and G.M. Doose, *Bull. Environ. Contam. Toxicol.*, 32 (1984) 651.
50 P.A. Greve and H.A.G. Heusinkveld, *Med. Fac. Landbouww. Rijksuniv. Gent*, 46 (1981) 317.
51 B.G. Luke, J.C. Richards and E.F. Dawes, *Bull. Environ. Contam. Toxicol.*, 32 (1984) 651.
52 B.G. Luke and J.C. Richards, *J. Assoc. Off. Anal. Chem.*, 67 (1984) 902.
53 J. Mes and D.J. Davies, *Int. J. Environ. Anal. Chem.*, 19 (1985) 203.
54 S.L. Head and V.W. Burse, *Bull. Environ. Contam. Toxicol.*, 39 (1987) 848.
55 J.C. Peterson and P. Robinson, *J. Res. Natl. Bur. Stand. (U.S.)*, 93 (1988) 343.
56 B.G. Luke, *Anal. Methods Pestic. Plant Growth Regul.*, 17 (1989) 75.
57 T. Wehner and J.N. Seiber, *ACS Extended Abstract, Pest. 007*, 182nd National ACS Meeting, New York, 1981.
58 T.C. Thomas and Y.A. Nishioka, *Bull. Environ. Contam. Toxicol.*, 35 (1985) 460.
59 I. Saito, N. Hisanga, M. Goto, T. Matsumoto and Y. Takeuchi, *Bull. Environ. Contam. Toxicol.*, 37 (1986) 664.
60 J.E. Woodrow, M.S. Majewski and J.N. Seiber, *J. Environ. Sci. Health, Part B*, B21 (1986) 143.
61 R.A. Moore and F.W. Karasak, *Int. J. Environ. Anal. Chem.*, 17 (1984) 187.
62 G.L. LeBel, D.T. Williams and F.M. Benoit, *Adv. Chem. Ser.*, 214 (1987) 309.
63 C. Leuenberger and J.F. Pankow, *Anal. Chem.*, 56 (1984) 2518.
64 J.F. Pankow, M.P. Ligocki, M.E. Rosen, L.M. Isabelle and K.M. Hart, *Anal. Chem.*, 60 (1988) 40.
65 G.A. Junk and J.J. Richard, *Anal. Chem.*, 60 (1988) 451.
66 G.A. Junk and J.J. Richard, *J. Res. Natl. Bur. Std. (U.S.)*, 93 (1988) 274.
67 P.W. Kohler and S.Y. Su, *Chromatographia*, 21 (1986) 531.

68 W.P. Cochrane, *Chem. Deriv. Anal. Chem.*, 1 (1981) 1.
69 M. Chiba, *Pestic. Sci. Biotechnol., Proc. Int. Cong. Pestic. Chem., 6th 1986 (Pub. 1987)*, p. 337.
70 K. Ramsteiner, *Int. J. Environ. Anal. Chem.*, 25 (1986) 49.
71 F.M. Gretch and J.D. Rosen, *J. Assoc. Off. Anal. Chem.*, 70 (1987) 109.
72 M.L. Hopper and K.R. Griffitt, *J. Assoc. Off. Anal. Chem.*, 70 (1987) 724.
73 I. Laws and N. Jones, *Int. Anal.*, 1 (1987) 30.
74 M.E.P. McNally and J.R. Wheeler, *J. Chromatogr., 435 (1988) 63.*
75 M.E.P. McNally and J.R. Wheeler, *J. Chromatogr.*, 447 (1988) 53.
76 P. Capriel, A. Haisch and Sh.U. Khan, *J. Agric. Food Chem.*, 34 (1986) 70.
77 W. Specht and H.P. Thier, Method S-10 and T. Stijve and H.P. Thier, Method S-9, cited in Reference 43.
78 J.L. Daft, *Anal. Chem.*, 56 (1984) 2687.
79 J.E. Froberg and G.M. Doose, *Anal. Methods Pestic. Plant Growth Regul.*, 14 (1986) 41.
80 F.I. Onuska, *J. High Resolut. Chromatogr. Chromatogr. Commun.*, 7 (1984) 660.
81 F.I. Onuska and S. Davies, *Chromatogr. Forum*, 1 (1986) 45.
82 L. Zenon-Roland, R. Agneessens, P. Nangniot and H. Jacobs, *J. High Resolut. Chromatogr. Chromatogr. Commun.*, 7 (1984) 480.
83 B.D. Ripley and H.E. Braun, *J. Assoc. Off. Anal. Chem.*, 66 (1983) 1084.
84 G. Zweig, *Anal. Methods Pestic. Plant Growth Regul.*, 14 (1986) 75.
85 N.V. Fehringer and S.M. Walters, *J. Assoc. Off. Anal. Chem.*, 67 (1984) 91.
86 N.V. Fehringer and S.M. Walters, *J. Assoc. Off. Anal. Chem.*, 69 (1986) 90.
87 H.J. Stan and H.M. Mueller, *J. High Resolut. Chromatogr. Chromatogr. Commun.*, 11 (1988) 140.
88 M.L. Hopper, *J. High Resolut. Chromatogr. Chromatogr. Commun.*, 10 (1987) 620.
89 H.J. Stan and H. Goebl, *J. Chromatogr.*, 314 (1984) 413.
90 H. Steinwandter, *Fresenius' Z. Anal. Chem.*, 304 (1980) 137.
91 O. Wust and W. Meier, *Z. Lebensm.-Unters.- Forsch.*, 177 (1983) 25.
92 H. Roseboom and H.A. Herbold, *J. Chromatogr.*, 202 (1980) 431.
93 S. Nitz, P. Moza and F. Korte, *J. Agric. Food Chem.*, 30 (1982) 593.
94 R. Deleu and A. Copin, *J. High Resolut. Chromatogr. Chromatogr. Commun.*, 3 (1980) 299.
95 K. Grob, Jr., *J. Chromatogr.*, 208 (1981) 217.
96 T.A. Wehner and J.N. Seiber, *J. High Resolut. Chromatogr. Chromatogr. Commun.*, 4 (1981) 348.
97 P.G. Baker and P. Bottomley, *Analyst (London)*, 107 (1982) 206.
98 T. Spitzer and G. Nickless, *J. High Resolut. Chromatogr. Chromatogr. Commun.*, 4 (1981) 151.
99 W. Gilsbach and H.P. Thier, *Z. Lebensm.-Unters.-Forsch.*, 175 (1982) 327.
100 M. Wolf, R. Deleu and A. Copin, *J. High Resolut. Chromatogr. Chromatogr. Commun.*, 4 (1981) 346.
101 H.J. Stan, *CLB, Chem. Labor. Betr.*, 35 (1984) 284.
102 H.J. Stan, *Lebensmittelchem. Gerichtl. Chem.*, 42 (1988) 31.
103 L.G.M.T. Tuinstra, W.A. Traag, A.J. van Munsteren and V. van Hese, *J. Chromatogr.*, 395 (1987) 307.
104 W.J. Trotter, *Int. J. Environ. Anal. Chem.*, 30 (1987) 299.
105 R. Deleu and A. Copin, *J. High Resolut. Chromatogr. Chromatogr. Commun.*, 7 (1984) 338.
106 J. Sherma, *Manual of Analytical Methods for the Analysis of Pesticides in Human and Environmental Samples*, EPA-600/8-80-038, June, 1980, available from NTIS.
107 J. Sherma, *Manual of Analytical Quality Control for Pesticides and Related Compounds*, EPA-600/2-81-059, April, 1981, available from NTIS.
108 S.M. Prinsloo and P.R. DeBeer, *J. Assoc. Off. Anal. Chem.*, 68 (1985) 1100.
109 S.M. Prinsloo and P.R. DeBeer, *J. Assoc. Off. Anal. Chem.*, 70 (1987) 878.
110 W.L. Saxton, *J. Chromatogr.*, 393 (1987) 175.
111 J.F. Suprock and J.H. Vinopal, *J. Assoc. Off. Anal. Chem.*, 70 (1987) 1014.

112 T. Cairns and E.G. Siegmund, *Anal. Methods Pestic. Plant Growth Regul.,* 14 (1986) 193.
113 L.O. Ruzo and W.M. Draper, *Anal. Methods Pestic. Plant Growth Regul.,* 14 (1986) 133.
114 R.E. Clement and H.M. Tosine, *Mass Spectrom. Rev.,* 7 (1988) 593.
115 E.A. Stemmler and R.A. Hites, *Anal. Chem.,* 57 (1985) 684.
116 J.J. Stamp, E.G. Siegmund and T. Cairns, *Anal. Chem.,* 58 (1986) 873.
117 J.P.G. Wilkins, A.R.C. Hill and D.F. Lee, *Analyst (London),* 110 (1985) 1045.
118 M. Oehme, S. Manoe and W. Thomas, *Fresenius' Z. Anal. Chem.,* 321 (1985) 655.
119 H. Prigge and K. Naumann, *Fresenius' Z. Anal. Chem.,* 320 (1985) 704.
120 R. Vaz and G. Blomkvist, *Chemosphere,* 14 (1985) 223.
121 D.L. Swackhamer, M.J. Charles and R.A. Hites, *Anal. Chem.,* 59 (1987) 913.
122 J.A. Laramee, B.C. Arbogast and M.L. Deinzer, *Anal. Chem.,* 60 (1988) 1937.
123 P.G. Nielsen, *Biomed. Mass Spectrom.,* 12 (1985) 695.
124 E.A. Stemmler and R.A. Hites, *Biomed. Mass Spectrom.,* 14 (1987) 417.
125 P. Begley and B.E. Foulger, *J. Chromatogr.,* 438 (1988) 45.
126 J.S.M. De Wit, C.E. Parker, K.B. Torner and J.W. Jorgenson, *Biomed. Environ. Mass Spectrom.,* 16 (1988) 47.
127 K. Jaklin, P. Krenmayr, K. Varmuza, W. Heegemann and W. Landvoigt, *Fresenius' Z. Anal. Chem.,* 330 (1988) 704.
128 Y. Tondeur, G.W. Sovocool, R.K. Mitchum, W.J. Niederhut and J.R. Donnelly, *Biomed. Environ. Mass Spectrom.,* 14 (1987) 733.
129 K.S. Kalasinsky, *Anal. Methods Pestic. Plant Growth Regul.,* 17 (1989) 101.
130 D.F. Gurka, M. Umana, E.D. Pellizzari, A. Moseley and J.A. De Haseth, *Appl. Spectrosc.,* 39 (1985) 297.
131 T.T. Holloway, B.J. Fairless, C.E. Freidine, H.E. Kimball, R.D. Kloepfer, C.J. Wurrey, L.A. Jonoby and H.G. Palmer, *Appl. Spectrosc.,* 42 (1988) 359.
132 D.F. Gurka and R. Titus, in W.R. Lang (Editor), *Anal. Chem. Instrum., Proc. Conf. Anal. Chem. Energy Technol., 28th Meeting, 1985,* published 1987, Lewis, Chelsea, MI, pp. 17-23.
133 J. Lipinski and H.J. Stan, *J. Chromatogr.,* 441 (1988) 213.
134 J.E. Dierkes, *J. Res. Natl. Bur. Stand. (U.S.),* 93 (1988) 298.
135 H.J. Stan, *Anal. Methods Pestic. Plant Growth Regul.,* 17 (1989) 167.
136 P. De Voogt, J.C. Klamer and H. Govers, *J. Chromatogr.,* 363 (1986) 407.
137 E.E. Hargesheimer, *J. Assoc. Off. Anal. Chem.,* 67 (1984) 1067.
138 D. Rinne and H. Groh, *Fresenius' Z. Anal. Chem.,* 322 (1985) 462.
139 United Kingdom Dept. of the Environment, *Methods Exam. Waters Assoc. Mater.,* 1985 (Determ. Organochlorine, Insec. Polychlorinated Biphenyls Sewage Sludges, Muds Fish 1978), 29-31.
140 M.P. Seymour and T.M. Jeffries, *Analyst (London),* 112 (1987) 427.
141 L. Torreli, A. Simonella, A. Falgiani, C. Filipponi and F. Gramenzi, *J. High Resolut. Chromatogr. Chromatogr. Commun.,* 10 (1987) 510.
142 M. Suzuki and M. Morimoto, *J. High Resolut. Chromatogr. Chromatogr. Commun.,* 9 (1986) 692.
143 E. Noroozian, F.A. Maris, M.W.F. Nielsen, R.W. Frei, G.J. de Jong and U.A.T. Brinkman, *J. High Resolut. Chromatogr. Chromatogr. Commun.,* 10 (1987) 17.
144 J.A. Ault, T.E. Spurgeon, D.S. Gillard and E.T. Mallinson, *J. Assoc. Off. Anal. Chem.,* 68 (1985) 941.
145 K. Boek, *Lebensmittelchem. Gerichtl. Chem.,* 38 (1984) 117.
146 H.P. Thier and T. Stijve, *Lebensmittelchem. Gerichtl. Chem.,* 40 (1986) 73.
147 H.P. Thier, *Lebensmittelchem. Gerichtl. Chem.,* 38 (1984) 111.
148 J.D. Millar, R.E. Thomas and H.J. Schattenberg, Report 1984, EPA-600/4-84-061; order number PB84-211358, 197 pp.
149 A.L. Alford-Stevens, J.W. Eichelberger and W.L. Budde, *Environ. Sci. Technol.,* 22 (1988) 304.
150 F.C. Garner, M.T. Homsher and J.G. Pearson, *ASTM Spec. Tech. Publ.,* 925 (1986) 132.

151 American Society for Testing and Materials, *ASTM Standard D 3086-85,* 1985.

152 T.A. Amin and R.S. Narang, *Anal. Chem.,* 57 (1985) 648.

153 R. Reding, *J. Chromatogr.,* 25 (1987) 338.

154 P.J. Marsden and J. Pearson, *ASTM Spec. Tech. Publ.,* 925 (1986) 198.

155 V. Lopez-Avila, S. Schoen, J. Milanes and W.F. Beckert, *J. Assoc. Off. Anal. Chem.,* 71 (1988) 375.

156 United Kingdom Dept. of the Environment, *Methods Exam. Waters Assoc. Mater.,* 1986 (Chlorobenzenes Water, Organochlorine Pestic. PCBs Turbid Waters, Halogenated Solvents Relat. Compd. Sewage Sludge Waters 1985), 44 pp.

157 W.A. Korfmacher, L.G. Rushing, P.H. Sitonen, C.J. Branscomb and C.L. Holder, *J. High Resolut. Chromatogr. Chromatogr. Commun.,* 10 (1987) 332.

158 F.M. Dunnivant and A.W. Elzerman, *J. Assoc. Off. Anal. Chem.,* 71 (1988) 551.

159 D.G. Patterson, Jr., L. Hampton, C.R. Lapeza, Jr., W.T. Belser, V. Green, L. Alexander and L.L. Needham, *Anal. Chem.,* 59 (1987) 2000.

160 R.A. Niemann, *J. Assoc. Off. Anal. Chem.,* 69 (1986) 976.

161 J.S. Stanley, T.M. Sack, Y. Tondeur and W.F. Beckert, *Biomed. Environ. Mass Spectrom.,* 16 (1988) 27.

162 R.W. Storherr, P. Ott and R.R. Watts, *J. Assoc. Off. Anal. Chem.,* 54 (1971) 513.

163 I. Levi and T.W. Nowicki, *J. Assoc. Off. Anal. Chem.,* 57 (1974) 924.

164 Committee for Analytical Methods for Residues of Pesticides and Veterinary Products in Foodstuffs and the Working Party on Pesticide Residues of the Ministry of Agriculture, Fisheries and Food (U.K.), *Analyst (London),* 102 (1977) 858.

165 M.C. Bowman, in H.A. Moye (Editor), *Chemical Analysis (Analysis of Pesticide Residues),* Volume 58, Wiley-Interscience, New York, 1981, p. 263.

166 G.J. Sharp, J.G. Bryan, S. Dilli, P.R. Haddad and J.M. Desmarchelier, *Analyst (London),* 113 (1988) 1493.

167 J.F. Lawrence, *Int. J. Environ. Anal. Chem.,* 29 (1987) 289.

168 K. Sasaki, T. Suzuki and Y. Saito, *J. Assoc. Off. Anal. Chem.,* 70 (1987) 450.

169 E.M. Lores, J.C. Moore and P. Moody, *Chemosphere,* 16 (1987) 1065.

170 M.L. Hopper, *J. Assoc. Off. Anal. Chem.,* 71 (1988) 731.

171 J.R. Ferreira and A.M.S. Silva Fernandes, *J. Assoc. Off. Anal. Chem.,* 63 (1980) 517.

172 K. Adachi, N. Ohokuni and T. Mitsuhashi, *J. Assoc. Off. Anal. Chem.,* 67 (1984) 798.

173 O. Nishijima, *Noyaku Kensasho Hokoku,* 24 (1984) 35.

174 Committee for Analytical Methods for Residues of Pesticides and Veterinary Products in Foodstuffs and the Working Party on Pesticide Residues of the Ministry of Agriculture, Fisheries and Food (U.K.), *Analyst (London),* 110 (1985) 765.

175 J. Kjoelholt, *J. Chromatogr.,* 325 (1985) 231.

176 J.J. Blaha and P.J. Jackson, *J. Assoc. Off. Anal. Chem.,* 68 (1985) 1095.

177 A. Neicheva, E. Kovacheva and G. Marudov, *J. Chromatogr.,* 437 (1988) 249.

178 H. Nijhuis and C. Ewers, *Milchwissenschaft,* 41 (1986) 133.

179 J.S. Warner, T.M. Engel and P.J. Mondron, *U.S. Eviron. Prot. Agency, Off. Res. Dev., [Rep.] EPA,* EPA/600/4-85/016, Apr. 1985, p. 66.

180 United Kingdom Dept. of the Environment (U.K.), *Methods Exam. Waters Assoc. Mater.,* 1986 (Organophosphorus Pestic. Sewage Sludge; Organophosphorus Pestic. River Drinking Water, Addit. 1985), 20 pp.

181 B.E. Wallbank, *J. Chromatogr.,* 208 (1981) 305.

182 S. Brauckhoff and H.P. Thier, *Z. Lebensm.- Unters.-Forsch.,* 184 (1987) 91.

183 G.M. Richardson and S.U. Qadri, *J. Agric. Food Chem.,* 35 (1987) 877.

184 R.C. Hall and D.E. Harris, *J. Chromatogr.,* 169 (1979) 245.

185 B.D. McGarvey, M. Chiba and T.H.A. Olthof, *J. Assoc. Off. Anal. Chem.,* 68 (1985) 753.

186 B.W. Wright and R.D. Smith, *J. High Resolut. Chromatogr. Chromatogr. Commun.,* 9 (1986) 73.

187 H.B. Lee, Y.D. Stokker and A.S.Y. Chau, *J. Assoc. Off. Anal. Chem.,* 69 (1986) 557.

188 A.J. Cessna, R. Grover, L.A. Kerr and M.L. Aldred, *J. Agric. Food Chem.,* 33 (1985) 504.

189 H.B. Lee and Y.D. Stokker, *J. Assoc. Off. Anal. Chem.,* 69 (1986) 568.

190 M. Grandet, L. Weil and K.E. Quentin, *Z. Wasser Abwasser Forsch.*, 21 (1988) 21.
191 J. Tekel, P. Farkas, K. Schultzova, J. Kovacicova and A. Szokolay, *Z. Lebensm.-Unters.-Forsch.*, 186 (1988) 319.
192 L. Ogierman, *Fresenius' Z. Anal. Chem.*, 320 (1985) 365.
193 L. Ogierman, *J. Assoc. Off. Anal. Chem.*, 69 (1986) 912.
194 T. Tamiri and S. Zitrin, *Biomed. Environ. Mass Spectrom.*, 14 (1987) 39.
195 W. Ebing, G. Richtarsky, K. Boeck, M. Eichner, K. Kypke-Hutter, B. Fatteroll, R. Oberdieck, W. Gilsbach and N. Thi Hanah, *Lebensmittelchem. Gerichtl. Chem.*, 39 (1985) 126.
196 S.M. Waliszewski and G.A. Szymczynski, *Fresenius' Z. Anal. Chem.*, 322 (1985) 510.
197 R. Agnessens, L. Zenon-Roland and P. Nangnoit, *Meded. Fac. Landbouwwet., Rijksuniv. Gent.*, 49 (1984) 1241.
198 H.A. Meemken, P. Rudolph and P. Fuerst, *Dtsch. Lebensm.-Rundsch.*, 83 (1987) 239.
199 D.F. Gurka, F.L. Shore, S.T. Pan and E.N. Amick, *J. Assoc. Off. Anal. Chem.*, 69 (1986) 970.
200 A.H. Ahmed, B. Sarrasin and V.N. Mallet, *Anal. Chem.*, 59 (1987) 1302.
201 American Society for Testing and Materials, *ASTM Standard*, D3478-85 (1985).
202 J.S. Warner, T.M. Engel and P.J. Mondron, *U.S. Environ. Prot. Agency, Off. Res. Dev., [Rep.] EPA*, EPA/600/4-85/017, Apr. 1985, p. 60.
203 E. Jensen, A. Ernsten and G. Sandberg, *Plant Growth Regul.*, 4 (1986) 55.
204 P.C. Bardalaye and W.B. Wheeler, *J. Assoc. Off. Anal. Chem.*, 68 (1985) 750.
205 P.C. Bardalaye, W.B. Wheeler, C.W. Meister and J.L. Templeton, *Food. Addit. Contam.*, 2 (1985) 283.
206 J.M. van der Poll, M. Vink and J.K. Quirijns, *Chromatographia*, 25 (1988) 511.
207 B.L. Worobey and W.F. Sun, *Chemosphere*, 16 (1987) 1457.
208 I. Ahmad, *J. Assoc. Off. Anal. Chem.*, 70 (1987) 745.
209 H. Steinwandter, *Fresenius' Z. Anal. Chem.*, 327 (1987) 363.
210 W.H. Newsome, *Anal. Methods Pestic. Plant Growth Regul.*, 11 (1980) 197.
211 E. Lemperle, *GIT Fachz.*, 31 (1987) 281.
212 T.M. Engel, J.S. Warner and W.M. Cooke, *Report 1985*, EPA/600/4-85/072, Order No. PB86-118726/GAR, 58 pp.
213 T.K. McGhie and P.T. Holland, *Analyst (London)*, 112 (1987) 1075.
214 J.S. Warner, T.M. Engel and P.J. Mondron, *U.S. Environ. Prot. Agency, Off. Res. Dev., [Rep.] EPA*, EPA/600/4-85/018, Apr. 1985, p. 60.
215 M.B. Taccheo, C. Spessotto, B. Bresin and L. Bagarolo, *Pestic. Sci.*, 15 (1984) 612.
216 M .T. Lafuente and J.L. Tadeo, *Fresenius' Z. Anal. Chem.*, 328 (1987) 105.
217 E.A. Hogendoorn, C.E. Goewie, H.H. van der Broek and P.A. Greve, *Meded. Fac. Landbouwwet., Rijkuniv. Gent*, 49 (1984) 1219.
218 R. Brennecke and K. Vogeler, *Pflanzenschutz- Nachr.*, 37 (1984) 46.
219 N.A. Smart, *Analyst (London)*, 112 (1987) 1559.
220 T. Clark and A.H.B. Deas, *J. Chromatogr.*, 329 (1985) 181.
221 R.S. Burden, A.H.B. Deas and T. Clark, *J. Chromatogr.*, 391 (1987) 273.
222 J. Chovancova, E. Matisova and V. Batora, *J. Assoc. Off. Anal. Chem.*, 68 (1985) 741.
223 H. Steinwandter, *Fresenius' Z. Anal. Chem.*, 321 (1985) 599.
224 J.J. Sullivan, J.D. Torkelson, M.M. Wekell, T.A. Hollingsworth, W.L. Saxton, G.A. Miller, K.W. Panaro and A.D. Uhler, *Anal. Chem.*, 60 (1988) 626.
225 J. Miyamoto, *Pure Appl. Chem.*, 53 (1981) 1967.
226 E. Papadopoulou-Mourkidou, *Residue Rev.*, 89 (1983) 179.
227 E. Papadopoulou-Mourkidou, *Anal. Methods Pestic. Plant Growth Regul.*, 16 (1988) 179.
228 M.A. Saleh, A.M. Marei and J.E. Casida, *J. Agric. Food Chem.*, 28 (1980) 592.
229 L. Ogierman and A. Slilowiecki, *Chromatographia*, 14 (1981) 459.
230 R.A. Simonaitis and R.S. Cail, *Chromatographia*, 18 (1984) 556.
231 S. Forbes and A.J. Dutton, *Pestic. Sci.*, 16 (1985) 404.
232 R. Mestres and H. Susilo, *Trav. Soc. Pharm. Montpellier*, 40 (1980) 277.

233 H.M. Akhtar, *J. Chromatogr.*, 246 (1982) 81.
234 R.S. Greenberg, *J. Agric. Food Chem.*, 29 (1981) 856.
235 S.J. Cave, *Pestic. Sci.*, 12 (1981) 156.
236 M.D. Awasthi, *J. Food Sci. Technol.*, 22 (1985) 4.
237 H.E. Braun and K. Stanek, *J. Assoc. Off. Anal. Chem.*, 65 (1982) 685.
238 T. Okadu, M. Uno, M. Nozawa and K. Tanigawa, *Shokuhin Eiseigaku Zasshi*, 24 (1983) 147.
239 R.A. Simonaitis and R.S. Cali, in M. Eliot (Editor), *ACS Symposium Series No. 42*, American Chemical Society, Washington, DC, 1977, p. 211.
240 B.S. Joia, L.P. Sarna and G.R.B. Webster, *Int. J. Environ. Anal. Chem.*, 21 (1985) 179.
241 S.H. Kennedy, *Int. Lab.*, 17 (1987) 44.
242 R.A. Chapman and C.R. Harris, *J. Chromatogr.*, 174 (1979) 369.
243 R.O. Lidgard, A.M. Duffield and R.J. Wells, *Biomed. Environ. Mass Spectrom.*, 13 (1986) 677.
244 K.A. Scudamore, *Anal. Methods Pestic. Plant Growth Regul.*, 16 (1988) 207.
245 J.W. DeVries, P.A. Larson, R.H. Bowers, H. Raymond, J.A. Keating, J.M. Broge, P.S. Wehling, H.H. Patel, H. Hasmukh and J.W. Zurawski, *J. Assoc. Off. Anal. Chem.*, 68 (1985) 759.
246 T.G. Alleman, R.A. Sanders and B.L. Madison, *J. Assoc. Off. Anal. Chem.*, 69 (1986) 575.
247 D.L. Heikes and M.L. Hopper, *J. Assoc. Off. Anal. Chem.*, 69 (1986) 990.
248 D.B. Page, W.H. Newsome and S.B. MacDonald, *J. Assoc. Off. Anal. Chem.*, 70 (1987) 446.
249 K.A. Scudamore and G. Goodship, *Pestic. Sci.*, 17 (1986) 385.
250 K.A. Scudamore, *Pestic. Sci.*, 18 (1987) 33.
251 L.G.M. Tuinstra, W.A. Traag and A.H. Roos, *J. High Resolut. Chromatogr. Chromatogr. Commun.*, 11 (1988) 106.
252 J.W. DeVries, J.M. Broge, J.P. Schroeder, R.H. Bowers, P.A. Larson and N.M. Burns, *J. Assoc. Off. Anal. Chem.*, 68 (1985) 1112.
253 J. Delventhal, *Lebensmittelchem. Gerichtl. Chem.*, 38 (1984) 42.
254 J. Gilbert, J.R. Startin and C. Crews, *Food Addit. Contam.*, 2 (1985) 55.
255 J.L. Daft, *J. Assoc. Off. Anal. Chem.*, 70 (1987) 734.
256 J.L. Daft, *J. Assoc. Off. Anal. Chem.*, 71 (1988) 748.
257 J.L. Daft, *J. Assoc. Off. Anal. Chem.*, 66 (1983) 228.
258 T. Stijve and H.P. Thier, *Lebensmittelchem. Gerichtl. Chem.*, 39 (1985) 125.
259 L.D. Sawyer and S.M. Walters, *J. Assoc. Off. Anal. Chem.*, 69 (1986) 847.
260 Association of Official Analytical Chemists, *Official Methods of Analysis*, 13th Edn., Sections 29.056-29.057, 1980.
261 K. Hunter, *Anal. Methods Pestic. Plant Growth Regul.*, 16 (1988) 119.
262 C.M. White and R.K. Houck, *J. High Resolut. Chromatogr. Chromatogr. Commun.*, 9 (1986) 3.
263 B.W. Wright and P.D. Smith, *J. High Resolut. Chromatogr. Chromatogr. Commun.*, 9 (1986) 73.
264 H. Frehse, *Pestic. Sci. Biotechnol., Proc. Int. Congr. Pestic. Chem., 6th 1986*, (1987) 293.
265 D.E. Knowles, B.E. Richter, M.B. Wygant, L. Nixon and M.R. Andersen, *J. Assoc. Off. Anal. Chem.*, 71 (1988) 451.
266 H.T. Kalinoski, H.R. Udseth, B.W. Wright and R.D. Smith, *J. Chromatogr.*, 400 (1987) 307.
267 H.T. Kalinoski, B.W. Wright and R.D. Smith, *Biomed. Environ. Mass Spectrom.*, 13 (1986) 33.
268 H.T. Kalinoski and R.D. Smith, *Anal. Chem.*, 60 (1988) 529.
269 A.C. Barefoot and R.W. Reiser, *J. Chromatogr.*, 398 (1987) 217.
270 A.J. Berry, D.E. Games, I.C. Mylchreest, J.A. Perkins and S. Pleasance, *Biomed. Environ. Mass Spectrom.*, 15 (1988) 105.
271 K.E. Markides, E.D. Lee, R. Bolick and M.L. Lee, *Anal. Chem.*, 58 (1986) 740.

272 J.R. Wheeler and M.E. McNally, *J. Chromatogr.*, 410 (1987) 343 and *Fresenius' Z. Anal. Chem.*, 330 (1988) 237.
273 S. Kennedy and R.J. Wall, *LC-GC*, 6 (1988) 930.
274 S.B. Hawthorne and D.J. Miller, *J. Chromatogr.*, 403 (1987) 63.
275 D.A. Laude, Jr., S.L. Pentoney, P.R. Griffiths and C.L. Wilkins, *Anal. Chem.*, 59 (1987) 2283.
276 R.C. Wiebolt, K.D. Kempfert, D.W. Later and E.R. Campbell, *J. High Resolut. Chromatogr. Chromatogr. Commun.*, 12 (1989) 106.
277 J.F. Lawrence, *Anal. Methods Pestic. Plant Growth Regul.*, Academic Press, New York, Volume 12, 1982.
278 L. Muszkat and N. Aharonson, *Anal. Methods Pestic. Plant Growth Regul.*, 14 (1986) 95.
279 K.F. Ivie, *Anal. Methods Pestic. Plant Growth Regul.*, 11 (1980) 55.
280 P.H. Cramer, A.D. Drinkwine, J.E. Going and A.E. Carey, *J. Chromatogr.*, 235 (1982) 489.
281 I. Ahmed, *J. Environ. Sci. Health*, B18 (1983) 207.
282 F.A. Maris, R.B. Geerdink, R.W. Frei and U.A.T. Brinkman, *J. Chromatogr.*, 323 (1985) 113.
283 M.W.F. Nielen, G.Koomen, R.W. Frei and U.A.T. Brinkman, *J. Liq. Chromatogr.*, 8 (1985) 315.
284 C.E. Goewie, P. Kwakman, R.W. Frei, U.A.T. Brinkman, W. Maasfeld, T. Seshadri and A. Kettrup, *J. Chromatogr.*, 284 (1984) 73.
285 C.E. Goewie and E.A. Hogendoorn, *J. Chromatogr.*, 410 (1987) 211.
286 R. Hammann and A. Kettrup, *Chemosphere*, 16 (1987) 527.
287 W.T. Kok, U.A.T. Brinkman, R.W. Frei, H.B. Hanekamp, F. Nooitgedacht and H. Poppe, *J. Chromatogr.*, 237 (1982) 357.
288 D. Chaput, *J. Assoc. Off. Anal. Chem.*, 69 (1986) 985.
289 C.E. Goewie and E.A. Hoogendoorn, *J. Chromatogr.*, 404 (1987) 352.
290 I. Fogy, E.R. Schmid and J.F.K. Huber, *Z. Lebensm.-Unters.-Forsch.*, 170 (1980) 194.
291 P.A. Hargreaves and K.J. Melksham, *Pestic. Sci.*, 14 (1983) 347.
292 R.W. Frei, M.W.F. Nielsen and U.A.T. Brinkman, *Int. J. Environ. Anal. Chem.*, 25 (1986) 3.
293 H. Roseboom and P.A. Greve, in J. Miyamoto and P.C. Kearney (Editors), *Pesticide Chemistry: Human Welfare and the Environment*, Pergamon Press, New York, 1982, Vol. 4, pp. 1-9.
294 H.A. Moye and P.A. St John, in J. Harvey and G. Zweig (Editors), *Pesticide Analytical Methodology*, ACS 136, American Chemical Society, Washington, DC, 1980, p. 89.
295 J.N. Seiber, M.M. McChesney, R. Kon and A.R. Leavitt, *J. Agric. Food Chem.*, 32 (1984) 681.
296 G. Petrick, D.E. Schulz and J.C. Duinker, *J. Chromatogr.*, 435 (1988) 241.
297 K.A. Ramsteiner, *J. Chromatogr.*, 393 (1987) 123.
298 T. Noy, E. Weiss, T. Herps, H. Van Crutchen and J. Rijks, *J. High Resolut. Chromatogr. Chromatogr. Commun.*, 11 (1988) 181.
299 L. Grant Rice, *J. Chromatogr.*, 317 (1984) 523.
300 J.S.M. De Wit, C.E. Parker, K.B. Tomer and J.W. Jorgenson, *Anal. Chem.*, 59 (1987) 2400.
301 C. Drossel and G. Perez Herrera, *Lebensmittelchem. Gerichtl. Chem.*, 41 (1987) 7.
302 V.L. McGuffin and M. Novotney, *Anal. Chem.*, 55 (1983) 580.
303 F.A. Maris, A. van der Vliet, R.B. Geerdink and U.A.T. Brinkman, *J. Chromatogr.*, 347 (1985) 75.
304 R. Giebelmann, *Pharmazie*, 42 (1987) 44.
305 P.G. Baker, *Pestic. Sci. Biotechnol., Proc. Int. Congr. Pestic. Chem., 6th*, (1987) 329.
306 R.A. Chapman, *J. Chromatogr.*, 258 (1983) 175.
307 G.R. Cayley and B.W. Simpson, *J. Chromatogr.*, 356 (1986) 123.
308 M.D. Osselton and R.D. Snelling, *J. Chromatogr.*, 368 (1986) 265.

309 J.F. Lawrence and D. Turton, *J. Chromatogr.*, 159 (1978) 207.
310 A.R. Hanks and B.M. Colvin, in K.G. Das (Editor), *Pesticide Analysis*, Dekker, New York, Chapter 3, 1981.
311 T.A. Bellar and W.L. Budde, *Anal. Chem.*, 60 (1988) 2076.
312 E. Popadopoulou, Y. Iwata and F.A. Gunther, *J. Agric. Food Chem.*, 31 (1983) 629.
313 K.A. Ramsteiner and W.A. Hormann, *J. Chromatogr.*, 104 (1975) 438.
314 C.E. Goewie and E.A. Hogendoorn, *Food Addit. Contam.*, 2 (1985) 217.
315 J.F. Lawrence, *Chromatographia*, 24 (1987) 45.
316 K. Oulakand and F. Jonas, *J. Chromatogr.*, 396 (1987) 433.
317 E.G. Cotterill and T.H. Byast, in J.F. Lawrence (Editor), *Conference Proceeding Liquid Chromatography in Environmental Analysis*, Humana Press, Clifton, NJ, 1984.
318 A.N. Wilson and R.J. Bushway, *J. Chromatogr.*, 214 (1980) 140.
319 A. Di Corcia, M. Marchetti and R. Samperi, *J. Chromatogr.*, 405 (1987) 357.
320 P.A. Greve and C.E. Goewie, *Int. J. Environ. Anal. Chem.*, 20 (1985) 29.
321 H. Jansen, U.A.T. Brinkman and R.W. Frei, *Chromatographia*, 20 (1985) 453.
322 R.T. Krause, *J. Chromatogr. Sci.*, 18 (1978) 281.
323 R.T. Krause, *J. Chromatogr.*, 185 (1979) 615.
324 K. Hunter, *J. Chromatogr.*, 270 (1983) 267.
325 U.A.T. Brinkman, *Chromatographia*, 24 (1987) 190.
326 L.G.M. Tuinstra and P.G.M. Kienhuis, *Chromatographia*, 24 (1987) 696.
327 C.J. Miles and H.A. Moye, *Chromatographia*, 24 (1987) 628.
328 C.J. Miles and H.A. Moye, *Anal. Chem.*, 60 (1988) 220.
329 F.A. Maris, M. Nijenhuis, R.W. Frei, G.J. de Jong and U.A.T. Brinkman, *Chromatographia*, 22 (1986) 235.
330 U.A.T. Brinkman and F.A. Maris, *Trends Anal. Chem.*, 4 (1985) 55.
331 J.A. Apffel, U.A.T. Brinkman and R.W. Frei, *J. Chromatogr.*, 312 (1984) 153.
332 X.D. Ding and I.S. Krull, *J. Agric. Food Chem.*, 32 (1984) 622.
333 G.J. Clark, R.R. Goodin and J.W. Smiley, *Anal. Chem.*, 52 (1985) 2223.
334 M.B. Thomas and P.E. Sturrock, *J. Chromatogr.*, 357 (1986) 318.
335 J.L. Anderson, K.K. Whiten, J.D. Brewster, Ou Tse-Yuan and W.K. Nonidez, *Anal. Chem.*, 57 (1986) 1366.
336 J.L. Anderson and D.J. Chesney, *Anal. Chem.*, 52 (1980) 2156.
337 Q.G. Von Nehring, J.W. Hightower and J.L. Anderson, *Anal. Chem.*, 58 (1986) 2777.
338 D.S. Owens and P.E. Sturrock, *Anal. Chim. Acta*, 188 (1986) 269.
339 D. Barcelo, *Chromatographia*, 25 (1988) 928.
340 K. Slais and D. Kourilova, *J. Chromatogr.*, 258 (1983) 57.
341 S.M. Walters, *J. Chromatogr.*, 259 (1983) 227.
342 M.J. Cope and A. Townshend, *Anal. Chim. Acta*, 134 (1982) 93.
343 V.L. McGuffin and M. Novotney, *Anal. Chem.*, 55 (1983) 2296.
344 J.C. Gluckman, D. Barcelo, G.J. de Jong, R.W. Frei, F.A. Maris and U.A.T. Brinkman, *J. Chromatogr.*, 367 (1986) 35.
345 A. de Kok, R.B. Geerdink and U.A.T. Brinkman, *J. Chromatogr.*, 252 (1982) 101.
346 F.A. Maris, R.B. Geerdink, R. van Delft and U.A.T. Brinkman, *Bull. Environ. Contam. Toxicol.*, 35 (1985) 711.
347 F.A. Maris, M. Nijenhuis, R.W. Frei, G.J. de Jong and U.A.T. Brinkman, *J. Chromatogr.*, 435 (1988) 297.
348 E.A. Woolson and N. Aharonson, *J. Assoc. Off. Anal. Chem.*, 63 (1980) 523.
349 J. Harvey, Jr., in J. Harvey, Jr. and G. Zweig (Editors), *Pesticide Analytical Methodology*, ACS 136, American Chemical Society, Washington, DC, 1980, pp. 1-14.
350 P.C. White, *Analyst (London)*, 109 (1984) 697.
351 S. Wachholz, H. Geissler, G. Perner and J. Bleck, *Fresenius' Z. Anal. Chem.*, 329 (1988) 768.
352 R.D. Voyksner and C. A. Haney, *Anal. Chem.*, 57 (1985) 991.
353 D. Barcelo, *LC-GC*, 6 (1988) 324.
354 R.D. Voyksner, J.T. Bursey and E.D. Pellizzari, *Anal. Chem.*, 56 (1984) 1507.
355 D. Barcelo, *Chromatographia*, 25 (1988) 295.

356 R.D. Voyksner, J.T. Bursey and E.D. Pellizzari, *Finnigan MAT Newsletter,* Number 205, San Jose, CA, 1984.
357 C.A. Parker, A.V. Geeson, D. E. Games, E.D. Ramsey, E.O. Abusteit, F.T. Corbin and K.B. Tomer, *J. Chromatogr.,* 438 (1988) 359.
358 R.D. Voyksner, in J.D. Rosen (Editor), *Applications of New Mass Spectrometry Techniques in Pesticide Chemistry,* John Wiley and Sons, New York, 1987, Chapter 11.
359 L.M. Shalaby, *Biomed. Mass Spectrom.,* 12 (1985) 262.
360 K. Levsen, K.H. Shafer and J. Freudenthal, *J. Chromatogr.,* 271 (1983) 51.
361 R.B. Geerdink, F.A. Maris, G.J. de Jong, R.W. Frei and U.A.T. Brinkman, *J. Chromatogr.,* 394 (1987) 51.
362 T. Cairns, E.G. Sigmund and G.M Doose, *Biomed. Mass Spectrom.,* 10 (1983) 21.
363 K.D. White, Z. Min, W.C. Brumley, R.T. Krause and J.A. Sphon, *J. Assoc. Off. Anal. Chem.,* 66 (1983) 1358.
364 R.T. Rosen and J.E. Dziedzic, *Chem. Anal. (N.Y.),* 91 (1987) 176.
365 T. Cairns, E.G. Siegmund and G.M. Doose, *Biomed. Mass Spectrom.,* 10 (1983) 24.
366 C.E. Parker, C.A. Haney and J.R. Haas, *J. Chromatogr.,* 237 (1982) 233.
367 C.E. Parker, C.A. Haney, J.R. Harvan and J.R. Haas, *J. Chromatogr.,* 242 (1982) 77.
368 D. Barcelo, F.A. Maris, R.B. Geerdink, R.W. Frei, G. de Jong and U.A.T. Brinkman, *J. Chromatogr.,* 394 (1987) 65.
369 R.D. Voyksner, W.H. McFadden and S.A. Lammert, in J.D. Rosen (Editor), *Applications of New Mass Spectrometry Techniques in Pesticide Chemistry,* John Wiley and Sons, New York, 1987, Chapter 17.
370 R.D. Voyksner and T. Cairns, *Anal. Methods Pestic. Plant Growth Regul.,* 17 (1988) 119.
371 R. Reupert and E. Ploeger, *Fresenius' Z. Anal. Chem.,* 331 (1988) 503.
372 R.A. Hoodless, J.A. Sidwell, J.C. Skinner and R.D. Treble, *J. Chromatogr.,* 166 (1978) 279.
373 B. Ohlin, *Var Foda,* 38 (1986) 111.
374 R.T. Krause and T. August, *J. Assoc. Off. Anal. Chem.,* 66 (1983) 234.
375 F.H. Funch, *Z. Lebensm.-Unters.-Forsch.,* 173 (1981) 95.
376 P. Cabras, P. Diana, M. Meloni, F.M. Pirisi and R. Pirisi, *J. Chromatogr.,* 256 (1983) 176.
377 M.P. Seymour, T.M. Jeffries and L.J. Notarianni, *Analyst (London),* 111 (1986) 1203.
378 K.L. McDonald, *J. Chromatogr. Sci.,* 22 (1984) 293.
379 C.E. Goewie, R.J. Berkhof, F.A. Maris, M. Treskes and U.A.T. Brinkman, *Int. J. Environ. Anal. Chem.,* 26 (1986) 305.
380 E.W. Zahnow, *LC-GC,* 4 (1984) 644.
381 C.M. Selavka, K.S. Jiao, I.S. Krull, P. Sheih, W. Yu and M. Wolf, *Anal. Chem.,* 60 (1988) 250.
382 R. Saleh, A. El-Ahraf and W.V. Willis, *Vet. Med. J.,* 34 (1986) 5.
383 D. Barcelo, *Biomed. Environ. Mass Spectrom.,* 17 (1988) 363, 622.
384 G.J. Clark, R.R. Goodin and J.W. Smiley, *Anal. Chem.,* 57 (1985) 2223.
385 G. Blaicher, W. Pfannhauser and H. Woidich, *Chromatographia,* 13 (1980) 438.
386 R.T. Krause, *J. Assoc. Off. Anal. Chem.,* 68 (1985) 726.
387 R.T. Krause, *J. Assoc. Off. Anal. Chem.,* 63 (1980) 1114.
388 D. Chaput, *J. Assoc. Off. Anal. Chem.,* 71 (1988) 542.
389 A. De Kok, A. Hiemstra and C.P. Vreeker, *Chromatographia,* 24 (1987) 469.
390 R.T. Krause, *J. Chromatogr.,* 442 (1988) 333.
391 T.D. Spittler and R.A. Marafioti, *J. Chromatogr.,* 255 (1983) 191.
392 Y.W. Lee and N.A. Westcott, *J. Agr. Food Chem.,* 28 (1980) 719.
393 W.P. Cochrane and M. Lanouette, *J. Assoc. Off. Anal. Chem.,* 64 (1981) 1032.
394 T.J. Spierenburg, M.B.H. Kemmeren-van Dijk and P.E.F. Zoun, *J. Chromatogr.,* 393 (1987) 137.
395 H. Mori, M. Kobavashi, K. Yagi, M. Takahashi, T. Gondo and N. Umetsu, *Nippon Noyaku Gakkaishi,* 12 (1987) 491.

396 P. Jandera, L. Svoboda, J. Kubat, J. Schvantner and J. Churacek, *J. Chromatogr.,* 292 (1984) 71.
397 H. Roseboom, H.A. Herbold and C.J. Berkhoff, *J. Chromatogr.,* 249 (1982) 323.
398 A. Di Corcia, M. Marchetti and R. Samperi, *J. Chromatogr.,* 405 (1987) 357.
399 Y. Xu, W. Lorenz, G. Pfister, M. Bahalir and F. Korte, *Fresenius' Z. Anal. Chem.,* 325 (1986) 377.
400 D.S. Owens and P.E. Sturrock, *Anal. Chim. Acta,* 188 (1986) 269.
401 C. Malan, J.H. Visser and H.A. Van de Venter, *S. Afr. J. Plant Soil,* 3 (1986) 82.
402 I.G. Ferris and B.M. Haigh, *J. Chromatogr. Sci.,* 25 (1987) 170.
403 P. Jandera, J. Churacek, P. Butzke and M. Marz, *J. Chromatogr.,* 387 (1987) 155.
404 G. Chiavari and C. Bergamini, *J. Chromatogr.,* 346 (1985) 369.
405 R.G. Luchtefeld, *J. Chromatogr.,* 23 (1985) 516.
406 R.G. Luchtefeld, *J. Assoc. Off. Anal. Chem.,* 70 (1987) 740.
407 M.W.F. Nielen, A.J. Valk, R.W. Frei, U.A.T. Brinkman, P. Mussche, R. de Nijs, B. Ooms and W. Smink, *J. Chromatogr.,* 393 (1987) 69.
408 A.J. Cessna, *J. Liq. Chromatogr.,* 11 (1988) 725.
409 G. Guinn, D.L. Brummett and R.C. Beier, *Plant Physiol.,* 81 (1986) 997.
410 J.M. Anderson, *J. Chromatogr.,* 330 (1985) 347.
411 W. Ternes and H.A. Ruessel-Sinn, *Fresenius' Z. Anal. Chem.,* 326 (1987) 757.
412 J.W. Dornseiffen and J.W. Verwaal, *Meded. Fac. Landbouwwet., Rijksuniv. Gent,* 50 (1985) 867.
413 R.T. Kon, L. Geissel and R.A. Leavitt, *Food Addit. Contam.,* 1 (1984) 67.
414 R. Binner and U. Banasiak, *Z. Gesamte Hyg. Ihre Grenzgeb.,* 31 (1985) 404.
415 D.R. Lauren and M.P. Agnew, *J. Chromatogr.,* 303 (1984) 206.
416 R. Binner, U. Banasiak and M. Giltschka, *Z. Gesamte Hyg. Ihre Grenzgeb.,* 31 (1985) 565.
417 E.W. Zahnow, *J. Agric. Food Chem.,* 35 (1987) 403.
418 E. Jenses, A. Crozier and A.M. Monteiro, *J. Chromatogr.,* 367 (1986) 377.
419 C.J. Miles, L.R. Wallace and H.A. Moye, *J. Assoc. Off. Anal. Chem.,* 69 (1986) 458.
420 J.E. Cowell, J.L. Kuntsman, P.J. Nord, J.R. Steinmetz and G.R. Wilson, *J. Agric. Food Chem.,* 34 (1986) 955.
421 L.G.M.T. Tuinstra and P.G.M. Kienhuis, *Chromatographia,* 24 (1987) 696.
422 M. Aklyama, N. Sakurai and S. Kuraishi, *Plant Cell Physiol.,* 24 (1983) 1431.
423 M.A. Alawi, *Fresenius' Z. Anal. Chem.,* 319 (1984) 524.
424 E. A. Queree, S.J. Dickson and S.M. Shaw, *J. Anal. Toxicol.,* 9 (1985) 10.
425 T. Nagayama, T. Maki, K. Kan, M. Iida and T. Nashima, *J. Assoc. Off. Anal. Chem.,* 70 (1987) 1008.
426 E.W. Zahnow, *J. Agric. Food Chem.,* 33 (1985) 1206.
427 A. Pena Heras and F. Sanchez-Rasero, *J. Chromatogr.,* 358 (1986) 302.
428 J.L. Prince and R.A. Guinivan, *J. Agric. Food Chem.,* 36 (1988) 63.
429 M. Arjmand, T.D. Spittler and R.O. Mumma, *J. Agric. Food Chem.,* 36 (1988) 492.
430 D.R. Lauren, H. Taylor and A. Rahman, *J. Chromatogr.,* 439 (1988) 470.
431 M. Schmidt, R. Hamann and A. Kettrup, *Int. J. Environ. Anal. Chem.,* 33 (1988) 1.
432 S.D. West and E. W. Day, Jr., *J. Agric. Food Chem.,* 36 (1988) 53.
433 M. Negre, M. Gennari and A. Cignetti, *J. Chromatogr.,* 387 (1987) 541.
434 M. Patumi, C. Marucchini, M. Businelli and C. Vischetti, *Pestic. Sci.,* 21 (1987) 193.
435 C.E. Parker, A.V. Geeson, D.E. Games, E.D. Ramsey, E.O. Abusteit, F.T. Corbin and K.B. Toner, *J. Chromatogr.,* 438 (1988) 359.
436 B.L. Worobey and B.J. Shields, *J. Assoc. Off. Anal. Chem.,* 70 (1987) 1021.
437 K. Isshiki, S. Tsumura and T. Watanabe, *J. Assoc. Off. Anal. Chem.,* 63 (1980) 747.
438 W. Van Haver, *Z. Lebensm.-Unters.-Forsch.,* 172 (1981) 1.
439 C.F. Aten, J.B. Bourke and R.A. Marafioti, *J. Agric. Food Chem.,* 30 (1982) 610.
440 K. Nakashimi, T. Nakagawa and S. Era, *Shokuhin Eiseigaku Zasshi,* 22 (1981) 233.
441 F. Burzi Fiori, A. Berri and M. Zacchetti, *Riv. Soc. Ital. Sci. Aliment.,* 10 (1981) 19.
442 B. Buttler and W. D . Hormann, *J. Agric. Food Chem.,* 29 (1981) 257.
443 P.G. Baker and P.G. Clarke, *Analyst (London),* 109 (1984) 81.
444 C. Paolo, *J. Chromatogr.,* 256 (1983) 176.

445 B. Luckas, *Z. Lebensm.-Unters.-Forsch.*, 184 (1987) 195.
446 G. Caccialanza, C. Gandini, C. Roggi and E. Zecca, *Farmaco*, 36 (1981) 73.
447 R.C. Massey, P.E. Key and J. McWeeny, *J. Chromatogr.*, 240 (1982) 254.
448 R.T. Krause and Y. Wang, *J. Liq. Chromatogr.*, 11 (1988) 349.
449 J.L. Tadeo and M.T. Lafuente, *J. Chromatogr.*, 391 (1987) 338.
450 F.G.P. Mullins and G.F. Kirkbright, *Analyst (London), 112 (1987) 701.*
451 N. Shiga, O. Matano and S. Goto, *J. Chromatogr.*, 396 (1987) 327.
452 R.M. Noble, D.J. Hamilton and W.J. Osborne, *Pestic. Sci.*, 13 (1982) 246.
453 R.B. Leidy, T.J. Sheets and L.A. Nelson, *Beitr. Tabakforsch. Int.*, 13 (1986) 191.
454 R.A. Chapman, *J. Chromatogr.*, 258 (1983) 175.
455 T. Doi, S. Sakaue and M. Horiba, *J. Assoc. Off. Anal. Chem.*, 68 (1985) 911.
456 E. Papadopoulou-Mourkidou, *Chromatographia*, 20 (1985) 376.
457 P.W. Lee, W.R. Powell, S.M. Stearns and O.J. McConnell, *J. Agric. Food Chem.*, 35 (1987) 384.
458 V. Batora, S.L.J. Vitorovic, H.P. Thier, M.A. Klisenko and R. Greenhalgh, *Pure Appl. Chem.*, 53 (1981) 1039.
459 L. Gyorfi, A. Ambrus and E. Bolygo, *Pestic. Sci. Biotechnol., Proc. Int. Congr. Pestic. Chem., 6th 1986*, (1987).
460 A. Ambrus, E. Hargital, G. Karoly, A. Fulop and J. Lantos, *J. Assoc. Off. Anal. Chem.*, 64 (1981) 743.
461 D. Lienig, K. Schaefer and G. Reichelt, *Acta Hydrochim. Hydrobiol.*, 13 (1985) 443.
462 A. Lawrerenz, H. Goralczyk and H. Hermenau, *Acta Hydrochim. Hydrobiol.*, 14 (1986) 121.
463 K. Fodor-Csorba and F. Dutka, *J. Chromatogr.*, 365 (1986) 309.
464 United Kingdom Department of the Environment (UK), *Methods Exam. Waters Assoc. Mater.*, (Organo-Phosphorus Pestic. Sewage Sludge; Organo-Phosphorus Pestic. River Drinking Water, Addit. 1985), 1986, 20 pp.
465 J. Sherma, *Anal. Methods Pestic. Plant Growth Regul.*, 7 (1973) 3.
466 J. Sherma, *Anal. Methods Pestic. Plant Growth Regul.*, 11 (1980) 79.
467 J. Sherma, *Anal. Methods Pestic. Plant Growth Regul.*, 14 (1986) 1.
468 J. Sherma, in J.C. Touchstone and J. Sherma (Editors), *Densitometry in Thin Layer Chromatography - Practice and Applications*, Wiley-Interscience, New York, 1979, Chapter 22.
469 S.N. Tewari, in J. Harvey, Jr. and G. Zweig (Editors), *Pesticide Analytical Methodology*, ACS Symposium Series No. 136, American Chemical Society, Washington, DC, 1980, Chapter 14.
470 M. Zakaria, M.F. Gonnord and G. Guiochon, *J. Chromatogr.*, 271 (1983) 127.
471 J. Sherma and J.L. Boymel, *J. Chromatogr.*, 247 (1982) 201.
472 E. Weller, *Physiol. Plant,* 54 (1982) 510.
473 K.M.S. Sundaram and R. Hindle, *J. Chromatogr.*, 194 (1980) 100.
474 Z. A. Leika and D.B. Girenko, *Gig. Sanit.*, (2) (1982) 77.
475 J. Sherma and T. Regan, *Pesticides*, 15 (1981) 21.
476 M. Franci, N. Andreoni and P. Fusi, *Bull. Environ. Contam. Toxicol.*, 26 (1981) 102.
477 A.E. Smith and L.J. Milward, *J. Chromatogr.*, 265 (1983) 378.
478 M.T. Giardi, M.C. Giardina and E. Brancaleoni, in A. Frigerio (Editor), *Chromatography in Biochemistry, Medicine and Environmental Research, 1*, Elsevier, Amsterdam, 1983, p. 53.
479 J. Sherma and N.T. Miller, *J. Liq. Chromatogr.*, 3 (1980) 901.
480 J. Sherma, P.J. Sielicki, Jr. and S. Charvat, *J. Liq. Chromatogr.*, 6 (1983) 2679.
481 A.A. Krasnykh, V.S. Shustov and M.F. Zelenina, *Gig. Sanit.*, (3) (1982) 53.
482 S. Traore and J.J. Aaron, *Talanta*, 28 (1981) 765.
483 J. Sherma and S. Stellmacher, *J. Liq. Chromatogr.*, 8 (1985) 2949.
484 S.P. Srivastava and Reena, *J. Liq. Chromatogr.*, 6 (1983) 139.
485 W. Hyde, H.M. Stahr, R. Moore, M. Donato and R. Pfeiffer, *Adv. Thin Layer Chromatography (Proc. Bienn. Symp.), 2nd 1980*, 439 (1982).
486 L.G. Aleksandrova and M.A. Klisenko, *J. Chromatogr.*, 247 (1982) 255.
487 J. Sherma and J.L. Boymel, *J. Liq. Chromatogr.*, 6 (1983) 1183.

488 S.V. Padalikar, S.S. Shinde and B.M. Shinde, *Analyst (London),* 113 (1988) 411.
489 J. Tancogne, *Ann. Tab.,* Sect. 2, Number Spec. (1987) 69.
490 M.H.K. Abdel-Kader, D.A. Stiles and M.T.H. Ragab, *Int. J. Environ. Anal. Chem.,* 18 (1984) 281.
491 I. Vukusic and B. Laskarin, *J. High Resolut. Chromatogr. Chromatogr. Commun.,* 4 (1981) 659.
492 T. Stijve, *Dtsch. Lebensm. Rundsch.,* 76 (1980) 234.
493 V.S. Vasilenko, *Gig. Sanit.,* 45 (1980) 54.
494 J. Brueggemann and H.D. Ocker, *Chem. Mikrobiol., Technol. Lebensm.,* 10 (1986) 113.
495 H. Rahmann, *Nahrung,* 25 (1981) 49.
496 H. Thieme and U. Kurzik-Dumke, *Pharmazie,* 37 (1982) 370.
497 J. Sherma and R. Slobodien, *J. Liq. Chromatogr.,* 7 (1984) 2735.
498 A.A. Krasynykh and L.G. Pavlova, *Gig. Sanit.,* (2) (1982) 76.
499 H. Thielemann, *Pharmazie,* 35 (1980) 329.
500 M.P. Kurhekar, F.C. D'Souza, M.D. Pundlik and S.K. Meghal, *J. Chromatogr.,* 209 (1981) 101.
501 R. Kumar and C.B. Sharma, *J. Liq. Chromatogr.,* 10 (1987) 3681.
502 J. Polak, *Chem. Listy,* 77 (1983) 306.
503 S.V. Mirashi, M.P. Kurhekar and F.D. D'Souza, *J. Chromatogr.,* 268 (1983) 352.
504 V.N. Kavetskii and G.G. Andrienko, *Fiziol. Biokhim. Kil't. Rast.,* 14 (1982) 89.
505 S.V. Mirashi, V.B. Patil and K.A. Ambade, *Curr. Sci.,* 54 (1985) 635.
506 H.N. Katkar and V.D. Joglekar, *Curr. Sci.,* 49 (1980) 350.
507 S.U. Bhaskar, *Talanta,* 29 (1982) 133.
508 P.P. Singh and R.P. Chawla, *J. Chromatogr.,* 450 (1988) 452.
509 D. Gottstein, D. Gross and H. Lehmann, *Z. Gesamte Hyg. Ihre Grenzgeb.,* 30 (1984) 620.
510 P.H. Sackett, *Anal. Chem.,* 56 (1984) 1600.
511 J. Kovac, M. Kurucova, V. Batora, J. Tekel and V. Strniskova, *J. Chromatogr.,* 280 (1983) 176.
512 J. Kovac, J. Tekel and M. Kurucova, *Z. Lebensm.-Unters.-Forsch.,* 184 (1987) 96.
513 H.S. Rathore and S. Gupta, *J. Liq. Chromatogr.,* 10 (1987) 3659.
514 J. Tekel, P. Farkas, J. Kovacicova and A. Szokolay, *Nahrung,* 32 (1988) 357.
515 R.W. Reiser, I.J. Belasco and R.C. Rhodes, *Biomed. Mass Spectrom.,* 10 (1983) 581.
516 A.N. Kadam and B.B. Ghatge, *Pesticides,* 18 (1984) 20.
517 M. Ozaki and S. Kuwatsuka, *Nippon Noyaku Gakkaishi,* 9 (1984) 769.
518 V.S. Shustrov, M.F. Zelenina and E.V. Kulintsova, *Khim. Sel'sk Khoz.,* (7) (1981) 57.
519 V.P. Lynch and H.R. Hudson, *Pestic. Sci.,* 12 (1981) 65.
520 U.S. Singh and R.K. Tripathi, *J. Chromatogr.,* 200 (1980) 317.
521 A. Riebel and C. Rellich, *Tagungsber. Akad. Landwirtschaftwiss. D.D.R.,* 187 (1981) 167.
522 J. Schneider, *Nahrung,* 30 (1986) 859.
523 H. Thielemann, *Z. Martin Luther Univ. Halle Wittenberg, Math. Naturwiss. Reihe.,* 36 (1987) 16.
524 K.R. Ziminski, T.J. Manning and L. Lukash, *Clin. Toxicol.,* 18 (1981) 731.
525 J. Lantos, U.A.T. Brinkman and R.W. Frei, *J. Chromatogr.,* 292 (1984) 117.
526 A. Deshpande, S.V. Padalikar and S.K. Meghal, *Curr. Sci.,* 50 (1981) 814.
527 S. Gocan, D.A. Trujillo, S. Kira and Y. Hrung, *Bull. Environ. Contam. Toxicol.,* 25 (1980) 824.
528 R. Pfeiffer and H.M. Stahr, *Adv. Thin Layer Chromatography (Proc. Bienn. Symp.), 2nd 1980,* 541 (1982).
529 J. Sherma, *J. Liq. Chromatogr.,* 11 (1988) 2121.
530 S.P. Srivastava and Reena, *Anal. Lett.,* 15 (1982) 39.
531 M.P. Kurhekar, M.D. Pundlik and S.K. Meghal, *J. Anal. Toxicol.,* 4 (1980) 322.
532 I. Fogy, G.M. Allmaier and E.R. Schmid, *Int. J. Mass Spectrom. Ion Phys.,* 48 (1983) 319.

533 J. Sherma and M. Getz, *Adv. Thin Layer Chromatography (Proc. Bienn. Symp.), 2nd 1980*, 483 (1980).
534 S. Renvall and M. Akerblom, *Var. Foeda.*, 34 (Suppl. 3) (1982) 240.
535 S.L. Vitorovic, *Pestic. Chem.: Hum. Welfare Environ., Proc. Int. Congr. Pestic. Chem., 5th 1982*, Pergamon, Oxford, 1983, Vol. 4, p. 101.
536 Z. Sonnenfeld and J. Paul, *Microchem. J.*, 32 (1985) 137.
537 C. Marutolu, C. Sarbu, M. Vlassa and C. Liteanu, *Analusis*, 14 (1986) 95.
538 J.A. Federici and J. Paul, *Microchem. J.*, 34 (1986) 211.
536 R. Kumar and C.B. Sharma, *J. Liq. Chromatogr.*, 10 (1987) 3637.
540 C. Marutoiu, M. Vlassa and C. Sarbu, *J. High Resolut. Chromatogr. Chromatogr. Commun.*, 10 (1987) 465.
541 V.B. Patil, S.V. Padalikar and G.B. Kawale, *Analyst (London)*, 112 (1987) 1765.
542 A.B. Wood and L. Kanagasabapathy, *Pestic. Sci. (Engl.)*, 14 (1983) 108.
543 S.V. Ohlsson and W.W. Hintze, *J. High Resolut. Chromatogr. Chromatogr. Commun.*, 6 (1983) 89.
544 M. Adinarayana, U.S. Singh and T.S. Dwivedi, *J. Chromatogr.*, 435 (1988) 210.
545 P. Tomboulain, S.M. Walters and K.K. Brown, *Mikrochim. Acta*, II (1-3) (1987) 11.
546 M. Van den Heede, J. Cordonnier, L. Van Bever and A. Heyndrickx, *Meded. Fac. Landbouwwet., Rijksuniv. Gent*, 47 (1982) 421.
547 J. Ikebuchi, I. Yuasa and S. Kotoku, *J. Anal. Toxicol.*, 12 (1988) 80.
548 G.S. Raju and K. Visweswariah, *Pesticides*, 18 (1984) 26.
549 S.U. Bhaskar and N.V.N. Kumar, *J. Assoc. Off. Anal. Chem.*, 64 (1981) 1312.
550 H. Breuer, *J. Chromatogr.*, 243 (1982) 183.
551 Z.H. Zhou and G.M. Pan, *Huan Ching K'o Hsueh*, 1 (1980) 5.
552 H.M. Stahr, *J. Liq. Chromatogr.*, 7 (1984) 1393.
553 H.J. Petrowitz and M. Wagner, *Holz Roh-Werkst.*, 42 (1984) 345.
554 M.A. Sattar, *J. Chromatogr.*, 209 (1981) 329.
555 G. Hermenau and K. Grahl, *Acta Hydrochim. Hydrobiol.*, 12 (1984) 685.
556 Y.D. Ha and K.G. Bergner, *Dtsch. Lebensm. Rundsch.*, 76 (1980) 390.
557 Y.D. Ha and K.G. Bergner, *Dtsch. Lebensm. Rundsch.*, 77 (1981) 102.
558 L. Ogierman and G. Brysz, *Fresenius' Z. Anal. Chem.*, 308 (1981) 463.
559 H. Sonobe, R.A. Carver, R.T. Krause and L.R. Kamps, *J. Agric. Food Chem.*, 30 (1982) 696.
560 G. Huber Kovacs, *J. Chromatogr.*, 303 (1984) 309.
561 I.A. Gentile and E. Passera, *J. Chromatogr.*, 236 (1982) 254.
562 A. Nelcheva, P. Vasileva-Aleksandrova and E. Kovacheva, *Mikrochim. Acta*, I (1984) 393.
563 M. Uno, T. Okada, T. Ohmae, I. Terada and K. Tanigawa, *Shokuhin Eiseigaku Zasshi*, 23 (1982) 191.
564 R. Sundararajan and R.P. Chawla, *J. Assoc. Off. Anal. Chem.*, 66 (1983) 1009.
565 K. Opong-Mensah and W.R. Porter, *J. Chromatogr.*, 455 (1988) 439.
566 S. Pavkov, A. Letic, M. Vojinovic and O. Stefanovic, *Hrana Ishrana*, 23 (1982) 81.
567 G. Becker, *Fresenius' Z. Anal. Chem.*, 318 (1984) 276.
568 T. Cserhati and F. Orsi, *Period Polytechn. Chem. Eng.*, 26 (1982) 111.
569 H. Thielemann, *Z. Gesamte Hyg. Ihre Grenzgeb.*, 25 (1979) 556.
570 H. Thielemann and H. Grahneis, *Z. Gesamte Hyg. Ihre Grenzgeb.*, 28 (1982) 324.
571 B. Wiemer, *Acta Hydrochim. Hydrobiol.*, 13 (1985) 527.
572 T. Cserhati and T. Bellay, *Acta Phytopathol. Entomol. Hung.*, 23 (1988) 257.
573 J. Sherma, *J. Liq. Chromatogr.*, 9 (1986) 3433.
574 J.F. Lawrence, *J. Assoc. Off. Anal. Chem.*, 64 (1980) 758.
575 N.I. Kiseleva and L.I. Kobylinskaya, *Gig. Sanit.*, (11) (1981) 48
576 K.K. Mazumdar, N. Samajpati and T. Chakrabarti, *Indian J. Exp. Biol.*, 20 (1982) 865.

Chapter 24

Environmental analysis

K. P. NAIKWADI and F. W. KARASEK

CONTENTS

24.1 INTRODUCTION

This chapter describes advances in the analysis of environmental volatile, semivolatile, and nonvolatile organic compounds and demonstrates the utility of the methods by examining a series of examples. Emphasis will be given to the most recently published methods, in particular the methods developed for the analysis of municipal solid-waste incinerator (MSWI) fly ash samples, polychlorinated biphenyl (PCB) fire samples, airborne particulate matter, water, and sediment/soil.

References on p. B579

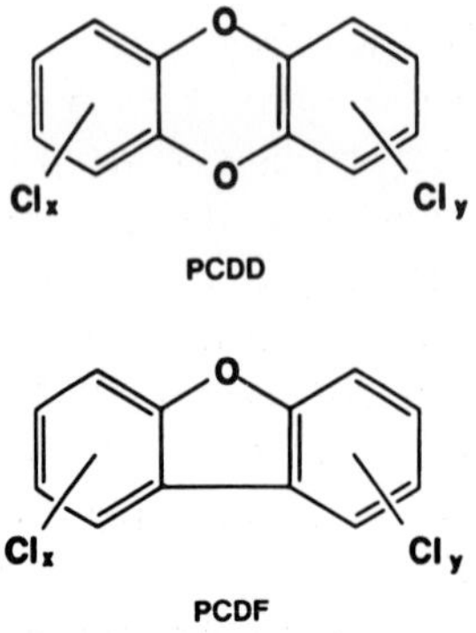

Fig. 24.1. Structures of PCDD and PCDF.

The structures of polychlorinated dibenzo-p-dioxins (PCDD) and polychlorinated dibenzofurans (PCDF) are shown in Fig. 24.1. Each of these structures represents a series of compounds having 1-8 chlorine atoms attached to the rings. There are 75 PCDD and 135 PCDF isomers. Analysis of total PCDD and PCDF, along with the most toxic isomers in each congener group, is difficult and challenging. Generally, attempts are made to quantify the total PCDD/PCDF in the environmental samples. Although analyses of the most toxic 2,3,7,8-substituted isomers of PCDD and PCDF in all environmental samples are desired, these are very difficult to perform by general analytical techniques. The sources, toxicity, mechanism of formation, and the analytical methods for PCDD/PCDF are reviewed in several articles and books [1-10].

Polycyclic aromatic hydrocarbons (PAHs) comprise the largest class of chemical carcinogens known to enter the environment through various sources. Natural sources include forest fires, volcanos, sediments, fossils, and fossil fuels. Anthropogenic sources include the burning of coal, oil, and wood; coal production, automobiles, and MSWIs. During the last three decades, many studies have been undertaken to characterize the PAH contents of various environmental samples. On this subject, several reviews and books have been published [11-15], covering the use of modern analytical techniques, such as HPLC, GC, GC/MS, and GC/FTIR for the analysis of PAHs.

Analysis of organics in soil and sediment is complicated, if the water content of the sample is high. Methods used for volatile organics in water samples are useful for wet sediments and soil after some modifications. A high-temperature purge-and-trap method for the analysis of semivolatile organics has been developed [16]. Extractions of semivolatile compounds in wet sediments and soil samples by traditional methods, such as Soxhlet, sonication, and stirring have been reported [17,18].

Analysis of water samples for organic pollutants is performed by designated methods. However, analytical method development work for the analysis of drinking, natural, and industrial waste water has not stopped. There is a continuous search for faster and easier analytical methods. GC/MS has been used successfully in the identification and quantitation of a large number of "volatile" organic pollutants in water [19-22]. HPLC is a particularly well-suited technique for analytical fractionation of the nonvolatile organics.

Many toxic or carcinogenic compounds are distributed in the environment at low levels. Determination of organics at parts-per-trillion (ppt) levels can be performed by combining sensitive and selective detection with sample preconcentration. Because concentrated samples are highly complex, several prefractionation steps may be required to isolate compounds of interest. A complete scheme for trace analysis of organic compounds generally consists of sampling, extraction, prefractionation, and analysis by GC or GC/MS. In addition to chlorinated organics, attention has been focused on PAHs and volatile organics in water, air, and soil samples. Selection of the best analytical approach is one of the most important factors influencing the reliability of results.

Environmental samples may be viewed in two perspectives: In the first, the sample is completely analyzed for organic volatile compounds by HPLC, GC, or GC/MS. In the second, there is a more specific analysis of target compounds, which are generally highly toxic, by means of highly sensitive and selective detectors in GC and HPLC. Complete analysis of an environmental sample, including analysis of target compounds, gives a realistic assessment of the toxicity of a particular environmental sample.

24.2 SAMPLE TREATMENT AND ANALYSIS

24.2.1 Municipal solid-waste incinerator fly ash samples

MSWIs generate thousands of tons of fly ash each year throughout the world. Approximately 1-2% of the fly ash is not trapped by an electrostatic precipitator (EP) and escapes with the flue gases. The fly ash collected at the EP contains high levels of dioxins compared to the bottom ash. Total dioxins in the environment originate from different sources, MSWI fly ash being a major contributor.

24.2.1.1 Sample collection

In a MSWI there are five main sampling regions: furnace, boiler, economizer, EP, and stack. Generally, 80-90% reduction in volume of municipal refuse has been observed after incineration. Out of the total ash produced, 70-75% is collected at the bottom of the furnace (called bottom ash) and contains small amounts of environmentally significant organic pollutants. However, 25-30% fly ash, which is light and much finer than bottom ash, escapes the boiler. This fly ash is collected at the EP by means of water sprays that settle the light particles. The fly ash accumulated at different points of the EP is generally combined by means of a moving belt. The combined fly ash sample is collected and transported to the laboratory for analysis. Although water is used in the EP, the later is at temperatures between 100 and 150°C. Hence, fly ash samples collected at the EP are always dry, and further drying prior to extraction is not needed.

References on p. B579

24.2.1.2 Sample extraction and cleanup

Routinely, 10 to 20 g of fly ash is extracted in a Soxhlet extractor with 350 ml of benzene or toluene for 48 h. For recovery estimates, fly ash is spiked with ^{13}C-labeled standards prior to extraction. The fly ash extract is then concentrated. Repeated sonication is much faster and less laborious than Soxhlet extraction [23]. Supercritical-fluid extraction (SFE) is a method of choice if requisite instrumentation is available [24].

Fly ash extract is a complicated matrix, containing many types of organic and inorganic compounds, which are present at ppb levels [25]. For the analysis of PCDD/PCDF by GC/MS with selected-ion monitoring, liquid chromatographic sample cleanup on a column, packed with silica or alumina, is highly recommended [26,27]. HPLC fractionation is essential for the complete analysis of a fly ash extract.

24.2.1.3 HPLC fractionation

PCDD and PCDF can be separated from other organic compounds by a single-step normal-phase HPLC fractionation on a 10-μm semipreparative Spherosorb silica column (250 x 9.4 mm ID) and elution with hexane, followed by dichloromethane and acetonitrile [25]. The first fraction consists of aliphatic hydrocarbons and the second fraction contains PCDD and PCDF. The latter has been found to be adequate for the analysis of total PCDD and PCDF in fly ash samples by GC/MS. Further separation of PCDD/PCDF into different congener groups can be achieved by reversed-phase HPLC [28]. For the analysis of the most toxic 2,3,7,8-tetrachlorodibenzo-*p*-dioxin (2,3,7,8-TCDD), and 2,3,7,8-tetrachloro-dibenzofuran (2,3,7,8-TCDF) by GC/MS and isomer-specific GC capillary columns, it is necessary to separate the tetra-congeners from the penta- to octa-congeners by HPLC.

24.2.1.4 GC and GC/MS analysis

Several hundred organic compounds have been identified in Ontario fly ash by HPLC, GC, and GC/MS techniques. The determination of total organic compounds and halogenated compounds in fly ash extracts was performed by GC/FID and GC/ECD. For complete identification of compounds in fly ash extract, HPLC was used to fractionate the extract, and the HPLC fractions were then analyzed by GC/FID and GC/ECD. The identification of compounds was performed by MS/library search/probability-based matching (PBM), retention indices, mass spectral data, and HPLC retention [25]. Comparison of PCDD/PCDF quantitative analyses by different techniques, such as GC/FID, GC/ECD, and GC/MS showed that GC/FID and GC/ECD give higher values than GC/MS/SIM [29]. Although GC/FID for combustible compounds and GC/ECD for electronegative compounds are highly sensitive, they are not reliable for the quantitation of compounds, such as PCDD/PCDF in fly ash, because these detectors respond to a variety of other compounds in the extracts. The GC/MS/SIM technique is highly selective and reliable for PCDD/PCDF analysis.

TABLE 24.1

IONS MONITORED FOR ^{13}C-LABELED AND UNLABELED PCDD

MI = Most intense ion in M, M + 2, and M + 4; 4CDD = tetrachlorodibenzo-*p*-dioxins, 5CDD = pentachlorodibenzo-*p*-dioxins, 6CDD = hexachlorodibenzo-*p*-dioxins, 7CDD = heptachlorodibenzo-*p*-dioxins, and 8CDD = octachlorodibenzo-*p*-dioxin.

Ions	4CDD	5CDD	6CDD	7CDD	8CDD
^{13}C-Labeled PCDD					
M	331.9	365.9	399.9	433.8	–
M + 2	333.9	367.9	401.9	435.8	469.7
M + 4	–	–	–	–	471.7
MI – COCl	269.9	303.9	337.9	371.8	407.7
Unlabeled PCDD					
M	319.9	353.9	387.9	421.8	–
M + 2	321.9	355.9	389.9	423.8	457.7
M + 4	–	–	–	–	459.7
MI – COCl	258.9	292.9	326.9	360.8	396.7

The combination of high resolution, achieved by capillary GC, and the high sensitivity and selectivity of the mass spectrometer results in a powerful and universal technique for environmental analysis. Analysis of PCDD/PCDF in fly ash extracts by the GC/MS/SIM technique involves an initial determination of the retention time windows. To determine such retention time windows, a sample or standard containing all PCDD/PCDF isomers or a sample containing the first-eluted and last-eluted isomers for each congener group needs to be injected, and two ions for each congener group should be monitored as a single group by GC/MS/SIM. Redrawing of mass chromatograms for a congener group provides the retention window for the elution of all isomers in that particular congener group. Once the retention time windows have been determined, three ions for each congener group in a particular retention window should be monitored, keeping all chromatographic conditions constant. The sensitivity for the detection of a compound depends on a variety of parameters, including the number of ions monitored for the specific compound and the dwell times of the ions monitored. Hence, it is necessary to monitor a minimum of characteristic ions for each congener group. Ions monitored for native and ^{13}C-labeled PCDD are shown in Table 24.1. For positive identification of PCDD/PCDF congeners the following criteria have to be satisfied:

(1) The PCDD/PCDF congeners in the sample must fall in the same retention window as that previously established for a sample containing the first-eluted and last-eluted isomers in each congener group. Retention times of the peaks in the sample must match

References on p. B579

the retention times of the peaks in the standard that contains all PCDD and PCDF isomers.

(2) The ratio of the relative intensities of the two major characteristic ions monitored for a particular congener group must correspond to that resulting for a congener group in the PCDD/PCDF standard, within ± 10 %.

(3) The signal-to-noise ratio must be greater than 3. Isomer-specific separation of 2,3,7,8-TCDD/TCDF requires special columns for both HPLC and GC. In GC, polar capillary columns, such as SP-2330 and SP-2331, are used for separating 2,3,7,8-TCDD from other TCDD isomers. Separation of 2,3,7,8-TCDF from other TCDF isomers is facilitated by LC on alumina columns, followed by GC/MS with a combination of capillary columns, such as a 60-m SP-2331 and a 12-m nonpolar DB-17 column [30]. For isomer-specific separations and quantitation of PCDD and PCDF isomers, in particular for the separation of 2,3,7,8-TCDD and 2,3,7,8-TCDF, laborious procedures are required, such as HPLC fractionation followed by GC and GC/MS [31,32]. Analytical methods required for quantitation of 2,3,7,8-TCDD/TCDF also depend upon the sample matrices.

Precise methods for measurements of dioxins in ambient air have been described [33-35]. A modified high-volume sampler and the sampling head module for a collection of environmental samples are illustrated diagramatically in Fig. 24.2 [33]. The method was designed to measure 0.1 to 1.0 pg/m^3 2,3,7,8-TCDD in order to obtain reliable measurements at the 5.5 pg/m^3 level as an estimated no-observed-effect level (NOEL). The method for sample preparation with a high-capacity aqueous-phase extractor was tested by Clement et al. [36]. The extraction of 200 l water at a time was achieved. The results of this large-scale extraction were compared with laboratory-scale liquid/liquid extraction techniques. Using this method, attempts were made to analyze only total PCDD/PCDF. Isomer-specific determination of PCDD/PCDF in several incinerator samples has been reported [37]. Sample preparation was performed by Soxhlet extraction, followed by normal- and reversed-phase LC. Isomer-specific separation and quantitation of PCDD/PCDF was carried out on a SP-2330 column for tetra- and penta-chlorodibenzo-p-dioxins. A study of PCDD/PCDF in sediments at different locations suggested that combustion was the source of these chemicals and implied that PCDD/PCDF are transported through the atmosphere [38]. Analysis of cow's milk from different locations in Switzerland for PCDD/PCDF showed the presence of the most toxic 2,3,7,8-substituted isomers of dioxins and furans at ppb levels. Multi-step liquid extraction procedures have been described for sample preparation [39]. The levels (ppt range) of 2,3,7,8-TCDD in human serum and plasma samples in the general population with no known exposure to 2,3,7,8-TCDD have been reported for the first time by Patterson et al. [40]. The analytical method initially developed for a 200-g sample of serum was further modified for analyzing a 10-g serum sample. In these methods 2,3,7,8-TCDD is quantified by the isotope dilution technique with $(^{13}C_{12})$2,3,7,8-TCDD as the internal standard by high-resolution GC/high-resolution MS (HRGC/HRMS).

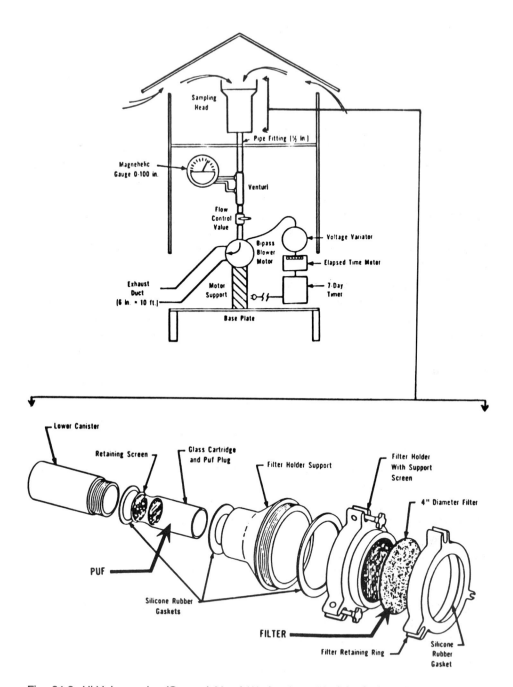

Fig. 24.2. Hi-Vol sampler (General Metal Works, Inc., Model PS-1) and sampling head module. (Reproduced from Ref. 33 with permission.)

References on p. B579

24.2.2 Polychlorinated dibenzo-*p*-dioxins in polychlorinated biphenyl fire samples

24.2.2.1 Sample cleanup and HPLC fractionation

PCB fire samples are usually very complicated, owing to the presence of high levels of PCBs and a variety of other organic compounds formed at high temperatures during accidental fires. Analysis of PCDD in these samples by GC/MS is very difficult due to strong interference by PCB. For example, the M + 2 and M + 4 ions of tetra- and penta-chlorodibenzo-*p*-dioxins have masses similar to the fragment ions of penta- to decachlorobiphenyls, which are unresolved on commercially available GC columns. Generally, analysis of dioxins in samples containing PCBs are performed by HRGC/HRMS [41]. More recently, open-column LC, usually on alumina columns, has been used for separating PCB from 2,3,7,8-TCDD [42]. A combination of solvents for the complete separation of PCBs from 2,3,7,8-TCDD by LC on an alumina column has been described [43,44]. Separation of PCBs from total PCDD/PCDF by normal- and reversed-phase HPLC has been reported [45].

An analytical-scale column (250 x 4.4 mm ID), packed with 5-μm alumina at 9000 psi by the slurry technique with a column packer (Shandon Southern Products) has been evaluated. Gradient elution, at a flowrate of 1 ml/min, consisted of 100% hexane for 10 min, programed to 100% dichloromethane in 30 min, 100% dichloromentane for 10 min, and back to 100% hexane in 10 min. PCBs were eluted within the first 16 min, PCDD/PCDF between 16 and 22 min, and polar compounds between 22 and 45 min. An internal standard, containing ^{13}C-PCDD isomers, was added to the PCB fire sample prior to separation by HPLC. The second HPLC fraction, containing PCDD, was further concentrated and analyzed by GC/MS.

24.2.2.2 GC and GC/MS analysis

A standard practice, applied by most laboratories analyzing for PCDD, involves the use of a limited ^{13}C-labeled dioxin mixture, containing one isomer for each congener group, as an internal standard or a mixture of unlabeled dioxins as an external standard. Quantitative estimation of dioxins can be carried out by comparing the response to known concentrations of dioxins in the standard to those of the dioxins present in the sample. The accuracy of quantitation by the use of an external standard can vary according to the run-to-run GC and GC/MS conditions and with variations in injection techniques. True quantitation of total amounts of PCDD by the use of an external standard cannot be assessed when extensive sample cleanup has been performed prior to GC/MS/SIM analysis. However, precision in PCDD analysis can be achieved by spiking the environmental sample with a known amount of ^{13}C-PCDD mixture before sample preparation and then analyzing the ^{13}C-PCDD along with the unlabeled PCDD in the environmental sample. The percent recovery of ^{13}C-PCDD thus provides a measure of the efficiency of the analytical procedures. Although ^{13}C-PCDD standards have a major advantage over unlabeled PCDD standards, only a few labeled PCDD isomers have been prepared and are gener-

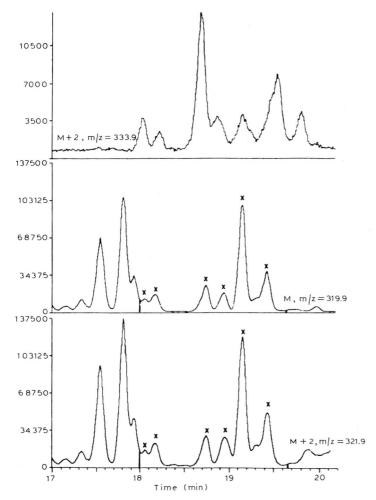

Fig. 24.3. Mass chromatograms of TCDDs in a PCB fire sample, M+2 (321.9) and M (319.9) ions (lower and middle tracings) and for unlabeled M+2 (333.9) ion spiked with ^{13}C-labeled TCDD (top tracing). Chromatographic conditions: DB-5, 30 m x 0.32 mm ID, fused-silica capillary column, temperature program: 80°C for 1 min, then to 230°C at 15°C/min, and to 300°C at 3°C/min, finally at 300°C for 20 min. Peaks marked "x" are dioxins.

ally available, and thus, the extent to which these procedures can be applied is limited. Accidental PCB fire samples are usually very complicated, due to the formation of traces of a variety of interfering compounds at higher temperatures. Often, the traditional criteria used for identification of PCDD [46,47] by GC/MS/SIM are not sufficient for analyzing PCDD in very complex samples because of interferences. In particular, analysis of a PCB fire sample for PCDD by these criteria can result in erroneous identification of interfering compounds as dioxins [48]. The mass chromatograms for the M and M+2 (319.9 and 321.9) ions of unlabeled TCDD in a PCB fire sample and the M+2 (333.9) ion of the ^{13}C-labeled TCDD reference standard are shown in Fig. 24.3. By applying the two major criteria of correct retention window and proper intensities, all peaks in the mass chromatogram for TCDD in the sample can be considered as positively identified. However, by comparison of retention times for known peaks in the mixture spiked with ^{13}C-PCDD (Fig. 24.3, upper tracing) it can be seen that the peaks marked "x" in the mass chromatograms for TCDDs in the PCB sample are the only dioxins. Other peaks, although having the correct retention window and intensities, are not dioxins. Similarly, based on the criteria of comparison of retention times of peaks in the retention window, several peaks of interfering compounds were observed in penta-, hexa-, and heptachlorodioxin windows that do not match the retention times of the peaks in the ^{13}C-PCDD mixture. It should be emphasized that the analysis of a complicated sample by the use of only one ^{13}C-labeled dioxin isomer for each congener group may also result in erroneous identification of interfering compounds as dioxins. Use of the ^{13}C-PCDD mixture and the additional criteria of comparison of retention times of individual peaks in the ^{13}C-PCDD mixture with that of peaks for the unlabeled PCDD in a PCB fire sample result in a more positive identification of dioxins [48].

24.2.3 Airborne particulate matter samples

24.2.3.1 Sample collection and cleanup

In most analytical schemes for organic compounds associated with airborne particulate matter, the samples are collected by high-volume filtration techniques, and the organic compounds are recovered by Soxhlet extraction. This technique was used to collect particulate emissions from a residential wood burning furnace [49]. Organic compounds were isolated by Soxhlet extraction or by solvent desorption when cartridges were used in sample collection. Sample extracts of airborne particulate matter are very complicated and need to be separated into three major groups: acidic, basic, and neutral by means of acid/base extraction. Acidic and basic compounds are analyzed, directly or after derivatization, by GC/MS. The neutral fraction consists of hundreds of compounds and hence must be fractionated by HPLC.

24.2.3.2 HPLC fractionation

A semipreparative 10-μm μBondapack amine column (250 x 9.4 mm ID) was used in HPLC separations. The gradient elution program consisted of 100% hexane for 30 min, then to 100% dichloromethane in 25 min, and then to 100% acetonitrile in 20 min, held there for 5 min, and back to the original conditions in 10 min. The flowrate was 2.5 ml/min. Ten fractions were collected, up to: 7, 10, 15, 20, 30, 37, 43, 48, 55, and 75 min. Each fraction was concentrated to 200 μl for GC and GC/MS analysis. The first three fractions

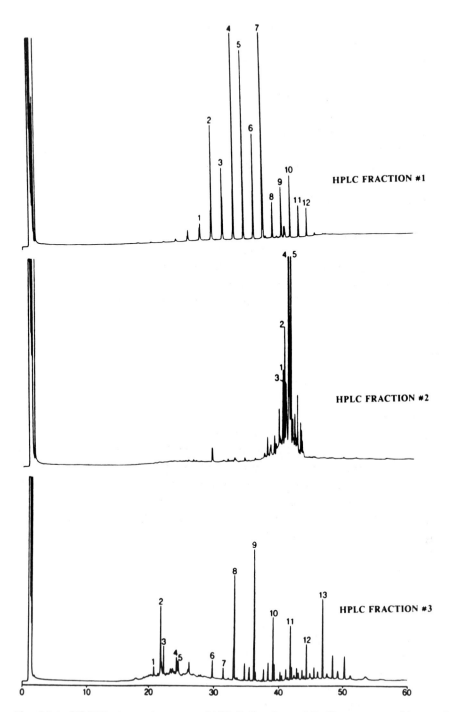

Fig. 24.4. GC/FID chromatograms of HPLC Fractions 1-3. Chromatographic conditions: DB-5, 30 m x 0.32 mm ID, fused-silica capillary column, temperature program: 80°C for 1 min, then to 300°C at 5°C/min, finally 20 min at 300°C. For identification of peaks see Table 24.2.

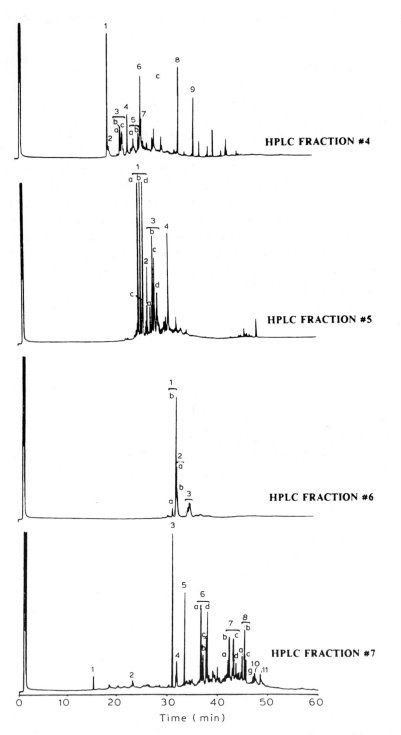

Fig. 24.5. GC/FID chromatograms of HPLC Fractions 4-7. For conditions see Fig. 24.4, for peak identification see Table 24.2.

contained aliphatic compounds, followed by two- to five-ring PAHs in the subsequent 4-7 fractions [49].

24.2.3.3 GC and GC/MS analysis

An airborne particulate extract was separated into three major fractions, containing acidic, basic, and neutral compounds [49]. The acidic and basic fractions were analyzed by GC/FID and GC/MS without further fractionation. However, because the neutral fraction contained numerous compounds, as shown by GC/FID screening, it was further fractionated by normal-phase HPLC (Section 24.2.3.2). All HPLC fractions were concentrated and analyzed by GC/FID and GC/MS. The GC/FID tracings of the first three HPLC fractions are shown in Fig. 24.4. These HPLC fractions were also analyzed by GC/MS and the compounds were identified by their mass spectra, PBM library search, retention indices, and comparison with the retention times of authentic compounds in a standard, analyzed under identical conditions. Most of the compounds identified in the first three fractions were aliphatic. The retention times of these compounds in the GC analysis were between 20 and 50 min. Under these conditions, PAHs are not separated from these compounds. Thus, HPLC separation of PAHs from aliphatic compounds prior to GC/MS analysis proved to be advantageous.

GC/FID tracings of the HPLC Fractions 4-7 are shown in Fig. 24.5. The normal-phase amine column yielded more selectivity than the silica column for the separation of PAHs, based on the number of rings. It is very difficult to separate compounds based on the number of rings, even with the amine column, when the sample contains a large number of PAHs. There is always some overlapping of compounds in consecutive HPLC fractions. Since the physical properties, such as vapor pressure and polarity, of PAHs are not consistent with the number of rings, several PAHs have the same retention time in GC. However, due to their different polarities and, consequently, different interactions with the amine column, their separation can be achieved by HPLC. In HPLC Fractions 4-7 numerous peaks were seen in the GC/FID tracings and GC/MS total-ion chromatograms. All the peaks numbered in Figs. 24.4 and 24.5 were identified by their mass spectra, molecular weights, and the PBM search information. However, even with this information, the identification of isomeric PAHs is very difficult. The identification of isomeric PAHs can be carried out by analysis of authentic standards under identical conditions and comparison of retention indices, both calculated and reported in the literature [50,51].

Air particulate extracts contain a large number of PAHs. It is beyond the capacity of any laboratory to keep all the PAH standards; moreover, several PAHs are not commercially available. However, it is possible to apply a common method of identification and to use available data for several standards in different laboratories. In particular, the retention index (R.I.) method developed by Lee et al. [51] was applied and retention indices were obtained for a mixture of 20 PAHs by using a R.I. program stored in a HP-1000 data system, which was accessible to the HP-5987A GC/MS system. The R.I.s obtained for a standard PAH mixture were identical to those reported [51], which demonstrated the applicability of R.I. data in different laboratories. Thus, using several methods, the peaks in

References on p. B579

TABLE 24.2

COMPOUNDS IDENTIFIED IN AN AIRBORNE PARTICULATE MATTER EXTRACT AFTER
SEPARATION OF A NEUTRAL FRACTION BY HPLC [49]

Peak No.	Compound	Retention indices		Identification method*
		calcd.	ref.	
Fraction 1:				
1	Docosane			a,e
2	Tricosane			a,e
3	Tetracosene			a,e
4	Pentacosane			a,e
5	Hexacosane			a,e
6	Heptacosane			a,e
7	Octacosene			a,e
8	Nonacosene			a,e
9	Triacontene			a,e
10	Hentriacontane			a,e
11	Dotricontane			a,e
12	Tritricontane			a,e
Fraction 2:				
1	Branched hydrocarbon $C_{28}H_{46}$			a,e
2	Branched hydrocarbon $C_{30}H_{48}$			a,e
3	Branched hydrocarbon $C_{30}H_{48}$			a,e
4	Branched hydrocarbon $C_{30}H_{48}$			a,e
5	Branched hydrocarbon $C_{29}H_{48}$			a,e
Fraction 3:				
1	1-Phenylnaphthalene	314.07	315.19	a,b,c,e,f
2	4H-Cyclopenta[def]phenanthrene	322.25	322.08	a,b,c,e,f
3	Methylhexadecanoate			a,b,e
4	C1-4H-Cyclopenta[def]phenanthrene	337.19		a,e
5	C1-4H-Cyclopenta[def]phenanthrene	341.75		a,e
6	Long-chain ester $C_{21}H_{42}O_2$			a,e
7	Methyl octadecanoate			a,b,e
8	Methyl heneicosanoate			a,b,e
9	Methyl tricosanoate			a,b,e
10	Long-chain ester $C_{27}H_{54}O_2$			a,e
11	Long-chain ester $C_{29}H_{58}O_2$			a,e
12	Long-chain ester $C_{31}H_{62}O_2$			a,e
13	Long chain ester $C_{33}H_{66}O_2$			a,e
Fraction 4:				
1	Phenanthrene	300.08	300.00	a,c,d
2	Anthracene	301.69	301.96	a,b,c,e,f
3a	3-Methylphenanthrene	319.58	319.46	a,b,c,e

TABLE 24.2 (continued)

Peak No.	Compound	Retention indices		Identification method*
		calcd.	ref.	
3b	2-Methylphenanthrene	320.52	320.17	a,b,c,e
3c	9- or 4- or 1-Methylphenanthrene	323.29	323.06	a,b,c
4	2-Phenylnaphthalene	331.76	332.59	a,b,c,e,f
5a	C2-(Anthracene/phenanthrene)	341.51		a,b,e
5b	C2-(Anthracene/phenanthrene)	349.21		a,b,e
6	Aceanthrylene	352.04		a,b,e
7	PAH, mol. wt. 204	354.09		a,e
8	Tricosanal			a,e
9	Pentacosanal			a,e
Fraction 5:				
1a	Fluoranthene	344.73	344.01	a,b,c,e,f
1b	Acephenanthrylene	348.27	347.67	a,b,c,f
1c	Aceanthrylene	351.26		a,b,e,f
1d	Pyrene	352.20	351.51	a,b,c,e,f
2	Long-chain ester			a,e
3a	C1-(Fluoranthene/pyrene/acephen-anthrylene/aceanthrylene)	363.13		a,b,e,f
3b	Benzo[a]fluorene	366.82	366.73	a,b,c,e,f
3c	Benzo[b]fluorene	369.47	369.39	a,b,c,e,f
3d	C1-(Fluoranthene/pyrene/acephen-anthrylene)	374.37		a,b,e,f
4	Benzo[ghi]fluoranthene	390.80	389.60	a,b,c,e,f
Fraction 6:				
1a	Unknown PAH, mol. wt. 226	392.92		a,e,f
1b	Unknown PAH, mol. wt. 226	397.64		a,e
2a	Benz[a]anthracene	398.90	398.50	a,b,c,e,f
2b	Chrysene/triphenylene	400.46	400.00	a,b,c,d,e,f
3	C1-(Benzanthracene/chrysene/tri-phenylene)	423.28		a,e
Fraction 7:				
1	Unidentified			a,e
2	9,10-Anthracenedione	333.25		a,e
3	Dioic ester			a,e
4	Triphenylene/chrysene	400.73	400.00	a,b,c,e,f
5	Phthalate ester			a
6a	Benzofluoranthene	443.92	442.56	a,b,c,e,f
6b	Benzo[e]pyrene	447.85	450.73	a,b,c,e
6c	Benzo[a]pyrene	453.33	453.44	a,b,c,e,f
6d	Perylene	455.07	456.22	a,b,e
7a	C1-(Benzopyrene/benzofluor-anthene/perylene)			a,e,f

(Continued on p. B570)

References on p. B579

TABLE 24.2 (continued)

Peak No.	Compound	Retention indices		Identification method*
		calcd.	ref.	
7b	C1-(Benzopyrene/benzofluoranthene/perylene)			a,e,f
7c	C1-(Benzopyrene/benzofluoranthene/perylene)			a,e,f
7d	C1-(Benzopyrene/benzofluoranthene/perylene)			a,e,f
8a	C2-(Benzopyrene/benzofluoranthene/perylene)			a,e,f
8b	C2-(Benzopyrene/benzofluoranthene/perylene)			a,e,f
8c	C2-(Benzopyrene/benzofluoranthene/perylene)			a,e,f
9	Dibenzopyrene/dibenzo[def,p]chrysene			a,e,f
10	Dibenzopyrene/dibenzo[def,p]chrysene			a,e,f
11	Coronene			a,e,f

* a = Electron-impact mass spectrum; b = probability-based matching; c = retention index; d = retention of standard reference compound; e = methane positive chemical ionization mass spectrum; f = methane negative chemical ionization mass spectrum.

TABLE 24.3

COMPOUNDS IDENTIFIED IN THE ACID FRACTION OF AIRBORNE PARTICULATE MATTER EXTRACT [49]

Peak No.	Compound	Identification method*
1	2,6-Dimethoxyphenol	a,b,e
2	4-Hydroxy-3-methoxybenzoic acid	a,e
3	Hydroxymethoxymethylbenzoic acid	a,e
4	2,6-Dimethoxy-4-(2-propenyl)-phenol	a,b,e
5	Hydroxyethoxymethylbenzoic acid	a,e
6	4-Hydroxy-3,5-dimethoxybenzaldehyde	a,b,e
7	4-Hydroxy-3,5-dimethoxybenzaldehyde	a,b,e
8	Hydroxymethoxymethylbenzoic acid	a,b,e
9	Hydroxymethoxydimethylbenzoic acid	a,b,e
10	Compound mol. wt. 274	a,e
11	Compound mol. wt. 344	a,e

* Refer to Table 24.2.

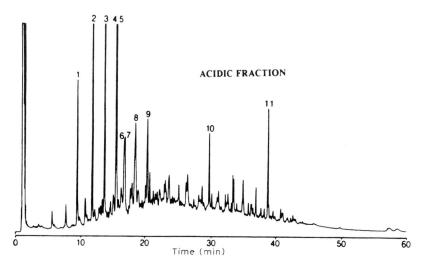

Fig. 24.6. GC/FID chromatogram of acidic fraction. For conditions see Fig. 24.4. For compound identification see Table 24.3.

HPLC Fractions 1-7 were identified and are reported in Table 24.2 (Figs. 24.4 and 24.5). HPLC Fractions 8-10 were also analyzed, and very few highly polar compounds, such as long-chain alcohols and aromatic ketones, were detected in those fractions. Compounds identified in the acidic fraction of the acid/base extraction are listed in Table 24.3 (Fig. 24.6). By correcting the amount of PAHs detected in different fractions for recoveries of standard PAHs under identical conditions, it was found that PAHs detected in the airborne particulate matter varied from 0.1 to 1.1 $\mu g/g$.

24.2.4 2,3,7,8-Tetrachlorodibenzo-*p*-dioxin

Based on the results of animal studies and in vitro assays, the symmetrically substituted 2,3,7,8-TCDD isomer appears to be the most toxic. The closely related 2,3,7,8-TCDD and 1,2,3,8-TCDD isomers differ in toxicity by a factor of 1000 to 10 000 [52]. Current techniques for the analysis of 2,3,7,8-TCDD in environmental samples are time-consuming and not completely satisfactory [53]. However, a liquid-crystal polysiloxane (LCP) capillary column separates 2,3,7,8-TCDD from all PCBs and from all other TCDD isomers [54], although retention times for OCDD and OCDF are very long [55]. To achieve the separation of 2,3,7,8-substituted PCDD/PCDF isomers along with analysis of total PCDD/PCDF in a single GC/MS analysis a more selective LCP column was developed [56]. Based on the separation mechanism of liquid-crystal stationary phases, for compounds with equal volatility the linear and symmetrical molecules will be retained longer than the bulkier molecules because their geometry is favorable to that of the liquid-crystal matrix. The 2,3,7,8-TCDD is the most symmetrical and linear of all 22 TCDDs,

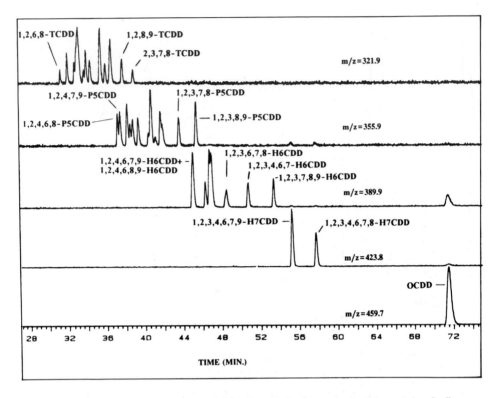

Fig. 24.7. Separation of tetra- to octachlorodibenzodioxins on a liquid-crystal polysiloxane capillary column by GC/MS. Chromatographic conditions: LCP, 25 m x 0.32-mm-ID fused-silica capillary column, temperature program: 100°C for 1 min, then to 210°C at 4°C/min, then to 260°C at 2°C/min, finally 30 min at 260°C.

and hence it is eluted after all other TCDD isomers. The separations achieved on the LCP column in GC/MS for a synthetic mixture that contains all isomers in the tetra- to octa-chlorodioxin congeners is shown in Fig. 24.7. The last peak in the mass chromatogram for TCDD was confirmed to be 2,3,7,8-TCDD. There are several advantages of the newly developed column for the analysis of environmental samples. It can be used for the analysis of the total PCDD/PCDF in addition to 2,3,7,8-TCDD in environmental samples. The most toxic 2,3,7,8-TCDD is eluted as the last isomer of all 22 TCDD isomers. More than 1 min retention time difference between 2,3,7,8-TCDD and the closest-eluted 1,2,8,9-TCDD isomer guarantees that there is no overlap in the case of minor column deterioration or a very high concentration of 1,2,8,9-TCDD.

24.2.5 Isomer-specific separation of polycyclic aromatic hydrocarbons

The separation of PAHs can also be performed on highly selective GC capillary columns. On the liquid-crystal stationary phases, isomeric PAHs separate on the basis of

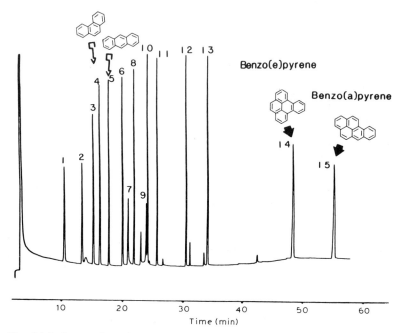

Fig. 24.8. Separation of isomeric PAHs on a liquid-crystal polysiloxane column. For chromatographic conditions see Fig. 24.7. Peaks: 1 = 9H-fluorene; 2 = 1-methyl-9H-fluorene; 3 = dibenzothiophene; 4 = phenanthrene; 5 = anthracene; 6 = 1-methylphenanthrene; 7 = 9-methylanthracene; 8 = fluoranthene; 9 = 2-methylanthracene; 10 = pyrene; 11 = 9,10-dimethylanthracene; 12 = 11H-benzo[b]fluorene; 13 = triphenylene; 14 = benzo[e]pyrene; 15 = benzo[a]pyrene.

their structural differences [57-61]. Recently, polymeric liquid-crystal stationary phases have been developed for use in GC capillary columns (Section 24.2.4) [62-64], and evaluated for the separation of PAHs in environmental samples [65,66]. The neutral fraction of airborne particulate matter was fractionated into aliphatic hydrocarbons and two fractions of PAHs on a normal-phase silica column. The GC/MS analysis of the PAH fractions showed that a large number of isomeric PAHs were separated on a liquid-crystal polysiloxane column that were difficult to separate and identify on conventional GC capillary columns [55]. It should be emphasized that the use of liquid-crystal columns reduces the time and effort required to use a specific column, such as the amine column in HPLC, for the separation of PAHs, based on the number of aromatic rings. HPLC fractionation by commonly used silica columns is sufficient prefractionation for the analysis of PAHs by a liquid-crystal column in GC/MS. The liquid-crystal polysiloxane capillary column showed the complete separation of benzo[a]pyrene from benzo[e]pyrene, anthracene from phenanthrene, and chrysene from triphenylene [56]. Separation of selected PAH standards on this column is shown in Fig. 24.8.

References on p. B579

24.2.6 Water samples

Organic analysis of natural water, drinking water, or industrial waste water determines specific organic compounds in μg/l or even ng/l (ppt) concentration. Several analytical procedures have been developed for the analysis of water. Each method is tailored for a few chemicals or classes of chemicals in the water samples. There are three types of organics present in the water samples: (1) volatile, (2) semivolatile, and (3) nonvolatile compounds. GC has been used for the analysis of volatile and some semivolatile compounds. HPLC is mainly used for nonvolatile and some semivolatile compounds. Usually, analytical chemists are not involved in sample collection techniques, but it is essential that the samples to be analyzed be representative of the total matrix studied. A detailed review of sampling techniques and related problems has been published [67].

24.2.6.1 Extraction and concentration techniques

Although some work has been reported on the direct analysis of water samples by GC, a concentration step prior to instrumental analysis is generally required. The water samples must be concentrated two to three orders of magnitude before analysis. For the semivolatile and nonvolatile organic compounds liquid/liquid extraction or sorbent-trapping techniques are being used extensively. In liquid/liquid extraction, typically 1 l of sample is extracted with two or three 100- to 150-ml portions of solvent. The combined extracts are concentrated to 1 ml or less for HPLC fractionation, followed by GC/MS analysis. Dichloromethane is the preferred solvent for the liquid/liquid extraction. Extraction of acidic and basic compounds at high and low pH values, respectively, separates these compounds. A Kuderna-Danish evaporator is recommended for concentration of the organic extract [68]. Sorbent-trapping techniques are used quite frequently for semivolatile and nonvolatile organics in water samples. The method involves two steps: the water is passed through a column containing a suitable adsorbent; and the adsorbed compounds are eluted for analysis. Numerous sorbents have been studied, among which organic resins, in particular XAD resins, are very effective. Desorption of compounds can be performed with organic solvents or by heating. The desorbed compounds can be analyzed directly by GC or GC/MS or they may be eluted by solvents and analyzed by HPLC or GC and GC/MS after concentration.

24.2.6.2 HPLC, GC, and GC/MS

Depending upon the complexity of the sample, HPLC fractionation may be useful to simplify samples prior to GC or GC/MS analysis. In a typical procedure, the organic extract of the environmental sample is screened by GC/ECD and GC/FID. If a large number of unresolved peaks are observed, the sample requires HPLC fractionation. It has been observed that a chemically bonded amine column is superior to silica for the prefractionation of complex samples. Detailed procedures for HPLC fractionation and for analysis of each fraction by GC and GC/MS have been reported. More than 100 compounds were identified in a St. Clair river sample [69]. The United States Environmental Protection Agency (EPA) has developed HPLC Method 605 for benzidines and Method 610 for PAHs [70]. Analysis of organics in water samples by HPLC only has been reported [71-73].

Nonvolatile acidic and basic compounds can be separated from water by ion chromatography. Ion-exchange resins are used to separate ionic nonvolatile compounds from the sample matrix in two separate sample aliquots. Nonvolatile strong acids, such as organic sulfonic acids, are separated from the water on anion-exchange resins, then eluted with HCl in methanol solution. Finally, the acids are methylated and analyzed by GC and GC/MS.

24.2.6.3 Purgeable organic compounds

Analysis of purgeable organics in water samples is a rapidly developing field. Because of their high vapor pressure, volatile compounds present difficult sampling, sample preparation, and analysis problems. The aim is to analyze trace organic volatile compounds quickly in a small volume of sample. Grob and Zorcher [74] designed a closed-loop stripping analysis (CLSA) system to remove compounds of intermediate molecular weight range from water samples. A schematic drawing of a modified CLSA system is shown in Fig. 24.9 [75]. The CLSA system has been used extensively in Europe and it is available from commercial sources. During the period of development of the CLSA method, Bellard and Lichtenberg [76] developed a purge-and-trap system for a wide range of purgeable organics in water samples (Fig. 24.10). In both systems the same principle of removing purgeable organics and trapping on adsorbents was used. The major difference is in the

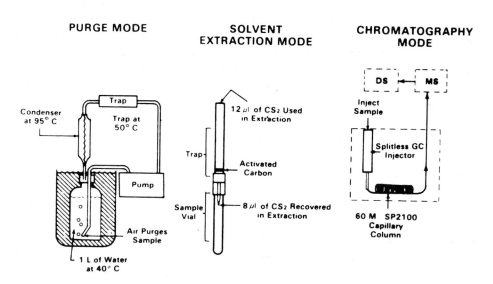

Fig. 24.9. Schematic drawing of the modified Grob CLSA system. (Reproduced from Ref. 75 with permission.)

References on p. B579

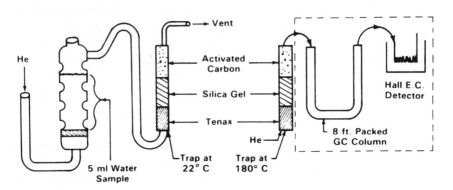

Fig. 24.10. Schematic drawing of the Bellard P and T method. (Reproduced from Ref. 75 with permission.)

desorption of the organics. In Grob's system CS_2 and in the Bellard and Lichtenberg system heat is used to desorb the organics. In spite of some disadvantages and limitations of both methods, both are complementary and widely used [77]. A comprehensive list of methods developed by the EPA for water analysis has been published [78].

24.2.7 Soil and sediments

Analysis of total organics in sediment and soil samples is greatly complicated by the sample matrix. Often, the methods used for volatile organics in water (Section 24.2.6) are applicable to the analysis of volatiles in wet sediments and soil. Sample cleanup and isolation of organics depends upon the sample matrix. Wet samples are more difficult to treat than dry soil and sediments and pose challenging problems. Methods for the analysis of purgeable organic pollutants in waste water treatment sludges have been described [79]. Analytical procedures for analyzing various complicated soil samples have been published in detail [80]. The analysis of semivolatile and nonvolatile organics in dry or wet soil is relatively easy. Traditional Soxhlet extraction, shaker extraction, or sonication is useful. More recently, sampling procedures, cleanup, and analysis of sediment and soil samples for PCB have been reported [81]. PCDD and PCDF are by far the most toxic common soil contaminants. Sample preparation, isolation, and analysis of PCDD/PCDF in soil and sediment samples have been described in detail [82]. Other important toxic contaminants are pesticides (Chapter 23). A complete analysis of soil samples for pesticides has been reported [83]. Analytical methods for soil and sediment extract are similar to those already described for fly ash and air-particulate matter.

24.3 SUPERCRITICAL-FLUID EXTRACTION AND SUPERCRITICAL-FLUID CHROMA-TOGRAPHY

Supercritical fluids (SF) have been used for the extraction of organics from various matrices for some time. Their application to environmental sample extraction and analysis is attracting increasing attention. The use of supercritical-fluid extraction (SFE) for environmental samples has proved to be superior to traditional methods of solvent extraction. SFE in combination with GC/MS has enabled a fast sample preparation and analysis of environmental samples. Even though GC and HPLC complement each other and are traditionally used for the separation of complicated environmental samples, they are inadequate to fill all requirements. In particular, nonvolatile or thermally labile analytes are not amenable to GC, and trace analysis of such compounds in complicated environmental samples by HPLC is difficult. Thus, supercritical-fluid chromatography (SFC) has shown great potential to fill the gap between GC and HPLC. Both packed and capillary column SFC techniques are being used for the analysis of environmental samples. SFC combined with MS provides universal instrumentation for the analysis of environmental samples. This section covers only environmental applications of SFC and SFE; for details of the method, see Chapter 8.

24.3.1 Supercritical-fluid extraction

The extraction and estimation of the recovery of organics from environmental solids is an essential step in sample preparation prior to analysis. Recent studies have shown that the use of SFs for analytical extractions provides a powerful alternative to traditional liquid solvent extraction methods. The use of SFE for extraction of environmental samples and comparison with liquid extraction methods have been reported in several recent studies [84-90]. SFE exploits the solvent power of various substances (gaseous or liquid at ambient temperatures and pressures) above their critical temperature and pressure. SFs have major advantages over liquid solvents: (1) They have solvent strengths approaching those of liquid solvents but much lower viscosities and higher solute diffusivities, resulting in faster extraction than with liquid solvents. (2) Very pure SFs are available, which are inert and have low critical temperatures, permitting the extraction of thermally labile compounds. (3) Generally used SFs, such as CO_2 and N_2O, are gaseous under ambient conditions, thus facilitating the concentration of the extracted analyte. (4) The solvent strength of SFs increases with increasing density, allowing the extraction conditions to be optimized for specific analytes by changing the extraction pressure. (5) The gaseous nature of generally used SFs under ambient conditions permits direct coupling of SFE with capillary GC or GC/MS.

The main components of a SFE system are a SF supply (usually a cylinder), the SF delivery system (pump), extraction cell, restrictor, and analyte collector (Fig. 24.11) [89]. Several research groups have developed their own SFE systems [89,91], which are less expensive than presently available commercial instruments. There are several benefits of SFE over traditional methods of liquid solvent extraction, the most notable of which with

References on p. B579

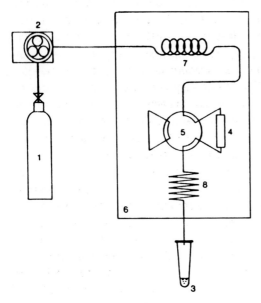

Fig. 24.11. Schematic drawing of an SFE system. 1 = Cylinder; 2 = pumping system; 3 = collection of extract; 4 = extractor; 5 = 6-way valve; 6 = oven; 7 = preheating coil. (Reproduced from Ref. 89 with permission.)

respect to environmental analysis is the use of nontoxic extraction media, some class-specific extraction selectivity, less sample handling, and direct coupling of SFE to GC [92-94] and GC/MS [95-97]. This is a new frontier in environmental sample preparation. SFE could replace traditional liquid extraction methods, such as Soxhlet and sonication, in the future.

The extraction of 2,3,7,8-TCDD from sediment samples by SFE has been reported for the first time by Onuska and Terry [89]. CO_2 and N_2O were tested, and complete extraction of the 2,3,7,8-TCDD was achieved within 30 min. The effect of moisture in sediment samples has also been studied, although the SFE efficiency decreases with increasing water content in sediment samples. Complete extraction of the sediment sample containing water was achieved in 60 min. This is much faster than the liquid solvent extraction methods. The SFE and recovery of PAHs from environmental samples, such as urban dust, fly ash, and river sediment, has been reported [84]. SFE efficiencies and recovery of PAHs have been compared with traditional methods of sonication and Soxhlet extraction. SFE has also been used for desorption and isolation of organic compounds from sorbents, such as Tenax, XAD, PUF, and polyimide resins.

24.3.2 Supercritical-fluid chromatography

Advances in micro-HPLC in the early Eighties have renewed the interest in the use of SFC. Initially, packed-column SFC had been studied by several researchers. Although greater sample capacities are obtained with packed columns than with capillary columns in SFC, the column length is a limiting factor, because the high pressure differential across the column results in poor chromatographic efficiency, Packed-column SFC is best for the rapid separation of less complex mixtures. Capillary column SFC can produce a greater number of theoretical plates, resulting in more efficient chromatographic separation. Therefore, capillary SFC is more useful in resolving very complex mixtures, particularly environmental samples. Since the first report of its use, steady progress has been made in instrumentation and column technology in SFC. Automated systems are currently available from several suppliers. Despite the interest in SFC, the environmental researchers must ask what problems SFC can tackle that have not already been solved by GC or HPLC. There are several environmentally important substances which are not volatile enough for GC or are thermally unstable. In complicated samples these substances cannot be detected well enough by HPLC. The ideal technique to fill the gap would have a low-temperature solvating mobile phase of programable strength and high chromatographic efficiency. In SFC, the diffusivity of the mobile phase is a function of density, which affects the column efficiency. Thus, in SFC, column efficiency is greatest for gas-like density conditions, and it decreases as liquid-like density conditions are approached. In general, resolution in SFC is greater than in HPLC and less than in GC. Hence, SFC is a suitable technique for nonvolatile, thermally labile, and highly polar organic pollutants in complex mixtures. In SFC, density programing is analogous to temperature programing in GC and gradient elution in HPLC. SFC has been increasingly used for the analysis of environmental samples because it is easily coupled with GC and GC/MS. Some recent developments in SFC and SFE for environmental analysis have been cited [98-107].

REFERENCES

1 C. Rappe, G. Choudhary and L.H. Keith (Editors), *Chlorinated Dioxins and Dibenzofurans in Perspective*, Lewis Publishers, Chelsea, MI, 1986.
2 G.G. Choudhary and O. Hutzinger (Editors), *Mechanistic Aspects of the Thermal Formation of Halogenated Organic Compounds Including Polychlorinated Dibenzo-p-dioxins*, Gordon and Breach, New York, 1983.
3 G. Choudhary, L.H. Keith and C. Rappe (Editors), *Chlorinated Dioxins and Dibenzofurans in the Total Environment*, Butterworth, London, 1983.
4 L.H. Keith, C. Rappe and G. Choudhary (Editors), *Chlorinated Dioxins and Dibenzofurans in the Total Environment II,* Butterworth, London, 1983.
5 M.A. Kamrin and P.W. Rodgers (Editors), *Dioxins in the Environment,* Hemisphere Publishing Corporation, Washington, 1985.
6 B.K. Afghan, J. Carron, P.D. Goulden, J. Lawrence, D. Leger, F. Onuska, J. Sherry and R. Wilkinson, *Can. J. Chem.,* 65 (1987) 1086.
7 F.W. Karasek and O. Hutzinger, *Anal. Chem.,* 58 (1986) 633A.
8 F.W. Karasek and F.I. Onuska, *Anal. Chem.,* 54 (1982) 309A.
9 C. Rappe, *Environ. Sci. Technol.,* 18 (1984) 78A.
10 R.E. Tucker, A.L. Young and A.P. Gray (Editors), *Human and Environmental Risks of Chlorinated Dioxins and Related Compounds*, Plenum Press, New York, 1983.
11 A. Bjorseth (Editors), *Handbook of Polycyclic Aromatic Hydrocarbons*, Dekker, New York, 1983.
12 K.D. Bartle, M.L. Lee and S.A. Wise, *Chem. Soc. Rev.,* 10 (1983) 113.

13 L.H. Keith (Editor), *Identification and Analysis of Organic Pollutants in Air,* Butterworth, London, 1984.

14 D.J. Futoma, S.R. Smith, J. Tanaka and T.E. Smith, *CRC Crit. Rev. Anal. Chem.,* 12 (1981) 69.

15 T. Vo-Dinh (Editor), *Chemical Analysis of Polycyclic Aromatic Compounds,* Wiley-Interscience, New York, 1989.

16 R.L. Spraggins, R.G. Oidham, C.L. Prescott and K.J. Baughman, in L.H. Keith (Editor), *Advances in Identification and Analysis of Organic Pollutants in Water, Volume 2,* Ann Arbor Science Publishers, Ann Arbor, MI, 1981, p. 747.

17 I. Temmerman, W. Mathys and D. Quaghebeur, in A. Bjorseth and G. Angeletti (Editors), *Proc. of the Fourth European Symposium, Organic Micropollutants in the Aquatic Environment,* Reidel, Dordrecht, 1986, p. 136.

18 U. Wahle, W. Kordel and W. Klein, *Int. J. Environ. Anal. Chem.,* 39 (1990) 121.

19 R. Otson and C. Chan, *Int. J. Environ. Anal. Chem.,* 30 (1987) 275.

20 I.M. Sayre, *J. Am. Water Work Assoc.,* 80 (1988) 53.

21 B. Dowty, L.E. Green and J.L. Laster, *J. Chromatogr. Sci.,* 14 (1976) 187.

22 Q.V. Thomas, J.R. Stork and S.L. Lammert, *J. Chromatogr. Sci.,* 18 (1980) 583.

23 A. Beard, K.P. Naikwadi and F.W. Karasek, *J. Chromatogr.,* submitted for publication.

24 S.B. Hawthorne and D.J. Miller, *Anal. Chem.,* 59 (1987) 1705.

25 H.Y. Tong, D.L. Shore and F.W. Karasek, *J. Chromatogr.,* 285 (1984) 423.

26 L.L. Lamparski and T.J. Nestrick, *Anal. Chem.,* 52 (1980) 2045.

27 L.L. Lamparski, T.J. Nestrick and R.H. Stehl, *Anal. Chem.,* 51 (1979) 1483.

28 H.Y. Tong, D.L. Shore and F.W. Karasek, *Anal. Chem.,* 56 (1984) 2442.

29 H.Y. Tong and F.W. Karasek, *Chemosphere,* 15 (1986) 1141.

30 H. Hagenmaier, H. Brunner, R. Haag and M. Kraft, *Fresenius Z. Anal. Chem.,* 323 (1986) 24.

31 D.R. Thielen and G. Olsen, *Anal. Chem.,* 60 (1988) 1332.

32 M. Swerev and K. Ballschmiter, *Anal. Chem.,* 59 (1987) 2536.

33 B.J. Fairless, D.I. Bates, J. Hudson, R.D. Kleopfer, T.T. Holloway and D.A. Morey, *Environ. Sci. Technol.,* 21 (1987) 550.

34 J.L. Hudson and D.A. Morey, *Chemosphere,* 18 (1989) 141.

35 R.M. Smith, P.W. O'Keefe, D.R. Hilker and K.M. Aldous, *Anal. Chem.,* 58 (1986) 2414.

36 R.E. Clement, S.A. Suter and H.M. Tosine, *Chemosphere,* 18 (1989) 133.

37 A. Yasuhara, H. Ito and M. Morita, *Environ. Sci. Technol.,* 21 (1987) 971.

38 J.M. Czuczwa and R.A. Hites, *Environ. Sci. Technol.,* 20 (1986) 195.

39 C. Rappe, M. Nygren, G. Lindstorm, H.R. Buser, O. Blaser and C. Wuthrich, *Environ. Sci. Technol.,* 21 (1987) 964.

40 D.G. Patterson, Jr., L. Hampton, C.R. Lapeza, Jr., W.T. Belser, V. Green, L. Alexander and L.L. Needham, *Anal. Chem.,* 59 (1987) 2000.

41 R.D. Kleopfer, R.L. Greenall, T.S. Viswanathan, C.J. Kirchmer, A. Gier and J. Muse, *Chemosphere,* 18 (1989) 109.

42 H. Hagenmaier, H. Brunner, R. Haag and M. Kraft, *Fresenius Z. Anal. Chem.,* 323 (1986) 24.

43 P.W. O'Keefe, R.M. Smith, D.R. Hilker, K.M. Aldous and W. Gilday, in L.H. Keith, C. Rappe and G. Choudhary (Editors), *Chlorinated Dioxins and Dibenzofurans in the Total Environment II,* Butterworth, London, 1983, p. 111.

44 R.E. Adams, M.M. Thomason, D.L. Strother, R.H. James and H.C. Miller, *Chemosphere,* 15 (1986) 1113.

45 F.W. Karasek, T.S. Thompson, D.H. Shalenburge and K.P. Naikwadi, *Proc. Technology Transfer Conference,* Ontario Ministry of the Environment, Toronto, 1986, 215 pp.

46 T.S. Thompson, K.P. Naikwadi and F.W. Karasek, *Proc. Technology Transfer Conference,* Ontario Ministry of the Environment, Toronto, 1987, 19 pp.

47 C.H. Williams, Jr., C.L. Prescott and P.B. Stewart, in L.H. Keith, C. Rappe and G. Choudhary (Editors), *Chlorinated Dioxins and Dibenzofurans in the Total Environment II*, Butterworth, London, 1983, p. 457.
48 K.P. Naikwadi and F.W. Karasek, *Int. J. Environ. Anal. Chem.*, 38 (1990) 329.
49 F.W. Karasek, G.M. Charbonneau and K.P. Naikwadi, Development of methodology for complete analysis of complex environmental samples, *Report No 216R*, Ontario Ministry of the Environment, Toronto, 1988.
50 K.P. Naikwadi, M. Charbonneau, F.W. Karasek and R.E. Clement, *J. Chromatogr.*, 398 (1987) 227.
51 M.L. Lee, D.L. Vassilaros, C.M. White and M. Novotny, *Anal. Chem.*, 51 (1979) 768.
52 A. Poland, E. Glover and A.S. Kende, *J. Chem. Biol.*, 251 (1976) 4926.
53 G. Choudhary, L.H. Keith and C. Rappe (Editors), *Chlorinated Dioxins and Dibenzofurans in the Environment*, Butterworth, London, 1983, p. 165.
54 K.P. Naikwadi and F.W. Karasek, *J. Chromatogr.*, 369 (1986) 203.
55 K.P. Naikwadi, A.M. Mcgovern and F.W. Karasek, *Can. J. Chem.*, 65 (1987) 970.
56 K.P. Naikwadi and F.W. Karasek, *Chemosphere*, 20 (1990) 1379.
57 K.P. Naikwadi, D.G. Panse, B.V. Bapat and B.B. Ghatge, *J. Chromatogr.*, 195 (1982) 309.
58 P.P. Pawar, K.P. Naikwadi, S.M. Likhite, B.V. Bapat and B.B. Ghatge, *J. Chromatogr.*, 254 (1982) 57.
59 J.M. Janini, G.M. Muschik and W. Zielinski, Jr., *Anal. Chem.*, 48 (1976) 809.
60 F. Janssen and T. Kalidin, *J. Chromatogr.*, 235 (1982) 323.
61 R.C. Kong, M.L. Lee, Y. Tominaga, R. Pratap, M. Iwao and R.N. Castle, *J. Chromatogr. Sci.*, 20 (1982) 502.
62 K.P. Naikwadi, A.L. Jadhav, S. Rokushika, H. Hatano and M. Ohshima, *Macromol. Chem.*, 187 (1986) 1407.
63 A.L. Jadhav, K.P. Naikwadi, S. Rokushika, H. Hatano and M. Ohshima, *J. High Resolut. Chromatogr. Chromatogr. Commun.*, 10 (1987) 77.
64 S. Rokushika, K.P. Naikwadi, A.L. Jadhav and H. Hatano, *J. High Resolut. Chromatogr. Chromatogr. Commun.*, 8 (1985) 480.
65 B.A. Jones, J.S. Bradshaw, M. Nishioka and M.L. Lee, *J. Org. Chem.*, 49 (1984) 4947.
66 K.E. Markides, M. Nishioka, B.J. Tarbet, J.S. Bradshaw and M.L. Lee, *Anal. Chem.*, 57 (1985) 1296.
67 L.H. Keith (Editor), *Principles of Environmental Sampling*, American Chemical Society, Salem, MA, 1988.
68 R.G. Webb, *Isolating Organic Water Pollutants: XAD Resins, Urethane Foams, Solvent Extraction*, EPA-66014-75-0035, U.S. EPA, Corvallis, OR, 1975.
69 J.H. Carey and J.H. Hart, *Water Pollut. Res. J. Can.*, 21 (3) (1986) 309.
70 Q.V. Thomas, J.R. Stork and S.L. Lammert, *J. Chromatogr. Sci.*, 18 (1980) 583.
71 K. Grob and A. Habich, *J. High Resolut. Chromatogr. Chromatogr. Commun.*, 6 (1983) 11.
72 *Method 610, Polycyclic Aromatic Hydrocarbons*, Federal Register (U.S.A.), 44 (233) 69514, Dec. 3, 1979.
73 T. Vo-Dinh (Editor), *Chemical Analysis of Polycyclic Aromatic Compounds*, Wiley-Interscience, New York, 1989, p. 98.
74 K. Grob and F. Zorcher, *J. Chromatogr.*, 117 (1976) 285.
75 R.G. Melton, W.E. Coleman, R.W. Slater, F.C. Kopfler, W.K. Allen, T.A. Aurand, D.E. Mitchell and S.J. Voto, in L.H. Keith (Editor), *Advances in Identification and Analysis of Organic Pollutants in Water, Volume 2*, Ann Arbor Science Publishers, Ann Arbor, MI, 1981, p. 597.
76 T.A. Bellar and J.J. Lichtenberg, *J. Am. Water Works Assoc.*, 66 (1974) 739.
77 R.D. Lingg, R.G. Melton, F.C. Melton, F.C. Kopfler, W.E. Coleman and D.E. Mitchell, *J. Am. Water Works Assoc.*, 69 (1977) 605.
78 L.H. Keith (Editor), *Advances in the Identification and Analysis of Organic Pollutants in Water, Volume 2*, Ann Arbor Science Publishing, Ann Arbor, MI, 1981.

79 C.L. Haile, Y.A. Shan, L.S. Malone and R.V. Northcutt, in L.H. Keith (Editor), *Advances in Identification and Analysis of Organic Pollutants in Water, Volume 2,* Ann Arbor Science Publishers, Ann Arbor, MI, 1981, p. 763.

80 K. Wolf, W.J. van den Brink and F.J. Colon (Editors), *Contaminated Soil'88, Vol. 2,* Kluwer Academic Publishers, Dordrecht, 1988.

81 A.L. Alford-Steven, J.W. Elchelberger and W.L. Budde, *Environ. Sci. Technol.,* 22 (1988) 304.

82 J.H. Exner, W.D. Keffer, R.O. Gilbert and R.R. Kinnison, in C. Rappe, G. Choudhary and L.H. Keith (Editors), *Chlorinated Dioxins and Dibenzofurans in Perspective,* Lewis Publishers, Chelsea, MI, 1986, p. 139.

83 F.I. Onuska, in F.W. Karasek, O. Hutzinger and S. Safe (Editors), *Mass Spectrometry in Environmental Chemistry,* Plenum Press, New York, 1984, p. 367.

84 S.B. Hawthorne and D.J. Miller, *Anal. Chem.,* 59 (1987) 1705.

85 M. Saito, T. Hondo and Y. Yamauchi, in R.M. Smith (Editor), *Supercritical Fluid Chromatography,* Royal Society of Chemistry, London, 1988, p. 203.

86 M.A. McHugh and V.J. Krukonis (Editors), *Supercritical Fluid Extraction,* Butterworth, London, 1986.

87 B.W. Wright, S.R. Frye, D.G. McMinn and R.D. Smith, *Anal. Chem.,* 59 (1987) 640.

88 S.B. Hawthorne and D.J. Miller, *J. Chromatogr.,* 403 (1987) 63.

89 F.I. Onuska and K.A. Terry, *J. High Resolut. Chromatogr. Chromatogr. Commun.,* 12 (1989) 357.

90 K.S. Nam, S. Kapila, D.S. Viswanath, T.E. Clevenger, J. Johansson and A.F. Yanders, *Chemosphere,* 19 (1989) 33.

91 K. Sugiyama, M. Saito, T. Honda and M. Senda, *J. Chromatogr.,* 332 (1985) 107.

92 S.B. Hawthorne, D.J. Miller and J.J. Langenfeld, *J. Chromatogr. Sci.,* 28 (1990) 1.

93 S.B. Hawthorne, D.J. Miller and M.S. Krieger, *J. Chromatogr. Sci.,* 27 (1989) 347.

94 S.B. Hawthorne and D.J. Miller, *J. Chromatogr. Sci.,* 24 (1986) 258.

95 S.B. Hawthorne, M.S. Krieger and D.J. Miller, *Anal. Chem.,* 61 (1989) 736.

96 J.R. Wheeler and M.E. McNally, *J. Chromatogr. Sci.,* 27 (1989) 534.

97 B.W. Wright, C.W. Wright and J.S. Fruchtler, *Energy Fuels,* 3 (1989) 474.

98 J.C. Fjeldsted and M.L. Lee, *Anal. Chem.,* 56 (1984) 619A.

99 R.D. Smith, B.W. Wright and C.R. Yonker, *Anal. Chem.,* 60 (1988) 1323A.

100 H.H. Hill and M. Morissey, *Spectra-Physics Chromatogr. Rev.,* 12 (1985) 2.

101 S. Rokushika, K.P. Naikwadi, A.L. Jadhav and H. Hatano, *Chromatographia,* 22 (1986) 209.

102 S. Rokushika, K.P. Naikwadi, A.L. Jadhav and H. Hatano, *J. High Resolut. Chromatogr. Chromatogr. Commun.,* 8 (1985) 480.

103 S.L. Pentoney, Jr., K.H. Shafer and P.R. Griffiths, *J. High Resolut. Chromatogr. Chromatogr. Commun.,* 24 (1986) 230.

104 H.T. Kalinoski, H.R. Udseth, B.W. Wright and R.D. Smith, *Anal. Chem.,* 58 (1986) 2421.

105 S.B. Hawthorne and D.J. Miller, *J. Chromatogr.,* 468 (1989) 115.

106 R.D. Smith, H.T. Kalinoski, H.R. Udseth and B.W. Wright, *Anal. Chem.,* 56 (1986) 2467.

107 R.D. Smith and H.R. Udseth, *Anal. Chem.,* 59 (1986) 13.

Chapter 25

Amines from environmental sources

H.A.H. BILLIET

CONTENTS

25.1 INTRODUCTION

Aliphatic and aromatic mono-, di-, and polyamines are used as raw materials or at an intermediate stage in the production of industrial chemicals, e.g., pharmaceuticals, polymers, pesticides, dyestuffs, and corrosion inhibitors. Many of them are suspected of being carcinogenic or mutagenic substances due to their adsorption tendency in tissues. Low-molecular-weight aliphatic amines also receive much attention as odorous substances in studies of air pollution. They can be present or develop through natural decomposition processes in a variety of systems, such as foods, cattle feed yards, and fish industries. Amines can be toxic of themselves, or more often, they can become toxic via chemical reactions, e.g., with nitriles, leading to the formation of nitrosamines.

Growing concern about the quality of air, and new evidence for health hazards as a consequence of exposure to amines in low concentrations have stimulated the development of sensitive analytical techniques, chromatography being an important one. Chromatography of free amines is difficult undoubtedly due to the tendency of these compounds to become adsorbed in the analytical system. It is generally more difficult to chromatograph aliphatic than aromatic amines, and the adsorption (interaction) tendency is in the order: primary > secondary > tertiary.

Amines are difficult to extract form water due to their high polarity. The determination of amines in water can be approached by different ways: (a) direct gas chromatography (GC) of aqueous samples; (b) concentration of the amines on an adsorbent followed by desorption, separation, and detection; and (c) precolumn derivatization of the amines, followed by separation and detection.

References on p. B593

25.2 METHODOLOGY

25.2.1 Gas chromatography

Several basic problems must be tackled when analyzing amines by GC: adsorption on the equipment, oxidation under chromatographic conditions, and the often very low levels of amines in the samples. One way to solve part of the problems is by derivatizing the amines in order to improve their chromatographic properties (peakshape and detection). For instance, introducing a halogen (fluorine)-containing group in the molecule enhances the response of the electron-capture detector (ECD). Another way to enhance the selectivity and sensitivity of a constant-current ECD [1] for the determination of polyaromatic amines is by addition of oxygen to the makeup gas. The use of the nitrogen/phosphorus detector (NPD) is an obvious choice for improving the detection of amines.

Prediction of the retention index of amines from the number of carbon and carbon-equivalent atoms in the molecule has been reported by Peng et al. [2]. The usefulness of that relationship lies in the tentative identification of a broad range of compounds, although positive identification by means of mass spectrometry (MS) will always be preferable.

High-temperature-stable ($<420°C$) stationary phases of the ladder-type siloxane polymer Lestosil have been effective in the GC of high-boiling polar compounds, including amines, without conversion to more volatile derivatives [3]. Direct analysis of aqueous samples minimizes sample preparation, thereby improving precision [4]. However, direct GC, even with an amine-deactivated column, is limited to the parts-per-million (ppm) level. Detection limits can be improved by concentrating the aqueous samples before measurement on a Cu(II) absorber column [5], on an XE-340 column [6], or by steam destillation [7]. These methods lower the detection limits but, as a sideeffect, also concentrate impurities leading to possible false positives.

Derivatization reactions, often selective for the amine type (primary, secondary, tertiary) have been used to improve detection. Because some of the derivatization reagents react with water (fluoroacetates [8], boron chelates [9], m-toluamides [10]), their use is limited to organic solvent extracts of amines. Direct derivatization of amines in water is feasible with reagents, such as o-phthalaldehyde [11], dinitrofluorobenzene [12], and 2-methoxy-2,4-diphenyl-3[2H]-furanone [13]. Pentafluorobenzaldehyde [14] is selective for primary amines, and derivatizations have been performed in ethanolic and aqueous samples, but not all primary amines will react with the same efficiency. After reaction, the imines are extracted with hexane and analyzed by capillary GC/MS.

Aromatic amines at very low concentrations can be determined by GC of the heptafluorobutyrate derivatives, with the ECD [15]. A generator has been described for the production of standard atmospheres of eleven aromatic amines at very low concentrations [15]. A method for trace analysis of free aromatic amines, applicable to the rubber industry, involves the separation of the amines by high-temperature glass-capillary GC on an OV-73 stationary phase by means of on-column injection and N-selective detection [16]. In earlier publications, the same authors described gas chromatographic methods, based on perfluoro fatty acid amides of aromatic amines, with the NPD [17] and ECD [18]. Using MS with electron impact and methane chemical ionization, the structure of the fluorobutyric acid amides was investigated [19]. The same methodology has been applied to the determination of isocyanates and amines in workplace atmospheres [20].

Aqueous acidic extraction was used to isolate basic nitrogen compounds from coal liquefaction products. The samples were analyzed by GC/MS [21]. Primary aromatic amines in coal liquefaction products were determined by GC/MS [22] after they were shown to be the predominant mutagenic species.

Because of their toxicity and mutagenicity, the determination of trace amounts of amines in workplaces [23] is very important. Enrichment and cleanup steps are necessary to reach parts-per-billion (ppb) levels. Direct analysis of free amines in salt solutions at sub-ppm levels by GC has been reported by Audunsson and Mathiasson [24]. Different parameters influencing the analysis were investigated: sample handling and storing, injector temperature, injector impregnation, pH of the solution, type of salt matrix, and detector response. The detection limit was of the order of 10 ppb for most amines.

Air sampling and transferring high-boiling compounds from sample adsorption tubes to capillary GC columns has been studied by Lawrence [25]. Results with diphenylamine, among other high-boiling compounds, were reported. The adsorption tubes used in this study were glass tubing containing a small amount of Tenax-GC material. Others used sampling trains, consisting of glass-fiber and silver membrane filters and support pad assemblies, followed by glass tubes containing Tenax, silica gel, or XAD-2 as sorbents for the analysis of aromatic amines in air [26]. Tenax was the sorbent of choice, since relatively poor and inconsistent recoveries were obtained with silica gel, and XAD-2 had a poor load capacity for amines. Amines in treated municipal water were adsorbed on an ion exchanger and, after elution with 1 M KOH solution and concentration by a purge-and-trap technique ($CuCl_2$ on Chromosorb W-AW), they were determined as heptafluoro-butyryl derivatives by GC on a fused-silica SPB-5 column and conventional or MS detection [27].

Stationary-phase selectivity for amines in gas chromatography can be optimized by using different types of liquid phases in different concentrations [28]. By varying the concentration of either UCON 50-HB-2000 or Igepal CO-880 liquid phase on Carbopack B packing, deactivated with 0.8% KOH, a column is selected for the separation of a mixture of aliphatic, aromatic, and cyclic amines. Other parameters to be optimized include the column length and the oven temperature conditions.

Capillary GC can also be used for the enantio-selective separation of amines and amino alcohols. The use of a particular cyclodextrin, like heptakis(3-O-acetyl-2,6-di-O-pen-tyl)-β-cyclodextrin as a chiral stationary phase in capillary GC for the separation of trifluoroacetylated chiral amines and amino alcohols was described by Konig et al. [29]. Enantioselectivity was obtained by the inclusion properties of the cyclodextrin derivatives. The chiral stationary phase exhibited a wide operating temperature range and was stable above 200°C.

References on p. B593

25.2.2 High-performance liquid chromatography

The separation of amines by liquid chromatography (LC) was formerly analogous to the classical amino acid analysis method of Stein and Moore [30,31]. This method was very popular in the early Seventies and is well documented in Chapter 30 of the book by Deyl et al. [32]. In the introduction to that chapter, the author makes the statement: "High-speed techniques and gel permeation chromatography are not popular at present, and it is likely that classical ion-exchange methods will dominate this area, as they give good and rapid separations". A typical example of cation-exchange chromatography can be found in a paper on the separation of mutagenic aromatic amines in coal conversion oil [33]. The chromatography of amines has moved away from the classical ion-exchange resin, and ion-interaction chromatography now dominates the field.

Nevertheless, a material specially suited for the separation of very basic compounds is alumina, almost forgotten as column material in HPLC [34]. Alumina under certain conditions acts as an ion exchanger, and due to its amphoteric character it can have cation- and anion-exchange properties. In acidic media, its cation-exchange capacity is small, and acidic alumina would be better suited as anion exchanger. In a basic environment, alumina acts predominantly as a cation exchanger. However, the variation with pH is gradual, and either cation- or anion-exchange properties can be observed over a fairly broad pH range. Alumina has also proven to be very stable at extreme pH values. Even pH 13 does not have an adverse effect: less than 1% of the column material is dissolved after 150 h of continuous operation. An example of the separation of moderately strong bases at pH 9.2 is given in Fig. 25.1. This procedure would not be possible on a silica-based ion exchanger. Lingeman et al. [35] made a comparison between the use of alumina and unmodified silica gel as cation-exchange material for the separation of amines with aqueous solvent mixtures. The chromatographic selectivity can be improved by addition of organic solvents to the aqueous mobile phase.

Besides ion exchange, ligand-exchange methods have often been used for the separation of amines, e.g., on Cu(II) stearate-coated columns. The columns are more stable than alkali-hydroxide-coated ones, and peaktailing is reduced [36]. A retention mechanism of the amines in that system has been proposed, and constants of complex formation between the amines and Cu(II) stearate in the gas phase in the presence of ammonia vapor have been obtained. The selectivity in ligand-exchange chromatography depends not only on the temperature but also on the concentration of the mobile-phase ligand.

Ligand-exchange chromatography on paper impregnated with Cu(II), sorbed on zinc silicate, has been studied for amines [37]. R_F values of 16 amines in 6 different solvent systems, and retention data on columns of Cu(II), sorbed on zinc silicate, have been published. No significant tailing was observed during the elution of the various amines, and the column was sufficiently stable to permit the same column to be used repeatedly.

Ligand exchange on Cu(II)-modified bonded silica gel as the stationary phase and hexane, containing ammonia and methanol, as the mobile phase showed good efficiency for separating sample amines [38]. Ammonia was found to be very effective in improving both the separation and the peakshape. Aliphatic amines were well separated.

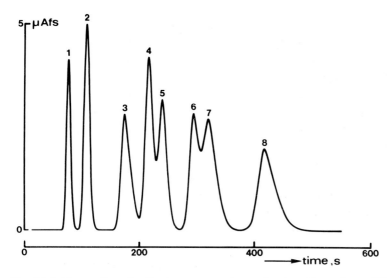

Fig. 25.1. Separation of cationic compounds on alumina. The column (25 cm x 4.6 mm ID) was packed with ALOX T10 (Batch EF 42) from Merck. The mobile phase contained tetraethylammonium borate buffer (pH 9.2), ionic strength 4×10^{-2} M. Amperometric detection was performed at +1.4 V with regard to the standard calomel electrode. So-lutes: 1 = 1,4-diaminobutane, 2 = triethanolamine, 3 = quinine, 4 = ephedrine, 5 = allylamine, 6 = piperidine, 7 = pyrrolidine, 8 = monoethanolamine. (Reprinted from Ref. 34 with permission.)

More fundamental studies on the chromatography of aromatic amines have been performed by Papp and Vigh [39]. The role of the buffer cations in RP-HPLC of low-, medium-, and high-pK aromatic amines was investigated in a reversed-phase system with methanol as modifier. The "silanophilic" (i.e. ion-exchange) contribution to the retention of amines by the modified silicas was large over the entire methanol concentration range. This ionic contribution can be largely suppressed by higher concentrations of inorganic or organic cations. Silanophilic interactions of aromatic amines and pyridine derivatives in reversed-phase (RP)LC have also been modified with borate or acetate salts of transition metals [40]. It was shown that selectivity and resolution can be improved.

Ion-interaction (ion-pairing) HPLC has been employed for the separation of some typical aliphatic and aromatic amines [41]. The effects on retention of the alkyl chain-length of the interaction ion and the eluent flowrate have been studied. UV and conduc-tometric detection showed an average sensitivity of 40 ng without pretreatment or derivatization.

Poor peakshapes of amines in RP-HPLC are frequently observed. In order to improve column efficiency, the mechanism of broadening must be understood. In addition to the silanol groups on the packing surface, the metal sites have been implicated as possible adsorption sites for amines. Stainless-steel frits appear to have a deleterious effect on peakshape by way of both mechanical and chemical interaction. The use of stainless-steel

References on p. B593

meshes or screens instead of frits is recommended. Remaining interactions are due to the silanol groups and have a complex nature [42].

The amine concentration in samples from environmental sources is often so small that very specific detection is needed. Reversed-phase HPLC with UV detection of some carcinogenic amines in the work atmosphere of the polyurethane industry has been described by Nieminen et al. [43]. Primary and secondary amines can also be derivatized with benzoyl chloride in sodium bicarbonate to form N-alkyl and N,N-dialkyl benzamides, showing very strong UV absorbance [44]. The derivatives are stable and easy to chromatograph on a reversed-phase column. The method is sensitive to nanomol amounts in the sample.

Derivatization and sorption of amines in air, sampled on a silica gel-packed tube, is a very elegant method for monitoring air quality [45]. The resulting m-toluoyl derivatives were analyzed by a RP-HPLC with UV detection. The absorptivity of m-toluoyl derivatives is more than sufficient to detect traces of aliphatic amines in air at levels of $< 1\ \mu g/m^3$. For the sensitive detection of amines in the picogram to nanogram range, fluorometric and electrochemical detection is used. Pre- and postcolumn derivatization methods have been used to improve separation selectivity and to enhance detection properties.

The classical detection method was the ninhydrin postcolumn reaction. Although this is a sensitive reaction, much lower detection limits are needed for the determination of amines in process fluids or environmental samples. Selectivity in electrochemical detection is often needed for the determination of trace amounts in complex samples. For this, pre- and postcolumn reagents have been developed, which have been reviewed by Kissinger et al. [46]. This technique has been applied by Shimada et al. [47]. N-(4-Anilinophenyl)isomaleimide (APIM) and N-(4-anilinophenyl)isophthalimide (APIP) were used to derivatize phenethylamine and piperidine as model compounds. The APIM derivatives were more responsive to the electrochemical detector than those formed with APIP. The detection limit for the phenethylamine/APIM adduct was ca. 0.1 pmol. Good linearity was observed in the low-ng range. Ferrocene reagent for precolumn labeling of amines with oxidative electrochemical detection, involving the aromatic amino and guaiacol group as electrophores, has been described by Tanaka et al. [48].

Precolumn derivatization of aliphatic amines with o-acetylsalicyloyl chloride results in 2-hydroxybenzamide after alkaline hydrolysis. The free phenolic group can then be detected electrochemically, following separation by HPLC [49]. Using a solid-phase reactor, containing a polymeric anhydride substituted with o-acetylsalicyl groups, amines can be derivatized to the corresponding o-acetylsalicylamides. They are usually detected spectrophotometrically, but the derivatives also respond in an electrochemical detector, despite the absence of a free phenolic group [50].

Instead of the classical o-phthalaldehyde/thiol reagent for the derivatization of primary amines, o-phthalaldehyde/sulfite, can be used to enhance electrochemical detection after LC. In the presence of sulfite, N-alkyl-1-isoindole sulfonates are formed rapidly and cleanly. They are easily oxidized at carbon electrodes [51]. The OPA/sulfite derivatives are relatively inert towards excess OPA and exhibit good in situ stability.

Trinitrobenzene sulfonic acid acts as a chromophore as well as an electrophore precolumn derivatizing agent in the HPLC of alkylamines [52]. The reaction is quantitative, and trinitrophenyl derivatives are amenable to UV and electrochemical detection, the latter providing lower detection limits.

Electrochemical detection of aromatic amines without derivatization has been reported by Concialini et al. [53]. The aromatic amines are easily oxidized electrochemically. The electrochemical behavior of a large number of aromatic amines of environmental interest was studied with emphasis on the selectivity of the detector.

Polyaromatic amines in synthetic fuel mixtures can be analyzed directly by LC with electrochemical detection [54]. Some selectivity among the different arylamines can be achieved by careful selection of the working electrode potential. A similar setup has been used to determine aromatic amines in sea water [55]. Detection limits of 15 and 1.5 nM for each aromatic amine after preconcentration were achieved with coulometric and amperometric electrochemical cells, respectively. Aliphatic amines separated by liquid chromatography can be detected in the picomol range by electrochemical generation and detection of excess bromine [56].

Aromatic amines in soil-core and river water samples were preconcentrated and electrochemically detected at the low-pg/ml range [57]. Synthetic fuel products contain primary aromatic amines which are suspected of mutagenicity. A monitor based on a liquid chromatograph with electrochemical detection was constructed to operate in the field [58].

Many different fluorigenic reagents for (primary and secondary) amines have been used to enhance detection selectivity and sensitivity [59]. Table 25.1 lists some of the most frequently used reagents for the pre- and postcolumn detection reactions of amines.

1,2-Naphthoylenebenzimidazole 6-sulfochloride compares favorable with dansyl (DNS) chloride as a fluorescent reagent for the derivatization of aliphatic amines with respect to sensitivity and detection limits. Chromatography on C_{18} silica columns and TLC on silica gel can be employed for the rapid separation of aliphatic amines after derivatization with this reagent [77]. Fluorescent derivatives of primary amines, analyzed on a RP column, can also be detected by a chemiluminescence detector [78]. Derivatization was performed with DNS chloride, 4-chloro-7-nitrobenzo-1,2,5-oxadiazole, and o-phthaldialdehyde. The DNS derivatives gave the lowest limits of detection (fmol). The corresponding values for the other two derivatives were several hundred fmol. Chemiluminescence was generated and detected with a specially designed system, using photon counting for measuring the emitted light. Actual detection limits achieved corresponded to ca. 10^{-10} mol of the amine in a spot after elution from a thin-layer plate and to ca. 5 x 10^{-14} mol of the amine in a sample volume of 10 μl, injected into a liquid chromatograph [77].

Tertiary amines were separated on a Nucleosil $5N(CH_3)_2$ resin column and detected by postcolumn derivatization, based on a reaction with a color reagent (acetic anhydride solution of citric acid) [79]. This reaction is specific; primary and secondary amines do not react or interfere.

References on p. B593

TABLE 25.1

SOME REAGENTS USED FOR PRE- AND POSTCOLUMN REACTION DETECTION OF AMINES

Reagent	Type of amines	Detection		Ref.
		method	limit	
Phenyl isocyanate	primary, secondary	UV	1-5 ng	60,61
1-Naphthyl isocyanate	primary	UV	1 ng	61
m-Toluoyl chloride		UV	1 ppm	62
2-Naphthyl chloroformate	tertiary	fluorescence	few ng per injection	63
o-Phthalaldehyde	primary	fluorescence	500 fmol	64,65
Fluorescamine	primary			66
Dansyl chloride, dabsyl chloride	primary, secondary	UV + fluores-cence		67-69
Different halogen nitro-benzofurazans				64,70,71
7-Chloro-4-nitrobenzo-2-oxa-1,3-diazole	secondary	fluorescence	few pmol	70
4-(2-Phthalimidyl)benzoyl chloride		fluorescence	10 pmol	72
3-(2-Phthalimidyl)benzoyl chloride				72
3-(2-Phthalimidyl)methoxy-benzoyl chloride				72
Polymeric activated 9-fluo-renyl methyl chloroformate		UV, fluores-cence		73,74
3-Benzoyl-2-quinoline carboxaldehyde		fluorescence		75
3,5-Dinitrobenzoyl chloride		UV		76

Polycyclic aromatic amines can be detected by chemiluminescence at 1 to 2 orders of magnitude lower concentration than by fluorescence in the same chromatographic system [80]. This high sensitivity allows these compounds to be determined in complex matrices without complete chromatographic isolation.

The determination of aromatic amines present as impurities in color additives for foods, drugs, and cosmetics is based on extraction from the pigment with chloroform, followed by diazotization and coupling of the diazonium salt with the disodium salt of 3-hydroxy-2,7-naphthalenedisulfonic acid or with 4,5-dihydro-5-oxo-1-(4-sulfophenyl)[1H]-pyrazole-3-carboxylic acid. The coupling products are then separated by HPLC and quantitated by measurement at 254 nm and 510 nm [81]. A similar method has been published by Bailey [82].

25.2.3 Thin-layer chromatography

Airaudo et al. [83] have reviewed the TLC of amine antioxidants and antiozonates used in elastomers. Data were given on 54 amines, chromatographed on silica gel thin-layer plates with benzene and benzene/ethyl acetate/acetone (100:5:1) as developing solvents and detected with N-chloro-2,6-dichloro-p-benzoquinone monoimine in buffered alkali as spray reagent. Some correlations between chemical structures, R_F values, and color reactions were established.

Aromatic amines in rubber products were rapidly screened at ng and pg detection limits by TLC and HPLC [84]. The thin-layer chromatograms were revealed by diazotization of the amines with nitrous oxide vapor, followed by coupling with N-(1-naphthyl)-ethylene diamine to produce spots with distinctly different colors.

Theoretical studies on the selectivity of particular mobile-phase systems in TLC with primary and secondary amines were carried out by Dzido and Soczewinski [85].

25.2.4 Supercritical-fluid chromatography

David and Sandra [86] applied SFC to amines and quaternary ammonium salts. With CO_2 as the mobile phase, problems due to the reactivity of CO_2 towards NH_2 groups would be expected. Carbon dioxide is known to react with amines at elevated temperatures and pressures to form carbamic acid derivatives, which are insoluble in supercritical CO_2. Acylation of the amines is easily and quantitatively performed. The acyl derivatives (trifluoroacetyl derivates) were prepared by heating the sample with trifluoroacetic acid at 60°C for 15 min. The direct analysis of quaternary ammonium salts by SFC is not an established technique. Analysis by high-temperature capillary GC after conversion to tertiary amines by off-line treatment with phosphoric acid at elevated temperatures is to be preferred.

The separation of model mixtures of nitrated diphenylamine and nitrated aniline by SFC on both capillary and packed columns was described by Ashraf-Khorassani and Taylor [87]. Highly deactivated phases could be used, and separation was performed with pressure-programed CO_2. On-line FTIR spectrometric detection was used for peak identification and to check peak purity.

A SFC system with nitrous oxide as the mobile phase and a nitrogen-sensitive gas chromatographic detector was evaluated for the determination of amines and their amide and carbamate derivatives by Mathiasson et al. [88]. The detector sensitivity for amines was comparable with that observed for amines when the same detector was used in GC. Nitrous oxide was chosen as the mobile phase instead of the commonly used carbon dioxide to eliminate the risk of reaction with primary or secondary amine functions.

25.3 PHENYLENE DIAMINES AND DIPHENYLAMINES

This group of aromatic amines is treated separately because of its importance. The nomenclature is somewhat confusing, but the easiest way is to name them starting from the three unsubstituted phenylene diamines. The application area of these compounds is very extensive. The most important is as intermediates in the synthesis of polyurethanes. Another important use is the application in oxidative hair dyes. Other applications are as antioxidants and antiozonates in plastics, in photographic developers, and as intermediates in fungicides, azo dyes, etc. The toxicity of these products is very high, and carcinogenic effects are known. Wastes containing them are environmental hazards.

The GC analysis of these amines is problematic due to their low volatility and high reactivity, although there has been some success in their determination at high levels [89]. The basic character of the compounds makes the use of C_{18} modified silicas difficult. According to the pK values of the compounds used in hair dyes, a pH of the mobile phase of 7.5 or higher is needed for reversed-phase systems.

A novel hydrophilic polystyrene packing has been used to separate diaminobenzene isomers by HPLC at pH 12 with a mobile phase of 30% aq. acetonitrile, containing 5% 2-hydroxyethylamine [90]. Good peakshapes were obtained for these highly basic compounds.

Phenylene diamines present in industrial waste water have been determined by HPLC on a C_{18} column with electrochemical (amperometric) detection [91]. Riggin and Howard [92] have shown that in most cases electrochemical detection is superior to UV detection. Chromatographic conditions and sample preparation procedures were described for many phenylene diamines of environmental significance.

p-Phenylenediamines and phenols are usually employed in gasoline as antioxidants. Precolumn concentration and micro-HPLC have been used to analyze these additives [93]. HPLC of o-phenylenediamine with electrochemical detection showed detection limits of ca. 10 ng/ml [94] in water used to rinse equipment without preconcentration. N,N'-Disubstituted derivatives of 1,4-phenylenediamine are the strongest and most widely used inhibitors of the autoxidation and ozonolysis of elastomers. Thin-layer and HPLC methods were used to identify and determine the antioxidants and antiozonants and their transformation products along with other additives in vulcanized rubbers [95,96].

The determination of hydrophobic constants of basic compounds from chromatographic experiments where silanol interaction cannot be excluded requires modifications, such as addition of pairing agents or rigorous endcapping of the stationary phase. Specific interactions can be excluded by using poly(butadiene)-coated alumina [97]. Such stationary phases can be operated in a wide pH range of eluents without any tendency for specific interaction with solutes, typical for silica-based RP materials.

Diphenylamine (DPA) is usually the stabilizer in single-base fire-arm propellants. The residual DPA may be identified by HPLC with sequential oxidative and reductive electrochemical detection [98]. Because of its low basicity, DPA adsorbed from acetonitrile onto a strongly acidic cation-exchange resin is readily displaced by small amounts of water. Microcolumns, filled with the resin, are used in the selective recovery of DPA from clothing debris and handswab extracts in forensic analysis.

REFERENCES

1 J.A. Campbell, E.P. Grimsrud and L.R. Hageman, *Anal. Chem.*, 55 (1983) 1335.
2 C.T. Peng, S.F. Ding, R.L. Hua and Z.C. Yang, *J. Chromatogr.*, 436 (1988) 137.
3 I.P. Yudina, G.N. Semina and K.I. Sakodynskii, *J. Chromatogr.*, 365 (1986) 19.
4 M.J. Avery and G.A. Junk, *Anal. Chem.*, 57 (1985) 790.
5 C.D. Chriswell and J.S. Fritz, *J. Chromatogr.*, 136 (1977) 371.
6 J.S. Smith and J.M. Hanrahan, *J. Chromatogr.*, 193 (1980) 271.
7 J.J. Richard and G.A. Junk, *Anal. Chem.*, 56 (1984) 1625.
8 M. Donike, *J. Chromatogr.*, 78 (1973) 273.
9 E. Hohaus, *Bunseki Kagaku*, 33 (1984) E55.
10 S.L. Wellons and M.A. Carey, *Chromatographia*, 10 (1978) 808.
11 R.L. Petty, W.C. Michel, J.P. Snow and K.S. Johson, *Anal. Chim. Acta*, 142 (1982) 299.
12 P. Koga, T. Akiyama and R. Shinohara, *Bunseki Kagaku*, 30 (1981) 745.
13 H. Nakamura, K. Takagi, Z. Tamura, R. Yoda and Y. Yamamoto, *Anal. Chem.*, 56 (1984) 919.
14 Y. Hoshika, *Anal. Chem.*, 49 (1977) 541.
15 D.W. Meddle and A.F. Smith, *Analyst (London)*, 106 (1981) 1082.
16 M. Dalene and G. Skarping, *J. Chromatogr.*, 331 (1985) 321.
17 G. Skarping, L. Renman and M. Dalene, *J. Chromatogr.*, 270 (1983) 207.
18 G. Skarping, L. Renman and B.E.F. Smith, *J. Chromatogr.*, 267 (1983) 315.
19 G. Skarping, B.E.F. Smith and M. Dalene, *J. Chromatogr.*, 303 (1984) 89.
20 G. Skarping, L. Renman and G. Sango, *J. Chromatogr.*, 346 (1985) 191.
21 P. Burchill, A.A. Herod and C.A. Mitchell, *Chromatographia*, 21 (1986) 67.
22 M.R. Guerin, C.H. Ho, T.K. Rao, B.R. Clark and J.L. Epler, *Environ. Res.*, 23 (1980) 42.
23 K. Kuwata, E. Akiyama, Y. Yamazaki, H. Yamasaki and Y. Kuge, *Anal. Chem.*, 55 (1983) 2199.
24 G. Audunsson and L. Mathiasson, *J. Chromatogr.*, 315 (1984) 299.
25 A.H. Lawrence, *J. Chromatogr.*, 395 (1987) 531.
26 R. Otson, J.M. Leach and L.T.K. Chung, *Anal. Chem.*, 59 (1987) 58.
27 F.E. Scully, G.D. Howell, H.H. Penn, K. Mazina and J.D. Johnson, *Environ. Sci. Technol.*, 22 (1988) 1186.
28 D.S. Treybig, *J. Chromatogr. Sci.*, 21 (1983) 310.
29 W.A. Konig, S. Lutz, G. Wenz and E. von der Bey, *J. High Resolut. Chromatogr. Chromatogr. Commun.*, 11 (1988) 506.
30 D.H. Spackman, W.H. Stein and S. Moore, *Anal. Chem.*, 30 (1958) 1190.
31 T.L. Perry and W.A. Schroeder, *J. Chromatogr.*, 12 (1963) 358.
32 Z. Deyl, in Z. Deyl, K. Macek and J. Janak (Editors), *Liquid Column Chromatography*, Elsevier, Amsterdam, 1975, Chap. 30.
33 D.A. Haugen, M.J. Peak and K.M. Suhrbier, *Anal. Chem.*, 54 (1982) 32.
34 C.J.C.M. Laurent, A. Reappreciation of Alumina in HPLC, *Thesis*, Delft University of Technology, Delft, 1983.
35 H. Lingeman, H.A. van Munster, J.H. Beynen, W.J.M. Underberg and A. Hulshoff, *J. Chromatogr.*, 352 (1986) 261.
36 K. Fujimura, M. Kitanaka, H. Takyanagi and T. Ando, *Anal. Chem.*, 54 (1982) 918.
37 D.K. Singh and A. Darbari, *Chromatographia*, 23 (1987) 93.
38 H. Takayanagi, H. Tokuda, H. Uehira, K. Fujimura and T. Ando, *J. Chromatogr.*, 356 (1986) 15.
39 E. Papp and Gy. Vigh, *J. Chromatogr.*, 282 (1983) 59.
40 G.A. Eiceman and F.A. Janecka, *J. Chromatogr. Sci.*, 21 (1983) 555.
41 M.C. Gennaro and E. Marengo, *Chromatographia*, 25 (1988) 603.
42 P.C. Sadek, P.W. Carr and L.W. Bowers, *J. Liq. Chromatogr.*, 8 (1985) 2369.
43 E.H. Nieminen, L.H. Saarinen and J.T. Laakso, *J. Liq. Chromatogr.*, 6 (1983) 453.

44 E. Solon Barreira, J. Paz Parente and J. Wilson de Alencar, *J. Chromatogr.,* 398 (1987) 381.

45 P. Simon and C. Lemacon, *Anal. Chem.,* 59 (1987) 480.

46 P.T. Kissinger, K. Bratin, G.C. Davis and L.A. Pachla, *J. Chromatogr. Sci.,* 17 (1979) 137.

47 K. Shimada, M. Tanaka and T. Nambara, *J. Chromatogr.,* 280 (1983) 271.

48 M. Tanaka, K. Shimada and T. Nambara, *J. Chromatogr.,* 292 (1984) 410.

49 R.M. Smith, A.A. Ghani, D.G. Haverty, G.S. Bament, A.Y. Chamsi and A.G. Fogg, *J. Chromatogr.,* 455 (1988) 349.

50 T.-Y. Chou, S.T. Colgan, D.M. Kao, I.S. Krull, C. Dorschel and B. Bidlingmeyer, *J. Chromatogr.,* 367 (1986) 335.

51 W.A. Jacobs, *J. Chromatogr.,* 392 (1987) 435.

52 W.L. Caudill and R.M. Wightman, *Anal. Chim. Acta,* 141 (1988) 269.

53 V. Conciliani, G. Chiavari and P. Vitali, *J. Chromatogr.,* 258 (1983) 244.

54 L.J. Felice, R.E. Schirmer, D.L. Springer and C.W. Veverka, *J. Chromatogr.,* 354 (1986) 442.

55 M.S. Varney and M.R. Preston, *J. Chromatogr.,* 348 (1985) 265.

56 K. Isaksson, J. Lindquist and K. Lundstrom, *J. Chromatogr.,* 324 (1985) 333.

57 J.R. Rice and P.T. Kissinger, *Environ. Sci. Technol.,* 16 (1982) 263.

58 T. Otagawa, I.R. Steter and S. Zaromb, *J. Chromatogr.,* 360 (1986) 252.

59 K. Imai, T. Toyo'oka and H. Miyano, *Analyst (London),* 109 (1984) 1365.

60 B. Bjorkqvist, *J. Chromatogr.,* 204 (1981) 109.

61 K. Anderson, C. Hallgren, J.-O. Levin and C.-A. Nilsson, *J. Chromatogr.,* 312 (1984) 482.

62 E.C.M. Chen and R.A. Farquharson, *J. Chromatogr.,* 178 (1979) 358.

63 G. Gubitz, R. Wintersteiger and A. Hartinger, *J. Chromatogr.,* 218 (1981) 51.

64 I.R.C. Whideside, P.J. Worsfold and E.H. McKerrell, *Anal. Chim. Acta,* 212 (1988) 155.

65 P. Kucera and H. Umagat, *J. Chromatogr.,* 255 (1983) 563.

66 D.L. Ingles and D. Gallimore, *J. Chromatogr.,* 325 (1985) 346.

67 P. Lehtonen, *J. Chromatogr.,* 314 (1984) 141.

68 M.-L. Henriks-Eckerman and T. Laijoki, *J. Chromatogr.,* 333 (1985) 220.

69 J.-K. Lin and C.-C. Lai, *Anal. Chem.,* 52 (1980) 630.

70 Y. Nishikawa and K. Kuwata, *Anal. Chem.,* 56 (1984) 1790.

71 G.M. Murray and M.J. Sepaniak, *J. Liq. Chromatogr.,* 6 (1983) 931.

72 Y. Tsuruta and K. Kohashi, *Anal. Chim. Acta,* 192 (1987) 309.

73 C.-X. Gao, I.S. Krull and Th.M. Trainor, *J. Chromatogr.,* 463 (1989) 192.

74 C.-X. Gao, T.-X. Chou, S.T. Colgan, I.S. Krull, C. Dorschel and B. Bidlingmeyer, *J. Chromatogr. Sci.,* 26 (1988) 449.

75 S.C. Beale, J.C. Savage, D. Wiesler, S.M. Wietstock and M. Novotny, *Anal. Chem.,* 60 (1988) 1765.

76 D.H. Neiderhiser and R.K. Fuller, *J. Chromatogr.,* 229 (1982) 470.

77 P. Jandera, H. Pechova, D. Tocksteinova, J. Churacek and J. Kralovsky, *Chromatographia,* 16 (1982) 275.

78 G. Mellbin and B.E.F. Smith, *J. Chromatogr.,* 312 (1984) 203.

79 M. Kudoh, I. Matoh and S. Fudano, *J. Chromatogr.,* 261 (1983) 293.

81 N. Richfield-Fratz, I.E. Bailey, Jr. and C.J. Bailey, *J. Chromatogr.,* 331 (1985) 109.

82 I.E. Bailey, Jr., *Anal. Chem.,* 57 (1985) 189.

83 Ch.B. Airaudo, A. Gayte-Sorbier, P. Aujoulat and V. Mercier, *J. Chromatogr.,* 437 (1988) 59.

84 A.S. Narang, D.R. Choudhury and A. Richards, *J. Chromatogr. Sci.,* 20 (1982) 235.

85 T. Dzido and E. Soczewiński, *J. Chromatogr.,* 395 (1987) 489.

86 F. David and P. Sandra, *J. High Resolut. Chromatogr. Chromatogr. Commun.,* 11 (1988) 897.

87 M. Ashraf-Khorassani and L.T. Taylor, *J. High Resolut. Chromatogr. Chromatogr. Commun.,* 12 (1989) 40.

89 L. Mathiasson, J.A. Jonsson and L. Karlsson, *J. Chromatogr.,* 467 (1989) 61.

89 E. Krasuska and W. Celler, *J. Chromatogr.*, 147 (1978) 470.
90 Y.-B. Yang and M. Verzele, *J. Chromatogr.*, 387 (1987) 197.
91 E.P. McGovern and J.H. Brewer, *J. High Resolut. Chromatogr. Chromatogr. Commun.*, 7 (1984) 709.
92 R.M. Riggin and C.C. Howard, *J. Liq. Chromatogr.*, 6 (1983) 1897.
93 A. Nakanishi, D. Ishii and T. Takeuchi, *J. Chromatogr.*, 291 (1984) 398.
94 L. Elrod, Jr. and S.G. Spanton, *J. High Resolut. Chromatogr. Chromatogr. Commun.*, 5 (1982) 628.
95 J. Rotschova, L. Taimr and J. Pospisil, *J. Chromatogr.*, 216 (1981) 251.
96 J. Rotschova and J. Pospisil, *J. Chromatogr.*, 211 (1981) 299.
97 R. Kaliszan, R.W. Blain and R.A. Hartwick, *Chromatographia,* 25 (1988) 5.
98 J.B.F. Lloyd, *Anal. Chem.,* 59 (1987) 1401.

Manufacturers and Dealers of Chromatography and Electrophoresis Supplies

(Reproduced in part from the *International Laboratory Buyers' Guide* with permission of International Scientific Information, Inc.)

Aabspec Instrumentation Ltd., 16 Rathmore Ave., Stillorgan, Dublin 18, IRELAND.
ABC Laboratories, P.O. Box 1097, Columbia, MO 65205, USA.
AC Analytical Controls BV, Postbus 374, NL-2600 AJ Delft, THE NETHERLANDS.
Accurate Chemical & Scientific Corp., 300 Shames Drive, Westbury, NY 11590, USA.
Ace Glass Inc., 1430 N.W. Boulevard, Vineland, NJ 08360, USA.
Ace Scientific Supply Co., 40-A Cotters La., East Brunswick, NJ 08816, USA.
Advanced Chemtech Inc., 2500 Seventh Street Rd., P.O. Box 1403, Louisville, KY 40201, USA.
Advanced Chromatographic Technologies, Blücher Strasse 22, D-1000 Berlin 61, GERMANY.
Advanced Separation Technologies (Astec), 37 Leslie Court, Box 297, Whippany, NJ 07981, USA.
AHL Inc., P.O. Box 742, Laurel, MD 20707, USA.
Ahlstrom Filtration Inc., 15 West High Street, Carlisle, PA 17013, USA.
Air Products and Chemicals Inc., P.O. Box 538, Allentown, PA 18105, USA.
Alameda Chemical and Scientific, 922 E. Southern Pacific Drive, Phoenix, AZ 85034, USA.
Alcott Chromatography, 5300 Oakbrook Pkwy., Ste. 100, Norcross, GA 30093, USA.
Aldrich Chemical Co., 940 W. St. Paul Ave., P.O. Box 355, Milwaukee, WI 53201, USA.
Alltech Associates, 2051 Waukegan Road, Deerfield, IL 60015, USA.
ALPHA Applied Research, 2355 McLean Blvd., Eugene, OR 97405, USA.
Alphagaz, Specialty Gases Div., Liquid Air Corporation, 2121 N. California Blvd., Walnut Creek, CA 94596, USA.
Altex Division, Beckman Instruments Inc., 4550 Norris Canyon Rd., San Ramon, CA 94583, USA.
AMBIS Systems Inc., 3939 Ruffin Road, San Diego, CA 92123, USA.
American Bionetics Inc., 21377 Cabot Blvd., Hayward, CA 94545, USA.
American Research Products Co., 30175 Solon Industrial Pkwy., Solon, OH 44139, USA.
American Scientific Products, 1430 Waukegan Rd., McGaw Park, IL 60085, USA.
Amersham Corp., 2636 S. Clearbrook Dr., Arlington Heights, IL 60005, USA.
Amicon Division, W.R. Grace & Co., 17 Cherry Hill Drive, Danvers, MA 01923, USA.
Analabs, 140 Water St., Norwalk, CT 06854, USA.
Analog & Digital Peripherals Inc., P.O. Box 499, Troy, OH 45373, USA.
Analtech Inc., 75 Blue Hen Drive, P.O. Box 7558, Newark, DE 19711, USA.
Analytical Bio-Chemistry Laboratories, P.O. Box 1097, Columbia, MO 65205, USA.
Analytical Chromatography Support, 10606 Brooklet, Ste. 202, Houston, TX 77099, USA.
Analytical Instruments Pty. Ltd., P.O. Box 215, Scarborough, Qld. 4020, AUSTRALIA.
Analytical Measuring Systems, London Rd., Pampisford, Cambridge CB2 4EF, GREAT BRITAIN.
Analytical Products Inc., 511 Taylor Way, Belmont, CA 94002, USA.
Analytichem International, 24201 Frampton Avenue, Harbor City, CA 90710, USA.
Analytic Parameters, P.O. Box 25035, Chicago, IL 60625, USA.
Anamed Instruments (Pvt.) Ltd., New Bombay 400 706, INDIA.
Angar Scientific Co., P.O. Box 538, Florham Park, NJ 07932, USA.
Anotec Separations Ltd., Wildmere Rd., Banbury, Oxon OX16 7JU, GREAT BRITAIN.
Anspec Co. Inc., 50 Enterprise Dr., P.O. Box 7730, Ann Arbor, MI 48107, USA.
Antek Instruments Inc., 6005 N. Freeway, Houston, TX 77076-3998, USA.

Applied Analytical Industries Inc., Rt. 6, P.O. Box 55, Wilmington, NC 28405, USA.
Applied Automation Inc., Pawhuska Rd., Bartlesville, OK 74004, USA.
Applied Biosystems Inc., 850 Lincoln Center Dr., Foster City, CA 94404, USA.
Applied Chromatography Systems Ltd., The Arsenal, Heapy Street, Macclesfield,
  Cheshire SK11 7JB, GREAT BRITAIN.
Applied Science Labs., P.O. Box 440, State College, PA 16804, USA.
Applied Separations, Box 6032/B, Franklin Tech Center, Bethlehem, PA 18001, USA.
Arnel Inc., 3141 Bordentown Ave., Parlin, NJ 08859, USA.
Asahi Chemical Ind. Co., 1-3-2 Yakoo, Kawasaki-ku, Kawasaki-shi, Kanagawa-ken 210,
  JAPAN.
Associated Laboratories, 806 N. Batavia, Orange, CA 92668, USA.
Atto Corp., 2-3, Hongo 7-chome, Bunkyo-ku, Tokyo 113, JAPAN.
Aura Industries Inc., P.O. Box 898, Staten Island, NY 10314, USA.
Autochrom Inc., P.O. Box 207, Milford, MA 01757-0207, USA.
Automated Microbiology Systems Inc., 3939 Ruffin Rd., San Diego, CA 92123, USA.
Automatic Switch Co., 60 Hanover Rd., Florham Park, NJ 07932, USA.
Axxiom Chromatography Inc., 23966 Craftsman Rd., Calabasas, CA 91302, USA.
Bacharach Inc., 625 Alpha Dr., Pittsburgh, PA 15238, USA.
Baekon Inc., 18866 Allendale Ave., Saratoga, CA 95070-5239, USA.
J.T. Baker Chemical Company, 222 Red School Lane, Phillipsburg, NJ 08865, USA.
Bal Seal Engineering Co. Inc., 620 W. Warner Ave., Santa Ana, CA 92707, USA.
Balston Inc., 703 Massachusetts Ave., Lexington, MA 02173, USA.
Barspec, P.O. Box 560, Rehovot 76103, ISRAEL.
Baxter Scientific Products, 1430 Waukegan Rd., McGaw Park, IL 60085, USA.
Beckman Instruments Inc., 2500 Harbor Blvd., Fullerton, CA 92634, USA.
Bendix Kansas City Division, Allied Bendix Aerospace, 2000 East 95th Street, Kansas
  City, MO 64131, USA.
Benson Polymerics, P.O. Box 12812, Reno, NV 89510, USA.
Laboratorium Prof. Dr. Berthold, Calmbacher Str. 22, D-7547 Wildbad, GERMANY.
J. C. Binzer Papierfabrik GmbH, Berleburger Strasse 71, Postfach 44, D-3559 Hatzfeld/
  Eder, GERMANY.
Bioanalytical Systems Inc., 2701 Kent Avenue, West Lafayette, IN 47906, USA.
Biocom, B.P. 53, F-91942 Les Ulis, FRANCE.
Bio-Fractionations, 1725 S. State Highway 89-91, Logan, UT 84321, USA.
Biomedical Enterprises Inc., P.O. Box 257, Irvington, NY 10533, USA.
Biomed Instruments Inc., 1020 S. Raymond Ave., Ste. B, Fullerton, CA 92631, USA.
Biometra Biomed. Analytik GmbH, Wagenstieg 5, D-3400 Göttingen, GERMANY.
Bio-Probe International, 14272 Franklin Ave., Tustin, CA 92680, USA.
Bioprocessing Ltd., 1 Industrial Estate, Cosett, Durham DH8 6TJ, GREAT BRITAIN.
Bio-Rad Chemical Div., 1414 Harbour Way South, Richmond, CA 94804, USA.
BIOS Corp., 291 Whitney Avenue, New Haven, CT 06511, USA.
Bioscan Inc., 4590 McArthur Blvd. NW, Washington, DC 20007, USA.
Biotec-Fischer GmbH, Daimlerstrasse 6, D-6301 Reiskirchen, GERMANY.
Biotech Instruments Ltd., 183 Cambord Way, Luton, Beds. LU3 3AN, GREAT BRITAIN.
Biotronik Wissenschaftliche Geräte GmbH, Benzstrasse 28, D-8039 Puchheim-Bahnhof,
  GERMANY.
Bischoff Analysentechnik und -geräte GmbH, Umler Str. 2, D-7250 Leonberg, GERMANY.
Bodman Chemicals, P.O. Box 2221, Aston, PA 19014, USA.
Boehringer-Mannheim GmbH, Sandhoferstrasse 116, D-6800 Mannheim 31, GERMANY.
Herman Bohlender, Postfach 1145, Bischofsheimer Weg 14, D-6970 Lauda-Königshofen,
  GERMANY.
Bomem Inc., 625 Marais, Vanier, Que., G1M 2Y2, CANADA.
John Booker & Co., 3825 Bee Cave Rd., Austin, TX 78746, USA.
Brinkmann Instruments Inc., Cantiague Road, Westbury, NY 11590, USA.
B. Brown Diessel Biotech GmbH, Postfach 120, D-3508 Meisungen, GERMANY.
Brownlee Labs. Inc., 2045 Martin Avenue, Santa Clara, CA 95050, USA.
Bruker Analytische Messtechnik GmbH, Silberstreifen, D-7512 Rheinstetten 4, GERMANY.
Büchi Laboratory Techniques Ltd., Meierseggstr. 40, CH-9230 Flawill, SWITZERLAND.
Buchler Instruments, 9900 Pflumm Rd., 17 Lenexa Business Center, Lenexa, KS
  66215-1223, USA.
Buck Scientific Inc., 58 Fort Point St., East Norwalk, CT 06855-1097, USA.

Burdick & Jackson, 1935 South Harvey Street, Muskegon, MI 49442, USA.
Burrell Corp., 2223 Fifth Ave., Pittsburgh, PA 15219, USA.
Cal Glass for Research Inc., 3012 Enterprise Ave., Costa Mesa, CA 92626, USA.
CAMAG, CH-4132 Muttenz, SWITZERLAND.
Camlab Ltd., Nuffield Road, Cambridge, Cambs. CB4 1TH, GREAT BRITAIN.
Carnegie Medicin AB, Roslagsvagen 101, S-104 05 Stockholm, SWEDEN.
Carolina Biological Supply Co., 2700 York Road, Burlington, NC 27215, USA.
Cavro Scientific Instruments Inc., 242 Humboldt Court, Sunnyvale, CA 94089, USA.
Cecil Instruments Ltd., Milton Industrial Estate, Cambridge Road, Milton, Cambs. CB4
    4AZ, GREAT BRITAIN.
Cera Inc., 14180 Live Oak Ave., Ste. F, Baldwin Park, CA 91706, USA.
C.G.A. Strumenti Scientifici S.p.A., Via del Della Robbia N. 38, I-50132 Firenze, ITALY.
Chemcon Inc., 34 Mann St., South Attleboro, MA 02703, USA.
Chemetron, Via Gustavo Modena 24, I-20129 Milano, ITALY.
Chemical Data Systems, 7000 Limestone Rd., Oxford, PA 19363, USA.
Chemical Dynamics Corp., P.O. Box 395, South Plainfield, NJ 07080, USA.
Chemical Research Supplies, P.O. Box 888, Addison, IL 60101, USA.
Chemtrix Inc., P.O. Box 1329, Hillsboro, OR 97123, USA.
Chromacol Ltd., Glen Ross House, Summers Row, London N12 0LD, GREAT BRITAIN.
Chromapon Inc., P.O. Box 4131, Whittier, CA 90607, USA.
ChromatoChem, Inc., 2837 Fort Missoula Road, Missoula, MT 59801, USA.
Chromatofield, Zila Valampe, F-13220 Châteauneuf-les-Martigues, FRANCE.
Chromatography Sciences Co., 5750 Vanden Abeele, Ville Saint-Laurent, Que. H4S 1R9,
    CANADA.
Chromatography Services Ltd., Carr Lane Industrial Estate, Hoylake, Merseyside, GREAT
    BRITAIN.
Chromatography Technology Services, 3301 W.134th St., Burnsville, MN 55337, USA.
Chromatronix, Inc., 2300 Leghorn Street, Mountain View, CA 94043, USA.
Chrompack International BV, Postbus 8033, NL-4330 EA Middelburg, THE
    NETHERLANDS.
Chrom Tech Inc., P.O. Box 24248, Apple Valley, MN 55124, USA.
Ciba Corning Analytical, Colchester Road, Halstead, Essex C09 2DX, GREAT BRITAIN.
CJB Developments Ltd., Airport Service Road, Portsmouth P03 5PG, GREAT BRITAIN.
Clarkson Chemical Co., P.O. Box 97, Williamsport, PA 17703-0097, USA.
Clontech, 4030 Fabian Way, Palo Alto, CA 94303, USA.
Cluzeau Info-Lab Sarl, B.P. 88, 35 rue Jean-Louis Fauré, F-33220 Ste. Foy-la-Grande,
    FRANCE.
P. J. Cobert Associates Inc., P.O. Box 12668, St. Louis, MO 63141, USA.
Cole-Parmer Instruments, 7425 N. Oak Park Ave., Chicago, IL 60648, USA.
Combined Sciences Corp., 433 Boston Post Rd., Darien, CT 06820, USA.
Computer Chemical Systems, P.O. Box 683, Rt. 41 & Newark Road, Avondale, PA 19311,
    USA.
Consort pvba, Parklaan 36, B-2300 Turnhout, BELGIUM.
Core Laboratories Inc., 1300 E. Rochelle Blvd., Irving, TX 7515-2053, USA.
Cortex Biochem Inc., 459 Hester St., San Leandro, CA 94677, USA.
Coulter Electronics Ltd., Northwell Dr., Luton, Beds. LU3 3RH, GREAT BRITAIN.
Crane Co., 175 Titus Ave., Warrington, NY 18976, USA.
Crescent Chemical Co., 1324 Motor Pkwy., Hauppague, NY 11788, USA.
Crown Glass, 990 Evergreen Dr., Somerville, NJ 08876, USA.
Cryogenic Rare Gas, 913 Commerce Circle, Hanahan, SC 29410, USA.
Cuno Inc., Life Science Div., 400 Research Parkway, Meriden, CT 06450, USA.
Curtin Matheson Scientific, 9999 Veterans Memorial Dr., Houston, TX 77038, USA.
Cyborg, 94 Bridge Street, Newton, MA 02158, USA.
Daicel Chemical Industries Ltd., 8-1 Kasumigaseki 3-chome, Chiyoda-ku, Tokyo 100,
    JAPAN.
Daiichi Pure Chemicals Co., 13-5 Nihombashi 3-chome, Chuo-ku, Tokyo 103, JAPAN.
Dani SpA, V. le Elvezia 42, I-20052 Monza (Milano), ITALY.
Data Translation Inc., 100 Locke Dr., Marlboro, MA 01752, USA.
Del Electronics Corp., 250 E. Sandford Blvd., Mount Vernon, NY 10550, USA.
Delsi Nermag, 15701 W. Hardy, Houston, TX 77060, USA.
Delta Technical Products Co., 7259 W. Devon, Chicago, IL 60631, USA.

DESAGA GmbH, Maass Strasse 26-28, Postfach 101969, D-6900 Heidelberg 1, GERMANY.
Detector Engineering & Technology Inc., 2212 Brampton Rd., Walnut Creek, CA 94598, USA.
Dexsil Chemical Corp., 1 Hamden Park Dr., Hamden, CT 06517, USA.
Digital Equipment Corporation, 4 Results Way, MR04-2/C16, Marlborough, MA 01752-9122, USA.
Dionex Corp., 1228 Titan Way, Sunnyvale, CA 94086, USA.
Diversified Biotech, 46 Marcellus Drive, Newton Centre, MA 02159, USA.
Domnick Hunter Filters Ltd., Durham Rd., Birtley Co., Durham DH3 2SF, GREAT BRITAIN.
Dorr-Oliver, 77 Havemeyer Lane, P.O. Box 9312, Stamford, CT 06904, USA.
Dracard Ltd., Wallis Ave., Park Wood, Maidstone, Kent ME15 9HE, GREAT BRITAIN.
Drew Scientific, 12 Barley Mow Passage, London W4 4PH, GREAT BRITAIN.
E. I. du Pont de Nemours & Co., Barley Mill Plaza, Wilmington, DE 19898, USA.
Duryea Assoc. Inc., 701 Alpha Dr., Pittsburgh, PA 15238, USA.
Dychrom, P.O. Box 70116, Sunnyvale, CA 94086, USA.
Dynamic Solutions Corp., 2355 Portola Rd., Ste. B, Ventura, CA 93003, USA.
Dyson Instruments Ltd., Hetton Lyons Industrial Estate, Hetton, Houghton-le-Spring, Tyne and Ware DH5 0RH, GREAT BRITAIN.
Eastman Kodak Co., P.O. Box 92894, Rochester, NY 14692-9939, USA.
E-C Apparatus Corp., 3831 Tyrone Blvd. North, St. Petersburg, FL 33709, USA.
E-D Scientific Specialties, P.O. Box 369, Carlisle, PA 17013, USA.
EG&G Princeton Applied Research Corp., CN 5206, Princeton, NJ 08543, USA.
Electro Biotransfer Inc., 790 Lucerne Drive, Sunnyvale, CA 94086, USA.
Electronic & Scientific Devices, 100 U.B. Jawahar Nagar, Delhi 110 007, INDIA.
EM Science, 11 Woodcrest Rd., Cherry Hill, NJ 08034-0395, USA.
Carlo Erba Strumentazione, Strada Rivoltana, I-20090 Rodano (Milano), ITALY.
Erma Inc., 2-4-5 Kajicho Chiyoda-ku, Tokyo 101, JAPAN.
ES Industries, 8 S. Maple Ave., Marlton, NJ 08053, USA.
ESA Inc., 45 Wiggins Ave., Bedford, MA 01730, USA.
ETPCORTEC Pty. Ltd., 31 Hope Street, Ermington, Sydney, N.S.W. 2115, AUSTRALIA.
Extrel Corporation, 240 Alpha Drive, P.O. Box 11512, Pittsburgh, PA 15238, USA.
E-Y Laboratories Inc., 105-127 N. Amphlett Blvd., San Mateo, CA 94401, USA.
F.E.R.O.S.A., Fabricación Espanola de Reactivos Organicos SA, C/la Jota 86, E-08016 Barcelona, SPAIN.
FFFractionation Inc., P.O. Box 8718, Salt Lake City, UT 84108, USA.
Fiatron Systems Inc., 510 S. Worthington St., Oconomowoc, WI 53066, USA.
Finnigan MAT, Barkhausenstrasse 2, Postfach 144062, D-2800 Bremen, GERMANY.
Fischer Labor- und Verfahrenstechnik GmbH, Industriepark Kottenforst, D-5309 Meckenheim, GERMANY.
Fisher Scientific Co., 711 Forbes Avenue, P.O. Box 1962, Pittsburgh, PA 15219, USA.
Fisher & Porter, Lab-Crest Scientific Div., East County Line Road, Warminster, PA 18974, USA.
Fisons Scientific Equipment, Bishop Meadow Road, Loughborough, Leicester LE11 ORG, GREAT BRITAIN.
Flow Laboratories SA, Via Campagna, Centro Nord-Sud, Bioggio, CH-6934 Lugano, SWITZERLAND.
Fluid Management Systems, 125 Walnut St., Watertown, MA 02172, USA.
Fluid Metering Inc., 29 Orchard St., P.O. Box 129, Oyster Bay, NY 11771, USA.
Fluka Chemie AG, Industriestrasse 25, CH-9470 Buchs, SWITZERLAND.
FMC Corp., 5 Maple St., Rockland, ME 04841, USA.
Foss Electric Ltd., Sandyford Industrial Estate, Foxrock, Dublin 18, IRELAND.
Fotodyne Inc., 16700 W. Victor Rd., New Berlin, WI 53151-4131, USA.
Foxboro Co., Bristol Park, Foxboro, MA 02035, USA.
Funakoshi, Pharmaceutical Co., 2-3 Surugadai, Kanda, Chiyoda-ku, Tokyo 101, JAPAN.
Galactic Industries Corp., 417 Amherst St., Nashua, NH 03063, USA.
Gallard-Schlesinger Industries Inc., 584 Mineola Ave., Carle Place, NY 11514, USA.
Gallenkamp, Belton Road West, Loughborough LE11 OTR, GREAT BRITAIN.
Gargya Research Inst., A-25, Rajouri Garden, Najafgarh Rd., New Delhi, Delhi 110 027, INDIA.
Gasukuro Kogyo Inc., 6-12-18 Nishi Shinjuku, Shinjuku-ku, Tokyo, JAPAN.

GAT Gamma Analysentechnik GmbH, Dionysiusstr. 6, D-2850 Bremerhaven, GERMANY.
Gelman Sciences Inc., 600 South Wagner Road, Ann Arbor, MI 48106, USA.
Geltech Inc., 934 Salem Pkwy., Salem, OH 44460-313, USA.
Genetic Research Instrumentation Ltd., Gene House, Dunnow Road, Felsted, Dunnow, Essex CM6 3LD, GREAT BRITAIN.
Genex Corporation, 16020 Industrial Drive, Gaithersburg, MD 20877, USA.
Genie Scientific, 17430 Mt. Cliffwood Circle, Unit A, Fountain Valley, CA 92708, USA.
Gibco Ltd., P.O. Box 35, Trident House, Renfrew Rd., Paisley, PA3 4EF, GREAT BRITAIN.
Gilson Medical Electronics Inc., 3000 West Beltline Hwy., Middleton, WI 53562, USA.
Gold Biotech, 10143 Paget, St. Louis, MO 63132, USA.
Gow-Mac Instruments Co., P.O. Box 32, Bound Brook, NJ 08805, USA.
Graphic Controls Corp., P.O. Box 1271, Buffalo, NY 14240, USA.
Griffin & George, Bishop Meadow Rd., Loughborough, Leicestershire LE11 ORG, GREAT BRITAIN.
Grupo Químico Industrial Ltd., Rua Jacurunta, 628 Penha, 21.020 Rio de Janeiro, BRAZIL.
Guelph Chemical Laboratories Ltd., 246 Silvercreek Parkway N., Guelph, Ont. N1H 1E7, CANADA.
Haake Buchler Instruments Inc., 244 Saddle River Road, Saddle Brook, NJ 07662-6001, USA.
Hach Co., P.O. Box 389, Loveland, CO 80539, USA.
Hamilton Company, 4970 Energy Way, Reno, NV 89502, USA.
Harrison Co., Palo Alto, CA 94303, USA.
Haskel Inc., 100-88 E. Graham Place, Burbank, CA 91502, USA.
Helena Laboratories, 1530 Lindbergh Dr., P.O. Box 752, Beaumont, TX 77704-0752, USA.
Hellma Cells Inc., P.O. Box 544, Borough Hall Station, Jamaica, NY 11424, USA.
HETP LC Components, 4 Victoria Road, Wilmslow, Cheshire SK9 5HN, GREAT BRITAIN.
Hewlett-Packard, P.O. Box 10301, Palo Alto, CA 94303-1501, USA.
Hichrom Ltd., 6 Chiltern Enterprise Ctr., Station Road, Theale, Berks. RG7 4AA, GREAT BRITAIN.
Hirschmann Gerätebau GmbH, Lohestrasse 5, D-8025 Unterhaching, GERMANY.
Hitachi Scientific Instruments, 460 E. Middlefield Rd., Mountain View, CA 94043, USA.
HNU Systems Inc., 160 Charlemont St., Newton, MA 02161, USA.
Hoefer Scientific Instruments, 654 Minnesota, P.O. Box 77387, San Francisco, CA 94107, USA.
HP Genenchem, 460 Point San Bruno Blvd., South San Francisco, CA 94080, USA.
HPLC Technology Ltd., Waterloo Street West, Macclesfield, Cheshire SK11 6PJ, GREAT BRITAIN.
HT Chemicals Inc., 4221 Forest Park Ave., St. Louis, MO 63108-2810, USA.
Iatron Laboratories Inc., Tokyo, JAPAN.
IBF Biotechnics Inc., 8510 Corridor Rd., Savage, MD 20763, USA.
IBF Reactifs, 35 Ave. Jean-Jaurès, F-92390 Villeneuve la Garenne, FRANCE.
IBM Instruments Inc., P.O. Box 3020, Wallingford, CT 06492, USA.
ICI Australia Operations Pty. Ltd., 5 Lake Dr., Redwood Gardens, Dingley, Vic. 3172, AUSTRALIA.
ICN Biochemicals Inc., 3300 Hyland Avenue, Costa Mesa, CA 92626, USA.
Idea Scientific Co., P.O. Box 2078, Corvallis, OR 97339, USA.
Ikemoto Scientific Technology Co., P.O. Box 14, Hongo, Bunkyo-ku, Tokyo 113, JAPAN.
Illinois Water Treatment Co., 4669 Shepherd Trail, Rockford, IL 61103, USA.
Immunetics Inc., 380 Green Street, Cambridge, MA 02139, USA.
Infometrix Inc., 2200 Sixth Ave., Ste. 833, Seattle, WA 98121, USA.
Ingold Electrodes Inc., 261 Ballardvale St., Wilmington, MA 01887, USA.
Innovative Chemistry Inc., P.O. Box 90, Marshfield, MA 02050, USA.
Institut für Chromatographie, Postfach 1141, D-6702 Bad Dürkheim 1, GERMANY.
Instrumentation Laboratory, 113 Hartwell Ave., Lexington, MA 02173, USA.
Integrated Separation Systems, 1 Westinghouse Plaza, Hyde Park, MA 02136, USA.
Interaction Chemicals, 1615 Plymouth Street, Mountain View, CA 94043, USA.
Interactive Microware Inc., P.O. Box 139, State College, PA 16804, USA.
Interfacial Dynamics Cop., P.O. Box 279, Portland, OR 97207, USA.
International Biotechnologies Inc., 25 Science Park, P.O. Box 9558, New Haven, CT 06535, USA.
International Equipment Co., 300 Second Avenue, Needham Heights, MA 02194, USA.

Introtek International LP, 120-C Jefryn Blvd., E. Dear Park, NY 11729, USA.
Invicta Biosystems Inc., 2225 Faraday Ave., Carlsbad, CA 92008, USA.
Ion Exchange Products Inc., 4834 S. Halsted Street, Chicago, IL 60609, USA.
Ionics Inc., 65 Grove St., Watertown, MA 02172, USA.
Isco Inc., P.O. Box 5374, Lincoln, NE 68505, USA.
Isolab Inc., P.O. Box 4350, Akron, OH 44321, USA.
Ithaca Laboratory Equipment Co., 305 W. Green St., Ithaca, NY 14850, USA.
Janus Laboratories Inc., 9307 Rock Canyon Way, P.O. Box 1406, Orangevale, CA 95662, USA.
Japan Analytical Industry Co., 208 Musashi, Mizuho, Nashitama, Tokyo 190-12, JAPAN.
JASCO, Japan Spectroscopic Co., 2967-5 Ishikawa-cho, Hachioji City, Tokyo 192, JAPAN.
Jaytee Biosciences Ltd., Kent Research & Development Centre, University of Kent, Canterbury, Kent CT2 7PD, GREAT BRITAIN.
JEOL Ltd., 1-3 Musashino 3-chome, Akishima City, Tokyo 196, JAPAN.
JJ's (Chromatography) Ltd., Hardwick Industrial Estate, Kings Lynn PE30 4JG, GREAT BRITAIN.
JM Science Inc., 5820 Main Street, Ste. 300, Buffalo, NY 14221, USA.
Jobin Yvon, 16-18 Rue du Canal, F-91160 Longjumeau, FRANCE.
Jones Chromatography Ltd., New Road, Hengoed, Mid Glamorgan CF8 8AU, GREAT BRITAIN.
Jookoo Co Ltd., 3-19-4 Hongo, Bunkyo-ku, Tokyo 113, JAPAN.
Jordan Scientific Co., 4315 S. State Rd. 446, Bloomington, IN 47401, USA.
Jordi Associates Inc., 26 Pearl St., Bellingham, MA 02019, USA.
Joyce-Loebl, Marquisway, Team Valley, Gateshead NE11 0QW, GREAT BRITAIN.
JPS Chimie, B.P. 343, CH-2022 Bevaix, SWITZERLAND.
Jule Biotechnologies Inc., 25 Science Park, Hew Haven, CT 06511, USA.
J & W Scientific Inc., 91 Blue Ravine Rd., Folsom, CA 95630, USA.
Kalex Scientific Co., 7 Mora Court, Manhasset, NY 11030, USA.
Keystone Scientific Inc., Penn Eagle Industrial Park, 320 Rolling Ridge Dr., Bellefonte, PA 16823, USA.
Keystone Valve, 9700 W. Gulf Bank Dr., Houston, TX 77040, USA.
Kinetek Systems Inc., 11802 Borman Dr., St. Louis, MO 63146, USA.
Kipp & Zonen BV, Mercuriusweg 1, NL-2624 Delft, THE NETHERLANDS.
Dr. Herbert Knauer KG, Heuchelheimer Strasse 9, D-6380 Bad Homburg, GERMANY.
Koch-Light Ltd., Edison House, 163 Dixons Hill Rd., North Mymms, Hatfield, Herts., GREAT BRITAIN.
Koken Co., 14-18, 3-chome Mejiro, Toshima-ku, Tokyo 171, JAPAN.
Konik Instruments S.A., Ctra. Cerdanyola 65-7, P.O. Box 136, Sant Cugat del Valles, E-08024 Barcelona, SPAIN.
Kontes Biotechnology, P.O. Box 729, Vineland, NJ 08360-2899, USA.
Kontron Instruments AG, Bernerstrasse Süd 169, CH-8010 Zürich, SWITZERLAND.
Kratos Analytical, Barton Dock Rd., Urmston, Lancas. M31 2LD, GREAT BRITAIN.
Kronus Inc., P.O. Box 1075, Ste. 312, Dana Point, CA 92629, USA.
Kupper & Co., Montalinweg 12, Postfach 55, CH-7402 Bonaduz, SWITZERLAND.
Labclear, 508 - 29th Ave., Oakland, CA 94601, USA.
Lab-Crest Scientific, E. County Line Rd., Warminster, PA 18974, USA.
Lab Glass Inc., 1172 NW Blvd., Vineland, NJ 08360, USA.
Labindustries Inc., 620 Hearst Ave., Berkeley, CA 94710-0128, USA.
Labomatic AG, Im Kirschgarten 30, CH-4124 Schönenbuch, SWITZERLAND.
Labotron Instruments AG, Forrlibuckstrasse 66, CH-8005 Zürich, SWITZERLAND.
Lachema, Research Institute for Pure Chemicals, Karásek 28, CS-621 33 Brno, CZECHO-SLOVAKIA.
LaSalle Scientific Inc., 103 Elmslie St., Lasalle, Que. HBR 1V4, CANADA.
Laser Precision Analytical, 17819 Gillette Ave., Irvine, CA 92714, USA.
LC Packings, Wilhelminastraat 118, NL-1054 WP Amsterdam, THE NETHERLANDS.
LC Resources, 3182 C Old Tunnel Road, Lafayette, CA 94549, USA.
LDC Analytical Inc., 3661 Interstate Industrial Park Road North, P.O. Box 10235, Riviera Beach, FL 33404, USA.
Lee Scientific Corp., 4426 S. Century Drive, Salt Lake City, UT 84123, USA.
Lida Manufacturing Corp., 9115 26th Ave., Kenosha, WI 53140, USA.

Life Science Laboratories Ltd., Sedgewick Rd., Luton LU4 9DT, GREAT BRITAIN.
Linear Instruments Corp., 500 Edison Way, P.O. Box 12610, Reno, NV 89510, USA.
Liquid Carbonic, 135 S. LaSalle, Chicago, IL 60603, USA.
LKB-Produkter AB, P.O. Box 305, S-161 26 Bromma, SWEDEN.
Logos Scientific Inc., 700 Sunset Rd., Henderson, NV 89015, USA.
Lurex Manufacturing Co., 1298 North West Blvd., Vineland, NJ 08360, USA.
MAC-MOD Analytical Inc., 127 Commons Court, Chadds Ford, PA 19317, USA.
Macherey-Nagel, Neumann-Neander Strasse, Postfach 307, D-5160 Düren, GERMANY.
Mallinckrodt Inc., 675 McDonell Blvd., St. Louis, MO 63134, USA.
M.A.L.T.A., srl, Via Gustavo Modena 24, I-20129 Milano, ITALY.
Malvern Instruments Ltd., Spring Lane South, Malvern, Worcester WR14 1AQ, GREAT
    BRITAIN.
Manville Filtration & Minerals, Ken-Caryl Ranch, Denver, CO 80217-5108, USA.
Markson Science, 10201 S. 51st Street, Ste. 100, Phoenix, AZ 85044, USA.
Marsh Biomedical Products, 274 N. Goodman St., Rochester, NY 14607, USA.
Mattson Instruments Inc., 1001 Fourier Court, Madison, WI 53717, USA.
May & Baker Ltd., Liverpool Rd., Eccles, Manchester M30 7RT, GREAT BRITAIN.
MCRA Applied Technologies Inc., P.O. Box 377, Rockaway, NJ 07866, USA.
Medatronics Corp., 3901 Clark St., Seaford, NY 11783, USA.
Medical Air Technology Ltd., Canto House, Wilton St., Denton, Manchester M34 3LZ,
    GREAT BRITAIN.
Memtek Corp., 28 Cook St., Billerica, MA 01821, USA.
E. Merck, P.O. Box 4119, D-6100 Darmstadt, GERMANY.
Metrohm Ltd., CH-9101 Herisau, SWITZERLAND.
Michrom BioResources, 4193 Sundown Road, Livermore, CA 94550, USA.
Micro Filtration Systems, 6800 Sierra Ct., Dublin, CA 94568, USA.
Micromeritrics, 1 Micromeritrics Drive, Norcross, GA 30093, USA.
Micron Separations, Inc., 135 Flanders Rd., Westborough, MA 01581, USA.
Microphoretic Systems Inc., 750 N. Pastoria Avenue, Sunnyvale, CA 94086, USA.
Microsensor Technology Inc., 41762 Christy St., Fremont, CA 94538, USA.
Midwest Scientific Inc., 228 Meramec Station Rd., Valley Park, MO 63088, USA.
Mikrolab Aarhus A/S, Axel Kiers Vej 34, DK-8270 Hojbjerk, DENMARK.
Miles Scientific, 30 W 475 N. Aurora Rd., Naperville, IL 60566, USA.
Milevac Scientific Glass Ltd., 38/40 Broton Dr. Trading Estate, Halstead, Essex C09 1HB,
    GREAT BRITAIN.
Millipore Corporation, 75C Wiggins Avenue, Bedford, MA 01730, USA.
Milton Roy LDC Div., P.O. Box 10235, Riviera Beach, FL 33404, USA.
Mitsui Toatsu Chemicals Inc., 2-5 Kasumigaseki, 3-chome, Chiyoda-ku, Tokyo 100,
    JAPAN.
Modchrom Inc., 8666 Tyler Blvd., Ste. 3, Mentor, OH 44060, USA.
Molecular Dynamics, 230 Santa Ana Circle, Sunnyvale, CA 94086, USA.
Molecular Instruments Co., P.O. Box 1652, Evanston, IL 60201, USA.
Mott Metalurgical Corp., Spring Lane, Farmington, CT 06032, USA.
The Munhall Company, 5655 N. High St., Worthington, OH 43085, USA.
Muromachi Kagaku Kogyo Kaisha Ltd., No. 3, 4-chome, Muromachi, Nihonbashi,
    Chuo-ku, Tokyo 103, JAPAN.
National Diagnostics Inc., 1013-7 Kennedy Blvd., Manville, NJ 08835, USA.
PE Nelson Systems, 10061 Bubb Road, Cupertino, CA 95014, USA.
Neslab Instruments Inc., P.O. Box 1178, Portsmouth, NH 03801, USA.
The Nest Group, 45 Valley Rd., Southboro, MA 01771, USA.
New Brunswick Scientific Co., 44 Talmadge Rd., P.O. Box 4005, Edison, NJ 08818-4005,
    USA.
Newman-Howells Assoc. Ltd., Wolvesey Palace, Winchester, Hamps. S023 9NB, GREAT
    BRITAIN.
Nicolet Instrument Corp., 5225-1 Verona Rd., Madison, WI 53711, USA.
Nihon Seimitsu Kagaku Co, 25-10 Futaba-cho, Itabashi-ku, Tokyo 173, JAPAN.
Nimbuchem SA, 115 Ch. de Charleroi, B-5900 Jodoigne, BELGIUM.
Nissei Sangyo America, 1701 Golf Rd., Ste. 401, Rolling Meadows, IL 60008, USA.
Noah Technologies Corp., Tool Fairgrounds Parkway, San Antonio, TX 78238, USA.
Nordion Instruments Oy Ltd., P.O. Box 1, SF-00371 Helsinki, FINLAND.
Nuclear Sources & Services Inc., P.O. Box 34042, Houston, TX 77234, USA.

Nuclide Corp., 1155 Zion Rd., Bellefonte, PA 16823, USA.
Ohio Valley Specialty Chemical Inc., 115 Industry Rd., Marietta, OH 45750, USA.
O.I. Analytical, P.O. Box 2980, College Station, TX 77841-2980, USA.
Olympus Clinical Instruments Div., 4 Nevada Dr., Lake Success, NY 11042, USA.
Omnifit Ltd., 51 Norfolk St., Cambridge, Cambs. CB1 2LE, GREAT BRITAIN.
Oncor, P.O. Box 870, Gaithersburg, MD 20884, USA.
On-Line Instruments Systems Inc., Route 2, P.O. Box 111, Jefferson, GA 30549, USA.
Owl Scientific Plastics Inc., P.O. Box 566, Cambridge, MA 02139, USA.
Oyster Bay Pump Works Inc., No. 1 Bay Ave., P.O. Box 96, Oyster Bay, NY 11771, USA.
Pall Biosupport Co., 77 Crescent Beach Rd., Glen Cove, NY 11542, USA.
Pallflex Inc., Kennedy Dr., Putnam, CT 06260, USA.
Parker Hannifin Corp., 9400 S. Memorial Pkwy., P.O. Box 4288, Huntsville, AL 35802, USA.
Particle Data Inc., 11 Hahn St., P.O. Box 265, Elmhurst, IL 60126, USA.
P.C. Inc., 11805 Kim Pl., Potomac, MD 20854, USA.
PCP Inc., 2155 Indian Rd., West Palm Beach, FL 33409-3287, USA.
Pen Kem Inc., 341 Adam St., Bedford Hills, NY 10507, USA.
Peris Industries Inc., P.O. Box 1008, State College, PA 16804-1008, USA.
Perkin-Elmer Corp., 761 Main Avenue, Norwalk, CT 06859-0090, USA.
Peptides International Inc., 10101 Linn Station Rd., Ste. 445, Louisville, KY 40223, USA.
PerSeptive Biosystems Inc., 222 Third Street, Ste. 0300, Cambridge, MA 02142, USA.
Petazon Co., Zug, SWITZERLAND.
Pharmacia LKB Biotechnology AB, S-751 82 Uppsala, SWEDEN.
Pharma-Tech Research Corp., 8800 Kelso Dr., Baltimore, MD 21221, USA.
Phase Separations Inc., 140 Water Street, Norwalk, CT 06854, USA.
Phenomenex, 2320 W 205 St., Torrance, CA 90501, USA.
Philips Analytical, NL-5600 MD Eidhoven, THE NETHERLANDS.
Phortran Inc., 344 Lekeside Drive, Foster City, CA 94404, USA.
Photometrics, 2010 N. Forbes Blvd., Tucson, AZ 85745, USA.
Photovac Inc., 134 Doncaster Ave., Thornhill, Ont. L3T 1L3, CANADA.
PI Technologies, 3182 C Old Tunnel Road, Lafayette, CA 94549, USA.
Pickering Laboratories Inc., 1951 Colony St., Ste. S, Mountain View, CA 94043, USA.
Pierce Chemical Co., P.O. Box 117, Rockford, IL 61105, USA.
Poly Labo Paul Block & Cie, 305 Route de Colmar, F-67100 Strasbourg, FRANCE.
Poly LC, 9052 Belwart Way, Columbia, MD 21045, USA.
Polymer Laboratories Inc., 160 Old Farm Road, Amherst, MA 01002, USA.
Polymicro Technologies Inc., 3035 N 33rd Dr., Phoenix, AZ 85017, USA.
Polysciences Inc., 400 Valley Rd., Warrington, PA 18976-2590, USA.
PQ Corporation, Conshohocken, PA 19428, USA.
PreComp Inc., 17 Barstow Rd., P.O. Box 461, Great Neck, NY 11021, USA.
Preiser Scientific Inc., 94 Oliver Street, St. Albans, WV 25177, USA.
Princeton Separations Inc., P.O. Box 300, Adelphia, NJ 07710, USA.
Princeton Testing Laboratory, P.O. Box 3108, Princeton, NJ 08543, USA.
Process Analyzers Inc., 8 Headley Pl., Fallsington, PA 19054, USA.
Prochrom SA, Chemin de Blanches Terres, B.P. 9, F-54250 Champigneulles, FRANCE.
Puregas/General Cable Co., P.O. Box 666, 5600 W. 88 Ave., Westminster, CO 80030, USA.
Philips Analytical Pye Unicam Ltd., York Street, Cambridge CB1 2PX, GREAT BRITAIN.
Quadrant Scientific Ltd., 36 Brunswick Rd., Gloucester GL1 1JJ, GREAT BRITAIN.
Quadrex Corp., P.O. Box 3881, New Haven, CT 06525, USA.
Radiomatic Instruments & Chemical Co., 5102 S. Westshore Blvd., Tampa, FL 33611, USA.
Radnoti Glass Technology Inc., 227 West Maple Ave., Monrovia, CA 91016, USA.
Rainin Instrument Co., Mack Road, Woburn, MA 01801, USA.
Realco Chemical Co., New Brunswick, NJ 08903, USA.
Reanal, P.O. Box 54, H-1441 Budapest 70, HUNGARY.
Regis Chemical Company, 8210 Austin Ave., P.O. Box 519, Morton Grove, IL 60053, USA.
Reichelt Chemietechnik GmbH, Englerstr. 18, D-6900 Heidelberg 1, GERMANY.
Reliable Scientific Inc., 881 Richland Dr., Memphis, TN 38116, USA.
Reliance Glass Works, 220 Gateway Rd., P.O. Box 825, Bensenville, IL 60106, USA.
Repligen Corp., 1 Kendall Square, Bldg. 700, Cambridge, MA 02139, USA.

Restek Corp., 110 Benner Circle, Bellefonte, PA 16823, USA.
Rheodyne Inc., P.O. Box 996, Cotati, CA 94931, USA.
Rhône-Poulenc Ltd., Liverpool Rd., Eccles, Manchester M30 7RT, GREAT BRITAIN.
Richard Scientific Inc., 250 Bel Marin Keys Blvd., Ste. D3, Novato, CA 94949, USA.
Riedel de Haen AG, Wunstorfer Str. 40, D-3016 Seelze, GERMANY.
Rockland Inc., P.O. Box 316, Gilbertsville, PA 19380, USA.
Rohm & Haas Company, 727 Norristown Road, Spring House, PA 19477, USA.
Ruska Laboratories Inc., P.O. Box 630009, Houston, TX 77263-0009, USA.
Russell pH Ltd., Station Rd., Auchtermuchty, Fife KY14 7DP, GREAT BRITAIN.
SAC Chromatography Ltd., Summerhouse Hill, Cardington, Bedford MK44 3SD, GREAT
    BRITAIN.
Sadtler Research Labs., 3316 Spring Garden Street, Philadelphia, PA 19104, USA.
Saitron SpA, 39, Via del Crocifisso, Ponte a Ema, I-50126 Firenze, ITALY.
Sanki Engineering Ltd., 2-16-10 Imazato, Nagaokakyo, Kyoto 617, JAPAN.
Serasep Inc., 1600 Wyatt Drive, Ste. 10, Santa Clara, CA 95054, USA.
Sargent-Welch Scientific Co., 7300 N. Linden Ave., Skokie, IL 60077, USA.
Sarstedt Inc., P.O. Box 4090, Princeton, NJ 08543, USA.
Sartorius GmbH, Weender Landstr. 94/108, D-3400 Göttingen, GERMANY.
Sartorius Filters Inc., 30940 San Clemente St., Hayward, CA 94544, USA.
Säulentechnik, Dr. Ing. H. Knauer GmbH, Am Schlangengraben 16, D-1000 Berlin 20,
    GERMANY.
Savant Instruments Inc., 110 Bi-County Blvd., Farmingdale, NY 11735, USA.
Scanivalve Corp., 10222 San Diego Mission Rd., San Diego, CA 92108, USA.
Schleicher & Schuell GmbH, Postfach 4, D-3354 Dassel, GERMANY.
Scientific Equipment & Instrument Co., 265 Houret Dr., Milpitas, CA 95035, USA.
Scientific Glass Engineering Co., 2007 Kramer La., Austin, TX 78758, USA.
Scientific Instrument Services Inc., Rt. 179, P.O. Box 593, Ringoes, NJ 08551, USA.
Scientific Logics Inc., 21910 Alcazar Ave., Cupertino, CA 95014, USA.
Scientific Systems Inc., 349 N. Science Park Road, State College, PA 16803, USA.
Scientific Technologies Inc., 3121 Glen Royal Rd., Raleigh, NC 27612, USA.
Scott Specialty Gases, Route 611, Plumsteadville, PA 18949, USA.
S-Cubed, P.O. Box 1620, La Jolla, CA 92038, USA.
SEAC srl, Via Carlo del Prete 139, I-50127 Firenze, ITALY.
Seamark Corp., 11618 Busy St., Richmond, VA 23236, USA.
Dr. R. Seitner Mess- und Regeltechnik GmbH, Mühlbachstrasse 20, D-8031 Seefeld/Obb.,
    GERMANY.
Separation Industries, P.O. Box 4338, 4 Leonard Street, Metuchen, NJ 08840, USA.
Separations Group Inc., 17434 Mojave Street, P.O. Box 867, Hesperia, CA 92345, USA.
Separations Technology, P.O. Box 352, Wakefield, RI 02880-0352, USA.
Sepragen Corp., 2126 Edison Ave., San Leandro, CA 94577, USA.
Serapine Corp., 821 Franklin Ave., Garden City, NY 11530, USA.
Serva Fine Biochemicals Inc., 200 Shames Dr., Westbury, NY 11590, USA.
Servomex Ltd., Crowborough, Sussex TN6 3DU, GREAT BRITAIN.
Severn Analytical Ltd., Unit 2B, St. Francis' Way, Shefford Industrial Park, Shefford, Beds.
    SG17 5DZ, GREAT BRITAIN.
SGE International Pty. Ltd., 2/76 Charles St., Ryde, NSW 2112, AUSTRALIA.
Shandon Southern Products Ltd., Chadwich Rd., Astmoor, Runcorn, Ches. WA7 1PR,
    GREAT BRITAIN.
Shimadzu Corporation, 1 Nishinokyo-Kuwabaracho, Nakagyo-ku, Kyoto 604, JAPAN.
Showa Denko K.K., 2-24-25 Tamagawa, Ohta-ku, Tokyo 146, JAPAN.
Siemens AG, Mess- u. Prozesstechnik, E 687 Postfach 21 1262, D-7500 Karlsruhe,
    GERMANY.
Sievers Research Inc., 1930 Central Ave., Ste. C, Boulder, CO 80301, USA.
Sigma Chemical Co., 3050 Spruce St., P.O. Box 14508, St. Louis, MO 63178, USA.
SKALAR Analytical BV, Spinveld 2, NL Breda, THE NETHERLANDS.
Small Parts Inc., P.O. Box 381966, Miami, FL 33238-1966, USA.
Sonntek/Knauer, P.O. Box 8589, Woodcliff Lake, NJ 07675, USA.
Sopar-Biochem, 124 rue J. Besme, B-1080 Bruxelles, BELGIUM.
Sota Chromatography, P.O. Box 693, Crompond, NY 10517, USA.
Southland Cryogenics, 2424 Lacy Lane, P.O. Box 110627, Carrollton, TX 75011, USA.
Spark Holland BV, P.O. Box 388, NL-7800 AJ Emmen, THE NETHERLANDS.

Spectra Gases Inc., 277 Coit St., Irvington, NJ 07111, USA.
Spectramass Ltd., Radnor Park Industrial Estate, Congleton, Cheshire, CW12 4XR,
    GREAT BRITAIN.
SpecTran Corp., 50 Hall Rd., Sturbridge, MA 01566, USA.
Spectra-Physics, 3333 N. First St., San Jose, CA 95134, USA.
Spectra-Tech Europe Ltd., Genesis Centre, Science Park South, Birchwood, Warrington
    WA3 7BH, GREAT BRITAIN.
Spectron Instrument Corp., 1342 W. Cedar Ave., Denver, CO 80223, USA.
Spectronics Corp., 956 Brush Hollow Rd., P.O. Box 483, Westbury, NY 11590, USA.
SpectroVision Inc., 25 Industrial Ave., Chelmsford, MA 01824, USA.
Spectrum Medical Industries Inc., 8430 Santa Monica Blvd., Los Angeles, CA 90069, USA.
Spectrum Scientific Inc., 15413 Vantage Pkwy, East Houston, TX 77032, USA.
Spellman High Voltage Electronics Corp., 7 Fairchild Ave., Plainview, NY 11803, USA.
Spinco Division, Beckman Instruments Inc., Palo Alto, CA 94304, USA.
SPIRAL Sarl, 3, rue des Mardors, Z.A. Couternon, F-21560 Arc-sur-Tille, FRANCE.
SRI Instruments, 548 South Gertruda Ave., Redondo Beach, CA 90277, USA.
Sterling Organics Ltd., Hadrian House, E. Gefield Ave., Fawdon, Newcastle-on-Tyne NE3
    3TT, GREAT BRITAIN.
Sterogene Bioseparations, 140 E. Santa Clara Street, Arcadia, CA 91006, USA.
St. John Associates Inc., 4805 Prince George's Ave., Beltsville, MD 20705, USA.
Stratagene Inc., 11099 N. Torrey Pines Rd., La Jolla, CA 92037, USA.
Strawberry Tree Inc., 160 S. Wolfe Rd., Sunnyvale, CA 94086, USA.
Sugiyama Shoji Co., Ato Bldg., 1-2-7 Shibadaimon, Minato-ku, Tokyo 105, JAPAN.
Sun Brokers Inc., P.O. Box 2230, Wilmington, NC 28405, USA.
Supelco Inc., Supelco Park, Bellefonte, PA 16823, USA.
Suprex Corporation, 125 William Pitt Way, Pittsburgh, PA 15238, USA.
Swagelok Co., 31400 Aurora Rd., Solon, OH 44139, USA.
Sycopel Scientific Ltd., The Laboratory, Station Road, E. Bolden, Tyne and Ware NE36
    OEB, GREAT BRITAIN.
Synchrom Inc., P.O. Box 310, Lafayette, IN 47902, USA.
Synthetic Peptides Inc., Dep. of Biochemistry, University of Alberta, Edmonton, Alta. T6G
    2P5, CANADA.
Systems Instrument Corp., America/SICA, 106 Centre St., Dover, MA 02030, USA.
Tecan U.S., P.O. Box 8101, Hillsborough, NC 27278, USA.
Techlab GmbH, Evessener Str. 2, D-3305 Erkerode, GERMANY.
Technical & Analytical Solutions, 49 Pelham Street, Ashton under Lyne, Lancashire OL7
    ODT, GREAT BRITAIN.
Technical Assoc., 7051 Eaton Ave., Canoga Park, CA 91303, USA.
Technicon Industrial Systems, 511 Benedict Ave., Tarrytown, NY 10591, USA.
Technimed Corp., 4987 NW 23rd Ave., Ft. Lauderdale, FL 33309, USA.
Technology Applications Inc., 26 West Martin Luther King Drive, Cincinnati, OH 45219,
    USA.
Tegal Scientific Inc., P.O. Box 5905, Concord, CA 94524, USA.
Tekmar Co., P.O. Box 371856, Cincinnati, OH 45222-1856, USA.
Teknivent Corp., 11684 Lilburn Park Rd., St. Louis, MO 63146, USA.
Tescom Corp., 12616 Industrial Blvd., Elk River, MN 55330, USA.
Tessek, 1070 Catleton Way, Sunnyvale, CA 94087, USA.
Texas Instruments, P.O. Box 655012, Dallas, TX 75240, USA.
Thermedics Inc., 470 Wildwood Street, Woburn, MA 01888-1799, USA.
Thermo Environmental Instruments Inc., 8 West Forge Parkway, Franklin, MA 02038,
    USA.
TM Analytic Inc., 303 E. Robertson, Brandon, FL 33511, USA.
Tohoku Electronic Industrial Co., 6-6-6 Shirakashidai, Rifu-cho, Miyagi 981-01, JAPAN.
Tokyo Kasei Kogyo Co., 3-9-4 Nihonbashi-Honcho, Chuo-ku, Tokyo 103, JAPAN.
Tokyo Rikakikai Co., Toei Bldg. 4-3, 4-chome, Muromachi, Nihombashi, Chuo-ku, Tokyo,
    JAPAN.
Toso Haas, Independence Mall West, Philadelphia, PA 19105, USA.
Tosoh Corporation, 4560 Tonda, Shinnanyo, Yamaguchi 746, JAPAN.
Toyo Soda Mfg. Co., 1-14-15 Akasak, Minato-ku, Tokyo 107, JAPAN.
Trace Analytical, P.O. Box 2523, Stanford, CA 94305, USA.
Tracor Instruments, 6500 Tracor La., Austin, TX 78725, USA.

Triangle Laboratories Inc., P.O. Box 13485, Research Triangle Park, NC 27709, USA.
Trivector Systems International Ltd., Sunderland Rd., Sandy, Beds. SG19 1RB, GREAT BRITAIN.
Tudor Scientific Glass Co., 555 Edgefield Rd., Belvedere, SC 29841, USA.
Tyler Research Corp., 6128 103 St., Edmonton, Alta. T6H 2H8, CANADA.
Ultra-Lum Inc., 217 E. Star of India La., Carson, CA 90746, USA.
Ultra Scientific Inc., 1 Main St., Hope, RI 02831, USA.
Ultra-Violet Products Ltd., Science Park, Milton Rd., Cambridge CB4 4BN, GREAT BRITAIN.
U-MicroComputers Ltd., Winstanley Industrial Estate, Long Lane, Warrington WA2 8PR, GREAT BRITAIN.
Unimetrics Corp., 501 Earl Rd., Shorewood, IL 60436, USA.
Union Carbide Industrial Gases, 200 Cottontail La., Somerset, NJ 08873, USA.
United States Biochemical Corp., P.O. Box 22400, Cleveland, OH 44122, USA.
Universal Absorbents Inc., 2801 Bankers Industrial Dr., Atlanta, GA 30360, USA.
Universal Biochemicals, 6 Sathya Sayee Bagar, Madurai, Tamil Nadu 625 003, INDIA.
Universal Scientific Inc., 2801 Bankers Industrial Dr., Atlanta, GA 30306, USA.
Upchurch Scientific Inc., 2969 N. Goldie Rd., Oak Harbor, WA 98277, USA.
USA Scientific Plastics, P.O. Box 3565, Ocala, FL 32678, USA.
UTI Instruments Co., 325 N. Mathilda Ave., Sunnyvale, CA 94063, USA.
UTI-tect Inc., 2233-F Northwestern Ave., Waukegan, IL 60087, USA.
UVP Inc., 5100 Walnut Grove Ave., San Gabriel, CA 91778, USA.
Vangard International, 1111-A Green Grove Rd., P.O. Box 308, Neptune, NJ 07754-0308, USA.
Varex Corporation, Burtonsville Commerce Center, 4000 Blackburn La., Burtonsville, MD 20866, USA.
Varian Instrument Group, 2700 Mitchell Drive, Walnut Creek, CA 94598, USA.
Verifio Corp., 250 Canal Blvd., Richmond, CA 94804, USA.
Vested Corp., 9299 Kirby Dr., Houston, TX 77054, USA.
Vetter Laborgeräte GmbH, Postfach 1348, Rudolf-Diesel-Str. 21, Bad Wurt/BRD, D-6908 Wiesloch 4 (Baiertal), GERMANY.
VG Analytical Ltd., Floats Road, Wythenshawe, Manchester M23 9LE, GREAT BRITAIN.
VG Masslab Ltd., Tudor Rd., Altrincham WA14 5RZ, GREAT BRITAIN.
VICI Valco Instruments Co., P.O. Box 55603, Houston, TX 77255, USA.
Vickers Instruments Inc., P.O. Box 99, 300 Commercial St., Malden, MA 02148, USA.
Violet, Via Giovanni Giorgi No. 22, I-00149 Roma, ITALY.
VWR Scientific, P.O. Box 7900, San Francisco, CA 94120, USA.
Waitaki International Biosciences, 55 Glen Scarlett Rd., Toronto, Ont. M6N 1P5, CANADA.
Wako Chemicals USA Inc., 12300 Ford Rd., Ste. 130, Dallas, TX 75234, USA.
Wakunaga Pharmaceutical Co., Shimokotachi 1624, Koda-cho, Takata-gun, Hiroshima 729-64, JAPAN.
Walden Precision Apparatus Ltd., The Old Station, Linton, Cambridge, CB1 6NW, GREAT BRITAIN.
Wale Apparatus Co., 400 Front St., P.O. Box D, Hellertown, PA 18055, USA.
Waters Chromatography, 34 Maple Street, Milford, MA 01757, USA.
Watson Products Inc., 1068 1/2 N. Allen Ave., Passadena, CA 91104, USA.
Wescan Instruments Inc., 3018 Scott Blvd., Santa Clara, CA 95054-0984, USA.
Whatman Ltd., Springfield Mill, Maidstone, Kent ME14 2LE, GREAT BRITAIN.
Wheaton Scientific, 1301 N. Tenth St., Millville, NJ 08332, USA.
Wilmad Glass Co., Route 40 & Oak Rd., Buena, NJ 08310, USA.
Wyatt Technology, P.O. Box 3003, Santa Barbara, CA 93130, USA.
Xavier Industries, 3627 W. Warner Ave., Santa Ana, CA 92704, USA.
Xerox Analytical Laboratories, 0114-42D Joseph C. Wilson Ctr. for Technology, Rochester, NY 14644, USA.
Xydex Corporation, 4 Alfred Cr., Bedford, MA 01730, USA.
YMC Inc., 51 Gibraltar Dr., Ste. 2D2, Morris Plains, NJ 07950, USA.
Yokagawa Corp., 200 W. Park Dr., Peachtree City, GA 30269, USA.
J. Young (Scientific Glassware) Ltd., 11 Colville Rd., London W3 8B3, GREAT BRITAIN.
Zeta-Meter Inc., 50-17 5th Street, Long Island City, NY 11101, USA.
Zinsser Analytic GmbH, Eschborner Landstr. 135, D-6000 Frankfurt 94, GERMANY.
Zymed Laboratories Inc., 52 S. Linden Ave., Ste. 4, South San Francisco, CA 94080, USA.

Subject Index

JOURNAL OF CHROMATOGRAPHY LIBRARY

A Series of Books Devoted to Chromatographic and Electrophoretic Techniques and their Applications

Although complementary to the *Journal of Chromatography*, each volume in the Library Series is an important and independent contribution in the field of chromatography and electrophoresis. The Library contains no material reprinted from the journal itself.

Other volumes in this series